Primate Adaptation and Evolution

Second Edition

Primate Adaptation and Evolution

Second Edition

John G. Fleagle

State University of New York, Stony Brook

Academic Press

An Imprint of Elsevier

San Diego London Boston New York Sydney Tokyo Toronto

Front cover photograph by Ron Garrison, Zoological Society of San Diego

This book is printed on acid-free paper. ♾

Academic Press
An Imprint of Elsevier
525 B Street, Suite 1900, San Diego, California 92101-4495, USA
http://www.academicpress.com

Academic Press
84 Theobald's Road, London WC1X 8RR, UK
http://www.academicpress.com

Library of Congress Catalog Card Number: 98-87186

International Standard Book Number: 0-12-260341-9

PRINTED IN THE UNITED STATES OF AMERICA
05 06 07 9 8 7 6

CONTENTS

TABLES & ILLUSTRATIONS

Tables

Illustrations

CHAPTER 1

CHAPTER 2

CHAPTER 3

CHAPTER 4

CHAPTER 7

CHAPTER 8

CHAPTER 9

CHAPTER 10

CHAPTER 11

CHAPTER 12

CHAPTER 13

CHAPTER 14

CHAPTER 15

CHAPTER 16

CHAPTER 17

CHAPTER 18

PREFACE TO SECOND EDITION

I would never have imagined that writing a second edition of *Primate Adaptation and Evolution* would be so much more difficult than writing the first one. However, as every author knows, once your words are typeset, they take on a life of their own and it is very difficult to think and write about the same topic in any other way. Nevertheless, our knowledge of past and present primates is growing by leaps and bounds and the first edition certainly needed updating.

In this edition, every chapter has been revised and rewritten. In some cases, the new information was so great that whole new chapters or major parts of chapters were required. Thus, Primate Lives (Chapter 3) now has an extensive introduction to Primate Life Histories, and Primate Origins (Chapter 11) has an expanded discussion about tree shrews, flying lemurs, and bats. I have added a new chapter on primate communities (Chapter 8) and have made separate chapters for early anthropoids (Chapter 13) and fossil platyrrhines (Chapter 14).

As in the first edition, I have provided estimates of body size for most of the living and fossil species to give the reader some general appreciation for the size of the animals being discussed. These are obviously estimates based on available information. The information of living species is drawn from an impressive recent compilation by Richard Smith and Bill Jungers (*Journal of Human Evolution* 1997, 32: 523–559). Body size estimates for extinct species are generally based on dental and cranial dimensions. I am especially grateful to Jay Norejko, Greg Gunnell, Laurie Godfrey, John Kappelman, and Eric Delson, who provided many of these estimates.

In addition to all those who helped with the first edition, many of my colleagues contributed to this revision with advice, data, illustrations, and help with proofreading. Among those I want to thank for so generously sharing their wisdom and skills with me are Kristina Aldridge, Kamla Alhuwalia, Friderun Ankel-Simons, Berhane Asfaw, Zelalem Assefa, Robert Barton, Chris Beard, Brenda Benefit, Tom Bown, Ric Charnov, Steve Churchill, Russ Ciochon, Kate Clark-Schmidt, Glenn Conroy, Bert Covert, Lisa Davis, Louis deBonis, Eric Delson, Brigitte Demes, Todd Disotell, Diane Doran, Stephane Ducrocq, Robin Dunbar, Robert Fajardo, Craig Feibel, Alexandra Fleagle, Snowball Fleagle, Susan Ford, John Flynn, Jorg Ganzhorn, Phil Gingerich, Laurie Godfrey, Michelle Goldsmith, Fred Grine, Greg Gunnell, Sharon Gursky, J.-L. Hartenberger, Walter Hartwig, Kristen Hawkes, Chris Heesy, Robert Hill, Ines Horovitz, Charles Janson, Andy Jones, William Jungers, Peter kappeler, John Kappelman, Rich Kay, Jay Kelley, Bill Kimbel, David Krause, Mike Lague, Susan Larson, Meave Leakey, Steve leigh, Pierre lemelin, Charles Lockwood, Anita Lubensky, Mary Maas, Laura MacLatchy, Curtis Marean, Bob Martin, Lawrence Martin, Judith Masters, Monte McCrossin, Scott McGraw, Kelly Mc-

Neese, Jeff Meldrum, Ellen Miller, Russ Mittermeier, Jay Mussell, Jay Norejko, Deborah Overdorff, Osbjorn Pearson, David Pilbeam, Mike Plavcan, Leila Porter, John Polk, Todd Rae, Tab Rasumussen, Denne Reed, Kaye Reed, Jennifer Reig, Brian Richmond, Ken Rose, Alfie Rosenberger, Callum Ross, Carolyn Ross, Henry Schwarcz, Natasha Shah, Liza Shapiro, John Shea, Mary Silcox, Elwyn Simons, Nancy Stevens, Caro-Beth Stewart, David Strait, Suzanne Strait, Randall Susman, Adan Tauber, Mark Teaford, Peter Ungar, Alan Walker, Chris Walker, Mireya West, Blythe Williams, Milford Wolpoff, Patricia Wright, Roshna Wunderlich, Carey Yaeger, Solomon Yirga, and Anne Yoder.

Bob Martin offered especially pointed comments and suggestions on the first edition that much improved this one. The tables, bibliographies, and index were made possible by Herculean effots from Ines Horovitz, Mary Maas, Leila Porter, and Kaye Reed. Joan Kelly typed so many versions of the manuscript that it drove her into early retirement, and we are forever lost without her. Although many people have generously shared ideas and information for this edition, Kaye Reed deserves special thanks for her major contributions to the chapter on primate communities and for unfailing support and encouragement.

Many institutions have supported various aspects of my work while this edition was being prepared, including the National Geographic Society, the National Science Foundation, and the Wenner–Gren Foundation. I am especially grateful to the L.S.B. Leakey Foundation, its chairman, Gordon Getty, and its president, Kay Woods, for their support in many forms over the past decade.

In my view, the illustrations are the heart of this book. New artwork for this edition was prepared by William Yee, Stephen Nash, and especially Ronald Futral and Luci Betti. The person who said that a picture was worth a thousand words had obviously never worked with artists as talented as these. The information conveyed by a single illustration from their drafting tables could not be captured in a thousand pages.

ONE

Adaptation, Evolution, and Systematics

Adaptation

Adaptation is a concept central to our understanding of evolution, but the term has proved very difficult to define in a simple phrase. One of the most succinct definitions has been offered by Vermeij (1978, p. 3): "An adaptation is a characteristic that allows an organism to live and reproduce in an environment where it probably could not otherwise exist." In the following chapters, we examine extant (living) and extinct (fossil) primates as a series of **adaptive radiations**—groups of closely related organisms that have evolved morphological and behavioral features enabling them to exploit different ecological niches. Adaptive radiations provide especially clear examples of evolutionary processes. The adaptive radiation of finches on the Galapagos Islands of Ecuador played an important role in guiding Darwin's views on the origin of species.

Adaptation also refers to the process whereby organisms obtain their adaptive characteristics. The primary mechanism of adaptation is natural selection. **Natural selection** is the process whereby any heritable features, anatomical or behavioral, that enhance the fitness of an organism relative to its peers, increase in frequency in the population in succeeding generations. Fitness, in an evolutionary sense, is reproductive success. It is important to remember that natural selection acts primarily through differential reproductive success of individuals within a population (Williams, 1966).

Evolution

Evolution is modification by descent, or genetic change in a population through time. Although biologists consider most evolution to be the result of natural selection, there are other, non-Darwinian mechanisms that can and do lead to genetic change within a population. **Genetic drift** is change in the genetic composition of a population from generation to generation due to chance sampling events independent of selection. **Founder effect** is a more extreme change in the genetic makeup of a population that occurs when a new population is established by only a few individuals. This new population may sample only a small part of the variation found in the ancestral population. Thus recessive alleles that are not expressed in the larger population may become more common or even fixed in the new population. In this way, the chance characteristics of a founder population can have dramatic effects on the subsequent evolution and adaptive diversity of a group of organisms.

Species and Speciation

The fact that the diversity of life is the result of evolution means that all organisms are related by virtue of sharing a common genetic ancestry in the distant past. However, it is also clear that living organisms are not a continuous spread of variation. The living world is composed of distinct kinds of organisms that we recognize as species. Although virtually all biologists recognize species as the natural units of life, defining exactly what a species is or how species form are more difficult problems. These are the problems that Darwin (1859) set out to explain, and they are still the subject of intense study and debate (e.g. Kimbel and Martin, 1993; Tattersall, 1992; deQueiroz, 1998).

Until recently, most biologists and anthropologists generally accepted the **Biological Species Concept (BSC),** in which species are defined as "groups of actually or potentially interbreeding natural populations that are reproductively isolated from other groups" (Mayrs, 1942). While appealing, in that it emphasizes the genetic and phyletic distinctiveness of species through reproductive isolation, the Biological Species Concept is obviously impossible to apply to fossils or allopatric populations of living animals and even difficult to apply to sympatric populations without detailed data on mating behavior and fertility (Tattersall, 1989, 1992). Many students of living organisms are more comfortable with a **Mate Recognition Concept** (Paterson, 1978, 1985; Masters, 1993). In this concept, species are defined as "the group of individuals sharing a common fertilization system" (Paterson, 1985). A particularly important aspect of this common fertilization system is a specific mate-recognition system. Members of a species recognize one another as potential mates through such behavior or morphological features as vocalizations, mating displays, or ornamentation. Like the Biological Species Concept, the Mate Recognition Concept is virtually impossible to apply to extinct organisms.

In contrast with the Biological Species Concept and the Mate Recognition Concept, which are based on information about reproductive behavior, many paleontologists prefer a species concept based on morphological differences. The **Phylogenetic Species Concept** (Cracraft, 1983) is commonly adopted by students of phylogenetic systematics. In this approach, a species is "the smallest diagnosable cluster of individuals within which there is a parental pattern of ancestry and descent". In this concept a species is defined on the basis of morphological distinctions from other taxa (Cracraft, 1983). In principle this could be based on a single feature. Other researchers, using what may be called a **Phenetic Fossil Species Concept** would argue that species defined morphologically should have approximately the same amount of metrical variation as extant populations (e.g., Gingerich and Schoeninger, 1979; Cope, 1993).

Most biologists agree that a species is a distinct segment of an evolutionary lineage, and many of the differences among species concepts reflect attempts to find criteria that can be used to identify species based on different types of information, such as behavioral observations of living populations or morphological information from teeth and skulls. In addition, some of the diversity of species concepts may be more useful in distinguishing species at different phases of their formation (de Quieroz, 1998). Paradoxically, the greatest challenge to species identification often comes not from incomplete information, but from those rare paleontological instances in which there is a continuous temporal sequence of populations undergoing directional selection (e.g. K. D. Rose and Bown, 1993). In this situation, the endpoints are clearly differentiable, but any species boundary is necessarily arbitrary (see Chapter 18).

Patterns of Evolution

Evolutionary change within a population can take place at different rates and can yield different results. There are many descriptive terms available to describe patterns of evolutionary change and numerous theories about how common these different patterns are in the history of life as well as about the causal factors underlying these different evolutionary patterns. The pattern of evolutionary change just described, in which a lineage undergoes gradual change over time, is called **anagenesis. Cladogenesis** is the division of a single lineage into two lineages. Gradual change in the morphology of a population of organisms through time, either anagenetic or cladogenetic, is often called **phyletic gradualism.** This type of evolutionary pattern is very common in the fossil record, and many biologists believe that most evolutionary change has been of this type.

The rates of evolutionary change that take place in populations through time may vary considerably, theoretically over several orders of magnitude. In addition, directional selection may shift over the course of relatively short time periods due to climatic fluctuations (see Carroll, 1997). Some paleontologists argue that the most common type of evolutionary change is **punctuated equilibrium,** in which the morphology of most species is essentially stable (in equilibrium) for long periods of time, and that speciation events are the result of rapid morphological and genetic shifts (or punctuations) over brief periods of evolutionary time so that intermediate forms are rarely recovered (e.g. Gould and Eldridge, 1993). Unfortunately, the punctuated equilibrium model is very difficult to distinguish from a discontinuous fossil record and is generally defined in a way that virtually precludes the possibility of determining how often it actually takes place.

Phylogeny

The branching pattern of successive species that results from numerous cladogenetic events is a **phylogeny.** Because this book deals with the adaptive radiations of primates, we are interested in reconstructing the evolutionary branching sequence, or phylogeny, of various primate groups to see how they are related to one another. Although some of us can trace our own genealogies (or those of our pets) through several generations, tracing the genealogical relationships among all primates is a much more daunting undertaking. The evolutionary radiation of primates has taken place over tens of millions of years of geological time and has involved thousands of species, millions of generations, and billions of individuals. Moreover, the records available for reconstructing primate phylogeny are meager, consisting of individuals of fewer than 300 living species and occasional bony remains of several hundred extinct species drawn from various parts of the world at various times during the past 65 million years.

The methods we use to reconstruct phylogenies are based primarily on identifying groups of related species through similarities in their morphology and in the molecular sequences of their genetic material. However, rather than just looking at overall similarity, most biologists agree that organisms should be grouped together on the basis of shared specializations (or shared-derived features) that distinguish them from their ancestors (e.g. Tassy, 1996; Forey et al., 1992). For example, body hair is a shared specialization that unites humans, apes, monkeys, and cats as mammals and distinguishes them from other types of vertebrates, whereas the common possession of a tail by many monkeys, lizards, and crocodiles is an ancestral feature that is of no particular value in assessing the evolutionary relationships among these organisms, since their

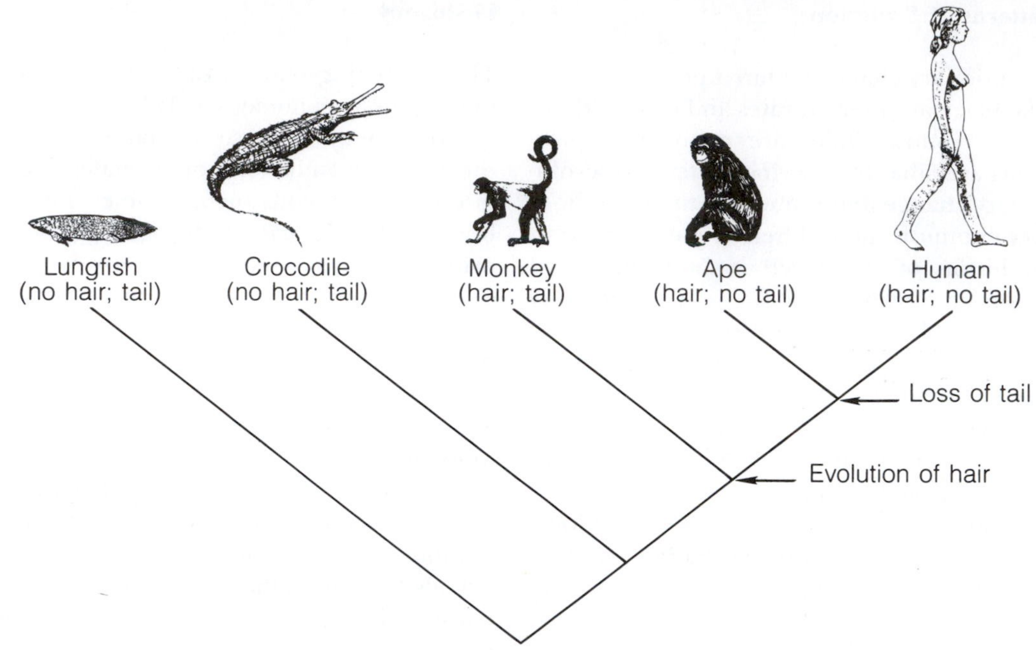

FIGURE 1.1 Shared specializations and ancestral features.

common ancestor had a tail. On the other hand, the absence of a tail in apes and humans represents a derived specialization that sets us apart from our ancestors (Fig. 1.1). The common possession of a group of specializations by a cluster of species or genera is interpreted as indicating that this cluster shares a unique heritage relative to other related species. Most current studies of primate phylogeny are based on extensive analysis of morphological features and/or amino acid sequences from parts of a species genome, using sophisticated computer programs to determine the pattern of shared derived similarities (Tassy, 1996; Kay et al., 1997).

Unfortunately, not all derived similarities among organisms are indicative of a unique heritage. Animals have frequently evolved morphological similarities independently. In addition to apes and humans, for example, a few monkey species and a few prosimians (as well as frogs) have lost all or most of their tail. The biologist's task in reconstructing phylogeny is to distinguish those specializations that are the result of a unique heritage from those that are not. The sameness of features that results from common ancestry is called **homology,** and identical features that are the result of a common ancestry are **homologous.** In contrast, the presence of similar features in different species that is not the result of a common inheritance is described as **homoplasy.**

Homoplasy is a very common phenomenon in evolution (Sanderson and Hufford, 1996). Most analyses of morphological evolution find that nearly half of all similarities are the result of homoplasy rather than common ancestry. There are several types of homoplasy and many factors that cause it to be such a common phenomenon. The evolution of superficially similar features in different lineages, such as the wings of bats and the wings of birds, is often called **convergence.** In contrast, **parallelism** refers to the independent evolution of identical features in closely related organisms, such as the independent evolution of white fur in many lineages of monkeys. **Reversal** refers to an evolutionary change in an organism that resembles the condition found in an earlier ancestor, such as the presence of hair on the fingers of humans descended from apes with hairless digits. In many cases homoplasy is the result of natural selection for similar functional adaptations. For example, larger animals may face similar mechanical problems. In addition, homoplasy may also reflect the fact that evolutionary pathways are constrained by development and available genetic potential. In molecular evolution, the potential for homoplasy is dictated by the limited number of amino acids making up the genetic code. Although frequently treated as an undesirable distraction in attempts to reconstruct phylogeny, homoplasy is an important evolutionary phenomenon that deserves greater study.

Taxonomy and Systematics

Taxonomy is a means of ordering our knowledge of biological diversity through a series of commonly accepted names for organisms. If scientists wish to communicate about animals and plants and to discuss their similarities and differences, they need a standard system of names both for individual types of organisms and for related groups of organisms. For example, the tufted capuchin monkey of South America, known to many people as the organ-grinder monkey, goes by over a dozen different names among the different tribal and ethnic groups of Suriname alone. To scientists around the world, however, this species is known by a single name, *Cebus apella.* The practice of assigning every biological species, living or fossil, a unique name composed of two Latin words was initiated by Carolus Linnaeus, a Swedish scientist of the eighteenth century whose system of biological nomenclature is universally followed today. Under the Linnean system, *Cebus* is the name of a **genus** (pl. *genera*), or group of animals, in this case all kinds of capuchin monkeys. (The name of a genus is always capitalized.) The word *apella,* the **species** name, refers to a particular type of capuchin monkey, the tufted capuchin monkey. (A species name always begins with a lowercase letter.) Each genus name must be unique, but species names need be unique only within a particular genus so that the combination of genus and species names is unique and refers to only one kind of organism. (The name is always written in italics—or underlined.) Somewhere in a museum there is a preserved skeleton (or skull, or skin) that has been designated as the **type specimen** for this species. The type specimen provides an objective reference for this species so that any scientist who thinks he or she may have discovered a different kind of monkey can examine the individual on which *Cebus apella* is based.

The Linnean system contains a hierarchy of levels for grouping organisms into larger and larger units (Table 1.1). Within the genus *Cebus,* for example, there are several species:

TABLE 1.1
A Classification of the Tufted Capuchin Monkey

Kingdom	Animal
Phylum	Chordata
Class	Mammalia (mammals)
Order	Primates (primates)
Suborder	Anthropoidea (higher primates)
Infraorder	Platyrrhini (New World monkeys)
Superfamily	Ceboidea (New World monkeys)
Family	Cebidae (capuchins, squirrel monkeys, and marmosets)*
Subfamily	Cebinae (capuchins and squirrel monkeys)*
Genus	*Cebus* (capuchins)
Species	*Cebus apella* (tufted capuchin monkey)

*Indicates only one of several common classifications (see Chapter 5).

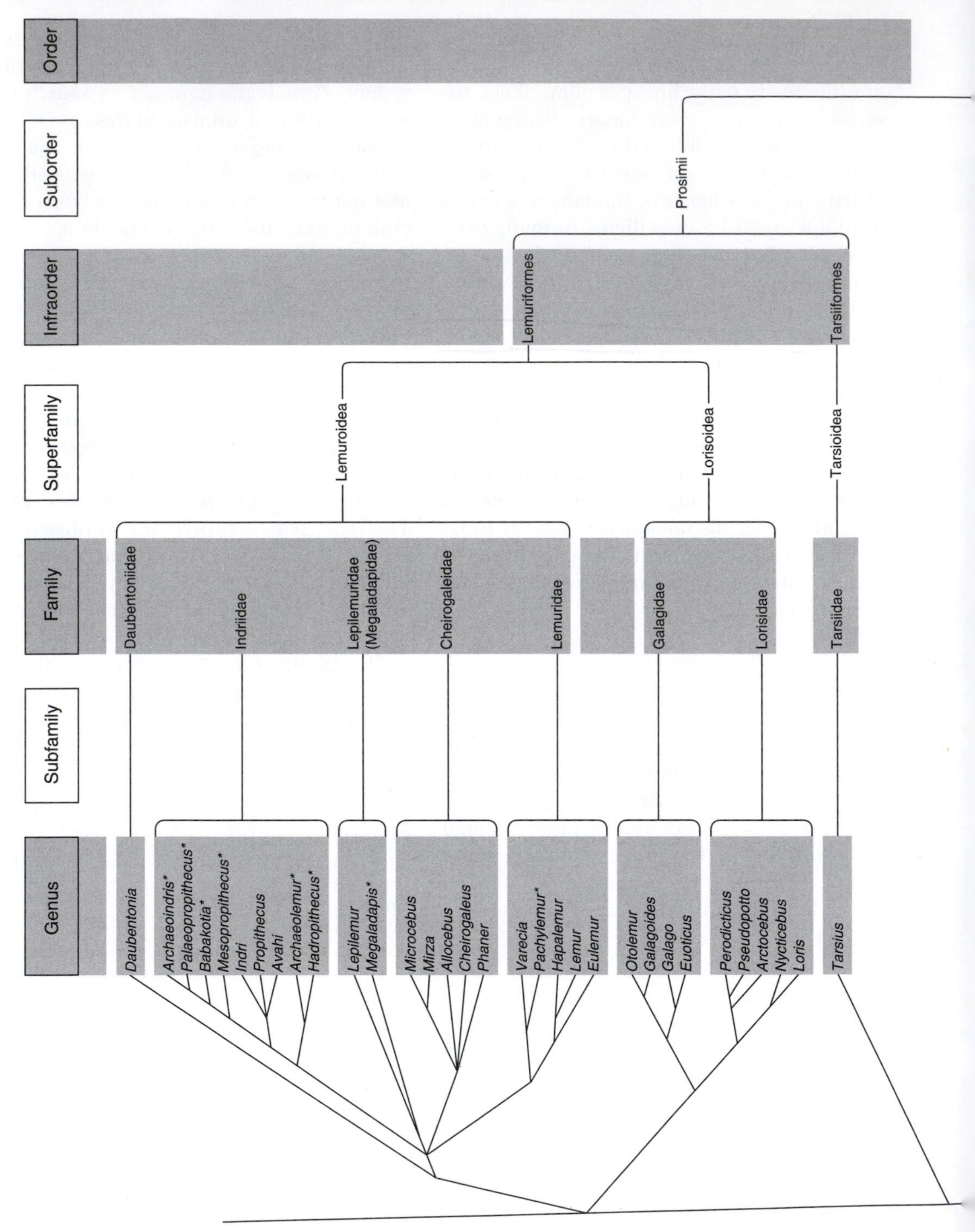

FIGURE 1.2 A classification of extant and subfossil (= recently extinct) (*) primate genera.

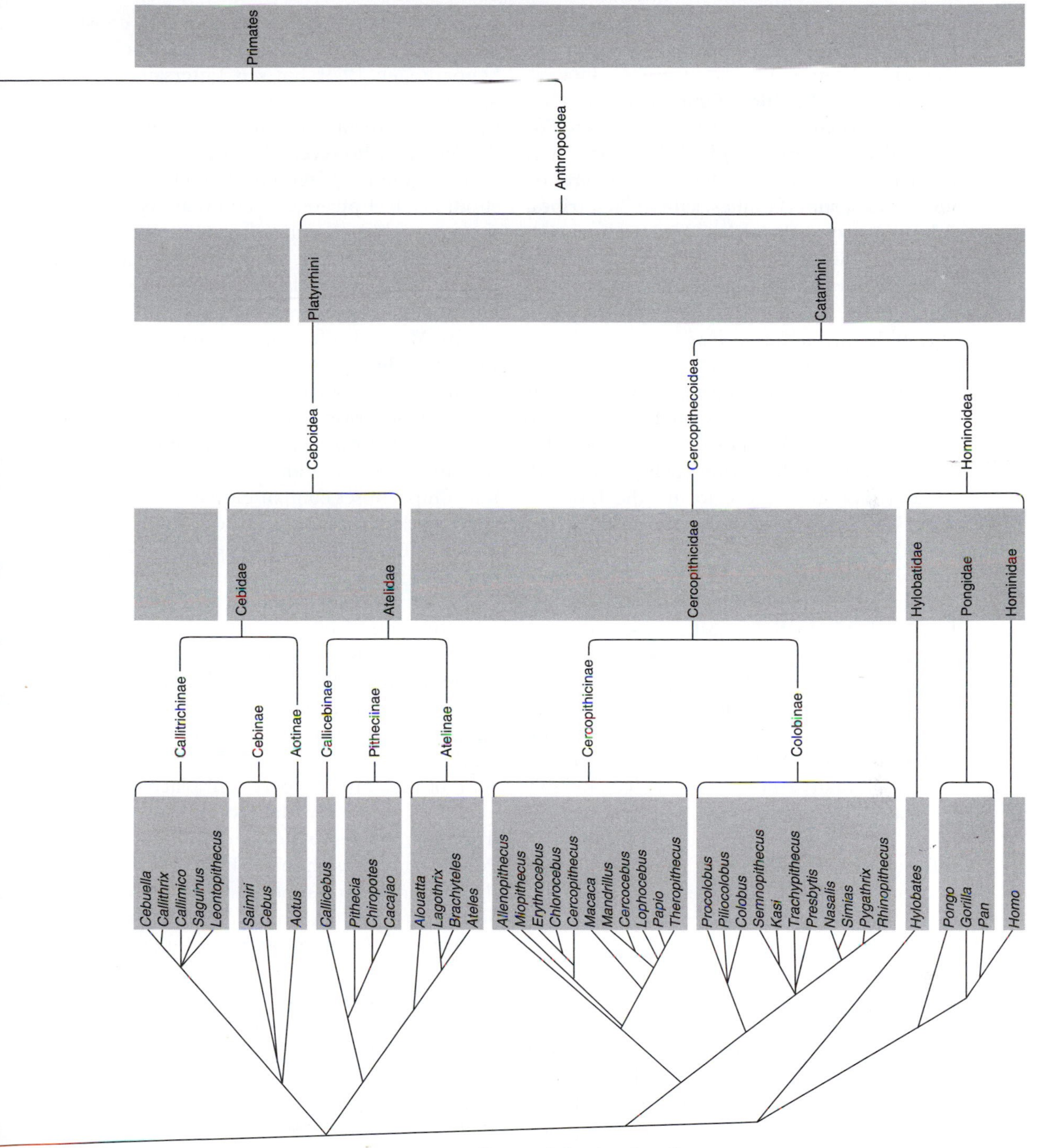

FIGURE 1.2 *(continued)*

Cebus apella, the tufted capuchin; *Cebus albifrons,* the white-fronted capuchin; *Cebus capucinus,* the capped capuchin; and others. Genera are grouped into **families,** families into **orders,** orders into **classes,** and classes into phyla. For particular lineages, these basic levels are often further subdivided or clustered into **suborders, infraorders, superfamilies, subfamilies, tribes, subgenera,** or **subspecies.** For convenience, names at different levels of the hierarchy are given distinctive endings. Family names usually end in *dae,* superfamily names in *oidea,* and subfamily names in *inae.*

In the science of classifying organisms, **systematics,** we attempt to apply the tidy Linnean system to the untidy, unlabeled world of animals. Figure 1.2, the classification used in this book, is the result of one such attempt. Although biologists agree to use the Linnean framework for naming organisms, they frequently disagree about the proper classification of particular creatures. They may disagree as to whether each of the gibbon types on different islands in Southeast Asia is a distinct species or only a subspecies of a single species. Some authorities may feel that gibbons and great apes should be placed in a single family, others that they should be placed in separate families. Once they have learned the Linnean hierarchy, many students are understandably frustrated and annoyed to find that textbooks often do not agree on the classification of different species. There are, however, usually good reasons for the disagreements about primate classification, as we see in the following chapters.

One reason for disagreement in primate classification is that the rules for distinguishing a genus, a family, or a superfamily are somewhat arbitrary. Scientists usually set their own standards. The only generally accepted rules are for species. However, as already discussed, there are many different ways proposed for distinguishing what a species is. Living species of mammals are remarkably consistent in their metric variability (Gingerich and Schoeninger, 1979; but see Tattersall, 1993), and we can use this standard to identify species in the fossil record. The limits for genera and families are, however, much more arbitrary.

It is generally agreed that classification should reflect phylogeny, and that taxonomic groups such as families, superfamilies, and suborders should be **monophyletic** groups; that is, that they should have a single common ancestor that gave rise to all members of the group. Many also feel that taxonomic groups should be **holophyletic** groups as well—they should contain all the descendants of their common ancestor, not just some of them. But it is often not practical or possible to achieve this unambiguously, and classifications are often compromises compatible with several possible phylogenies. In addition, many biologists feel that classification should reflect not only phylogeny but also major adaptive differences, even among closely related species. For example, most biologists now agree that humans are much more closely related to chimpanzees and gorillas than to orangutans. Thus a true phyletic taxonomy would group humans with the African apes in a single family and the orangutan in a separate family. In spite of this, many experts still place humans in a separate family, the Hominidae, and all living great apes in a common family, the Pongidae, because humans have departed further from the common ancestor of humans and great apes than have chimpanzees and gorillas. The family Pongidae is called a **paraphyletic** grouping because some of its members (chimpanzees and gorillas) are more closely related to a species (humans) placed in another family than they are to other members (orangutans) of their own family (Fig. 1.3). The taxonomy used in this book (Fig. 1.2) contains several such departures from a strictly phyletic classification. In all

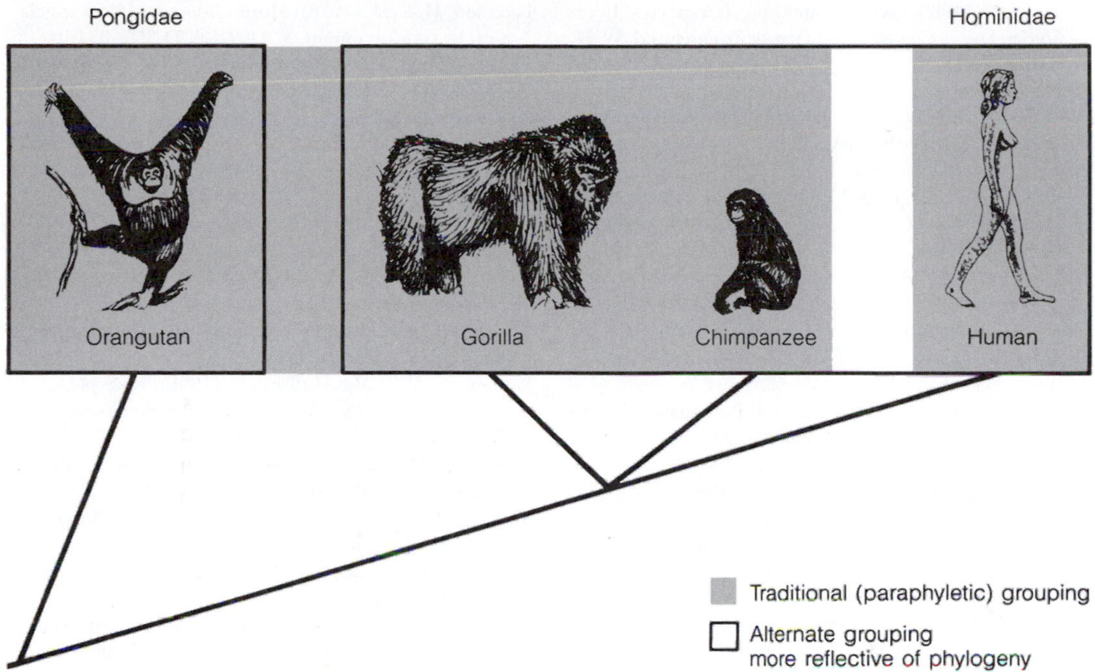

FIGURE 1.3 A strictly phyletic classification recognizes that humans, chimpanzees, and gorillas are more closely related to each other than any of them are to orangutans; the latter are thus grouped separately as the only pongids. A more traditional classification recognizes adaptive differences; in this case, chimpanzees and gorillas are classified with orangutans (pongids), and humans are grouped separately (hominids) because of the great degree of adaptation that distinguishes humans from even their closest primate relatives.

cases, the evolutionary relationships are discussed in the text.

BIBLIOGRAPHY

Carroll, R. L. (1997). *Patterns and Processes of Vertebrate Evolution.* New York: Cambridge University Press.

Cope, D. (1993). Measures of dental variation and indicators of multiple taxa in samples of sympatric Cercopithecus species. In *Species, Species Concepts, and Primate Evolution,* ed. W. H. Kimbel and L. B. Martin, pp. 211–238. New York: Plenum Press.

Cracraft, J. (1983). Species concepts and speciation analysis. *Curr. Ornithol.* **1**:159–187.

——— (1989). Species concepts and the ontology of Evolution. Biol. Philkos. 2:63–90.

Cracraft, J., and Eldridge, N., eds. (1979). *Phylogenetic Analysis and Paleontology.* New York: Columbia University Press.

Darwin, C. (1859). *On the Origin of Species by Means of Natural Selection, or the Preservation of Favoured Races in the Struggle for Life.* London: John Murray.

Dawkins, R. (1976). *The Selfish Gene.* Oxford: Oxford University Press.

——— (1987). *The Blind Watchmaker.* New York: W. W. Norton.

deQueiroz, K. (1998). The general lineage concept of species, specias criteria, and the process of speciation: A conceptual unification and terminological recommendations. In *Endless Forms: Species and Speciation.* Oxford: Oxford University Press.

Eggleton, P., and Vane-Wright, R. I. (1994). *Phylogenetics and Ecology.* Linnean Society Symposium Series No. 17. San Diego: Academic Press.

Eldridge, N. (1993). What, if anything, is a species? In *Species, Species Concepts, and Primate Evolution,* ed. W. H. Kimbel and L. B. Martin, pp. 3–20. New York: Plenum Press.

Eldredge, N., and Stanley, S. M., eds. (1984). *Living Fossils.* New York: Springer Verlag.

Forey, P. L., Humphries, C. J., Kitching, I. J., Scotland, R. W., Siebert, D. J., and Williams, D. M. (1992). *Cladistics: A Practical Course in Systematics.* Oxford: Oxford University Press.

Gingerich, P. D. (1993). Quantification and comparison of evolutionary rates. *Am. J. Science* **293A**:453–478.

Gingerich, P. D., and Schoeninger, M. J. (1979). Patterns of tooth size variability in the dentition of Primates. *Am. J. Phys. Anthropol.* **51**:457–566.

Gould, S. J., and Eldridge, N. (1993). Punctuated equilibrium comes of age. *Nature* **366**:223–227.

Grine, F. E., Martin, L. B., and Fleagle, J. G. (1987). *Primate Phylogeny.* San Diego: Academic Press.

Hall, B. K., ed. (1994). *Homology: The Hierarchical Basis of Comparative Biology.* San Diego: Academic Press.

Harvey, P. H., Leigh Brown, A. J., Maynard Smith, J., and Nee, S., eds. (1996). *New Uses for New Phylogenies.* New York: Oxford University Press.

Harvey, P. H., and Pagel, M. D. (1991). *The Comparative Method in Evolutionary Biology.* Oxford: Oxford University Press.

Kay, R. F., Ross, C., and Williams, B. A. (1997). Anthropoid origins. *Science* **275**:797–804.

Kimbel, W. H., and Martin, L. B. (eds.). (1993). *Species, Species Concepts, and Primate Evolution.* New York: Plenum Press.

Martins, E. P., ed. (1996). *Phylogenies and the Comparative Method in Animal Behavior.* New York: Oxford University Press.

Masters, J. C. (1993). Primates and paradigms: Problems with the identification of genetic species. In *Species, Species Concepts, and Primate Evolution,* ed. W. H. Kimbel and L. B. Martin, pp. 43–64. New York: Plenum Press.

Mayr, E. (1942). *Systematics and the Origin of Species.* New York: Columbia University Press.

McHenry, H. M. (1996). Homoplasy, clades, and hominid phylogeny. In *Contemporary Issues in Human Evolution,* ed. W. E. Meikle, F. C. Howell, and N. G. Jablonski, pp. 77–89. Wattis Symposium Series in Anthropology, Memoir 21. San Francisco: California Academy of Sciences.

Paterson, H. E. H. (1978). More evidence against speciation by reinforcement. *S. Afr. J. Sci.* **74**:369–371.

——— (1985). The recognition concept of species. In *Species and Speciation,* ed. E. S. Vrba. *Transvaal Museum Monogr.* **4**:21–29.

Purvis, A. (1995). A composite estimate of primate phylogeny. *Phil. Trans. R. Soc.* **B348**:405–421.

Rose, K. D., and Bown, T. M. (1993). Species concepts and species recognition on Eocene primates. In *Species, Species Concepts, and Primate Evolution,* ed. W. H. Kimbel and L. B. Martin, pp. 299–330. New York: Plenum Press.

Rose, M. R., and Lauder, G. V., eds. (1996). *Adaptation.* San Diego: Academic Press.

Sanderson, M. J., and Hufford, L. (1996). *Homoplasy: The Recurrence of Similarity in Evolution.* San Diego: Academic Press.

Schoch, R. M. (1986). *Phylogeny Reconstruction in Paleontology.* New York: Van Nostrand–Reinhold.

Simpson, G. G. (1953). *The Major Features of Evolution.* New York: Columbia University Press.

——— (1961). *Principles of Animal Taxonomy.* New York: Columbia University Press.

Smith, A. B. (ed.). (1994). *Systematics and the Fossil Record: Documenting Evolutionary Patterns.* Oxford: Blackwell Scientific Publications .

Tassy, P. (1996). Grades and clades: A paleontological perspective on phylogenetic issues. In *Contemporary Issues in Human Evolution,* ed. W. E. Meikle, F. C. Howell, and N. G. Jablonski, pp.55–76. Wattis Symposium Series in Anthropology, Memoir 21 San Francisco: California Academy of Sciences.

Tattersall, I. (1989). The roles of ecological and behavioral observation in species recognition among primates. *Hum. Evol.* **4**:117–124.

——— (1992). Species concepts and species identification in human evolution. *J. Hum. Evol.* **22**:341–349.

——— (1993). Speciation and morphological differentiation on the genus *Lemur.* In *Species, Species Concepts, and Primate Evolution,* W. H. Kimbel and L. B. Martin (eds.), pp.163–176. New York: Plenum Press.

Vermeij, G. J. (1978) *Biogeography and Adaptation: Patterns of Marine Life.* Cambridge, Mass.: Harvard University Press.

Williams, C. C. (1966). *Adaptation and Natural Selection: A Critique of Some Current Evolutionary Thought.* Princeton, N. J.: Princeton University Press.

TWO

The Primate Body

PRIMATE ANATOMY

Fossil and living primates are an extraordinarily diverse array of species. Some are among the most generalized and primitive of all mammals; others show morphological and behavioral specializations unmatched in any other mammalian order. This diversity in structure, behavior, and ecology is our topic of study in this book. The purpose of this chapter is to establish an anatomical frame of reference—a survey of features common to all (or almost all) primates. This chapter, then, provides pictures and descriptions of primate anatomy and preliminary indications of those anatomical features that have undergone the greatest changes in primate evolution.

Compared to most other mammals, we primates have retained relatively primitive bodies. Some of us are specialized in that we have lost our tails, and many have a relatively large brain. But no primates have departed so dramatically from the common mammalian body plan as bats, whose hands have become wings; as horses, whose fingers and toes have reduced to a single digit; or as baleen whales, who have lost their hindlimbs altogether, adapted their tails into flippers, and replaced their teeth with great, hairlike combs. The anatomical features that distinguish the bones and teeth of primates from those of many other mammals are the result of subtle changes in the shape and proportion of **homologous** elements rather than major rearrangements, losses, or additions of body parts. We generally find the same bones and teeth in all species of primates, with only minor differences, reflecting different diets or locomotor habits. The fact that humans are constructed of the same bony elements as other primates (and generally other mammals) is a major piece of evidence demonstrating our evolutionary origin.

Size

Size is a basic aspect of an organism's anatomy and plays a major role in its ecological adaptations. It is a feature that can be readily compared, both among living species and between living and fossil primates. Adult living primates range in size from mouse lemurs and pygmy marmosets, with a mass less than 100 g, to male gorillas, with a body mass of over 200 kg (Fig. 2.1). The fossil record provides evidence of a few extinct primates from early in the age of mammals that were much smaller (probably as small as 20 g) and at least one, *Gigantopithecus blacki* from the Pleistocene of China (see Chapter 16), that was much larger (probably over 300 kg). In their range of body sizes, primates are one

FIGURE 2.1 A mouse lemur (*Microcebus*) and a gorilla (*Gorilla*), the smallest and largest living primates.

of the more diverse orders of living mammals. As a group, however, primates are rather medium-size mammals (Fig. 2.2)—larger than most insectivores and rodents and smaller than most ungulates, elephants, and marine mammals.

Cranial Anatomy

The anatomy of the head, or cranial region, plays a particularly important role in studies of primate adaptation and evolution. Many of the anatomical features that have traditionally been used to delineate the systematic relationships among primates are cranial features, and most of our knowledge of fossil primates is based on this region.

Bones of the Skull

The adult primate skull (Figs. 2.3, 2.4) consists of many different bones that together form a hollow, bony shell that houses the brain and special sense organs and also provides a base for the teeth and chewing muscles. Only the lower jaw, or mandible, and the three bones of the middle ear are separate, movable elements; the others are fused into a single unit, the **cranium.** This unit can be roughly divided into two regions: a more posterior braincase, or **neurocranium,** and a more anterior facial region, or **splanchnocranium.**

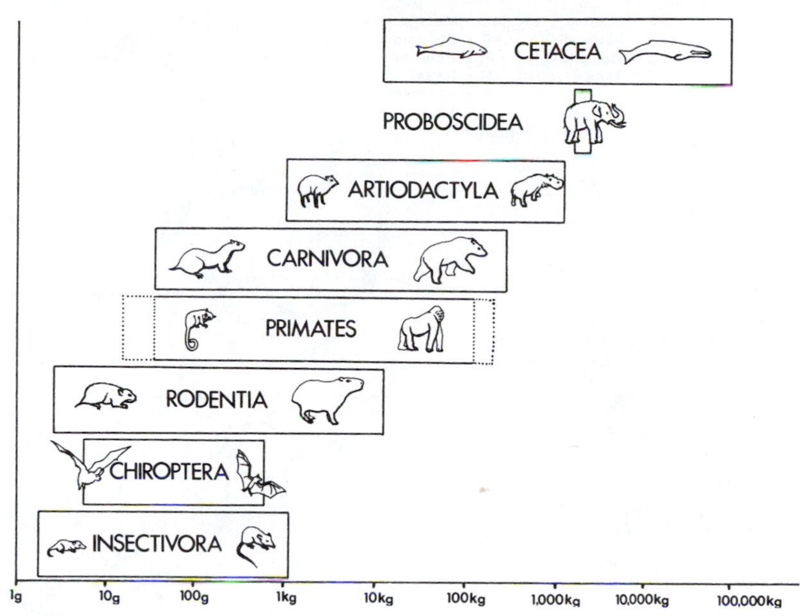

FIGURE 2.2 Size ranges for various orders of mammals, including primates. The solid lines include all living species; the dotted lines include all known living and fossil species.

The braincase serves as a protective bony case for the brain, a housing for the auditory region, and an area of muscle attachment for the larger chewing muscles and the muscles that move the head on the neck. Three paired flat bones—the **frontal, parietal,** and **temporal** bones—make up the top and sides of the braincase. (The temporal bone is a relatively complicated bone with several distinct parts.) The posterior and inferior surfaces of the braincase are formed by a single bone, the **occipital,** which also has a number of distinct parts. A complex, butterfly-shaped bone, the **sphenoid,** forms the anterior surface of the braincase and joins it with the facial region.

The facial region is formed by the **maxillary** and **premaxillary bones,** which contain the upper teeth; the **zygomatic bone,** which forms the lateral wall of the **orbit,** or eye socket; the **nasal bones,** which form the bridge of the nose; and numerous small bones that make up the orbit and the internal nasal region. The lower jaw, or **mandible,** contains the lower teeth. In many mammals, and in most prosimian primates, the two halves of the mandible are loosely connected anteriorly in such a way that they can move somewhat independently of one another. This joint is called the **mandibular symphysis.** In higher primates, including humans, the two sides of the lower jaw are fused to form a single bony unit.

Although all primate skulls are made up of these same components, they can have very different appearances, depending on the relative size and shape of individual bones (Figs. 2.3; 2.4). The skull functions as a base and structural framework for the first part of the digestive system and as a housing for the brain and special sense organs of sight, smell, and hearing. Much of the diversity in primate skull shape reflects the need for this single bony structure to serve numerous, often conflicting

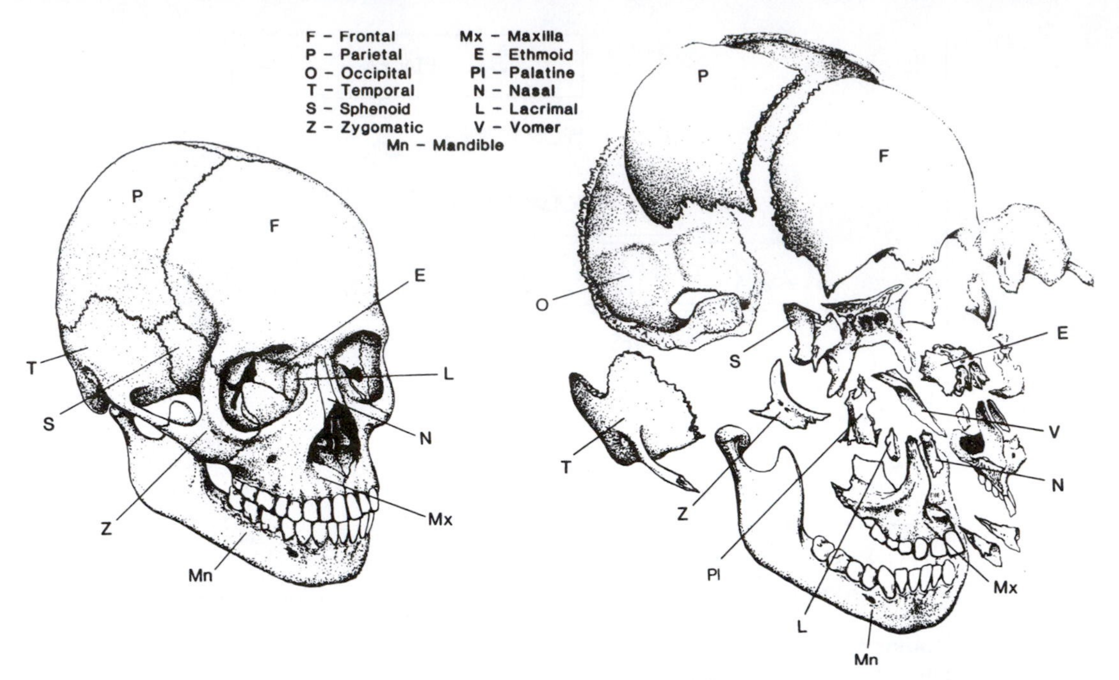

FIGURE 2.3 The human skull.

functions. For example, although the size of the orbits is most directly related to the size of the eyeball and to whether a species is active during the day or night, it influences the shape and position of the nasal cavity and the space available for chewing muscles.

Teeth and Chewing

Many parts of the head and face are important in the acquisition and initial preparation of food. The lips, cheeks, teeth, mandible, tongue, **hyoid bone** (a small bone suspended in the throat beneath the mandible), and muscles of the throat all participate in this complex activity, and many of these same parts also play a role in communication and sound production. The two parts of the skull that can be linked most clearly to dietary habits are the teeth and the chewing muscles that move the lower jaw.

Teeth, more than any other single part of the body, provide the basic information underlying much of our understanding of primate evolution. Because of their extreme hardness and compact shape, teeth are the most commonly preserved identifiable remains of most fossil mammals. But teeth are more than just plentiful; they are also complex organs that provide considerable information about both the phyletic relationships and the dietary habits of their owners. Because of the importance of teeth in evolutionary studies, there is an extensive but fairly simple terminology for dental anatomy.

All primates have teeth in both the upper jaw (**maxilla**) and the lower jaw (**mandible**), and, like most features of the primate skeleton, primate teeth are **bilaterally symmetrical**—the teeth on one side are mirror images of those on the other. Each primate jaw normally contains four types of teeth (Fig. 2.5). These

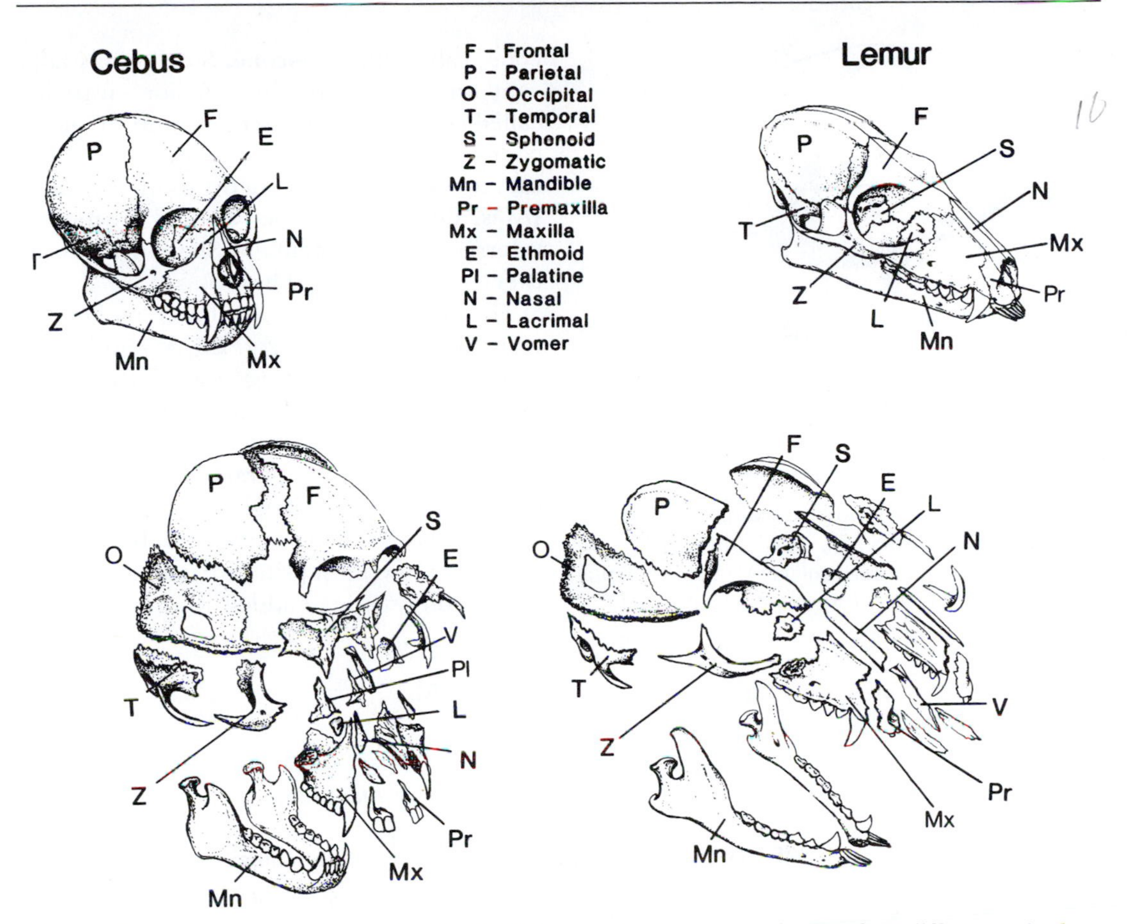

FIGURE 2.4 Skulls of a capuchin monkey (*Cebus*) and a lemur (*Lemur*) showing how differences in the size and shape of individual bones contribute to overall differences in skull form.

are, from front to back, **incisors, canines, premolars,** and **molars.** The number of teeth a particular species possesses is usually expressed in a **dental formula.** The human dental formula is 2.1.2.3./2.1.2.3., indicating that we normally have two incisors, one canine, two premolars, and three molars on each side of both the upper and the lower jaw for a total of thirty-two adult teeth. In most primate species, formulae for the upper and lower dentition are the same. In addition to **adult** (or **permanent**) **teeth,** primates have an earlier set of teeth, the **milk** (or **deciduous**) **dentition,** which precedes the adult incisors, canines, and premolars and occupies the same positions in the jaws. The human milk dentition, for example, contains two deciduous incisors, one deciduous canine, and two deciduous **premolars** (often called "milk molars") in each quadrant, for a total of twenty deciduous teeth.

The three main cusps of an upper molar (Fig. 2.6) are the **paracone,** the **metacone,** and the **protocone.** The triangle formed by these cusps is called the **trigon.** Many primates have evolved a fourth cusp distal to the proto-

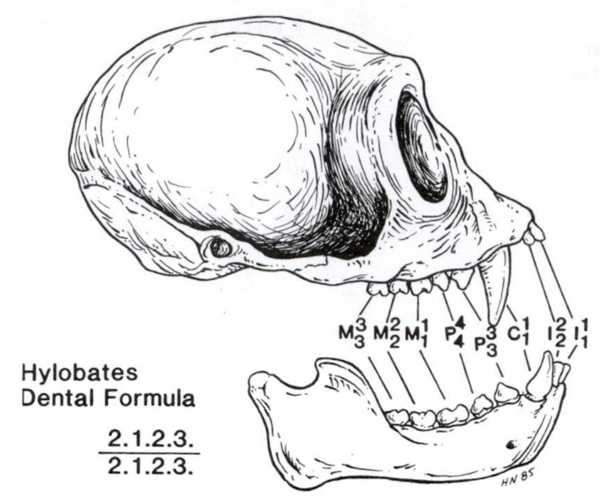

FIGURE 2.5 The dentition of a siamang (*Hylobates*) showing two incisors (I), one canine (C), two premolars (P), and three molars (M) in each dental quadrant for a dental formula of 2.1.2.3./2.1.2.3.

cone, called the **hypocone.** Small cusps adjacent and lingual to these major cusps are called **conules** (the **paraconule** and the **metaconule**). Accessory folds of enamel on the **buccal** (cheek-side) surface of the tooth are called **styles,** and an enamel belt around the tooth is referred to as a **cingulum.** Shallow areas between crests are called **basins.**

The basic structure of a lower molar in a generalized mammal (Fig. 2.6) is another triangle, this one pointing toward the cheek side. The cusps have the same names as those of the upper molars, but with the suffix *-id* added (**protoconid, paraconid,** and **metaconid**). This basic triangle in the front of a lower molar is called the **trigonid.** In primates and all but the most primitive mammals there is an additional area added to the distal end

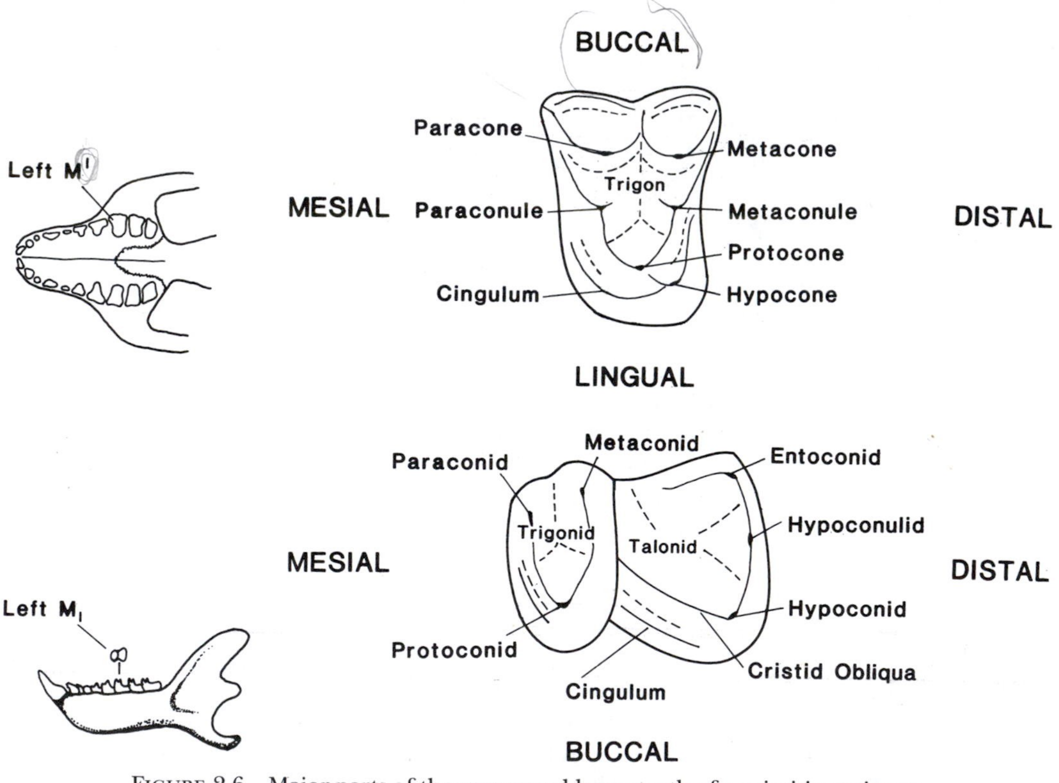

FIGURE 2.6 Major parts of the upper and lower teeth of a primitive primate.

of this primitive trigonid. This extra part, the **talonid,** is formed by two or three additional cusps: the **hypoconid** on the buccal side, the **entoconid** on the lingual (tongue) side, and, in many species, a small, distalmost cusp between these two, the **hypoconulid.**

Primate dentitions are involved in two different aspects of feeding. The anterior part of the dentition, the incisors and often the canines (together with the lips and often the hands), is primarily concerned with ingestion—the transfer of food from the outside world into the oral cavity in manageable pieces that can then be further prepared by the cheek teeth (the molars and premolars). The subsequent breakdown of food items by chewing is called mastication.

The molars and premolars of primates break down food mechanically in three ways: (a) by puncture-crushing or piercing the food with sharp cusps, (b) by shearing the food into small pieces, that is, by trapping particles between the blades of enamel that are formed by the crests that link cusps, and (c) by crushing or grinding food in mortar-and-pestle fashion between rounded cusps and flat basins. Different types of food require different types of dental preparation before swallowing, and it is possible to relate the various characteristics of both the anterior teeth (for obtaining and ingesting objects) and the cheek teeth (for puncturing, shearing, or grinding) to diets with different consistencies (as we discuss in Chapter 9).

The movement of the lower jaw relative to the skull in both ingestion and chewing (mastication) is brought about by four major chewing muscles that originate on the skull and insert on different parts of the lower jaw (Fig. 2.7). The largest is the **temporalis,** which has a fan-shaped origin on the side of the skull and inserts onto the **coronoid process** of the mandible. The second large muscle is the **masseter,** which originates from the **zygomatic arch** and inserts on the lateral surface of the mandible. Both of these muscles close the jaw when they contract. There are two smaller muscles on the inside of the jaw: the **medial** and **lateral pterygoids.** Much of the bony development of the primate skull seems to be related to the size and shape of these muscles and to the magnitude and direction of the forces generated in the skull during chewing. These muscular differences have in turn evolved to meet mechanical demands associated with dietary differences.

Tongue and Taste

In primates, as in all mammals, the tongue forms the floor of the oral cavity. Primate tongues vary considerably in shape (Fig. 2.8).

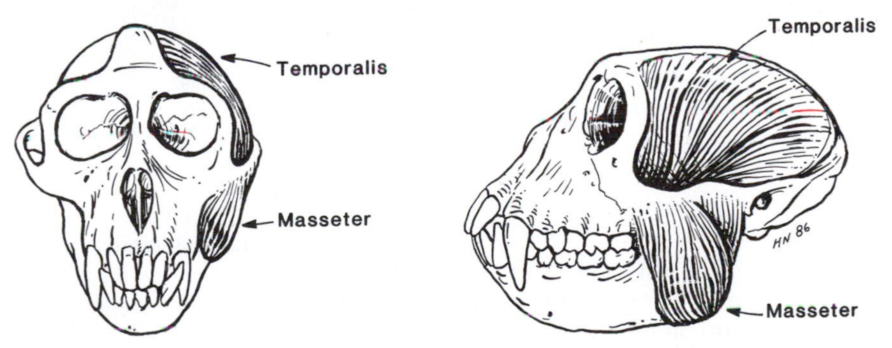

FIGURE 2.7 Anterior and lateral views of a primate skull showing the major chewing muscles.

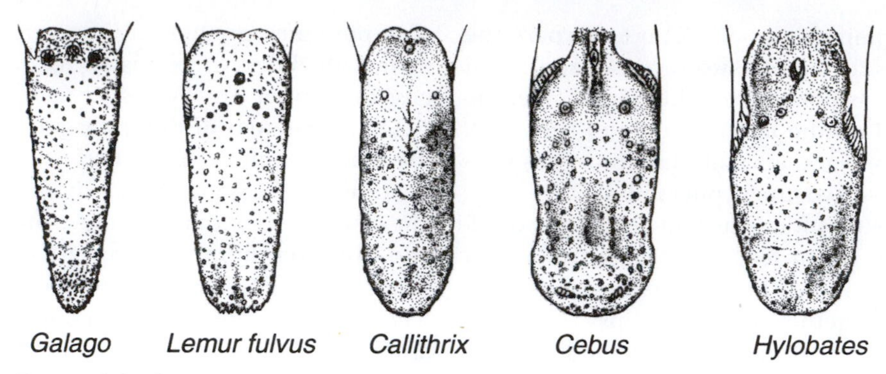

FIGURE 2.8 Superior view of the tongues of five primates showing differences in overall shape and in the distribution of taste papillae.

Part of this variation reflects differences in the shape of the oral cavity itself and is determined by the relative breadth of the anterior dentition. However, part of this variation is related to differences in the function of the tongue as an organ of taste, touch, and vocalization. For example, in all strepsirhines, there is a distinctive projection from the underside of the tongue, the **lingula,** that serves to clean the tooth comb. The brown lemur, *Lemur fulvus,* has a series of highly sensitive conical structures at the tip of its tongue, and feathery projections have been described on the tip of the tongue of *Lemur rubriventor* that seem associated with its habit of licking nectar from flowers (Hofer, 1981; Overdorff, 1992), and the fork-marked lemur, *Phaner,* has a long, muscular tongue that it uses while feeding on exudates (gums and resins). Among higher primates, humans have a very unusual tongue, in that it extends very far posteriorly and plays a major role in vocalization by changing the shape of the throat. Although all primate tongues contain the taste buds (**papillae**) that are responsible for the sense of taste, the size, shape, and number of different taste sensors vary considerably from species to species (Fig. 2.8 ; see also Hladik and Simmen, 1997).

Muscles of Facial Expression

Another aspect of cranial anatomy in primates that deserves special consideration is facial musculature (Fig. 2.9). Among primates, and especially in humans, the muscles of facial expression are more highly developed and differentiated into separate units than among any other groups of mammals (Huber, 1931).

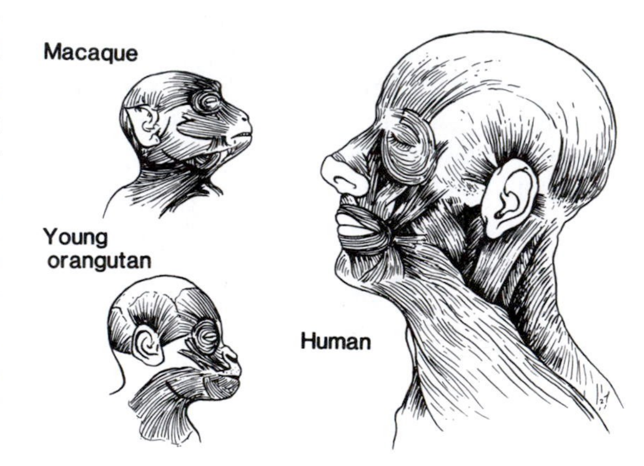

FIGURE 2.9 Muscles of facial expression in a macaque (*Macaca*), a young orangutan (*Pongo*), and a human (*Homo sapiens*). Note the increasing differentiation of individual muscles, which enables finer control of expressions.

It is these muscles that make possible the range of visual expressions that characterizes and facilitates the complex social behavior of primates.

The Brain and Senses

The structural shape of the skull—the development of bony buttresses and crests as well as the relative positioning of the face and the neurocranium—seems to be greatly influenced by the size and functional requirements of the masticatory system. However, the relative sizes of many parts of the skull, such as the neurocranium and the orbits, as well as the size and position of various openings in the skull, seem more directly related to the skull's role in housing the brain and the sense organs responsible for smell, vision, and hearing.

The Brain

The brain is the largest organ in the head, and its relative size is an important determinant of skull shape among primates. Relative to body weight, primates have the largest brains of any terrestrial mammals; only marine mammals are comparably brainy. There are, however, differences in relative brain size among primates. Lemurs, lorises, and *Tarsius* all have relatively smaller brains, compared to monkeys and apes, and human brains are relatively enormous. Still, the brain is a complex organ with many parts, and although some parts of primate brains are relatively large by mammalian standards, others are relatively small. In gross morphology, a primate brain can be divided into three parts (Fig. 2.10)—the **brainstem,** the **cerebellum,** and the **cerebrum.** Each part has very different functions, and each, in turn, is made up of many different functionally distinct sections.

The brainstem forms the lower surface and base of the brain. It is an enlarged and modified continuation of the upper part of the spinal cord and is the part of the primate brain that differs least from that found among other mammals and lower vertebrates. The brainstem is concerned with basic physiological functions such as reflexes, control of heartbeat and respiration, and temperature regulation, as well as the integration of sensory input before it is relayed to "higher centers" in the cerebrum. Many of the cranial nerves, which are responsible for innervation of such things as the organs of sight and hearing and the muscles of the orbit and face, arise from the brainstem. Very little of the primate brainstem is visible in either a lateral or superior view; it is covered by two areas that have become large and specialized: the cerebellum and the cerebral hemispheres.

The **cerebellum,** which lies between the brainstem and the posterior part of the cerebrum, is a developmental outgrowth of the caudal region of the brainstem. It is concerned primarily with control of movement and with motor coordination. Compared with other mammals, primates have a relatively large cerebellum.

The paired **cerebral hemispheres** are the part of the brain that has undergone the greatest change during primate evolution. It is in this part that we find the greatest differences between primates and other mammals and the greatest differences among living primates (Fig. 2.10). Gigantic cerebral hemispheres are one of the hallmarks of human evolution. Anatomically, this part of the brain is divided into **lobes** named for the bones immediately overlying them—**frontal, parietal, temporal,** and **occipital.** In most primates the surface of the cerebral hemispheres is cov-

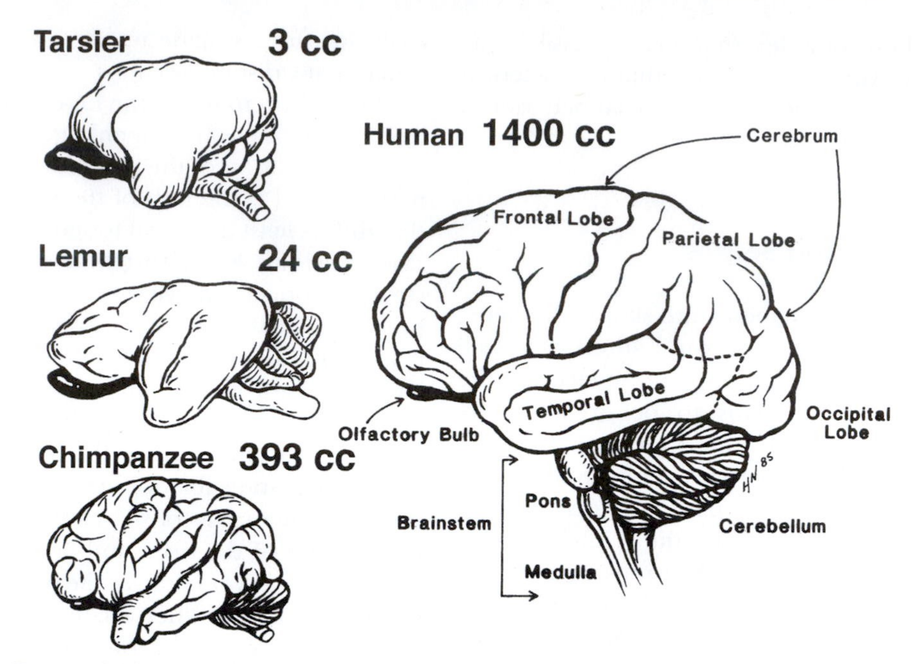

FIGURE 2.10 Lateral view of the brains of a lemur (*Lemur*), a tarsier (*Tarsius*), a chimpanzee (*Pan*), and a human (*Homo sapiens*), showing differences in relative size of the parts of the brain. Note especially the differences in size of the olfactory bulb and the size and development of convolutions on the cerebral hemispheres.

ered with convolutions made up of characteristic folds, or **gyri,** which are separated by grooves, or **sulci.** The development of these convolutions is most apparent in larger species and reflects the fact that the most functionally significant part of the cerebrum, the **gray matter,** lies at the surface. The convolutions or foldings of the brain surface provide a greater increase in the surface area of the cerebral hemispheres relative to brain or body volume than would be provided by a smooth spherical surface.

Overall, the cerebral hemispheres are involved with recognition of sensations, with voluntary movements, and with mental functions such as memory, thought, and interpretation. Different regions of the cerebrum (i.e., specific gyri) can be related to particular functions (Fig. 2.11). The central sulcus, for example, separates an anterior gyrus related to voluntary movement from a more posterior gyrus concerned with sensation. Within each of these areas, it is possible to identify more specific regions concerned with voluntary movement or sensation of particular parts of the body. In addition, there are other parts of the cerebral hemispheres, called **association areas,** which are related to the integration of input from several different senses (such as hearing and vision) and to specific tasks, such as language and speech. Two particularly well-developed association areas in the human brain are those related to language, **Broca's area** in the frontal lobe

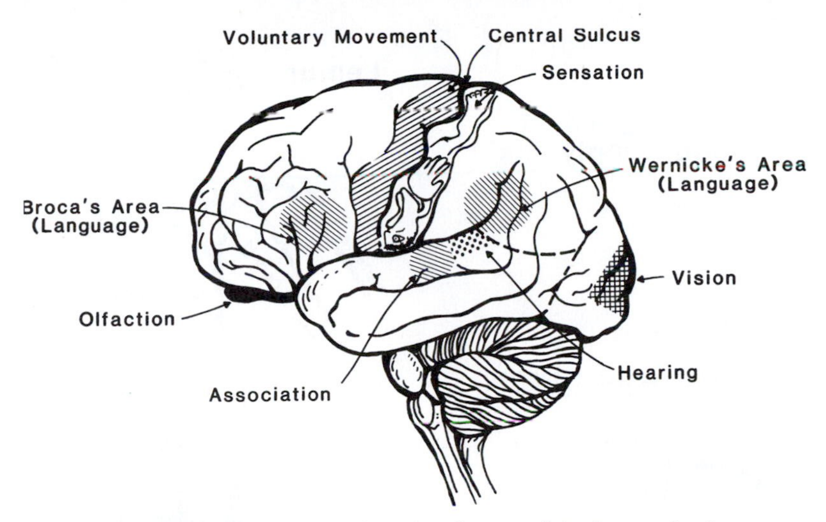

FIGURE 2.11 Important functional areas of the human brain.

and **Wernike's area** in the parietal and temporal lobes.

Although the brain is a soft structure, primate brains often leave their mark on the bony morphology of the skull. Size (in particular, volume) is an obvious feature of a primate brain that can be determined from a skull. Furthermore, in many species, sulci and gyri also leave impressions on the internal surface of the cranium. Such impressions on fossil skulls can provide limited information about the development of different functional regions on the cerebral hemispheres of extinct primates.

The nerves that take signals to and from the brain enter and leave the cranial cavity through various holes, called **foramina,** in the skull bones. The largest of these holes is the **foramen magnum,** through which the spinal cord passes. The many smaller foramina vary considerably in size and position among living primates and are widely used in systematics. In a few cases it seems possible to correlate the size of a foramen carrying a specific nerve to the development of a particular function or anatomical region.

Cranial Blood Supply

Foramina also serve as passages for the arteries that supply blood to the brain and other cranial structures and for the veins that drain those same structures. The pathway of the blood supply to the brain shows a number of distinctly different patterns among living primates (Fig. 2.12). Although we know little about the functional significance of these differences, they have proved useful in sorting the phyletic relationships among many living and fossil primate species. The major blood supply to the head in primates comes from two branches of the common carotid artery at the base of the neck. The **external carotid** is responsible primarily for supplying structures in the neck and face; the **internal carotid** (along with the vertebral arteries) supplies the brain and the eye. The internal carotid artery enters the cranial cavity as two distinct arteries, a **stapedial artery** passing through the stapes bone and a **promontory artery** that generally lies medial to the stapedial artery and crosses the **promontorium,** a raised sur-

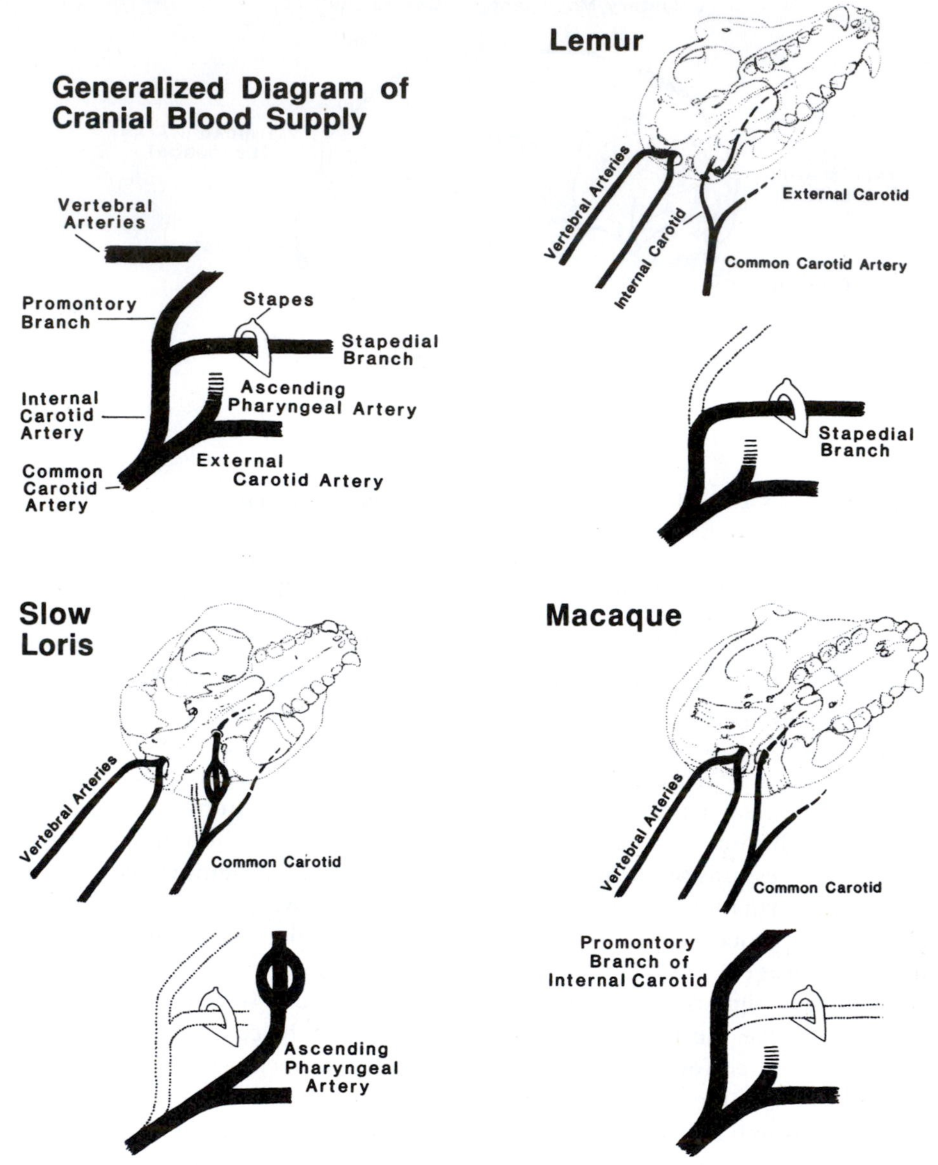

FIGURE 2.12 Cranial blood supply in several types of living primates. In all living primates, the vertebral arteries supply blood to the brain; however, species differ considerably in the relative contributions of the stapedial and promontory branches of the internal carotid artery and of the ascending pharyngeal branch of the external carotid artery. In the lemur (*Lemur*), the stapedial branch provides the major arterial supply to the anterior part of the brain; in a slow loris (*Nycticebus*), the blood to the front of the brain comes from a large ascending pharyngeal artery; in a macaque (*Macaca*) and all higher primates, the promontory branch of the internal carotid provides the major arterial blood supply to the anterior part of the brain.

face in the middle ear, to enter the cranial cavity farther anteriorly. In most lemurs, for example, the stapedial is the larger artery; in tarsiers, New World monkeys, Old World monkeys, apes, and humans, the stapedial is generally absent in adults and promontory provides most of the blood supply to the brain. In lorises, galagos, and cheirogaleids, a branch of the external carotid artery, the **ascending pharyngeal,** provides the major blood supply to the brain (Fig. 2.12).

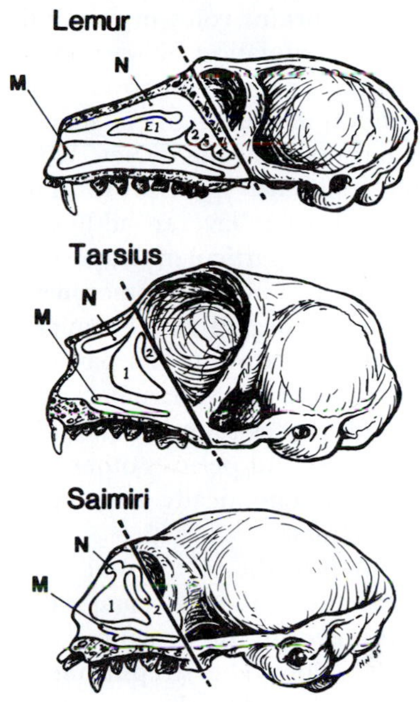

FIGURE 2.13 Structure of the interior nasal region of a lemur (*Lemur*), a tarsier (*Tarsius*), and a squirrel monkey (*Saimiri*). Note the reduction in number and relative size of the turbinates in *Tarsius* and *Saimiri*. M, maxilloturbinate; n, nasoturbinate; E, ethmoturbinates (numbered).

Olfaction

In many mammals, smell is the dominant sensory mode. It provides much of the information on which animals rely to find their way around, locate their food, locate potential predators, communicate with their kin and neighbors, and determine the sexual status of potential mates. Among more **diurnal** (active during the day) higher primates, smell seems to be less important for some of these functions than are other senses, such as vision. But even for these species this most basic of senses has not been abandoned. It still plays an important, but relatively poorly understood, role in reproduction, communication, and food evaluation in most primate species.

The sensation of smell is carried by the **olfactory nerves,** which end in paired swellings, the **olfactory bulbs** that lie under the large frontal lobes of most primates (Fig. 2.10). The olfactory nerves receive their input from the special sensory membranes lining the scroll-like **turbinates** of the internal nasal cavity. The development of the nasal part of the olfactory system and its position with respect to the orbits shows two distinctly different arrangements among primates (Fig. 2.13). In lemurs and lorises, as well as in most other mammals, the nerves responsible for olfaction pass between the orbits from the internal cavity to the brain. Within the nasal cavity, large numbers of turbinates are attached to several different bones, including several derived from the ethmoid bone that lies in a special cul-de-sac, the **sphenoid recess.** In tarsiers, monkeys, apes, and humans, the structure of this region is greatly simplified. The olfactory nerves pass over the interorbital septum, rather than between the orbits, and the sphenoid recess and posteriormost two turbinates are missing or greatly reduced (Maier, 1993). In apes and humans this region is even further reduced.

Although primate noses and the tissue-lined passages that make up their internal structure are associated primarily with olfaction, they also play important roles in respiration and temperature regulation by warming and humidifying the air that passes over them.

In addition to their sense of smell, lemurs, lorises, tarsiers, and many New World monkeys (but apparently not Old World monkeys, apes, or humans) have an additional sense that seems to be particularly important in sexual communication. The **vomeronasal organ** (or **Jacobson's organ**) is a chemical-sensing organ that lies in the anterior part of the roof of the mouth in many mammals. It is stimulated by substances found in the urine of female primates, and permits other individuals to determine chemically the reproductive status of a female.

In addition to these "internal" differences in nasal structure, there is a major dichotomy among primates in the structure of the external nasal region. The strepsirhine primates (lemurs, lorises, and galagos) are so named because of the median cleft and moist region that extends from the base of the nasal opening to the inside of the upper lip (Fig. 2.14), as in many mammals, such as dogs and cats. It has been generally argued that the strepsirhine condition is related to the function of the Jacobson's organ. In contrast, tarsiers and anthropoids, the haplorhines, have a continuous upper lip and often a hairy region between the lip and the base of the nasal opening (Fig. 2.14). Concordantly, the importance of the Jacobson's organ seems reduced, but not absent, in most higher primates.

Vision

Primates rely extensively on vision to understand the world around them. Nevertheless, there are considerable differences among primate species in many aspects of their visual systems, both in the bony structure of the orbit and in the soft anatomy of the eye and the parts of the brain related to sight. Primate eyes vary strikingly in relative size. Nocturnal (active during the night) species have relatively larger eyes and bony orbits than do diurnal species.

In addition to their differences in size, the bony orbits of primate skulls show important differences in construction (Fig. 2.15). In most mammals, and among the primitive, primate like plesiadapiforms, each eye lies nestled in a pocket of tough but flexible connective tissue on the side of the skull, medial to the zygo-

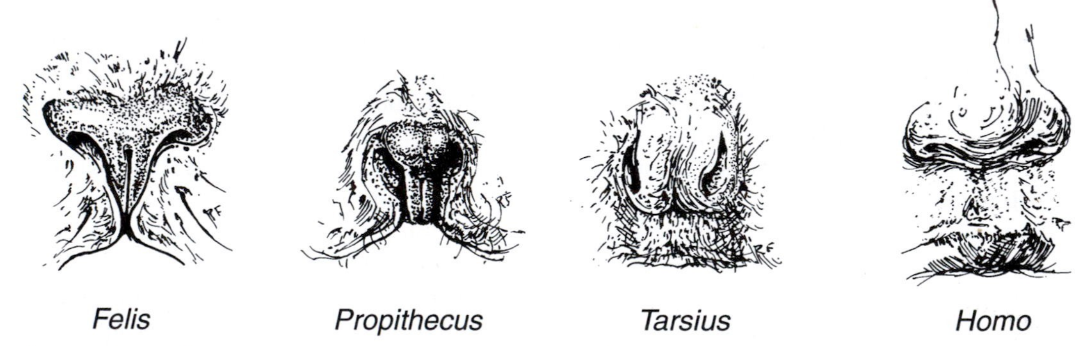

FIGURE 2.14 The nostrils and upper lip of two strepsirhine mammals—a cat and a lemur—and two haplorhine primates—a tarsier and a person. Note the midline cleft in the lip and nose of the cat and the lemur.

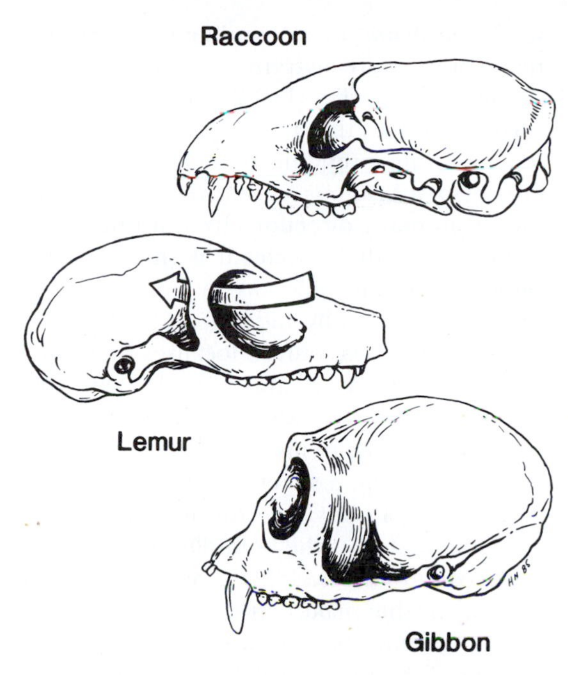

FIGURE 2.15 The bony structure of the orbit in a raccoon, a lemur, and a gibbon. In the raccoon, the orbit is open laterally. In the lemur, the orbit is surrounded by a bony ring but is open posteriorly. In the gibbon, the posterior opening of the orbit is closed off so that the eye is surrounded by a bony cup.

matic bone. The lateral side of the orbit is formed by a fibrous ligament rather than by bone. In all living primates, however, the zygomatic bone and the frontal bones join to form a lateral strut, or **postorbital bar,** so that the eye is surrounded by a complete bony ring. In higher primates, and to a lesser extent in *Tarsius,* the orbit is further walled off behind by a bony partition, the **postorbital plate**; thus the eyeball lies within a bony cup. This condition is described as **postorbital closure.** The functional significance of the postorbital bar and postorbital plate is regularly debated, with no clear solution. The most important function seems to be in isolating the orbital contents from the effects of chewing muscles. In addition to these major differences in the mechanical structure of primate orbits, there is considerable variation among primate species in the arrangement of the mosaic of small bones forming the medial wall of the orbit and in the size of the eyeball relative to the size of the bony orbit.

The overall structure of most primate eyeballs is similar; the main differences lie in the structure of the **retina,** the filmlike sheet of light-sensitive cells that lines the back of the eye. Two types of cells make up the retina in most primates: **rods,** which are very sensitive to light but do not distinguish color, and **cones,** which are sensitive to color. In many nocturnal primates, the retina is composed predominantly of rods. Furthermore, in lemurs and lorises we find an additional feature characteristic of many nocturnal mammals: the retina contains an extra layer that reflects light. This layer, the **tapetum lucidum,** seems to reduce visual acuity but enhances an animal's ability to see at night by "recycling" incoming light. In tarsiers, monkeys, apes, and humans (all of which lack a tapetum lucidum), we find a different modification of the retina—a specialized area of the retina, called the **fovea,** in which the light-sensitive cells are packed extremely close together, allowing very good visual acuity.

While most diurnal primates have color vision, there is diversity in the color vision abilities found among different species and among individuals within species (Jacobs, 1995). All Old World monkeys and apes that have been tested have color vision similar to that found in most humans, with three types of cones, each sensitive to a different part of the visual spectrum. However, New World monkeys are much more diverse; some species and individuals have three types of cones, whereas others have only two. The nocturnal owl monkey, *Aotus,* has only a single type of cone, as does the greater bushbaby, *Otolemur.* The significance of this diversity in color vision remains to be explained.

Hearing

Hearing plays an important role in many aspects of primate life. Many species, especially those active at night, use hearing to locate insect prey, and most use their ears to listen for approaching predators and to receive the vocal signals emitted by their family and neighbors. Although we know much about the anatomy of the auditory system, the physiological significance of many anatomical differences among primate ear regions is poorly understood.

Anatomically, the primate ear can be divided into three parts—the **outer ear,** the **middle ear,** and the **inner ear** (Fig. 2.16). The outer ear is composed of the **external ear,** or **pinna,** and a tube leading from that structure to the **eardrum,** or **tympanic membrane.** Primate pinnae are extremely variable in size, shape, and mobility (Fig. 2.17). In many nocturnal primates that rely extensively on hearing to locate prey, the outer ear is often a large, membranous structure that can be moved in many directions by a distinct set of muscles. In other species it is smaller, often only slightly movable (as in humans), and may even be totally hidden under fur. The outer ear collects sounds, localizes them with respect to direction, and funnels them into the auditory canal, where they set the tympanic membrane in motion.

The eardrum, a sheet of connective tissue spread over a bony ring formed by the tympanic bone, forms the boundary between the outer ear and the middle ear and changes the moving air that makes up sound into mechanical movements that are passed along the three ossicles of the middle ear (the **malleus, incus,** and **stapes**). The last of these, the stapes, transfers this motion to the fluid-filled inner ear.

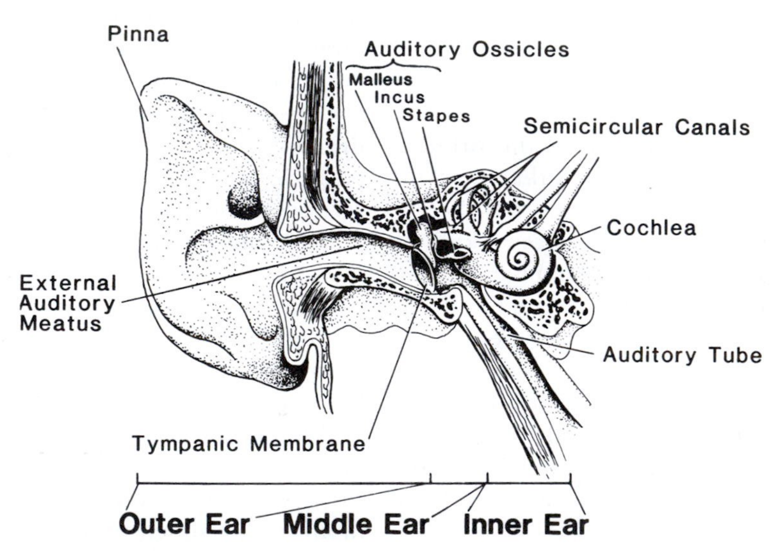

FIGURE 2.16 A primate ear, showing the three major parts and the individual elements in each part.

FIGURE 2.17 External ears (pinnae) of several primates.

The inner ear contains three functionally different parts within the petrous portion of the temporal bone. The part concerned with hearing is the **cochlea,** a coiled, snail-shaped bony tube. Within the fluid-filled cochlea is a pressure-sensitive organ that registers the movement of the fluid and sends impulses to the brain through the acoustic nerve. The other two parts of the inner ear, three **semicircular canals** and two other fluid-filled chambers (the **utricle** and **saccule**), are responsible for sensing movement and for orientation with respect to gravity.

Apart from differences in relative sensitivity to particular frequencies, all primate ears seem to function in much the same way. There are, however, considerable architectural differences in the way the bony housing of the middle ear is constructed (Fig. 2.18). In all living primates, the inferior surface of the middle ear is covered by a thin sheet of bone, called the **auditory bulla,** derived from the petrous part of the temporal bone. In some primates, this bulla is inflated, or balloonlike, and is often divided into many compartments; in others, it is flatter. The physiological significance of the different types of ear architecture is poorly understood. The inflated auditory bullae of many small nocturnal primates seem to increase perception of low-frequency sounds and may be associated with nocturnal predation on flying insects.

The spatial relationship between the tympanic ring and the auditory bulla differs considerably among major groups of living primates (Fig. 2.18). In lemurs the ring lies within the cavity formed by the bulla, in lorises the ring is attached to the inside wall of the bulla, and in New World monkeys it is attached to the outside wall of the bulla. In tarsiers and catarrhines the ring is also attached to the wall of the bulla, but it extends laterally to form a bony tube, the **external auditory meatus.**

The Trunk and Limbs

Whereas the skull is concerned primarily with sensing the environment, with communication, and with the ingestion and preparation of food, the part of the skeleton behind the skull, the **postcranial skeleton,** as it is often called, serves quite different functions. Obviously it provides support and protection for the organs of the trunk, but its primary functions and those that seem to account best for the major differences in skeletal shape are related to locomotion. In this capacity, the postcranial skeleton provides both a structural support and a series of attachments and levers to aid in movement. The primate postcranial skeleton (Figs. 2.19; 2.20) is relatively generalized by mammalian standards. Primates have

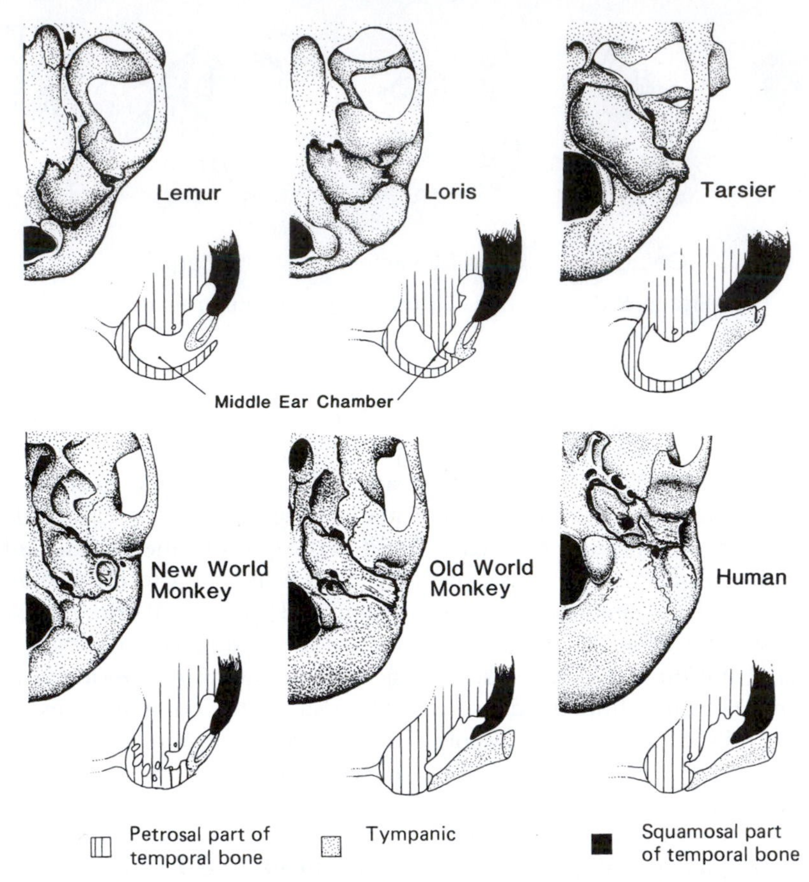

FIGURE 2.18 The structure of the tympanic (ectotympanic) bone surrounding the eardrum and its position in relation to the bones surrounding the middle ear cavity vary considerable among living primate species (inferior view above, cross-sectional view of the middle ear below). In a lemur (*Lemur*), the tympanic bone is ring-shaped and is suspended within the bony bullar cavity. In lorises, the tympanic bone lies at the edge of the middle ear cavity and is connected to the wall of the bulla; note, also, that the bulla cavity is divided. In tarsiers (*Tarsius*), the tympanic bone is elongated to form a bony tube at the lateral edge of the bullar cavity. In New World monkeys, the tympanic bone is a ringlike structure fused against the lateral wall of the bulla. In catarrhines, represented by an Old World monkey and a human, the tympanic bone is extended laterally to form a bony tube.

retained many bones from their early mammalian ancestors that other mammals have lost. For example, most primates have a primitive limb structure, with one bone in the upper (or **proximal**) part of each limb (the humerus or femur), a pair of bones in the lower (**distal**) part (the radius and ulna or tibia and fibula), and five digits on the hands

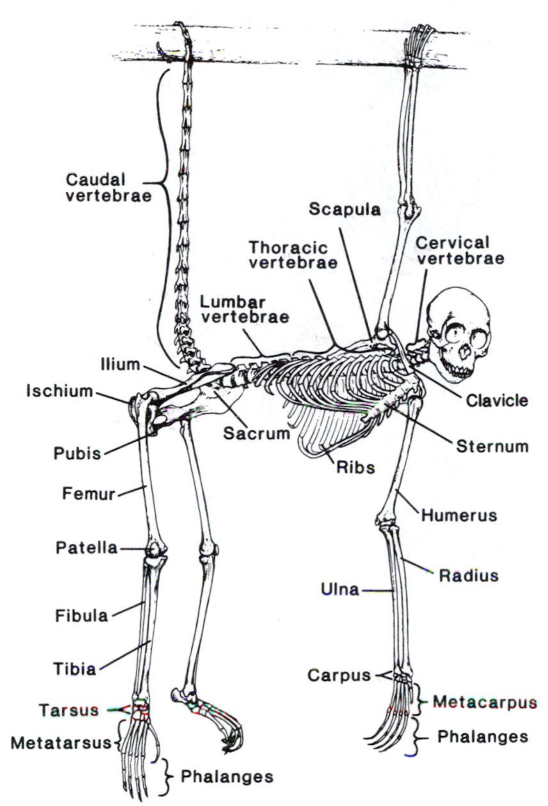

FIGURE 2.19 The skeleton of a spider monkey (*Ateles*). This species is unusual in having a very small thumb and a prehensile tail.

and feet. Primate skeletons can be divided into three parts: the **axial skeleton** (the backbone and ribs), the **forelimbs,** and the **hindlimbs.** To facilitate descriptions of anatomical features, we use a standard terminology for directions with respect to an animal's body (Fig. 2.21).

Axial Skeleton

The backbone, which is made up of individual bones called **vertebrae,** is divided into four regions (Fig. 2.19). The **cervical,** or neck region, contains seven vertebrae in almost all mammals. The first two vertebrae, the **atlas** and the **axis,** are specialized in shape and serve as a support and pivot for the skull. The second region of the backbone is the **thorax.** Most of the rotational movements of the trunk involve movements between thoracic vertebrae. Primates have between nine and thirteen thoracic vertebrae, each of which is attached to a rib. The ribs are connected anteriorly with the sternum to enclose the thoracic cage, within which lie the heart and lungs. The outside of the thorax is covered by the muscles of the upper limbs. The thoracic vertebrae are followed by the lumbar vertebrae. There are no ribs attached to the lumbar vertebrae, but there are very large **transverse** processes for the attachment of the large back muscles that extend the back. In most primates, these transverse processes arise from the body of the vertebrae; however, in apes, humans, and some fossil lemurs, they arise from parts of the vertebral arch. Most of the flexion and extension of the back takes place in the lumbar region, and the length of this part of the back varies considerably in association with differences in locomotion and posture, as discussed in later chapters. The next lower region of the backbone is the **sacrum,** a single bone composed of three to five fused vertebrae. The **pelvis,** or hipbone, is attached to the sacrum on its two sides, and the tail joins it distally. The last region of the spine, the **caudal region,** or tail, varies from a few tiny bones fused together (the **coccyx,** in humans) to a long, grasping organ of as many as thirty bones in some species (Fig. 2.19).

Upper Limb

The primate upper limb, or **forelimb,** is divided into four regions, most of which contain several bones. The most proximal part, nearest the trunk, is the **shoulder girdle,** which is composed of two bones—the **clavicle** anteriorly and the **scapula** posteriorly. All primates

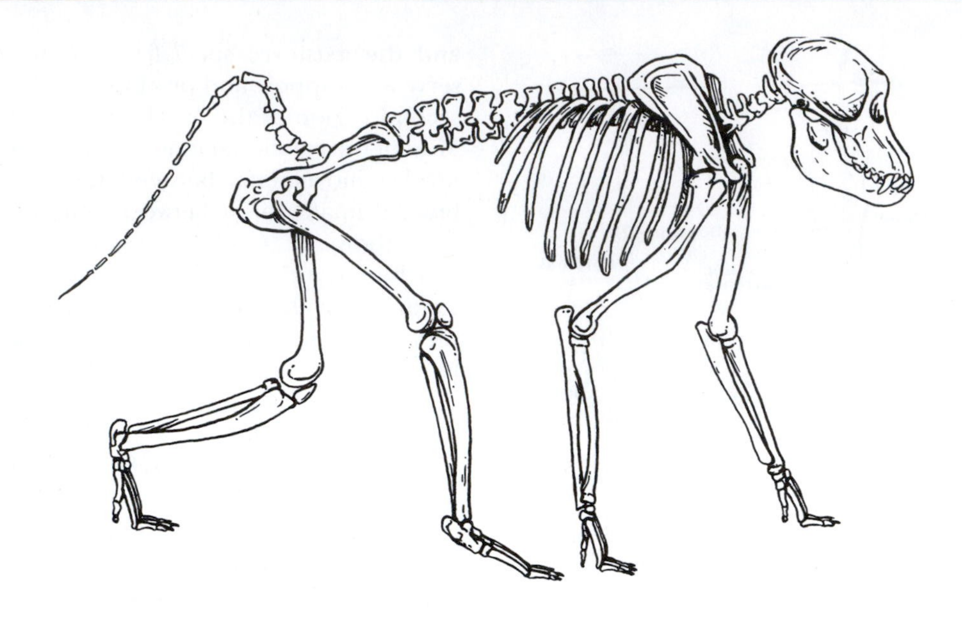

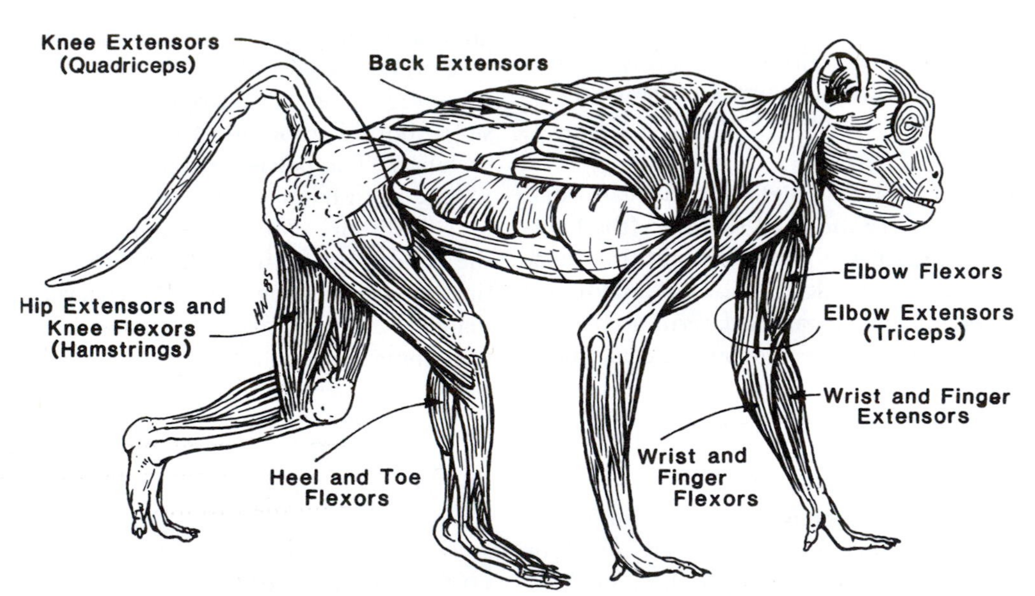

FIGURE 2.20 The skeleton of a baboon (*Papio*) and the superficial limb musculature of the same species showing the major muscle groups responsible for locomotion.

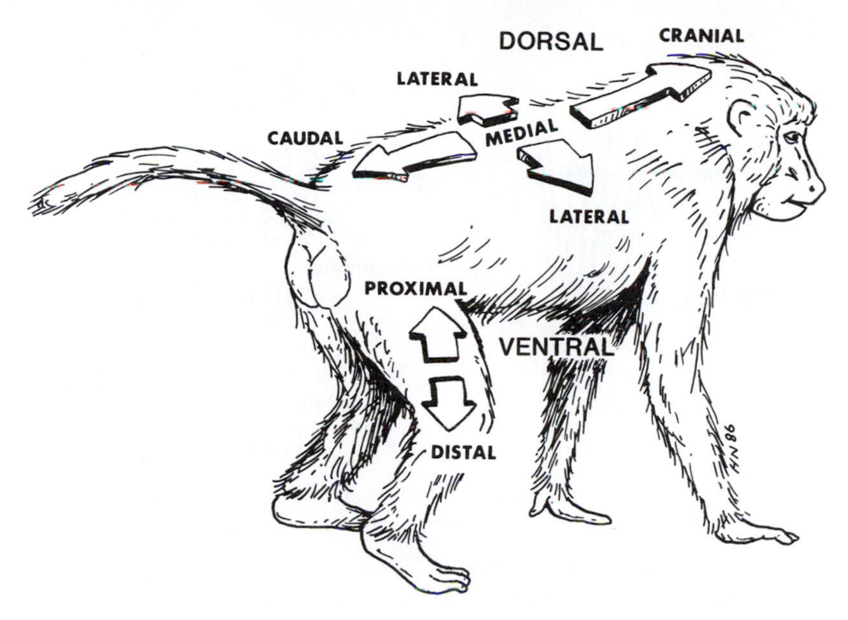

FIGURE 2.21 Terminology for anatomical orientation.

have a clavicle, in contrast to many other mammals—particularly, fast terrestrial runners such as dogs, cats, horses, and antelopes, which have lost this bone. The clavicle is one of the primitive skeletal characteristics of primates. This small S-shaped bone, attached to the sternum anteriorly and to the scapula posteriorly, provides the only bony connection between the upper limb and the trunk.

The flat, triangular scapula is attached to the thoracic wall only by several broad muscles. It articulates with the single bone of the upper arm, the **humerus,** by a very mobile ball-and-socket joint. Most of the large propulsive muscles of the upper limb originate on the chest wall or the scapula and insert on the humerus. The muscles responsible for flexing and extending the elbow originate on the humerus (or just above, on the scapula) and insert on the forearm bones.

There are two forearm bones that articulate with the humerus—the **radius,** on the lateral or thumb side, and the **ulna,** on the medial side. The elbow joint is a complex region that involves the articulation of these three bones. The articulation between the ulna and the humerus is a hinge joint that functions as a simple lever. The radius forms a more complex joint; this rodlike bone not only flexes and extends but also rotates about the end of the humerus. There are two articulations between the radius and the ulna, one at the elbow and one at the wrist. Because of its rotational movement, the radius can roll over the ulna. The movement of the radius and ulna is called **pronation** when the hand faces down and **supination** when the hand faces up. The muscles responsible for movements at the wrist and for flexion and extension of the fingers originate on the distal end of the humerus and on the two forearm bones. Distally, the radius and the ulna articulate with the bones of the wrist. The radius forms the larger joint between the forearm and the

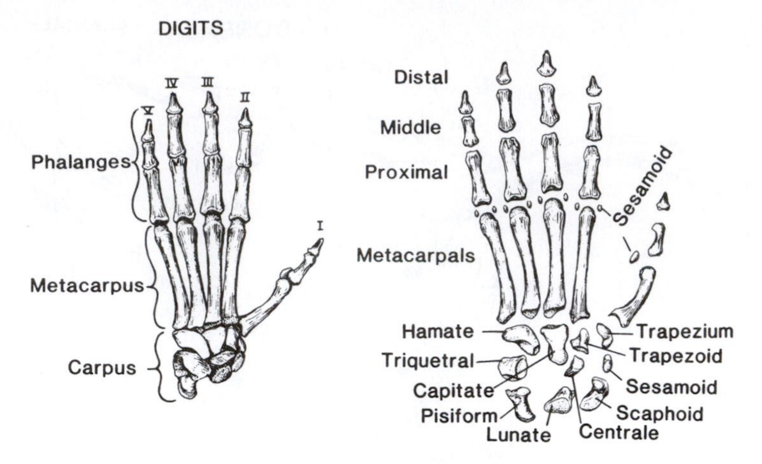

FIGURE 2.22 Dorsal view of the bony skeleton of a baboon left hand.

wrist, and in some primates (lorises, humans, and apes) the ulna does not even contact the wrist bones.

Primate hands (Figs. 2.22; 2.23) are divided into three regions—the **carpus,** or wrist, the **metacarpus,** and the **phalanges** (singular = **phalanx**). The wrist is a complicated region consisting of eight or nine separate bones aligned in two rows. The proximal row articulates with the radius, and the distal row articulates with the metacarpals of the hand. Between the two rows of bones is another composite joint, the **midcarpal joint,** which has considerable mobility in flexion, extension, and rotation.

The five rodlike metacarpals form the skeleton of the palm and articulate distally with the phalanges, or finger bones, of each digit. The joints at the base of most of the metacarpals are formed by two flat surfaces, offering little mobility, but the joint at the base of the first digit, the **pollex,** or thumb, is more elab-

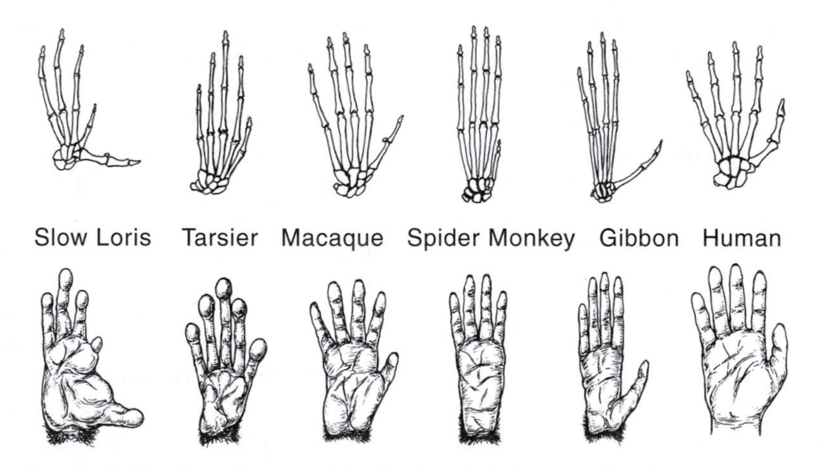

FIGURE 2.23 Dorsal views of the left hand skeleton and palmar views of the right hand of six primate species.

orate and allows the more complex movements associated with grasping. The joints between the metacarpal and the proximal phalanx of each finger allow mainly flexion and extension and a small amount of side-to-side movement (**abduction** and **adduction**) for spreading the fingers apart. There are three phalanges (proximal, middle, and distal) for each finger except the thumb, which has only two (proximal and distal). The interphalangeal joints are purely flexion and extension joints.

As just noted, the muscles mainly responsible for flexing and extending the fingers and thumb lie within the forearm and send long tendons into the hand that insert on the middle and distal phalanges. The only muscles that lie completely within the hand are those forming the ball of the thumb, which are responsible for fine movements of that digit, a smaller group forming the other side of the palm, and a series of small muscles within the palm that aid in complex movements of the digits. The **palmar** (relating to the palm) surfaces of primate hands and feet are covered with **friction pads,** a special type of skin covered with **dermatoglyphics** (fingerprints), and sweat glands. In most living primates, the tips of the distal phalanges have flattened nails, in contrast with the claws on the digits of most primitive mammals or the hooves of ungulates. A few primates have specialized claws on some of their digits.

Although primate hands usually have approximately the same numbers of bones, the relative sizes of the hand elements can vary greatly in conjunction with particular needs for locomotion or manipulation (Fig. 2.23). The slow-climbing loris, for example, has a robust thumb and long lateral digits for grasping branches; the more suspensory, hanging primates, such as the gibbon and spider monkey, have very long, slender fingers. Primates that use their hands for manipulating food, such as mountain gorillas and macaques, or tools, such as humans, have well-developed thumbs that can be opposed to the fingers.

Lower Limb

The primate lower limb, or **hindlimb,** can be divided into four major regions: pelvic girdle, thigh, leg, and foot. These regions are comparable to the shoulder girdle, arm, forearm, and hand of the forelimb.

The primate **pelvic girdle** is composed of three separate bones on each side (the **ilium, ischium,** and **pubis**) that fuse to form a single rigid structure, the **bony pelvis.** In contrast with the pectoral girdle, which is quite mobile and loosely connected to the trunk, the pelvic girdle is firmly attached to the backbone through a nearly immobile joint between the sacrum and the paired ilia. The primate pelvis, like that of all mammals, serves many roles. Forming the bottom of the abdominopelvic cavity, the internal part supports and protects the pelvic viscera, including the female reproductive organs, the bladder, and the lower part of the digestive tract. The bony pelvis also forms the birth canal through which the newborn must pass. In conjunction with this requirement, many female primates (including women) have a bony pelvis that is relatively wider than that of males. Finally, the pelvis plays a major role in locomotion; it is the bony link between the trunk and the hindlimb bones, and it is the origin for many large hindlimb muscles that move the lower limb.

The ilium is the largest of the three bones forming the bony pelvis. A long, relatively flat bone in most primates, it lies along the vertebral column and is completely covered with large hip muscles, primarily those responsible for flexing, abducting, and rotating the hip joint. The rodlike ischium lies posterior to the ilium; the **hamstring** muscles responsible for extending the hip joint and flexing the knee arise from its most posterior surface, the **ischial tuberosity.** This tuberosity is the primate sitting bone. In Old World monkeys it is expanded and covered by a tough fatpad, the ischial callosity. The pubis lies anterior to the other two bones and gives rise to many of the muscles that adduct the hip joint. The ischium and pubis join inferiorly and surround a large opening, the **obturator foramen.** The relative sizes and shapes of the ilium, ischium, and pubis vary considerably among different primate species in conjunction with different locomotor habits.

The hip joint is a ball-and-socket joint that allows mobility in many directions. The socket part of the bony pelvis, the **acetabulum,** that articulates with the head of the femur, lies at the junction of the three bones.

The single bone of the thigh is the **femur.** The prominent features of this long bone are a round **head** that articulates with the pelvis, the **greater trochanter** where many hip extensors and abductors insert, the shaft, and the **distal condyles,** which articulate with the tibia to form the knee joint. Most of the surface of the femur is covered by the quadriceps muscles, which are responsible for extension of the knee. Attached to the tendon of this set of muscles is the third bone of the knee, the small **patella.**

Two bones make up the lower leg, the **tibia** medially and the **fibula** laterally. The tibia is larger and participates in the knee joint; distally, it forms the main articulation with the ankle. The fibula is a slender, splintlike bone that articulates with the tibia both above and below and also forms the lateral side of the ankle joint. In tarsiers, the tibia and fibula are fused to form a single element in the distal half of the leg; and in some small prosimians and New World monkeys, much of the distal part of the fibula is appressed against the shaft of the tibia. Arising from the surfaces of the tibia and fibula (and also from the distalmost part of the femur) are the large muscles responsible for movements at the ankle and those that flex and extend the toes during grasping.

Like the hand, the primate foot (Figs. 2.24; 2.25) is made up of three parts: tarsus, metatarsus, and phalanges. The most proximal two tarsal bones are part of the ankle—the **talus** above and the **calcaneus** below. The head of the talus articulates with the **navicular bone.** This boat-shaped bone articulates with three small **cuneiform bones,** which in turn articulate with the first three metatarsals. The body of the talus sits roughly on the center of the calcaneus, the largest of the tarsal bones. The tuberosity of the calcaneus extends well posterior of the rest of the ankle and forms the heel process, to which the Achilles tendon from the calf muscles attaches. This process acts as a lever for the entire foot. Anteriorly, the calcaneus articulates with the **cuboid,** which in turn articulates with the metatarsals of the fourth and fifth digits.

In nonhuman primates, the digits of the foot resemble those of the hand (Figs. 2.21, 2.23). Each of the four lateral digits has a long metatarsal with a flat base and a rounded head, followed by three phalanges. The shorter first digit, the **hallux,** is opposable, like the thumb, and has a mobile joint at its base for grasping. Primate feet, like primate hands, show considerable differences from species to species in the relative proportions of different elements in association with different locomotor abilities. The climbing loris has a grasping foot, the tarsier has a very long ankle region for rapid leaping, and the suspensory gibbon and spider monkey have long, slender digits for hanging. With their short phalanges and lack of an opposable hallux, human feet are stiff, propulsive levers most suitable for walking on flat surfaces. The feet of primates, like our hands have friction pads on the sole and digits and nails (or claws) on the distal phalanges.

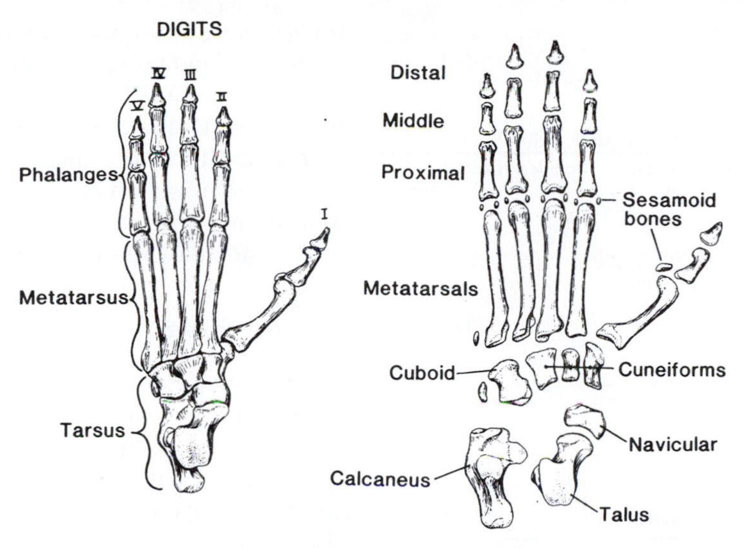

FIGURE 2.24 Dorsal view of the skeleton of a baboon left foot.

Limb Proportions

Primates vary dramatically in their overall body proportions. Some species have forelimbs longer than hindlimbs; others have hindlimbs longer than forelimbs. Some have limbs relatively long for the length of their trunk; others have relatively short limbs. These proportional differences are often described by a **limb index,** a ratio of the length of one part to the length of another part of the same animal. Table 2.1 gives the formula for some of the most commonly used indices. Of these, the intermembral index, a ratio of forelimb

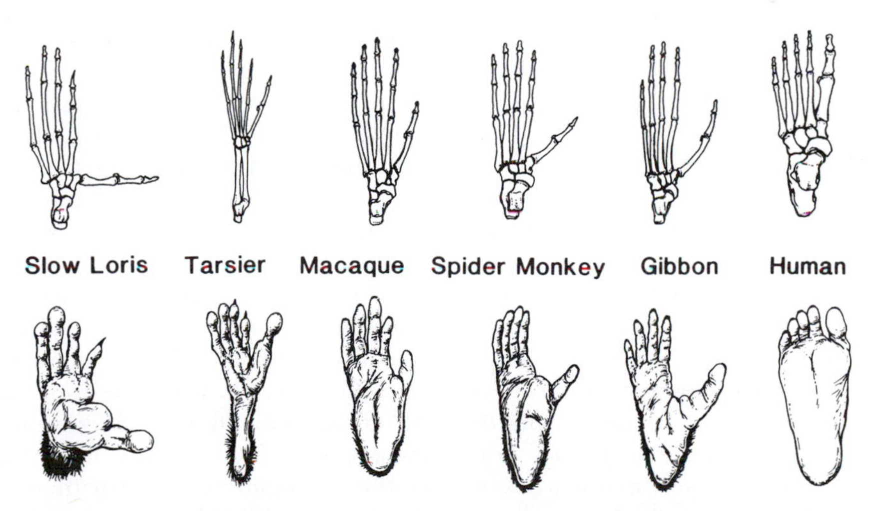

FIGURE 2.25 Dorsal views of the left foot skeleton and plantar views of the right foot of six primate species.

TABLE 2.1
Skeletal Proportions

Intermembral index

$$\frac{\text{Humerus length} + \text{radius length}}{\text{Femur length} + \text{tibia length}} \times 100$$

Humerofemoral index

$$\frac{\text{Humerus length}}{\text{Femur length}} \times 100$$

Brachial index

$$\frac{\text{Radius length}}{\text{Humerus length}} \times 100$$

Crural index

$$\frac{\text{Tibia length}}{\text{Femur length}} \times 100$$

length to hindlimb length, is especially useful for describing the body proportions of a species and also seems to be correlated with locomotor differences in many primates (see Chapter 9). In general, leapers have a low intermembral index (longer hindlimbs), suspensory species have a high intermembral index (longer forelimbs), and quadrupedal species have intermediate indices (forelimbs and hindlimbs similar in size).

Soft Tissues

Primates are composed of more than just bones and teeth, but these are the parts usually preserved in the fossil record and in most museum collections of extant primate species. For extinct species, our knowledge of other aspects of anatomy must be based on inferences derived from our knowledge of the relationships between bony anatomy and the softer structures associated with that bony anatomy. For example, we can often reconstruct details of muscular attachments in extinct species from scars on bones. However, for understanding the adaptations and phylogenetic relationships of living primates, details of "soft" anatomy are often very important. In addition to primate musculature, several other primate organ systems have been well studied and provide insight into the evolution and adaptations of living primates.

Digestive System

In previous sections of this chapter we discussed the first part of the digestive system, the dentition and structures of the oral cavity; these cranial parts are involved in ingestion and the initial mechanical and chemical preparation of food items. The remainder of the digestive system (Fig. 2.26) lies primarily in the abdominal cavity and is concerned with further chemical preparation of food, absorption of nutrients, and excretion of wastes. The primate digestive system, like that of all vertebrates, is basically a long tube with some enlarged areas (stomach and large intestine), some coiled loops (small intestine), one cul-de-sac (caecum), and two developmental outgrowths of the digestive tract, the liver and pancreas, which produce various digestive enzymes. Although there is considerable variation among primate species in the relative size, shape, number of separate parts of individual organs in this system, largely associated with their different diets (see Chapter 9), the organs themselves and their functions are relatively similar throughout the order.

After food is prepared in the oral cavity, it is passed through the **esophagus,** a narrow, muscular tube that traverses the thoracic cavity, into the abdominal cavity, where it empties into the **stomach.** Here the food undergoes chemical preparation by digestive juices. The most specialized primate stomachs are those of the colobine monkeys; in these primates, this organ is divided into several sections that

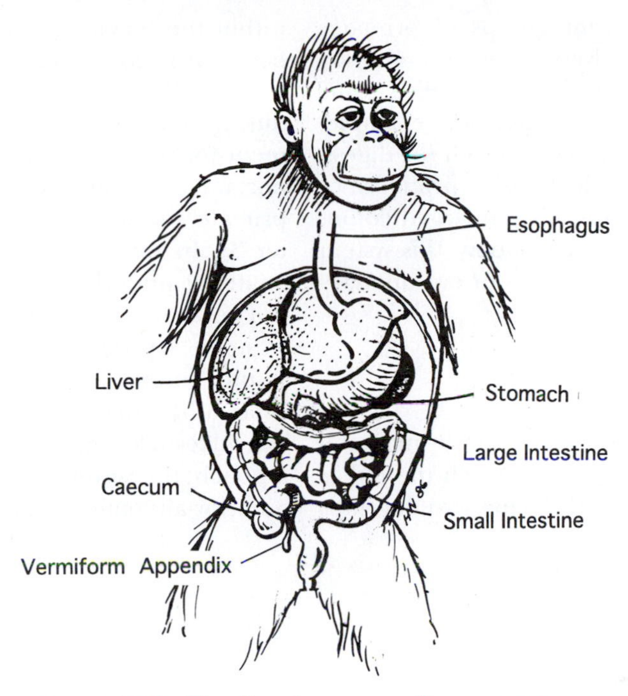

FIGURE 2.26 The digestive system of an orangutan.

function as fermenting chambers in which bacterial colonies break down cellulose (see Chapter 9).

From the stomach, the food passes to the **small intestine,** where further chemical preparation takes place. Here digestive juices from the liver and pancreas are mixed with the food, and much of the nutrient absorption takes place in this part of the gut. The small intestine is normally the longest part of the digestive tract. In most primates, it is several times as long as the animal's body and is usually folded into a series of loops within the abdominal cavity. At the end of the small intestine, the unabsorbed food and wastes are passed to the large intestine.

The **large intestine** is larger in diameter than the small intestine but usually shorter in length. It is involved primarily with further absorption of nutrients and water and, in its final parts, with excretion of solid wastes. At the beginning of the large intestine is the **caecum,** a cul-de-sac that varies considerably in size among different primate species and serves several special digestive functions. Like the colobine stomach, this out-of-the-way segment of the digestive tract is an ideal place for harboring the bacteria used to break down food items that primates cannot normally digest, such as leaves or gums. The remainder of the large intestine, the **colon,** is usually divided into several parts on the basis of position within the abdominal cavity. From this last part of the large intestine, solid wastes leave the body through the **rectum** and **anus.**

Many of the adaptive differences in the digestive system of living primates are discussed further in Chapter 9. It is worth noting here, however, that different groups of primates have frequently evolved quite different visceral adaptations for similar digestive functions. Leaf-eating colobines, for example, have evolved an enlarged stomach for digesting leaves, whereas leaf-eating primates of Madagascar have evolved an enlarged colon. Like all parts of primate anatomy, this system often reflects the interaction of evolutionary history and adaptation.

Reproductive System

All primates have a characteristically mammalian reproductive system, in which the egg is fertilized internally and the embryo develops within the female's uterus for many months before it is born. This basic mammalian pattern of extensive investment by the mother during development, and of infant nourishment for months or years after birth, has important implications for the evolution of primate social behavior (discussed in Chapter 3). In this chapter we briefly review the anatomy underlying primate reproduction.

The anatomical structures associated with primate reproduction are similar to those found among other mammals (Fig. 2.27). Like other mammals, male primates have paired **testicles** that normally lie suspended in a pouch, the **scrotum,** at the lower end of the anterior abdominal wall. Male primates differ

FIGURE 2.27 The male and female reproductive organs in gorillas.

from species to species in the position of the scrotum, which is usually behind the penis but may bc in front, and in the timing of the descent of the testes from their fetal position within the abdomen to the reproductive position in the scrotum. There are also considerable differences in the relative size of the testes, which are related to mating systems (see Chapter 9), and in the size and external appearance of the penis. In most nonhuman primates there is a bone in the penis, the **baculum.**

Like other mammals, female primates have paired **ovaries** and paired **fallopian,** or **uterine, tubes** extending laterally toward the ovaries from the midline uterus (Fig. 2.27). There is considerable variation among primate species in the relative size of the fallopian tubes and the body of the uterus. Among lemurs and lorises, the fallopian tubes are large relative to the body of the uterus, a condition normally found among mammals that have multiple births. Among tarsiers, monkeys, apes, and humans, the fallopian tubes are relatively slender and the body of the uterus is much larger, the condition normally found among mammals that give birth to single offspring.

The **vagina** lies below the uterus and opens onto the **perineum,** where the external genitalia are found. The external genitals of female primates generally consist of two sets of **labia,** on either side of the vaginal opening, and the **clitoris,** anterior to the vagina. The clitoris of female primates varies in size and shape: in some species it is small and hidden beneath a hood; in others it is large and pendulous, in some cases larger than the male's penis. In addition, many female primates have areas of sexual skin surrounding the external genitalia that change color and size during the sexual cycle. In some species, such as baboons and chimpanzees, these sexual swellings are extremely large and provide a rather spectacular advertisement of an individual's reproductive condition.

Primates vary considerably in the periodicity of their reproductive physiology. At the extremes are Malagasy lemurs, in which reproductive activity in both males and females is limited to one day per year, and most higher primates, in which male sperm production seems to be relatively constant throughout the year and female ovulation occurs regularly at approximately monthly intervals. There are also numerous intermediate species in which both male and female reproductive activity (sperm production and ovulation) is limited to one or two seasons each year, often in response to environmental cues such as food availability and day length.

One aspect of primate reproductive anatomy that shows considerable differences among living primate species is the form of the placenta and other structures associated with the developing fetus within the mother's womb (Fig. 2.28). In most lemurs and lorises, the placental membranes are spread diffusely throughout the uterine cavity, and fetal circulation is separated from maternal circulation by several tissue layers: this is an **epitheliochorial placenta.** In *Tarsius* and in all higher primates, the placenta is localized into one or two discrete disks, and there is a much closer approximation between fetal and maternal blood supplies: this is a **hemochorial placenta.** In great apes and humans, the intimacy of fetal and maternal circulation reaches its greatest degree and provides the most efficient transfer of nutrients to the fetus.

Growth and Development

In this chapter we have discussed primate bodies as if they were fixed entities. Obviously our bodies are constantly changing. During the early part of an individual's life

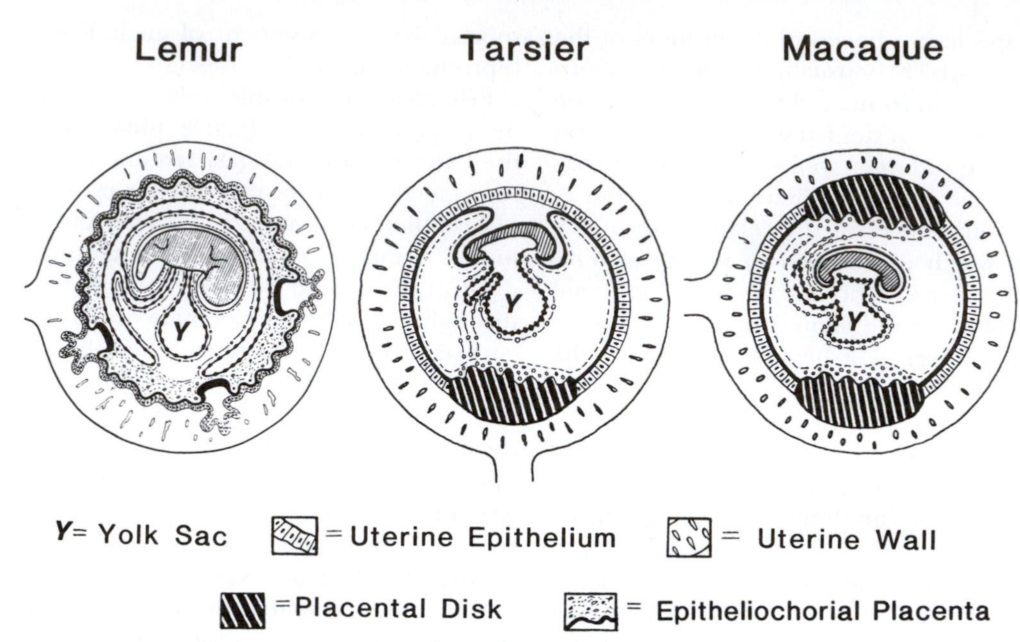

FIGURE 2.28 Fetal membranes in three primates. In the lemur, several layers of tissue separate the uterus of the mother from the diffuse epitheliochorial membrane of the fetus. In the tarsier and the macaque, the developing embryo forms one or two placental disks that invade the lining of the uterus to become embedded in the uterine wall, providing a more intimate interchange between fetal and maternal circulation.

span it undergoes both an increase in size (**growth**) and changes in shape and function of many organs (**development**). Together, these changes are referred to as ontogeny. Compared to most other mammals of a similar size, primates are characterized by a long period of growth and development. An extreme in this regard are the great apes and humans, in which individuals reach adult size and become sexually mature only after ten or more years of growth (Fig. 2.29). In most species, rapid growth during infancy is followed by a long childhood in which growth is relatively slow; just prior to sexual maturity, there is a phase of rapid growth, called the adolescent growth spurt; (Fig. 2.30). While many primates have an **adolescent growth spurt,** the timing of and magnitude varies considerably both among different species and between males and females. Indeed, differences in the timing and rates of growth are responsible for different patterns of sexual dimorphism (Fig. 2.31; Leigh, 1992).

Systematically collected data on primate growth and development are available for only a few species, and most of our knowledge of primate growth and development comes from isolated, anecdotal observations and the many works of Adolph Schultz (e.g. Schultz, 1956). While there is information on growth in body weight for a number of species, information on skeletal maturation has been studied in detail only in humans, macaques, chimpanzees, and capuchins. As a result,

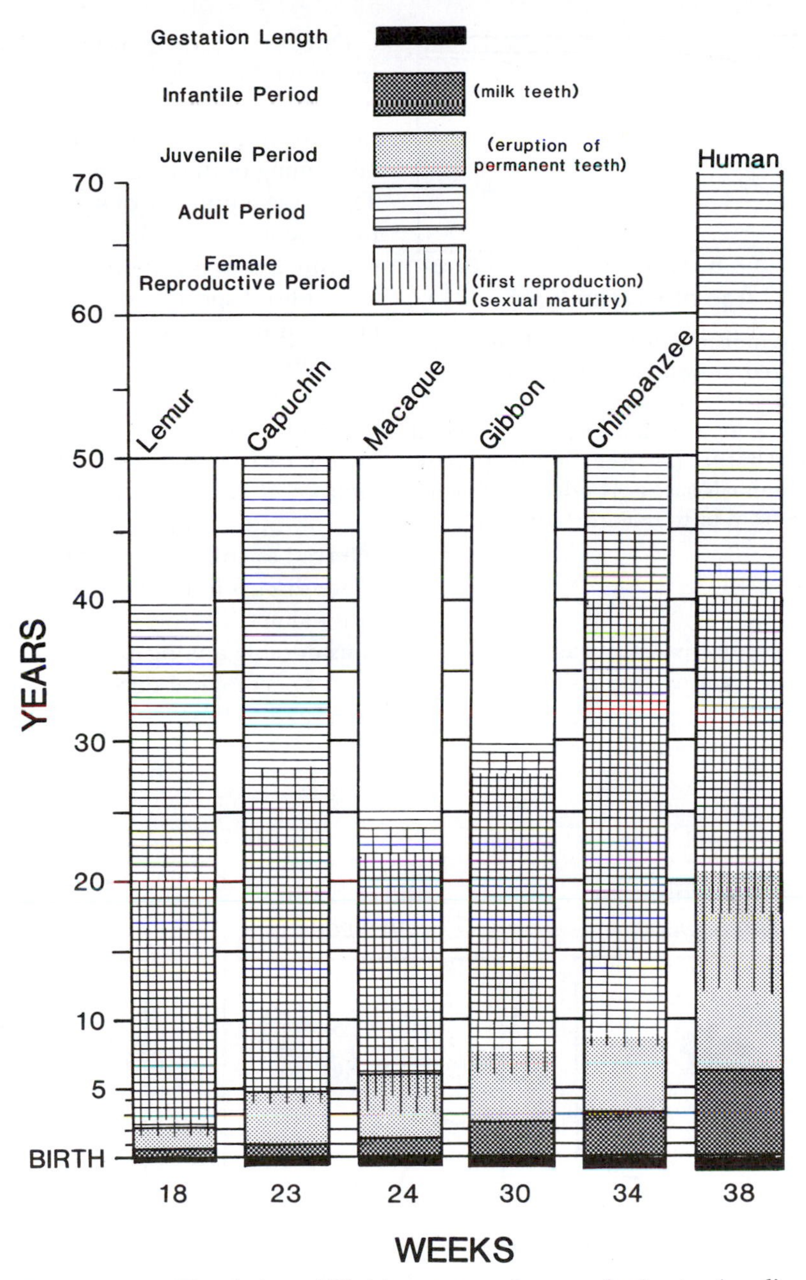

FIGURE 2.29 The timing of life history events in several primates (modified from Schultz, 1960).

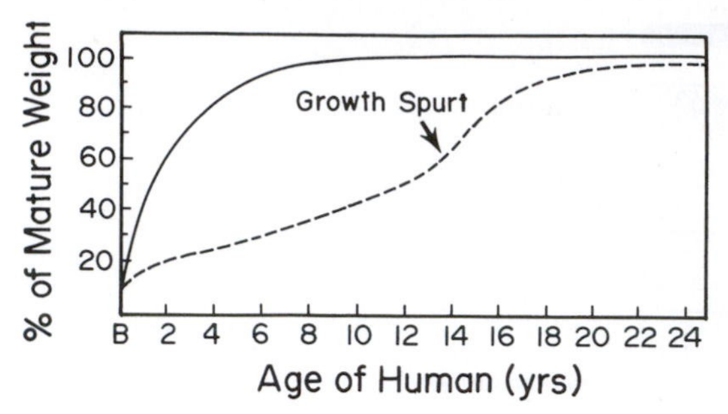

FIGURE 2.30 A human growth curve (dashed line) and a generalized growth curve of a nonprimate mammal (solid line). In humans and other primates, there is a long period of slow childhood growth followed by the adolescent growth spurt. In contrast, most animals have a growth curve that decreases in rate from birth onward (modified from Watts, 1986).

primate-wide comparisons on rates of skeletal maturation are almost impossible to make (Watts, 1986).

One aspect of growth and development which has been studied in many primates and has played an important role in studies of primate and human evolution is the sequence and timing of the eruption of teeth (Fig. 2.32). There are a number of significant differences in the sequence in which the permanent teeth of primates appear in the jaw. In general, larger, longer-lived species are characterized by later eruption of the last two molar teeth, and humans are unusual in the early eruption of their canine teeth. These differences have been important in providing clues to the lifestyle of our early human ancestors (Smith, 1992; Smith *et al.*, 1994; Macho and Wood, 1995).

Growth, development, and aging are often referred to collectively as **life history features.** As we discuss in the next chapter, these maturational and reproductive characteristics vary considerably among primates and can be related to ecological differences among species (e.g. Ross, 1998).

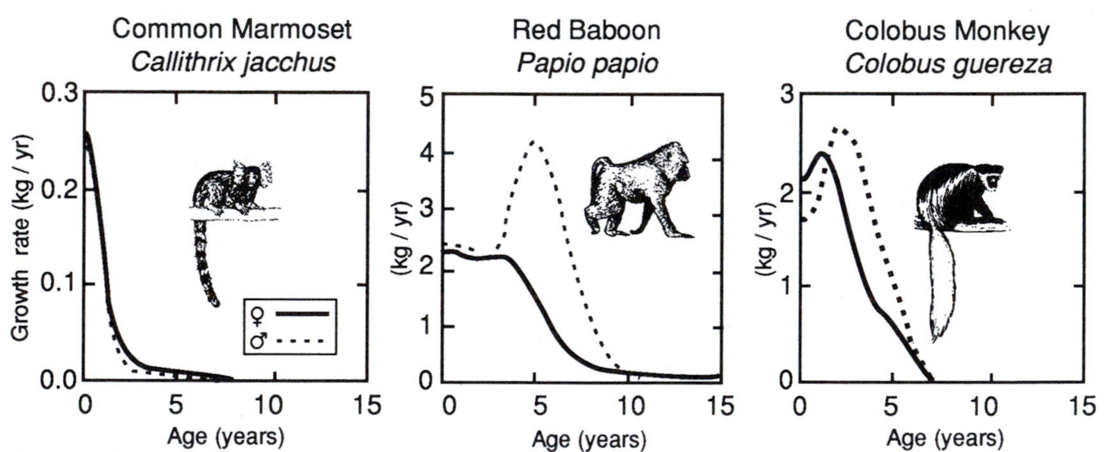

FIGURE 2.31 Primate species show striking differences in patterns of somatic growth, both between species and between sexes of a single species. Marmosets show no adolescent growth spurt; baboons show considerable dimorphism in the growth spurt; and colobus monkeys show a spurt in both sexes. Courtesy of S. Leigh.

FIGURE 2.32 Dental eruption sequences in a variety of primates. Arrows highlight eruption of the molar teeth to illustrate the sequence differences between genera (redrawn and modified from Schultz, 1960). C, canine; I, incisor; M, molar; P, premolar.

BIBLIOGRAPHY

GENERAL

Aiello, L., and Dean, C. (1992). *Human Evolutionary Anatomy.* London: Academic Press.

Gebo, D. L. , ed. (1993). *Postcranial Adaptation in Nonhuman Primates.* DeKalb: Northern Illinois Press.

Hartman, C. G., and Straus, W. L., Jr. (1933). *The Anatomy of the Rhesus Monkey.* Baltimore: Williams and Wilkins.

Hershkovitz, P. (1977). *Living New World Monkeys (Platyrrhini), With an introduction to Primates,* vol. 1. Chicago: University of Chicago Press.

Hill, W. C. O. (1953–1970). *Primates,* vols. 1–8. Edinburgh: Edinburgh University Press.

______ (1972). *Evolutionary Biology of the Primates.* New York: Academic Press.

Hofer, H., Schultz, A. H., and Starck, D. (1956–1973). *Primatologia,* vols. 1–4. Basel, Switzerland: Karger.

Iwamoto, M., ed. (1994). Morphology of the Japanese macaque. *Anthropological Science* 102(suppl):1–205

Le Gros Clark, W. E. (1959). *The Antecedents of Man.* Edinburgh: Edinburgh University Press.

Martin, R. D. (1990). *Primate Origins and Evolution.* Princeton, N. J.: Princeton University Press.

Schultz, A. H. (1969). *The Life of Primates.* New York: Universe Books.

Strasser, E., and Dagosto, M. (1988). *The Primate Postcranial Skeleton.* New York: Academic Press

Swindler, D. R., and Erwin, J., eds. (1986). *Comparative Primate Biology,* vol. 1: *Systematics, Evolution, and Anatomy.* New York: Alan R. Liss.

SIZE

Eisenberg, J. F. (1981). *The Mammalian Radiations.* Chicago: University of Chicago Press.

Jungers, W. L. (1985). *Size and Scaling in Primate Biology.* New York: Plenum Press.

Smith, R. J, and Jungers, W. L. (1997). Body mass in comparative primatology. Amer. *J. Human Evol.* 523–559.

CRANIAL ANATOMY

Gregory, W. K. (1922). *Origin and Evolution of the Human Dentition.* Baltimore: Williams and Wilkins.

Hiiemae, K. M., and Kay, R. F. (1973). Evolutionary trends in the dynamics of primate mastication. *Symp. Fourth Int. Cong. Primatol.* **3**:28–64.

Hladik, C. M., and Simmen, B. (1997). Taste perception and feeding behavior in nonhuman primates and human opulations. *Evol. Anthropol.* **5**:58–71.

Hofer, H. (1981). Microscopic anatomy of the apical part of the tongue of *Lemur fulvus* (Primates, Lemuriformes). *Gegenbaurs morph. Jahrb.* **127**:343–363.

Huber, E. (1931). Evolution of facial musculature and cutaneous field of trigeminus, pt. II. *Q. Rev. Biol.* **5**:389–437.

Kay, R. F. (1975). The functional adaptations of primate molar teeth. *Am. J. Phys. Anthropol.* **43**:195–216.

——— (1984). On the use of anatomical features to infer foraging behavior in extinct primates. In *Adaptation for Foraging in Nonhuman Primates,* ed. P. S. Rodman and J. G. H. Cant, pp. 21–53. New York: Columbia University Press.

Overdorff, D. (1992). Differential patterns in flower feeding by *Eulemur fulvus rufus* and *Eulemur rubriventer* in Madagascar. *Amer. J. Primatol.* **28**:191–204.

Moore, W. J. (1981). *The Mammalian Skull.* Cambridge: Cambridge University Press.

Schneider, R. (1958) Zunge und weicher Gaumen. In *Primatologia,* ed. H. Hofer, A. H. Schultz, and D. Starck, pp. 61–126. Basel, Switzerland: Karger.

The Brain

Armstrong, E., and Falk, D., eds. (1982). *Primate Brain Evolution: Methods and Concepts.* New York: Plenum Press.

Falk, D. (1982). Primate neuroanatomy: An evolutionary perspective. In *A History of American Physical Anthropology,* ed. F. Spencer, pp. 75–103. New York: Academic Press.

Noback, C. R., and Montagna, W., eds. (1970). *The Primate Brain.* New York: Appleton-Century-Crofts.

Radinsky, L. B. (1975). Primate brain evolution. *Am. Sci.* **63**:656–663.

Steklis, H. D., and Erwin, J. (1988). *Comparative Primate Biology,* vol. 4: *Neurosciences.* New York: Alan R. Liss.

Cranial Blood Supply

Bugge, J. (1980). Comparative anatomical study of the carotid circulation in New and Old World primates: Implications for their evolutionary history. In *Evolutionary Biology of the New World Monkeys and Continenal Drift,* ed. R. L. Ciochon and A. B. Chiarelli, pp. 293–316. New York: Plenum Press.

Conroy, G. C. (1982). A study of cerebral vasculature evolution in primates. In *Primate Brain Evolution,* ed. E. Armstrong and D. Falk, pp. 246–261. New York: Plenum Press.

MacPhee, R. D. E., and Cartmill, M. (1986). Basicranial structures and primate systematics. In *Comparative Primate Biology,* vol. 1: *Systematics, Evolution, and Anatomy,* ed. D. R. Swindler and J. Erwin, pp. 219–275. New York: Alan R. Liss.

Olfaction

Cave, A. J. (1973). The primate nasal fossa. *J. Linn. Soc.* **53**:377–387.

Fobes, J. L., and King, J. E. (1982). Auditory and chemoreceptive sensitivity in primates. In *Primate Behavior,* ed. J. L. Fobes and J. E. King, pp. 245–270. New York: Academic Press.

Maier, W. (1993). Zur evolutiven und funktionellen Morphologie des Gesichtsschadels der Primates. *Z. Morph. Anthrop.* **79**:279–299.

Vision

Cartmill, M. (1980). Morphology, function, and evolution of the anthropoid postorbital septum. In *Evolutionary Biology of the New World Monkeys and Continental Drift,* ed. R. L. Ciochon and A. B. Chiarelli, pp. 243–274. New York: Plenum Press.

Jacobs, G. H. (1995). Variations in primate color vision: Mechanisms and utility. *Evol. Anthropol.* **3**:196–205.

Hearing

MacPhee, R. D. E., and Cartmill, M. (1986). Basicranial structures and primate systematics. In *Comparative Primate Biology,* vol. 1: *Systematics, Evolution, and Anatomy,* ed. D. R. Swindler and J. Erwin, pp. 219–275. New York: Alan R. Liss.

THE TRUNK AND LIMBS

Larson, S. G. (1993). Functional morphology of the shoulder in primates. In *Postcranial Adaptation in Nonhuman Primates,* ed. D. L. Gebo, pp. 45–69. DeKalb: Northern Illinois Press.

Schultz, A. H. (1969). *The Life of Primates.* New York: Universe Books.

Schultz, M. (1986). The forelimb of the Colobinae. In *Comparative Primate Biology,* vol. 1: *Systematics, Evolution, and Anatomy,* ed. D. R. Swindler and J. Erwin, pp. 559–670. New York: Alan R. Liss.

Shapiro, L. (1993). Functional anatomy of the vertebral column in primates. In *Postcranial Adaptation in Nonhuman Primates,* ed. D. L. Gebo, pp. 121–149. DeKalb: Northern Illinois Press.

Sigmon, B. A., and Farslow, D. L. (1986). The primate hindlimb. In *Comparative Primate Biology*, vol. 1: *Systematics, Evolution, and Anatomy*, ed. D. R. Swindler and J. Erwin, pp. 671–718. New York: Alan R. Liss.

Stern, J. T., Jr. (1971). Functional myology of the hip and thigh of cebid monkeys and its implications for the evolution of erect posture. *Bibl. Primatol.* **14**:1–318.

SOFT TISSUES

Digestive System

Chivers, D. J., and Hladik, C. M. (1980). Morphology of the gastrointestinal tract in primates: Comparisons with other mammals in relation to diet. *J. Morphol.* **166**: 337–386.

Hill, W. C. O. (1958). Pharynx, oesophagus, stomach, small and large intestine: Form and position. *Primatologia* **III/1**:139–147.

——— (1972). *Evolutionary Biology of the Primates*. New York: Academic Press.

Hladik, C. M. (1867). Surface relative du tractus digestif de quelques primates, morphologie des villosities intestinales et correlations avec le regime alimentaire. *Mammalia* **31**:120–147.

Reproductive System

Dukelow, W. R., and Erwin, J., eds. (1986). *Comparative Primate Biology*, vol. 3: *Reproduction and Development*. New York: Alan R. Liss.

Hill, W. C. O. (1972). *Evolutionary Biology of the Primates*. New York: Academic Press.

Luckett, W. P. (1974). *Contributions to Primatology*, vol. 3: *Reproductive Biology of the Primates*. Basel, Switzerland: Karger.

——— (1975). Ontogeny of the fetal membranes and placenta. In *Phylogeny of the Primates*, ed. W. P. Luckett and F. S. Szalay, pp. 157–182. New York: Plenum Press.

GROWTH AND DEVELOPMENT

Hartwig, W. C. (1996). Perinatal life history traits in New World monkeys. *Am. J. Primatol.* **40**:99–130

Leigh, S. (1992). Patterns of variation in the ontogeny of primate body size dimorphism. *J. Hum. Evol.* **23**:27–50.

Macho, G. A. and Wood, B. A. (1995). The role of time and timing in hominid dental evolution. *Evol. Anthropol.* **4**:17–31.

Ravosa, M. J. and Gomez, A. M. eds. (1992). *Ontogenetic Perspectives on Primate Evoluting Biology*. London: Academic Press; also *Journal of Human Evolution* **23**:1–307.

Ross, C. (1998). Primate Life Histories. *Evol. Anthropol.* **6**: 54–63.

Ross, C. (1994). The craniofacial evidence for anthropoid and tarsier relationships. In *Anthropoid Origins*, ed. J. G. Fleagle and R. F. Kay, pp. 469–547. New York: Plenum.

Schultz, A. H. (1956). Postembryonic age changes. *Primatologia* **1**:887–964.

Smith, B. H. (1992). Life History and the Evolution of Human maturation. *Evol. Anthropol.* **1**:134–142.

Smith, B. H., Crummett, T. L., and Brandt, K. I. (1994). Ages of Eruption of Primate teeth: A Compendium for Aging Individuals and Comparing Life Histories. *Yrbk. Phys. Anthropol.* **37**:177–231.

Watts, E. S. (1986). Skeletal development. In *Comparative Primate Biology*, vol. 3: *Reproduction and Development*, ed. W. R. Dukelow and J. Erwin, pp. 415–439. New York: Alan R. Liss.

THREE

Primate Lives

BEHAVIOR, ECOLOGY, AND LIFE HISTORY

In the previous chapter we discussed physical characteristics of primates. The purpose of this chapter is to enliven the primate body by introducing the basic features of primate behavior, ecology and life history—where primates live, what they eat, how they move, and how they organize their social life and lives in general. In later chapters we see how these parameters vary from species to species; in this one we introduce terminology and general principles.

Primate Habitats

Nonhuman primates today are found naturally on five of the seven continents (Fig. 3.1). There are no living primates other than humans on either Antarctica or Australia and no evidence that primates ever inhabited either continent before the relatively recent arrival of humans. Although primates occupy only marginal areas of Europe (Gibraltar) and North America (Central America and southern Mexico), they were formerly much more widespread on both continents. For the present, however, Africa, Asia, South America, and their nearby islands are the home of most living nonhuman primates.

A few hardy primate species live in temperate areas, where the winters are cold such as Nepal, and Japan, but these are exceptional. The vast majority of primate species and individuals are found in tropical climates, where daily fluctuations in temperature between day and night far exceed the changes in average temperature from season to season. In these climates, seasonal changes in rainfall have a much greater effect on the vegetation and on the primates than do any seasonal differences in temperature or day length.

Forest Habitats

Within their geographic range, living primates are found in a variety of habitats, ranging from deserts to tropical rain forests. Only a few hardy types such as chimpanzees, baboons, and Senegal bushbabies manage to successfully ply their primate trade year after year in drier, more poorly vegetated areas. The majority of primate species and individuals live in tropical forests of one sort or another. The forests come in many shapes, with variations in climate, altitude, topography, and soil type as well as in the characteristic flora and fauna. A few of the more distinctive forest types are illustrated in Figure 3.2.

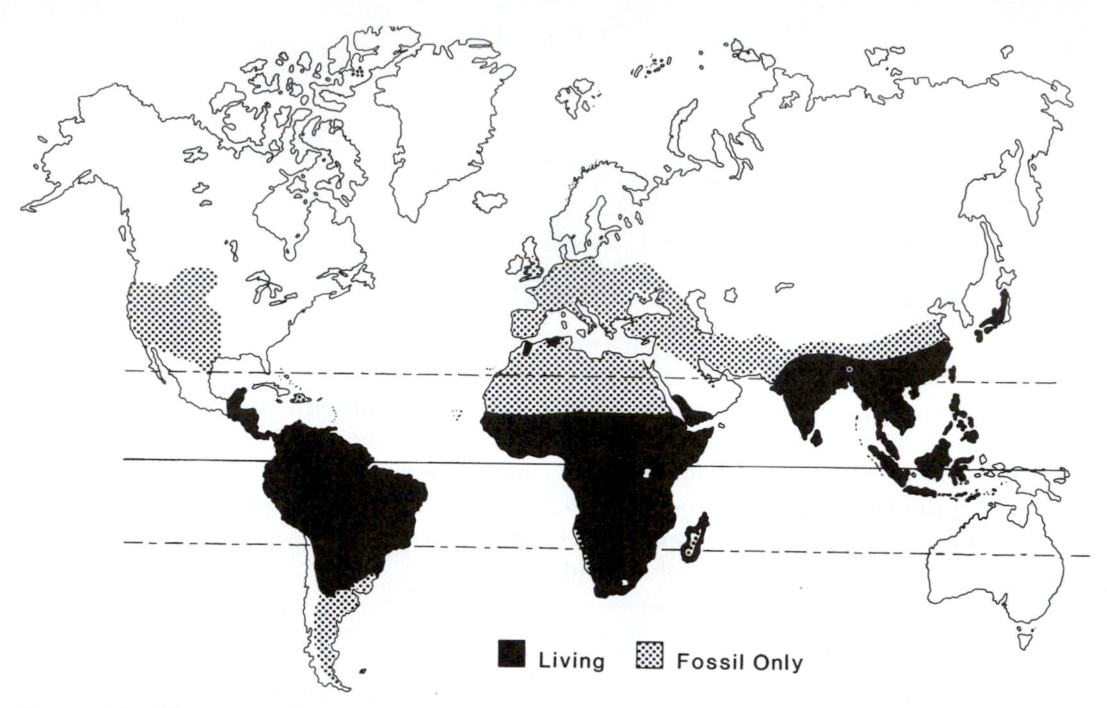

FIGURE 3.1 The geographical distribution of extant nonhuman primates and extinct primate species.

Primary rain forests are usually characterized by the height of the trees (up to 80 m) and the relatively continuous canopy that results from intense competition between many tree species for access to light. The dark understories of primary rain forests, which are usually quite open, are made up primarily of trunks and vines. The canopies of these rain forests are punctuated by occasional emergent trees, which stand above the rest, and by gaps resulting from tree falls. It is through these gaps that light reaches the forest floor, enabling the forests to renew themselves.

Secondary forests, like the areas around tree falls, are characterized by denser, more continuous vegetation because of the availability of light. The canopy structure is less distinct and is often characterized by an abundance of vines and short trees. Because of the high levels of light, leaves and fruit can be very abundant in secondary forests.

African **woodlands** are made up of relatively shorter, often deciduous trees. Between individual trees are continuous growths of grasses and low bushes. As the trees become more sparse, woodland gives way to **bushlands, scrub forests,** and **savannah.**

In many relatively dry tropical regions, forests are concentrated around rivers. These **gallery forests** can contrast strongly with surrounding areas in the types of animals they support.

There are other ways of categorizing forests. We find highland rain forests and lowland rain forests, as well as swamp forests, montane forests, and bamboo forests. Each of these environments presents a primate with a different array of trees on which to move, different places to sleep, and different things to

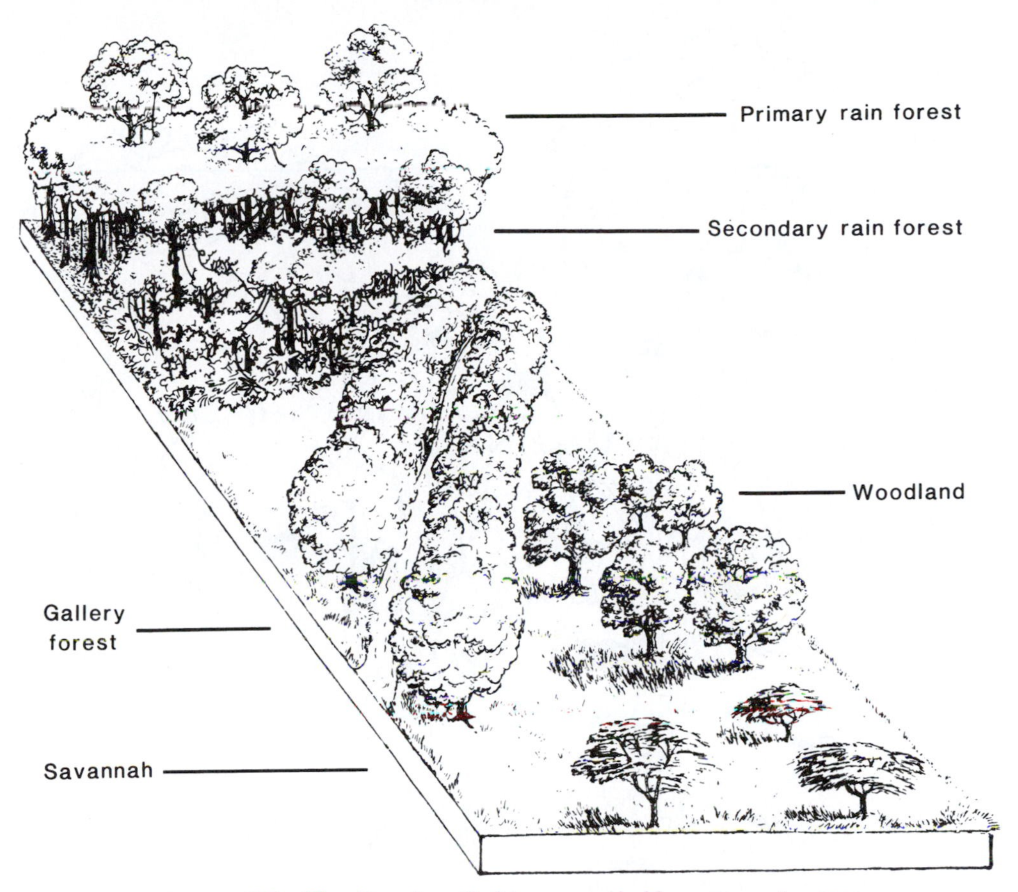

FIGURE 3.2 The diversity of habitats occupied by extant primates.

eat from season to season. The primates that inhabit these forests must meet these different demands. Many of the behavioral differences among living primate populations reflect adaptations to this diversity of habitats.

Habitats Within the Forest

Equally diverse as the types of forests primates inhabit are the different niches primate species may occupy within a single forest (Fig. 3.3). In a tropical forest, which reaches heights of 80 m, the temperature and humidity, the shapes of the branches, the kinds of plant foods, and the types of other animals a species encounters are usually quite different on the ground level, 20 m above the ground, or 40 m up in the canopy. Near the ground, there is little light, there are many vertical supports (such as small lianas and young trees), and there are terrestrial predators. Higher in the canopy, there are more horizontally continuous supports that provide convenient highways for arboreal travel and a greater abundance of leaves and fruits. Still higher, in the emergent layer, the canopy again becomes discontinuous, heat from the sun may be quite intense, and individuals are exposed to aerial predators. Primates, as

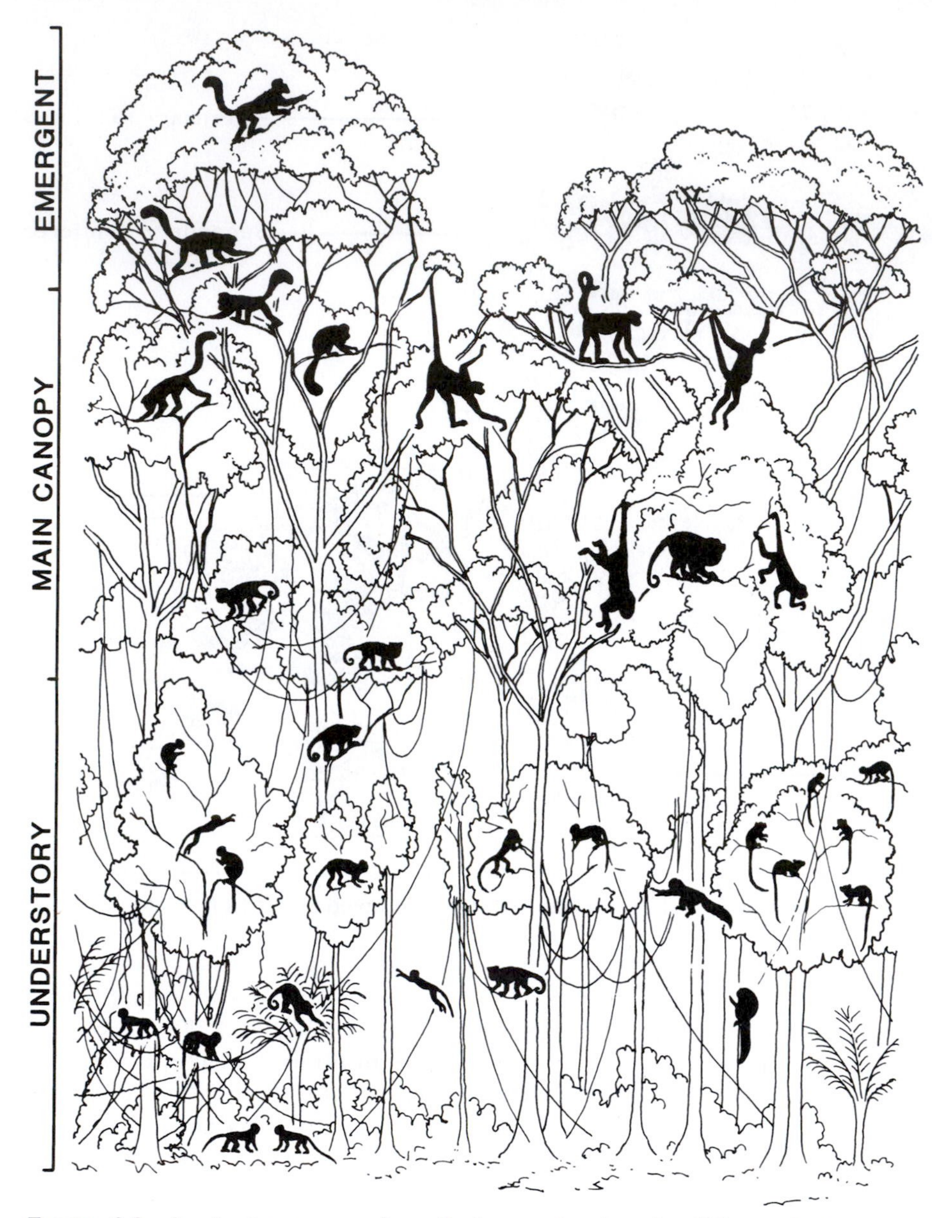

FIGURE 3.3 A rain forest scene from Suriname showing the different levels of a tropical forest, each with different types of substrates and each occupied by different primate species.

well as many other arboreal animals, often move and feed in specific forest levels; they have adapted to these different demands and opportunities.

Primates also specialize on different types of trees within a forest—trees that may have distinctive structures or produce foods with unique characteristics. Some species rely on bamboo patches within the forest, others on palms, and still others on vines. Primates often seem to specialize on trees with characteristic sizes and productivity; some species seem to feed primarily on small trees that produce small quantities of fruit, while others concentrate on the forest giants that produce huge bonanzas of fruit. In sum, there are many different niches within any single forest habitat, each of which offers a slightly different way of making a living for a primate.

Primates in Tropical Ecosystems

The tropical forests inhabited by most primates are the most complex ecosystems on earth, often containing thousands of plant species, hundreds of vertebrates, and innumerable insects and other invertebrates. It is important to remember that the evolution of living primates has taken place in conjunction with the evolution of other members of these complex environments. Plants, for example, are not the passive structural elements of the forest they might appear to be. Natural selection in plants has led to the evolution of elaborate and complex mechanisms for obtaining needed resources of light and nutrients, including defending leaves from herbivores, attracting pollinators, and ensuring that seeds are adequately dispersed and prepared for germination. The brightly colored, sweet, juicy fruits that form the diet of many primates have probably evolved those attributes for the purpose of attracting primates: once the fruits have been eaten, the seeds they contain will be scattered about the forest (e.g., Chapman, 1995). At the same time that many plants have evolved ways of enticing animals to help them disperse their seeds, they have also evolved mechanisms to protect their leaves and immature fruits and seeds from predators; they may cover them with spines, for example, fill them with indigestible materials or toxic substances, or encase them in hard shells. Primates, in turn, have evolved ways to overcome many of these plant defenses. Thus the dietary and foraging adaptations of living primates have evolved hand in hand with features in the tropical plants that affect their dietary choices (Sussman, 1995; Chapman, 1995).

In their roles as competitors, predators, and prey, the other animals of the forest have also had an important influence on the evolutionary history of primates. In Manu National Park, a pristine rain forest environment in the upper Amazon basin of Peru, **frugivorous** (fruit-eating) monkeys account for only about one-third of the biomass of frugivorous vertebrates (Terborgh, 1983). Birds, bats, various carnivores, and numerous rodents eat many of the same fruits as the primates and are often found in the same trees at the same time. There has certainly been competition among these different animals for access to the various food items in the forest. Many of the animals that inhabit the same forests as primates are interested not in the monkeys' food, but in the monkeys as food. Large felids (including lions, tigers, leopards, jaguars, and pumas) prey on primates, as do many large birds (Figure 4.8) and snakes. The presence of predators has exerted an important influence on the evolution of many aspects of primate ecology and behavior, including activity patterns, social organization, choice of sleeping sites, vocalizations, and coloration patterns (Isbell, 1994).

Only one primate, the Asian tarsier, relies exclusively on other animals for its food; but many primates, especially the smaller species, include various invertebrates and small vertebrates such as lizards as a regular part of their diet. As we shall see, primate species have evolved a number of unique predation strategies to exploit different types of prey in distinct parts of the forest structure. Capturing flying insects requires keen eyes and quick hands, and locating cryptic insects that live beneath the bark of trees or in leaf litter requires a keen sense of smell or hearing. Often such cryptic prey can be reached only by gnawing through the bark with specialized teeth, by ripping it open with strong hands, or by probing in crevices with slender fingers. Again, the evolution of primate adaptations reflects an interaction with the evolutionary history of other organisms in the forest.

Land Use

Primates live in a complex environment with many constantly changing variables. One way groups of primates deal with this complexity is to restrict their activities to a limited area of forest that they know well. Thus we find that primates are very conscious of real estate. In contrast with many birds or other mammals that have seasonal migrations, most primates spend their days, years, and often most of entire lives in a single, relatively restricted area of forest. To exploit this area effectively, they must know many things about it—the different food trees and their seasonal cycles, the best pathways for moving, the best water sources, and the safest places for sleeping. Many researchers have suggested that it is this need for knowledge of their environment that is responsible for the evolution of primate mental abilities (Milton, 1981). Nevertheless, while most researchers are convinced that primate groups have a very detailed knowledge of the distribution of resources in their area of forest, we still know very little about what type of "mental map" they may have of the geographical and temporal patterning of food resources or how this information is maintained or communicated.

There is a standard terminology used to describe the normal patterns of land use by primates and other animals (Fig. 3.4). The distance an individual or group moves in a single day (or night) is called a **day range** or **daily path length** (Fig. 3.4a, arrows). If we map all the day ranges for a primate group, we can see the total area of land used over a longer period of time, for example, a year. This area of land (or forest) is called the **home range** (Fig. 3.4b, dashed line). Often a group uses one part of its home range intensively, with only occasional, usually seasonal, forays into other parts. This heavily used area is called the **core area** (Fig. 3.4b, dotted line). Frequently the home ranges of neighboring groups of the same species overlap. In other instances there is almost no overlap and adjacent groups actively defend the boundaries of their home ranges with actual fighting or vocal battles. Such defended areas are called **territories.**

Activity Patterns

Most primates limit their activities to one particular segment of each twenty-four-hour day. Most mammals are **nocturnal**; they are active primarily at night and sleep during the day. By contrast, most birds are **diurnal**; they are active during the hours of light and sleep when it is dark. Some mammals are **crepuscular**; they are most active in the hours around dawn and dusk, when the light is at low levels. Nocturnality seems to be the primitive condition for primates; nearly three-quarters of the more primitive prosimians (lemurs, lorises, and tarsiers) are nocturnal, but there is only one nocturnal monkey. The majority of living

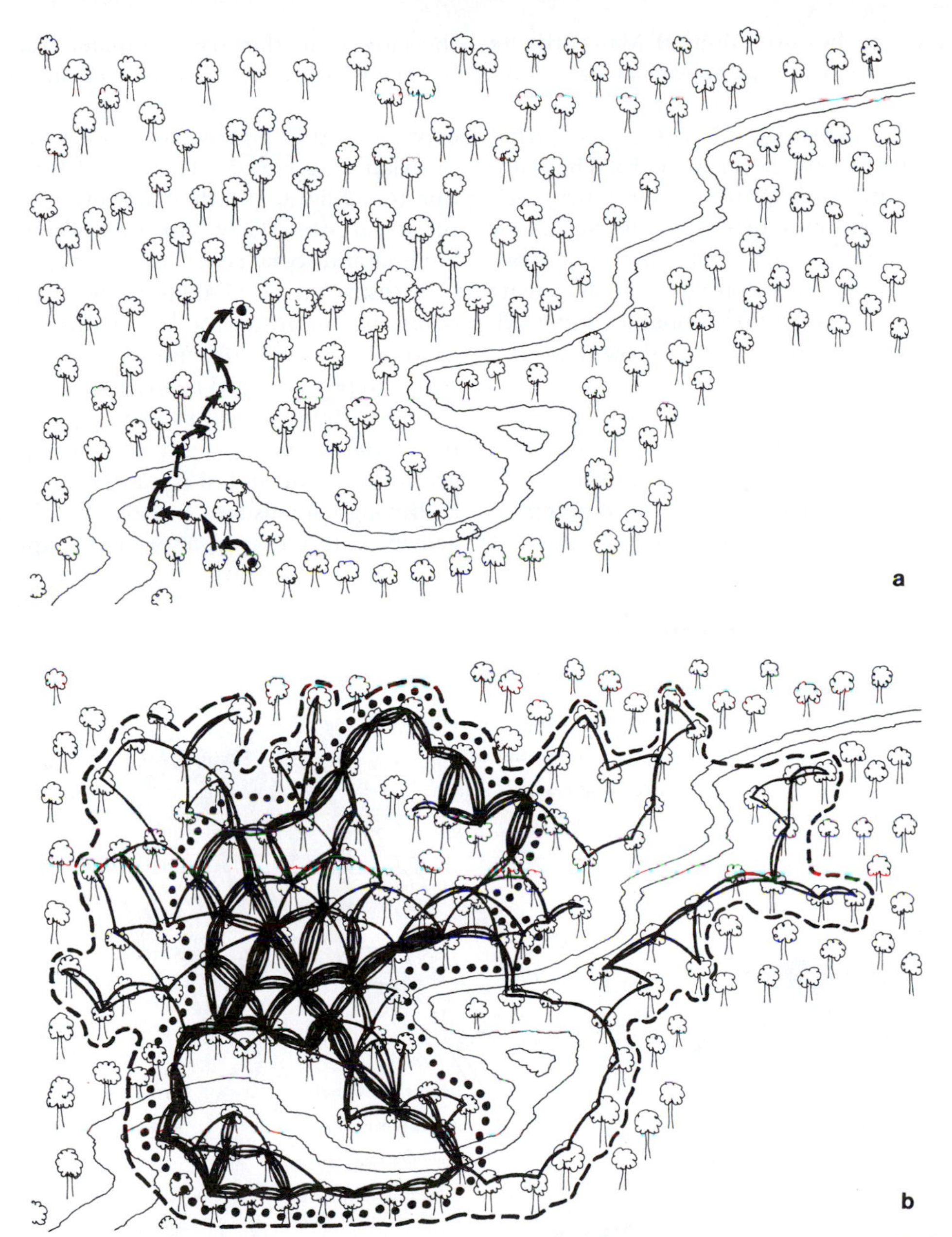

FIGURE 3.4 Primate land use: (a) The path an individual or group travels in a day is called a *day range* or *daily path length.* (b) If all day ranges are combined, the total area utilized by the group is its *home range* (dashed line). The part of the home range that is most heavily used is called the *core area* (dotted lines).

primates are, however, diurnal. Many primate species show peaks of activity at dawn and dusk and have a rest period at either midday, for diurnal species, or midnight, in nocturnal species; few, if any, are crepuscular. There are also primates with quite variable activity patterns. Rather than being strictly diurnal or nocturnal, they seem to be active at intervals throughout a twenty-four-hour day, an activity pattern that is called **cathemeral** (Tattersall, 1988). Several lemur species show this cathemeral activity pattern.

Each of these ways of life has its advantages and disadvantages (Fig. 3.5). Diurnal species presumably have a better view of where they are going, of available food, and of potential mates, friends, competitors, and predators. At the same time, they have a greater risk of being seen by predators. Nocturnal species are better concealed from many predators, except owls, and they have fewer direct primate or avian competitors. They avoid heat stress due to sunlight, and they may even avoid diurnal parasites. They have the difficulties in feeding and social communication associated with restricted visual abilities, but their vocal communication may be better during hours of darkness, and olfactory communication seems to be enhanced by the humid night air (Wright, 1996). It is thus not surprising that nocturnal primates tend to live in small groups or alone and to communicate primarily through smells and sounds. A cathemeral activity pattern enables a species to exploit the

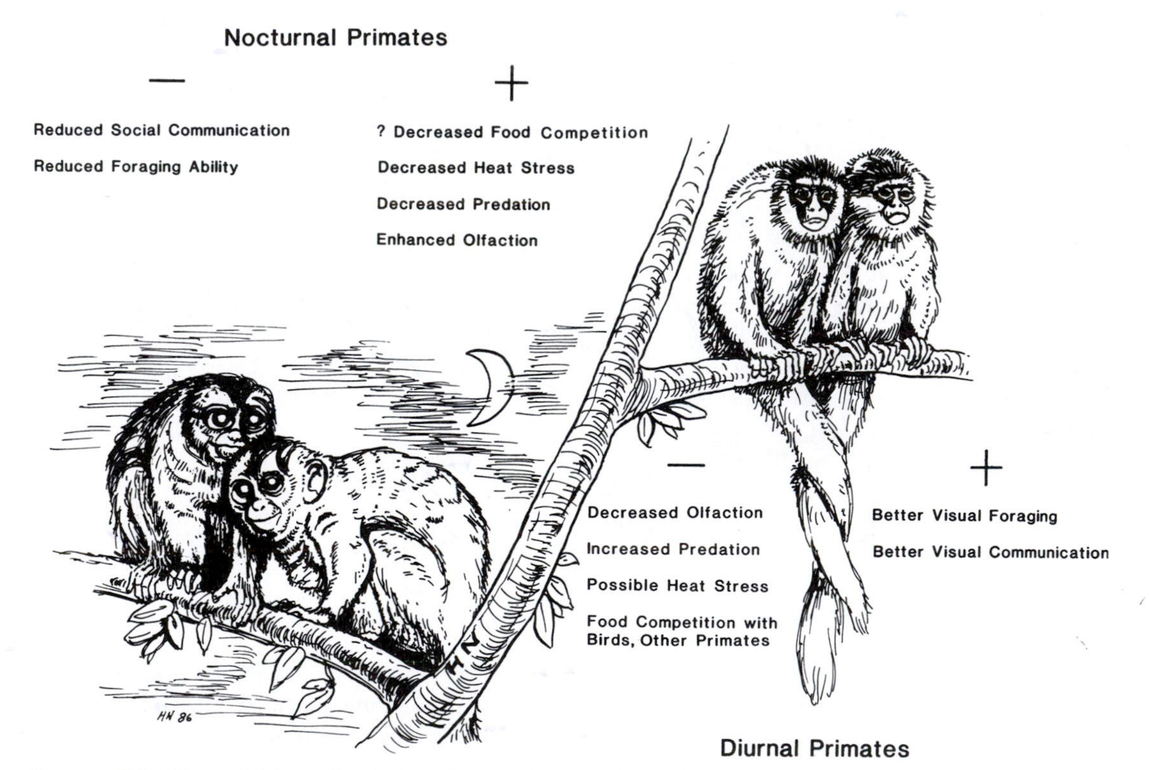

FIGURE 3.5 Potential benefits (+) and costs (−) of diurnality and nocturnality for two New World monkeys—the dusky tit monkey (*Callicebus*) and the owl monkey (*Aotus*).

advantages of both diurnality and nocturnality in conjunction with changes in temperature or food availability. The mongoose lemur, for example, is most active during daylight hours for the part of the year in which it feeds on fruits and new leaves; in the dry season, however, when these food items are scarce, it becomes more active at night and feeds on nectar (see Chapter 4).

A Primate Day

In addition to such drastic differences as diurnality and nocturnality, primates show differences in the way they spend each day or night. For most primates the day is generally divided among three main activities: feeding, moving, and resting. Activities such as sex, grooming, and territorial displays usually occupy a relatively small part of each day. There are exceptions, of course. During their short breeding season, males of some lemur species may spend half of their waking hours engaged in fighting with other males for the opportunity of mating with females during their brief period of sexual activity. For most primates, however, these activities are just occasional punctuations in long sequences of resting, feeding, and travel.

The distribution of activities throughout the day is usually not random (Fig. 3.6). Many diurnal primates generally travel early and late in the day and rest in the middle of the day. Most begin and end each day with a long feeding period. Gibbons, and many other species, show a temporal pattern in food preference; they eat fruit in the morning and leaves in the evening. This preference for fruit early in the day reflects a need for the quick energy available in fruits because of their high sugar and low fiber composition. Their choice of leaves in the evening perhaps reflects an attempt to maximize available digestion time (overnight) and to maximize various nutrient contents of leaves (e.g., Chapman and Chapman, 1991). Because plants cannot photosynthesize in the dark, the protein level in leaves increases during the morning and the sugar levels increase throughout the day (Ganzhorn and Wright, 1994).

Primate Diets

Variation in the choice of foods on a daily, seasonal, and yearly basis is one of the greatest differences among living primates and one that has far-reaching effects on virtually all aspects of their life and morphology. Primate diets have generally been divided into three main food categories: fruit, leaves, and fauna (usually insects and arachnids). Species that specialize on one of these dietary types are sometimes referred to as **frugivores, folivores,** and **insectivores** (or **faunivores**), respectively. These dietary categories accord well with the structural and nutritional characteristics of primate foods, and thus frugivores, folivores, and faunivores have characteristic features of teeth and guts that enable them to process their different diets (see Chapter 9).

In addition to particular nutritional and mechanical features, primate foods may vary considerably in their distribution and availability in both space and time (Fig. 3.7). Foods may be found in small patches that can feed one or two small animals or may be found in large patches that contain enough food to satiate many individuals of a large species. Similarly, in each habitat some foods may be densely distributed or relatively sparsely distributed. In addition to differences in spatial distribution, foods may show many types of temporal availability. Some foods may be available all year in some environments, others may be seasonal, and still others may be available only every two or three years. The distri-

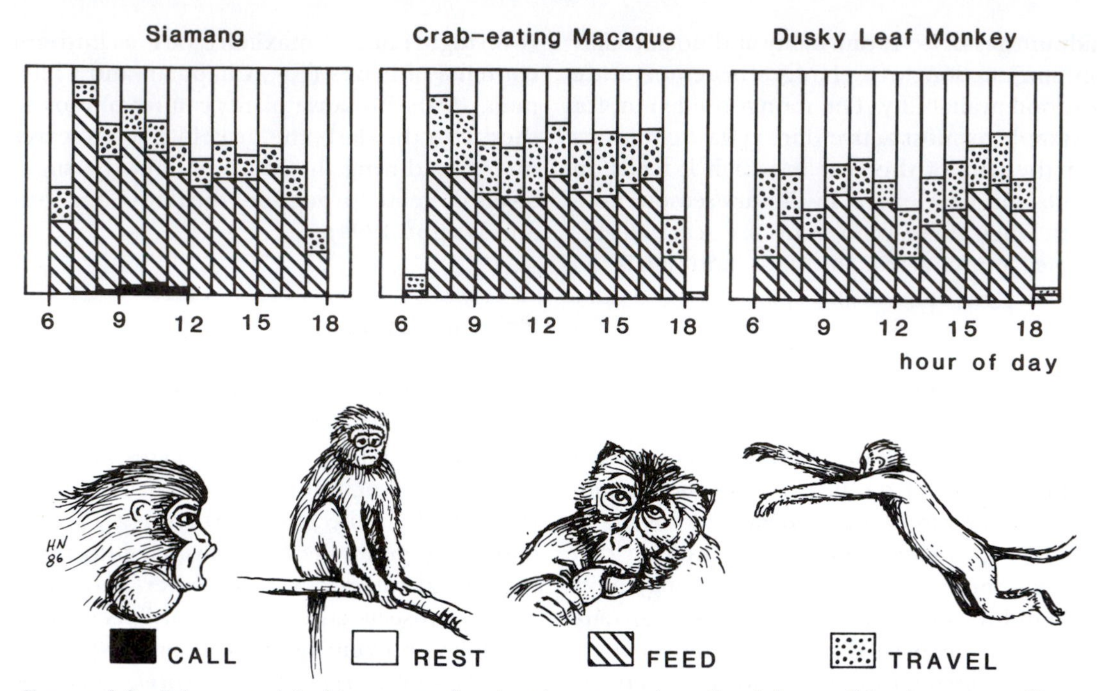

FIGURE 3.6 Primate activity histograms showing the proportion of each hour of the day spent calling, resting, feeding, and traveling by three Asian primates.

bution of food items is often correlated with primate grouping and ranging patterns. Folivores, for example, tend to have smaller home ranges and day ranges than frugivores, because leaves are more uniformly distributed in both time and space than are fruits. (see Chapter 9).

Like any categorization, these gross descriptions gloss over many subtle differences in the types of foods primates eat and the different problems they must overcome to obtain a balanced diet from day to day. For example, new leaves and mature leaves often have very different chemical, textural, and nutritional compositions and may be available during different seasons of the year (Ganzhorn, 1989). Some fruits appear in large clumps; others are more evenly scattered in small numbers over a large area. As already noted,

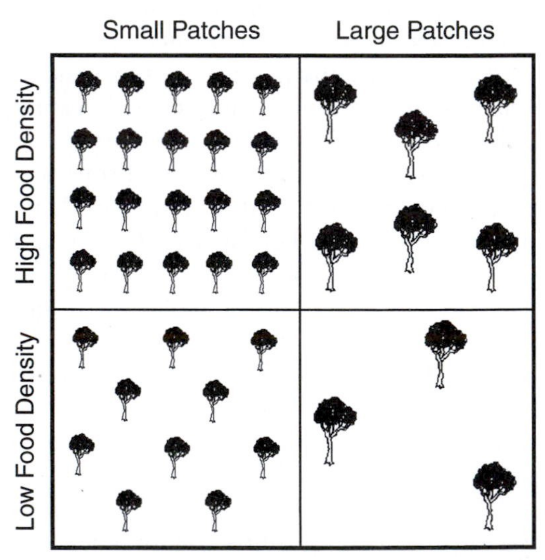

FIGURE 3.7 Different patterns of spatial distribution of potential food resources.

flying insects must be hunted differently than burrowing insects. In addition, feeding on foods such as gums (**gummivory**), seeds (**gramnivory**), or nectar (**nectivory**) is an important aspect of the dietary behavior of many primates and often requires unique adaptations (Nash, 1986; Roosmalen, *et al.*, 1988; Overdorff, 1992), but such behaviors do not fit easily into these three categories. Finally, there are many different ways of evaluating the importance of different food items in a primate's diet that may yield very different views of its dietary specializations. For example, the food that is eaten most often throughout most of the year may seem to be the most important from an energetic perspective, but other foods may contain rare nutritional elements not otherwise available. In many habitats, food availability may be highly seasonal, and the food that a species eats during the worst part of the year may be the most critical item. These limiting foods are called **keystone resources** (Terborgh, 1992).

The many intricate ways primates obtain their food are usually referred to as **foraging strategies.** They are called "strategies" because many factors are involved, and the behavior of any species is probably the result of compromises and decisions among an array of potential behaviors, each with unique costs and benefits. Thus, within any one dietary category, such as frugivory, different species may have quite different foraging strategies. One species specializes on fruits that are regularly available in small amounts throughout the forest, while another species may specialize on fruits that are found in more irregularly spaced, but larger, clumps. We would expect two such species to be similar in their dentition and digestive system but to have very different ranging patterns. Many of the descriptions of individual primate species in later parts of this book emphasize the subtle differences in foraging strategies that have been found among primate species within the same general habitat. These subtle differences in feeding habits demonstrate the richness of primate adaptations that have evolved over the past 60 million years.

Locomotion

A major aspect of the foraging strategy of any species, and an aspect of behavior that shows considerable variation among primates, is **locomotion,** the way animals move. No other order of mammals displays the diversity of locomotor habits seen among primates. Like diet, primate locomotor habits can be crudely divided into several major categories (Fig. 3.8), each characterized by different patterns of limb use: leaping, arboreal and terrestrial quadrupedalism, suspensory behavior, and bipedalism. Each of these ways of moving may provide a primate with better access to a particular type of forest structure or may be more efficient for traveling on a particular type of substrate.

Leaping (**saltation**) allows arboreal species to move between discontinuous supports, for example, between separate trees or between tree trunks in the understory. **Arboreal quadrupedalism** is more suitable for movement on a continuous network of branches and is probably less hazardous than leaping, especially for larger species. **Terrestrial quadrupedalism** enables a primate to move rapidly on the ground. **Suspensory behavior** allows larger species to spread their weight among small supports and also to avoid the problem of balancing their body above a support. Finally, **bipedalism** allows a species to progress on a continuous, level substrate while freeing the hands for other tasks.

As with dietary categories, the locomotor categories do considerable injustice to the actual diversity of primate movements. Some

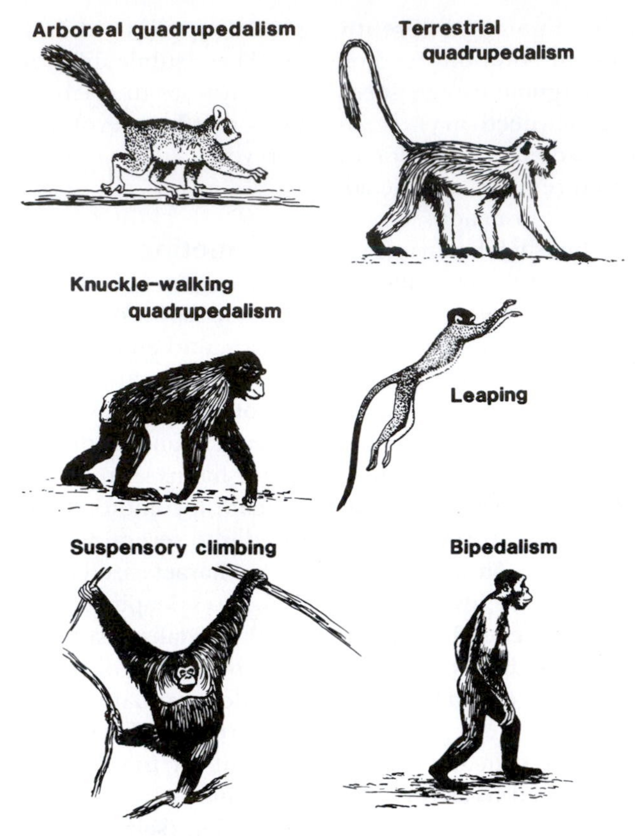

FIGURE 3.8 Examples of primate locomotor behavior.

species leap from a vertical clinging position, others from more horizontal supports. Quadrupedal walking and running may involve different gaits in the trees and on the ground. Suspensory behavior includes many different activities, including **brachiation** (swinging by two arms), climbing, and bridging. As with dietary groups this lumping of behaviors is mainly for the purpose of examining general patterns in either morphology or ecology (see Hunt et al., 1996, for a more detailed classification).

In addition to locomotion, primatologists also pay careful attention to differences in primate postures—the way primates sit, hang, cling, or stand while they obtain their food, rest, or sleep (Fig. 3.9) (McGraw, 1998). In many instances, feeding postures may be as important in the evolution of the species as locomotion. Primates that feed on gums or other tree exudates, for example, often must cling to the side of a large trunk. This clinging ability may be more important than the method by which the tree is reached. Likewise, the suspensory locomotion of many primates may be just a by-product of their need to hang below supports to feed on food sources at the end of small branches.

FIGURE 3.9 Examples of primate feeding postures.

Social Life

The size and composition of the groups in which primates carry on their daily activities and the methods they use to explore the area of land they inhabit are the most extensively studied aspects of primate behavior and ecology. All primates are social animals; they interact regularly with other members of their species (Charles-Dominique, 1971). Most diurnal species and some nocturnal ones are also **gregarious**—they feed, travel, and sleep in groups. The composition of primate social groups differs considerably, however, from species to species. Several distinct types of groups are particularly common (Fig. 3.10).

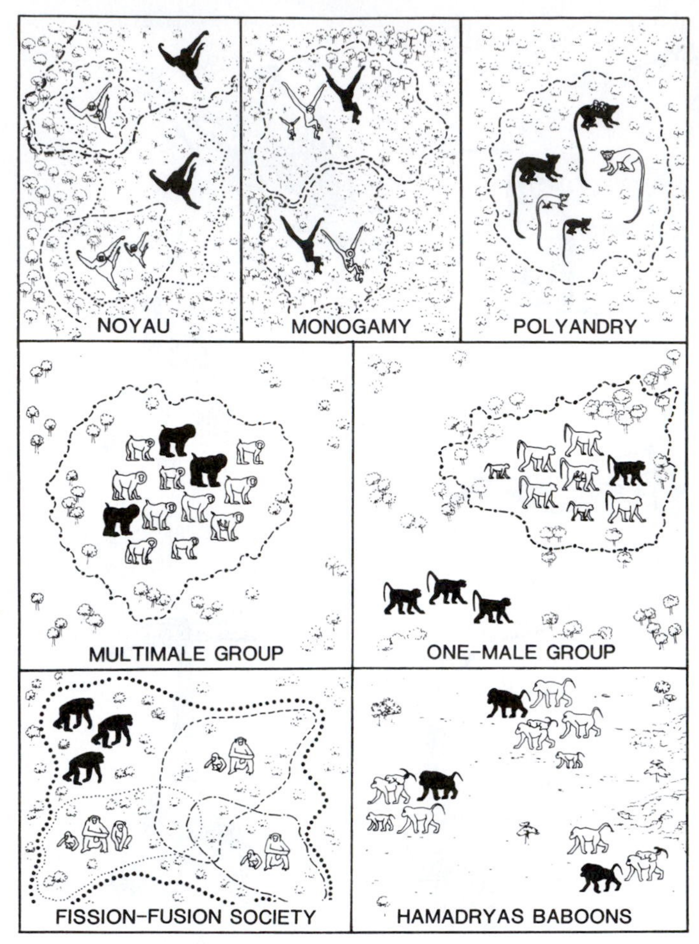

FIGURE 3.10 Common types of primate social groups. Light figures are females; dark figures are males.

The simplest, and certainly most primitive, social group is the **noyau,** which seems to characterize most primitive, nocturnal mammals (Charles-Dominique, 1983). The basic unit of this arrangement is the individual female and her offspring. In the noyau, adult males and females do not form permanent mixed-sex groups; rather, individual males have ranges that overlap several different female ranges. Thus, even though the two sexes do not travel together regularly, they interact often enough for males to monitor the reproductive status of the females and for females to have a choice of potential mates. This type of spatial arrangement may be associated with several different types of mating systems and patterns of intrasexual competition, depending on the synchrony of female receptivity (Kappeler, 1997).

The next simplest grouping, at least in terms of numbers, is the **monogamous family,** consisting of one adult female, one male, and their offspring. Nonhuman primates that live in these families have generally been thought to mate for life (but see Palombit, 1994), and there is usually intense territorial competition between adjacent groups. Most of this competition appears to be intersexual—males compete to exclude other males and females compete to exclude other females. The actual role of males in monogamous primate groups has recently come under considerable scrutiny and is the subject of considerable debate. Although males play a prominent role carrying infants in some species (owl monkeys, titi monkeys, siamang), males in other monogamous species seem superficially to contribute very little to infant care. One possibility is that resident males help deter infanticide by other males seeking to take over access to the female (van Schaik and Dunbar, 1990). Alternatively, the male's presence may be largely a mating strategy rather than directly related to child care (van Schaik and Paul, 1997).

Under naturalistic conditions, tamarins (small New World monkeys) live in **polyandrous** groups consisting of a single reproducing female and several sexually active males (Goldizen *et al.*, 1996). In these groups, several of the males, as well as many other group members, participate in the care of offspring, and some authors have suggested that these groups are better viewed as a "communal breeding system" (Garber, 1997).

Many primate species live in groups consisting of a single adult male along with several females and their offspring. These are usually based on **matrilineal** social systems and often called **female-bonded groups.** Adult males not living with females band together to form separate **all-male groups** in some species, and in others they live alone. In many species characterized by **one-male groups,** outside males regularly attack the troop, oust the resident male, kill dependent infants, and mate with the females (Hausfater and Hrdy, 1984). Although the resident male in the one-male groups of Asian langurs seem to be defending reproductive access to the females in the troop (van Schaik et al., 1992), in some cercopithecine species the mating season is characterized by a large influx of "extra males," and resident males seem to fare no better than the "visitors" (e.g., Rowell, 1991). Clearly, there can be considerable diversity in social relationships and reproductive activity within similar demographic patterns.

In contrast to these one-male groups, many species live in large bisexual groups that include several adult males, numerous females, and offspring, all of which forage as a group. Such groups are characterized by complex intratroop politics and competition. The distinction between one-male groups and such **multimale groups** is very difficult to make for many species of living primates. As the young males mature, many one-male groups seem to become multimale groups. Also, the composition of primate groups within a species may depend on factors such as troop size and population density. Because of this blurry distinction, Eisenberg and his colleagues (1972) have introduced an intermediate type of social group, the **age-graded group.**

Among spider monkeys and chimpanzees, the social units are even more fluid—being what is called a **fission–fusion society** (Chapman et al., 1993). The size and composition of subgroups within a community may vary from day to day, depending on availability of resources and on individual relationships. In many species, however, adult females tend to travel with their offspring and males tend to travel together. However, these subgroups frequently join to feed at a particularly rich food source. Although this fission–fusion behavior is especially characteristic of chimpanzees and spider monkeys, many other primates regularly divide into smaller foraging groups in some seasons or when feeding on some types of food (Kinzey and Cunningham, 1994).

There are many primate species for which social organization cannot be characterized so easily in terms of the numbers of males and females, because the groupings change for different activities or perhaps in different seasons. Hamadryas baboons, for example, forage all day in small one-male harem groups consisting of one male, one or a few females, and their offspring. Each evening, however, dozens of these small groups congregate on a single sleeping cliff, and sometimes the troop travels as a unit from one area to another.

Finally, there is increasing evidence from many species, including mandrills, golden monkeys, proboscis monkeys, and some New World monkeys, of separate troops coming together amicably on a regular basis to form **supertroops,** either during foraging or resting. The significance of these supertroops is not altogether clear. They may represent "family reunions" of related individuals, or they may represent some type of foraging strategies to maximize individual knowledge of food resources, to preempt competition, or to exploit superabundant resources.

These categories of social organization, like all such classifications, provide us mainly with a convenient framework for comparing different species. The ultimate goal of such classification is to facilitate investigation of the factors that have given rise to this diversity in social organization. Primate social groupings are the result of many selective factors, each of which influences in a different way the size, composition, and dynamics of the social group. It is the dynamics of interindividual interaction, and the genealogical relationships among individuals and groups of individuals, rather than just the numbers of males and females, that provide the real clues to understanding primate social systems (e.g., Harcourt, 1992). For example, single-male social groups of most Old World monkeys are composed of closely related females and their offspring and an unrelated adult male. In contrast, gorilla single-male groups usually consist of unrelated females attached to an adult male.

Individuals, Groups, and Communities

Although most primates are regularly found to live in species-specific types of social units, most individuals do not spend their entire lives in the same group. In most Old World monkeys and in *Lemur catta,* females generally stay in their natal group and males emigrate to other groups. In chimpanzees, males stay in their natal group and females migrate. In many other species (howling monkeys, tamarins), both sexes migrate from group to group. Interestingly, this last pattern can give rise to a situation in which a single primate group continues to occupy the same home range from year to year even though there is no continuity in the group membership (Goldizen et al., 1996). The relative importance of factors such as inbreeding avoidance, resource distribution, and "phylogenetic background" in determining the patterns of immigration and emigration in different species is far from elucidated but is one of the most intriguing problems in understanding primate life histories (e.g., Pusey and Packer, 1986; Moore, 1992).

From the perspective of an individual primate, its current social group is just one level in a hierarchical series of relationships that includes kin relationships, intergroup relationships with unrelated individuals, and relationships with members of nearby groups of either kin or unrelated individuals (Rowell, 1993; Dolhinow, 1994). The behavior of each individual in a group, and consequently the behavior of the group as a whole, results from the complex combination of all of these relationships (e.g., Fleagle and Wright, 1993). For example, the ranging behavior of a primate group frequently seems to involve both foraging behavior, to meet the needs of group members, and monitoring the whereabouts and the individual composition of adjacent groups. Likewise it is a commonly observed phenomenon that members of one group may have special affiliations and mating interactions with members of other groups.

From the perspective of the individuals that comprise them, primate social groups are not stable, permanent units; they are highly complex and dynamic networks that continually change as individuals are born, mature, emigrate, immigrate, mate, reproduce, and die. Selective factors affect individual group members in different ways. Factors that are of critical importance to one individual may be less significant to another of a different gender, age, or kinship or to the same individual from month to month or year to year as its reproductive status or its relationships with other individuals change.

Most of our early knowledge of primate social behavior came from studies that lasted only one or two years. There are, however, increasing numbers of long-term field studies as well as studies of genetic relationships among individuals that provide us with a more long-term and biologically realistic perspective on primate social behavior. Rather than yielding simple answers, this increase in information is demonstrating the many factors involved and their potential interactions in determining the social interactions among primates. This dynamic nature of primate groups reinforces the view that primate social groups are the result of many selective factors acting on each individual and that to understand the diversity of behavior we find within our order, we have to consider all of the potential costs and benefits of group living to an individual as well as the phylogenetic history of each species (DiFiori and Rendall, 1994).

Why Primates Live in Groups

Compared with most other types of mammals, primates are extremely social animals. This behavior is evident, not only in the diverse types of social groups just described, but also in the elaborate systems of scents, postures, facial expressions, and vocalizations that primates have evolved for communicating with their conspecifics. Primate social behavior has evolved through natural selection. Like all other primate adaptations, social behavior can be viewed as the result of a complex and dynamic balance of selective advantages and disadvantages. From an evolutionary perspective, the fitness of an individual animal, or its evolutionary success, is equivalent to its **reproductive success**—the number of reproductively successful offspring it contributes to the next generation. Thus, in evaluating the importance of different factors that might select for group living in primates, we must try to determine how group living would affect the fitness of an individual.

From the point of view of the individuals that make up primate groups, there are four potential advantages to group living: greater protection from predators, improved access to food, better access to mates, and assistance in caring for offspring. Each of these potential advantages is likely to have greater selective value for some individuals than for others, depending on the individual's gender and age, the reproductive physiology of the species, and the ecological environment. Each potential advantage also must be balanced against the potential disadvantages of group living:

greater visibility to predators, and increased competition with other individuals for these same resources of food, mates, and assistance in rearing offspring. The behavioral and physiological adaptations individual primates have evolved for maximizing their survival and that of their offspring in this maze of advantages and disadvantages are referred to as **reproductive strategies.**

Protection from Predators

Predator avoidance seems to be the major factor that selects for group living in diurnal primates (Janson, 1992; van Schaik, 1983; Dunbar, 1988), even though actually measuring the influence of predation on living populations is extremely difficult (Cheney and Wrangham, 1986; Isbell, 1994; Boinski and Chapman, 1995; but see Goodman et al., 1993). Individuals living in groups gain increased protection in several ways. Most significantly, each individual benefits from the eyes, ears, and warning calls of other individuals; thus individuals are safer, and each can spend less time in vigilance behavior. Groups of individuals can gang up on a potential predator. Should the group have to flee, any individual in the group is less conspicuous and more difficult for a predator to isolate and attack. Finally, it has been argued that members of groups may have a better knowledge of their home range and the likely whereabouts of predators than non–group-living individuals.

In contrast, it has also been suggested that group living may, in some cases, make an individual more susceptible to predation. A group may be easier for a predator to locate than an isolated monkey, and some predators may have large appetites that lead them to prefer "clumped prey." Thus some species may adopt a **cryptic strategy** to avoid predators.

Improved Access to Food

The reproductive success of any individual depends ultimately on its ability to obtain enough food for itself and its offspring. As noted earlier, the ways animals obtain and select their diets from the array of potential food sources are often referred to as **foraging strategies.** Foraging strategies are, in a larger sense, just one part of reproductive strategies of individuals. Overall, it would seem most likely that group living would be detrimental to the foraging success of an individual because each individual would have increased competition for food resources from other group members. Thus individuals living in groups should need to find larger food sources and/or travel farther and visit more food sources each day.

The most important factor in determining the size of groups in which primates live seems to be the distribution of food resources in time and space (Fig. 3.7) (van Schaik, 1989; Isbell, 1991). Primate species relying on foods that are found in small, evenly scattered patches, such as gums or many small forest fruits, usually live in small groups. Those that specialize on foods such as figs, which are usually found in gigantic but erratically spaced patches, tend to live in large groups. It is easy to see how the distribution of food resources can limit the size of groups that are able to feed on any single resource patch.

However, there are also several ways in which group living may provide individuals with better access to food than they might be able to obtain by foraging alone. Many primate species actively defend food sources—in some cases individual food trees, in others the troop's entire range. In general, disputes over food resources are often resolved by group size; larger groups can displace smaller groups in preferred food trees or in preferred areas. By joining a group, individuals gain access to its resources. There is obviously a fine balance between a group size that is small enough for the group to subsist on a particular resource or set of resources and one that is still large enough for the group to defend those resources from other groups. It should not be surprising that many primate groups that defend their food resources are composed of closely related individuals, usually females and their offspring (Wrangham, 1980). Living in groups may also help primates locate food. Individuals may benefit in several ways from communal knowledge about the location of food sources, either through the memory of other individuals or through food calls given by other group members who are foraging semi-independently. There are also suggestions that primates feeding on insects may benefit from the disturbance caused by other troop members who inadvertently flush out insects as they move.

As already discussed, primate foods may vary considerably in both their patchiness and their density. Accordingly, foods with different patterns of distribution should select for different types of grouping and ranging patterns.

Access to Mates

Sexual reproduction requires that each reproductively successful male and female find a mate of the opposite sex. The reproductive strategies of males and females are, however, quite different for virtually all sexually reproducing animals. A critical aspect of primate reproduction that influences individual reproductive strategies is the marked asymmetry in the roles played by males and females during the early development of offspring. Female primates, like all female mammals, nourish and carry developing young for many months before birth and also provide milk for the infant for months or years after birth. In contrast, the investment by a male primate to its offspring during this part of development is much less—and theoretically could be as little as a single sperm cell.

There are several consequences of this dramatic difference in the time and energy required of male and female primates. First, because of the time required by gestation, the maximum number of potential offspring a female primate can have in a lifetime is far less than the number that can be sired by a male, and the female's offspring must necessarily be more evenly spaced in time. With unlimited food resources, a single female with a twenty-year period of reproductive fertility, a litter size of one, and a six-month gestation period can theoretically (but not actually) produce forty offspring in her lifetime. To achieve this reproductive success, she will (again theoretically) need to associate with a male for mating purposes only briefly every six months. In many primate species, a male theoretically could father the same number of offspring in a week (or even a day) if that number of receptive, fertile females is available. Thus, in the number of offspring they can physiologically produce, females are limited primarily by time and food resources, whereas males are limited by their access to females.

There are other consequences of this asymmetry in required investment by males and females in early reproductive investment. One is that females are always sure that the offspring they bear are their own. An individual male, on the other hand, can never be sure that he is the father of a newborn. Only by limiting the access of his mates to other males can he increase the likelihood that the offspring they produce are his own.

From these physiological differences in the relative minimal investment required to produce offspring and the relative certainty of parentage, we can predict that the theoretically optimal strategies of males and females for maximizing their reproductive success will be very different. The most successful male is the one that mates with the greatest number of females and excludes other males from mating with these same females in order to ensure that all offspring are his own progeny. Females, on the other hand, have fewer obvious strategies for producing greater numbers of offspring. Female reproductive strategies seem to emphasize the quality rather than the quantity of offspring. Because every offspring involves such a large investment in time and energy, female strategies are concerned with ensuring that the male that sires the offspring is likely to engender healthy, strong progeny through paternal investment in such forms as protection from predators and access to food resources.

From these considerations we would expect male reproductive behavior to involve more intensive competition with other males for access to reproductively active females. The relatively greater intensity of male–male competition for access to mates over that expected in females is generally regarded as a major cause of sexual dimorphism in overall body size and in the size of canine teeth, features that are important in fighting and in dominance displays (see Chapter 9). Thus it seems that access to mates plays an important role in the reproductive strategies of males and a major selective factor in males joining groups. In general, it is the distribution of available females that determines male ranging behavior (e.g. Wrangham, 1979).

Access to many potential mates seems to be a less important factor favoring group living for females of many primate species. Indeed, adult females outnumber males in most primate groups. However, there is certainly competition among females for access to mates, and females frequently mate with multiple males in species that live in groups with numerous males. For females, a major important factor is the contribution that one or more of these males can make toward the survival of the offspring.

Assistance in Protecting and Rearing Offspring

Mating is only the first step in successful reproduction. An individual's reproductive success is determined by the number of offspring that live to reproduce themselves, not by the number of conceptions. Offspring that do not survive to successfully reproduce are, from an evolutionary perspective, a wasted effort. For primates that give birthto relatively helpless young that require a relatively long time to reach adulthood, parental investment in the growing offspring is a particularly important aspect of reproductive behavior.

Because of their greater initial investment in offspring and the certainty of maternity, females always make a substantial contribution toward the upbringing of infants in a primate group. Milk is expensive to produce and females may eat twice as much food when they are lactating. Thus it is not surprising that female primates usually solicit and receive help in raising offspring from other troop members. There is, however, considerable variability among primates in the contributions of males, females, and other, less closely related troop members in the care and rearing of immature animals. Investment in infants and dependent young seems to be correlated with the degree to which individuals are likely to be related to the offspring. In some monogamous species, such as titi monkeys and owl monkeys, males often contribute as much or more to the care of infants as do females. In larger, more complex social groups, adults of both sexes often assist the mother in caring for infants. In many primate societies, the adult females in the group are probably related, so infants are the "nieces" and "nephews" of other troop members. In addition, female primates have evolved many behavioral strategies to ensure assistance in rearing infants. By mating with several males, for example, females in multimale troops can confuse the issue of paternity and perhaps elicit some investment from all of the males, since none can exclude the possibility that an infant is his offspring.

Perhaps more significantly, a male's willingness or ability to care for offspring may be a prerequisite for future matings. Baboon females are more likely to mate with males who have helped care for her offspring in the previous year (Smuts, 1985), and some researchers have suggested that because paternity is so uncertain for most male primates, male care is probably best seen as a mating strategy, even in monogamous species (van Schaik and Paul, 1997). Access to help in rearing offspring from other individuals of all genders and ages is probably a more important factor favoring group living by females than is access to mates.

In addition to direct care through infant carrying or baby sitting, males of many species may provide critical assistance to female consorts through deterring infanticide by other males. It has been argued that this infanticide protection is actually more important than any infant carrying in many monogamous species (van Schaik and Dunbar, 1990) and perhaps especially in Malagasy primates that seem to live in groups with a relatively high number of males to females (Kappeler, 1996).

Phylogeny

Although the details of grouping behavior and social interactions among primates are undoubtedly the result of natural selection for maximum individual fitness, evolution by natural selection has not operated on a blank slate for each species. Rather, each species has evolved from a particular phylogenetic background. There may be more than one satisfactory solution to particular ecological problems, and it is quite evident that closely related taxa are often similar to one another in their social behavior and that different primate radiations often show distinctive patterns of social organization (Di Foiri and Rendall, 1994: Rendall and Di Fiori, 1995). Thus many of the most revealing studies of the factors determining social adaptations involve comparisons of closely related species or even of the same species in different environments. In these studies it is possible to control for phylogenetic differences and tease out the ecological parameters responsible for the social differences.

Polyspecific Groups

In addition to living regularly in social groups composed of members of their own species, some primates are commonly found in groups composed of several species. These **polyspecific groups,** or mixed-species groups, are particularly common among species of *Cercopithecus* (Fig. 6.13) and *Saguinus* (Fig. 5.22). In many cases these groups containing members of several different species may forage and sleep together for several days, in other instances the interactions may be for only a few hours. It seems likely that two of the factors that select for intraspecific grouping behavior—predator avoidance and shared information about food resources—are responsible for these polyspecific associations as well (Gautier-Hion et al., 1983; Norconk, 1990).

Primate Life Histories

In the previous sections we have discussed primate lives and activities from the relatively static perspective of an adult individual at a single point in time. What does it do in the course of a 24-hour day? What does it eat? How does it move? What type of social group does it live in? In addition, primate lives, like those of all organisms, have a temporal dimension over which the features of their life take place. Many aspects of the lives of individual primates, and the way in which they expend their energies, are constantly changing from conception to birth to sexual maturation to death. Moreover the timing of developmental and reproductive events varies dramatically from species to species (Fig. 2.29). Consider, for example the timing of reproduction in the smallest living primate, the mouse lemur, and in the largest, the gorilla. In the 10 years it takes a female gorilla to reach sexual maturity, a female mouse lemur born on the same day could theoretically have left 10 million descendants (Harvey, 1992). Why do gorillas take so long to grow up and produce offspring so slowly? Why do some species produce twins twice a year and others a single offspring every four or five years?

Analyses of the timing of events in the lives of individuals of different species are called **life history studies** (Stearns, 1992; DeRousseau, 1990). Many temporal aspects of primate biology discussed in this and the previous chapter fall under the category of life history features, including growth and development, and patterns of immigration and emigration of individuals over the course of their life span. However, most studies of life history tend to focus specifically on the allocation of energy and time towards reproduction. The patterns that characterize individual species or different sexes within a species are called life history strategies because they involve a balance of diverse selective factors. Life history studies are one of the most exciting areas in evolutionary biology today.

Life History Diversity

Primates show considerable diversity in many aspects of life history (Fig. 3.11; e.g. Harvey and Clutton-Brock, 1985; Harvey, 1992; Ross, 1998). Gestation length may vary from about 60 days to over 250 days and birth mass varies from less than 10 grams in the smallest species to more than 2000g in apes and humans. Most primate species, like humans, give birth to one offspring at a time, but for many species twins are the norm, and a few regularly produce triplets. Once they are born, primates show very diverse ways of caring for their infants. In most species, the offspring are carried around after birth by their parents or other relatives. However in some species, newborns are kept in nests, and in others they are parked on nearby branches while the mother feeds. The time primate mothers spend nursing their offspring varies from less than two months in some species to more than four years in others (Lee, 1997). In some species an individual may reach sexual maturity in less than a year while in others it may take over ten years, with males and females following very different schedules. Finally, the expected life span of an individual is less than 10 years in some species, but over fifty in ourselves (Hill, 1993).

There are numerous problems in trying to make sense of all this diversity. Many aspects of life history, such as life span or weaning age, are poorly known for most species, or only known for captive individuals in others. There are considerable analytical problems involved in comparing features among members of closely related species (Harvey and Pagel, 1991). Perhaps most importantly, there are just too many potential variables to consider at the same time. Nevertheless, studies of primate life history have been tremendously productive during the last decade and there are a number of general patterns in the way

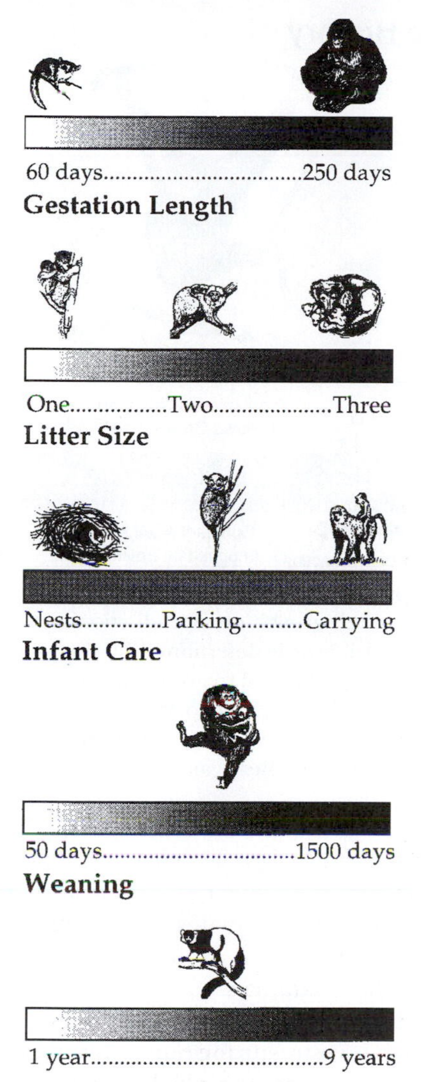

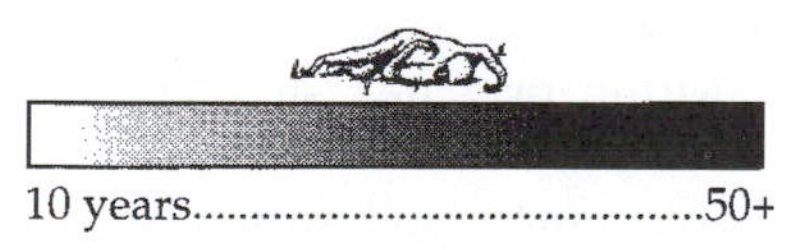

FIGURE 3.11 Primates show striking interspecific differences in many aspects of their life histories.

primate lives are organized. In some aspects of our life histories, primates are very similar to other mammals. For example, in primates, as among virtually all mammals, weaning weight seems to be fixed at 4 times birth weight regardless of the timing of growth or the size of the species (Lee *et al.,* 1991). However, there are also ways in which primates are unusual as an order or individual primate families or species have their own unique life history features.

General Patterns—Small and Large / Fast and Slow

The most obvious correlate of life history variation is body size. In general, larger species have longer gestation, fewer and larger infants, longer weaning ages, and longer time to sexual maturity. Their lives are lived at a much slower reproductive pace than those of smaller animals. Primates follow these trends found in mammals as a class. However, the factors related to reproduction seem to be more highly correlated with one another than with size alone. Thus, even when body size is held constant, analyses of life history differences still show a fast–slow continuum between and among primate species, as well as among mammals as a group, suggesting that overall variation in life history features is determined by environmental factors such as age-specific mortality rates (Promislaw and Harvey, 1990).

Primate Patterns

In a broad comparison of life history variables, Charnov (1993, Charnov and Berrigan, 1993) has shown that the relative amount of energy an individual of a species devotes to either growth or to production of offspring seems to be a constant function of body size for growing juveniles and for reproducing adults (Fig. 3.12). Thus, the amount of energy that a mother can expend on production of offspring is determined by her size. However, the relative contribution allocated to growth

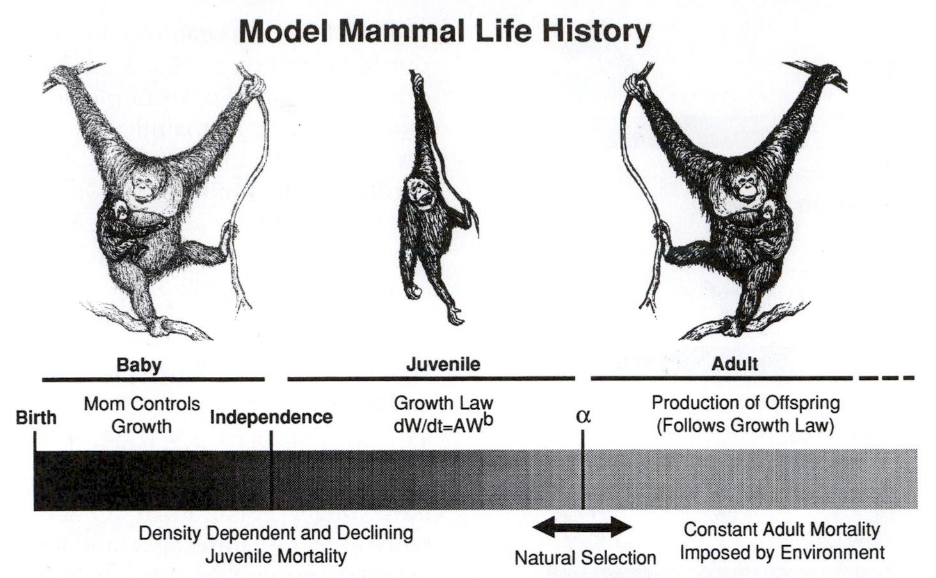

FIGURE 3.12 Mammalian life histories can be represented as a production model in which individuals allocate a portion of their resources to growth or reproduction as a function of their body weight. The growth rate of an infant is determined by the resource allocation of the mother; after weaning the energy devoted to growth is proportional to body size; at adulthood, the growth energy is allocated to reproduction. According to this model, primates contribute a relatively small amount of their energy budgets to either growth or reproduction. (From Charnov and Berrigan, 1993).

and reproduction varies considerably among groups of vertebrates.

Compared to other mammals, we primates invest remarkably little of our daily energy budgets in growth and reproduction throughout our lives (Charnov and Berrigan, 1993). For our size, we primates have smaller litters, take longer to reach sexual maturity, and live longer than members of other orders (Fig. 3.13). These differences are effected through our lower rates of prenatal and postnatal growth. Correlates of our slow rate of postnatal growth are the relatively low-nutritional content of primate milk compared with that of many other mammalian mothers (e.g. Oftedal, 1991) and our long juvenile growth period (e.g. Janson and van Schaik, 1994). The ultimate reasons for the slow rate of primate lives are less clear. One possibility is the high energetic cost of our large brains; another is that primates suffer relatively low rates of juvenile and adult mortality compared with other mammals.

Within this general trend of low-level reproductive investment for our order, there are a number of patterns of life history variation associated with particular phylogenetic groups and a few broad associations between life history variation and aspects of ecology. For example, within primates, strepsirhines have a much lower neonatal mass and larger litters than haplorhines; and within strepsirhines, lorises have long gestation times and slow growth rates while cheirogaleids have short gestation times and fast growth rates compared with lemurs (Rasmussen and Izard,

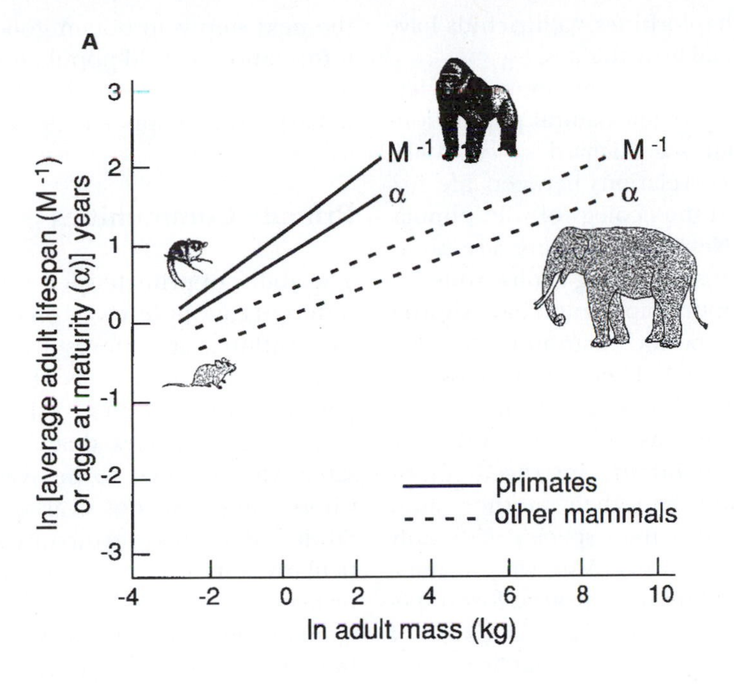

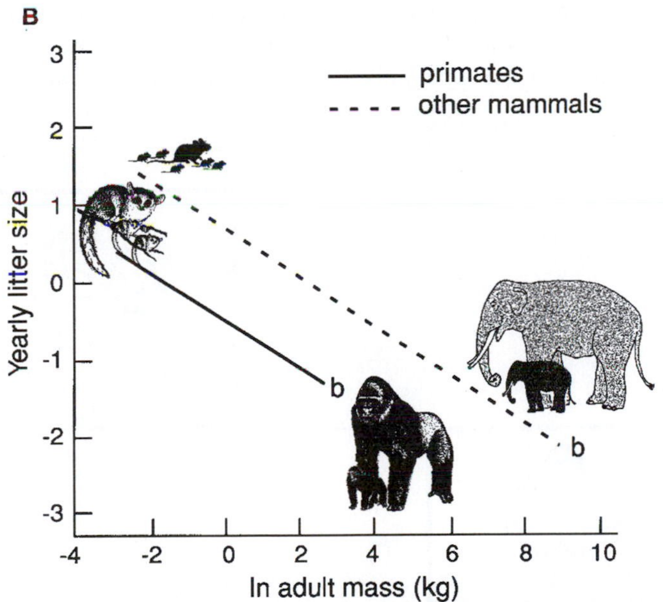

FIGURE 3.13 Compared with other mammals, primates have a longer adult lifespan and higher age of maturity (A). Primates also produce smaller litters than other mammals (B). (From Charnov and Berrigan, 1993).

1988). Among haplorhines, callitrichids have very high postnatal growth rates.

Because demographic and mortality data are available for very few natural populations of primates (but see Richard *et al.*, 1991; Wright, 1995), correlations between life history patterns and the ecology of wild populations are few. Nevertheless there are some broad associations emerging. Folivorous anthropoids grow more rapidly and have shorter postnatal growth periods than nonfolivorous species (Leigh, 1994). There are also associations between growth rates and social interactions; species, such as colobines and callitrichids, in which infants are cared for by individuals other than their mother show higher growth rates than species with only maternal care (Fig. 3.14). Also, within radiations of Old World monkey groups, forest species have slower growth rates and later ages of first reproduction than species that live in secondary forests or more open habitats (Ross, 1992, 1998). As life history variables become increasingly well documented for primates, the next step is to obtain good demographic information on wild populations and to relate life history variation to details of ecology and behavior (e.g. Kappeler, 1996).

Primate Communities

The ability of primate species to specialize on different canopy levels or on different types of food within a single forest habitat—and to do so during different parts of each day—often permits several species to thrive in the same habitat. Two species that are found in the same area are said to be **sympatric**; species whose ranges do not overlap are **allopatric.** Studies of groups of sympatric species are particularly important for our understanding of primate adaptations because they allow direct comparison of ecological variables (locomotion, diet, social organization) within one environment. Such studies of primate communities are essentially natural experiments in which the climate, the forest, and the com-

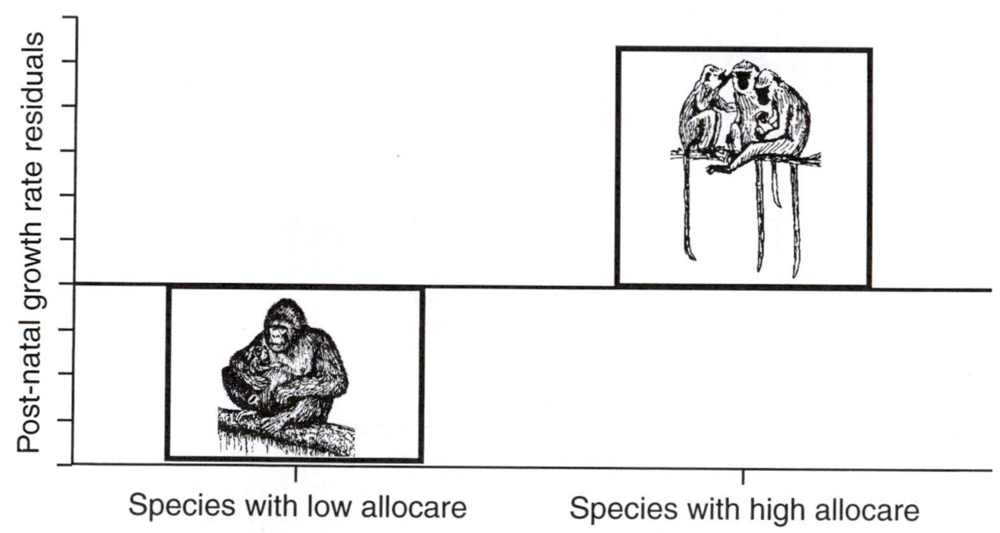

FIGURE 3.14 Primates with high amounts of nonmaternal care show relatively higher rates of infant growth than species with low amounts of nonmaternal care. Courtesy of Caroline Ross.

peting species (such as predators, parasites, or other arboreal mammals or birds) are held constant. Thus the observer can see how changes in one ecological variable are correlated with changes in other variables.

In the following chapters we generally discuss primate species individually, but we often also compare related species that are sympatric. In this. way we not only highlight the diversity of related animals but also see how differences in one parameter, such as diet, change in relation to other parameters, such as locomotion or social organization. Many of this book's illustrations show the ecological features of several sympatric species, enabling comparisons both within selected communities and across various communities. In each taxonomic chapter we discuss the diversity of adaptations within a particular evolutionary radiation, and in Chapter 8 we compare the primate communities within and between major continental areas. Only by looking at primates from several perspectives can we appreciate how adaptation and evolution have produced the diversity of species we see today.

BIBLIOGRAPHY

PRIMATE HABITATS

Chapman, C. A. (1995). Primate seed dispersal: Coevolution and conservation implications. *Evol. Anthropol.* **4**: 74–82.

Cheney, D. L., and Wrangham, R. W. (1987). Predation. In *Primate Societies,* ed. B. B. Smuts, D. L. Cheney, R. M. Seyfarth, R. W. Wrangham, and T. T. Struhsaker, pp. 227–239. Chicago: University of Chicago Press.

Isbell, L. A. (1994). Predation on primates: Ecological patterns and evolutionary consequences. *Evol. Anthropol.* **3**(2):61–71.

Napier, J. R. (1966). Stratification and primate ecology. *J. Anim. Ecol.* **35**:411–412.

Napier, J. R., and Napier, P. H. (1967). *A Handbook of Living Primates.* New York: Academic Press.

Sussman, R. W. (1995). How primates invented the rainforest and vice versa. In *Creatures of the Dark: The Nocturnal Prosimians,* ed. L. Alterman, G. A. Doyle, and M. K. Izard, pp.1–10. New York: Plenum Press.

Terborgh, J. (1983). *Five New World Primates.* Princeton, N. J.: Princeton University Press.

______. (1992). *Diversity and the Tropical Rain Forest.* New York: Scientific American.

Whitmore, T. C. (1990). *An Introduction to Tropical Rain Forests.* Oxford: Oxford University Press.

Wolfheim, J. H. (1983). *Primates of the World: Distribution, Abundance, and Conservation.* Seattle: University of Washington Press.

LAND USE

Chevalier-Skolnikoff, S., Galdikas, B. M. F., and Skolnikoff, A. Z. (1982). The adaptive significance of higher intelligence in wild orangutans: A preliminary report. *J. Hum. Evol.* **11**:639–652.

Dunbar, R. (1992). Neocortex size as a constraint on group size in primates. *J. Hum. Evol.* **20**:469–493.

Milton, K. (1981). Distribution patterns of tropical plant foods as an evolutionary stimulus to primate mental development. *Am. Anthropol.* **83**:534–548.

Mitani, J. C., and Rodman, P. S. (1979). Territoriality and the relation of ranging patterns and home range size to defendability, with an analysis of territoriality among primate species. *Behav. Ecol. Sociobiol.* **5**:241–251.

Oates, J. F. (1987). Food distribution and foraging behavior. In *Primate Societies,* ed. B. B. Smuts, D. L. Cheney, R. M. Seyfarth, R. W. Wrangham, and T. T. Struhsaker, pp. 197–209. Chicago: University of Chicago Press.

Pollock, J. I. (1974). Spatial distribution and ranging behavior in lemurs. In *The Study of Prosimian Behavior,* ed. G. A. Doyle and R. D. Martin, pp. 359–409. New York: Academic Press.

ACTIVITY PATTERNS

Chapman, C. A., and Chapman, L. (1991). The foraging itinerary of spider monkeys: When to eat leaves. *Folia primatol.* **56**:162–166.

Charles-Dominique, P. (1975). Nocturnality and diurnality: An ecological interpretation of these two modes of life by an analysis of the higher vertebrate fauna in tropical forest ecosystems. In *Phylogeny of the Primates: A Multidisciplinary Approach,* ed. W. P. Luckett and F. S. Szalay, pp. 69–88. New York: Plenum Press.

Ganzhorn, J. U., and Wright, P. C. (1994). Temporal patterns in primate leaf eating: The possible role of leaf chemistry. *Folia Primatol.* **63**:203–208.

Kappeler. P. M. (1995). Life history variation among nocturnal prosimians. In *Creatures of the Dark: The Nocturnal Prosimians,* ed. L. Alterman, G. A. Doyle, and M. K. Izard, pp. 75–92. New York: Plenum Press.

Overdorff, D., and Rasmussen, M. (1995). Determinants of nighttime activity in "diurnal" lemuroid prosimians. In *Creatures of the Dark: The Nocturnal Prosimians,* ed. L. Alterman, G. A. Doyle, and M. K. Izard, pp. 61–74. New York: Plenum Press.

Raemaekers, J. (1978). Changes through the day in the food choice of wild gibbons. *Folia Primatol.* **30**: 194–205.

Sussman, R. W., and Tattersall, I. (1976). Cycles of activity, group composition, and diet of *Lemur mongoz mongoz* Linnaeus 1766 in Madagascar. *Folia Primatol.* **26**: 270–283.

Tattersall, I (1988). Cathemeral activity in primates: A definition. *Folia Primatol.* **49**:200–202.

Wright, P. C. (1996). The Neotropical primate adaptation to nocturnality: Feeding in the night (*Aotus nigriceps* and *A. azarae*). In *Adaptive Radiations of Neotropical Primates.* eds. M. A. Norconk, A. L. Rosenberger, and P. A. Garber, pp. 369–382. New York: Plenum Press.

PRIMATE DIETS

Barton, R. A., Whiten, A., Strum, S. C., Byrne, R. W., and Simpson, A. J. (1992). Habitat use and resource availability in baboons. *Anim. Behav.* **43**:831–844.

Barton, R. A, Whiten, A., Byrne, R. W., and English, M. (1993). Chemical composition of baboon plant foods: Implications for the interpretation of intra- and interspecific differences in diet. *Folia Primatol.* **61**:1–20.

Chivers, D. J., and Hladik, C. M. (1980). Morphology of the gastrointestinal in primates: Comparisons with other mammals in relation to diet. *J. Morphol.* **166**: 337–386.

Chivers, D. J., Wood, B. A., and Bilsborough, A. (1984). *Food Acquisition and Processing in Primates.* New York: Plenum Press.

Ganzhorn, J. U. (1988). Food partitioning among Malagasy primates. *Oecology* **75**:436–450.

Ganzhorn, J. H. (1989). Primate species separation in relation to secondary plant chemicals. *Human Evol.* **4**: 125–132.

Hladik, C. M. (1988). Seasonal variation in food supply for wild primates. In *Coping with Uncertainty in Food Supply,* ed. I. de Garine and G. A. Harrison, pp. xx-yy. Oxford: Clarendon Press.

Kay, R. F. (1984). On the use of anatomical features to infer foraging behavior in extinct primates. In *Adaptation for Foraging in Nonhuman Primates: Contributions to an Organismal Biology of Prosimians, Monkeys and Apes,* ed. P. S. Rodman and J. G. H. Cant, pp. 21–53. New York: Columbia University Press.

Milton, K. (1978). The quality of diet as a possible limiting factor on the Barro Colorado Island howler monkey population. In *Recent Advances in Primatology,* vol. I: *Behavior,* ed. D. J. Chivers and K. A. Joysey, pp. 387–389. London: Academic Press.

Nash, L. T. (1986). Dietary, behavioral and morphological aspects of gummivory in primates. Yrbk. Phys. Anthropol **29**:113–137.

Oates, J. F. (1987). Food distribution and foraging behavior. In *Primate Societies,* ed. B. B. Smuts, D. L. Cheney, R. M. Seyfarth, R. W. Wrangham, and T. T. Struhsaker, pp. 197–209. Chicago: University of Chicago Press.

Overdorff, D. J. (1992). Differential patterns of flower feeding by *Eulemur fulvus rufus* and *Eulemur* in *Amer. J. Primatol.* **28**:191–204.

Rodman, P. S., and Cant, J. G. H., eds. (1984). *Adaptations for Foraging in Nonhuman Primates: Contributions to an Organismal Biology of Prosimians, Monkeys and Apes.* New York: Columbia University Press.

Roosmalen, M. G. M., Mittermeier, R. A., and Fleagle, J. G. (1988). Diet of the northern bearded saki (*Chiropotes satanas chiropotes*): A neotropical seed predator. *Am. J. Primatol.* **14**:11–35.

Terborgh, J. (1992). *Diversity and the Tropical Rain Forest.* New York: Scientific American.

LOCOMOTION

Cant. J. G. H. (1992). Positional behavior and body size of arboreal primates: A theoretical framework for field studies and an illustration of its application. *Amer. J. Phys. Anthropol.* **76**:29–37.

Fleagle, J. G. (1979). Primate positional behavior and anatomy: Naturalistic and experimental approaches. In *Environment, Behavior and Morphology: Dynamic Interactions in Primates,* ed. M. E. Morbeck, H. Preuschoft, and N. Gomberg, pp. 313–325. New York: Gustav Fischer.

Fleagle, J. G., and Mittermeier, R. A. (1980). Locomotor behavior, body size and comparative ecology of seven Surinam monkeys. *Am. J. Phys. Anthropol.* **52**:301–322.

Hunt, K. D., Cant, J. G. H., Gebo, D. L., Rose, M. D., Walker, S. E., and Youlatos, D. (1996). Standardized descriptions of primate locomotor and postural modes. *Primates* **37**:363–387.

McGraw, W. S. (1998). Posture and support use of Old World monkeys (Cercopithecidae): The influence of foraging strategies, activity patterns, and the spatial distribution of preferred food items. *Am. J. Primatol.* **44.**

Mittermeier, R. A., and Fleagle, J. G. (1976). The locomotor and postural repertoires of *Ateles geoffroyi* and *Colobus guereza,* and a reevaluation of the locomotor category semibrachiation. *Am. J. Phys. Anthropol.* **45**(2): 235–251.

Prost, J. (1965). A definitional system for the classification of primate locomotion. *Am. Anthropol.* **67**: 1198–1214.

Ripley, S. (1967). The leaping of langurs: A problem in the study of locomotor adaptation. *Am. J. Phys. Anthropol.* **26**:149–170.

Rose, M. D. (1974). Postural adaptations in New and Old World monkeys. In *Primate Locomotion,* ed. F. A. Jenkins, pp. 201–222. New York: Academic Press.

SOCIAL LIFE

Altmann, S. A., and Altmann, J. (1979). Demographic constraints on behavior and social organization. In *Primate Ecology and Human Origins,* ed. I. S. Bernstein and E. O. Smith, pp. 47–63. New York: Gartland STPM Press.

Chapman, C. A., White, F. J., and Wrangham, R. W. (1993). Defining subgroup size in fission–fusion societies. *Folia Primatol.* **61**:31–34.

Chapman, C. A., Wrangham, R. W., and Chapman, L. J. (1995). Ecological constraints on group size: An analysis of spider monkey and chimpanzee subgroups. *Behav. Ecol. Sociobiol.* **36**:59–70.

Charles-Dominique, P. (1971). Sociologie chez les lemuriens. *La Recherche* **15**:780–781.

______. (1983). Ecology and social adaptations in didelphid marsupials: Comparison with eutharians of similar ecology. In *Advances in the Study of Mammalian Behavior,* ed. J. E. Eisenberg and D. Weiman, pp.: 395–422. CITY: American Society of Mammalogists, Special Publication no. 7.

Di Fiore, A., and Rendall, D. (1994). Evolution of soical organization: A reappraisal by using phylogenetic methods. *Proc. Natl. Acad. Sci. USA* **91**:994–995.

Dolhinow, P. (1994). Social systems and the individual. *Evol. Anthropol.* **3**:73–74.

Eisenberg, J. F., Muckenhirn, N. A., and Rudran, R. (1972). The relationship between ecology and social structure in primates. *Science* **176**:863–874.

Fleagle, J., and Wright, P. C. (1993), Primate groups and primate populations. *Amer. J. Phys. Anth.* **16**(suppl.):87.

Garber, P. A. (1984). A preliminary study of the moustached tamarin monkey (*Saguinus mystax*) in northeastern Peru: Questions concerned with the evolution of a communal breeding system. *Folia Primatol.* **42**:17–32.

______. (1991). Primate behavioral ecology. In *Encyclopedia of Human Biology* **6**:127–133.

______. P. A. (1997). One for all and breeding for one: Cooperation and competition as a calltirtrichine reproductive strategy. *Ev. Anth.* (in press).

Goldizen, A., Mendelson, J., van Vlaardingen, M., and Terborgh, J. (1996). Saddle-back tamarin (*Saguinus fuscicollis*) reproductive strategies: Evidence from a 13-year study of a marked population. *Am. J. Primatol.* **38**:57–83.

Harcourt, A. H. (1989). Environment, competition and reproductive performance of female monkeys. *Tree* **4**(4):101–105.

______. (1992). Coalitions and alliances: Are primates more complex than non-primates? In *Coalitions and Alliances in Humans and Other Animals,* ed. A. H. Harcourt and F. B. MF. B. de Waal, pp. 445–471. Oxford: Oxford University Press.

Hrdy, S. B. (1984). Assumptions and evidence regarding the sexual selection hypothesis: A reply to Boggess. In *Infanticide: Comparative and Evolutionary Perspectives,* ed. G. Hausfater, and S. B. Hrdy, pp. 315–330. Hawthorne, N. Y.: Aldine.

Hoof, J. A., and Schaik, C. P. van. (1992). Cooperation in competition: The ecology of primate bonds. In *Coalitions and Alliances in Humans and Other Animals,* ed. A. H. Harcourt and F. de Waal, pp. 357–389. Oxford: Oxford University Press.

Isbell, L., and van Vuren, (1996). Differential costs of locational and social dispersal and their consequences for female group-living primates. *Behavior* **133**:1–36.

Jay, P. C. (1968). *Primates: Studies in Adaptation and Variability.* New York: Holt, Rinehart and Winston.

Kappeler, P. M. (1993). Variation in social structure: The effects of sex and kinship on social interactions in three lemur species. *Ethology* **93**:125–145.

Kappeler, P. M. (1997). Intrasexual selection in *Mirza cocquereli*; evidence for scramble competition polygyny in a solitary primate. *Behav. Ecol. Sociobiol.* **45**: 115–127.

Kinzey, W. G., and Cunningham, E. P. (1994). Variability in Platyrrhine social organization. *Am. J. Primatol.* **34:** 185–198.

Leighton, D. R. (1987). Gibbons: Territoriality and monogamy. In *Primate Societies,* ed. B. B. Smuts, D. L. Cheney, R. M. Seyfarth, R. W. Wrangham, and T. T. Struhsaker, pp. 135–145. Chicago: University of Chicago Press.

McFarland, M. J. (1986). Ecological determinants of fission–fusion sociality in *Ateles* and *Pan.* In *Primate*

Ecology and Conservation, ed. J. G. Else and P. C. Lee, pp. 181–190. Cambridge: Cambridge University Press.

McFarland Symington, M. (1987). Sex ratio and maternal rank in wild spider monkeys: When daughters disperse. *Behav. Ecol. Sociobiol.* **20**:421–425.

———. (1990). Fission–fusion social organization in *Ateles* and *Pan. Int. J. Primatol.* **11**(1):47–61.

Moore, J. (1992). Dispersal, nepotism, and primate social behavior. *Int. J. Primatol.* **13**:361–379.

———. (1984). Female transfer in primates. *Int. J. Primatol.* **5**(6) 537–589.

Palombit, R. (1994). Dynamic pair bonds in Hylobatids: Implications regarding monogamous social systems. *Behavior* **128** 65–101.

Pusey, A. E., and Packer, C. (1987). Dispersal and philopatry. In *Primate Societies,* ed. B. B. Smuts, D. L. Cheney, R. M. Seyfarth, R. W. Wrangham, and T. T. Struhsaker, pp. 250–267. Chicago: University of Chicago Press.

Rowell, T. E. (1987). On the significance of the concept of the harem when applied to animals. *Social Science Information* **26**(3) 649–669.

———. (1991). What can we say about social structure? In *The Development and Integration of Behavior. Essays in Honor of Robert Hinde,* ed. P. Bateson, pp.255–270. Cambridge: Cambridge University Press.

———. (1993). Reification of Social Systems. *Evol Anthropol.* **2**:135–137.

Rubenstein, D. I., and Wrangham, R. W. (1986). *Ecological Aspects of Social Evolution.* Princeton, N. J.: Princeton University Press.

Runciman, W. C., Maynard Smith, J., and Dunbar, R. J. M. ed. (1996). *Evolution of Social Behavior Patterns in Primates and Man. Proc. Boston Acad.* **88**:1–297.

Schaik, C. P. van, and Dunbar, R. I. M. (1990). The evolution of monogamy in large primates: A new hypothesis and some crucial tests. *Behavior* **115**:30–62.

Schaik, C. P. van, and Paul, A. (1997). Male care in primates: Does it ever reflect paternity? *Evol. Anth.* **5**:152–156.

Smuts, B. B., Cheney, D. L., Seyfarth, R. M., Wrangham, R. W., and Struhsaker, T. T., eds. (1987). *Primate Societies.* Chicago: University of Chicago Press.

Strum, S. C. (1987). *Almost Human.* New York: Random House.

Sussman, R. W., and Garber, P. A. (1987). A new interpretation of the social organization and mating system of the Callitrichidae. *Int. J. Primatol.* **3**:73–92.

Terborgh, J., and Goldizen, A. Wilson. (1985). On the mating system of the cooperatively breeding saddle-backed tamarin (*Saguinus fuscicollis*). *Behav. Ecol. Sociobiol.* **16**:293–299.

WHY PRIMATES LIVE IN GROUPS

Alexander, R. K. (1974). The evolution of social behavior. *Ann. Rev. Ecol. Syst.* **5**:325–383.

Barton, R. A., Byrne, R. W., and Whiten, A. (1996). Ecology, feeding competition and social structure in baboons. *Behav. Ecol. Sociobiol.* **38**:321–329.

Hoof, J. A., and Schaik, C. P. van. (1992). Cooperation in competition: The ecology of primate bonds. In *Coalitions and Alliances in Humans and Other Animals,* ed. A. H. Harcourt and F. B. M. de Waal, pp.357–389. Oxford: Oxford University Press.

Henzi, S. P. (1988). Many males do not a multimale troop make. *Folia Primatol.* **51**:165–168.

Janson, C. H., and Goldsmith, M. L. (1995). Predicting group size in primates: Foraging costs and predation risks. *Behav. Ecol.* **6**:326–336.

Kappeler, P. M. (1997). Determinants of primate social organization: Comparative evidence and new insights from Malagasy lemurs. *Biol. Rev.* **72**:111–151.

Schaik, C. P. Van. (1983). Why are diurnal primates living in groups? *Behavior* **87**:120–144.

———. C. P. (1989) The ecology of social relationships amongst female primates. In *Comparative Socioecology,* ed. V. Standen and R. A. Foley, pp. 195–218. Oxford: Blackwell.

Schaik, C. P. van, and Dunbar, R. I. M. (1990). The evolution of monogamy in large primates: A new hypothesis and some crucial tests. *Behavior:* **115**:30–62.

Schaik, C. P. van, and Hoof, J. A. R. A. M. van.(1983). On the ultimate causes of primate social systems. *Behavior* **85**:91–117.

Terborgh, J. (1983). *Five New World Primates.* Princeton, N. J.: Princeton University Press.

Terborgh, J., and Janson, C. H. (1986). The socioecology of primate groups. *Ann. Rev. Ecol. Syst.* **17**:111–135.

Wilson, E. O. (1975). *Sociobiology, the New Synthesis.* Cambridge, Mass.: Belknap Press.

Wrangham, R. W. (1979). On the evolution of ape social systems. *Social Science Information* **18**:335–368.

———. (1980). An ecological model of female-bonded primate groups. *Behavior* **75**:262–300.

———. (1987). Evolution of social structure. In *Primate Societies,* ed. B. B. Smuts, D. L. Cheney, R. M. Seyfarth, R. W. Wrangham, and T. T. Struhsaker, pp. 282–296. Chicago: University of Chicago Press.

Protection from Predators

Boinski, S., and Chapman, C. A. (1995). Predation on Primates: Where are we and what's next? *Evol. Anth.* **4**:1–2.

Cheney, D. L., and Wrangham, R. W. (1986). Predation. In *Primate Societics,* ed. B. B. Smuts, D. L. Cheney, R. M. Seyfarth, R. W. Wrangham, and T. T. Struhsaker, pp. 227–239. Chicago: University of Chicago Press.

Dunbar, R. I. (1988). *Primate Social Systems.* Ithaca, N. Y.: Cornell University Press.

Goodman, S. M., O'Conner, S., and Klangrand, O. (1993). A review of predation on lemurs: Implications for the evolution of social behavior in small nocturnal primates. In *Lemur Societies and their Ecological Basis,* ed. P. M. Kappeler and J. U. Ganzhorn, pp. 51–66. New York: Plenum Press.

Isbell, L. A. (1994). Predation on primates: Ecological patterns and evolutionary consequences. *Evol. Anthropol.* **3**(2):61–71.

Janson, C. H. (1992). Evolutionary ecology of primate social structure. In *Evolutionary Ecology and Human Behavior,* 5–1, ed. E. A. Smith and B. Winterhalder, pp. 95–130. New York: Aldine.

Schaik, C. P. van, and Horstermann (1994). Predation risk and the number of adult males in a primate group. *Behav. Ecol. Sociobiol.* **35**:261–272.

Schaik, C. P van, van Noordwijk, M. A., Warsono, B., and Sutriono, E. (1983). Party size and early detection of predators in Sumatran forest primates. *Primates* **24**(2): 211–221.

Terborgh, J. (1986). The social systems of New World primates: An adaptationist view. In *Primate Ecology and Conservation,* ed. J. G. Else and P. C. Lee, pp. 199–212. Cambridge: Cambridge University Press.

Improved Access to Food

Charles-Dominique, P. (1995). Food distribution and reproductive constraints in the evolution of social structure: Nocturnal priamtes and other mammals. In *Creatures of the Dark: The Nocturnal Prosimians,* ed. L. Alterman, G. A. Doyle, and M. K. Izard, pp. 425–438. New York: Plenum Press.

Isbell, L. A. (1991). Contest and scramble competition: Patterns of female aggression and ranging behavior among primates. *Behavioral Ecology* **2**:143–155.

Janson, C. H., and C. P. van Schaik, eds. (1988). Primate feeding competition. *Behavior* **105**:1–186.

Overdorff, D. J. (1998). Are *Eulemur* species pair-bonded? Social organization and mating strategies in *Eulemur fulvus rufus* from 1988–1995 in southeastern Madagascar. *Amer. J. Phys. Anthropol.* **105**:153–166.

Schaik, C. P. van. (1989). The ecology of social relationships amongst female primates. In *Comparative Socioecology,* ed. V. Standen and R. A. Foley, pp.195–218. Oxford: Blackwell.

Wrangham, R. W. (1980). An ecological model of female-bonded primate groups. *Behavior* **75**:262–300.

Access to Mates

Milton, K. (1985). Mating patterns of woolly spider monkeys, *Brachyteles arachnoides:* lmplications for female choice. *Behav. Ecol. Sociobiol.* **17**:53–59.

Schaik, C. P., Assink, P. R., and Salafsky, (1992) Territorial behavior in Southeast Asian langurs: Reseouce defense or mate defense? *Am. J. Primatol.* **26**:233–242.

Wrangham, R. W. (1979). On the evolution of ape social systems. *Social Science Information* **18**:335–368.

Assistance in Rearing Offspring

Janson, C. H. (1984). Female choice and mating system of the brown capuchin monkey *Cebus apella* (Primates: Cebidae). *Z. Tierpsychol.* **65**:177–200.

Kappeler, P. M. (1997). Determinants of primate social organization comparative evidence and new insights from Malagasy lemurs. *Biol Rev.* **72**:111–151.

Leutenegger, W. (1980). Monogamy in callitrichids: A consequence of phyletic dwarfism? *Int. J. Primatol.* **13**: 63–176.

Schaik, C. P. van, and Dunbar, R. I. M. (1990). The evolution of monogamy in large primates: A new hypothesis and some crucial tests. *Behavior* **115**:30–62.

Schaik, C. P. van, and Paul, A. (1997). The paradox of male care in primates. *Evol. Anth.* **5**:152–156.

Smuts, B. B. (1985). *Sex and Friendship in Baboons.* Hawthorne, N. Y.: Aldine.

Weiman, D. (1977). Monogamy in mammals. *Q. Rev. Biol.* **52**:39–69.

Whitten, P. L. (1987). Infants and adult males. In *Primate Societies,* ed. B. B. Smuts, D. L. Cheney, R. M. Seyfarth, R. W. Wrangham, and T. T. Struhsaker, pp. 343–357. Chicago: University of Chicago Press.

Wright, P. C. (1993). Variations in male–female dominance and offspring care in non-human primates. In *Sex and Gender Hierarchies,* ed. B. D. Miller, pp.127–145. Cambridge: Cambridge University Press.

______ . (1996). Patterns of paternal care in primates. *Int. J. Primatol.* **11**:89–102.

Phylogeny

Di Fiore, A., and Rendall, D. (1994). Evolution of social organization: A reappraisal by using phylogenetic methods. *Proc. Natl. Acad. Sci. USA* **91**:9941–9945.

Rendall, D., and Di Fiori, A. (1995). The road less travelled: Phylogenetic perspectives in primatology. *Evol. Anthropol.* **4**:43–52.

Polyspecific Groups

Gautier-Hion, A., Quris, R., and Gautier, J. P. (1983). Monospecific vs polyspecific life: A comparative study of foraging and antipredatory tactics in a community of Cercopithecus monkeys. *Behav. Ecol. Sociobiol.* **12**: 325–335.

Norconk, M., ed. (1990). Special issue: Ecological and behavioral correlates of polyspecific primate troops. *Am. J. Primatol.* **21**:81–170.

LIFE HISTORY

Charnov, E. L. (1993). *Life History Invariants.* Oxford: Oxford Univ. Press.

Charnov, E. L. and Berrigan, D. (1993). Why do female primates have such long lifespans and so few babies? or life in the slow lane. *Evol. Anthrop.* **(6):**191–193.

DeRousseau, C. J. ed. (1990). Primate Life History and Evolution Monographs in *Primatology* **14**:1–366 New York: Wiley-Liss.

Harvey, P. H. (1990). Life-History Variation: Size and Mortality Patterns. In *Primate Life History and Evolution,* ed. DeRousseau, C. J. pp. 81–88. New York: Wiley-Liss.

———. (1992). Life-History Patterns. In *Encyclopedia of Human Evolution,* ed. Jones, S., Martin, R., and Pilbeam, D., pp. 95–97.

Harvey, P. H. and Clutton-Brock, T. H. (1985). Life History Variation in Primates. *Evolution* **39**(3):559–581.

Harvey, P. H. and Pagel, M. (1991). The comparative method in evolutionary biology. Oxford: Oxford University Press.

Hill, K. (1993). Life History Theory and Evolutionary Anthropology. *Evol. Anthropol.* **2**:78–88.

Janson, C. H. and Van Schaik, C. P. (1993). Ecological Risk Aversion in juvenile primates: Slow and steady wins the race. In Periera, M. E. and Fairbanks, L. A., ed. *Juvenile Primates: Life History, Development and Behavior,* pp. 57–74. Oxford: Oxford University Press.

Kappeler, P. M. (1996). Causes and Consequences of Life History Variation Among Strepsirhine Primates. *Am. Nat.* **148**:868–891.

Lee, P. C. (1997). The Meanings of Weaning: Growth, Lactation, and Life History. *Evol. Anthropol.* **5**:87–96.

Lee, P. C., Majluf, P. and Gordon, I. J. (1991). Growth, weaning and maternal investment from a comparative perspective. *J. Zool. Lond.* **225**:99–114.

Leigh, S. R. (1992). Patterns of variation in the ontogeny of primate body size dimorphism. *J. Hum. Evol.* **23**: 27–50.

———. (1994). Ontogenetic correlates of diet in anthropoid primates. *Am. J. Phys. Anthrop.* **94**:499–522.

Martin, R. D. and MacLarnon, A. M. (1985). Gestation period, neonatal size and maternal investment in placental mammals. *Nature* **313**:220–223.

———. (1990). Reproductive Patterns in Primates and Other Mammals: The Dichotomy Between Altrical and Precocial Offspring. In *Primate Life History and Evolution.* ed. DeRousseau, C. J. pp. 47–79. New York: Wiley-Liss.

Oftedal, O. T. (1991). The nutritional consequences of foraging in primates: The relationship of nutrient intakes to nutrient requirements. *Phil. Trans. R. Soc. Lond. B* **334**:161–170.

Partridge, L. and Harvey, P. H. (1988). The ecological context of life history and evolution. *Science* **241**: 1449–1455.

Promislow, D. E. L. and Harvey, P. H. (1990). Living fast and dying young: A comparative analysis of life-history variation among mammals. *J. Zool. Lond.* **220**: 417–437.

Rasmussen, D. T. and Izard, M. K. (1988). Scaling of growth and life history traits relative to body size, brain size, and metabolic rates in lorises and galagos (Lorisidae, Primates) *Amer. J. Phys. Anthropol.* **75**: 357–367.

Richard, A. F., Rakotomanga, P., and Schwartz, M. (1991). Demography of *Propithecus verreauxi* at Beza Mahafaly, Madagascar: sex ratio, survival, and fertility. *Amer. J. Phys. Anthropol.* **84**:307–322.

Read, A. F. and Harvey, P. H. (1989). Life history differences among the eutherian radiations. *J. Zool. Lond.* **219**:329–353.

Roberts, M. (1994). Growth, development and perental care in the western tarsier (*Tarsius bancanus*) in captivity: Evidence for "slow" life history and nonmonogamous mating system. *Int. J. Primatol.* **15**:1–28.

Ross, C. (1988). The intrinsic rate of natural increase and reproductive effort in primates. *J. Zool. Lond.* **214**: 199–219.

———. (1991). Life History Patterns of New World Monkeys. *Intl. J. Primatol.* **12**(5):481–502.

———. (1992). Life History Patterns and Ecology of Macaque Species. *Primates* **33**(2):207–215.

———. (1992). Environmental correlates of the intrinsic rate of natural increase in primates. *Oecologia* **90**: 383–390.

———. (1998). Primate Life Histories *Evol. Anthropol.* **6**: 54–63.

Shea, B. T. (1990). Dynamic Morphology. Growth, Life History, and Ecology in Primate Evolution. In *Pri-*

mate Life History and Evolution, ed. DeRousseau, C. J. pp. 325–352. New York: Wiley-Liss.

______. (1992). Developmental perspective on size change and allometry in evolution. *Evol. Anthrop.* **1**(4): 125–134.

Smith, B. H. (1992). Life history and the evolution of human maturation. *Evol. Anthrop.* **1**(4):134–142.

Stearns, S. (1992). *The Evolution of Life Histories.* New York: Oxford University Press.

Wright, P. C. (1990). Patterns of Paternal Care in Primates. *Int. J. Primatol.* **11**:89–102.

______. (1995). Demography and life history of free-ranging *Propithecus diadema edwardsi* in Ranomafana National Park, Madagascar. *Int. J. Primatol.* **16**:835–854.

FOUR

Prosimians

SUBORDER PROSIMII

In many respects, the living members of the primate order seem to form a natural ladder from primitive to more advanced, or specialized, types. This remarkable array provides us with living species that preserve many features that characterized primates of earlier epochs. This diversity of form can give us some idea of the pathways of primate evolution. Unfortunately, this series of more primitive and more advanced groups is not easily classified, especially when we are not certain of the true phylogeny.

The order Primates is commonly divided into two major groups, or suborders, Prosimii (lemurs, lorises, and tarsiers) and Anthropoidea (monkeys, apes, and humans). Although specialized in many respects, living prosimians generally retain more primitive features than do anthropoids; and in many aspects of the teeth, skulls, and limbs, they preserve a morphology similar to that found in primates of the Eocene epoch, 50 to 40 million years ago. Their English name, pro-simian ("before apes"), suggests this primitive nature, and the German *halbaffen* ("half-ape") is even more evocative.

There are eight families of living prosimians, all from the Old World (Fig. 4.1). Five of these are from the island of Madagascar: cheirogaleids, lemurids, indriids, and two families that contain only a single living genus each, the lepilemurids (or megaladapids) and the daubentoniids. Closely related to the Malagasy families are the lorises of Africa and Asia and the galagos of Africa. Finally, there are the tarsiers of Southeast Asia. The diverse groups contained in the suborder Prosimii are united by their retention of primitive primate features and their lack of the features characteristic of the other suborder, the Anthropoidea, or higher primates.

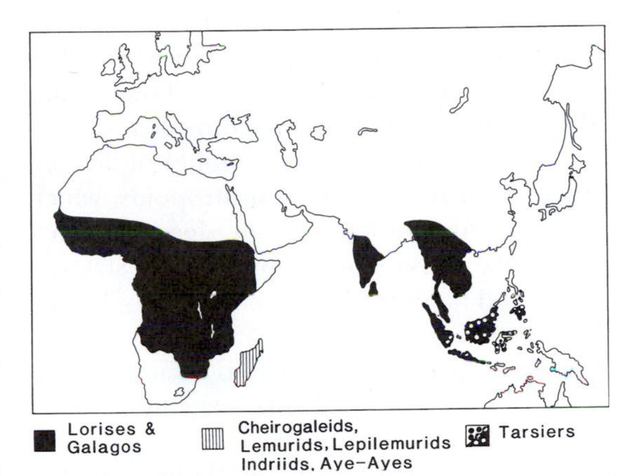

FIGURE 4.1 Geographic distribution of extant prosimians.

This division of living primates into two suborders (Fig. 4.2) is a **gradistic,** or **horizontal,** classification; that is, the two groups, prosimians and anthropoids, are grades (or, in a rough sense, stages) of evolution. Such a classification provides no indication of which group of living prosimians may be closer to the origin of anthropoids, nor does it emphasize the derived characteristics that may be used to group prosimians. It simply expresses the fact that prosimians are primitive primates that lack anthropoid features.

An alternate grouping of living primates is by presumed lines of descent into a **phyletic,** or **vertical,** classification (Fig. 4.2). In this approach, the first seven groups of prosimians just mentioned (cheirogaleids, lemurids, lepilemurids, indriids, daubentoniids, galagids, and lorisids) are grouped together in one suborder, Strepsirhini. Tarsiers are then grouped with anthropoids as the Haplorhini, since they share a number of derived anatomical features with anthropoids, which suggests that anthropoids are derived from a tarsierlike prosimian. In this scheme, strepsirhines and haplorhines are both monophyletic groups.

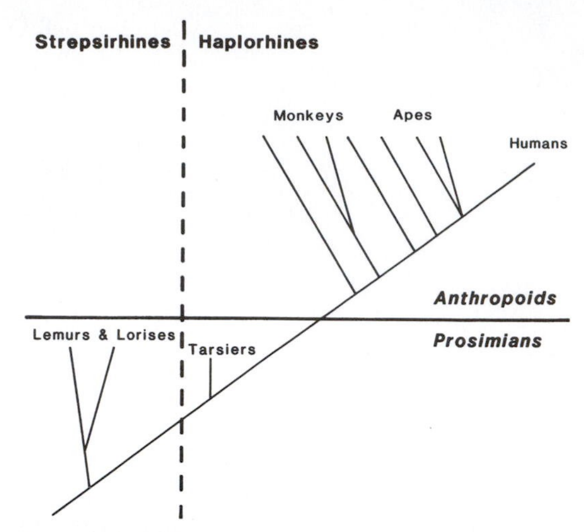

FIGURE 4.2 The gradistic division of primates into prosimians and anthropoids contrasted with the phyletic division into strepsirhines and haplorhines.

Neither grouping scheme is totally satisfactory, especially when fossil primates are included in the classification. Although most students of living primates believe that a strepsirhine–haplorhine division expresses the phyletic relationships among the living taxa, many paleoanthropologists are less comfortable with a tarsier–anthropoid relationship, as discussed in later chapters. This book follows the traditional gradistic classification, even though the strepsirhine–haplorine division is currently the best assessment of the true phylogeny.

Strepsirhines

Living strepsirhines are united by at least three specialized features of "hard anatomy" that can be identified in fossils: their unusual dental tooth comb (and associated small upper incisors), the laterally flaring talus, and the grooming claw on the second digit of their feet (Figs. 4.3, 4.4). Their skull (Fig. 4.5) is characterized by the retention of primitive primate features such as a simple postorbital bar (without postorbital closure), a relatively small braincase, and a primitive mammalian nasal region with a sphenoid recess. Many of the distinctive soft structures of the strepsirhine cranial region, such as the well-developed rhinarium, are primitive features found in many other groups of mammals. The reflecting tapetum lucidum in the eye is a more complicated feature. Although a tapetum is a common feature in many mammalian groups, the tapetum of strepsirhines seems to involve different chemicals than that of other mammals, suggesting that the strepsirhine tapetum may be a derived feature of that group (Martin, 1995).

The reproductive system of all strepsirhines is characterized by at least two pairs of nipples,

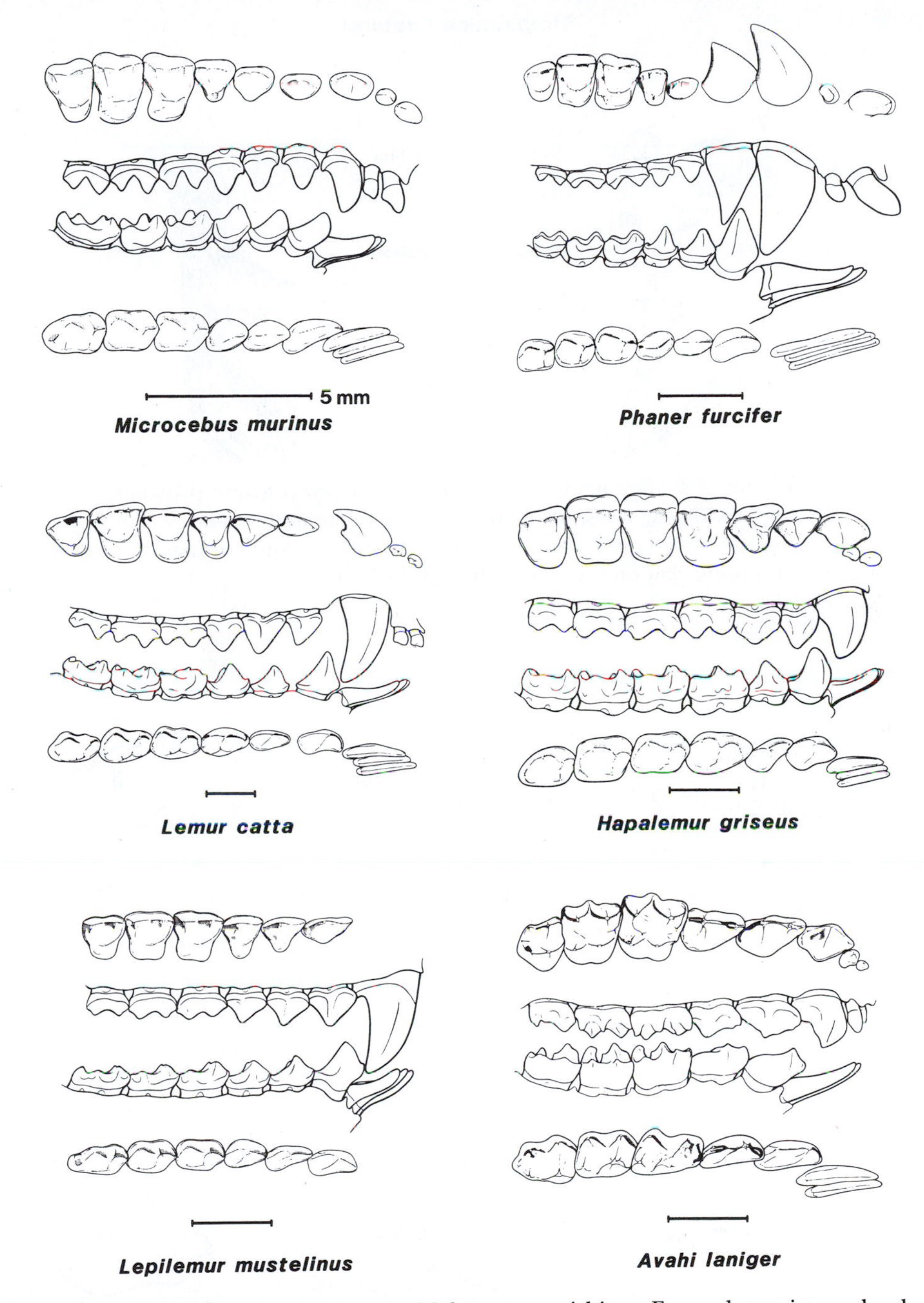

FIGURE 4.3 Dentition of representative Malagasy strepsirhines. For each species, occlusal view of upper dentition (above); lateral view of upper and lower dentition (center); and occlusal view of lower dentition (below) (from Maier, 1980).

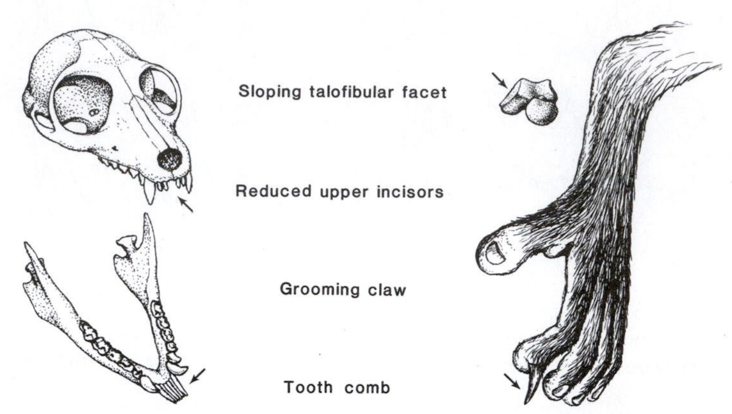

FIGURE 4.4 Distinctive skeletal features of strepsirhine primates: laterally flaring talus, small upper incisors separated by a large cleft, dental tooth comb composed of lower incisors and canines, and grooming claw on the second digit of the foot.

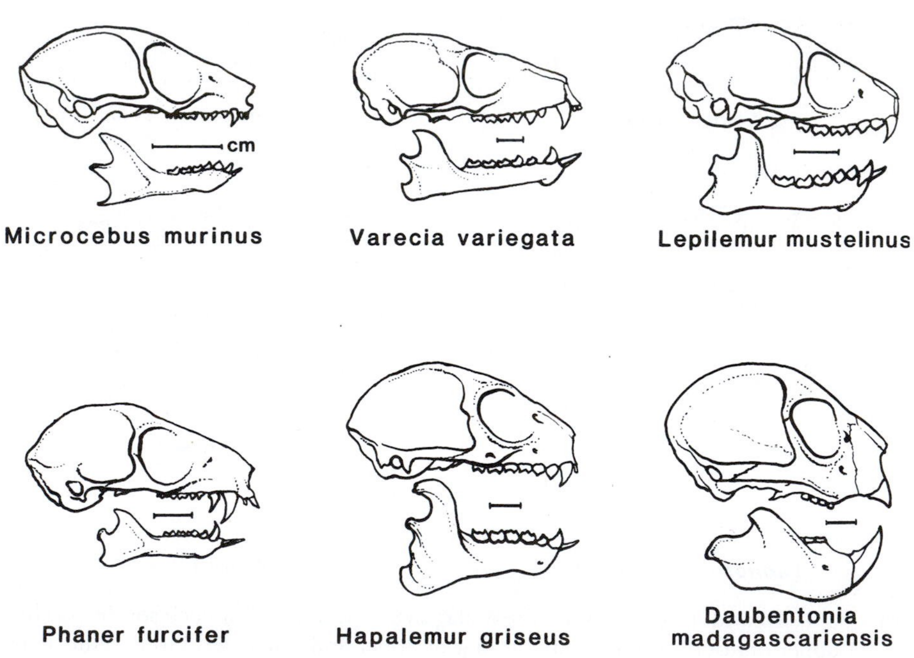

FIGURE 4.5 Skulls of a variety of extant strepsirhines.

a bicornate uterus, and an epitheliochorial type of placentation. Although these features distinguish strepsirhines from other primates, they too are probably primitive primate characteristics, for they are found in many other groups of mammals.

Malagasy Strepsirhines

The greatest abundance and diversity of extant strepsirhines occur on the island of Madagascar, off the eastern coast of southern Africa (Fig. 4.6). The world's fourth largest island, Madagascar has an area approximately that of California and Oregon combined and lies totally in the southern hemisphere. There is tremendous regional diversity in the flora, with tropical rain forests along the east coast and in the northwest, mountain regions in the north, dry forests and spiny deserts in the west and south, and a heavily cultivated central plateau that has been almost totally denuded of natural vegetation (Fig. 4.6). Madagascar has been separated from the African mainland for over 100 million years and has an unusual and relatively limited mammalian fauna, with none of the large carnivores or ungulates found in other parts of the world, and the only primates are strepsirhines; there are no monkeys or apes. However, the radiation of strepsirhines, or lemurs in the broad sense, is one of the three major radiations of living primates and Madagascar is home to fourteen genera of primates—25% of the total in the world today. The Malagasy lemurs are usually divided into five families: cheirogaleids, lemurids, lepilemurids (or megaladapids), indriids, and daubentoniids.

FIGURE 4.6 A map of Madagascar showing the distribution of different forest types.

Cheirogaleids

The smallest strepsirhines and perhaps the most primitive of the Malagasy families are the cheirogaleids (Fig. 4.7, Table 4.1). They are all nocturnal, nest-building animals with a mass of less than 500 g. They share with lemurids the primitive strepsirhine dental formula of 2.1.3.3. and, in all but one genus, a primitive ear structure, with the tympanic ring lying free within the bulla (see Fig. 2.18). The arrangement of the cranial blood supply in cheirogaleids shows the same unique pattern as that of the lorises and galagos. In both groups, the ascending pharyngeal artery enters the skull near the center of the cranial base to form the internal carotid artery supplying the brain (see Fig. 2.12). The reproductive system is unusual among primates in that females have three pairs of nipples and normally give birth to twins. There are five extant genera.

The **mouse lemurs** (*Microcebus*) are among the smallest of all living primate species. They have a fairly short, pointed snout and large membranous ears. Their limbs are short relative to the length of their trunk, and their forelimbs are slightly shorter than their hindlimbs. Their hands are very humanlike in proportion. Their tail is approximately the same length as their body.

There are three species of mouse lemurs (Atsalis et al., 1996). The large gray species (*M. murinus*) is found throughout the drier forests of the western, northern, and southern coastal areas, while the smaller brown species (*M. rufus*) lives in the more humid forests of the east coast as well as in patches of humid forest in the north and on the central plateau. A tiny red species (*M. myoxinus*) with a body mass of approximately 30 g has recently been discovered in the dry forests of the west (Schmid and Kappeler, 1994). Mouse lemurs are particularly abundant in secondary forests and in the undergrowth and lower levels of virtually all forest types, including cultivated areas, and seem to be subject to extraordinary predation pressure from a wide range of carnivores, snakes, and birds (Fig. 4.8). A recent study in the southwest of Madagascar, at Beza-Mahafaly, indicates that a population of 2000 individuals may lose as many as 500 individuals per year to predation by a single owl species (Goodman et al., 1991, 1993). Mouse lemurs are arboreal quadrupeds that move primarily by walking and running along very small branches and leaping between terminal twigs.

Mouse lemurs are the most faunivorous of the cheirogaleids and of all Malagasy primates (Fig. 4.9). They eat invertebrates as well as small vertebrates (tree frogs, chameleons), which they catch by quick hand grasps. Although they are mainly arboreal, they frequently prey on terrestrial insects by leaping on them from low perches. Their diet also contains various fruits, flowers (nectar), buds, and leaves. In feeding, they use a wide range of postures, including hindlimb suspension.

Mouse lemurs are nocturnal and seem to be most active just after nightfall and before sunrise. During the day they sleep in leaf nests, which they make among small branches or in hollow trees. On the west coast, the activity of *M. murinus* shows considerable seasonal variation. In the relatively lush wet season they increase their weight from 50 to 80 g, largely by storing fat in their tail, which increases fourfold in volume. During the long dry season their activity decreases, and they may spend several days without feeding, using the stored fat for sustenance. In captivity, mouse lemurs show a seasonal dietary preference, with a relatively greater protein intake during the more active season.

Mouse lemurs are basically solitary foragers with individuals foraging in separate overlapping home ranges. Adult females occupy small, distinct ranges, but neighboring individuals frequently interact with one another prior to nesting and often nest together in common sleeping sites, or dormitories. The

FIGURE 4.7 Six genera of cheirogaleids: upper left, a mouse lemur (*Microcebus*) reaches for a dragonfly while a hairy-eared dwarf lemur (*Allocebus trichotis*) rests on a branch; to their right, Coquerel's dwarf lemur (*Mirza coquereli*) reaches for a piece of fruit; just below, a fat-tailed dwarf lemur (*Cheirogaleus medius*) also reaches for fruit; below it, a greater dwarf lemur (*Cheirogaleus major*) walks along a branch; on the right, a fork-marked lemur (*Phaner furcifer*) clings to a tree and licks exudates.

TABLE 4.1
Infraorder Lemuriformes
Family CHEIROGALEIDAE

Common Name	Species	Intermembral Index	Mass (g)	
			F	M
Gray mouse lemur	*Microcebus murinus*	72	63	59
Brown mouse lemur	*M. rufus*	71	49	46
Western rufous mouse lemur	*M. mioxinus*	—	30	31
Coquerel's dwarf lemur	*Mirza coquereli*	70	326	304
Greater dwarf lemur	*Cheirogaleus major*	72	362	438
Fat-tailed dwarf lemur	*C. medius*	68	282	283
Fork-marked lemur	*Phaner furcifer*	68	460	
Hairy-eared dwarf lemur	*Allocebus trichotis*	—	84	92

home ranges of male mouse lemurs are much larger and usually overlap with several (about four) female ranges in a noyau type of social structure. Young, probably nonreproductive, males appear to occupy adjacent, less optimal areas that do not overlap with female ranges.

Like most Malagasy species, mouse lemurs are seasonal breeders. Individual females are receptive for only one day at the end of the dry season (September–October), and their birth season coincides with the wet season (November–February). Female mouse lemurs usually have litters of two or three infants, which remain in the nest while the mother forages. There are indications from a captive study that the sex ratio of individual litters is influenced by the presence or absence of other females. Isolated females give birth to more daughters than sons, while the reverse is true of females nesting together. Infant mouse lemurs do not cling to the mother's fur. When moving them, she carries them in her mouth.

Coquerel's dwarf lemur (*Mirza coquereli*) is a substantially larger species. Males are significantly heavier than females. It shares a number of dental features with the dwarf lemurs (*Cheirogaleus*) but has a very long tail and limb proportions similar to those of the mouse lemurs. Like mouse lemurs, *Mirza* has a pointed snout and large, membranous ears. *Mirza* is found only on the western and northwestern

FIGURE 4.8 In southwestern Madagascar, mouse lemurs (*Microcebus murinus*) are preyed on extensively by barn owls (*Tyto alba*).

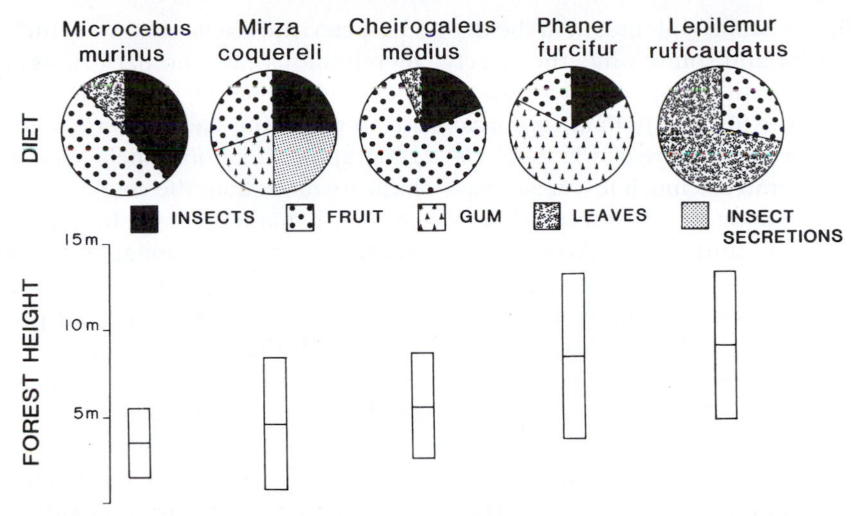

FIGURE 4.9 Diet and forest height preference for five sympatric prosimians in the dry forest of western Madagascar (data from Hladik *et al.*, 1980).

coasts of Madagascar. It is sympatric with *Microcebus murinus* but seems to prefer thicker and taller forests near rivers or ponds and is found in slightly higher parts of the canopy (Fig. 4.9). Coquerel's dwarf lemur moves mainly by quadrupedal running, with some leaping.

Mirza feeds on insects and vertebrate prey, fruits, nectar, and some gums but seems to specialize on secretions produced by the larvae of colonial insects. During the dry season, these insect secretions account for up to 60 percent of all feeding time. Like mouse lemurs, *Mirza* uses a variety of feeding postures, including clinging on the trunks of trees.

Coquerel's dwarf lemurs construct large, elaborate circular nests of leaves for their daytime resting. After leaving their nests at nightfall, these primates seem to devote the first half of the night to feeding and the second half to social interaction with conspecifics. They live in a noyau social system. The female and male ranges appear to be similar in size for most of the year although female ranges overlap while males have nonoverlapping ranges. In the mating season male ranges expand fourfold in size to overlap with numerous females as well as other males. Male testes enlarge considerably during the mating season and the mating system of this species is best described as scramble competition polygyny (Kappeler, 1997). In the dry season, male–female interactions (including contact calls, grooming, play, and chasing) are very common and take place in the central areas of overlapping ranges. These lemurs show no indication of reduced activity during the dry season. Like mouse lemurs, they give birth to twins or triplets, which stay in the nests during the first three weeks of life.

There are two allopatric species of **dwarf lemurs,** which differ mainly in size: *Cheirogaleus medius,* the fat-tailed dwarf lemur, and *Cheirogaleus major,* the greater dwarf lemur. Both have pointed snouts and moderate-size ears that are often partly hidden by their fur. In both species the tail is slightly shorter than the long trunk and the arms are much shorter than the legs.

The smaller *C. medius* is abundant in the dry forest of the west and south, while the larger *C. major* is found in the more humid forests of the east and on the plateau. Both are arboreal quadrupeds that move more slowly than *Microcebus* or *Mirza* and are much less agile leapers. *Cheirogaleus* major is subject to predation by both carnivores and snakes (Wright and Martin, 1995).

Cheirogaleus medius is predominantly frugivorous but opportunistically eats small amounts of insects, small vertebrates, gums, and nectar. *Cheirogaleus major* is also predominantly frugivorous (86 percent of observations) but relies heavily on nectar during the dry months of October and November (Wright and Martin, 1995). Dwarf lemurs are less versatile than mouse lemurs in their feeding postures, and move almost exclusively by quadrupedal walking and running. During the night they intersperse periods of activity with periods of rest.

Dwarf lemurs adapt to the dry seasons of Madagascar by hibernating for six to eight months of each year. During this time, they metabolize the enormous fat reserves stored in their tails during the wet season. In western Madagascar, the mean adult body weight of *C. medius* drops by one-third between March and November (Hladik et al., 1980). Similarly, *C. major* hibernates between May and late September and may halve its weight during hibernation. Lesser dwarf lemurs have been reported to nest in hollow trees, during both normal daytime sleeping and hibernation. *Cheirogaleus major* builds leaf nests. All that is known about the social grouping of dwarf lemurs is that several individuals are frequently found hibernating in the same nest (Kappeler, 1997). In *C. medius,* mating is from September to October, and litters of two or three are born in December or January.

The **fork-marked lemur** (*Phaner furcifer*) is one of the largest and ecologically most specialized of the cheirogaleids (Fig. 4.7). Its characteristic facial features include large membranous ears and dark rings around the eyes that join on the top of the skull to form a stripe down the back. There is considerable geographic variation in pelage color that may indicate more than the single recognized species. Fork-marked lemurs have relatively long hindlimbs and a very long, bushy tail.

Fork-marked lemurs are widely but discontinuously distributed over many parts of Madagascar, although they are most common in the west. These lemurs forage in all levels of the forest and specialize on gum. They have a number of distinctive anatomical adaptations commonly found among primates with this unusual diet. They have very large hands and feet with expanded digital pads, and their fingernails are keeled like claws, for clinging to the trunks of trees. For obtaining gums, they have relatively long procumbent upper and lower incisors, long canines and anterior upper premolars (see Fig. 4.3), and a long and narrow tongue. Their gut is characterized by a large caecum in which the gums are chemically broken down. Their locomotion is rapid quadrupedal walking and running interspersed with leaps from branch to branch.

In contrast to other cheirogaleids, male and female fork-marked lemurs seem to live in more or less permanent groups, many of which contain one male and one female. The male follows between 1 and 10 m behind the female during the night, and the two stay in constant vocal contact. They forage one at a time at the gum sites, with the females appearing to have first choice. During the day, a pair of fork-marked lemurs normally sleeps in a tree hole or in a nest built by *Mirza coquereli.* Pairs commonly groom one another.

Until recently, the **hairy-eared dwarf lemur** (*Allocebus trichotis*) was known only from a handful of osteological specimens and was thought to be possibly extinct. However, it has recently been "rediscovered" in a rain forest in northeast Madagascar (Meier and Albignac, 1991). *Allocebus* is slightly larger than *Microcebus murinus* and has ear tufts (Fig. 4.7). The most distinctive feature of the genus, which separates it from other cheirogaleids, is the construction of its auditory region. Rather than having a free tympanic ring within the auditory bulla, *Allocebus* resembles lorises and galagos in having a tympanic ring fused to the wall of the bulla. Nothing is known of its natural behavior, but the dental similarities to *Phaner* and the presence of keeled nails on its digits suggest a diet of gums. Molecular studies indicate that it is more closely related to mouse lemurs and dwarf lemurs than to *Phaner* (Rumpler et al., 1994).

Lemurids

The lemurids (Fig. 4.10, Table 4.2) are the typical Malagasy lemurs. They share the same dental formula with the cheirogaleids, 2.1.3.3, and the tympanic ring lies free within the auditory bulla as in most cheirogaleids and indriids (Fig. 2.18). Their cranial blood supply to the anterior part of the brain, however, is largely through the stapedial artery rather than through the ascending pharyngeal artery as in cheirogaleids. Most are medium-sized (1–4 kg), diurnal or cathemeral, group-living prosimians that do not build nests. Traditionally, most species of lemurids (except bamboo lemurs) were placed in a single genus, *Lemur*. However, in recent years this genus has been limited to the ring-tailed lemur, with most other species placed in the genus *Eulemur* (Simons and Rumpler, 1988) and the ruffed lemurs separated into their own genus, *Varecia*.

Lemur catta, the **ring-tailed lemur** (Fig. 4.10), is one of the larger species of the genus. It is a gray animal with a long, striped tail; sexes look alike. Ring-tailed lemurs are diurnal, live in the dry south of Madagascar, and feed both on the ground and in the trees. They are the most terrestrial of living strepsirhines, spending 30 percent of each day and 65 percent of their traveling time on the ground. They are primarily quadrupedal walkers and runners. Their diet contains large amounts of both fruit and leaves, and varies from region to region, depending on both habitat and competition from other lemur species.

Ring-tailed lemurs live in large social groups of about twenty individuals that contain approximately equal numbers of males and females. The groups travel almost a kilometer a day and occupy a home range averaging between 10 and 32 ha (hectares), depending on the richness of habitats. *Lemur catta* societies are strikingly similar to those of many Old World monkeys, such as macaques, in that they are centered around one or more female matrilines, with males normally emigrating from their natal troop and changing troops approximately every three years (Sussman, 1992). As in other strepsirhines, females are dominant over the males in the troop. One or more of the adult immigrant males in a ring-tailed lemur troop usually occupies a central position and is frequently associated with the core of females, while the other males, both natal and immigrant, occupy more peripheral positions. There are consistent dominance hierarchies among both females and males.

Eulemur fulvus, the **brown lemur,** is similar in size to *L. catta* (Fig. 4.10), but different in many aspects of ecology. The many subspecies of *E. fulvus* form a ring around the island of Madagascar and can be very different ecologically. Despite a considerable variability in chromosome number, all seem to be capable of interbreeding. Many of the subspecies of *E. fulvus* are sexually dichromatic; that is, males and females have different pelage patterns. All have a cathemeral activity pattern. In contrast to ring-tailed lemurs, brown lemurs are totally arboreal. They move primarily by quadrupedal walking and running and

FIGURE 4.10 Three lemurid species from different parts of Madagascar: above, a pair of brown lemurs (*Eulemur fulvus*); center, a pair of ruffed lemurs (*Varecia variegata*); below, three ring-tailed lemurs (*Lemur catta*), on the ground.

TABLE 4.2
Infraorder Lemuriformes
Family LEMURIDAE

Common Name	Species	Intermembral Index	Mass (g)	
			F	M
Ring-tailed lemur	*Lemur catta*	70	2210	2210
Brown lemur	*Eulemur fulvus*[a]	72	2250	2180
Mongoose lemur	*E. mongoz*	72	1560	1410
Black lemur	*E. macaco*	71	1760	1880
Red-bellied lemur	*E. rubriventer*	68	1940	1980
Crowned lemur	*E. coronatus*	69	1080	1280
Ruffed lemur	*Varecia variegata*	72	3520	3630
	Pachylemur insignis[†]	98	10,000	
	P. julliyi[†]	94	12,800	
Gray bamboo lemur or gentle lemur	*Hapalemur griseus*	64	670	748
Greater bamboo lemur	*H. simus*	65	1300	2150
Golden bamboo lemur	*H. aureus*	—	1390	1520

[a]Simplified systematics.
[†]Extinct. Estimated mass from Godfrey *et al.*, 1997.

by leaping. Their diet consists of leaves, fruit, and flowers, with percentages varying considerably from habitat to habitat and season to season. The two subspecies that have been most thoroughly studied are the brown lemur (*E. fulvus fulvus*), from the southwest, and the rufous lemur, *E. fulvus rufous,* from the eastern rain forest.

Compared with *L. catta, E. fulvus* lives in somewhat smaller groups, averaging seven to twelve individuals, with roughly equal numbers of males and females. Unlike many other strepsirhine species, at least one subspecies, *E. fulvus rufous,* shows no evidence of female dominance over males (Pereira *et al.*, 1990) or male dominance over females. In the southwest of Madagascar, brown lemurs travel less than 150 m each day within their tiny (less than 1-ha) home ranges. In contrast, *Eulemur fulvus rufus,* in the eastern rain forest, has a much larger day range of nearly one kilometer and a home range of approximately 100 ha. Moreover, in time of food scarcity it regularly makes long foraging trips outside of its normal home range (Overdorff, 1993).

The **red-bellied lemur,** *Eulemur rubriventer,* is found sympatric with *E. fulvus rufus* in the eastern rain forests and has been extensively studied in recent years at Ranomafana (e.g., Overdorff, 1996). Red-bellied lemurs are similar in size to rufous lemurs. They are also cathemeral. Red-bellied lemurs move about half the time by leaping and half by quadrupedal walking and running. Their diet varies considerably from season to season, but overall consists of about 80 percent fruit, with lesser amounts of leaves and nectar from flowers.

Red-bellied lemurs live in small, monogamous groups, usually with only a single offspring at a time. Males assist in caring for the infant. They have relatively small home ranges, averaging 19 ha. Day ranges average less than 450 m, with considerable seasonal variability (Overdorff, 1993, 1996).

The **mongoose lemur** (*Eulemur mongoz*) is a most unusual primate behaviorally and ecologically. These small lemurs live in forested areas both in the north of Madagascar and on the nearby Comoro Islands of Anjouan and

Moheli. They are exclusively arboreal. Mongoose lemurs show extreme variability in their activity pattern, both between different populations and, in at least one area, from season to season. Tattersall (1976) found them active at night on the island of Moheli and in the lowlands of Anjouan. Yet in the cold, wet highlands of Anjouan they are active in the daytime. On Madagascar, Tattersall and Sussman (1975) found mongoose lemurs active only in the night in July and August (the dry season), whereas Harrington (1978) saw the same population active during the day in the rainy months of February through July. The field observations suggest that mongoose lemurs, at least, tend to be nocturnal in dry conditions and diurnal in cold, wet conditions.

Oddly enough, only the nocturnal diet and ranging patterns of mongoose lemurs are well known. When they were nocturnal, during the dry season in Madagascar, they fed 80 percent of the time on a single species and from a total of only five species during the entire observation period. The major component of their diet was flowers and nectar (and some fruit), for which their main competitors were bats (Sussman, 1978). They ate virtually no leaves.

The social organization of this dichromatic species seems to be as flexible as its activity period, but the two are not clearly correlated. Most populations live in monogamous family groups composed of an adult male, an adult female, and their offspring. Other groups seem to have more adults.

The ecology and behavior of other species of the genus *Eulemur* (*E. macaco* and *E. coronatus*) are at present known only from preliminary results (e.g., Colquhoun, 1993; Freed, 1993).

The beautiful **ruffed lemur** (*Varecia variegata*), from the east coast forests, is the largest lemurid and seems to be the most primitive in many aspects of its behavior (Fig. 4.10). Ruffed lemurs have a long, doglike snout, thick fur, and a long tail. Compared with other lemurs, they have relatively short limbs. Ruffed lemurs are totally arboreal and restricted to forests with a continuous canopy of large trees. They are reported to be almost exclusively quadrupedal. In captivity they frequently adopt hindlimb suspensory postures. They are almost totally frugivorous and are effective seed dispersers for many fruits.

The social behavior of ruffed lemurs has been studied in a wide range of both seminatural and wild conditions. Some populations have been found to live in small, monogamous groups (White, 1991), while larger communities of about a dozen individuals are found in other populations (Moreland, 1991) and in seminatural groups where dispersion is more limited than food resources. All observations indicate considerable seasonal changes in group cohesion. In many features of their reproductive behavior, ruffed lemurs resemble cheirogaleids more than other lemurids. As in cheirogaleids, female *Varecia* have three pairs of nipples and regularly give birth to twins. They build nests where newborns are kept. At several weeks of age, the infants are carried by the mother, in her mouth, and parked on branches while she forages. Infant *Varecia* receive considerable attention and care from other group members, suggesting some type of cooperative breeding system among the adults of larger groups. Like other Malagasy species, ruffed lemurs are seasonal breeders, and females are fertile on only one day per year. The single study of reproductive behavior in the wild reported very heavy infant mortality, with less than half of the infants born surviving the first three months.

Hapalemur (Fig. 4.11) is a distinctive genus of lemurids that has a unique dietary and habitat specialization on bamboo plants. There are three species: a small one, the **gray bamboo lemur** or **gentle lemur** (*H. griseus*), with a fairly broad distribution; and two larger, rarer

FIGURE 4.11 A family of gentle bamboo lemurs (*Hapalemur griseus*) in a typical bamboo habitat.

species, the **greater bamboo lemur** (*H. simus*) and the **golden bamboo lemur** (*H. aureus*), both from restricted areas in the southeastern rain forest (Fig. 4.12). All *Hapalemur* species have relatively short faces and small, hairy ears. They have short arms and long legs compared with other lemurids (Jungers, 1979). All show a preference for bamboo stands within the forests and live almost entirely on bamboo shoots and leaves, but each of the bamboo lemurs seems to specialize on different parts of the plant. *Hapalemur griseus* eats primarily new shoots of several bamboo lianas (Overdorff et al., 1997); the larger *H. simus* can break open and eat the pith from mature culms; and *H. aureus* specializes on the growing shoots of a species containing very high levels of cyanide (Glander et al., 1989). The locomotor and postural behavior of bamboo lemurs in this vertical habitat is mainly clinging and leaping, but they also move quadrupedally along bamboo branches when feeding. All three species are diurnal in their activity. *Hapalemur griseus* and *H. aureus* are regularly found in small (presumably family) groups and have single births. Groups of the larger *H. simus* have more than one breeding female.

Lepilemurids (or Megaladapids)

Lepilemur, the **sportive,** or **weasel lemurs** (Fig. 4.13, Table 4.3), are often grouped with the lemurids or indriids, but this genus is distinctive enough among living lemurs to

FIGURE 4.12 Three sympatric bamboo lemurs from Ranomafana National park in southeastern Madagascar: above, the gentle bamboo lemur (*Hapalemur griseus*); middle, the golden bamboo lemur (*Hapalemur aureus*); below, the greater bamboo lemur (*Hapalemur simus*).

FIGURE 4.13 Several sportive lemurs (*Lepilemur mustelinus*) in a dry forest of Didiereaceae bush.

deserve its own family (which it shares with the giant, extinct *Megaladapis*). This small, drab-colored lemur is characterized by a lack of permanent upper incisors (Fig. 4.3), an unusual articulation between the mandible and the skull, large digital pads on its hands and feet, and a large caecum. There is considerable variability in the chromosomes among the widespread populations of this genus, and most authorities recognize up to seven distinct species. Because *Lepilemur* shares many of its unusual cranial features with the large extinct *Megaladapis* the family is more properly called Megaladapidae than Lepilemuridae.

Sportive lemurs are found in all types of forests throughout Madagascar and are often locally abundant. They prefer vertical postures and travel mainly by leaping (Ganzhorn, 1993). Lepilemurs are predominantly nocturnal and folivorous (Fig. 4.9). Like other folivorous primates, they are unable to digest cellulose, the main structural component of leaves, so they rely on the bacteria in their digestive tract for this task. In sportive lemurs, as in horses and rabbits, the bacteria live in the caecum, at the base of the large intestine; therefore the cellulose is broken down very near the end of the digestive tract. *Lepilemur* (like rabbits) have been reported to reingest their feces (containing the broken-down cellulose) during the day. The basal metabolic rate of *Lepilemur* is one of the lowest recorded for any primates, this is both an energy conservation strategy and perhaps an adaptation for digesting toxic chemicals (Schmid and Ganzhorn, 1996).

Like many of the cheirogaleids, sportive lemurs live as solitary individuals with overlapping

TABLE 4.3
Superfamily Lemuroidea
Family LEPILEMURIDAE

<table>
<tr><th rowspan="2">Common Name</th><th rowspan="2">Species</th><th rowspan="2">Intermembral Index</th><th colspan="2">Mass (g)</th></tr>
<tr><th>F</th><th>M</th></tr>
<tr><td>Subfamily Lepilemurinae</td><td></td><td></td><td></td><td></td></tr>
<tr><td>Weasel lemur</td><td>Lepilemur mustelinus</td><td>65</td><td colspan="2">770</td></tr>
<tr><td>Red-tailed sportive lemur</td><td>L. ruficaudatus</td><td>63</td><td>779</td><td>771</td></tr>
<tr><td>Gray-backed sportive lemur</td><td>L. dorsalis</td><td></td><td colspan="2">550</td></tr>
<tr><td>White-footed sportive lemur</td><td>L. leucopus</td><td></td><td>594</td><td>617</td></tr>
<tr><td>Small-toothed weasel lemur</td><td>L. microdon</td><td>60</td><td></td><td>970</td></tr>
<tr><td>Milne-Edward's sportive lemur</td><td>L. edwardsi</td><td>63</td><td>934</td><td>908</td></tr>
<tr><td>Northern sportive lemur</td><td>L. septentrionalis</td><td></td><td colspan="2">750</td></tr>
<tr><td>(!)Subfamily Megaladapinae</td><td></td><td></td><td></td><td></td></tr>
<tr><td></td><td>Megaladapis edwardsi†</td><td>120</td><td colspan="2">75,000</td></tr>
<tr><td></td><td>M. grandidieri†</td><td>115</td><td colspan="2">65,000</td></tr>
<tr><td></td><td>M. madagascariensis†</td><td>114</td><td colspan="2">40,000</td></tr>
</table>

(†), Extinct. Estimated mass from Godfrey *et al.*, 1997.

home ranges in a noyau arrangement or in small groups (especially in the mating season). Because of the low nutritional content of leaves and their small size, these lemurs are on a particularly tight energy budget, and their nightly activity pattern reflects this condition. Individuals have very small home ranges (0.2–0.5 ha), and they are intensely territorial, as their common name suggests. Males give loud crowlike calls throughout the night and have been reported to engage in fisticuffs in defense of their tiny, leafy domains. Most of their evening, however, is spent in resting and guarding their territory, presumably letting their leaves digest. Sportive lemurs have single births.

Indriids

The living indriids (Fig. 4.14, Table 4.4) are a very uniform family, consisting of three similar genera that differ most obviously in size and activity pattern. They have a reduced dental formula, with only two premolars in each quadrant and four rather than six teeth in their tooth comb (Fig. 4.3). Indriids are remarkably uniform in their cranial morphology (Fig. 4.15) and, like lemurids, have a tympanic ring that lies free in the bulla and a large stapedial artery. All extant indriids are specialized leapers with long hindlimbs and a short, dorsally oriented ischium. Their digestive tract has an enlarged caecum and large intestine. Indriids tend to have single births.

The smallest of the indriids is the **woolly lemur** (*Avahi*), a mottled brown, white, and beige, thickly furred animal with a long tail. It is the only nocturnal indriid. There are two species. *Avahi laniger* is common in the wet forests of eastern Madagascar, and *A. occidentalis* is from the moister areas of the northwest and west. Woolly lemurs move mainly by leaping between vertical supports in the understory of the forest. Their diet is almost

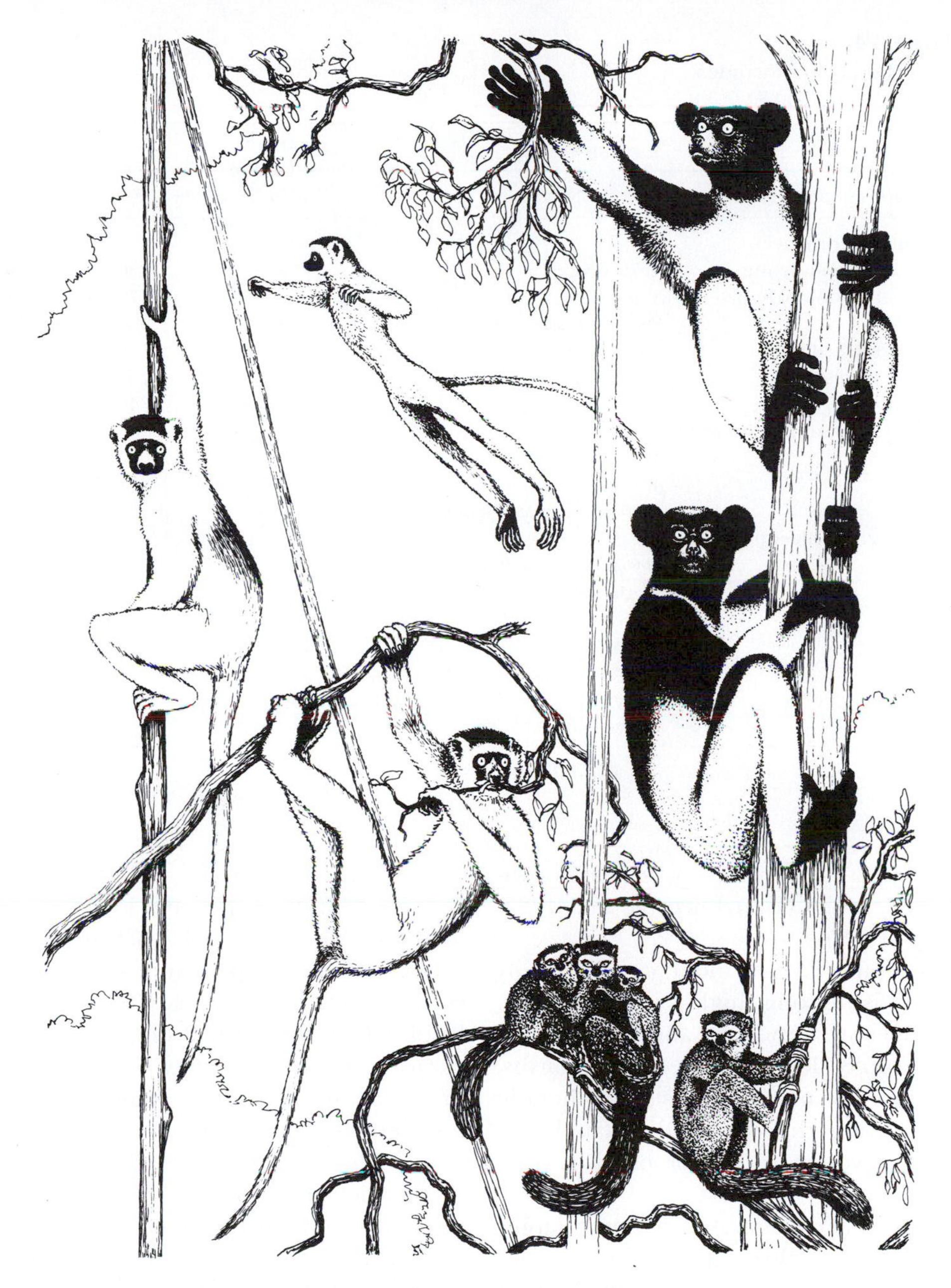

FIGURE 4.14 Three indriids from different parts of Madagascar: left, the sifaka (*Propithecus verreauxi*); right, a pair of indris (*Indri indri*); below, a family of the nocturnal woolly lemurs (*Avahi laniger*) in a typical sleeping posture.

TABLE 4.4
Superfamily Lemuroidea
Family INDRIIDAE

Common Name	Species	Intermembral Index	Mass (g) F	Mass (g) M
Subfamily Indriinae				
Eastern woolly lemur	*Avahi laniger*	58	1320	1030
Western woolly lemur	*A. occidentalis*		777	814
Sifaka	*Propithecus verreauxi*	59	2950	3250
Tattersall's sifaka	*P. tattersalli*		3590	3390
Diademed sifaka	*P. diadema*	64	6260	5940
Indri	*Indri indri*	64	6840	5830
(†)Subfamily Archaeolemurinae				
	Archaeolemur edwardsi†	92	22,000	
	A. majori†	92	17,000	
	Hadropithecus stenognathus†	100	28,000	
(†)Subfamily Palaeopropithecinae†				
	Mesopropithecus pithecoides†	99	11,000	
	M. globiceps†	97	10,000	
	M. dolichobrachion†	113	12,000	
	Babakotia radofilai†	119	15,000	
	Palaeopropithecus ingens†	138	45,000	
	P. maximus†	144	54,000	
	Archaeoindris fontoynonti†	—	200,000	

(†), Extinct. Estimated mass from Godfrey *et al.*, 1997.

exclusively leaves, but they avoid leaves containing alkaloids (Ganzhorn *et al.*, 1985; Ganzhorn, 1989).

Woolly lemurs live in monogamous family groups and issue long, high-pitched whistles. Home ranges are between 1 and 2 ha. Like all indriids, woolly lemurs have single births. Woolly lemurs do not build nests; they huddle together during the day among tangles of vines and leaves in the lower parts of trees (Fig. 4.14).

The **sifaka** (*Propithecus*) is a much larger indriid. There are three species: the smaller *P. verreauxi,* from the dry forest on the west coast; the larger diademed sifaka, *P. diadema,* from the rain forests in the east; and the recently described *P. tattersalli,* a small species from a restricted area in the northeast (Simons, 1988). Like *Avahi,* species of *Propithecus* have very long limbs, especially long legs, and a long tail. All are diurnal. They are found in a variety of forest types, and all are primarily vertical clingers and leapers and travel by leaping between vertical supports. When they come to the ground, they progress by means of bipedal hops. During feeding they also use a variety of suspensory postures, often hanging upside down among small branches.

The diet of *P. verreauxi* varies from locality to locality but always includes large amounts of fruit (particularly in the wet season) and leaves (especially in the dry season). All *Propithecus* species have similar day ranges, but *P. diadema* has a home range fifteen times the size of that used by the dry forest species, *P. verreauxi.* Sifakas live in moderate-size groups

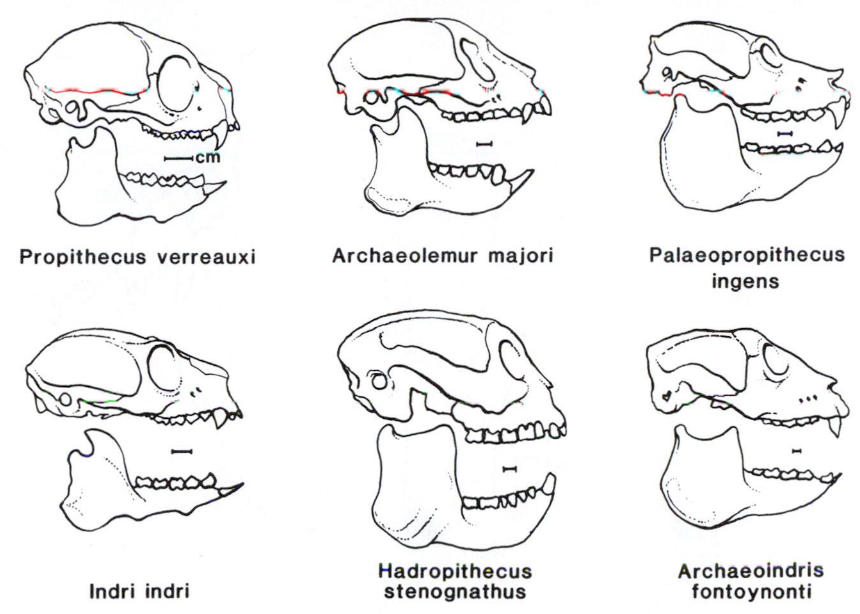

FIGURE 4.15 Skulls of a variety of extant and recently extinct indriids.

(3–9 individuals) often with more than one breeding female. Females generally lead group progressions and are dominant over males in access to food. Sifakas normally give birth to one infant every two years. Males regularly change groups, and there is intense male–male competition in the breeding season. In a nine-year study of *P. diadema* in eastern Madagascar, Wright (1995) found that 67 percent of the young died before the age of reproduction and that all males emigrate from their natal group before breeding. There were two deaths by infanticide. Diademed sifakas are regularly preyed on by the fossa, a large carnivorous mustelid (Wright, 1995).

The **indri** (*Indri indri*) is a diurnal species from the hilly rain forests of the east coast. Like *Propithecus, Indri* has a short face and thick fur. They have body weights that are similar to those of the largest individuals of *P. diadema.* These creatures have extremely long hands, long feet, and slender arms as well as very long legs, but no tail (Fig. 4.16). They move primarily by leaping in lower levels of the forest but also hang below more horizontal supports, especially during feeding. Like sifakas, indris eat both fruit and leaves, with the proportions varying seasonally. They seem to avoid mature leaves, however, and specialize on young leaves and shoots. Like *Avahi* they seem to avoid leaves containing alkaloids. The small family groups, averaging about four individuals, defend relatively large territories from other groups. They are extremely vocal and give long, haunting morning calls to advertise their presence to neighboring groups, which answer sequentially.

Daubentoniids

The **aye-aye** (*Daubentonia madagascariensis*) is about as improbable a primate as one could imagine (Fig. 4.17, Table 4.5). This moderate-size (3 kg), black animal with coarse, shaggy

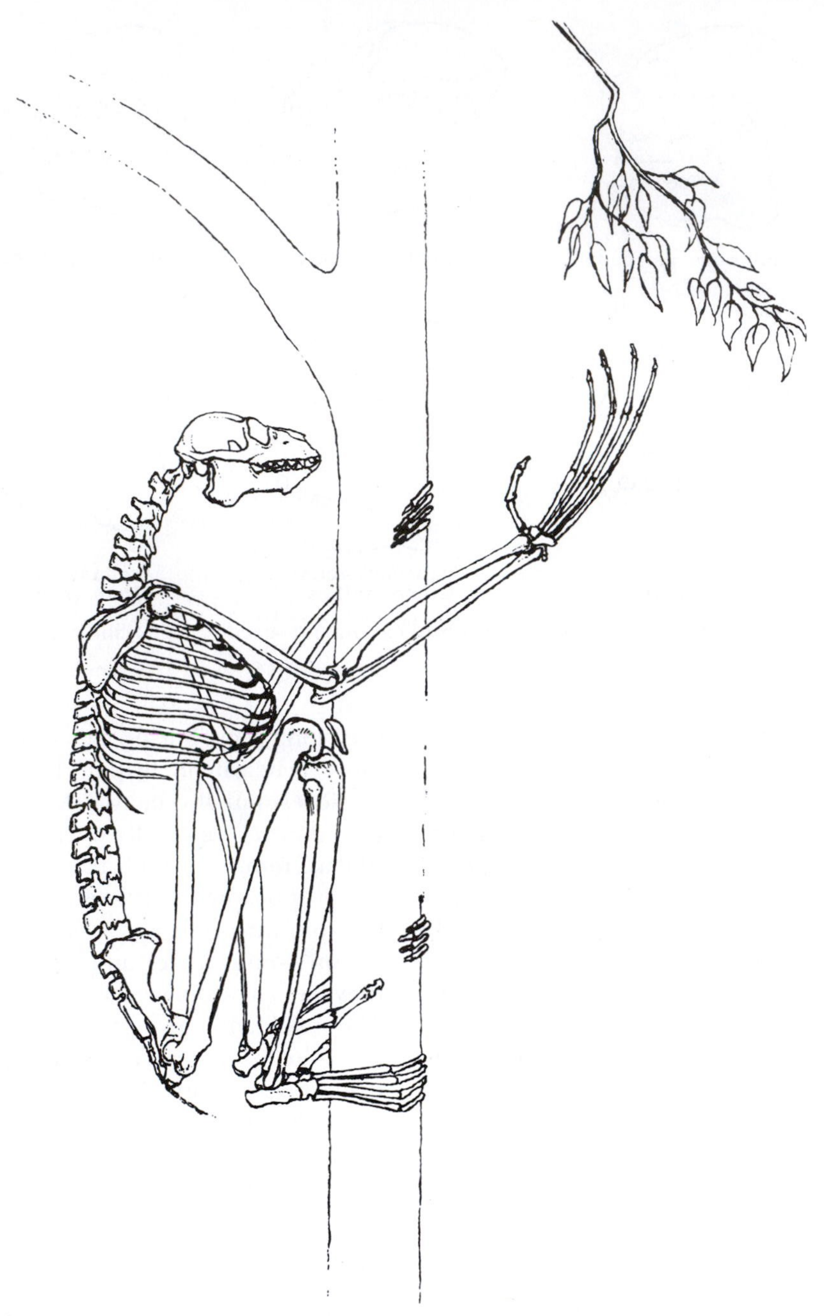

FIGURE 4.16 The skeleton of an indri (*Indri indri*). Note the extremely long hindlimbs, the slender forelimbs, and the reduced tail.

FIGURE 4.17 The solitary and nocturnal aye-aye (*Daubentonia madagascariensis*) probing the bark of a tree with its slender third digit. In the background is an aye-aye nest made of leaves.

fur, enormous ears, and a large bushy tail has more extreme morphological specializations than any other living primate (Oxnard, 1981) but retains many features that clearly link it with lemurs (Tattersall, 1982b). The aye-aye has a greatly reduced dental formula of 1.0.1.3. The most distinctive feature of its dentition, indeed of the entire skull (Fig. 4.5), is the pair of large, ever-growing, rodentlike incisors. The skull has a relatively large, globular

TABLE 4.5
Infraorder Lemuriformes
Family DAUBENTONIIDAE

Common Name	Species	Intermembral Index	Mass (g) F	Mass (g) M
Aye-aye	*Daubentonia madagascariensis*	71	2,490	2,620
	D. robusta†	85	10,000	

(†), Extinct. Estimated mass from Godfrey *et al.*, 1997.

braincase compared with other lemurs, but the auditory bulla and the cranial arteries are like those found in lemurids and indriids. Aye-ayes have relatively large, clawed digits on both hands and feet (except for the hallux), and the third digit of each hand is extremely long and slender.

Aye-ayes are nocturnal and seem to have a limited diet consisting of fruits, seeds, plant parts, insect larvae, and flowers—most of which are obtained through the use of their gnawing teeth and probing finger. They have been described as specialists on structurally defended foods, both fruits and insects, that are unavailable to other animals (Sterling, 1993, 1994). Captive studies have shown that aye-ayes locate insect prey within a log by tapping with their specialized digits and listening to the response (Erikson, 1991, 1994). They then obtain the food by gnawing away the bark with their large incisors and by using the slender finger as a probe. There are no woodpeckers on Madagascar, but the aye-aye has been described as a woodpecker avatar (Cartmill, 1974).

Aye-ayes are mainly quadrupedal, and they are solitary foragers. They live in a noyau-type spatial arrangement, with very large overlapping male ranges and smaller nonoverlapping female ranges. Both sexes have very large night ranges averaging about one and a half kilometers. In contrast with most Malagasy species, aye-ayes seem to be nonseasonal breeders, with females coming into estrus at different times during the year, each pursued by several males (Sterling, 1993, 1994). They have single births and build large round nests of sticks and leaves in forks of tree trunks.

Subfossil Malagasy Prosimians

Despite their diversity, the living strepsirhines of Madagascar are a small part of the primate fauna that inhabited the island in the very recent past. An extensive fauna of larger genera and species is known from fossil deposits as young as one thousand years old (Fig. 4.18), and there are reports of large lemurs still living as recently as three hundred years ago (Flacourt, 1661, in Tattersall, 1982). Bones of the extinct species have been found in conjunction with human artifacts or in sites indicating human activity, suggesting that the first appearance of humans on Madagascar approximately 2000 years ago is largely responsible for the demise of the lemurs, either directly through hunting or indirectly through habitat modification. However, the persistence of some species for over 1000 years after the appearance of humans indicates that it took many centuries for the lemurs to become extinct (Simons et al., 1995). Most of the extinct species were large and probably diurnal; some were partly terrestrial, while many were suspensory—two locomotor behaviors that are rare on Madagascar today. In discussing the adaptive radiation of the Malagasy strepsirhines, these subfossil taxa are most appropriately considered with the living species, since their demise is a very recent and perhaps ongoing event. Perhaps the most striking evidence that the large extinct Malagasy forms are an integral part of the present radiation is that all of the extinct species are related to the living families and are often found in association with fossil remains of living species and genera (Godfrey et al., 1997). The extinct taxa are not spread evenly among families. There are no known extinct cheirogaleids, and the fossil aye-aye species was much larger but probably ecologically similar to the living one. The extinct members of other families greatly extend our knowledge of the adaptive radiation of those groups. This is most dramatic in the case of the indriids, whose extinct relatives include six genera in two very distinct families.

FIGURE 4.18 Artistic reconstruction of the fossil site of Ampazambazimba, Madagascar (ca. 8000–1000 B.P.), showing a variety of subfossil prosimians from that locality. At the upper left, a *Megaladapis* feeds on leaves while clinging to a trunk. Below are two individuals of *Pachylemur insignis* and to the right a family of slothlike *Palaeopropithecus*. On the ground, an *Archaeoindris* feeds in the background while another individual of *Megaladapis* ambles along to another tree. In the foreground, a group of *Archaeolemur* feed on tamarind pods while a group of *Hadropithecus* wander in from the right.

Palaeopropithecines, or Sloth Lemurs

Among the most remarkable discoveries in primate evolution during recent years has been the increasing documentation of an extensive radiation of large, folivorous suspensory indriid relatives that inhabited the forests of Madagascar in recent millennia. The **sloth lemurs,** as they are called, share with the living indriids a reduced dental formula, with two premolars in each quadrant and only four anterior teeth. However, they are generally placed in a separate subfamily, the Palaeopropithecinae, with four genera that exhibit an increasing adaptation to slothlike behavior with increasing size. Like extant Malagasy primates, they apparently lacked any sexual size dimorphism (Godfrey et al., 1993). Although the limb proportions and skeletal adaptations in many of these extinct lemurs were extraordinary, suspensory feeding postures are not unusual in the larger living indriids and provide evidence for a behavioral continuity between the living leapers and this extinct radiation.

The genus ***Mesopropithecus,*** with three species, is the smallest and most primitive paleopropithecine and the fossil indriid most similar to the living genera (Table 4.4). Dentally and cranially *Mesopropithecus* is similar to *Propithecus,* but it is slightly larger in size and more robust and has larger upper incisors. Like the living indriids, it seems to have been largely folivorous. However, skeletal remains of this genus, and especially those of a new species from the north of Madagascar, show that *Mesopropithecus* is a basal member of the sloth lemurs, with several of the distinctive adaptations for suspension found in the larger paleopropithecines (Simons et al., 1995). *Mesopropithecus* was a slow arboreal quadruped with some suspensory abilities.

Babakotia is a newly discovered sloth lemur from northern Madagascar (Godfrey et al., 1990; Jungers et al., 1991) that has an estimated body mass of just over 15 kg. Its dentition suggests a folivorous diet. Its limbs show more indications of suspensory habits than any of the living indriids or *Mesopropithecus* but less than the larger sloth lemurs described next. For moving about it seems to have relied on a combination of vertical climbing and suspension, without the leaping behavior that characterizes living indriids.

Palaeopropithecus is a much larger genus, with an estimated weight, based on limb size, of about 50 kg (Godfrey et al., 1995). *Palaeopropithecus* is dentally similar to *Propithecus,* with long, narrow molars and well-developed shearing crests, but it has small, vertical lower incisors with no tooth comb. It was almost certainly folivorous. The skull is similar to that of living indriids (Fig. 4.15) but more robustly built, with a longer snout and a heavily buttressed nasal region, suggesting more prehensile lips. The auditory region is superficially quite different from that of living indriids in that it has a tubular meatus extending laterally from the tympanic ring, apparently related to the extreme development of the mastoid region.

Unlike the living indriids, which have relatively long legs and extraordinary leaping abilities, *Palaeopropithecus* has considerably longer forelimbs than hindlimbs (Fig. 4.19). It has very long, curved phalanges and very mobile joints. It was the most suspensory of all known strepsirhines, with locomotor abilities that have generally been compared with those of sloths or orangutans (Jungers et al., 1997).

The genus ***Archaeoindris*** is closely related to *Palaeopropithecus* but was substantially larger, with an estimated weight of approximately 200 kg, the size of a male gorilla. Dentally and cranially (Fig. 4.14) *Archaeoindris* is similar to *Palaeopropithecus* and was probably folivorous. The few limb bones attributed to this giant indriid, together with its great size, suggest that

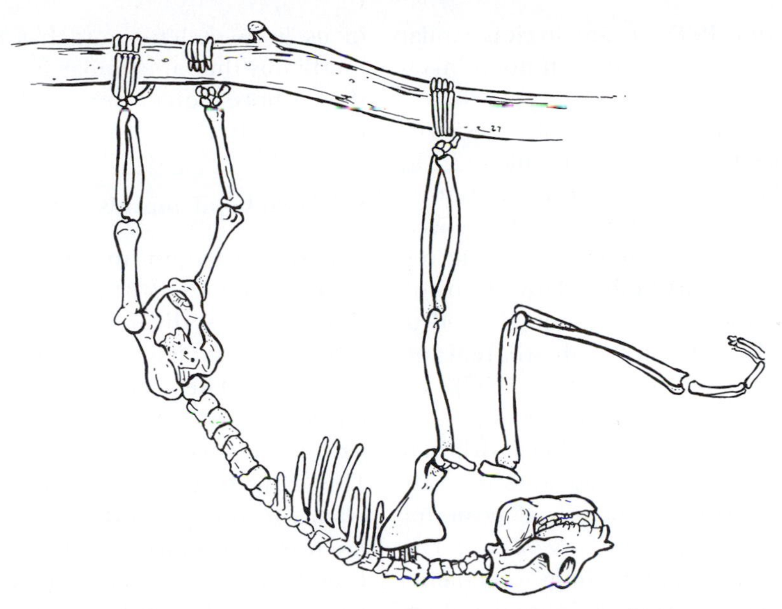

FIGURE 4.19 A skeleton of *Palaeopropithecus ingens* restored in a suspensory feeding position.

it was probably more terrestrial (Fig. 4.18). Its limb bones suggest that the closest locomotor analogues for this giant prosimian are the extinct ground sloths of North and South America.

Whereas the sloth lemurs seem to have been the Malagasy equivalents of orangutans or sloths, the other subfamily of extinct indrids, the Archaeolemurinae evolved remarkable morphological similarities to cercopithecoid monkeys such as macaques. Although small compared with the largest of the sloth lemurs (Table 4.4), they were larger than the living genera. Archaeolemurines, like lemurs and cheirogaleids, have one more premolar than either indriids or palaeopropithecids, and they have large upper central incisors, expanded but slightly procumbent lower incisors and canines, and a fused mandibular symphysis.

Archaeolemur is one of the best-known fossil Malagasy primates, with hundreds of bones from nearly twenty localities. Traditionally, *Archaeolemur* has been divided into two species, the larger *A. edwardsi* and the smaller *A. majori*. However, the geographical distribution of specimens suggests a broad cline of allopatric species, with larger forms in moister habitats and smaller ones in dryer regions (Godfrey et al., 1990). Regardless of the systematics, the dental and skeletal adaptations of this widespread and distinctive genus are well known. In *Archaeolemur*, the anterior premolar is caniniform, and the entire premolar row forms a long cutting edge. The broad molars have low, rounded cusps arranged in a bilophodont pattern similar to that characterizing Old World monkeys. The dental similarities between the extinct indriid and living Old World monkeys such as macaques and baboons suggest that *Archaeolemur* probably had a diverse diet that included fruit, seeds, and invertebrates. It has been suggested that it ate the seeds of baobab trees and facilitated seed dispersal (Baum, 1995). Cranially, it is similar to living indriids (Fig. 4.15), with no monkey-like features.

In skeletal anatomy *Archaeolemur* has striking similarities to Old World monkeys in its limb proportions and in the configuration of individual limb elements and joint surfaces. While many details of its limbs suggest a terrestrial habitat, the short limbs relative to trunk length are characteristic of arboreal quadrupeds. It probably exploited both arboreal and terrestrial supports (Godfrey et al., 1997).

The single species of ***Hadropithecus*** was the most specialized of the monkeylike indriids. It has smaller incisors, reduced anterior premolars, and expanded, molarized posterior premolars compared with *Archaeolemur.* The molars have additional foldings of enamel that form a complex array of dentine and enamel crests and develop extremely flat wear. Because of similarities to the graminivorous (grass-feeding) gelada baboons, several authors have suggested that *Hadropithecus* fed on small grass seeds (Fig. 4.18). The reduced anterior dentition has been suggested as evidence that in *Hadropithecus,* as in gelada baboons, the hands played a major role in feeding (Jolly, 1970). *Hadropithecus* has a very short face and a robust skull with well-developed sagittal and nuchal crests (Fig. 4.15). Limb bones of *Hadropithecus* are relatively rare, but a recent study (Godfrey et al., 1997) has shown that *Hadropithecus,* like *Archeolemur,* had relatively short limbs for its trunk length and probably moved both on the ground and in the trees.

Subfossil Lemurids

Pachylemur insignis (Fig. 4.18, Table 4.2), a subfossil lemurid, is dentally similar to the living ruffed lemur, *Varecia variegata,* and probably had a frugivorous diet (Seligsohn and Szalay, 1974). This fossil genus is more robustly built in its limb skeleton and has forelimbs and hindlimbs that are similar in length. It was a slow, arboreal quadruped with even less leaping ability than the living *Varecia.*

Subfossil Lepilemurids

One of the largest and most unusual of the extinct lemurs was ***Megaladapis,*** known from three species divided into two subgenera (Table 4.3; Vuillaume-Randriamanantena et al., 1992). The largest species had an estimated body size of about 70 kg, based on limb size, but had teeth of an animal twice as large. This extinct giant shares several dental and cranial features with the living *Lepilemur,* and the two are usually placed in the same family (Lepilemuridae or Megaladapidae). Like *Lepilemur, Megaladapis* has long, narrow molars with well-developed crests, indicating a folivorous diet, and no upper incisors. The lack of incisors suggests to some authors that *Megaladapis* may have had a pad covering the premaxillary region (as in living artiodactyls) that occluded with the lower incisors for cropping herbivorous foods. In addition, the pronounced nasal region suggests the possibility of large prehensile lips, perhaps for cropping tough, thorny vegetation. The expanded mandibular condyle on *Megaladapis* is like that of *Lepilemur* and distinct from other strepsirhines. Like many of the larger fossil lemurs, *Megaladapis* has a fused mandibular symphysis. It has a long, flat cranium, a long snout, huge frontal sinuses, and a small braincase.

Megaladapis has long forelimbs relative to the size of its hindlimbs, and a very long trunk. In contrast to the slender limbs of *Palaeopropithecus,* those of *Megaladapis* are extremely robust and short. The hands and feet are enormous (longer than its femur), with moderately curved phalanges (Wunderlich et al., 1996). The locomotor behavior of this genus seems to have been that of a vertical clinger and climber similar to the living koala of Australia. *Megaladapis* probably clung to vertical trunks while cropping vegetation with its snout (Fig. 4.18). On the ground *Megaladapis* no doubt moved quadrupedally between trees. Differences in the skeletal morphology of the different subgenera suggest that the larger *M. edwardsi* was more terrestrial than the other species.

ADAPTIVE RADIATION OF MALAGASY PRIMATES

Like the Galapagos finches, the Malagasy strepsirhines were a natural experiment in evolution. Isolated from repeated faunal invasions and from ecological competition with other primates (until humans arrived) and many other groups of mammals, this lineage evolved an extremely diverse array of species with dietary and locomotor adaptations for exploiting a wide range of ecological conditions (Fig. 4.20). In size, these animals ranged from the smallest living primate, a mouse lemur (30 g), to *Archaeoindris,* which must have weighed as much as living gorillas. In their gross dietary preferences there are species specializing on insects, on gums, on fruits, on nectar, on leaves, on bamboo plants, and probably on seeds, each showing dental adaptations in accordance with these dietary differences. In locomotor abilities they include prodigious leapers, arboreal and probably terrestrial quadrupeds, long-armed suspensory species, and some, such as the koala-like *Megaladapis* and the ground sloth–like *Archaeoindris,* which have no analogue among other primates (e.g. Godfrey et al., 1997). Their limb proportions and many other aspects of the musculature and skeletal anatomy show considerable anatomical diversity that is functionally related to the behavioral differences.

The Malagasy radiation includes numerous diurnal, nocturnal, and cathemeral species.

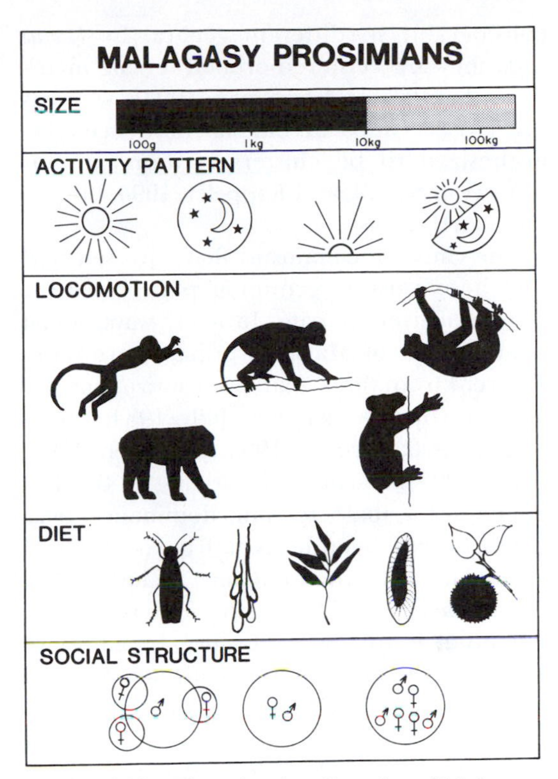

FIGURE 4.20 The adaptive diversity of Malagasy prosimians, including both extant and subfossil species.

The remarkably flexible activity period of the mongoose lemur is not uncommon among lemurs, but it is not known for other primates. Generalizations about the social organization of the Malagasy primates are more difficult, both because many aspects are poorly understood and because we have no information about the behavior of the larger (diurnal and in some cases probably terrestrial) subfossil species. Many species (notably the nocturnal ones) have very primitive, relatively simple noyau social structures; others live in monogamous pairs; and still others live in larger groups with numerous males and females. The organization of the larger groups varies considerably in different genera. In *Lemur catta* they are centered around female matrilines (Sauther and Sussman, 1994), whereas the large groups in *Eulemur* have been hypothesized to be clusters of monogamous units (van Schaik and Kappeler, 1994; but see Overdorff, 1998).

The causal mechanisms that have selected for these various grouping patterns are a source of lively debate. In many ways, social relationships of Malagasy primates are quite different from those that characterize most living anthropoids (see, e.g., Jolly, 1984; Young et al., 1990; Wright, 1993; Kappeler, 1994, 1996, 1997). Particularly striking is the frequency of female feeding dominance over males. Several hypothesis have been advanced to explain this phenomenon, including low metabolic rate and low-quality food resources, but none is totally satisfactory. In addition, Malagasy primates are frequently dichromatic, but in contrast with most anthropoids there is rarely sexual dimorphism in either body mass or canine size. Both males and females have large canines, suggesting considerable intraspecific competition in both sexes. The uniformity of these features among all of the Malagasy species, despite their great adaptive diversity and the striking habitat diversity of the island, is remarkable and difficult to understand (Kappeler, 1990, 1991). Studies of this extraordinary radiation continue at a rapid pace (Crompton and Harcourt, 1998; Goodman and Patterson, 1997).

Galagos and Lorises

In addition to the diverse radiation of strepsirhines on Madagascar, there is a smaller, mainland radiation represented by the galagos in Africa and the lorises in Africa and Asia (Figs. 4.21, 4.22). Galagos and lorises share with the Malagasy families the strepsirhine characteristics of a dental tooth comb and a grooming claw on the second digit and a flared talus (Fig. 4.4), but they also have cranial features that separate them from most of the prosimians of Madagascar. They share with cheirogaleids a unique blood supply to the anterior part of the brain through the ascending pharyngeal artery (Fig. 2.12) rather than through the stapedial. In the ear region, the tympanic ring is fused to the lateral wall, as in *Allocebus,* rather than being suspended within the bulla as in most Malagasy forms. In both of these features, the lorises and galagos show similarities with cheirogaleids. The overall cranial morphology of lorises and galagos is very similar and distinct from that seen in indriids and most lemurids. Galagos and lorises are nocturnal and arboreal, but the two families are extremely different in their postcranial morphology and locomotor behavior. The former are primarily leapers; the latter are slow climbers.

Galagids

The galagos (Table 4.6) are a far more diverse group than many early systematists realized, and the exact arrangement of species and genera is a subject of considerable disagreement (Nash et al., 1987; Masters et al., 1994). Most current authorities recognize three or four genera and at least ten species. There are undoubtedly more species remaining to be described and many unresolved problems in galago systematics (Bearder, 1995; Masters, 1998).

There are two species of *Otolemur,* the greater galagos. *Otolemur crassicaudatus,* the **thick-tailed bushbaby,** or **large-eared greater galago,** is the largest of the galagos. It has a wide distribution throughout eastern and southern Africa. A slightly smaller species, *O. garnettii,* has a more limited distribution on the coast of east Africa. Like all galagos,

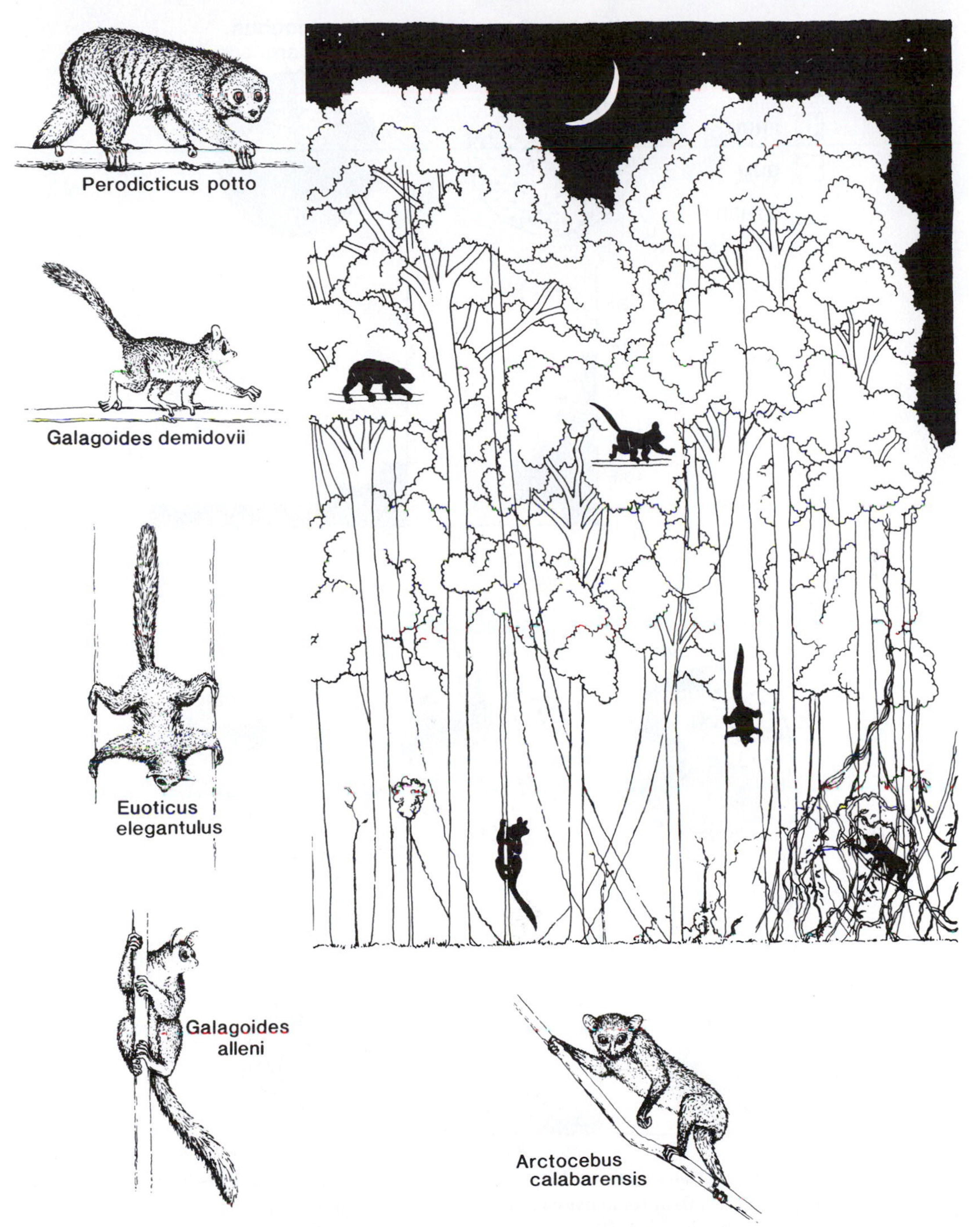

FIGURE 4.21 Five sympatric lorises and galagos from Gabon.

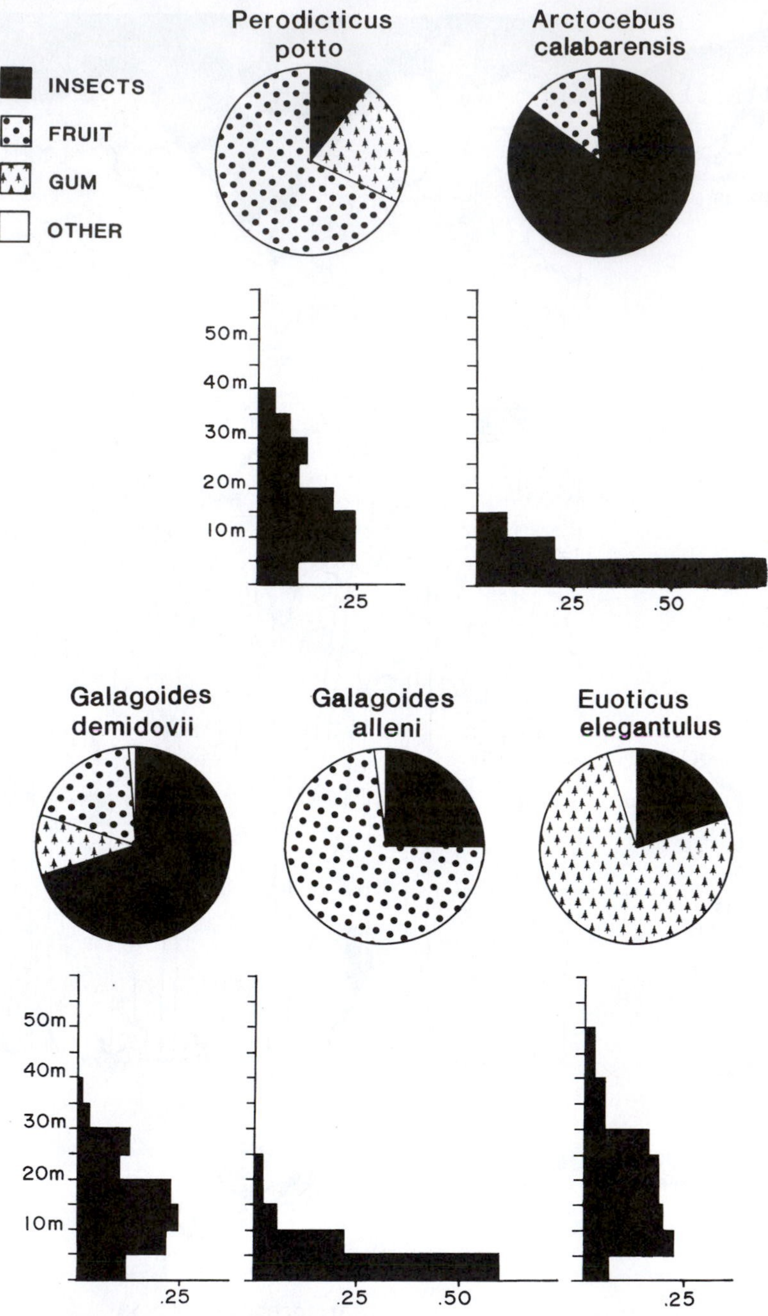

FIGURE 4.22 Diet and forest height preference for five sympatric lorises and galagos from Gabon (data from Charles-Dominique, 1977).

TABLE 4.6
Infraorder Lemuriformes
Family GALAGIDAE

Common Name	Species	Intermembral Index	Mass (g)	
			F	M
Thick-tailed bushbaby	*Otolemur crassicaudatus*	70	1110	1190
Small-eared greater bushbaby	*O. garnettii*	69	734	794
Senegal bushbaby	*Galago senegalensis*	52	199	227
Somali bushbaby	*G. gallarum*	—		200
South African lesser bushbaby	*G. moholi*	54	173	187
Needle-clawed bushbaby	*Euoticus elegantulus*	64	300	
Matschie's needle-clawed bushbaby	*E. matschiei*	60	212	207
Demidoff's bushbaby	*Galagoides demidoff*	68	60	63
Thomas' bushbaby	*G. thomasi*	67	130	103
Zanzibar bushbaby	*G. zanzibaricus*	60	137	149
Allen's bushbaby	*G. alleni*	64	269	277

thick-tailed bushbabies have large ears, a long tail, relatively long lower limbs, and an elongated calcaneus and navicular. These proportions of both the limbs and the ankle in *Otolemur* are less extreme than in other galagos. In both species of *Otolemur*, males are significantly heavier than females (Kappeler, 1991).

Like all galagos, thick-tailed bushbabies are nocturnal. They are found in relatively low forests between 6 and 12 m high and move mainly by quadrupedal walking and running, less often by leaping. Their diet consists primarily of fruits and gums and varies considerably from season to season.

Like cheirogaleids, thick-tailed bushbabies are solitary foragers that live in a noyau social organization. Females build leaf nests for their twin or even triplet offspring and carry their infants in their mouth if they must move them.

There are several species of lesser bushbabies placed in the genus *Galago*. However, the precise number of species in this group and their actual phylogenetic relationships are not resolved (Bearder, 1995; Masters, 1997). *Galago senegalensis*, the **Senegal bushbaby,** is the most widespread species, extending from Senegal in the west across central Africa to eastern Africa. *G. moholi,* the **South African lesser galago,** has a broad distribution over much of southern Africa. *G. gallarum,* the **Somali galago,** has more restricted distributions in eastern Africa. These lesser bushbabies inhabit a wide range of forests, savannahs, open woodlands, and isolated thickets. They are all smaller than *Otolemur,* have relatively long legs and ankle bones, and are spectacular leapers that occupy the opposite extreme of the galago locomotor spectrum from *Otolemur* (McArdle, 1981). They prefer the lower forest levels and smaller supports and travel almost exclusively by leaping. Their diet is comprised mainly of insects, but gum is a major component in the dry season. During gum feeding, these small galagos cling to the rough bark of acacia trees by grasping.

The best studied species of the lesser bushbabies is *G. moholi.* In this species individuals forage separately in large individual home ranges, but they often group together in daytime sleeping nests. Among the adult males there appears to be a social dominance hierarchy related to age and weight. Because of the harsh, unpredictable nature of their environment and high mortality rates, these primates seem to have a "boom or bust" reproductive strategy, with females capable of having up to two litters of twins per year. They are not particular about their choice of nest sites, frequently choosing tangles as well as holes in trees.

The dwarf galagos include two closely related species: **Demidoff's galago,** *Galagoides demidoff,* (Fig. 4.21) and **Thomas' galago,** *G. thomasi.* Dwarf galagos resemble *Microcebus* in many behavioral features (Charles-Dominique and Martin, 1970). Their range extends in a band across central Africa from west to east. Throughout this region, they are very common in dense vegetation of either the canopy of primary forests or the understory of secondary forests. They are less specialized leapers than either *Galago senegalensis* or *Galagoides alleni* but move mainly by quadrupedal walking and running, with short leaps between branches. Their diet (Fig. 4.22) in western Africa is predominantly insects (70 percent), with lesser amounts of fruit (19 percent) and gums (10 percent). Their social structure is a noyau system with overlapping male and female ranges, and day sleeping nests are shared by groups of females and occasional visiting males. They seem to have single births once a year in some parts of their range, but they frequently have twins in other areas. The **Zanzibar galago** (*G. zanzibaricus*) is also usually placed in this genus.

Allen's bushbaby (*Galagoides alleni*) is a medium-sized galago from West Africa whose affinities are subject to some debate. Traditionally it has been placed with the lesser bushbabies; however, some authors group it with the dwarf galagos; others place it with the greater galagos (Masters et al., 1994). Its behavior has been studied by Pierre Charles-Dominique in Gabon (Fig. 4.21). *Galagoides alleni* moves by leaping between small vertical supports in the understory and between these small trees and the ground. Its diet varies considerably at different sites and perhaps seasonally. In the primary forests of Gabon, Allen's bushbabies eat 25 percent animal matter, 75 percent fruit, much of which is from the ground, and some gums (Fig. 4.22) (Charles-Dominique, 1977); but in a secondary forest locality they eat a much higher number of insects (Molez, 1976). Individuals forage alone, and males have large home ranges overlapping the ranges of several females. Allen's bushbabies are much less prolific than Senegal bushbabies. In their relatively stable rain forest habitat, females give birth to only a single infant per year.

The most specialized of the galagos is *Euoticus elegantulus,* the **needle-clawed galago** (Fig. 4.21). This medium-size species resembles the cheirogaleid *Phaner* in having numerous morphological specializations such as procumbent upper incisors, caniniform upper anterior premolars, and laterally compressed, clawlike nails related to its gum-eating habits. These galagos use all levels of the canopy and move both quadrupedally and by leaping. They are particularly adept at clinging to large trunks and branches, which are the source of their main food—gums (Fig. 4.22). Most of their foraging is solitary, and their social behavior is largely unknown. The poorly known *E. matschiei* from central Africa is often grouped with this species.

Lorisids

In contrast with the rapid running and leaping of galagos, lorises (Table 4.7) are best known for their slow, stealthy habits. They have smaller ears than galagos, their forelimbs and hind limbs are more similar in length, and they lack long tails. Despite their slow-

TABLE 4.7
Infraorder Lemuriformes
Family LORISIDAE

Common Name	Species	Intermembral Index	Mass (g)	
			F	M
Potto	*Perodicticus potto*	88	836	830
Martin's false potto	*Pseudopotto martini*	85		
Angwantibo,	*Arctocebus calabarensis*	89	306	312
Golden angwantibo	*Arctocebus aureus*			210
Slender loris	*Loris tardigradus*	90	269	264
Common slow loris	*Nycticebus coucang*	88	626	679
Bengal loris	*N. coucang bengalensis*		1110	1310
Pygmy slow loris	*N. pygmaeus*	91	307	

ness, they have the broadest geographic distribution (between endpoints) of any prosimian family, with two or three genera in Africa and two in Asia. The number of lorisid species is debated, and there is considerable morphological diversity within recognized taxa (Schwartz and Beutel, 1995; Smith and Jungers, 1997).

The **potto** (*Perodicticus potto*) is the largest of the African lorises and the most widespread (Fig. 4.21). Its range extends from Liberia in the west to Kenya in the east. Pottos prefer the main continuous canopy of both primary and secondary forests and move on relatively large supports. Their diet in western Africa has been reported to be largely frugivorous and to include much smaller amounts of animal matter and gums (Fig. 4.22). In eastern Africa, however, they reportedly eat much more gum (60 percent) and less fruit (10 percent) and animal material. The potto has been described as an animal that specializes in olfactory foraging, and its most characteristic activity is slow climbing along large supports, with its nose to the branch.

Like galagos, pottos seem to be solitary foragers and to have overlapping male and female ranges. The females have single births. Pottos do not build nests but rather sleep curled up on branches. During the night, females do not carry their infant but "park" it for the evening and return for it later. The infants are white at birth and turn a drab brown within a few months.

A new genus and species of African loris, *Pseudopotto martini* (**Martin's false potto**), has recently been described from Cameroon in west Africa (Schwartz, 1996). It is slightly smaller than *Perodictus potto* and is distinguished by its reduced P^3, tiny third molar, and moderate-sized tail. Nothing is known of its habits.

The angwantibos are smaller, more slender African lorisids, with a restricted distribution in west central Africa. There are two species, *Arctocebus calabarensis* (**angwantibo**), and *A. aureus* (**golden angwantibo**). Angwantibos prefer the understory of primary and secondary forests, where they are usually found less than 5 m above the ground. They move by slow quadrupedal climbing on very small branches and lianas. They are predominantly insectivorous, specializing on noxious caterpillars. Socially, they appear to be very similar to pottos.

Asian lorises (Fig. 4.23), like their African counterparts, come in two shapes, thin and plump, but the two genera are not sympatric. The **slender loris,** *Loris tardigradus,* is from Sri Lanka and southern India. This species has

FIGURE 4.23 The two Asian lorisids: left, a slow loris (*Nycticebus coucang*) from Southeast Asia; right, a slender loris (*Loris tardigradus*) from India and Sri Lanka.

been most accurately described as a banana on stilts. It is found in the understory of dry forests and in the canopy of wetter forests, where it moves about among the fine twigs. It is mainly insectivorous. Virtually nothing is known of its social behavior.

The **slow loris** (*Nycticebus*), from Southeast Asia, is the stockier Asian genus and contains two species: *N. coucang* (with four subspecies) and *N. pygmaeus*. The subspecies of *N. coucang* have body masses ranging from about 400 g to over 1100 g. The small, generally dimorphic *N. pygmaeus* from Vietnam and Laos is sympatric with the largest subspecies of *N. coucang* (Ravosa, 1998). Slow lorises are found primarily in the main canopy but seem to show no preference for primary, secondary, or deciduous forests (Barrett, 1981). Like all lorises, they move mainly by slow quadrupedal climbing and prefer larger supports, 3–6 cm in diameter. All reports indicate that they are frugivorous. There are no detailed studies of their social organization.

ADAPTIVE RADIATION OF GALAGOS AND LORISES

Compared with that of their Malagasy relatives, the adaptive radiation of galagos and lorises is limited. They are all small and nocturnal, a likely correlate of their geographic overlap with the larger, diurnal catarrhine primates throughout their range. Nevertheless, on the basis of recent systematic and ecological studies we are coming to appreciate more of the diversity of these nocturnal animals (Rasmussen and Nekaris, 1998). Within their nocturnal habits, the galagos and lorises have evolved sufficient adaptive diversity to permit as many as five sympatric species and an increasing array of allopatric taxa (Figs. 4.21, 4.22: Charles-Dominique, 1977; Bearder et al., 1995). Both families include some species that specialize on fruits, gums, or insects, as well as many that show seasonal specializations but a more diverse diet over a full year. Even though the lorises seem to have a rather stereotyped, stealthy locomotor behavior, their different sizes permit sympatric species (in Africa) to use different types of supports in different parts of the canopy. The galagos are more speciose and show more diversity in diet and locomotor abilities, with extremely saltatory species, largely quadrupedal species, and specialized, clawed trunk scramblers. In contrast with the Malagasy strepsirhines, the nocturnal lorises and galagos appear to show little diversity in social organization. They are all solitary foragers and vary in their tendencies to sleep in groups (Bearder, 1987).

PHYLETIC RELATIONSHIPS OF STREPSIRHINES

There is fairly general, though not universal, agreement that lemurids, indriids, lepilemurids, cheirogaleids, and the aye-aye are distinct groups of strepsirhines, and that the first three are more closely related to one another than to any other group of strepsirhines. The phyletic relationships of the aye-aye and cheirogaleids to the other families of large Malagasy strepsirhines are debated (Fig. 4.24). Dental and cranial studies place the aye-aye with indriids, but molecular studies suggest that this odd primate is more distantly related and is the outgroup to other Malagasy taxa (e.g., Yoder, 1997). The difficulty in reconstructing the phyletic relationships of strepsirhines is the same one that bedevils all phylogenetic studies—the considerable amount of parallel evolution that occurred during the radiation of the group. Between

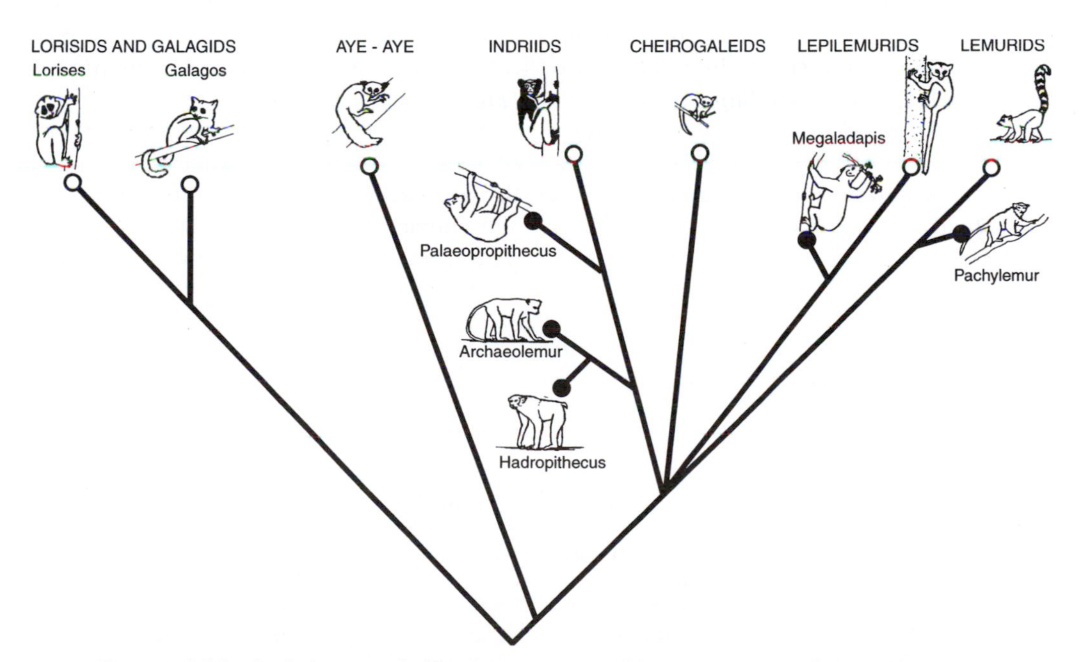

FIGURE 4.24 A phylogeny of strepsirhines, including representative subfossil genera.

20 and 40 percent of all features used to distinguish the different taxonomic groups must have evolved independently in at least two radiations to account for their distribution within living taxa (Eaglen, 1983). Thus, depending on which anatomical characters are used, we can reconstruct a different phylogenetic branching sequence for many of the groups.

A particularly vexing issue in strepsirhine phylogeny concerns the relationship of the cheirogaleids to the other Malagasy families and to the lorises and galagos. Historically, the cheirogaleids have been grouped with the other Malagasy families rather than with the galagos and lorises, a division that is supported by the structure of the ear region (Fig. 2.18) as well as by some molecular studies. Cheirogaleids, galagos, and lorises are, however, unique among living primates in their cranial arterial system (Fig 2.12), and one genus of cheirogaleid (*Allocebus*) has a galago-like ear region, so many authorities have argued that cheirogaleids should be grouped with the galagos and lorises. The relationship between Malagasy and mainland strepsirhines raises a number of interesting evolutionary and biogeographical questions. Are the mainland galagos and lorises descended from a cheirogaleid, or are the cheirogaleids evolved from galagos that found their way to Madagascar? To what extent do small cheirogaleids and galagos represent primitive strepsirhines? There is evidence from both morphological and molecular studies to support each of these alternatives, but the most recent molecular evidence supports grouping the cheirogaleids with other Malagasy families (Fig. 4.24; Yoder, 1994, 1997).

Tarsiers

The **tarsiers** (genus *Tarsius*) of Southeast Asia (Table 4.8) are among the smallest and most unusual of all living primates. They show a mixture of prosimian and anthropoid features. However, their similarities to other prosimians are primitive features: an unfused mandibular symphysis, molar teeth with high cusps, grooming claws on their second (and third) toes, multiple nipples, and a bicornate uterus. Their similarities to higher primates seem to be derived specializations indicative of a phyletic relationship. In addition, tarsiers have many distinctive features all their own (Fig. 4.25).

A most striking feature of a tarsier is the size of its eyes, each of which is actually larger than the animal's brain, not to mention its stomach. Tarsiers differ from other nocturnal primates, and resemble all diurnal primates in

TABLE 4.8
Infraorder Tarsiiformes
Family TARSIIDAE

Common Name	Species	Intermembral Index	Mass (g)	
			F	M
Philippine tarsier	*Tarsius syrichta*	58	117	104
Bornean tarsier	*T. bancanus*	52	117	128
Spectral tarsier	*T. spectrum*	—	108	125
Diane's tarsier	*T. dianae*	—	107	104
Pygmy tarsier	*T. pumilus*	—		<100

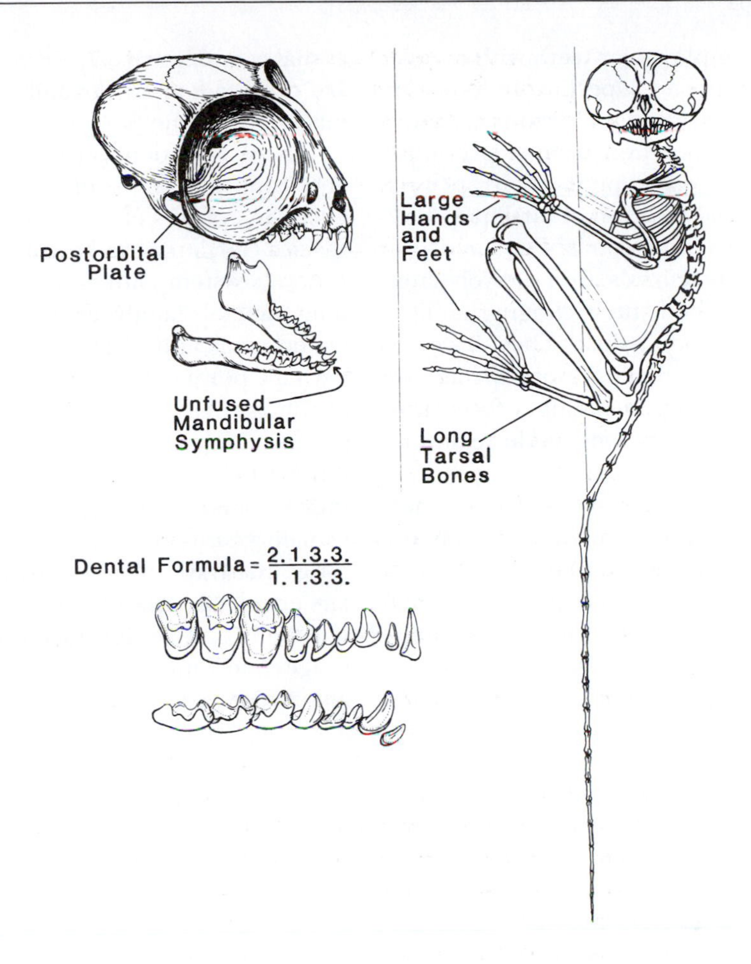

FIGURE 4.25 The skull, dentition, and skeleton of *Tarsius*, showing some of the distinctive features of the genus.

having a retinal fovea and lacking the reflective tapetum found in all lemurs and lorises as well as many other groups of mammals. Their large eyes are protected by a bony socket that is partly closed posteriorly, similar to higher primates. The nose of tarsiers resembles that of higher primates as well, both externally in the lack of an attached upper lip with a median fold, and internally in the greatly reduced turbinates and in the absence of a sphenoid recess. In tarsiers, as in higher primates, the major blood supply to the brain comes through the promontory branch of the internal carotid (Fig. 2.12). The tympanic ring lies external to the auditory bulla and extends laterally to form a bony tube, the external auditory meatus (Fig. 2.18). The tarsier dental formula, 2.1.3.3/1.1.3.3, is unique among primates, but tarsier teeth resemble those of anthropoids in overall proportions, with large upper central incisors, small lower incisors, and large canines. Their high-cusped, simple molar teeth with conules on the upper molars superficially look very primitive, but there are indications that they may have been modified from a previously more complex molar type in their ancestry.

The postcranial skeleton of tarsiers is striking in many of its proportions. The hands and feet of these tiny animals are relatively enormous, reflecting both their clinging abilities and their predatory habits. They have extremely long legs and many more specific adaptations for leaping, including a fused tibia and fibula and the very long ankle region responsible for their name.

In their reproductive physiology, tarsiers show several similarities to higher primates. They have a hemochorial placenta like that of monkeys, apes, and humans rather than the epitheliochorial type found in lemurs, and they produce relatively large offspring. Female tarsiers undergo monthly sexual cycles, with swellings reminiscent of some Old World monkeys.

There are five distinct species of *Tarsius. Tarsius syrichta* is from the Philippines, and *T. bancanus* is from Borneo, Sumatra, and allegedly Java. There are three tarsiers described from Sulawesi (formerly called Celebes): *T. spectrum* (Fig. 4.26), *T. dianae,* and the smaller *T. pumilus* from montane mossy forests. Both *T. bancanus* and *T. spectrum* seem to be most common in the lower levels of all types of forests, but they are particularly abundant in secondary forests and scrub. They are totally nocturnal and spend their days sleeping in the grass or on vines. Both the Bornean and spectral tarsiers travel and feed very near the ground. They move mainly by rapid leaps of up to 3 m. In Sulawesi, other activities, such as calling and resting, take place higher in the canopy. *Tarsius pumilus* is unusual in having keeled nails for clinging to moss-covered trees. Tarsiers are totally faunivorous; they eat insects, arachnids, and small vertebrates such as snakes and lizards. *T. syrichta* and *T. spectrum* are reported to be sexually dimorphic with males larger than females.

There is evidence from both naturalistic and captive studies of considerable differences in social behavior among and within species. Tarsiers have been reported to live in a noyau system with individual but overlapping ranges, family groups consisting of a mated pair and their offspring, or larger, perhaps polygynous groups. Spectral tarsiers live in families or perhaps small polygynous groups, sleep together, give complex territorial duet calls early every evening, and then forage together all night. Males and females actually chase other tarsiers out of their territories (Gursky, 1995, 1998). In the Bornean tarsier (*T. bancanus*), on the other hand, males and females have separate, overlapping ranges and forage independently (Crompton and Andau, 1987). All tarsiers have a remarkably long, six-month gestation period and one of the lowest rates of fetal growth among mammals (Roberts et al., 1994). The single infants weigh up to 30 percent of the mother's weight at birth and suckle for two months. During that time, they seem to develop adult prey-catching patterns. Females are largely responsible for care of the infant and usually park the infants while foraging (Gursky, 1994). In contrast to many monogamous species, there is no indication of male infant care in tarsiers.

PHYLETIC RELATIONSHIPS OF TARSIERS

The relationship of tarsiers to other primates, both living and fossil, has been a source of debate for many years and remains a topic of lively controversy and widely divergent views today. On the basis of their many primitive primate features, tarsiers are often grouped with the strepsirhines as prosimians. But most authorities feel that *Tarsius* and the higher primates share a large enough number of distinctive cranial, dental, and reproductive features to justify grouping them in a single lineage, the haplorhines. Although some researchers (e.g., Rasmussen, 1994) argue that the features shared by tarsiers and anthropoids are not homologous or are the result of parallel evolution, a haplorhine clade is strongly supported by cranial morphology (Ross, 1994) and by DNA systematics (Miyamoto and Goodman, 1990; Shoshani et al., 1996).

Nevertheless, the evolutionary history of tarsiers is far from being resolved, and increasing evidence of morphological and behavioral differences among tarsier species fur-

FIGURE 4.26 A pair of tarsiers (*Tarsius spectrum*) from the island of Sulawesi.

ther complicates the picture. In the past, most researchers have regarded tarsiers as a group of prosimians that gave rise to anthropoids. Still, many of their similarities to anthropoids, such as their lack of a tapetum, their partial postorbital closure, and their very large teeth for their body size, are equally compatible with a view of tarsiers as a group of anthropoids that have "reinvaded" the "prosimian adaptive zone" (Cartmill, 1980). Perhaps tarsiers are dwarfed anthropoids. The phyletic history of tarsiers and the adaptive story behind their similarities with anthropoids are important issues in primate evolution that remain unresolved at present. We discuss these topics again in later chapters when we examine fossil prosimians.

BIBLIOGRAPHY

GENERAL

Alterman, L., Doyle, G. A., and Izard, M. K. (1995). *Creatures of the Dark: the Nocturnal Prosimians.* New York: Plenum Press.

Bourliere, E. (1974). How to remain a prosimian in a simian world. In *Prosimian Biology,* ed. R. D. Martin, G. A. Doyle, and A. C. Walker, pp. 17–22. London: Duckworth.

Cartmill, M. (1982). Basic primatology and prosimian evolution. In *Fifty Years of Physical Anthropology in North America,* ed. F. Spencer, pp. 147–186. New York: Academic Press.

Doyle, G. A. and Martin, R. D., eds. (1979). The Study of Prosimian Behavior. New York: Academic Press.

Harcourt, C., Crompton, R. H., and Feistner, A. T. C., eds. (1998). The Biology and Conservation of Prosimians. *Folia Primatol.* **69**(Supplement 1):1–417.

Maier, W. (1980). Konstruktion morphologishe Untersuchungen am Gebis der rezenten. Prosimiae (Primates). *Abh. semckenb. naturforsch. Ces.* **538:**1–158.

Martin, R. D. (1979). Phylogenetic aspects of prosimian behavior. In *The Study of Prosimian Behavior,* ed. G. A. Doyle and R. D. Martin, pp. 45–78. New York: Academic Press.

——— (1995). Prosimians: From obscurity to extinction. In *Creatures of the Dark: The Nocturnal Prosimians,* ed. L. Alterman, G. A. Doyle, and M. K. Izard, pp. 535–563. New York: Plenum Press.

Martin, R. D., Doyle, G. A., and Walker, A. C., eds. (1974). *Prosimian Biology.* London: Duckworth.

Petter, J. J., and Petter-Rousseaux, A. (1979). Classification of the prosimians. In *The Study of Prosimian Behavior,* ed. G. A. Doyle and R. D. Martin, pp. 1–44. New York: Academic Press.

Walker, A. (1979). Prosimian locomotor behavior. In *The Study of Prosimian Behavior,* ed. G. A. Doyle and R. D. Martin, pp. 543–566. New York: Academic Press.

STREPSIRHINES

Charles-Dominique, P., and Martin, R. D. (1970). Evolution of lorises and lemurs. *Nature* **227**:257–260.

Conroy, G. G., and Packer, D. J. (1981). The anatomy and phylogenetic significance of the carotid arteries and nerves in strepsirhine primates. *Folia Primatol.* **55**: 237–247.

Hill, W. C. O. (1952). *Primates,* vol. 1: *Strepsirhini.* Edinburgh: University of Edinburgh Press.

Jouffroy, F. K. (1975). Osteology and myology of the lemuriform postcranial skeleton. In *Lemur Biology,* ed. I. Tattersall and R. W. Sussman, pp. 149–192. New York: Plenum Press.

Jouffroy, F. K., and Lessertisseur, J. (1979). Relationships between limb morphology and locomotor adaptations among prosimians: An osteometric study. In *Environment, Behavior and Morphology: Dynamic Interactions in Primates,* ed. M. E. Morbeck, H. Preuschoft, and N. Gomberg, pp. 143–181. New York: Gustav Fischer.

Jungers, W. L. (1979). Locomotion, limb proportions and skeletal allometry in lemurs and lorises. *Folia Primatol.* **52**:8–28.

Schwartz, J. H., and Tattersall, I. (1985). Evolutionary relationships of living lemurs and lorises (Mammalia, Primates) and their potential affinities with European Eocene Adapidae. *Anthropol. Papers Am. Mus. Nat. Hist.* **60**(1):1–100.

Yoder, A. (1997). Back to the future: A synthesis of strepsirhine systematics. *Evol. Anthrop.* **6**:11–22.

MADAGASCAR

Albrecht, G. H., Jenkins, P. D. and Godfrey, L. R. (1990). Ecogeographic size variation among the living and subfossil prosimians of Madagascar. *Am. J. Primatol.* **22**: 1–50.

Ganzhorn, J. U., and Sorg, J.-P., eds. (1996). Ecology and Economy of a tropical dry forest in Madagascar. *Primate Report* 46-1. Gottingen.

Glander, K. E., Wright, P. C., Daniels, P. S. and Merelender, A. M. (1992). Morphometrics and testicle size of rain forest lemur species from southeastern Madagascar. *J. Hum. Evol.* **22**:1–17.

Goodman, S. M., O'Connor, S., and Langrand, O. (1993). A review of predation on lemurs: Implications for the evolution of social behavior in small, nocturnal primates. In *Lemur Social Systems and Their Ecological Basis,* ed. P. M. Kappeler and J. U. Ganzhorn), pp. 51–66. New York: Plenum Press.

Harcourt, C., and Thornback, J. (1990). *Lemurs of Madagascar and Comoros.* Glanz: IUCN Publication Services.

Hladik, A. (1980). The dry forest of the west coast of Madagascar: Climate, phenology and food available for prosimians. In *Nocturnal Malagasy Primates,* ed. P. Charles-Dominique, H. M. Cooper, A. Hladik, C. M. Hladik, E. Pages, G. E Pariente, A. PetterRousseaux, and A. Schilling, pp. 3–40. New York: Academic Press.

Jouffroy, F. K. (1962). La musculature des membres chez les lemuriens de Madagascar. Etude descriptive et comparative. *Mammalia* **26** (supple. 2): 1–326.

Kappler, P. M. (1990) The evolution of sexual size dimorphism in prosimian primates. *Am. J. Primatol.* **21**: 201–214.

______ (1991) Patterns of sexual dimorphism in body weight among prosimian primates. *Folia Primatol.* **57**: 132–146.

______ (1993). Sexual selection and lemur social systems. In: *Lemur Social Systems and Their Ecological Basis,* ed. P. M. Kappeler and J. U. Ganzhorn, pp. 223–240. New York: Plenum Press.

______ (1997). Determinants of Primate Social Organization: Comparative Evidence and New Insights from Malagasy lemurs. Biol. Rev. **72**:111–151.

Kappeler, P. M., and Ganzhorn, J. U. (1993a). The evolution of primate communities and societies in Madagascar. *Evol. Anthrop.* **2**(6):159–171.

______ , eds. (1993b). *Lemur Social Systems and their Ecological Basis.* New York: Plenum Press.

Martin, R. D. (1972). Adaptive radiation and behavior of the Malagasy lemurs. *Philos. Trans. R. Soc. London B.* **264**:295–352.

Mittermeier, R. A., Tattersall, I., Konstant, W. R., Meyers. D. M., and R. B. Mast. (1994). *Lemurs of Madagascar.* Washington, D. C.: Conservation International.

Oxnard, C. E., Crompton, R. H., and Lieberman, S. S. (1990). *Animal Lifestyles and Anatomies.* Seattle: Universtiy of Washington Press.

Petter, J. J., Albignac, R., and Rumpler, Y. (1977). *Faune de Madagascar,* vol. 44: *Mammiferes lemuriens* (*Primates, Prosimiens*). Paris: Orstrom, CNRS.

Richard, A. (1986). Malagasy prosimians: Female dominance. In *Primate Societies,* ed. B. B. Smuts, D. L. Cheney, R. M. Seyfarth, R. W. Wrangham, and T. T. Struhsaker, pp. 25–33. Chicago: University of Chicago Press.

Richard, A., and Dewar, R. E. (1991). Lemur ecology. *Annu. Rev. Ecol. Syst.* **22**:145–175.

Schaik, C. P. van, and Kappeler, P. M. (1993). Life history, activity period and lemur social systems. In *Lemur Social Systems and Their Ecological Basis,* ed. P. M. Kappeler and J. U. Ganzhorn, pp. 241–260. New York: Plenum Press.

Sussman, R. W. (1982). The adaptive array of lemurs of Madagascar. *Mongr. Syst. Bot. Missouri Bot. Gard.* **25**: 215–226.

Tattersall, I. (1982). *The Primates of Madagascar.* New York: Columbia University Press.

Tattersall, I., and Sussman, R. W. (1975). *Lemur Biology.* New York: Plenum Press.

Wright, P. C. (1992). Primate ecology, rainforest conservation, and economic development: Building a national park in Madagascar. *Evol. Anthropol.* **1**:25–33.

Cheirogaleids

Atsalis, S., Schmid, J., and Kappeler, P. M. (1996). Metrical comparisons of three species of mouse lemur. *J. Human Evol.* **31**:61–68.

Cartmill, M. (1975). Strepsirhine basicranial structures and the affinities of the Cheirogaleidae. In *Phylogeny of the Primates: A Multidisciplinary Approach,* ed. W. P. Luckett and F. S. Szalay, pp. 313–356. New York: Plenum press.

______ (1982). Assessing tarsier affinities: ls anatomical description phylogenetically neutral? ln *Phylogenie et paleobiogeographic,* ed. E. Buffetaut, P. Janvier, J. C. Rage, and P. Tassy, pp. 279–287. Geobios, Mem. Spec. 6.

Charles-Dominique, P. (1977). *Ecology and Behavior of Nocturnal Primates.* New York: Columbia University Press.

Charles-Dominique, P., and Petter, J. J. (1980). Ecology and social life of *Phaner furcifer.* In *Nocturnal Malagasy Primates,* ed. P. Charles-Dominique, H. M. Cooper, A. Hladik, C. M. Hladik, E. Pages, G. E Pariente, A. Petter-Rousseaux, and A. Schilling, pp. 75–96. New York: Academic Press.

Groves, C. P., and Tattersall, I. (1991). Geographical variation in the fork-marked lemur, *Phaner furcifer* (Primates, Cheirogaleidae). *Folia Primatol.* **56**:39–49.

Harcourt, C. (1987). Brief trap/retrap study of brown mouse lemur (*Microcebus rufus*). *Folia primatol.* **49**: 209–211.

Hladik, C. M., Charles-Dominique, P., and Petter, J. J. (1980). Feeding strategies of five nocturnal prosimians in the dry forest of the west coast of Madagascar. In *Nocturnal Malagasy Primates,* ed. P. Charles-Dominique, H. M. Cooper, A. Hladik, C. M. Hladik, E. Pages, G. E Pariente, A. Petter-Rousseaux, and A. Schilling, pp. 41–74. New York: Academic Press.

Martin, R. D. (1972a). A preliminary field study of the lesser mouse lemur (*Microcebus murinus* J. F. Miller, 1777). Z. *Comp. Ethol. Suppl.* **9**:43–89.

——— (1972b). Adaptive radiation and behavior of the Malagasy lemurs. *Philos. Trans. R. Soc. London B.* **264**: 295–352.

——— (1973). A review of the behavior and ecology of the lesser mouse lemur (*Microcebus murinus* J. F. Miller, 1777). In *Comparative Ecology and Behavior of Primates,* ed. R. P. Michael and J. H. Crook, pp. 1–68. London: Academic Press.

McCormick, S. A. (1980). Oxygen consumption and torpor in the fat-tailed dwarf lemur (*Cheirogaleus medius*): Rethinking prosimian metabolism. *Comp. Biochem Physiol.* **68A**:605–610.

Meier, B., and Albignac, R. (1991). Rediscovery of *Allocebus trichotis* Gunther, 1875 (primates) in northeast Madagascar. *Folia Primatol.* **56**:57–63.

Pages, E. (1980). Ethoecology of *Microcebus coquereli* during the dry season. In *Nocturnal Malagosy Primates,* ed. P. Charles-Dominique, H. M. Cooper, A. Hladik, C. M. Hladik, E. Pages, G. E Pariente, A. Petter-Rousseaux, and A. Schilling, pp. 3–40. New York: Academic Press.

Pages-Feuillade, E. (1988). Modalites de l'occupation de l'espace et relations interindividuelles chez un prosimien nocturne malgache (*Microcebus murinus*). *Folia Primatol.* **50**:204–220.

Perret, M. (1990). Influence of social factors on sex ratio at birth, maternal investment and young survival in a prosimian primate. *Behavioral Ecol. Sociobiol.* **27**: 447–454.

Petter, J. J., Schilling, A., and Pariente, G. (1971). Observations eco-ethologiques sur deux lemuriens malgaches nocturnes: *Phaner furcifer* et *Microcebus coquereli. Terre Vie* **25**:287–327.

Rumpler, Y., Crovell, S., and Montagnon, D. (1994) Systematic relationships among Cheirogaleidae (Primates, Strepsirhini) determined from analysis of highly repreated DNA. *Folia Primatol.* **63**:149–155.

Schmid, J., and Kappeler, P. M. (1994). Sympatric mouse lemurs (*Microcebus* spp.) in western Madagascar. *Folia primatol.* **63**:162–170.

Schwartz, J. H., and Tattersall, I. (1985). Evolutionary relationships of living lemurs and lorises (Mammalia, Primates) and their potential affinities with European Eocene Adapidae. *Anthropol. Papers Am. Mus. Nat. Hist.* **60**(1):1–100.

Szalay, F. S., and Katz, C. C. (1973). Phylogeny of lemurs, galagos and lorises. *Folia Primatol.* **19**:88–103.

Tattersall, I. (1982). *The Primates of Madagascar.* New York: Columbia University Press.

Wright, P. C., and Martin, L. B. (1993). Predation, pollination and torpor in two nocturnal primates, *Cheirogaleus major* and *Microcebus rufus,* in the rain forest of Madagascar. In *Creatures of the Dark: The Nocturnal Prosimians,* ed. L. Alterman, G. A. Doyle, and M. K. Izard, pp. 45–60. New York: Plenum Press.

Yoder, A. D. (1994). Relative position of the Cheirogaleidae in strepsirhine phylogeny: A comparison of morphological and molecular methods and results. *Am. J. Phys. Anthrop.* **94**:25–46.

Lemurids

Conley, J. M. (1975). Notes on the activity pattern of *Lemur fulvus. J. Mammal.* **56**:712–715.

Colquhoun, I. C. (1993). The socioecology of *Eulemur macaco.* In *Lemur Social Systems and Their Ecological Basis,* ed. P. M. Kappeler. and J. U. Ganzhorn, pp. 11–24. New York: Plenum Press.

Doyle, G. A. (1989). Speciation and the evolution of behavior in prosimians. *Human Evol.* **4**:97–104.

Freed, B. Z. (1993). Differences in habitat use of crowned lemurs (*Lemur coronatus*) and Sanford's lemurs (*Lemur fulvus sanfordi*) in Madagascar. *Am. J. Phys. Anthrop. Supp.* **16**:88.

Glander, K. E., Wright, P. C., Seigler, D. S., Randrianasolo, V., and Randrianasolo, B. (1989). Consumption of cyanogenic bamboo by a newly discovered species of bamboo lemur. *Am. J. Prim.* **19**:119–124.

Harrington, J. (1974). Olfactory communication in Lemur-fulvus. In *Prosimian Biology,* ed. R. D. Martin, G. A. Doyle, and A. C. Walker, pp. 331–346. London: Duckworth.

——— (1975). Field observation of social behavior of *Lemur fulvus fulvus* E. Geoffroyi, 1812. In *Lemur Biology,* ed. I. Tattersall and R. W. Sussman, pp. 259–279. New York: Plenum Press.

——— (1978). Diurnal behavior of *Lemur mongoz* at Ampijoroa, Madagascar. *Folia Primatol.* **29**:291–302.

Jolly, A. (1966). *Lemur Behavior—A Madagascar Field Study.* Chicago: University of Chicago Press.

Jungers, W. L. (1979). Locomotion, limb proportions and skeletal allometry in lemurs and lorises. *Folia Primatol.* **32**:8–28.

Kappeler, P. M. (1990). Female Dominance in *Lemur catta:* More Than Just Female Feeding Priority? *Folia Primatol.* **55**:92–95.

Macedonia, J. M., and Stanger, K. F. (1994). Phylogeny of the lemuridae revisited: Evidence from communication signals *Folia primatol.* **63**:1–43.

Meier, B., Albignac, R., Peyrierns, A., Rumpler, Y., and Wright, P. (1987). A new species of *Hapalemur* (Primates) from southeast Madagascar. *Folia Primatol.* **48**: 211–215.

Morland, H. S. (1991). Preliminary report on the social organization of ruffed lemurs (*Varecia variegata variegata*) in a northeast Madagascar rain rorest. *Folia Primatol.* **56**:157–161.

______ (1993a). Reproductive activity of ruffed lemurs (*Varecia variegata variegata*) in a Madagascar rain forest. *Am. J. Phys. Anthrop.* **91**:71–82.

______ (1993b). Seasonal behavioral variation and its relationship to thermoregulation in ruffed lemurs (*Varecia variegata variegata*). In *Lemur Social Systems and Their Ecological Basis,* ed. P. M. Kappeler. and J. U. Ganzhorn, pp. 193–204. New York: Plenum Press.

Overdorff, D. J. (1993). Ecological and reproductive correlates to range use in red-bellied lemurs (*Eulemur rubriventer*) and rufous lemurs (*Eulemur fulvus rufus*). In *Lemur Social Systems and Their Ecological Basis,* ed. P. M. Kappeler and J. U. Ganzhorn, pp. 167–179. New York: Plenum Press.

______ (1996). Ecological correlates to social structure in two lemur species in Madagascar. *Am. J. Phys. Anthrop.* **100**:487–506.

______ (1998). Are *Eulemur* species pair-bonded? Social organization and mating strategies in *Eulemur fulvus rufus* from 1988–1995 in southeast Madagascar. *Amer. J. Phys. Anthrop.* **105**:153–166.

Overdorff, D. J., Strait, S. G., and Telo, A. (1997). Seasonal variation in activity and diet in a small-bodied folivorous primate, *Hapalemur griseus,* in Southeastern Madagascar. *Amer. J. Primatol.* **43**:211–223.

Pereira, M. E. (1993). Seasonal adjustment of growth rate and adult body weight in ringtailed lemurs. In *Ecological Bases of Lemur Social Organization,* ed. P. M. Kappeler and J. U. Ganzhorn, pp. 205–222. New York: Plenum Press.

Pereira, M. E., and Weiss, M. L. (1991). Female mate choice, male migration, and the threat of infanticide in ringtailed lemurs. *Behav. Ecol. Sociobiol.* **28**: 141–152.

Pereira, M. E., Kaufman, R., Kappeler, P. M., and Overdorff, D. J. (1990). Female dominance does not characterize all of the Lemuridae. *Folia Primatol.* **55**:96–103.

Petter, J. J., and Peyrieras, A. (1975). Preliminary notes on the behavior and ecology of *Hapalemur griseus.* In *Lemur Biology,* ed. I. Tattersall and R. W. Sussman, pp. 281–286. New York: Plenum Press.

Petter, J. J., Albignac, R., and Rumpler, Y. (1977). *Faune de Madagascar,* vol. 44: *Mammiferes lemuriens* (*Primates, Prosimiens*). Paris: Orstom, CNRS.

Pollack, J. I. (1979). Spatial distribution and ranging behavior in lemurs. In *The Study of Prosimian Behavior,* ed. G. A. Doyle and R. D. Martin, pp. 359–410. New York: Academic Press.

______ (1986). A note of the ecology and behavior of *Hapalemur griseus. Primate Conservation* **6**:97–101.

Sauther, M. L. (1989). Antipredator behavior in troops of free-ranging *Lemur catta* at Beza Mahafaly Special Reserve, Madagascar. *Int. J. of Primatol.* **10**;595–605.

______ (1991). Reproductive behavior of free-ranging *Lemur catta* at Beza Mahafaly Special Reserve, Madagascar. *Am. J. Phys. Anthrop.* **84**:463–477.

______ (1993). Resource competition in wild populations of ringtailed lemurs (*Lemur catta*). In *Lemur Social Systems and Their Ecological Basis,* ed. P. M. Kappeler. and J. U. Ganzhorn, pp. 135–152. New York: Plenum Press.

Sauther, M., and Sussman, R. W. (1993). A new interpretation of the social organization and mating system of the ringtailed lemur (*Lemur catta*). In *Lemur Social Systems and Their Ecological Basis,* ed. P. M. Kappeler. and J. U. Ganzhorn, pp. 111–122.New York: Plenum Press.

Simons, E. L., and Rumpler, Y. (1988). *Eulemur:* New generic name for species of *Lemur* other than *Lemur catta. C. R. Acad. Sci. Paris, t.* **307**:547–551.

Sussman, R. W. (1974). Ecological distinctions in sympatric species of lemur. In *Prosimian Biology,* ed. R. D. Martin, G. A. Doyle, and A. C. Walker, pp. 75–108. London: Duckworth.

______ (1975). A preliminary study of the behavior and ecology of *Lemur fulvus fulvus* Audebert, 1880. In *Lemur Biology,* ed. I. Tattersall and R. W. Sussman, pp. 237–258. New York: Plenum Press.

______ (1978). Nectar feeding by prosimians and its evolutionary implications. In *Recent Advances in Primatology,* vol. 3, ed. D. J. Chivers and K. A. Joysey, pp. 119–125. London: Academic Press.

______ (1982). Male life history and intergroup mobility among ringtailed lemurs. *Int. J. Primatol.* **13**:395–413.

Sussman, R. W., and Tattersall, I. (1976). Cycles of activity, group composition and diet of *Lemur mongoz mongoz* (Linnaeus, 1766) in Madagascar. *Folia Primatol.* **26**: 270–283.

Tattersall, I. (1976). Group structure and activity rhythm in *Lemur mongoz* (Primates, Lemuriformes) on Anjouan and Moheli Islands, Comoro Archipelago. *Anthropol. Papers Am. Mus. Nat. Hist.* **53**:367–380.

______ (1978). Behavioral variation in *Lemur mongoz* (= *L. m. mongoz*). In *Recent Advances in Primatology,* vol. 3, ed. D. J. Chivers and K. A. Joysey, pp. 127–132. London: Academic Press.

——— (1982). *The Primates of Madagascar.* New York: Columbia University Press.

Tattersall, I., and Schwartz, J. H. (1992). Relationships within the Malagasy primate subfamily Lemurinae. In *Topics in Primatology,* vol. 3: *Evolutionary Biology, Reproductive Endocrinology and Virology,* ed. S. Matano, R. H. Tuttle, H. Ishida, and M. Goodman, pp. 103–112. Tokyo: University of Tokyo Press.

Tattersall, I., and Sussman, R. W. (1975). Observations on the ecology and behavior of the mongoose lemur, *Lemur mongoz mongoz* Linnaeus (Primates, Lemuriformes) at Ampyoroa, Madagascar. *Anthropol. Papers Am. Mus. Nat. Hist.* **52**:193–216.

Tilden, C. D. (1990). A study of locomotor behavior in a captive colony of red-bellied lemurs (*Eulemur rubriventer*). *Am. J. Primatol.* **22**:87–100.

White, F. J. (1991). Social organization, feeding ecology, and reproductive strategy of ruffed lemurs, *Varecia variegatta.* In *Primatology Today,* ed. A. Ehara, T. Kimura, O. Takenaka, and M. Iwamoto, pp. 81–84. Tokyo: Tokyo Press.

White, F. J., Balko, E. A., and Fox, E. A. (1993). Male transfer in captive ruffed lemurs, *Varecia variegata variegata.* In *Lemur Social Systems and Their Ecological Basis,* ed. P. M. Kappeler. and J. U. Ganzhorn, pp. 41–50. New York: Plenum Press.

Wright, P. C. (1987). The greater bamboo lemur in Madagascar. *From the Forest* **2**(1):1–4.

——— (1988). A lemur's last stand. *Animal Kingdom* **91**: 12–25.

Lepilemurids

Charles-Dominique, P., and Hladik, C. M. (1971). Le *Lepilemur* du sud de Madagascar: Ecologie, alimentation et vie sociale. *Terre Vie* **25**:3–66.

Ganzhorn, J. U. (1993). Flexibility and constraints of Lepilemur ecology. In *Lemur Social Systems and Their Ecological Basis,* ed. P. M. Kappeler. and J. U. Ganzhorn, pp. 153–166. New York: Plenum Press.

Hladik, C. M., and Charles-Dominique, P. (1974). The behavior and ecology of the sportive lemur (*Lepilemur mustelinus*) in relation to its dietary peculiarities. In *Prosimian Biology,* ed. R. D. Martin, G. A. Doyle, and A. C. Walker, pp. 24–38. London: Duckworth.

Hladik, C. M., Charles-Dominique, P., and Petter, J. J. (1980). Feeding strategies of five nocturnal prosimians in the dry forest of the west coast of Madagascar. In *Nocturnal Malagasy Primates,* ed. P. Charles-Dominique, H. M. Cooper, A. Hladik, C. M. Hladik, E. Pages, G. E. Pariente, A. PetterRousseaux, and A. Schilling, pp. 41–74. New York: Academic Press.

Hladik, C. M., Charles-Dominique, P., Valdebouze, P., Delort-Lavale, J., and Flanzy, J. (1971). La caecotrophie chez un primate phyllophage du genre *Lepilemur* et les correlations avec les particularites de son appareil digestif. *C. R. Acad. Sci. (Paris)* **272**:3191–3194.

Nash, L. T. (1998). Vertical clingers and sleepers: seasonal influences on the activities and substrate use of *Lepilemur leucopus* at Beza Mahafaly Special Reserve. *Folia Primatol.* **69**(Suppl 1):204–217.

Petter, J. J., Albignac, R., and Rumpler, Y. (1977). *Faune de Madagascar,* vol. 44: *Mammiferes lemuriens (Primates, Prosimiens).* Paris: Orstom, CNRS.

Schmid, J., and Ganzhorn, J. U. (1996). Resting metabolic rates of *Lepilemur ruficaudatus. Am. J. Primatol.* **38**: 169–174.

Tattersall, I., and Schwartz, J. H. (1974). Craniodental morphology and the systematics of the Malagasy lemurs (Primates, Prosimii). *Anthropol. Papers Am. Mus. Nat. Hist.* **52**(3):141–192.

Indriids

Albignac, R., and Dorst, J. (1981). Zoologies des vertebres variabilite dans l'organisation territoriale et l'ecologie de *Avahi laniger* (Lemurien nocturne de Madagascar). *C. R. Acad. Sci. (Paris)* **292**(ser. III): 331–334.

Demes, B., Jungers, W. L., and Selpien, K. (1991). Body size, locomotion, and long bone cross-sectional geometry in indriid primates. *Am. J. of Phys. Anthrop.* **86**: 537–547.

Fleagle, J. G., and Anapol, F. C. (1992). The indriid ischium and the hominid hip. *J. Hum. Evol.* **22**: 285–305.

Ganzhorn, J. U. (1989). Primate species separation in relation to secondary plant chemicals. *Hum. Evol.* **4**: 125–132.

Ganzhorn, J. U., Abraham, J. P., and Razanahoera-Rakotomalala, M. (1985). Some aspects of the natural history and food selection of *Avahi laniger. Primates* **25**: 452–463.

Gebo, D. L., and Dagosto, M. (1988). Foot anatomy, climbing, and the origin of the Indriidae. *J. Hum. Evol.* **17**:135–154.

Glander, K. E. , Wright, P. C., Daniels, P. C., and Merenlender, A. S. (1992). Morphometrics and testicle size of rain forest lemur species from southeastern Madagascar. *J. Human Evol.* **22**:1–17.

Jolly, A. (1966). *Lemur Behavior—A Madagascar Field Study.* Chicago: University of Chicago Press.

Jungers, W. L. (1979). Locomotion, limb proportions and skeletal allometry in lemurs and lorises. *Folia Primatol.* **52**:8–28.

Meyers, D. M., and Wright, P. C. (1993). Resource tracking: Food availability and *Propithecus* seasonal reproduction. In *Lemur Social Systems and Their Ecological Basis,* ed. P. M. Kappeler and J. U. Ganzhorn, pp. 179–192. New York: Plenum Press.

Petter, J. J., Albignac, R., and Rumpler, Y. (1977). *Faune de Madagascar,* vol. 44: *Mammiferes lemuriens (Primates, Prosimiens)*. Paris: Orstom, CNRS.

Pollock, J. I. (1975). Field observations on *Indri indri:* A preliminary report. In *Lemur Biology,* ed. I. Tattersall and R. W. Sussman, pp. 287–312. New York: Plenum Press.

——— (1979). Spatial distribution and ranging behavior in lemurs. In *The Study of Prosimian Behavior,* ed. G. A. Doyle and R. D. Martin, pp. 359–410. New York: Academic Press.

——— (1986). The song of the indris (*Indri indri;* Primates: Lemuroidea): Natural history, form, and function. *Int. J. Primatol.* **7**:225–264.

Ravosa, M. J., Meyers, D. J. and Glander, K. E. (1993). Relative growth of the limbs and trunk in Sifakas: Heterochronic, ecological, and functional considerations. *Am. J. Phys. Anthrop.* **92**:499–520.

Richard, A. E. (1974). Patterns of mating in *Propithecus verreauxi.* In *Prosimian Biology,* ed. R. D. Martin, G. A. Doyle, and A. C. Walker, pp. 49–74. London: Duckworth.

——— (1978). *Behavioral Variation: Case Study of a Malagasy Lemur.* Lewisburg, Pa.: Bucknell University Press.

Richard, A. E, and Heimbuch, R. (1975). An analysis of the social behavior of three groups of *Propithecus verreauxi.* In *Lemur Biology,* ed. I. Tattersall and R. W. Sussman, pp. 313–334. New York: Plenum Press.

Richard, A. E., Rakotomanga, P., and Schwartz, M. (1993). Dispersal by *Propithecus verreauxi* at Beza Mahafaly, Madagascar: 1984–1991. *Am. J. Primatol.* **30**:1–20.

Simons, E. L. (1988). A new species of *Propithecus* (Primates) from northeast Madagascar. *Folia Primatol.* **50**: 143–151.

Thalmann, U., Geissmann, T., Simona, A., and Mutschler, T. (1993). The Indris of Anjanaharibe-Sud, northeastern Madagascar. *Int. J. Primatol.* **14**:357–381.

Wright, P. C. (1995). Demography and life history of free-ranging *Propithecus diadema edwardsi* in Ranomafana National Park, Madagascar. *Int. J. Primatol.* **16**:835–854.

Daubentonids

Cartmill, M. (1974). *Daubentomia, Dactylopsila,* woodpeckers and kinorhynchy. In *Prosimian Biology,* ed. R. D. Martin, G. A. Doyle, and A. C. Walker, pp. 655–672. London: Duckworth.

Erikson, C. J. (1991). Precussive foraging in the aye-aye, *Daubentonia madagascariensis. Anim. Behav.* **41**:793–801.

——— (1994). Tap-scanning and extractive foraging in aye-ayes, *Daubentonia madagascariensis. Folia Primatol.* **62**:125–135.

Feistner, A. T. C., and Sterling, E. J. eds. (1994). The aye-aye: Madagascar's most puzzling primate. *Folia Primatol.* **62**:1–180.

Oxnard, E. (1981). The uniqueness of *Daubentonia. Am. J. Phys. Anthropol.* **54**:1–21.

Petter, J. J., and Peyrieras, A. (1970). Nouvelle contribution a l'etude d'un lemurien malgache, le aye-aye (*Daubentonia madagascarensis* E. Geoffroyi). *Mammalia* **54**:167–193.

Petter, J. J., Albignac, R., and Rumpler, Y. (1977). *Faune de Madagascar,* vol. 44: Mammiferes Lemuriens (Primates, Prosimins). Paris: Orstom, CNRS.

Sterling, E. J. (1993). Patterns of range use and social organization in aye-ayes (*Daubentona madagascariensis*) on Nosy Mangabe. In *Lemur Social Systems and Their Ecological Basis,* ed. P. M. Kappeler and J. U. Ganzhorn, pp. 1–10. New York: Plenum Press.

——— (1994). Evidence for Nonsexual Reproduction in Wild Aye-Ayes (*Daubentonia madagascarensis*). *Folia Primatol.* **62**:46–53.

Tattersall, I. (1982a). *The Primates of Madagascar.* New York: Columbia University Press.

——— (1982b). Two misconceptions of phylogeny and classification. *Am. J. Phys. Anthropol.* **57**:13.

SUBFOSSIL MALAGASY PROSIMIANS

Dewar, R. E. (1984). Recent extinctions in Madagascar: The loss of the subfossil fauna. In *Quaternary Extinctions: A Prehistoric Revolution,* ed. P. S. Martin and R. G. Klein, pp. 574–593. Tucson: University of Arizona Press.

Flacourt, E. de. (1661). Histoire de la grande lsle Madagascar. Avec une relation de ce qui s'est passe des annees 1655, 1656 et 1657, non encor veue Par la premiere impression. pp. 1–202 (Histoire); 203–471 (Relation). Troyes, N. Oudot: Paris, Pierre L' Amy.

Godfrey, L. R., Sutherland, M. R., Paine, R. R., Williams, F. L., Boy, D. S., and Vaillaume-Randriamanantena, (1995). Limb joint surface areas and their ratios in Malagasy lemurs and other mammals. *Amer. J. Phys. Anthropol.* **97**:11–36.

Godfrey, L. R., Jungers, W. L., Reed, K. E., Simons, E. L., and Chatrath, P. S. (1997). Subfossil lemurs: Inferences about past and present primate communities in Madagascar. In *Natural Change and Human Impact in Madagascar,* eds. S. M. Goodman and B. D. Patterson pp. 218–256. Washington: Smithsonian Institution.

MacPhee, R. D. E. (1986). Environment, extinction, and Holocene vertebrate localities in southern Madagascar. *National Geographic Research* **2**:441–455.

MacPhee, R. D. E., and Burney, D. A. (1991). Dating of modified femora of extinct dwarf *Hippopotamus* from southern Madagascar: Implications for constraining human colonization and vertebrate extinction events. *J. Archaeolog. Sci.* **18**:695–706.

MacPhee, R. D. E., Burney, D. A., and Wells, N. A. (1985). Early Holocene chronology and environment of Ampazambazimba, a Malagasy subfossil lemur site. *Int. J. Primatol.* **6**:463–489.

Simons, E. L. (1994). The giant aye-aye *Daubentonia robusta. Folia primatol.* **62**:14–21

Simons, E. L., Burney, D. A., Chatrath, P. S., Godfrey, L. R., Jungeres, W. L., and Rakotosamimanana, B. (1995). AMS ^{14}C dates for extinct lemurs from caves in the Ankarana Massif, northern Madagascar. *Quaternary Res.* **43**:249–254.

Tattersall, I. (1982). *The Primates of Madagascar.* New York: Columbia University Press.

Vuillaume-Randriamanantena, M., Godfrey, L. R., and Sutherland, M. R. (1985). Revision of *Hapalemur* (*Prohapalemur*) *gallieni* (Standing 1905). *Folia Primatol.* **45**: 89–116.

Walker, A. (1967). Patterns of extinction among the subfossil Madagascan lemuroids. In *Pleistocene Extinctions: The Search for a Cause,* ed. P. S. Martin and H. E. Wright, Jr., pp. 425–532. New Haven, Conn.: Yale University Press.

——— (1974). Locomotor adaptations in past and present prosimians. In *Primate Locomotion,* ed. E. A. Jenkins, pp. 349–381. New York: Academic Press.

Mesopropithecus

Baum, D. (1996). The ecology and conservation of the baobabs (Adansonia-Bombacaceae) in Madagascar. In *Economy and Ecology of a Tropical Dry Forest,* ed. J. Ganzhorn and J.-P. Sorg, *Primate Report* **46**(1): 311–327.

Simons, E. L., Godfrey, L. R., Jungers, W. L., Chatrath, P. S., and Ravaoarisoa, (1995). A new species of *Mesopropithecus* (Primates, Paleopropithecidae) from Northern Madagascar. *Int. J. Primatol.* **16**:653–682.

Vuillaume-Randriamanantena, M. (1982). Contribution a l'etude des os longs des lemuriens subfossiles malagaches: Leurs particularites au niveau des proportions. Dissertation de 3ed cycle, Universite de Madagascar, Antananarivo.

Babakotia

Godfrey, L. R., Simons, E. L., Chatrath, P. S., and Rakotosamimanana, B. (1990). A new fossil lemur (*Babakotia,* Primates) from northern Madagascar. *C. R. Acad. Sci. Paris* **310**(Serie II):81–87.

Jungers, W. L., Godfrey, L. R., Simons, E. L., Chatrath, P. S., and Rakotosamimanana, B. (1991). Phylogenetic and functional affinities of *Babakotia* (Primates), a fossil lemur from northern Madagascar. *Proc. Natl. Acad. Sci.* **88**:9082–9086.

Simons, E. L., Godfrey, L. B., Jungers, W. L., Chatrath, P. S., and Rakotosamimanana, B. (1992). A new giant subfossil lemur, *Babakotia,* and the evolution of the sloth lemurs. *Fol. Primatol.* **58**:197–203.

Palaeopropithecus

Jungers, W. L., Godfrey, L. R., Simons, E. L., and Chatrath, P. S. (1997). Phalangeal curvature and positional behavior in extinct sloth lemurs (Primates, Palaeopropithecidae). *Proc. Natl. Acad. Sci. USA* **94**: 11998–12001.

Saban, R. (1975). Structure of the ear region in living and subfossil lemurs. In *Lemur Biology,* ed. I. Tattersall and R. W. Sussman, pp. 83–109. New York: Plenum Press.

Tattersall, I. (1975). Notes on the cranial anatomy of the subfossil Malagasy lemurs. In *Lemur Biology,* ed. I. Tattersall and R. W. Sussman, pp. 111–124. New York: Plenum Press.

Archaeoindris

Carlton, A. (1936). The limb bones and vertebrae of the extinct lemurs of Madagascar. *Proc. Zool. Soc. London* **110**:281–307.

Lamberton, C. (1936). Fouilles faites en 1936. *Bull. Acad. Malgache, n.s.* **19**:1–19.

Archaeolemur

Godfrey, L. R., Sutherland, M. R., Petto, A. J., and Boy, D. S. (1990). Size, space and adaptation in some subfossil lemurs from Madagascar. *Am. J. Phys. Anthrop.* **81**:45–66.

Jolly, C. J. (1970). The seed-eaters: A new model of hominid differentiation based on baboon analogy. *Man* **5**: 5–26.

Tattersall, I. (1973). Cranial anatomy of the *Archaeolenurinae* (Lemuroidea, Primates). *Anthropol. Papers Am. Mus. Nat. Hist.* **52**:1–110.

——— (1974). Facial structure and mandibular mechanics in *Archaeolemur.* In *Prosimian Biology,* ed. R. D. Martin, G. A. Doyle, and A. C. Walker, pp. 563–578. London: Duckworth.

Hadropithecus

Godfrey, L. R., Jungers, W. L., Wunderlich, R. E. and Richmond, B. G. (1997). Reappraisal of the postcranium of *Hadropithecus* (Primates, Indroidea). *Am. J. Phys. Anthrop.* **103**:529–556.

Jolly, C. (1970). *Hadropithecus,* a lemuroid small-object feeder. *Man n.s.* **5**:525–529.

Tattersall, I. (1973). Cranial anatomy of the *Archaeolemurines* (Lemuroidea, Primates). *Anthropol. Papers Am. Mus. Nat. Hist.* **52**:1–110.

Pachylemur

Jouffroy, E. K., and Lessertisseur, J. (1979). Relationships between limb morphology and locomotor adaptations among prosimians: An osteometric study. In *Environment, Behavior and Morphology: Dynamic Interactions in Primates,* ed. M. E. Morbeck, H. Preuschoft, and N. Gomberg, pp. 143–181. New York: Gustav Fischer.

Seligsohn, D., and Szalay, F. S. (1974). Dental occlusion and the masticatory apparatus in *Lemur* and *Varecia:* Their bearing on the systematics of living and fossil primates. In *Prosimian Biology,* ed. R. D. Martin, G. A. Doyle, and A. C. Walker, pp. 543–562. London: Duckworth.

Megaladapis

Jungers, W. L. (1977). Hindlimbs and pelvic adaptations to vertical climbing and clinging in *Megaladapis,* a giant subfossil prosimian from Madagascar. *Yrbk. Phys. Anthropol.* **20**:508–524.

Major, C. J. E. (1893). On *Megaladapis madagascariensis,* an extinct gigantic lemuroid from Madagascar. *Philos. Trans. R. Soc. London* **185**:15–38.

Tattersall, I. (1972). The functional significance of airorhynchy in *Megaladapis. Folia Primatol.* **18**:20–26.

Vuillaume-Randriamanantena, M., Godfrey, L. R., Jungers, W. L. Jr., and Simons, E. L. (1992). Morphology, taxonomy and distribution of *Megaladapis*—giant subfossil lemur from Madagascar. *C. R. Acad. Sci. Paris.* **315**(Serie II):1835–1842.

Wunderlich, R. E., Simons, E. L., and Jungers, W. L. (1996). New pedal remains of *Megaladapis* and their functional significance. *Am. J. Phys. Anthrop.* **100**: 115–138.

ADAPTIVE RADIATION OF MALAGASY PRIMATES

Dagosto, M. (1994).

——— (1995). Seasonal variation in positional behavior of Malagasy lemurs. *Int. J. Primatol.* **16**:807–834.

Ganzhorn, J. U. (1989). Niche separation of seven lemur species in the eastern rainforest of Madagascar. *Oecologia* **79**:279–286.

Godfrey, L. R., Jungers, W. L., Reed, K. E., Simons, E. L., and Chatrath, P. S. (1997). Subfossil lemurs: Inferences about past and present primate communities in Madagascar. In *Natural Change and Human Impact in Madagascar,* eds. S. M. Goodman and B. D. Patterson pp. 218–256. Washington: Smithsonian Institution.

Hawkins, A. F. A., Chapman, P., Ganzhorn, J. U. (1990). Vertebrate conservation in Ankarana Special Reserve, northern Madagascar. *Biological Conservation* **54**:83–110.

Jolly, A. (1984). The puzzle of female feeding priority. In *Female Primates: Studies by Women Primatologists,* ed. M. E Small, pp. 197–215. New York: Alan R. Liss.

Kappeler, P. M. (1990a). Female dominance in *Lemur catta:* More than just female feeding priority? *Folia Primatol.* **55**:92–95

——— (1990b). The evolution of sexual size dimorphism in prosimian primates. *Am. J. Primatol.* **21**: 201–214.

——— (1991). Patterns of sexual dimorphism in body weight among prosimian primates. *Folia Primatol.* **57**: 132–146.

——— (1996). Causes and consequences of life-history variation among strepsirhine primates. *Am. Nat.* **148**: 868–889.

Kappeler, P. M., and Ganzhorn, J. U. (1993). The evolution of primate communities and societies in Madagascar. *Evol. Anthrop.* **2**(6):159–171.

——— (1993). *Lemur Social Systems and their Ecological Basis.* New York: Plenum Press.

——— (1993). Female dominance in primates and other mammals. In *Perspectives in Ethology. Vol. 10. Behavior and Evolution,* eds. P. P. G. Bateson, P. H. Klopfer, and N. S. Thompson. pp. 143–158. New York: Plenum.

——— (1997). Determinants of primate social organization: comparative evidence and new insights from Malagasy lemurs. *Biol. Rev.* **72**:111–151.

Martin, R. D. (1972). Adaptive radiation and behavior of the Malagasy lemurs. *Philos. Trans. R. Soc. London B.* **264**:295–352.

Overdorff, D. J. (1998). Are *Eulemur* species pair-bonded? Social organization amd mating strategies in *Eulemur fulvus rufus* from 1988–1995 in southeast Madagascar. *Amer. J. Phys. Anthrop.* **105**:153–166.

Oxnard, C. E., Crompton, R. H., and Lieberman, S. S. (1990). *Animal Lifestyles and Anatomies.* Seattle: University of Washington Press.

Richard, A. (1986). Malagasy prosimians: female dominance. In *Primate Societies,* ed. B. B. Smuts, D. L. Cheney, R. M. Seyfarth, R. W. Wrangham, and T. T. Struhsaker, pp. 25–33. Chicago: University of Chicago Press.

Richard, A., and Dewar, R. E. (1991). Lemur ecology. *Ann. Rev. Ecol. Syst.* **22**:145–175.

Schaik, C. P. van, and Kappeler, P. M. (1994). Life history, activity period and lemur social systems. In *Lemur Social Systems and Their Ecological Basis,* ed. P. M. Kappeler and J. U. Ganzhorn, pp. 243–263. New York: Plenum Press.

Seligsohn, D. (1977). Analysis of species-specific molar adaptations in strepsirhine primates. In *Contributions to Primatology,* vol. 11. Basel, Switzerland: S. Karger

Tattersall, I. (1992). Systematic versus ecological diversity: The example of the Malagasy primates. In *Systematics, Ecology and the Biodiversity Crisis,* ed. N. Eldredge, pp. 25–39. New York: Columbia University Press.

——— (1993). Madagascar's lemurs. *Sci. Amer.* **268**(1): 110–117.

GALAGIDS

Bearder, S. K. (1986). Lorises, bushbabies, and tarsiers: Diverse societies in solitary foragers. In *Primate Societies,* ed. B. B. Smuts, D. L. Cheney, R. M. Seyfarth, R. W. Wrangham, and T. T. Struhsaker, pp. 11–24. Chicago: University of Chicago Press.

Bearder, S. K., Honess, P. E., and Ambrose, L. (1995). Species diversity among Galagos, with special reference to mate recognition. In *Creatures of the Dark: The Nocturnal Prosimians,* ed. L. Alterman, G. A. Doyle, and M. K. Izard, pp. 331–352. New York: Plenum Press.

Bearder, S. K., and Doyle, G. A. (1974). Ecology of bushbabies, *Galago senegalensis* and *C. crassicaudatus,* with some notes on their behavior in the field. In *Prosimian Biology,* ed. R. D. Martin, G. A. Doyle, and A. C. Walker, pp. 109–130. London: Duckworth.

Charles-Dominique, P. (1971). Eco-ethologie des prosimiens du Gabon. *Biol. Gabonica* **7**:121–228.

——— (1972). Ecologie et vie sociale de *Galago demidovii* (Fischer 1808, Prosimii). *Z. Tierpsychol. Beih.* **9**:7–41.

——— (1977a). *Ecology and Behavior of Nocturnal Primates.* New York: Columbia University Press.

——— (1977b). Urine marking and territoriality in *Galago alleni* (Waterhouse 1837—Lorisoidea, Primates), a field study by radio telemetry. *Z. Tierpsychol.* **45**: 113–138.

Charles-Dominique, P., and Martin, R. D. (1970). Evolution of lorises and lemurs. *Nature (London)* **227**:257–260.

Crompton, R. H. (1983). Age differences in locomotion of two subtropical Galaginae. *Primates* **24**(2):241–259.

——— (1984). Foraging, habitat structure and locomotion in two species of *Galago.* In *Adaptations for Foraging in Non-human Primates,* ed. P. S. Rodman and J. G. H. Cant, pp. 73–111. New York: Columbia University Press.

Harcourt, C. S. (1983). Bright eyes at night, bushbabies in Diani Forest. *Swara* **6**:26–27.

Harcourt, C. S., and Nash, L. T. (1986). Social organization of galagos in Kenyan coastal forests: I, *Galago zanzibaricus. Am. J. Primatol.* **10**:339–356.

Jungers, W. L., and Olson, T. R. (1984). Relative brain size in galagos and lorises. *Fortschr. Zool.* **30**:6–9.

Martin, R. D., and Bearder, S. K. (1979). Radio bushbaby. *Nat. Hist.* **88**(8):77–81.

Masters, J. (1988). Speciation in the greater galagos (Prosimii: Galaginae): Review and synthesis. *Biol. J. Linnean Soc.* **34**:149–174.

——— (1998). Speciation in the lesser galagos. *Folia Primatol.* **69**(Suppl 1):357–370.

Masters, J., and Lubinsky, D. (1988). Morphological clues to genetic species: Multivariate analysis of Greater Galago sibling species. *Am. J. Phys. Anthrop.* **75**:37–52.

Masters, J., Rayner, R. J., Ludewick, H., Zimmermann, E., Molez-Verriere, N., Vincent, F., and Nash, L. T. (1994). Phylogenetic relationships among the galaginae as indicated by ethrocytic allozymes. *Primates* **35**: 177–190.

McArdle, J. E. (1981). Functional morphology of the hip and thigh of the lorisiformes. *Contributions to Primatology,* vol. 7, pp. 1–32. Basel, Switzerland: S. Karger.

Molez, N. (1976). Adaptation alimentaire du Galago D' Allen aux milieux forestiers secondaires. *Terre Vie* **30**: 210–228.

Nash, L. T., and Harcourt, C. S. (1986). Social organization of galagos in Kenyan coastal forests: II, *Galago garnettii. Am. J. Primatol.* **10**:357–370.

Nash, L. T., Bearder, S. K., and Olson, T. R. (1989). Synopsis of *Galago* species characteristics. *Int. J. Primatol.* **10**:57–80.

LORISIDS

Amerasinghe, E. B., Van Cuylenberg, B. W. B., and Hladik, C. M. (1971). Comparative histology of the alimentary tract of Ceylon primates in correlation with diet. *Ceylon J. Sci., Biol. Sci.* **9**:75–87.

Barrett, E. (1981). The present distribution and status of the slow loris in peninsular Malaysia. *Malaysian Applied Biol.* **10**(2):205–211.

Bearder, S. K. (1986). Lorises, bushbabies, and tarsiers: Diverse societies in solitary foragers. In *Primate Societies,* ed. B. B. Smuts, D. L. Cheney, R. M. Seyfarth, R. W. Wrangham, and T. T. Struhsaker, pp. 11–24. Chicago: University of Chicago Press.

Charles-Dominique, P. (1971). Eco-ethologie des prosimiens du Gabon. *Biol. Gabonica* **7**:121–228.

______ (1977). *Ecology and Behavior of Nocturnal Primates.* New York: Columbia University Press.

Charles-Dominique, P., and Bearder, S. K. (1979). Field studies of lorisid behavior: Methodological aspects. In *The Study of Prosimian Behavior,* ed. G. A. Doyle and R. D. Martin, pp. 567–630. New York: Academic Press.

Demes, B., and Jungers, W. L. (1989). Functional differentiation in long bones in lorises. *Folia Primatol.* **52**:58–69.

Demes, B., Jungers, W. L. and U. Nieschalk. (1990). Size- and Speed-related Aspects of Quadrupedal Walking in Slender and Slow Lorises. In: *Gravity, Posture and Locomotion in Primates.* (Jouffroy, F. K, Stack, M. H. and Niemitz, C., eds.). Il Sedicesimo, Firenze, pp. 175–197.

Eliot, O., and Eliot, M. (1967). Field notes on the slow loris in Malaysia. *J. Mammal.* **48**:497–498.

Gomez, A. M. (1992). Primitive and derived patterns of relative growth among species of lorisidae. *J. Hum. Evol.* **23**(3):219–234.

Hladik, C. M. (1975). Ecology, diet and social patterning in Old and New World primates. In *Socioecology and Psychology of Primates,* ed. R. H. Tuttle, pp. 3–35. The Hague: Mouton.

______ (1979). Diet and ecology of prosimians. In *The Study of Prosimian Behavior,* ed. G. A. Doyle and R. D. Martin, pp. 307–358. New York: Academic Press.

Izard, M. K., and Rasmussem, D. T. (1985) Reproduction in the slender loris (*Loris tardigradus malabaricus*). *Am. J. Primatol.* **8**:153–165.

Jolly, C. J., and Gorton, C. (1974). Proportions of the extrinsic foot muscles in some lorisid prosimians. In *Prosimian Biology,* ed. R. D. Martin, G. A. Doyle, and A. C. Walker, pp. 801–816. London: Duckworth.

Jungers, W. L., and Olson, T. R. (1984). Relative brain size in galagos and lorises. *Fortschr. Zool.* **30**:6–9.

Oates, J. E. (1984). The niche of the potto, *Perodicticus potto. Int. J. Primatol.* **5**(1):51–61.

Ravosa, M. J. (1998). Cranial allometry and geographic variation in slow lorises (*Nycticebus*) *Am. J. Primatol.* **44**.

Rasmussen, D. T. and Nekaris, K. A. (1998). The evolutionary history of lorisiform primates. *Folia Primatol.* **69**(Suppl. 1):250–285.

Schwartz, J. H. (1992). Phylogenetic relationships of African and Asian lorisids. In *Topics in Primatology,* vol. 3, *Evolutionary Biology, Reproductive Endocrinology, and Virology,* ed. S. Matano, R. H. Tuttle, H. Ishida, and M. Goodman, pp. 65–81. Tokyo: University of Tokyo Press.

______ (1996). *Pseudopotto martini:* A new genus and species of extant lorisiform primate. *Anthropol. Papers Amer. Mus. Nat. Hist.* **78**:1–14.

Walker, A. (1969). The locomotion of the lorises, with special reference to the potto. *E. Afr. Wildlife J.* **8**:1–5.

PHYLETIC RELATIONSHIPS OF STREPSIRHINES

Cartmill, M. (1975). Strepsirhine basicranial structures and the affinities of the Cheirogaleidae. In *Phylogeny of the Primates: A Multidisciplinary Approach,* ed. W. P. Luckett and F. S. Szalay, pp. 313–356. New York: Plenum Press.

Eaglen, R. H. (1983). Parallelism, parsimony and the phylogeny of the Lemuridae. *Int. J. Primatol.* **4**(3): 249–273.

Sarich, V. M., and Cronin, J. E. (1976). Molecular systematics of the primates. In *Molecular Anthropology,* ed. M. Goodman and R. Tashkan, pp. 141–170. New York: Plenum Press.

Tattersall, I., and Schwartz, J. H. (1985). Evolutionary relationships of living lemurs and lorises (Mammalia, Primates) and their potential affinities with European Eocene Adapidae. *Anthropol. Papers Am. Mus. Nat. Hist.* **60**:1–110.

Yoder, A. D. (1994). Relative position of the Cheirogaleidae in strepsirhine phylogeny: A comparison of morphological and molecular methods and results. *Am. J. Phys. Anthrop.* **94**:25–46.

Yoder, A. D. (1997). Back to the Future: A synthesis of strepsirhine systematics. *Evol. Anthrop.* **6**:11–22.

Yoder, A., Cartmill, M., Ruvolo, M., Smith, K., and Vilgalys, R. (1996). Ancient single origin for Malagasy primates. *Proc. Natl. Acad. Sci. USA* **93**:5122–5126.

TARSIERS

Aiello, L. C. (1986). The relationships of the tarsiformes: A review of the case for the Haplorhini. In *Major Topics in Primate and Human Evolution,* ed. B. Wood, L. Martin, and P. Andrews, pp. 47–65. Cambridge: Cambridge University Press.

Cartmill, M. (1980). Morphology, function and evolution of the anthropoid postorbital septum. In *Evolutionary Biology of the New World Monkeys and Continental Drift,* ed. R. L. Ciochon and A. B. Chiarelli, pp. 243–274. New York: Plenum Press.

______ (1982). Assessing tarsier affinities—Is anatomical description phylogenetically neutral? In *Phylogenie et Paleobiogeographie,* ed. E. Buffetaut, P. Janvier, J. C. Rage, and P. Tassy, pp. 279–287. *Geobios, Mem. Spec.* **6.**

Crompton, R. H., and Andau, P. M. (1986). Locomotion and habitat utilization in free-ranging *Tarsius bancanus:* A preliminary report. *Primates* **27**:337–355.

——— (1987). Ranging, activity rhythms, and sociality in free-ranging *Tarsius bancanus:* A preliminary report. *Int. J. Primatol.* **8**:43–72.

Fogden, M. (1974). A preliminary field study of the western tarsier, *Tarsius bancanus* Horsefield. In *Prosimian Biology,* ed. R. D. Martin, G. A. Doyle, and A. C. Walker, pp. 151–166. London: Duckworth.

Gebo, D. L. (1987). Functional anatomy of the tarsier foot. *Am. J. Phys. Anthropol.* **73**:9–31.

Gingerich, P. D. (1978). Phylogeny reconstruction and the phylogenetic position of *Tarsius.* In *Recent Advances in Primatology,* vol. 3, ed. D. J. Chivers and K. A. Joysey, pp. 249–255. London: Academic Press.

Gursky, S. (1994). Infant care in the spectral tarsier (*Tarsius spectrum*) Suluwesi, Indonesia. *Int. J. Primatol.* **15**: 843–854.

——— (1995). Group size and composition in the spectral tarsier, *Tarsius spectrum:* Implications for social organization. *Tropical Biodiversity* **3**:57–62.

——— (1998). Conservation status of the spectral tarsier, *Tarsius spectrum:* population density and home range size. *Folia Primatol.* **69**(Suppl. 1):191–203.

Hubrecht, A. A. W. (1908). Early ontogenetic phenomena in mammals and their bearing on our interpretation of the phylogeny of the vertebrates. *Quan. J. Microsc. Sci.* **53**:1–181.

Luckett, W. P., and Maier, W. (1982). Development of deciduous and permanent dentition in *Tarsius* and its phylogenetic significance. *Folia Primatol.* **37**:1–36.

Mackinnon, J., and Mackinnon, K. (1980). The behavior of wild spectral tarsiers. *Int. J. Primatol.* **1**:361–379.

Martin, R. D. (1975). Ascent of the primates. *Nat. Hist.* **74**(3):52–61.

Miyamoto, M., and Goodman, M. (1990). DNA systematics and the evolution of primates. *Ann. Rev. Ecol. Syst.* **21**:197–220.

Musser, G. G., and Dagosto, M. (1987). The identity of *Tarsius pumilus,* a pygmy species endemic to the montane mossy forests of central Sulawesi. *Am. Mus. Nov.,* no. 2867, pp. 1–53.

Niemitz, C. (1977). Zur functionellen Anatomie der Papillarleisten und ihrer Muster bei *Tarsius bancanus borneanus* Horsefield, 1821. *Z. Saugetierk.* **42**:321–346.

——— (1979). Outline of the behavior of *Tarsius bancanus.* In *The Study of Prosimian Behavior,* ed. G. A. Doyle and R. D. Martin, pp. 631–660. New York: Academic Press.

——— (1984). *Biology of Tarsiers.* New York: G. Fischer Verlag.

——— (1991) *Tarsius dianae:* A new primate species from central Sulawesi (Indonesia). *Folia Primatol.* **56**:105–116.

Pollock, J. I., and Mullin, R. J. (1987). Vitamin C biosynthesis in prosimians: Evidence for the anthropoid affinity of *Tarsius. Am. J. Phys. Anthropol.* **75**:65–70.

Rasmussen, D. T. (1994). The different meanings of a tarsoid-anthropoid clade. In *Anthropoid Origins.* ed. J. Fleagle and R. F. Kay, pp. 335–360. New York: Plenum Press.

Ross, C. (1994). The craniofacial evidence for anthropoid and tarsier relationships. In *Anthropoid Origins,* ed. J. Fleagle and R. F. Kay, pp. 469–547. New York: Plenum Press.

Roberts, M. (1994). Growth, development, and parental care in the western tarsier (*Tarsius bancanus*) in captivity: Evidence for a "slow" life-history and nonmonogamous mating system. *Int. J. Primatol.* **15**(1):1–28.

Schwartz, J. (1984). What is a tarsier? In *Living Fossils,* ed. N. Eldredge and S. M. Stanley, pp. 38–49. New York: Springer Verlag.

Shoshani, J., Groves, C. P., Simons, E. L., and Gunnell, G. F. (1996). Primate phylogeny: Morphological vs molecular results. *Molecular Phylogenetics and Evolution* **5**:102–154.

Wright, P. C., Izard, M. K., and Simons, E. L. (1986). Reproductive cycles in *Tarsius bancanus. Am. J. Primatol.* **11**:207–215.

FIVE

New World Anthropoids

PRIMATE GRADES AND CLADES

Primates have traditionally been divided into two suborders, Prosimii and Anthropoidea. This is a **gradistic** classification, because prosimians are a grade; they are identified by their retention of primitive anatomical features, not by unique derived features that distinguish them from other primates. In contrast to prosimians, anthropoids are also a **clade,** or natural phyletic unit, because the features that distinguish anthropoids from other primates are unique to anthropoids and are derived with respect to other primates. While we are certain from both morphological and molecular evidence that living anthropoids are a natural unit with a single common ancestry, there is much less agreement about where anthropoids came from, as we shall see in later chapters.

There are three major radiations of extant anthropoids, or higher primates: the platyrrhines, or New World monkeys, from South and Central America; and two groups of catarrhines from Africa, Europe, and Asia—the cercopithecoids, or Old World monkeys, and the hominoids, or apes and humans. The origins and evolutionary divergences of these groups, a topic of current debate, is discussed in later chapters. In this chapter and in Chapters 6 and 7 we consider the evolutionary diversity of the living higher primates. After a general characterization of the anthropoid suborder, we begin with the platyrrhines of the New World.

Anatomy of Higher Primates

Several anatomical features, primarily in the skull, distinguish anthropoids as a group from prosimians (Figs. 5.1, 5.2). Although some of these features are found in various other mammals (and quite a few in tarsiers), the suite as a whole is diagnostic of higher primates and presumably reflects a unique ancestry for the various groups of primates belonging to that suborder.

The dentition of anthropoids is more conservative than that of extant prosimians in both dental formula and the shape of the teeth. All higher primates have two incisors in each quadrant. Although these may be somewhat procumbent, as in pitheciines, they are never markedly so, as in strepsirhine prosimians. The anthropoid canine is always larger in caliber and usually taller in height than the incisors. There is considerable diversity in the size and shape of anthropoid canines, both among different species and between sexes within many species, but the canine is never absent, drastically reduced, or markedly procumbent, as in strepsirhines, and it usually

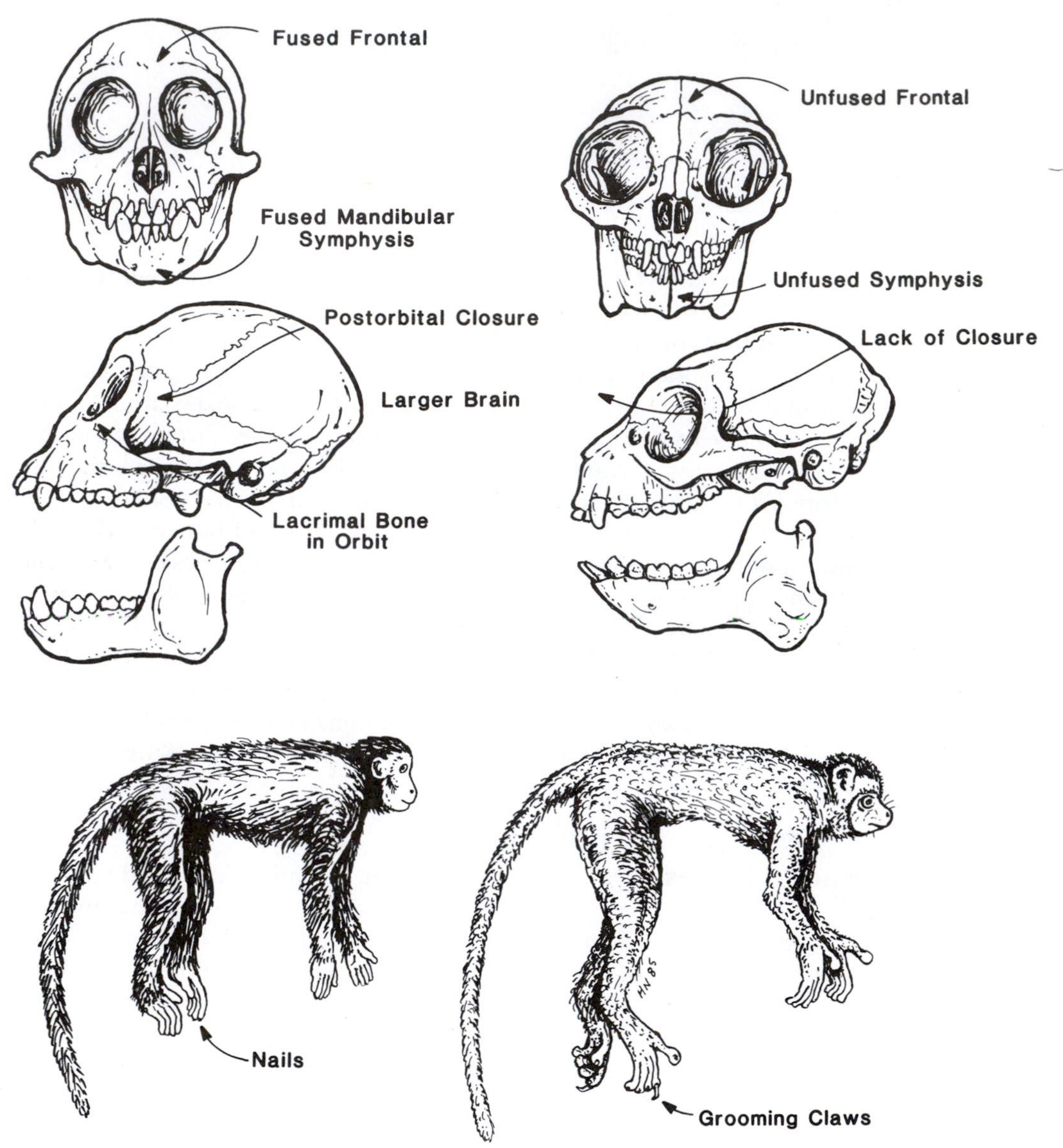

FIGURE 5.1 Anatomical features characteristic of anthropoids and prosimians.

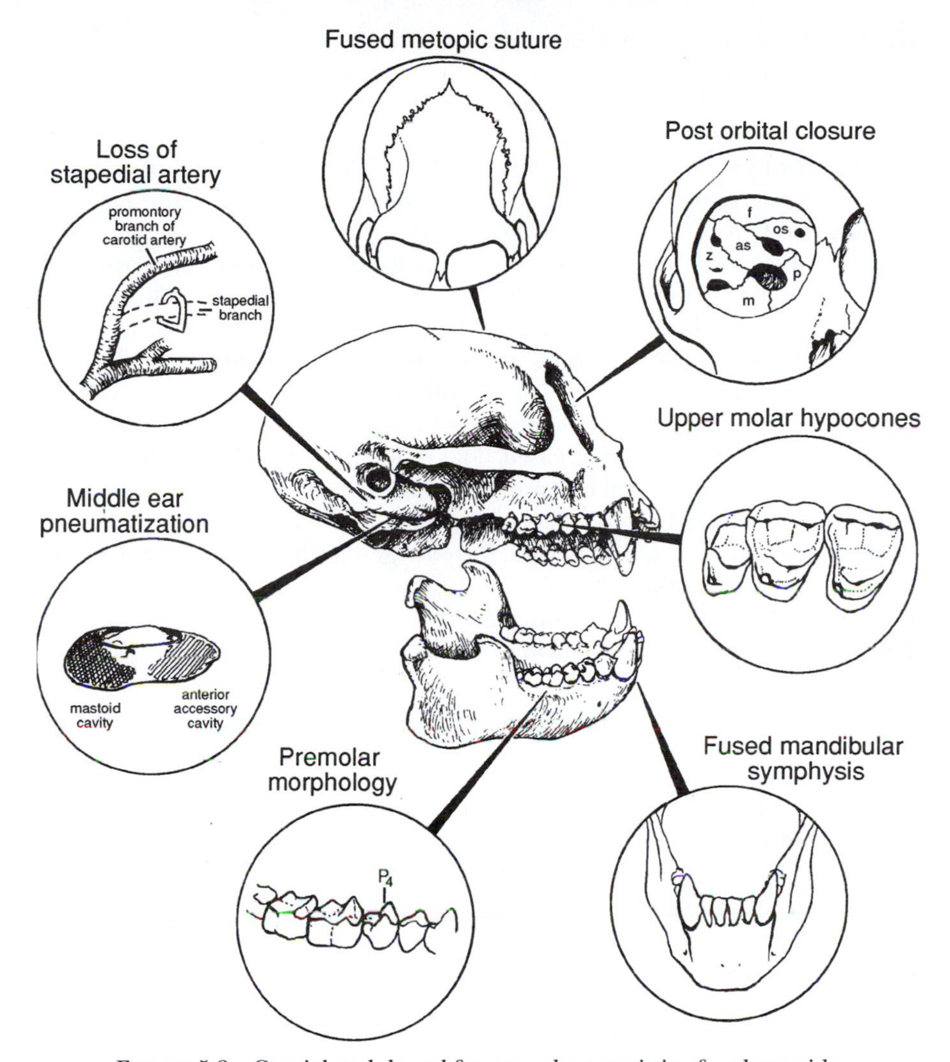

FIGURE 5.2 Cranial and dental features characteristic of anthropoids.

retains a large root, even when the crown is reduced, as in *Homo sapiens*.

Old World higher primates have two premolars, and New World monkeys have three. The morphology of higher primate premolars varies considerably, but the anteriormost premolar is often a simple tooth with a broad buccal surface that sharpens the upper canine, and the posteriormost premolar is always a semimolariform tooth with a differentiated trigonid and talonid and two subequal cusps on the trigonid (Fig. 5.2).

All living higher primates have three molars, except for marmosets and tamarins, which have two. Higher primate upper molars usually have a moderate-sized hypocone, making a relatively square tooth. The lower molars of higher primates are relatively broad with reduced trigonids (usually no paraconid) and an expanded talonid basin.

In most prosimians, and in placental mammals in general, the two halves of the mandible are joined at the front by a mobile symphysis that permits some degree of independent movement by the two halves of the jaw during chewing. In higher primates the two halves of the mandible are fused together. This condition seems to be functionally associated with the development of vertical lower incisors in anthropoids.

In conjunction with their diurnal habits, higher primates as a group have decreased their reliance on smell and chemical communication and have emphasized their reliance on sight. This change is reflected in several diagnostic characteristics of skull morphology (Figs. 5.1; 5.2). In prosimians the frontal bone joins with the zygomatic bone to form a postorbital bar and complete a bony ring around the orbit. In higher primates the frontal, zygomatic, and sphenoid bones expand the postorbital bar to form a bony cup that surrounds the eye and separates it from the temporal fossa behind, a condition known as **postorbital closure.** In anthropoids the paired frontal bones usually fuse into a single unit early in ontogeny.

The orbits of anthropoids face forward in conjunction with their well-developed stereoscopic vision. Like tarsiers, all higher primates lack a reflecting tapetum and have a retinal fovea on the posterior surface of their eyeball; most higher primates have color vision, although there is considerable variation in sensitivity to different parts of the spectrum (Jacobs, 1996). Most higher primates have a relatively short snout, and the lacrimal bone lies within the orbit rather than external to it on the snout. Internally, the nasal region of anthropoids is reduced, and there are fewer turbinates, as in tarsiers.

In higher primates, as in tarsiers, the blood supply to the brain is primarily through the promontory branch of the internal carotid artery; the stapedial artery is usually absent in adult anthropoids (Fig. 2.11). The tympanic ring is fused to the lateral wall of the auditory bulla in all anthropoids, but the shape of the tympanic bone differs among major living groups (Fig. 2.15).

In skeletal anatomy, anthropoids share few diagnostic features that clearly characterize them as a group. In general, most anthropoids are larger than most extant prosimians and have relatively shorter trunks. Anthropoid forelimbs and hindlimbs are more similar in length, or the forelimbs are longer. With our relatively long legs, we humans have very unusual higher primate proportions. Anthropoids generally do not have grooming claws.

Platyrrhines

The living anthropoids of the tropical areas of Central and South America (Fig. 5.3), the platyrrhines, have an evolutionary history extending back nearly 30 million years and are a much more diverse group than their common name, New World monkeys, suggests. They are "monkeys" only in that they are not apes (tailless, close relatives of humans). Evolving in South America in the absence of other primates (including prosimians), platyrrhines have evolved some species that are prosimian-like in habits, some that are apelike, and many that have no close analogy among other groups of living primates.

Several distinctive anatomical features distinguish platyrrhines from other primates. Living New World monkeys are all small- to medium-size primates, ranging from just under 100 g to just over 10 kg. Their name is derived from the broad, flat shape of their external nose, which often, but not always, distinguishes them from the Old World anthropoids, which often have narrow, or catarrhine, nostrils. In dental and cranial anatomy

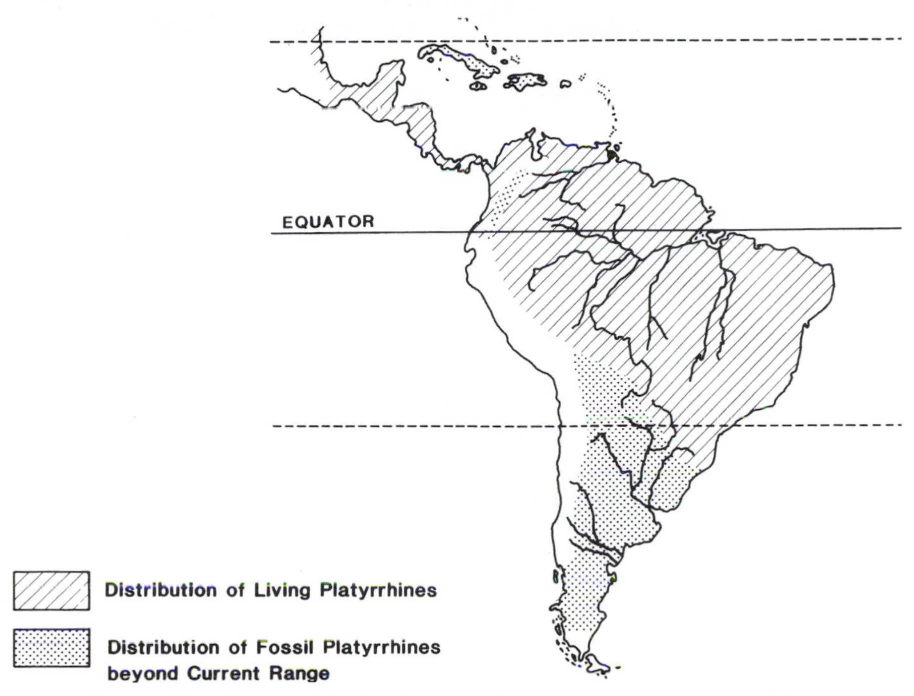

FIGURE 5.3 Geographic distribution of extant and extinct platyrrhines.

(Figs. 5.4, 5.5), platyrrhines have many primitive features that have been lost in the evolution of Old World catarrhines. For example, they have three premolars, and the tympanic ring is fused to the side of the auditory bulla but does not extend laterally as a bony tube. They also share some unique specializations of their own. The first two lower molar teeth of living platyrrhines usually lack hypoconulids, and on the lateral wall of the skull (the pterion region) the parietal and zygomatic bones join to separate the frontal bone above from the sphenoid below. In addition, the cranial sutures of platyrrhines fuse relatively late, and many species have relatively long, narrow skulls (Fig. 5.6).

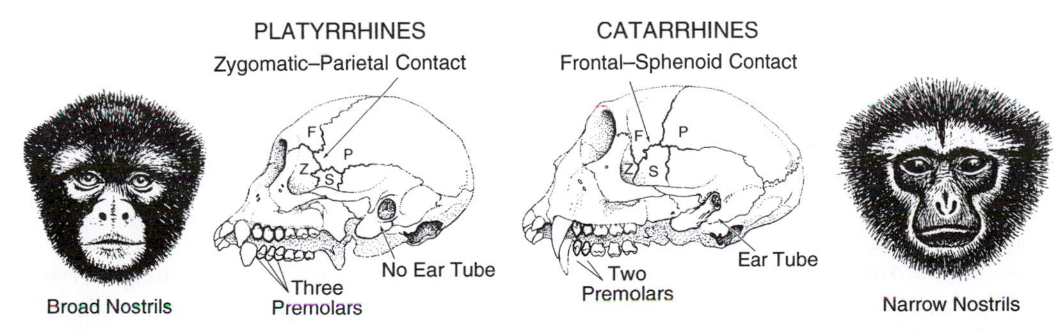

FIGURE 5.4 Skulls of a platyrrhine and a catarrhine, showing some of the features distinguishing these two major groups of anthropoids.

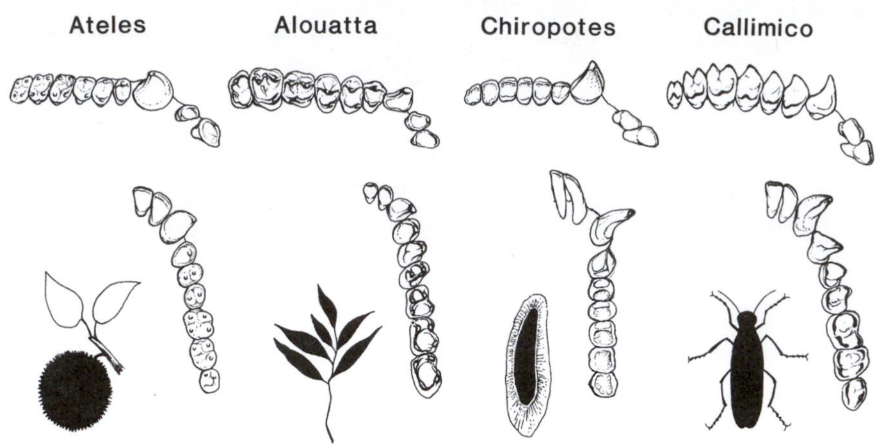

FIGURE 5.5 Upper and lower dentitions of four platyrrhines, illustrating structural differences in the teeth associated with ingesting and processing different types of food.

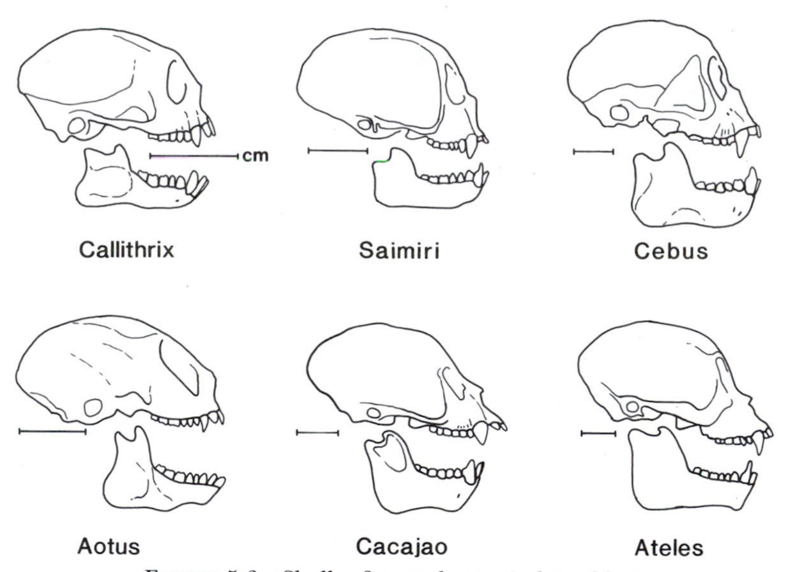

FIGURE 5.6 Skulls of several extant platyrrhines.

Platyrrhine limb proportions are relatively conservative, with intermembral indices ranging between 70 and 100. Platyrrhines lack the extremely low intermembral indices found among many prosimians or the high indices of apes. Most have a relatively short forearm, and most lack an opposable thumb. All have a tail of some sort, and in five genera the tail is a prehensile, fifth limb.

The sixteen genera of extant platyrrhines are a very diverse group, usually placed in many separate subfamilies. Six subfamilies are distinguished here: pitheciines, callicebines, atelines, cebines, callitrichines, and aotines. Each of these probably represents a relatively old lineage, and each is characterized by distinctive ecological (and morphological) adaptations. Some of the ecological diversity of platyrrhines is illustrated and characterized ecologically in Figures 5.8 and 5.9.

Pitheciines

The Pitheciines, a distinctive subfamily of platyrrhines, comprise three genera of medium-size monkeys: *Pithecia, Chiropotes,* and *Cacajao* (Table 5.1). The latter two genera are more closely related to one another than either is to *Pithecia.* Pitheciines are characterized by unusual dental specializations for processing relatively soft fruits and seeds encased in tough outer coverings that are generally too hard for other monkeys to bite through (Kinzey and Norconk, 1992). These dental specializations include large procumbent incisors, robust canines, and relatively small, square premolar and molar teeth with low cusps (Fig. 5.5, *Chiropotes*). In association with their dental specializations, they have a slightly prognathic snout, a narrow, U-shaped palate, and enlarged nasal bones (Fig. 5.6, *Cacajao*). Their pelage varies considerably from species to species. Pitheciines live in a wide variety of habitats, including rain forest, mountain savannah forest, flooded forests, and liane forest. They are found in extremely diverse social groups. However, many pitheciines seem to have a fluid, fission–fusion social structure in which individuals or small groups may forage separately or together, depending on available food resources. The subfamily ranges throughout Amazonia and parts of the Guianas, but not into Central America to the North or beyond the Amazonian ecosystem to the south.

With their broad noses, bushy fur, and long fluffy tails, **sakis** (*Pithecia*) are among the most distinctive of all New World primates (Fig. 5.7). They are the smallest pitheciine, averaging about 2 kg. They have a slightly more gracile skull and jaw than other pitheciines, a relatively longer trunk, and longer legs. There are five species of sakis, divided into two species groups. The best known is *Pithecia pithecia,* the white-faced saki, from the Guianas and northeastern Brazil, where it is frequently found sympatric with *Chiropotes.* White-faced sakis are dichromatic; that is, males are black with a white face, while females are a more uniform agouti color. In addition, there are four closely related Amazonian species, *P. albicans, P. irrorata, P. monachus,* and *P. aequatorialis* (Hershkovitz, 1987; Emmons, 1997). None of these is dichromatic like the white-faced saki, but *P. albicans* is a spectacular monkey with a white cape.

Species of *Pithecia* have been reported in a wide range of forest types, and the available data provide no clear evidence of habitat preference. White-faced sakis are most commonly seen in the understory and lower canopy levels, where they move primarily by leaping. They are among the most saltatory of all New World monkeys and frequently move by spectacular leaps. Other *Pithecia* species seem to be more quadrupedal and are found in higher parts of the forest (Peres, 1993). Like all pitheciines, sakis have unique dental adaptations that reflect their specialization on fruit and especially on seeds encased in a hard outer covering. However, compared with *Chiropotes* and *Cacajao, Pithecia* seems to eat fruits with relatively softer outer coverings and has a more diverse diet. They use their robust incisors and canines to open seed cases and their broad, flat molars for crushing relatively soft fruits and seeds. They rarely eat insects.

Sakis have generally been reported to live in monogamous family groups that travel and sleep together but often separate while feeding during the day. There are also reports of

TABLE 5.1
Infraorder Platyrrhini
Subfamily PITHECINAE

Common Name	Species	Intermembral Index	Mass (g) M	Mass (g) F
White-faced saki	*Pithecia pithecia*	75	1,940	1,580
Bald-faced saki	*P. irrorata*	—	2,250	2,070
Monk saki	*P. monachus*	77	2,610	2,110
White saki	*P. albicans*	—	3,000	—
Equatorial saki	*P. aequatorialis*	—	2,250	
Black-bearded saki	*Chiropotes satanas*	83	2,900	2,580
White-nosed bearded saki	*C. albinasus*	—	3,150	2,490
Bald uakari	*Cacajao calvus*	83	3,450	2,880
Black-headed uakari	*C. melanocephalus*	—	3,160	2,710

larger groups and of these small groups coming together. Saki social organization appears to be more a fission–fusion structure than one of strict monogamy. Estimated home range sizes vary considerably, from 10 ha or less in *Pithecia pithecia* to as much as 200 ha for *P. albicans*. There are no good estimates of day range. Sakis have single births, and the young seem to be cared for primarily by the females.

The **bearded saki** (*Chiropotes*) is larger than *Pithecia* but lacks any obvious sexual dichromatism. The skull and jaw of *Chiropotes* are more robust than those of *Pithecia*, and the limbs are more similar in length (Figs. 5.7, 5.10). There are two species: *Chiropotes satanas*, a smaller (3 kg), chocolate brown species with a black beard and bouffant hairdo from the Guianas and northeastern Brazil, and a larger species, *C. albinasus*, with black fur and a white nose, from Brazil south of the Amazon. *Chiropotes* prefers high rain forests (also mountain savannah forests in Suriname), where it is usually found in the middle and upper levels of the main canopy (Fig. 5.9). Bearded sakis are primarily arboreal quadrupeds that frequently use hindlimb suspension in feeding. They feed on hard, often unripe fruits and on seeds with very hard shells, which they open with their large canines. They occasionally ingest insects.

Bearded sakis live in large groups, with numerous males and females that split up into smaller foraging units during the day. They seem to have extremely large home ranges (over 200 ha), and groups may travel several kilometers in a day. They have single births and show no evidence of paternal care.

Uakaries (*Cacajao*) are the largest pitheciines and are easily distinguished from the other genera by their large size and short tail. There are three types: *C. melanocephalus*, the rare black uakari from black water rivers of the upper Amazon of Brazil, eastern Colombia, and southern Venezuela, and two subspecies of *C. calvus*, the bald uakari from the whitewater rivers of the upper Amazon (Solimoes) drainage. Both subspecies of *C. calvus* have scarlet faces and bald heads. In one subspecies, the long, shaggy fur is red; in the other, it is white. In contrast with bearded sakis, which are found in dry forests, uakaris are found in flooded forests, and the ranges of the two genera do not overlap (Ayres, 1989; Burnett & Brandon-Jones, 1997).

The ecology of uakaris has been the subject of several studies in recent years. Like other

FIGURE 5.7 Two pitheciine primates that live sympatrically in Suriname and other areas of northeastern South America: above, the bearded saki (*Chiropotes satanas*); below, the white-faced saki (*Pithecia pithecia*).

FIGURE 5.8 Seven sympatric platyrrhine species in a Surinam rain forest, showing typical locomotor and postural behavior as well as use of different heights in the forest. At the highest levels are the bearded saki (*Chiropotes satanas*) and the black spider monkey (*Ateles paniscus*); below them are tufted capuchins (*Cebus apella*) on the left and red howling monkeys (*Alouatta seniculus*) on the right; in the lower levels are squirrel monkeys (*Saimiri sciureus*) on the left and golden-handed tamarins (*Saguinus midas*) and white-faced sakis (*Pithecia pithecia*) on the right.

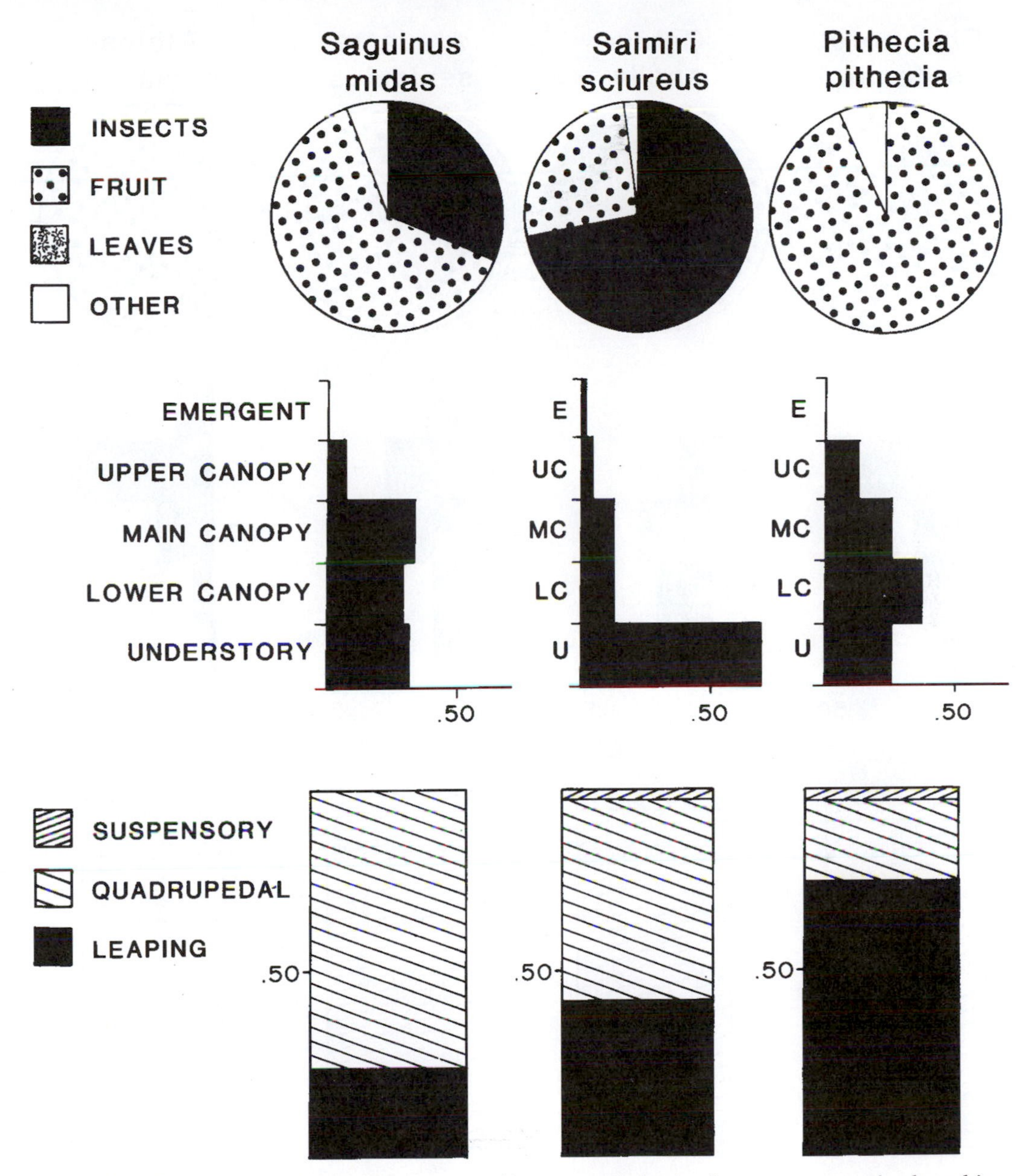

FIGURE 5.9 Diet, forest height preferences, and locomotor behavior for seven sympatric platyrrhines from Surinam. *(Continued on next page.)*

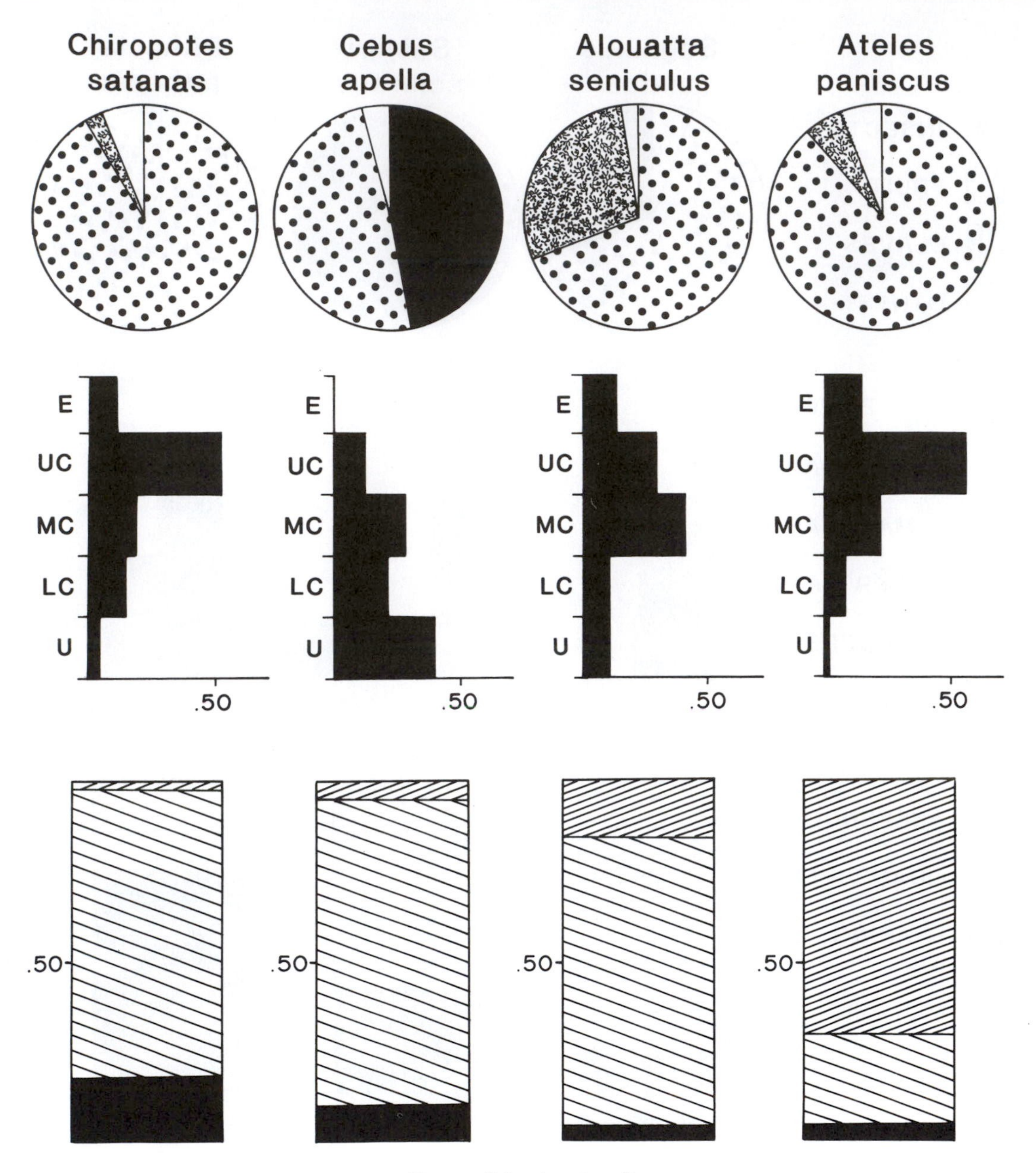

FIGURE 5.9 *(continued)*

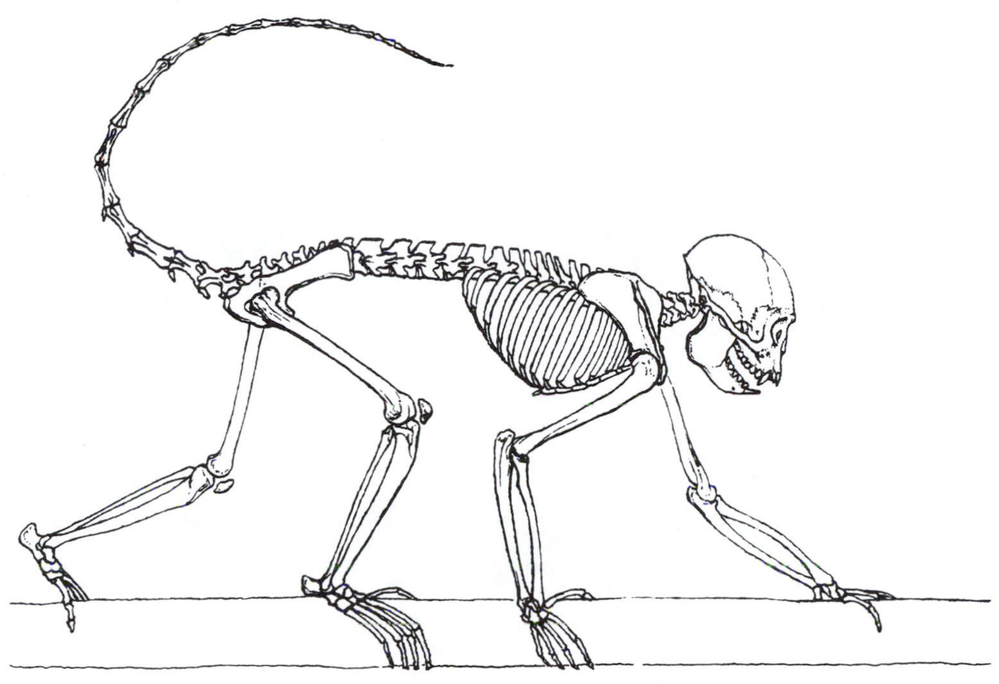

FIGURE 5.10 The skeleton of a bearded saki (*Chiropotes satanas*).

pitheciines, they specialize on the fruits with hard outer shells and immature seeds in their flooded forest habitats and move primarily by quadrupedal walking and running, both in the trees and on the ground. Like *Chiropotes,* uakaris often feed by hindlimb suspension (Walker and Ayres, 1996).

Like *Chiropotes,* both black uakaris and white uakaris live in large multimale, multifemale groups of 20–30 individuals or more, with even larger home ranges. Groups of white uakaris frequently divide into very small foraging parties, often for several days.

Callicebinae

Titi monkeys (*Callicebus;* Figs. 3.5, 5.11; Table 5.2) are frequently placed near the ancestry of pitheciines (e.g., Rosenberger, 1992; Kinzey, 1992; Schneider *et al.,* 1993). Other analyses identify them as a distinct subfamily of persistently primitive platyrrhines that has been separate from other platyrrhines since the beginning of the radiation (e.g., Kay, 1990). Titi monkeys have been traditionally divided into three species: *C. torquatus,* the yellow-handed titi; *C. moloch,* the dusky titi, from Amazonian areas of Brazil, Venezuela, Colombia, and Peru; and the beige, masked titi, *C. personatus,* from southeastern Brazil. However, Hershkovitz suggests that the "moloch group" should be divided into many more species (Table 5.2; Hershkovitz, 1988, 1990; see also Emmons, 1997). Titi monkeys have short faces, fluffy bodies with long, fluffy tails, and long legs. Compared with other platyrrhines, they have very short canine teeth. Titi monkeys have relatively short snouts and long skulls, with simple molar teeth compared to other platyrrhines There is slight dental sexual dimorphism in

FIGURE 5.11 The titi monkey (*Callicebus*).

C. torquatus, but none is reported for other species. Their skeletal anatomy is very similar to that of *Aotus* and has been considered most comparable to that of the ancestral platyrrhine (Ford, 1988).

Species of titi monkeys frequently show distinctive habitat preferences. In Peru *Callicebus torquatus* prefers high ground, and mature high forest, and is most frequently sighted in the main canopy, whereas *C. moloch* favors the understory and low levels in lowland forest and bamboo thickets (Kinzey, 1981, Defler. 1994). They are both quadrupeds and leapers, but *C. moloch* leaps more frequently than does *C. torquatus.* Both occasionally use vertical clinging positions when feeding, but *C. moloch* does so more frequently.

Titi monkeys are mainly frugivorous, but *C. moloch* eats large numbers of leaves (and bamboo shoots), whereas *C. torquatus* supplements its fruity diet with less foliage and more insects. Titi monkeys live in monogamous family groups that advertise their presence with elaborate dawn duets, lasting five to fifteen minutes, that are answered sequentially by neighboring pairs. In addition, they actively defend their rela-

TABLE 5.2
Infraorder Platyrrhini
Subfamily CALLICEBINAE

Common Name	Species	Intermembral Index	Mass (g) M	Mass (g) F
Dusky titi monkey	*Callicebus moloch*	74	1,020	956
	C. brunneus		854	805
	C. cupreus	77	1,020	1,120
	C. caligatus		880	
	C. donacophilus	75	991	909
	C. hoffmanni		1,090	1,030
	C. modestus		—	—
	C. olallae		—	—
	C. oenanthe		—	—
Yellow-handed titi monkey	*C. torquatus*	79	1,280	1,210
Masked titi monkey	*C. personatus*	73	1,270	1,380

tively small territories. All three species have relatively small day ranges (600 m), and they seem to specialize on fruits that are found in small patches, which they can harvest on a regular basis. All species are diurnal, with feeding peaks in the early morning and late afternoon and a long resting peak in late morning. They have single births, and the young are carried by the male after their first week.

Aotines

Aotus, the **owl monkeys,** night monkeys, or douracoulis (Figs. 3.5, 5.12, Table 5.3), are the only nocturnal higher primates. Their phylogenetic relationship to other platyrrhines is very obscure. They have been traditionally linked with *Callicebus,* because the two are very similar in many aspects of cranial and postcranial anatomy. However, recent biomolecular studies group them with marmosets and tamarins (Schneider *et al.,* 1993). Owl monkeys are found throughout South America, from Panama to northern Argentina, but they are absent from the Guianas and southeastern Brazil. Over this broad range there are many allopatric species that differ from one another in the relative size of their dentition, the coloration of their neck and tail, and often in their karyotypes (Hershkovitz, 1983; Emmons, 1997; Ford, 1994). All are of medium size (about 1 kg), with no marked sexual dimorphism. They have relatively long legs and a long tail. They have large digital pads on their hands and feet, a slightly opposable thumb, and often a compressed, clawlike grooming nail on the second digit of each foot. They have very large upper central incisors and small third molars. Their basal metabolic rate is lower than that of other platyrrhines, a feature common among nocturnal mammals (Wright, 1989). The most distinctive feature of their cranial anatomy, associated with their nocturnal habits, is the large size of their orbits, which are the largest of any anthropoid (see Fig. 5.6). Despite their nocturnal habits, *Aotus,* like other anthropoids, have no tapetum lucidum. Instead they have a tapetum fibrosum. The retina of *Aotus* has only a single type of cone, so they lack color vision (Jacobs *et al.,* 1996). *Aotus* almost certainly evolved from diurnal ancestors.

FIGURE 5.12 The night monkey, or owl monkey (*Aotus trivirgatus*), the only nocturnal higher primate.

TABLE 5.3
Infraorder Platyrrhini
Subfamily AOTINAE

Common Name	Species	Intermembral Index	Mass (g)	
			M	F
Night monkeys, owl monkeys	*Aotus*	74		
	A. azarae		1,180	1,230
	A. infulatus		1,190	1,240
	A. lemurinus		918	874
	A. miconax			
	A. nancymae		794	780
	A. nigriceps		875	1,040
	A. trivirgatus		813	736
	A. vociferans		708	698

Owl monkeys are found in a variety of forest habitats, and there are no indications that they prefer any particular canopy level. They are predominantly quadrupedal but are adept leapers. Their diet is primarily frugivorous and is supplemented by both foliage and insects. They feed in small, evenly dispersed trees that produce a small number of fruits on a regular basis. Unlike other small, monogamous platyrrhines, they also feed in larger trees at night when more dominant species are asleep.

Owl monkeys live in monogamous families of two to four individuals that occupy home ranges of 6–10 ha. Night ranges are long in the wet season and very short (250 m) in the dry season, when available food is more clumped in its distribution. During the southern winters of Paraguay and northern Argentina, owl monkeys are active during both day and night. Each family has several daytime sleeping nests consisting of tree holes, vine tangles, or open branches. Solitary individuals give low, owl-like hoots, perhaps to attract mates.

In contrast with most monogamous monkeys, adult owl monkeys rarely groom one another. Nevertheless, pairs stay in close contact throughout the night and sleep huddled together. Females give birth to single offspring annually. During the first week of life the infant is increasingly entrusted to the male, who, for the remainder of the infant's dependency, carries it throughout much of the night and sleeps with it during the day. The infant returns to its mother only to nurse.

Atelines

The four ateline genera (Table 5.4) are the largest platyrrhines; the largest individuals of each genus weigh approximately 10 kg. All atelines have a long, prehensile tail with friction ridges similar to fingerprints on the distal

TABLE 5.4
Infraorder Platyrrhini
Family ATELINAE

Common Name	Species	Intermembral Index	Mass (g)	
			M	F
Red howler	*Alouatta seniculus*	97	6,690	5,210
Black-and-red howler	*A. belzebul*	—	7,270	5,520
Brown howler	*A. fusca*	—	6,730	4,350
Mantled howler	*A. palliata*	98	7,150	5,350
Mexican black howler	*A. pigra*		11,400	6,430
Black howler	*A. caraya*	97	6,420	4,330
Common woolly monkey	*Lagothrix lagotricha*	98	7,280	7,020
Yellow-tailed woolly monkey	*L. flavicaudata*	—	10,000	
Woolly spider monkey	*Brachyteles arachnoides*	104	9,610	8,070
Black spider monkey	*Ateles paniscus*	105	9,110	8,440
Black-handed spider monkey	*A. geoffroyi*	105	7,780	7,290
Brown-headed spider monkey	*A. fusciceps*	103	8,890	9,160
Long-haired spider monkey	*A. belzebuth*	109	8,290	7,250

part of its ventral surface. In many aspects of limb and trunk anatomy and in their use of suspensory behavior, they show similarities to the extant apes (Erikson, 1963). In dental and cranial anatomy, as well as diet and social structure, they are quite diverse. Of the four genera, howling monkeys (*Alouatta*) are the most distinct genus and are often placed in a separate subfamily.

The **howling monkeys** (*Alouatta*) have a broad distribution, ranging from southern Mexico to northern Argentina (Fig. 5.13). There are six allopatric species (Emmons, 1997). All are large (6–10 kg) and sexually dimorphic in size and have prehensile tails, but individual species vary dramatically in color, ranging from red to brown, black, or blond. In some species both sexes are the same color; in others the sexes look strikingly different.

Howling monkeys have relatively small incisors and large sexually dimorphic canines (Fig. 5.5). The lower molars have a narrow trigonid and a large talonid; the upper molars are quadrate, with well-developed shearing crests characteristic of folivorous primates.

The skull of *Alouatta* is distinguished by its relatively small cranial capacity and lack of cranial flexion. The mandible is quite large and deep, and the hyoid bone is expanded into a very large, hollow resonating chamber. Howling monkeys have forelimbs and hindlimbs that are similar in length and a long, prehensile tail. Like many platyrrhines, howling monkeys have a poorly differentiated thumb and usually hold objects and branches between their second and third digits, a grasp known as **schizodactyly** ("between fingers").

Howling monkeys are found in a variety of habitats, including primary rain forests, secondary forests, dry deciduous forests, montane forests, and llanos habitats containing patches of relatively low trees in open savannah. Their distribution ranges from sea level to altitudes above 3200 m. Within these diverse forest habitats, most species seem to prefer the main canopy and emergent levels (Fig. 5.9), but several species that live in drier areas (especially *A. caraya*) regularly come to the ground and cross open areas between patches of forest. Howlers are slow, quadrupedal monkeys that rarely leap (Fig. 5.9). During feeding, and less often during travel, they use suspensory locomotion, primarily climbing, in which all five limbs grasp supports opportunistically. During feeding they frequently use their tail in suspension. Howling monkeys are one of the most folivorous of all New World monkeys (Figs. 5.5, 5.9). There is considerable variation in their diet from month to month, but leaves, especially new leaves, constitute half or more of the yearly diet. Fruits and flowers are the next most common components.

Most howling monkeys live in groups containing several adult males, several adult females, and their offspring, but the normal group composition seems to vary by species, by habitat richness and as a function of the age of a troop (Crockett, 1987). Groups of *A. palliata* may contain from twelve to thirty individuals, but in *A. seniculus* and *A. caraya* troops are usually smaller. Howler day ranges are very small, often less than 100 m, because of the howlers' ability to subsist on both a diversity of food items and on relatively common foods such as leaves. Home ranges generally vary from 4 to 20 ha, depending on group size, a small home range for New World monkeys of their biomass. In keeping with their small home ranges, howling monkeys spend little time traveling each day and have long periods of resting and digesting.

As their name indicates, howling monkey groups regularly advertise their presence with loud, lionlike roars given by both males and females. Vocal battles are often reinforced by actual physical combat between individuals,

FIGURE 5.13 A troop of red howling monkeys (*Alouatta seniculus*).

usually males, but there is disagreement regarding the extent to which howlers actively defend their territories or merely their daily positions. While many *Alouatta* troops have several adult males, most mating is by a single male, and there is considerable competition among males for access to a troop and the females within it. Males taking over a troop have been reported to kill the dependent infants that were probably the offspring of the ousted male. An unusual feature of howling monkey demography is that in some species, both males and females regularly transfer among troops, with neither sex consistently remaining in the natal troop. Howlers have single births. Infants are frequently cared for by females other than their mother.

There are two species of **woolly monkeys,** *Lagothrix* (Table 5.4). The more widespread Humbolt's woolly monkey, *L. lagotricha* (Fig. 5.14), is from the upper Amazon regions and adjacent areas in western Brazil, Peru, Colombia, Equador, Bolivia, and other parts of Colombia and Venezuela. The very rare yellow-tailed woolly monkey, *L. flavicaudata,* is restricted to the cloud forests of Peru. Like howlers, woolly monkeys are very sexually dimorphic in body size, with the largest males weighing over 10 kg (Peres, 1994).

Woolly monkeys seem to inhabit primarily high rain forests, but extend into gallery forests in Colombia and Venezuela. They are primarily arboreal quadrupeds and climbers that rely less extensively on forelimb suspension or leaping. Mature fruit pulp is the major item in their diet, but there is considerable seasonal variation, and some populations rely extensively on insects, a rare dietary item for such a large monkey. In the dry season, when fruits are less available, the primary item in their diet changes to new leaves, young seeds, or legume pod secretions, depending on availability.

Woolly monkeys live in large, loosely coherent groups ranging between 20 and 50 individuals, depending on habitat richness and food distribution. Home ranges are very large, as much as 860 h, with day ranges of over 3000 m. Woolly monkeys are particularly susceptible to hunting pressure.

The four allopatric species of **spider monkeys** (*Ateles*) range from the Yucatan peninsula of Mexico through Amazonia and range in color from beige to solid black (Table 5.4). They are large, graceful, long-limbed monkeys with a long, prehensile tail (Fig. 5.15). Male and female spider monkeys are virtually identical in size and color in every species, and this monomorphism is strengthened by the long, pendulous clitoris in the females, which is often mistaken for a penis.

The dentition of spider monkeys (Fig. 5.5) is characterized by relatively large, broad incisors and small molars with low, rounded cusps. The skull has large orbits, a globular braincase, and a relatively shallow mandible (Fig. 5.6). Spider monkeys have relatively long, slender limbs (see Fig. 2.16) that resemble those of gibbons and other suspensory species in many features. Their fingers and toes are long and slender (see Figs. 2.23, 2.25), and most species lack an external thumb.

Spider monkeys are restricted largely to high primary rain forests, where they prefer the upper levels of the main canopy (Fig. 5.9). They have extremely diverse locomotor abilities. During travel, they use both arboreal quadrupedalism and suspensory behavior, including brachiation and climbing. They move bipedally in the trees and occasionally leap. During feeding they are almost totally suspensory, and they use all five limbs to utmost advantage. Spider monkeys feed primarily on ripe fruit, but in some seasons they eat large amounts of new leaves (Fig. 5.9).

Spider monkey social organization is a fission–fusion pattern similar to that of chimpanzees. Groups are generally large, comprising a dozen or more individuals of both sexes and all ages. During the day, the large social

FIGURE 5.14 A troop of woolly monkeys (*Lagothrix lagotricha*).

FIGURE 5.15 A group of black spider monkeys (*Ateles paniscus*).

group generally breaks down into smaller foraging units of two to five individuals which give loud, barking contact calls. These units are most frequently either adult females and their offspring or groups of young males. Limited long-term observations indicate that males generally remain in their natal group while females transfer out. Spider monkeys have single births, and the young are cared for by the mother.

Brachyteles arachnoides, the **woolly spider monkey,** or **muriqui** (Fig. 5.16), is probably the largest nonhuman primate in the neotropics. Reliable body weights are rare and show a tremendous range, but seem to average between 8 and 10 kg. *Brachyteles* is one of the primate species closest to extinction. Its range is restricted to the few remaining patches of rain forest in southeastern Brazil. Although *Brachyteles* resembles the spider monkey in limb proportions and in the lack of a thumb, the dentition is superficially more like that of *Alouatta,* with numerous shearing crests on the molar teeth. The canines of both sexes are small. However, there are indications of differences in canine size and possibly also social organization in different subspecies of muriqui.

In their disappearing habitat, *Brachyteles* are totally restricted to high forest areas, where they prefer main canopy levels. Like *Ateles,* they are arboreal quadrupeds that rely extensively on suspensory behavior during travel and especially during feeding. Leaves make up the greater part of their diet, followed by fruit. Their social organization seems most comparable to that of spider monkeys, but perhaps more cohesive. Large multimale, multifemale groups split into smaller foraging units during the day, but usually remain in vocal contact. Larger, more cohesive groups have been found in the best-studied northern population (Strier, 1992) and smaller groups in the southern population (Milton, 1984). Mating is promiscuous, with females frequently copulating with several males successively. Males have extremely large testicles, suggesting that competition among males is largely by sperm competition rather than by interindividual aggression. Male muriquis remain in their natal group, females disperse.

Cebinae

This subfamily contains capuchin monkeys and squirrel monkeys, the two most omnivorous platyrrhines. Although the two genera have long been distinct, and some phylogenetic studies place them in separate subfamilies, recent molecular studies (Schneider et al., 1993) confirm that cebinae is a natural group.

Cebus, the **capuchin,** or organ-grinder, monkeys, are among the most well known of New World monkeys. Formerly common at street corners and circuses, they are now frequently seen in movies, television programs, and clothing advertisements. These medium-size, sexually dimorphic primates all have large premolars and square molar teeth with thick enamel, which they use to open hard nuts. Their forelimbs and hindlimbs are more similar in size than those of many platyrrhines. Capuchins have a relatively short, fur-covered, prehensile tail. They have short fingers and an opposable thumb and are the most dexterous platyrrhines. For their body size, capuchins seem to have a "slow" life history schedule with long lactation periods, long interbirth intervals and a long life span—up to 47 years (Fedigan and Rose, 1995).

The genus is usually divided into four species, in two species-groups (Table 5.5). *Cebus apella,* the black-capped, or tufted, capuchin (Fig. 5.17), is a relatively robust species found throughout most of the neotropics from northern South America to northern Argentina. Three allopatric species of untufted capuchins are sympatric with *C. apella* throughout

FIGURE 5.16 A troop of muriquis, or woolly spider monkeys (*Brachyteles arachnoides*), from southeastern Brazil, one of the most endangered living primates.

TABLE 5.5
Infraorder Platyrrhini
Family CEBINAE

Common Name	Species	Intermembral Index	Mass (g)	
			M	F
Tufted capuchin	*Cebus apella*	81	3,650	2,520
White-fronted capuchin	*C. albifrons*	81	3,180	2,290
White-throated capuchin	*C. capucinus*	81	3,680	2,540
Wedge-capped capuchin	*C. olivaceus*	83	3,290	2,520
Squirrel monkey	*Saimiri sciureus*[a]	80	779 −911	668 −711
Red squirrel monkey	*S. oerstedii*	80	897	680

[a]Simplified systematics.

much of South America. *Cebus olivaceus,* the wedge-capped, or brown, capuchin, is found in Venezuela, the Guianas, and parts of Brazil, and *C. albifrons,* the white-fronted capuchin, is from the upper Amazon. *Cebus capucinus,* the white-throated capuchin, is the only species from Central America.

Capuchins are found in virtually all types of neotropical forest. They seem to prefer the main canopy levels (Fig. 5.9), but they frequently descend to the understory or to the ground during both travel and feeding. All four species are arboreal quadrupeds that use their prehensile tails mainly during feeding. The capuchin diet includes many types of fruit and animal matter (Fig. 5.9). Tufted capuchins are opportunistic monkeys that use their manipulative abilities and their strength to obtain foods that are unavailable to other species. They forage destructively for invertebrates in bark and leaf litter and are also able to break open hard palm nuts and the hard shells covering immature flowers. In contrast, the more gracile *C. albifrons* specializes on superabundant fruit sources in the same forests as *C. apella.*

All capuchins live in social groups of eight to thirty individuals, with several adult males, several adult females, and offspring. There is greater sexual size dimorphism and a more obvious dominance hierarchy among males of *C. apella* than of *C. albifrons.* Capuchins have large home ranges and relatively long day ranges. They have single births.

Saimiri (Fig. 5.17; Table 5.5), the **squirrel monkeys,** are generally allied with *Cebus* on the basis of their dental proportions and some general behavioral similarities, such as their insectivorous diet and quadrupedal locomotion.

Squirrel monkeys are less than a kilogram in average body mass. They are commonly used as laboratory animals and were once common as pets. Squirrel monkeys have a distinctive cranial morphology with a very long occipital region and a foramen magnum that lies well under the skull base (Fig. 5.6). The orbits are so close together that the interorbital septum is perforated by a large opening. Squirrel monkeys have relatively broad, quadrate upper molars, with a large lingual cingulum and very small last molars. Their cheek teeth have very sharp cusps, which are associated with an insectivorous diet (Fig. 5.9). The canines are sexually dimorphic, with those of males larger than those of females. Squirrel monkeys are very precocious in their development compared

FIGURE 5.17 Two platyrrhines that live sympatrically throughout much of South America: above, the tufted capuchin (*Cebus apella*); below, the squirrel monkey (*Saimiri sciureus*).

with *Cebus,* with more prenatal brain growth and early motor development (Hartwig, 1995). However there is considerable variation among different populations in rates of postnatal development.

The postcranial skeleton of squirrel monkeys (Fig. 5.18) is characterized by a relatively long trunk, long hindlimbs, and a long tail that is prehensile in infants but not in adults. Their hands have comparatively short fingers and a short but relatively unopposable thumb. There is frequently a small amount of fusion between the tibia and fibula distally, making the ankle joint more restricted to simple hinge movements than in other platyrrhine species.

There are numerous allopatric populations of squirrel monkeys from Panama and Costa Rica through Amazonian Brazil, Colombia, Ecuador, Peru, and Bolivia, the Guianas, and southern Venezuela. All are small, gray to yellow monkeys with short fur and a long, thin tail. Hershkovitz (1984) divides them into two species groups based on the shape of the facial fur. I follow a more simple geographic arrangement, recognizing two species, *S. oerstedii* for the Central American red squirrel monkey and *S. sciureus* for the South American populations (Thorington, 1985).

Squirrel monkeys occupy a variety of rain forest habitats but show a preference for riverine and secondary forests, where they are commonly found in the lower levels (Fig. 5.9). They are arboreal quadrupeds that frequently leap, especially when traveling in lower forest levels. During feeding, they are almost totally quadrupedal and occasionally come to the ground.

Squirrel monkeys are frugivores and insectivores that specialize on large fruit trees throughout the year. As they travel between fruit trees, they forage for insects (for almost half of each day) and frequently catch large arthropods on the wing. *Saimiri sciureus* often travel and forage in conjunction with *Cebus apella* (Fig. 5.17).

Squirrel monkeys live in large, continuously active groups that range from 20 to over 50 individuals, with numerous adults of both sexes and offspring. They communicate fre-

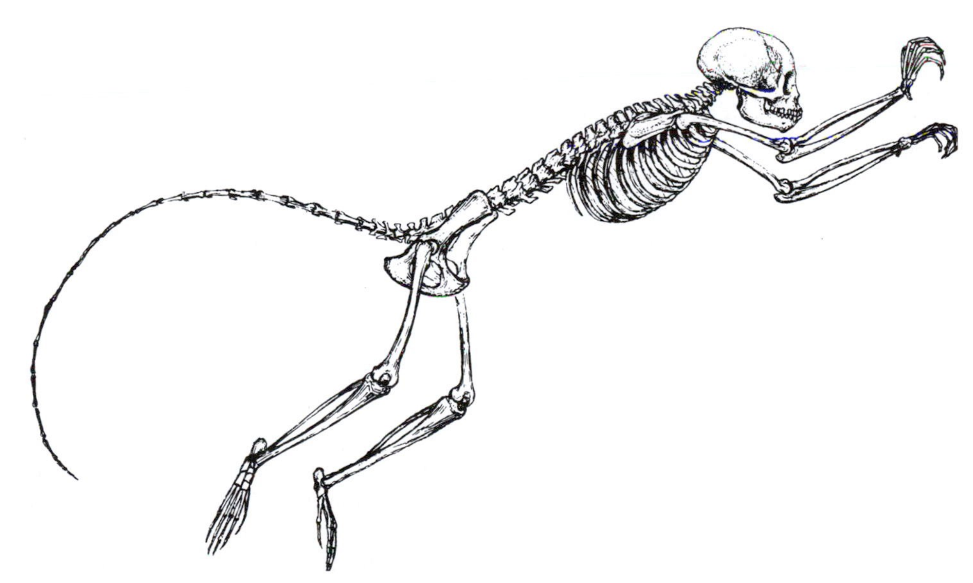

FIGURE 5.18 The skeleton of a squirrel monkey (*Saimiri sciureus*).

quently throughout the day by means of high-pitched whistles and chatter; a group is usually heard well before it comes into view. The estimated size of the overlapping home ranges varies dramatically from troop to troop, but is often greater than 200 ha. Squirrel monkeys occupy large day ranges (2.5–4 km), in keeping with the large group sizes and extensive insect foraging. There is considerable diversity in squirrel monkey social organization among the few populations that have been studied (Fedigan and Boinski, 1996). *Saimiri sciureus* in Peru tend to forage in relatively larger parties than *S. oerstedi* in Costa Rica. In addition, female interactions are strikingly different in the two species, in association with the use of differently distributed fruit resources. In *S. oerstedi,* fruit resources are found in small, evenly distributed patches. There is little direct feeding competition, and females form loose associations and frequently transfer among groups. However, in Peru, *S. sciureus* feed more frequently on larger fruit resources that generate more direct feeding competition. In this species, there is a strict dominance hierarchy among females, and transfer among troops (Mitchell *et al.,* 1991).

In both Costa Rica and Peruvian squirrel monkeys, as in many prosimian species, adult females are dominant over males for most of the year and also spend more time feeding. Males tend to be peripheral members of troops. However, in the mating season, males put on extra fat and become more aggressive. Mating success seems to be correlated with male size, as a result of both intrasexual competition and female choice (Boinski, 1987). The *Saimiri* male seems to be an integral part of groups and to be dominant to females (Fedigan and Boinski, 1996). Squirrel monkeys have restricted breeding seasons, with all births in a single troop occurring within a week. Females give birth to relatively large single offspring at yearly intervals. In contrast with the males of most small New World monkeys, male squirrel monkeys do not play an important role in the care of infants; rather, infants are cared for by several females in addition to their mother.

Callitrichines

Callitrichines are the smallest and most distinctive New World anthropoids (Table 5.6). There are three separate groups among the callitrichines: Goeldi's monkey (*Callimico*), tamarins (*Saguinus* and *Leontopithecus*), and marmosets (*Callithrix* and *Cebuella*). They are all small (100–750 g), often brightly colored monkeys with little if any sexual dimorphism in size or pelage coloration.

Marmosets and tamarins have a unique dentition (Fig. 5.19), with a dental formula of 2.1.3.2. They have lost their third molars from the primitive platyrrhine condition. They have simple, tritubercular upper molars, with no hypocone, that differ from the tritubercular molars of most primitive mammals in also lacking conules. *Callimico* has a more primitive dentition, with three molars and a tiny hypocone. All callitrichines have very short snouts and long braincases (Fig. 5.6).

Callitrichines have skeletons with relatively long trunks, tails, and legs. All digits except the great toe end in **tegulae,** or claws, rather than the nails characteristic of other higher primates (Fig. 5.19), an adaptation that enables them to cling to the sides of large tree trunks to feed on gums, saps, and insects.

Ecologically, callitrichines seem to be characterized by abilities to exploit marginal and disturbed habitats; dietary preferences for fruit, insects, and exudates; and locomotor adaptations for quadrupedal walking and running and leaping. Many species are adept at clinging to large, vertical supports . Within these general outlines, the subfamily exhibits considerable ecological diversity (e.g., Garber, 1992; Ferrari, 1993; Rylands, 1993).

TABLE 5.6
Infraorder Platyrrhini
Subfamily CALLITRICHINAE

<table>
<tr><th rowspan="2">Common Name</th><th rowspan="2">Species</th><th rowspan="2">Intermembral Index</th><th colspan="2">Mass (g)</th></tr>
<tr><th>M</th><th>F</th></tr>
<tr><td>Goeldi's monkey</td><td>Callimico goeldii</td><td>69</td><td>499</td><td>468</td></tr>
<tr><td>Black-and-red tamarin</td><td>Saguinus nigricollis</td><td>78</td><td>468</td><td>484</td></tr>
<tr><td>Saddle-back tamarin</td><td>S. fuscicollis</td><td>79</td><td>343</td><td>358</td></tr>
<tr><td>Tripartite tamarin</td><td>S. tripartitus</td><td>80</td><td></td><td></td></tr>
<tr><td>Moustached tamarin</td><td>S. mystax</td><td>74</td><td>510</td><td>539</td></tr>
<tr><td>White-lipped tamarin</td><td>S. labiatus</td><td>73</td><td>490</td><td>539</td></tr>
<tr><td>Emperor tamarin</td><td>S. imperator</td><td>75</td><td>474</td><td>475</td></tr>
<tr><td>Golden handed tamarin</td><td>S. midas</td><td>77</td><td>515</td><td>575</td></tr>
<tr><td>Inustus tamarin</td><td>S. inustus</td><td>—</td><td>585</td><td>?803</td></tr>
<tr><td>Barefaced tamarin</td><td>S. bicolor</td><td>—</td><td>428</td><td>430</td></tr>
<tr><td>Cottontop tamarin</td><td>S. oedipus</td><td>74</td><td>418</td><td>404</td></tr>
<tr><td>Geoffroy's tamarin</td><td>S. goeffroyi</td><td>76</td><td>482</td><td>502</td></tr>
<tr><td>White-footed tamarin</td><td>S. leucopus</td><td>74</td><td>494</td><td>490</td></tr>
<tr><td>Golden Lion tamarin</td><td>Leontopithecus rosalia</td><td>89</td><td>620</td><td>598</td></tr>
<tr><td>Gold-headed lion tamarin</td><td>L. chrysomelas</td><td>—</td><td>620</td><td>535</td></tr>
<tr><td>Black-faced lion tamarin</td><td>L. caissara</td><td>—</td><td></td><td>575</td></tr>
<tr><td>Golden-rumped lion tamarin</td><td>L. chrysopygus</td><td>—</td><td>620</td><td>535</td></tr>
<tr><td>Silvery marmoset</td><td>Callithrix argentata</td><td>76</td><td>333</td><td>360</td></tr>
<tr><td>Snethlage's marmoset</td><td>C. emiliae</td><td></td><td>313</td><td>330</td></tr>
<tr><td>Black-headed marmoset</td><td>C. nigriceps</td><td></td><td>370</td><td>390</td></tr>
<tr><td>Black-tailed marmoset</td><td>C. melanura</td><td></td><td></td><td></td></tr>
<tr><td>Tassel-eared marmoset</td><td>C. humeralifer</td><td></td><td>475</td><td>472</td></tr>
<tr><td>Golden-white tassel-eared marmoset</td><td>C. chrysoleuca</td><td></td><td></td><td></td></tr>
<tr><td>Maues marmoset</td><td>C. mauesi</td><td>—</td><td>345</td><td>398</td></tr>
<tr><td>Common marmoset</td><td>C. jacchus</td><td>76</td><td>317</td><td>324</td></tr>
<tr><td>Buffy tufted-ear marmoset</td><td>C. aurita</td><td>74</td><td colspan="2">429</td></tr>
<tr><td>Buffy-headed marmoset</td><td>C. flaviceps</td><td>—</td><td colspan="2">406</td></tr>
<tr><td>White-faced marmoset</td><td>C. geoffroyi</td><td>—</td><td colspan="2">359</td></tr>
<tr><td>Black tufted-ear marmoset</td><td>C. penicillata</td><td>76</td><td>344</td><td>307</td></tr>
<tr><td>Wied's marmoset</td><td>C. kuhlii</td><td>—</td><td colspan="2">375</td></tr>
<tr><td>Pygmy marmoset</td><td>Cebuella pygmaea</td><td>83</td><td>110</td><td>122</td></tr>
</table>

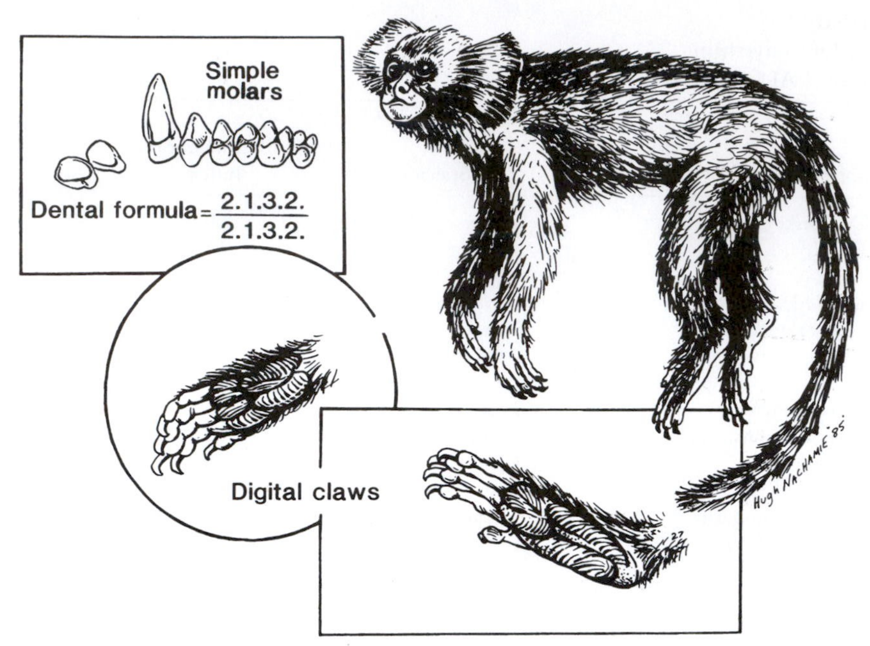

FIGURE 5.19 The unusual features of callitrichines.

Despite having a simple unicornate uterus and a single pair of nipples, features usually found among mammals characterized by single births, marmosets and tamarins typically give birth to dizygotic twins, which share a common placenta. The more primitive *Callimico* has single births. In all callitrichines, males play a major role in the care of infants and are primarily responsible for transporting them.

Although captive callitrichines regularly live in stable monogamous groups, in natural situations they exhibit a variety of larger social groupings and mating patterns (e.g., Savage and Baker, 1996). These include monogamy (*Cebuella*), polyandry (*Saguinus* and *Callithrix*), and polygynandry (*Saguinus*). Most commonly, there is a single breeding female and several adult males that mate with the female. Most groups also contain other, nonbreeding adult males, adult females, and offspring of various ages, all of whom assist in caring for infants. The genetic relationship among these various helpers differs among the genera. In *Callithrix* these groups are probably extended families (Digby and Barreto, 1996); in *Saguinus,* groups normally seem to be composed of unrelated adults living in a polyandrous or a communal breeding system (Garber, 1977). Both sexes normally disperse from their natal group in callitrichines, but there is considerable variability in the timing and frequency of dispersion. There is also considerable debate over the origin of the callitrichine social organization. Some researchers believe it evolved from a monogamous pattern as found in titi monkeys and owl monkeys (Goldizen, 1992); others postulate that callitrichine social systems evolved from large, multimale, multifemale groups (Garber, 1994).

The evolutionary history of the unique callitrichine morphological features is a topic of much discussion. Hershkovitz (1977) has argued that the small size, claws, and simple molars of callitrichines are primitive features indicating an independent origin of platyrrhines from a very primitive (nonanthropoid) primate ancestor. Others (e.g., Ford, 1980; Leutenegger, 1980; Rosenberger, 1984; Sussman and Kinzey, 1984; Garber, 1992; Martin, 1992) have argued—more convincingly in my opinion—that the unique anatomical features of callitrichines are derived specializations related to their small size and/or unusual ecological adaptations for insectivory and exudate eating. For example, among platyrrhines, and primates in general, smaller species generally have larger infants (relative to the size of the mother). This results in considerable problems for the female in birthing and in early postnatal care, which marmosets and tamarins have overcome by giving birth to twins (rather than a single large infant) and by extensive caring for infants by many group members (Leutenegger, 1980). However, as many authors (e.g., Martin, 1992) have subsequently noted, twinning has such extreme ramifications for callitrichine life history and ecology that it is probably more than just a simple solution to an obstetric problem. It is probably also related to the heavy predation pressure suffered by callitrichines (e.g., Caine, 1993) and their adaptations as colonizing species. Likewise, their small size permits both a high-energy diet and a reduced dentition, compared with that needed by larger, more folivorous or frugivorous species (see Chapter 9). Ecologically their small size has yielded a time-minimizing foraging strategy due to the benefits of a specialization on energy-rich insects and fruits found in small patches and the cost of high predation risk (Ferrari, 1993). However, the sequence in which many of the unique callitrichine features appeared and the ecological context of their evolution are far from resolved. Indeed, it has been argued that many of the morphological features, including third molar loss and small size, may have evolved several times within the radiation (Kay, 1994).

Goeldi's Monkey

Goeldi's monkey, *Callimico goeldii* (Fig. 5.20; Table 5.6), is a tufted, silky black monkey from the upper Amazon regions of Colombia, Ecuador, Peru, and Bolivia. It is intermediate in several anatomical features between other callitrichines and other platyrrhines. This species has a more primitive dentition than other callitrichines, with a full set of three molars and a small hypocone on the upper molars (Fig. 5.5). *Callimico* also has single births, the primitive condition for all New World monkeys. In other respects, such as claws and limb proportions, *Callimico* resembles tamarins and marmosets. It is slightly smaller than the squirrel monkey. Males and females are virtually indistinguishable in size and coloration.

Goeldi's monkeys have a very limited distribution, and occur very patchily throughout Amazonia. There is conflicting evidence regarding habitat preferences. They have been reported in primary forests and secondary forests, but most commonly in low bushes and bamboo thickets of the understory (see Christen and Geissmann, 1994). They move mainly by leaping from trunk to trunk a few meters off the ground. The diet of *Callimico* includes large amounts of invertebrates and also fruits. There is no evidence that they ever feed on exudates.

In captivity, these monkeys live in family groups of one adult male and one adult female, but under natural conditions groups contain more adults of both sexes. Their social organization has not been studied in detail. Goeldi's monkeys have relatively large home ranges and day ranges, and they frequently form polyspecific associations with tamarins. They have single offspring twice a year. As in

FIGURE 5.20 A group of Goeldi's monkeys (*Callimico goeldii*) in a bamboo habitat.

Callicebus and *Aotus*, males are largely responsible for carrying infants after birth.

Tamarins

The **tamarins,** *Saguinus,* are the most widespread and diverse of callitrichines, with ten to twelve distinct species displaying an extraordinary array of pelage color patterns and elaborations of facial hair (Fig. 5.21; Table 5.6). All are small, with relatively long trunks, legs, and tail. Like most other anthropoid primates, tamarins have canines that are much larger than their incisors. Although many tamarins eat exudates, they seem to lack the marmosets' ability to gnaw holes in tree bark. Compared with other callitrichines, tamarins are characterized by small, relatively unstable social groups, large home ranges, a single annual birth peak, and lower population densities (Ferrari and Lopes Ferrari, 1989).

Saguinus has the widest distribution of any callitrichine genus and is found throughout Amazonia, on both banks of the Amazon, into the Guianas, Colombia, and even Central America. Tamarins are commonly divided into three species groups, based on pelage and aspects of skeletal morphology. However, tamarin species show ecological groupings that often cut across the traditional species groups (Garber 1993). There have been many behavioral and ecological studies of tamarins in recent years (e.g., Garber, 1997; Rylands, 1993), only a few of which can be discussed here.

Saguinus oedipus, the **crested** or **cotton-top tamarin** (Fig. 5.21), lives in low secondary forests of Panama and Colombia. These monkeys

FIGURE 5.21 The faces of four tamarins.

move primarily by quadrupedal walking and running on medium-size supports and less frequently by leaping between vertical trunks. Fruit (40 percent) and animal material (40 percent) seem to make up the bulk of their diet, and exudates are an important third component. In foraging for fruits, they range from the middle of the canopy to the ground. They forage for insects in the shrub layer and feed on exudates by clinging to the sides of relatively large trunks.

Cotton-top tamarin groups have a mean size of six individuals, but this varies considerably. The relationships among individuals in these groups are not known from present studies. Group composition changes frequently, and there is considerable immigration and emigration between groups.

The **golden-handed tamarin** (*Saguinus midas*) has been studied in both eastern Colombia and Suriname. In Suriname these tamarins are most common in primary forest, but they prefer the edge habitats between forest types (Fig. 5.9). They spend most of their time in the middle levels of the forest, where they move primarily by quadrupedal walking and running along medium-size supports and by leaping between the ends of branches. They seem to be largely frugivorous; insects and exudates are less important components of their diet. Social groups of golden-handed tamarins average about six individuals, with a considerable range (2–12). There are no detailed studies of their composition or social dynamics.

Saguinus fuscicollis (Fig. 5.22), the **saddle-back tamarin** (named for the distinctive patterns on its back), has over a dozen subspecies throughout Amazonian Colombia, Peru, Bolivia, and Brazil. It has been studied in several localities, often in conjunction with other tamarins. These small monkeys move and feed in the lower levels of the forest. Their locomotion involves frequent leaps between large vertical tree trunks and from branches to trunks. During the wet season they are primarily frugivorous, specializing on small, widely dispersed fruits. In the dry season, when fruits are rare, the herbivorous portion of their diet consists almost totally of nectar. Insect foraging accounts for nearly half of the daily feeding time in this species, which specializes on relatively large, cryptic insects that it locates by probing in the hollows, crevices, and bases of trees.

Saddle-back tamarins live in small groups of three to eight individuals and actively defend their territories against neighboring groups of the same species. The most common group structure is a polyandrous mating system with a single breeding female and two breeding males. Monogamous and polygynous groups are less common, and all monogamous groups

FIGURE 5.22 Two sympatric tamarins from Bolivia: above, the white-lipped tamarin (*Saguinus labiatus*); below, the saddle-back tamarin (*Saguinus fuscicollis*).

studied failed to raise their offspring successfully without a second male caretaker. Although group territories remain stable from year to year, there is considerable turnover of individuals in a group as a result of predation and births as well as emigration and immigration. Groups often range more than a kilometer a day.

Throughout their distribution, saddle-back tamarins are normally found in association with one of the moustached tamarins. They seem to be the more parasitic members of this association, for they rarely participate in defense of the common territories, and gain considerable foraging benefit from insects flushed by moustached tamarins as well as the antipredator benefits of the larger groups. (Peres, 1992). There are three members of the moustached group whose behavior is well documented: the white-lipped tamarin, the emperor tamarin, and the moustached tamarin.

Saguinus labiatus, the **white-lipped tamarin** (Figs. 5.21, 5.22), has a relatively small distribution in the middle Amazonian region of western Brazil and eastern Bolivia, where it lives in sympatry with *S. fuscicollis.* Although similar in size to the saddle-back species, white-lipped tamarins differ in several aspects of behavior and ecology. They are found at higher levels in the forest, most commonly in the middle levels of the canopy, where they move by quadrupedal running and short leaps between branches. Like saddle-backs, they eat both fruits and insects. In contrast with the actively foraging saddle-back tamarins, these monkeys spend much of their time visually scanning for prey, which they find among the leaves and terminal branches within the main canopy.

White-lipped tamarins and saddle-back tamarins have home ranges of similar size (30 ha) for their small groups (2–4 individuals). Most striking is the fact that groups of the two species overlap almost completely in their daily ranging behavior, not only when traveling and feeding but also during resting and sleeping. This seems to be primarily an adaptation for predator detection.

Saguinus imperator, the **emperor tamarin** (Fig. 5.21), was named for Emperor Franz Joseph of Austria because of its sweeping moustache. This medium-size tamarin from the upper Amazonian region of Peru and Bolivia is similar in many aspects of its ecology to the white-lipped tamarin. It relies on fruit from small trees throughout the year as its dietary staple, but it takes a substantial amount of nectar during the dry season, when fruits are less abundant. Like white-lipped tamarins, emperor tamarins forage for visible insects among the leaves and small branches of the forest canopy.

The small family groups of emperor tamarins share their moderate territories (30 ha) with a group of saddle-backed tamarins of a similar size, but they defend it against conspecifics. They often travel as much as a kilometer a day. Twin offspring are normally born at the beginning of the rainy season.

The **moustached tamarin** (*S. mystax*), from the middle Amazon region of northern Peru and western Brazil, has been the subject of several recent studies undertaken in conjunction with the trapping and transfer of wild populations onto a natural island in northern Peru. These studies provide some experimental evidence on habitat preferences and group formation. Among the available habitats, these tamarins seem to prefer the drier, upland forest and to avoid flooded forests. They travel and feed most commonly on thin, flexible supports.

Moustached tamarins live in groups of three to eight individuals, which usually contain a single breeding female, up to three presumably reproductively active adult males, and several nonreproductive females and subadults. In the newly formed groups, there was frequent emigration and immigration of adults of both sexes. Usually more than one adult male participated in the care of infants, and census data on numerous groups show a correlation between the number of male helpers and the number of surviving infants.

Lion Tamarins

Lion tamarins (*Leontopithecus*), from southeastern Brazil (Fig. 5.23), are the largest callitrichines, with an adult body mass of over 600 g (Table 5.6). There are four allopatric species: *L. rosalia,* the golden lion tamarin; *L. chrysomelas,* the golden-headed lion tamarin; *L. chrysopygus,* the golden-rumped or black lion tamarin; and *L. caissara,* the black-faced lion tamarin. All are on the verge of extinction in the wild because of extensive habitat destruction. Of the four, *L. chrysomelas* is the most distinctive in dental and cranial features, with particularly large anterior teeth. *Leontopithecus rosalia* has the most gracile anterior dentition (Rosenberger and Coimbra-Filho, 1984).

Lion tamarins are confined largely to the lowland primary rain forest of southeast Brazil and seem to fare poorly in secondary forests. All species are found primarily in the main canopy levels, where they move in a quadrupedal fashion. Lion tamarins feed on a wide range of invertebrates and small vertebrates as well as on fruit, but they have never been observed eating exudates. They use their long fingers to extract insects from holes and crevices, beneath tree bark, and especially inside bromeliad plants. It has been suggested that the enlarged anterior teeth of *L. chrysomelas* are used to gnaw through bark to expose insects. Despite the similarities among these species, each species seems to have a distinct ecology, with the inland *L. chrysopygus* occupying the most distinctive habitat (Rylands, 1993).

Lion tamarins live in relatively small social groups more similar in size to those of *Saguinus* than to the larger groups of *Callithrix.* As in all callitrichines, the social groups of lion tamarins usually have a single breeding female, but group composition and mating patterns are quite variable, with monogamy, polyandry, and polygyny all reported for wild populations. Lion tamarin females generally breed once a year and give birth to twins that are cared for by other group members. Lion tamarins have the largest home ranges of all callitrichines, ranging from 40 to over 100 ha; likewise, day ranges are very large for most populations. All lion tamarins use holes in trees for sleeping.

Marmosets

The **marmosets** (Table 5.6) are the smallest platyrrhines and have the most specialized dentition (Fig. 5.24). They are distinguished from tamarins and other platyrrhines by their uniquely enlarged incisors, which are similar in height to their canines. Furthermore, these large incisors have only a thin layer of enamel on the lingual surface, which quickly wears away and causes the teeth to assume a chisel-like shape similar to the incisors of rodents. Marmosets use these chisel-like incisors for biting holes in trees to elicit the flow of gums, saps, and resins. Compared with tamarins. marmosets tend to occupy drier, more seasonal habitats, to have larger, more stable social groups and smaller home ranges, and to exhibit bimodal annual birth peaks.

Callithrix species (Fig. 5.25) are primarily from south of the Amazon and Madeira rivers in Brazil and the edge of Bolivia. They are divided into three species groups: *C. humeralifer,* the **tassel-eared marmoset** (and the closely related *C. emiliae*) from Amazonia; *C. argentata,* the **silvery marmoset,** from Amazonia, western Brazil, and eastern Bolivia; and *C. jacchus,* the **common marmoset,** a superspecies that includes five closely related allopatric species (or subspecies) from southeastern Brazil. *Callithrix* species seem to be found most com-

FIGURE 5.23 A group of lion tamarins (*Leontopithecus rosalia*), one of the most beautiful and most endangered primates.

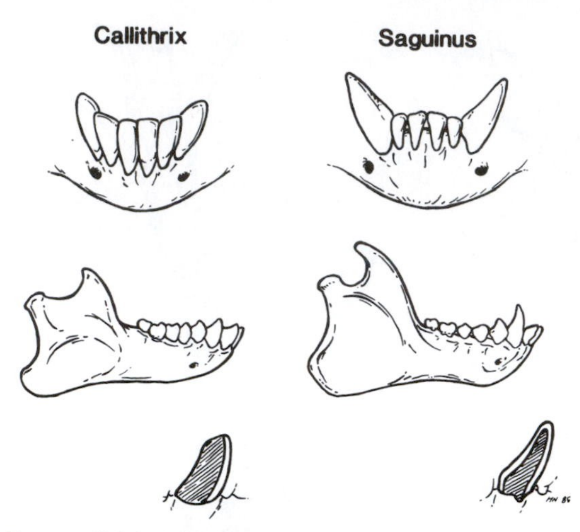

FIGURE 5.24 The lower jaw and teeth of a marmoset (left) and a tamarin (right), showing the differences in proportion of the canines and incisors and the thickness of the enamel on the lower incisors.

monly in dry rather than flooded forests, and especially in edge habitats and secondary forest environments. They forage for insects in the vine tangles of the understory and move by a combination of quadrupedal walking and running as well as by leaping. They use clinging postures on tree trunks when eating exudates. *Callithrix* species all eat fruits, insects, and exudates, with the latter being preferred especially at the end of the wet season, when fruits are scarce. However, there is considerable variability in the dietary specializations of marmosets, with the Amazonian species (*C. humeralifer, C. argentata*) being more frugivorous and those from more seasonal habitats in northeast and central Brazil (*C. jacchus, C. penicillata*) more dependent on exudates throughout the year (Rylands and deFaria, 1993).

Like other callitrichines, marmosets live in large groups, normally, but not always, with a single breeding female but several other adult males and females. In contrast with tamarin groups, which commonly seem to be made up of recently dispersed, unrelated individuals, marmoset groups seem to be composed of extended families. Marmosets give birth to twins at six-month intervals, and the young are carried by the male from the first week of life. At night *Callithrix* often sleep in tree holes.

Cebuella pygmaea, the **pygmy marmoset** (Fig. 5.26), has an adult body weight of approximately 100 g. It is the smallest marmoset, the smallest platyrrhine, and the smallest anthropoid and only a bit larger than *Microcebus* and *Galagoides demidoff,* the smallest living primates. Because of its small size, Hershkovitz (1977) has argued that the pygmy marmoset is the most primitive platyrrhine. However, because of their unusual diet and the dental specializations they share with *Callithrix* (Fig. 5.24), most other primatologists consider pygmy marmosets to be extremely specialized platyrrhines. They are found in the Amazonian regions of Colombia, Peru, Ecuador, Brazil, and Bolivia, where their distribution and density seem to be linked to the abundance of special feeding trees.

Feeding on tree exudates occupies 67 percent of pygmy marmoset feeding time, part of which is devoted to actual feeding from a primary exudate tree and part of which is devoted to preparing new trees by gnawing holes in their bark. The remainder of the monkeys' feeding time consists of foraging for insects and occasional fruits. Because they are so dependent on tree exudates, pygmy marmosets tend to be found in the lower levels of the forest. Their insect foraging takes place in vine tangles. During exudate eating, they frequently adopt clinging positions on the large trunks and move by leaping between vertical supports (Kinzey *et al.,* 1975).

Pygmy marmosets live primarily in small groups with a single adult male, a single adult female, and offspring of various ages. They

FIGURE 5.25 The faces of four marmosets, showing the diversity in facial ornamentation.

have tiny home ranges centered around whatever the main food tree is at the time. Because these primary exudate trees change from year to year, so do the home ranges of pygmy marmoset groups, and individual groups seem to be widely spaced from one another. Pygmy marmosets give birth to dizygotic twins at approximately six-month intervals. The young are carried most frequently by the adult male. During the night, pygmy marmosets sleep in vine tangles or in tree holes.

ADAPTIVE RADIATION OF PLATYRRHINES

Like the Malagasy prosimians, the platyrrhines of the neotropics arrived on an island continent tens of millions of years ago and have

FIGURE 5.26 A family of pygmy marmosets (*Cebuella pygmaea*).

evolved into a diverse radiation with no competition from other groups of primates. The extent of their adaptive diversity (Fig. 5.27) is indicated by the presence of numerous sympatric species throughout most of South America and up to thirteen species at some Amazonian sites (see Figs. 5.8, 5.9).

In size, platyrrhines are small- to medium-sized primates; living species range from approximately 100 g for the pygmy marmoset to approximately 10 kg for several of the atelines. The recently extinct ateline *Protopithecus* was several times larger (Chapter 14). All platyrrhine genera but one are diurnal, but the single nocturnal genus (*Aotus*) is very widespread. Platyrrhines show remarkable intraspecific variations in color vision. In some species color vision is commonly present in only one sex (Jacobs, 1995).

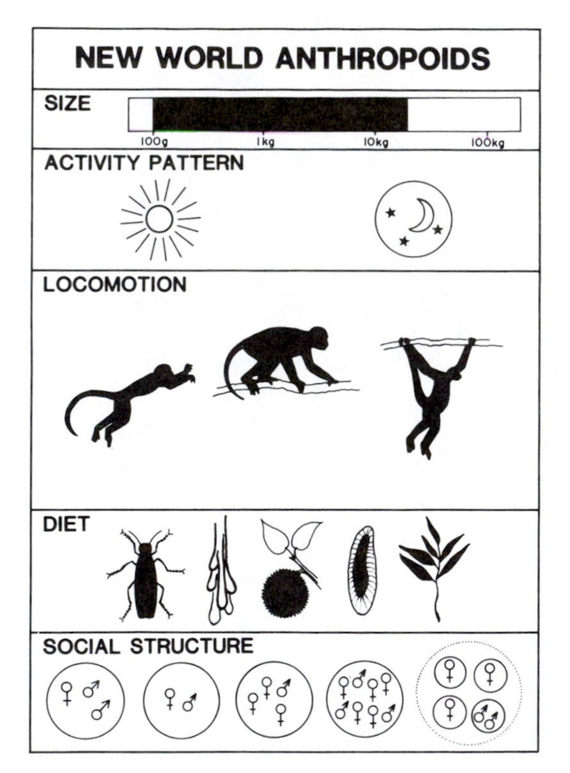

FIGURE 5.27 The adaptive diversity of the platyrrhines.

Although they lack the extremes in limb proportion or skeletal specialization seen in many other groups of primates, platyrrhines show a wide range of locomotor abilities. Some species are excellent leapers, many are arboreal quadrupeds, and the larger species frequently use suspensory postures. Platyrrhines are the only primates to have evolved a prehensile tail, an organ that adds considerably to the locomotor abilities of five genera. In addition to locomotor specializations, New World monkeys have evolved unique postural adaptations, such as the abilities of the clawed callitrichines to cling to vertical trunks, an adaptation for feeding on exudates and cryptic prey on trunks and in tree holes.

A striking feature of the platyrrhine radiation is the absence of terrestrial species. A few species (*Alouatta caraya, Cebus apella, Cacajao calvus,* and *Saimiri sciureus*) occasionally forage on the ground or travel for short distances between trees, but none spend a large portion of each day feeding on the ground.

The New World anthropoids include species that specialize on gums, on fruits, on leaves, and on seeds. Some of the smaller species rely heavily on nectar during the dry periods of the year. There are only two predominantly folivorous genera, *Alouatta* and *Brachyteles.* Only the woolly spider monkey seems to rely almost exclusively on leaves.

Platyrrhine social organization is much more diverse than that found among any other major radiation of primates (Di Fiori and Rendall, 1994). Many New World monkeys live in monogamous groups (*Aotus, Callicebus,* and *Pithecia pithecia*) that seem to be stable for several years. In several genera (*Ateles, Brachyteles,* and *Chiropotes*), the normal social structure is a large group of many adult males and females that fragments into smaller foraging units, depending on the distribution of food resources (Kinzey and Cunningham, 1994). One genus (*Alouatta*) lives in single-male groups in some environments and multimale groups at different sites or when the population density changes. *Cebus* and *Saimiri* live in more complex groups of several adults of each sex. The social organization of *Saimiri* is reminiscent of that of Malagasy lemurs, with female dominance throughout the year and intense male–male competition for a few brief weeks in the breeding season. Social organization in the callitrichines is much more complex than the simple monogamy suggested by studies of captive monkeys. Social groups of many species in natural environments usually contain a single breeding adult female with several reproductively active males and often several nonbreeding adults as well as younger animals. These seem to be polyandrous mating systems in *Saguinus* (Garber 1997). In retrospect, it is perhaps not the platyrrhine diversity that is so unusual, but the uniformity of Old World monkeys (Di Fiori and Rendall, 1994). In any case, its is this platyrrhine diversity that gives us our best clues to understanding the ecological causalities underlying primate social behavior.

PHYLETIC RELATIONSHIPS OF PLATYRRHINES

New World anthropoids have traditionally been divided into two families, the callitrichids (marmosets and tamarins) and the cebids (everything else). *Callimico* has always been a problem genus that does not fit cleanly into either group. Although the marmosets and tamarins are certainly the most distinctive group of platyrrhines, such a division does not resolve the relationships of the remaining genera, nor does it offer any insight into the problem of which other group of platyrrhines is most closely related to the callitrichids. Likewise, division of platyrrhines into numerous subfamilies (e.g. Anderson and Jones, 1984) emphasizes the morphological distinctiveness of the extant platyrrhine groups and accords with suggestions that each is the result of an early evolutionary diversification, but it provides no insight into the evolutionary history of the radiation. There have been numerous attempts to reconstruct the phyletic relationships of platyrrhines using a variety of morphological and molecular data (Rosenberger 1984, 1992; Ford, 1986; Kay, 1990; Schneider *et al.,* 1993; Schneider *et al.,* 1996).

The results of recent molecular studies by Schneider and colleagues (Schneider et al., 1993, 1996; Schneider and Rosenberger, 1996) can be used to summarize the current issues in platyrrhine systematics (Fig. 5.28). Virtually all studies agree in recognizing the integrity and distinctiveness of three groups of platyrrhines—pitheciines, atelines, and callitrichines. The areas of disagreement have traditionally concerned the relationships of these groups to one another and to the taxa *Aotus, Callicebus, Cebus,* and *Saimiri.* The molecular results generally agree with earlier morphological studies of Rosenberger (e.g., 1981, 1992) in placing *Callicebus* as a close relative of the pitheciines (*Pithecia, Chiropotes,* and *Cacajao*), and in grouping *Cebus* and *Saimiri* together as close relatives of callitrichines.

The most surprising result of the molecular studies is the grouping of *Aotus* with *Cebus, Saimiri* and callitrichines rather than with *Callicebus.* However, this result is consistent with Goldizen's (1990) suggestion that callitrichine social behavior was derived from an *Aotus*-like ancestor, and current molecular results can not resolve whether *Aotus, Cebus* or *Saimiri* is the closest relative of callitrichines.

While most researchers would group atelines and pitheciines together to the exclusion

FIGURE 5.28 A summary phylogeny of New World monkeys.

of other New World monkeys, the molecular results are equivocal on this relationship and sometimes cluster pitheciines (plus *Callicebus*) with callitrichines (plus *Cebus, Saimiri,* and *Aotus*). Thus while some researchers would divide platyrrhines into two families (Atelidae and Cebidae), others would recognize three families (Atelidae, Cebidae, and Pithecidae).

Within the individual subfamilies there remain several unresolved relationships—particularly those among callitrichines. Surprisingly, the molecular studies place *Callimico* well within the callitrichine clade, suggesting that many features, such as loss of the last molar and twin births, evolved independently several times, or that *Callimico* reverted to the primitive condition for these features. The position of *Brachyteles* closest to *Lagothrix* is also surprising and suggests that many of the postcranial similarities between *Brachyteles* and *Ateles* evolved in parallel.

However compelling it may seem to sort out the phylogeny of platyrrhines, this is, in fact just the first step in obtaining an understanding of the evolution of the group. Using this phylogeny as a base, we can then start to inquire as to how and when the unique adaptations of the different clades evolved and the morphological and ecological transformations that were involved (Ford, 1990, 1992; Garber and Kinzey, 1992; Rosenberger, 1992; Horovitz and Meyer, 1997). It is the origin of adaptations that should be the goal in understanding platyrrhine phylogeny.

BIBLIOGRAPHY

GENERAL

Coimbra-Filho, A. F, and Mittermeier, R. A., eds. (1981). *Ecology and Behavior of Neotropical Primates.* Rio de Janeiro: Academia Brasiliera de Ciencias.

Emmons, L. H. (1997). *Neotropical Rainforest Mammals: A Field Guide,* Second edition. Chicago: University of Chicago Press.

Ford, S. M. (1986). Systematics of the New World monkeys. In *Comparative Primate Biology,* vol. 1: *Systematics,*

Evolution, and Anatomy, ed. D. R. Swindler and J. Erwin, pp. 73–135. New York: Alan R. Liss.

Ford, S. M, and Davis, L. C. (1992). Systematics and body size: Implications for feeding adaptations in New World monkeys. *Am. J. Phys. Anthropol.* **88**:45–468.

Garber, P. H., and Kinzey, W. G., eds. (1992). Feeding adaptations in New World primates: An evolutionary perspective. *Am. J. Phys. Anthropol.* **88**(4): 411–562.

Hershkovitz, P. (1977). *Living New World Monkeys (Platyrrhini), with an Introduction to the Primates,* vol. 1. Chicago: University of Chicago Press.

Horovitz, I., and Meyer, A. (1997). Evolutionary trends in the ecology of New World monkeys inferred from a combined, phylogenetic analysis of nuclear, mitochondrial, and morphological data. In *Molecular Evolution and Adaptive Radiation,* eds. T. J. Givnish and K. J. Sytsma, pp. 189–224. Cambridge: Cambridge University Press.

Kinzey, W. G. ed. (1997). *New World Monkeys: Ecology, Evolution and Behaviors.* New York: Aldine de Gruyter.

Mittermeier, R. A., Rylands, A. B., Coimbra-Filho, A. F., and daFonseca, G. A. B., eds. (1988). *Ecology and Behavior of Neotropical Primates,* vol. 2. Washington, D. C.: World Wildlife Fund.

Norconk, M. A., Rosenberger, A. L., and Garber, P. A. eds. (1997). *Adaptive Radiations of Neotropical Primates,* New York: Plenum.

Rosenberger, A. L., and Fleagle, J. G., eds. (1989). Special Issue, "New World Monkeys." *J. Human Evol.* **18**(7):595–717.

PITHECIINES

Kay, R. F. (1990). The phyletic relationships of extant and fossil Pitheciinae (Platyrrhini, Anthropoidea). *J. Human Evol.* **19**:175–208.

Kinzey, W. G. (1992). Dietary and dental adaptations in the Pitheciinae. *Am. J. Phys. Anthrop.* **88**(4):499–514.

Walker, S. E. (1997). Evolution of positional behavior in the saki/uakari (*Pithecia, Chiropotes, Cacajao*) In *Adaptive Radiations of Neotropical Primates,* eds. M. Norconk, A. L. Rosenberger, and P. A. Garber pp. 335–382. New York: Plenum.

Sakis

Buchannon, D. B., Mittermeier, R. A., and van Roosmalen, M. G. M. (1981). The saki monkeys, genus Pithecia. In *Ecology and Behavior of Neotropical Primates,* ed. A. E. Coimbra-Filho and R. A. Mittermeier, pp. 391–417 Rio de Janeiro: Academia Brasiliera de Ciencias.

Fleagle, J. G., and Meldrum, D. J. (1988). Locomotor behavior and skeletal morphology of two sympatric pithecine monkeys, *Pithecia pithecia* and *Chiropotes satanas. Am J. Primatol.* **16**:227–249.

Happel, R. E. (1982). Ecology of *Pithecia hirsuta* in Peru. *J. Human Evol.* **11**:581–590.

Hershkovitz, P. (1979). The species of sakis, genus *Pithecia* (Cebidae, Primates) with notes on sexual dimorphism. *Folia Primatol.* **31**:1–22.

______ (1986). The piebald saki. *Field Museum Natural History Bulletin* **57**(2):24–25.

______ (1987). The taxonomy of South American sakis, genus *Pithecia* (Cebidae, Platyrrhini): A preliminary report and critical review with the description of a new species and a new subspecies. *Am. J. Primatol.* **12**: 387–468.

Izawa, K. (1975). Foods and feeding behavior of monkeys in the Upper Amazon Basin. *Primates* **16**:295–316.

______ (1976). Group sizes and composition of monkeys in the Upper Amazon Basin. *Primates* **17**:367–399.

Johns, A. (1986). Notes on the ecological current status of the buffy saki, *Pithecia albicans. Primate Conservation* **7**:26–29.

Kinzey, W. G., and Norconk, M. A. (1990). Hardness as a basis of fruit choice in two sympatric primates. *Am. J. Phys. Anthropol.* **81**:5–15.

______ (1993). Physical and chemical properties of of fruit and seeds eaten by *Pithecia* and *Chiropotes* in Surinam and Venezuela. *Int. J. Primatol.* **14**:207–228

Mittermeier, R. A., and van Roosmalen, M. G. M. (1981). Preliminary observations on habitat utilization and diet in eight Surinam monkeys. *Folio Primatol.* **36**:1–39.

Oliveira, J. M. S., Guerreiro de Lima, M., Bonvincino, C., Ayres, J. M., and Fleagle, J. (1985). Preliminary notes on ecology and behavior of the white-faced saki (*Pithecia pithecia,* Linnaeus, 1766; Cebidae, Primates). *Acta Amazonica* **15**:249–263.

Peres, C. (1993). Notes on the ecology of buffy saki monkeys (*pithecia albicans,* Gray 1860): A canopy seed-predator. *Am. J. Primatol.* **31**:129–140.

Bearded Sakis

Ayres, J. M. (1981). Observacoes sobre Ecologia e o Compartamento das Cuxius (*Chiropotes albinasus* and *Chiropotes satanus.:* Cebidae, Primates). Consuelho Nacional de Desenvolvimento Cientifico E Technologico Instituto Nacional de Pesquisas da Amazonia. Manaus: Fundacao Univ. do Amazonas.

______ (1989). Comparative feeding ecology of the uakari and bearded saki, *Cacajao* and *Chiropotes. J. Human Evol.* **18**(7):697–716. (Special Issue "New World Monkeys".)

Fleagle, J. G., and Mittermeier, R. A. (1980). Locomotor behavior, body size and comparative ecology of seven Surinam monkeys. *Am. J. Phys. Anthropol.* **52**: 301–314.

Hershkovitz, P. (1985). A preliminary taxonomic review of the South American bearded saki monkeys genus *Chiropotes* (Cebidae, Platyrrhini) with the description of a new subspecies. *Fieldiana* **27** (NS):1–46.

Kinzey, W. G., and Norconk, M. A. (1990). Hardness as a basis of fruit choice in two sympatric primates. *Am. J. Phys. Anthropol.* **81**:5–15.

——— (1993). Physical and chemical properties of of fruit and seeds eaten by *Pithecia* and *Chiropotes* in Surinam and Venezuela. *Int. J. Primatol.* **14**:207–228

Mittermeier, R. A., and van Roosmalen, M. G. M. (1981). Preliminary observations on habitat utilization and diet in eight Surinam monkeys. *Folia Primatol.* **36**:1–39.

van Roosmalen, M. G. M., Mittermeier, R. A., and Milton, K. (1981). The bearded sakis, genus *Chiropotes*. In *Ecology and Behavior of Neotropical Primates,* ed. A. E. Coimbra-Filho and R. A. Mittermeier, pp. 419–441. Rio de Janeiro: Academia Brasiliera de Ciencias.

van Roosmalen, M. G. M., Mittermeier, R. A., and Fleagle, J. G. (1988). Diet of the bearded saki (*Chiropotes satamas chiropotes*): A neotropical seed predator. *Am. J. Primatol.* **14**:11–35.

Uakaris

Ayres, J. M. (1986). *Uakaries and Amazonian Flooded Forest.* Ph.D. Dissertation. Cambridge University.

——— (1989). Comparative feeding ecology of the uakari and bearded saki, *Cacajao* and *Chiropotes. J. Human Evol.* **18** (7):697–716. (Special Issue "New World Monkeys".)

Barnett, A. A. and Brandon-Jones, D. (1997). The ecology, biogeography and conservation of the uakaris, *Cacajao* (Pitheciinae). *Folia primatol.* **68**:223–235.

Fontaine, R. (1981). The uakaris, genus *Cacajao.* In *Ecology and Behavior of Neotropical Primates,* ed. A. F. Coimbra-Filho and R. A. Mittermeier, pp. 443–493. Rio de Janeiro: Academia Brasiliera de Ciencias.

Hershkovitz, P. (1987). Uacaries, New World monkeys of the genus *Cacajao* (Cebidae, Platyrrhini): A preliminary taxonomic review with the description of a new subspecies. *Am. J. Primatol.* **12**:1–54.

Lehman, S. M., and Robertson, K. L. (1994). Preliminary survey of *Cacajao melanocephalus melanocephalus* in southern Venezuela. *Int. J. Primatol.* **15**:927–934.

Walker, S. E. and Ayres, J. M. (1996). Positional behavior of the white uakari (*Cacajao calvus calvus*) *Amer. J. Phys. Anthropol.* **101**:161–172.

Titi Monkeys

Hershkovitz, P. H. (1987). The titi. *Field Museum of Natural History Bulletin* **5**:11–15

——— (1988). Origin, speciation, and distribution of South American titi monkeys, genus *Callicebus* (Family Cebidae, Platyrrhine). *Proc. Acad. Nat. Sci. Philadelphia* **140**:240–272.

——— (1990). Titis, New World monkeys of the genus *Callicebus* (Cebidae, Platyrrhini): A prelininary taxonomic review. *Fieldiana* (Zoology, New Series, No. 55). **1410**:1–109.

Kinzey, W. G. (1981). The titi monkey, genus *Callicebus.* In *Ecology and Behavior of Neotropical Primates,* ed. A. E. Coimbra-Filho and R. A. Mittermeier, pp. 241–276. Rio de Janeiro: Academia Brasiliera de Ciencias.

Müller, A.-H. (1996). Diet and feeding ecology of masked titis (*Callicebus personatus*) in *Adaptive Radiations of Neotropical Primates,* eds. M. A. Norconk, A. L. Rosenberger, and P. A. Garber. pp. 383–401. New York: Plenum.

Robinson, J. G., Wright, P. C., and Kinzey, W. G. (1986). Monogamous cebids and their relatives: Intergroup calls and spacing. In *Primate Societies,* ed. B. B. Smuts, D. L. Cheney, R. M. Seyfarth, R. W. Wrangham, and T. T. Struhsaker, pp. 44–53. Chicago: University of Chicago Press.

Wright, P. C. (1984). Biparental care in *Aotus trivirgatus* and *Callicebus moloch.* In *Female Primates: Studies by Women Primiatologists,* ed. M. E. Small, pp. 59–75. New York: Alan R. Liss.

——— (1986). Ecological correlates of monogamy in *Aotus* and *Callicebus.* In *Primate Ecology and Conservation,* ed. J. G. Else and P. C. Lee, pp. 159–167. Cambridge: Cambridge University Press.

Owl Monkeys

Ford, S. (1994). Taxonomy and distribution of the owl monkey. In *Aotus: The Owl Monkey.* ed. J. F. Baer and R. E. Weller, pp. 1–57. New York: Academic Press.

Hershkovitz, P. (1983). Two new species of night monkeys, genus *Aotus* (Cebidae, Primates): A preliminary report on *Aotus* taxonomy. *Am. J. Primatol.* **4**:209–243.

Jacobs, G. H. (1977). Visual sensitivity: Significant within-species variations in non-human primate. *Science* **197**: 499–500.

Jacobs, G. H., Neitz, M., and Neitz, J. (1996). Mutations in S-cone pigment genes and the absence of color vision in two species of nocturnal primate. *Proc. R. Soc. Lond. B* **263**:705–710.

Moynihan, M. (1976). *The New World Primates—Adaptive Radiation and the Evolution of Social Behavior, Language and Intelligence.* Princeton, N. J.: Princeton University Press.

Wright, P. C. (1978). Home range, activity pattern and agonistic encounters of a group of night monkeys (*Aotus trivirgatus*) in Peru. *Folio Primatol.* **29**:43–55.

______ (1981). The night monkeys, genus *Aotus.* In *Ecology and Behavior of Neotropical Primates,* ed. A. F. Coimbra-Filho and R. A. Mittermeier, pp. 214–240. Rio de Janeiro: Academia Brasiliera de Ciencias.

______ (1984). Biparental care in *Aotus trivirgatus* and Callicebus moloch. In *Female Primates: Studies by Women Primatologists,* ed. M. F. Small, pp. 59–75. New York: Alan R. Liss.

______ (1985). *The Costs and Benefits of Nocturnality for Aotus trivirgatus (the Night Monkey).* Ph.D. Dissertation, City University of New York.

______ (1986). Ecological correlates of monogamy in *Aotus* and *Callicebus.* In *Primate Ecology and Conservation,* ed. J. G. Else and P. C. Lee, pp. 159–167. Cambridge: Cambridge University Press.

______ (1989). The nocturnal primate niche in the New World. *J. Human Evol.* **18**(7):635–658. (Special Issue "New World Monkeys".)

______ (1994). The behavioral ecology of the owl monkey. In *Aotus: The Owl Monkey,* ed. J. F. Baer and R. E. Weller, pp. 97–112. New York: Academic Press.

ATELINES

Anthony, M. R. L., and Kay, R. F. (1993). Tooth form and diet in ateline and alouattine primates: Reflections on the comparative method. *Am. J. Sci.* **293A**:356–382.

Cant, J. G. H. (1986). Locomotion and feeding postures of spider and howling monkeys: Field study and evolutionary interpretation. *Folia primatol.* **46**:1–14.

Glander, K. E., Fedigan, L. M., Fedigan, L., and Chapman, C. (1991). Field methods for capture and measurement of three monkey species in Costa Rica. *Folia Primatol.* **57**:70–82.

Rosenberger, A. L., and Strier, K. B. (1989). Adaptive radiation of the ateline primates. *J. Human Evol.* **18**:717–750.

Howling Monkeys

Altmann, S. A. (1959). Field observations on howling monkey society. *J. Mammal.* **40**:317–330.

Carpenter, C. R. (1934). A field study of the behavior and social relations of howling monkeys. *Comp. Psychol. Monogr.* **10**:1–168.

Chivers, D. J. (1969). On the daily behavior and spacing of howling monkey groups. *Folio Primatol.* **10**:48–102.

Crockett, C. M., and Eisenberg, J. E. (1986). Howlers: Variations in group size and demography. In *Primate Societies,* ed. B. B. Smuts, D. L. Cheney, R. M. Seyfarth, R. W. Wrangham, and T. T. Struhsaker, pp. 54–68. Chicago: University of Chicago Press.

DaSilva, E. C., Jr. (1981). A preliminary survey of brown howler monkeys (*Alouatta fusca*) at the Cantareira Reserve (Sao Paulo, Brazil). *Rev. Brasil. Biol.* **41**(4): 897–909.

Erikson, G. E. (1963). Brachiation in New World monkeys and in anthropoid apes. *Symp. Zool. Soc. London* **10**:135–164.

Fleagle, J. G., and Mittermeier, R. A. (1980). Locomotor behavior, body size and comparative ecology of seven Surinam monkeys. *Am. J. Phys. Anthropol.* **52**:301–314.

Gaulin, S. J. C., and Gaulin, C. K. (1982). Behavioral ecology of *Alouatta seniculus* in Andean Cloud Forest. *Int. J. Primatol.* **3**(1):1–52.

Gebo, D. L. (1992). Locomotor and postural behavior in *Alouatta palliata* and *Cebus capucinus. Am. J. Primatol.* **26**: 277–290.

Glander, K. E. (1975). Habitat description and resource utilization: A preliminary report on mantled howling monkey ecology. In *Socioecology and Psychology of Primates,* ed. R. H. Tuttle, pp. 37–57. The Hague: Mouton.

______ (1978). Howling monkey feeding behavior and plant secondary compounds: A study of strategies. In *The Ecology of Arboreal Folivores,* ed. G. G. Montgomery, pp. 561–573. Washington, D. C.: Smithsonian Institution Press.

______ (1981). Feeding behavior in mantled howling monkeys. In *Foraging Behavior: Ecological, Ethological and Psychological Approaches,* ed. A. C. Kamil and T. D. Sargent, pp. 231–257 New York: Garland Press.

______ (1992). Dispersal patterns in Costa Rican mantled howling monkeys. *Int. J. Primatol.* **13**(4):415–436.

Leighton, M., and Leighton, D. R. (1982). The relationship of size of feeding aggregate to size of food patch: Howler monkeys (*Alouatta polliata*) feeding in *Trithelia cipo* fruit trees on Barro Colorado Island. *Biotropica* **14**(2):81–90.

Malinow, M. R., Pope, B., Depaoli, J. R., and Katz, S. (1968). Laboratory observations on living howlers. In *Biology of the Howler Monkey,* ed. A. Caraya, pp. 224–230. Basel: S. Karger.

Mendel, F. (1976). Postural and locomotive behavior of *Alouatta palliata* on various substrates. *Folia Primatol.* **26**: 36–53.

Milton, K. (1980a). Food choice and digestive strategies of two sympatric primate species. *Am. Naturalist* **177**: 496–505.

——— (1980b). *The Foraging Strategies of Howler Monkeys: A Study in Primate Economics.* New York: Columbia University Press.

Mittermeier, R. A., and van Roosmalen, M. G. M. (1981). Preliminary observations on habitat utilization and diet in eight Surinam monkeys. *Folia Primatol.* **36**:1–39.

Pope, T. R. (1990). The reproductive consequences of male cooperation in the red howler monkey: Paternity exclusion in multimale and single-male troops using genetic markers. *Behav. Ecol. and Sociobiol.* **27**:439–446.

Rockwood, L. L., and Glander, K. E. (1979). Howling monkeys and leaf-cutting ants: Comparative foraging in a tropical deciduous forest. *Biotropica* **11**(1):1–10.

Rudran, R. (1979). The demography and social mobility of a red howler (*Alouatta seniculus*) population in Venezuela. In *Vertebrate Ecology in the Northern Neotropics,* ed. J. Eisenberg, pp. 107–126. Washington, D. C.: Smithsonian Institution Press.

Schon, M. A. (1968). *The Muscular System of the Red Howling Monkey.* Bulletin no. 273. Washington, D. C.: Smithsonian Institution Press.

Sekulic, R. (1982). Daily and seasonal patterns of roaring and spacing in four red howler (*Alouatta seniculus*) troops. *Folia Primatol.* **39**:22–48.

Woolly Monkeys

Defler, T. R. (1993). Genus *Lagothrix,* E. Geoffoy St. Hilaire (Atelinae, Cebidae, Platyrrhini): *Lagothrix lagotricha* in the N. W. Amazon of Colombia. *Int. J. Primatol.* **8**:420.

——— (1996). Aspects of the ranging pattern in a group of wild woolly monkeys (*Lagothrox lagotricha*) *Am. J. Primatol.* **38**:289–302.

Fooden, J. (1963). Revision of the woolly monkeys (genus *Lagothrix*). *J. Mammal.* **44**(2):213–247.

Klein, L. L., and Klein, D. (1975). Social and ecological contrasts between four taxa of neotropical primates. In *Socioecology and Psychology of Prinates,* ed. R. H. Tuttle, pp. 59–86. The Hague: Mouton.

Mittermeier, R. A., de Macedo-Ruiz, H., and Luscombe, A. (1975). A woolly monkey rediscovered in Peru. *Oryx* **13**(1):41–46.

Mittermeier, R. A., de Macedo-Ruiz, H., Luscombe, A., and Cassidy, J. (1977). Rediscovery and conservation of the Peruvian yellow-tailed woolly monkey (*Lagothrix flavicaudata*). In *Primate Conservation,* ed. HSH Prince Ranier III and G. H. Bourne, pp. 95–115. New York: Academic Press.

Nishimura, A. (1990). A sociological and behavioral study of woolly monkeys, *Lagothrix lagotricha,* in the upper Amazon. *Sci. Eng. Rev. Doshisha University* **31**:1–121.

Peres, C. A. (1991). Humboldt's woolly monkeys decimated by hunting in Amazonia. *Oryx* **25**(2):89–95.

——— (1994). Diet and feeding ecology of gray woolly monkeys (*Lagothrix lagotricha cana*) in central Amazonia: Comparisons with other atelines. *Int. J. Primatol.* **15**(3):333–372.

——— (1996). Use of space, spatial group structure, and foraging group size of gray woolly monkeys (*Lagothrix lagotricha cana*) at Urucu, Brazil. In *Adaptive Radiations of Neotropical Primates,* eds. M. A. Norconk, A. L. Rosenberger, and P. A. Garber. pp. 467–488. New York: Plenum.

Stevenson, P. R., Quinones, M. J., and Ahunmada, J. A. (1994). Ecological strategies of woolly monkeys (*Lagothrix lagotricha*) at Tinigua National Park, Colombia. *Am.J. Primatol.* **32**:123–140.

Spider Monkeys

Cant, J. G. H. (1977). Ecology, locomotion, and social organization of spider monkeys (*Ateles geoffroyi*). Ph.D. Dissertation. University of California, Davis, Calif.

——— (1986). Locomotion and feeding postures of spider and howling monkeys: Field study and evolutionary interpretation. *Folio Primatol.* **46**:1–14.

——— (1990). Feeding ecology of spider monkeys (*Ateles geoffroyi*) at Tikal, Guatemala. *Human Evol.* **5**(3): 269–281).

Chapman, C. A. (1987). Flexibility in diets of three species of Costa Rican primates. *Folia Primatol.* **49**:90–105.

——— (1989). Multiple central place foraging by spider monkeys: Travel consequences of using many sleeping sites. *Oecologia* **79**:506–511.

——— (1990). Association patterns of spider monkeys: The influence of ecology and sex on social organization. *Behav. Ecol. Sociobiol.* **26**:408–414.

Chapman, C. A., and Chapman, L. J. (1991). The foraging itinerary of spider monkeys: When to eat leaves? *Folia Primatol.* **56**:162–166.

Chapman, C. A., Chapman, L. J., and Lefebvre, L. (1989). Spider monkey alarm calls: Honest advertisement or warning kin? *Animal Behav.* **39**(1):197–198.

Eisenberg, J. F., and Kuehn, R. E. (1966). The behavior of *Ateles geoffroyi* and related species. *Smithson. Misc. Coll.* **151**:1–63.

Fedigan, L. M., Fedigan, L., Chapman, C., and Glander, K. E. (1988). Spider monkey home ranges: A comparison of radio telemetry and direct observation. *Am. J. Primatol.* **16**:19–29.

Klein, L. L., and Klein, D. (1975). Social and ecological contrasts between four taxa of neotropical primates. In *Socioecology and Psychology of Primates,* ed. R. H. Tuttle, pp. 59–86. The Hague: Mouton.

______ (1976). Neotropical primates: Aspects of habitat usage, population density and regional distribution in La Macarena, Colombia. In *Neotropical Primates: Field Studies and Conservation,* ed. R. W. Thorington and P. G. Heltne, pp. 70–78. Washington, D. C.: National Academy of Sciences.

McFarland, M. J. (1986). Ecological determinants of fission–fusion sociality in *Ateles* and *Pan.* In *Primate Ecology and Conservation,* ed. J. G. Else and P. C. Lee, pp. 181–190. Cambridge: Cambridge University Press.

Mittermeier, R. A. (1978). Locomotion and posture in *Ateles geoffroyi* and *Ateles paniscus. Folio Primatol.* **30**:161–193.

Mittermeier, R. A., and van Roosmalen, M. G. M. (1981). Preliminary observations on habitat utilization and diet in eight Surinam monkeys. *Folia Primatol.* **36**:1–39.

Norconk, M. A., and Kinzey, W. G. (1994). Challenge of neotropical frugivory: Travel patterns of spider monkeys and bearded sakis. *Am. J. Phys. Anthropol.* **34**: 171–183.

Symington, M. McFarland. (1987a). Food competition and foraging party size in the black spider monkey (*Ateles paniscus chamek*). *Behaviour* **105**:117–134.

______ (1987b). Sex ratio and maternal rank in wild spider monkeys: when daughters disperse. *Behav. Ecol. Sociobiol.* **20**:421–425.

______ (1988a). Demography, ranging patterns, and activity budgets of black spider monkeys (*Ateles paniscus chamek*) in the Manu National Park, Peru. *Am. J. Primatol.* **15**:45–67.

______ (1988b). Food composition and foraging party size in the black spider monkey (*Ateles paniscus chamek*). *Behavior* **105**:117–134.

______ (1990). Fission–fusion social organization in *Ateles* and *Pan. Int. J. Primatol.* **11**:47–61.

van Roosmalen, M. G. M. (1980). Habitat preference, diet, feeding strategy and social organization of the black spider monkey (*Ateles paniscus paniscus,* Linnaeus 1758) in Surinam. Ph.D. Dissertation, Agricult. Univ. Wageningen.

White, F. (1986). Census and preliminary observations on ecology of the black-faced spider monkey (*Ateles paniscus chamek*) in Manu National Park, Peru. *Am. J. Primatol.* **11**:125–32.

Woolly Spider Monkeys

Milton, K. (1984). Habitat, diet and activity patterns of free-ranging woolly spider monkeys (*Brachyteles arachnoides,* E. Geoffroy, 1806). *Int. J. Primatol.* **5**(5):491–514.

______ (1985). Mating patterns of woolly spider monkeys, *Brachyteles arachnoides:* Implications for female choice. *Behav. Ecol. Sociobiol.* **17**:53–59.

Mittermeier, R. A., Coimbra-Filho, A. F., Constable, I. D., Rylands, A. B., and Valle, C. (1982). Conservation of primates in the Atlantic forest region of east Brazil. *Int. Zoo Yrbk.* **22**:2–17.

Nishimura, A. (1979). In search of woolly spider monkeys. *Kyoto Univ. Primate Res. Inst., Reports of New World Monkey,* pp. 21–37.

Strier, K. B. (1985). Reproduao De *Brachyteles arachnoides* (Primates, Cebidae). *A Primatologia No Brasil* **2**: 163–175.

______ (1987a). Activity budgets of woolly spider monkeys, or muriquis (*Brachyteles arachnoides*). *Am. J. Primatol.* **13**:385–395.

______ (1987b). Ranging behavior of woolly spider monkeys, or muriquis, *Brachyteles arachnoides. Int. J. Primatol.* **8**:575–591.

______ (1990). New World primates, new frontiers: Insights from the woolly spider monkey, or muriquis. *Int. J. Primatol.* **11**:7–19.

______ (1989). Effects of patch size on feeding associations in muriquis (*Brachyteles arachnoides*). *Folia Primatol.* **52**:70–77.

______ (1991). Demography and conservation of an endangered primate, *Brachyteles arachnoides. Amer. J. Phys. Anthropol.* **81**:302–303.

______ (1991). Diet in one group of woolly spider monkeys *Am. J. Primatol.* **23**:113–126.

______ (1992a). Causes and consequences of nonaggression in the woolly spider monkey, or muriqui (*Brachyteles arachnoides*). In *Aggression and Peacefulness in Humans and Other Primates,* ed. J. Silverberg and J. P. Gray, pp. 100–116. New York: Oxford University Press.

______ (1992b). *Faces in the Forest: The Endangered Muriqui Monkeys of Brazil.* New York: Oxford University Press.

Capuchins

Fedigan, L. M. and Rose, L. M. (1995). Interbirth interval variation in the sympyhic species of neotropical monkey. *Am. J. Primatol.* **37**:9–24.

Freese, C., and Oppenheimer, J. R. (1981). The capuchin monkeys, genus *Cebus.* In *Ecology and Behavior of Neotropicol Primates,* ed. A. E. Coimbra-Filho and R. A. Mittermeier, pp. 331–390. Rio de Janeiro: Academia Brasiliera de Ciencias.

Gebo, D. L. (1992). Locomotor and postural behavior in *Alouatta palliata* and *Cebus capucinus. Am. J. Primatol.* **26**: 277–290.

Janson, C. H. (1984). Female choice and mating system of the brown capuchin monkey, *Cebus apella* (Primates: Cebidae). *Z. Tierpsychol.* **65**:177–200.

______ (1986a). Capuchin counterpoint. *Nat. Hist.* **95**(2): 45–53.

——— (1986b). The mating system as a determinant of social evolution in capuchin monkeys (*Cebus*). In *Primate Ecology, and Conservation,* ed. J. G. Else and P. C. Lee, pp. 169–179. Cambridge: Cambridge University Press.

Janson, C. H., and Boinski, S. (1992). Morphological and behavioral adaptations for foraging in generalist primates: The case of the cebines. *Am. J. Phys. Anthrop.* **88**: 483–498.

Jungers, W. L., and Fleagle, J. G. (1981). Postnatal growth allometry of the extremities of *Cebus albifrons* and *Cebus apella:* A longitudinal and comparative study. *Am. J. Phys. Anthropol.* **53**:471–478.

Robinson, J. G. (1981). Spatial structure in foraging groups of wedge-capped capuchin monkeys, *Cebus nigrivittatus. Anim. Behav.* **29**:1036–1056.

Robinson, J. G., and Janson, C. H. (1986). Capuchins, squirrel monkeys, and atelines: Socioecological convergence with Old World monkeys. In *Primate Societies,* ed. B. B. Smuts, D. L. Cheney, R. M. Seyfarth, R. W. Wrangham, and T. T. Struhsaker, pp. 69–82. Chicago: University of Chicago Press.

Terborgh, J. (1983). *Five New World Primates.* Princeton, N. J.: Princeton University Press.

Thorington, R. W., Jr. (1967). Feeding and activity of *Cebus* and *Saimiri* in a Colombian forest. In *Progress in Primatology,* ed. D. Starck, R. Schneider, and H. J. Kuhn, pp. 180–184. Stuttgart, Germany: Gustav Fischer.

Squirrel Monkeys

Baldwin, J. D., and Baldwin, J. I. (1981). The squirrel monkey, genus *Saimiri.* In *Ecology and Behavior of Neotropical Primates,* ed. A. E Coimbra-Filho and R. A. Mittermeier, pp. 277–330. Rio de Janeiro: Academia Brasiliera de Ciencias.

Boinski, S. (1987a). Birth synchrony in squirrel monkeys: A strategy to reduce neonatal predation. *Behav. Ecol. Sociobiol.* **21**:393–400.

——— (1987b). Mating patterns in squirrel monkeys (*Saimiri oerstedi*): Implications for seasonal sexual dimorphism. *Behav. Ecol. Sociobiol.* **21**:13–21.

——— (1988). Sex differences in the foraging behavior of squirrel monkeys in a seasonal habitat. *Behav. Ecol. Sociobiol.* **21**:177–187

Boinski, S., and Flower, M. L. (1989). Seasonal patterns in a tropical lowland forest. *Biotropica* **21**: 223–233.

Boinski, S., and Mitchell, C. L. (1992). The ecological and social factors affecting adult female squirrel monkey vocal behavior. *Ethology* **92**:316–330.

——— (1994). Male residence and association patterns in Costa Rican squirrel monkeys (*Saimiri oerstedi*). *Am. J. Primatol.* **34**:157–169.

Fedigan, L. M. and Boinski, S. (1997). Behavior and ecology issues in *Cebus* and *Saimiri* in *Adaptive Radiations of Neotropical Primates,* eds. M. A. Norconk, A. L. Rosenberger, and P. A. Garber. pp. 221–228. New York: Plenum.

Fleagle, J. G., Mittermeier, R. A., and Skopec, A. L. (1981). Differential habitat use by *Cebus apella* and *Saimiri sciureus* in central Surinam. *Primates* **22**(3): 361–367

Hartwig, W. C. (1995). Effect of life history on the squirrel monkey (*Platyrrhini Saimiri*) cranium. *Am. J. Phys. Anthropol.* **97**:435–449.

(Hershkovitz, P. (1984). Taxonomy of squirrel monkeys, genus *Saimiri* (Cebidae, Platyrrhini): A preliminary report with description of a hitherto unnamed form. *Am. J. Primatol.* **7**:155–210.

Janson, C. H., and Boinski, S. (1992). Morphological and behavioral adaptations for foraging in generalist primates: The case of the cebines. *Am. J. Phys. Anthrop.* **88**(4):483–498.

Mitchell, C. L., Boinski, S., van Schaik, C. P. (1991). Competitive regimes and female bonding in two species of squirrel monkey (*Saimiri oerstedi* and *S. sciureus*). *Behav. Ecol. Sociobiol.* **28**:55–60.

Terborgh, J. (1983). *Five New World Primates.* Princeton, N. J.: Princeton University Press.

Thorington, R. W., Jr. (1967). Feeding and activity of *Cebus* and *Saimiri* in a Colombian forest. In *Progress in Primatology,* ed. D. Starck, R. Schneider, and H. J. Kuhn, pp. 180–184. Stuttgart, Germany: Gustav Fischer.

——— (1968). Observations of squirrel monkeys in a Colombian forest. In *The Squirrel Monkey,* ed. L. A. Rosenblum and R. W. Cooper, pp. 69–85. New York: Academic Press.

Williams, L., Gibson, S., McDaniel, M., Bazzel, J., Barnes, S., and Abee, C. (1994). Allomaternal interactions in the Bolivian squirrel monkey (*Saimiri boliviensis boliviensis*). *Am. J. Primatol.* **34**:145–156.

CALLITRICHINES

Ferrari, S. F. (1993). Ecological differentiation in the Callitrichidae. In *Marmosets and Tamarins: Systematics, Behavior, and Ecology,* ed. A. B. Rylands, pp. 314–328. Oxford: Oxford University Press.

Ferrari, S. F., and Ferrari, M. A. L. (1989). A re-evaluation of the social organization of the Callitrichidae, with reference to the ecological differrences between genera. *Folia Primatol.* **52**:132–147.

Ford, S. M. (1980). Callitrichids as phyletic dwarfs and the place of the Callitrichidae in Platyrrhini. *Primates* **21**(1):31–43.

Garber, P. A. (1992). Vertical clinging, small body size, and the evolution of feeding adaptations in the callitrichinae. *Am. J. Phys. Anthrop.* **88**(4):469–482.

______ (1994). Phylogenetic approach to the study of tamarin and marmoset social systems. *Am. J. Primatol.* **34**:199–219.

Garber, P. A., Encarnación, and Pruetz, J. D. (1993). Demographic and reproductive patterns in moustached tamarin monkeys (*Saguinus mystax*): Implications for reconstructing platyrrhine mating systems. *Am. J. Primatol.* **29**:235–254.

Goldizen, A. W. (1986). Tamarins and marmosets: Communal care of offspring. In *Primate Societies,* ed. B. B. Smuts, D. L. Cheney, R. M. Seyfarth, R. W. Wrangham, and T. T. Struhsaker, pp. 34–43. Chicago: University of Chicago Press.

______ (1990). A comparative perspective of the evolution of tamarin and marmoset social systems. *Int. J. Primatol.* **11**:63–83.

Hershkovitz, P. (1977). *Living New World Monkeys (Platyrrhini), with an Introduction to the Primates,* vol. 1. Chicago: University of Chicago Press.

Kleiman, D. G. (1977). *The Biology and Conservation of the Callitrichidae.* Washington, D. C.: Smithsonian Institution Press.

Leutenegger, W. (1980). Monogamy in callitrichids: A consequence of phyletic dwarfism? *Int. J. Primatol.* **1**(1):95–98.

Martin, R. D. (1992). Goeldi and the dwarfs: The evolutionary biology of the small New World monkeys. *J. Human Evol.* **22**:367–393.

Rosenberger, A. L. (1984). Aspects of the systematics and evolution of the marmosets. In *A Primotologica No Brazil,* ed. M. T. de Mello, pp. 159–180 Angis do 1. Congresso Brasiliero de Primatologia, Sociedad de Primatologica.

Rylands, A. B., ed. (1993). *Marmosets and Tamarins: Systematics, Behavior, and Ecology.* Oxford: Oxford University Press, 396 pp.

Rylands, A. B., Coimbra-Filho, A. F., and Mittermeier, R. A. (1993). Systematics, geographic distribution, and some notes on the conservation status of the Callitrichidae. In *Marmosets and Tamarins: Systematics, Behavior, and Ecology,* ed. A. B. Rylands, pp. 11–77. Oxford: Oxford University Press.

Snowdon, C. T. (1993). A vocal taxonomy of the callitrichids. In Marmosets and Tamarins: Systematics, Behavior, and Ecology, ed. A. B. Rylands, pp. 78–94. Oxford: Oxford University Press.

Sussman, R. W., and Garber, P. A. (1987). A new interpretation of the social organization and mating system of the Callitrichidae. *Int. J. Primatol.* **8**:73–92.

Sussman, R. W., and Kinzey, W. G. (1984). The ecological role of the Callitrichidae: A review. *Am. J. Phys. Anthropol.* **64**(4):419–449.

Tardif, S. D., and Garber, P. A., eds. (1994). Social and reproductive patterns in neotropical primates: Relation to ecology, body size, and infant care. *Am. J. Primatol.* **34**(2):111–244.

Terborgh, J., and Goldizen, A. W. (1985). On the mating system of the cooperatively breeding saddle-backed tamarin (*Saguinus fuscicollis*). *Behav. Ecol. Sociobiol.* **16**: 293–299.

Goeldi's Monkey

Buchanan-Smith, H. (1991). Field observations of Goeldi's monkey, *Callimico goeldii,* in northern Bolivia. *Folia Primatol.* **57**:102–105.

Christen, A., and Geissmann, T. (1994). A primate survey in northern Bolivia, with special reference to Goeldi's monkey, *Callimico goeldii. Int. J. Primatol.* **15**:239–273.

Heltne, P. G., Wojcik, J. E, and Pook, A. G. (1981). Goeldi's monkey, genus *Callimico.* In *Ecology and Behavior of Neotropical Primates,* ed. A. F. Coimbra-Filho and R. A. Mittermeier, pp. 169–209. Rio de Janeiro: Academia Brasiliera de Ciencias.

Hershkovitz, P. (1977). *Living New World Monkeys (Platyrrhini), with an Introduction to the Primates,* vol. 1. Chicago: University of Chicago Press.

Martin, R. D. (1992). Geoldi and the dwarfs: The evolutionary biology of the small New World monkeys. *J. Human Evol.* **22**:367–393.

Moynihan, M. (1976). *The New World Primates—Adaptive Radiation amd the Evolution of Social Behavior, Language and Intelligence.* Princeton, N. J.: Princeton University Press.

Pook, A. G., and Pook, G. (1981). A field study of the socio-ecology of the Goeldi's monkey (*Callimico goeldii*) in northern Brazil. *Folia Primatol.* **35**:288–312.

______ (1982). Polyspecific associations between *Saguinus fuscicollis, Saguinus labiatus, Callimico goeldii* and other primates in northwestern Bolivia. *Folio Primatol.* **38**:196–216.

Tamarins

Dawson, G. A. (1978). Composition and stability of social groups of the tamarin, *Saguinus oedipus geoffroyi,* in Panama: Ecology and behavioral implications. In *The*

Biology and Conservation of the Callitrichidae, ed. D. G. Kleiman, pp. 23–38. Washington, D. C.: Smithsonian Institution Press.

Fleagle, J. G., and Mittermeier, R. A. (1980). Locomotor behavior, body size and comparative ecology of seven Surinam monkeys. *Am. J. Phys. Anthropol.* **52**: 301–314.

Garber, P. A. (1984a). Proposed nutritional importance of plant exudates in the diet of the Panamanian tamarin, *Saguinus oedipus geoffroyi. Int. J. Primatol.* **5**(1):1–15.

——— (1984b). Use of habitat and positional behavior in a neotropical primate, *Saguinus oedipus.* In *Adaptations for Foraging in Non-human Primates,* ed. P. S. Rodman and J. G. H. Cant, pp. 112–133. New York: Columbia University Press.

——— (1984). Proposed nutritional importance of plant exudates in the diet of the Panamanian tamarin, *Saguinus oedipus geoffroyi. Int. J. Primatol.* **5**(1):1–15.

——— (1989). Role of spatial memory in primate foraging Patterns: *Saguinus mystax* and *Saguinus fuscicollis. Am. J. Primatol.* **19**:203–216.

——— (1991). A comparative study of positional behavior in three species of tamarin monkeys. *Primates* **32**(2):219–230.

——— (1993a). Feeding ecology and behavior of the genus *Saguinus.* In *Marmosets and Tamarins: Systematics, Behavior, and Ecology,* ed. A. B. Rylands, pp. 273–295. Oxford: Oxford University Press.

——— (1993b). Seasonal patterns of diet and ranging in two species of tamarin monkeys: Stability versus variability. *Int. J. Primatol.* **14**(1):1–22.

Garber, P. A., Moya, L., and Malaga, C. (1984). A preliminary field study of the moustached tamarin monkey (*Saguinus mystaix*) in northeastern Peru: Questions concerned with the evolution of a communal breeding system. *Folia Primatol.* **42**:17–32.

Izawa, K., and Yoneda, M. (1981). Habitat utilization of non-human primates in a forest of the West Pando, Brazil. *Kyoto Univ. Primate Res. Inst.,* Reports of New World Monkeys, pp. 13–21.

Janson, C. H., Terborgh, J., and Emmons, L. H. (1981). Non-flying mammals as pollinating agents in the Amazonian forest. *Biotropica. Reprod. Botany Suppl.* 1–6.

Mittermeier, R. A., and van Roosmalen, M. G. M. (1981). Preliminary observations on habitat utilization and diet in eight Surinam monkeys. *Folia Primatol.* **36**:1–39.

Nehman, P. F. (1978). Aspects of the ecology and social organization of free ranging cotton-tamarins (*Saguinus oedipus*) and the conservation status of the species. In *The Biology and Conservation of the Callitrichidae,* ed. D. G. Kleiman, pp. 39–72. Washington, D. C.: Smithsonian Institution Press.

Norconk, M. A. (1990). Mechanisms promoting stability in mixed *Saguinus mystax* and *S. fuscicollis* troops. *Am. J. Primatol.* **21**:159–170.

Peres, C. A. (1992a). Consequences of joint-territoriality in a mixed-species group of tamarin monkeys. *Anim. Behav.* **123**(3–4):220–246.

——— (1992b). Prey-capture benefits in a mixed-species group of Amazonian tamarins, *Saguinus fuscicollis* and *S. mystax. Behav. Ecol. Sociobiol.* **31**:339–349.

Price, E. C. (1991). Stability of wild callitrichid groups. *Folia Primatol.* **57**:111–114.

——— (1992a). The benefits of helpers: effects of group and litter size on infant care in tamarins (*Saguinus oedipus*). *Am. J. Primatol.* **26**:179–190.

——— (1992b). The costs of infant carrying in captive cotton-top tamarins. *Am. J. Primatol.* **26**:23–33.

——— (1992c). Sex and helping: Reproductive strategies of breeding male and female cotton-top tamarins, *Saguinus oedipus. Anim. Behav.* **43**:717–728.

Soini, P. (1987). Ecology of the saddle-back tamarin *Saguinus fuscicollis illigeri* on the Ro Pacaya, Nnortheastern Peru. *Folia primatol.* **49**:11–32.

Terborgh, J. (1983). *Five New World Primates.* Princeton, N. J.: Princeton University Press.

Terborgh, J., and Goldizen, A. W. (1985). On the mating system of the cooperatively breeding saddle-backed tamarin (*Saguinus fuscicollis*). *Behav. Ecol. Sociobiol.* **16**: 293–299.

Yoneda, M. (1981). Ecological studies of *Saguinus fuscicollis* and *Saguinus labiatus* with reference to habitat segregation and height preference. Kyoto Univ. Primate Res. Inst., Reports of New World Monkeys, 43–50.

——— (1984). Comparative studies on vertical separation, foraging behavior and traveling mode of saddle-backed tamarins (*Saguinus fuscicollis*) and red-chested moustached tamarins (*Saguinus labiatus*) in northern Bolivia. *Primates* **25**(4):414–422.

Lion Tamarins

Coimbra-Filho, A. E, and Mittermeier, R. A. (1973). Distribution and ecology of the genus *Leontopithecus* (Lesson 1840) in Brazil. *Primates* **14**(1):47–66.

Dietz, J. M., and Baker, A. J. (1993). Polygyny and female reproductive success in golden lion tamarins (*Leontopithecus rosalia*). *Anim. Behav.* **46**:1067–1078.

Dietz, J. M., Baker, A. J., and Miglioretti, D. (1994). Seasonal variation in reproduction, juvenile growth, and adult body mass in golden lion tamarins (*Leontopithecus rosalia*). *Amer. J. Primatol.* **34**:115–132.

Kleiman, D. G., Hoage, R. J., and Green, K. M. (1988). The lion tamarins. In *Ecology and Behavior of Neotropical*

Primates, vol. 2, ed. R. A. Mittermeier, A. B. Rylands, and A. F. Coimbra-Filho, pp. 199–347. Washington, D. C.: World Wildlife Fund.

Peres, C. A. (1989). Cost and benefits of territorial defense in wild golden lion tamarins, *Leontopithecus rosalia. Behav. Ecol. Sociobiol.* **25**:227–233.

Rosenberger, A. F., and Coimbra-Filho, A. F. (1984). Morphology, taxonomic status and affinities of the lion tamarin, *Leontopithecus* (Callitrichinae, Cebidae). *Folia Primatol.* **42**:149–179.

Rylands, A. B. (1993). The ecology of the lion tamarins, *Leontopithecus:* Some intrageneric differences and comparisons with other callitrichids. In Marmosets and Tamarins: Systematics, Behavior, and Ecology, ed. A. B. Rylands, pp. 296–313. Oxford: Oxford University Press.

Marmosets

Coimbra-Filho, A. E, and Mittermeier, R. A. (1978). The gouging, exudate eating and the short-tusked condition in *Callithrix* and *Cebuella.* In *The Biology and Conservation of the Callitrichidae,* ed. D. G. Kleiman, pp. 105–117. Washington, D. C.: Smithsonian Institution Press.

Digby, L. J., and Barreto, C. E. (1996). In *Adaptive Radiations of Neotropical Primates,* eds. M. A. Norconk, A. L. Rosenberger, and P. A. Garber, pp. 173–183. New York: Plenum.

Digby, L. J., and Ferrari, S. F. (1993). Multiple breeding females in free-ranging groups of *Callithrix jacchus. Int. J. Primatol.* **15**(3):389–387.

Ferrari, S. F. (1992). The care of infants in a wild marmoset (*Callithrix flaviceps*) group. *Am. J. Primatol.* **26**:109–118.

Goldizen, A. W. (1990). A comparative perspective on the evolution of tamarin and marmoset social systems. *Int. J. Primatol.* **11**(1):63–83.

Koenig, A., and Rothe, H. (1991). Infant carrying in a polygynous group of common marmosets (*Callithrix jacchus*). *Primates* **32**:183–195.

Lacker, T. E., Jr., Bouchardet de Fonseca, G. A., Alves, C., Jr., and Magalhaes-Castro, B. (1984). Parasitism of trees by marmosets in a central Brazilian gallery forest. *Biotropica* **16**(3):202–209.

Rosenberger, A. L. (1978). Loss of incisor enamel in marmosets. *J. Mammal.* **59**:207–208.

Rylands, A. B. (1981). Preliminary field observations on the marmoset, *Callithrix humeralifer intermedius* (Hershkovitz, 1977) at Dardanelos, Rio Aripuana, Mato Grosso. Primates 22:46–59.

______ (1984). Tree gouging and exudate feeding in marmosets (Callitrichidae, Primates). In *Tropical Rainforest: Ecology and Management Supplemental Reports,* ed. S. L. Sulton, T. Whitmore, and A. C. Chadwick. Proceedings of Leeds Philosophical and Literature Society, Leeds, London.

Rylands, A. B., and deFaria, D. S. (1993). Habitats, feeding ecology, and home range size in the genus *Callithrix.* In *Marmosets and Tamarins: Systematics, Behavior, and Ecology,* ed. A. B. Rylands, pp. 262–272. Oxford: Oxford University Press.

Pygmy Marmosets

Hernandez-Camacho, J., and Cooper, R. W. (1976). The nonhuman primates of Colombia. In *Neotropical Primates: Field Studies and Conservation,* ed. R. W. Thoringon and P. G. Heltne, pp. 35–69. Washingon, D. C.: National Academy of Sciences.

Kinzey, W. G., Rosenberger, A. L., and Ramirez, M. (1975). Vertical clinging and leaping in a neotropical anthropoid. *Nature* **255**:327–328.

______ (1976). Notes on the ecology and behavior of the pygmy marmoset, Cebuella pygmaea, in Amazonian Colombia. In *Neotropical Primates. Field Studies and Conservation,* ed. R. W. Thorington and P. G. Heltne, pp. 79–84. Washington, D. C.: National Academy of Sciences.

Ramirez, M. F., Freese, C. H., and Revilla, C. J. (1978). Feeding ecology of the pygmy marmoset, *Cebuello pygmaea,* in northeastern Peru. In *The Biology and Conservation of the Callitrichidae,* ed. D. G. Kleiman, pp. 91–104. Washington, D. C.: Smithsonian Institution Press.

Soine, P. (1982). Ecology and population dynamics of the pygmy marmoset, *Cebuella pygmaea. Folia Primatol.* **39**:1–21.

______ (1993). The ecology of the pygmy marmoset, *Cebuella pygmaea:* Some comparisons with two sympatric tamarins. In *Marmosets and Tamarins: Systematics, Behavior, and Ecology.* ed. A. B. Rylands, pp. 257–261. Oxford: Oxford University Press.

Terborgh, J. (1983). *Five New World Primates.* Princeton, N. J.: Princeton University Press.

ADAPTIVE RADIATION OF PLATYRRHINES

Di Fiori, A., and Rendall, D. (1994). Evolution of social organization: A reappraisal for primates by using phylogenetic methods. *Proc. Natl. Acad. Sci. USA* **91**: 9941–9945.

Ford, S. M. (1994). Evolution of sexual dimorphism in body weight in platyrrhines. *Am. J. Phys. Anthropol.* **34**: 221–244.

Ford, S. M., and Davis, L. C. (1992). Systematics and body size: Implications for feeding adaptations in New World monkeys. *Am. J. Phys. Anthrop.* **88**(4):415–468.

Garber, P. A. (1997). One for all and breeding for one: Cooperation and competition as a tamarin reproductive strategy. *Evol. Anthropol.* **5**:187–199.

Garber, P. A., and Kinzey, W. G. (1992). Feeding adaptations in New World primates: An evolutionary perspective: Introduction. *Am. J. Phys. Anthrop.* **88**(4): 411–414.

Jacobs, G. H. (1995). Variations in primate color vision: Mechanisms and utility. *Evol. Anthropol.* **3**:196–205.

Kay, R. F., Plavcan, J. M, Glander, K. E., and Wright, P. C. (1988). Sexual selection and canine dimorphism in New World monkeys. *Am. J. Phys. Anthrop.* **77**:385–397.

Kinzey, W. G. (1992). Distribution of primates and forest refuges. In *Biological Diversification in the Tropics,* ed. G. T. Prance, pp. 455–481. New York: Columbia University Press.

Kinzey, W. G., and Cunningham, E. P. (1994). Variability in platyrrhine social organization. *Am. J. Primatol.* **34**: 185–198.

Peres, C. A. (1993). Structure and spatial organization of an Amazonian terra firme forest primate community. *J. Tropical Ecol.* **9**:259–276.

Plavcan, J. M., and Kay, R. F. (1988). Sexual dimorphism and dental variability in platyrrhine primates. *Int. J. Primatol.* **9**(3):169–177.

Rosenberger, A. L. (1992). Evolution of feeding niches in New World monkeys. *Am. J. Phys. Anthropol.* **88**:525–562.

Rosenberger, A. L., and Fleagle, J. G. , eds. (1989). *J. Human Evol., Special Issue, "New World Monkeys."* **18**(7): 595–717.

Strier, K. (1994). Myth of the typical primate. Yrbk. *Phys. Anthrop.* **37**:233–271.

PHYLETIC RELATIONSHIPS OF PLATYRRHINES

Ciochon, R. L., and Chiarelli, A. B. (1980). *Evolutionary Biology of the New World Monkeys and Continental Drift.* New York: Plenum Press.

Ford, S. M. (1986). Systematics of the New World monkeys. In *Comparative Primate Biology,* vol. 1: *Systematics, Evolution, and Anatomy,* ed. D. R. Swindler and J. Erwin, pp. 73–135. New York: Alan R. Liss.

——— (1990). Locomotor adaptations and fossil, platyrrhines. *J. Human Evol.* **19**:141–173.

Goldizen, A. W. (1990). A comparative perspective on the evolution of tamarin and marmoset social systems. *Int. J. Primatol.* **11**:63–83.

Horovitz, I., and Meyer, A. (1997). Evolutionary trends in the ecology of New World monkeys inferred from a combined, phylogenetic analysis of nuclear, mitochondrial, and morphological data. In *Molecular Evolution and Adaptive Radiation,* eds. T. J. Givnish and K. J. Sytsma, pp. 189–224. Cambridge: Cambridge University Press.

Kay, R. F. (1990). The phyletic relationships of extant and fossil Pitheciinae (Platyrrhine, Anthropoidea). *J. Human Evol.* **19**:175–208.

Rosenberger, A. L. (1981). Systematics: The higher taxa. In *Ecology and Behavior of Neotropical Primates,* ed. A. E. Coimbra-Filho and R. A. Mittermeier, pp. 9–27. Rio de Janeiro: Academia Brasiliera de Ciencias.

——— (1984). Aspects of the systematics and evolution of the marmosets. In *A Primatologica No Brosil,* ed. M. T. de Mello, pp. 159–180. Angis do 1. Congresso Brasiliero de Primatologia, Sociedad de Primatologica.

——— (1992). Evolution of feeding niches in New World monkeys. *Am. J. Phys. Anthropol.* **88**:525–562.

Sarich, V. M., and Cronin, J. E. (1980). South American mammal molecular systems, evolutionary clocks, and continental drift. In *Evolutionary Biology of the New World Monkeys and Continental Drift,* ed. R. L. Ciochon and A. B. Chiarelli, pp. 399–421. New York: Plenum Press.

Schneider, H., and Rosenberger, A. L. (1996). Molecules, morphology, and platyrrhine systematics. In *Adaptive Radiations of Neotropical Primates,* eds. M. A. Norconk, A. L. Rosenberger, and P. A. Garber, pp. 3–19. New York: Plenum.

Schneider, H., Schneider, M. P. C., Sampaio, I., Harada, M. L., Stanhope, M., Czelusniak, J., and Goodman, M. (1993). Molecular phylogeny of the New World monkeys (Platyrrhini, Primates). Molec. Phylogenet. Evol. **2**(3):225–242.

Schneider, H., Sampaio, J., Hnyuda, M. L., Barroso, C. M. L., Schneider, M. D. C., Czelusniak, J., and Goodman, M. (1996). Molecular phylogery of the New World Monkeys (Platyrrhini, Primates). Based on two unlinked nuclear genes IRBP Intron 1 and E-globin sequences. *Amer. J. Phys. Anthropol.* **100**:153–180.

Szalay, F. S., and Delson, E. (1979). *Evolutionary History of the Primates.* New York: Academic Press.

Thorington, R. W., Jr., and Anderson, S. (1984). Primates. In *Orders and Families of Recent Mammals of the World,* ed. S. Anderson and J. Knox Jones, Jr., pp. 187–217. New York: Wiley.

SIX

Old World Monkeys

INFRAORDER CATARRHINI

The platyrrhine monkeys are the only primates in the Neotropics, and they fill a diverse array of ecological niches there. In the Old World, there are two very distinct radiations of higher primates that make up the infraorder Catarrhini—the Old World monkeys (Cercopithecoidea) and the hominoids (Hominoidea) As we shall see in later chapters, the evolutionary history of these two groups is quite different, as is their current diversity. The hominoids are restricted to a few species from the tropical forests of Africa and Asia and one cosmopolitan species—humans; they are the subject of Chapter 7. There are many more species and genera of Old World monkeys than of hominoids; it is Old World monkeys that are the subject of the present chapter.

Catarrhine Anatomy

Catarrhines are characterized by numerous anatomical specializations that set them apart from the more primitive platyrrhines. The name is derived from the shape of their nostrils, which are usually narrow and facing downward rather than round and facing laterally as in most New World monkeys. In their dentition, catarrhines have two rather than three premolars in each quadrant, for a dental formula of 2.1.2.3. On the external surface of the side wall of their skull, the frontal bone contacts the sphenoid bone and separates the zygomatic bone anteriorly from the parietal bone posterioriy. In the auditory region, the tympanic bone extends laterally to form a tubular external auditory meatus (see Fig. 5.4). All Old World monkeys, all gibbons, and some chimpanzees have expanded ischial tuberosities and well-developed sitting pads, and all lack an entepicondylar foramen on the humerus. In general, the living catarrhines are much larger than the living platyrrhines, and they include more folivorous and terrestrial species.

Cercopithecoids

Old World monkeys, Cercopithecoidea, are the more taxonomically diverse and numerically successful of living catarrhines. Although Old World monkeys have traditionally been viewed as primitive catarrhines, it has become increasingly evident in recent years that this is a very specialized radiation of Old World primates, one that is not only very different from living apes but also quite derived. We now know that both Old World monkeys and apes are quite specialized in different ways with respect to the earliest catarrhines. We return to this issue in later chapters as we deal with catarrhine evolution.

Cercopithecoid monkeys have several anatomical features that distinguish them from apes and humans (Fig. 6.1). Most characteristic are the specialized molar teeth in which the anterior two cusps and the posterior two cusps are aligned to form two ridges, or lophs. Teeth with this structure are described as **bilophodont.** Most Old World monkeys have daggerlike canines in the males and smaller ones in the females. In both males and females, the canines are sharpened by a narrow anterior lower premolar. In cranial anatomy, Old World monkeys have relatively narrow nasal openings and narrow tooth rows compared with apes.

The limbs of Old World monkeys are characterized by a very narrow elbow joint with a reduced medial epicondyle and a relatively long olecranon process on the ulna. Sitting pads on the expanded ischial tuberosities are a distinctive feature of the group, and many have a long tail.

Cercopithecoid monkeys are found throughout Africa and Asia (Fig. 6.2). In Europe they are found only on the Gibraltar headland, but in the recent past they had a much more extensive distribution on that continent. Old World monkeys are found in a wider range of latitudes, climates, and vegetation types than any other group of living primates except people. There are two very different groups of Old World monkeys, placed in separate subfamilies: the cercopithecines, or cheek-pouch monkeys, and the colobines, or leaf-eating monkeys. Both have undergone extensive adaptive radiations and are represented by numerous genera and species.

The two subfamilies, Cercopithecinae and Colobinae, are distinct in many aspects of their anatomy (Figs. 6.3, 6.4). Many of their differences are related to basic dietary adaptations. The colobines are predominantly leaf and seed eaters, whereas the cercopithecines are predominantly fruit eaters. Cercopithecines have cheek pouches, broader incisor teeth, and molar teeth with high crowns and relatively low cusps, whereas colobines have no cheek pouches, narrower incisors, and molar teeth with high cusps. Colobines have a large, complex ruminant-like stomach. In cranial anatomy, cercopithecines have a narrow interorbital region; in colobines, the interorbital region is broader. In general, cercopithecines have longer snouts and shallower mandibles than do colobines. Most cercopithecines have longer thumbs and shorter fingers than colobines, which often lack a thumb. Cercopithecine forelimbs and hindlimbs tend to be similar in size, whereas colobines usually have longer hindlimbs.

Cercopithecines

The cercopithecines are a predominantly African group. Only a single, very successful genus, *Macaca,* is found in Asia or Europe. Cercopithecines range in size from the tiny (just over 1 kg) arboreal talapoin monkey of western Africa to the large (as much as 50 kg), mostly terrestrial baboons found throughout the African continent (Tables 6.1–6.5).

Macaques

Macaques (*Macaca*) are medium-size cercopithecines (Fig. 6.5) and are relatively generalized in many aspects of their anatomy compared with other members of the subfamily. Macaques are characterized by moderately long snouts, high-crowned molar teeth with very low cusps, and long third molars (Fig. 6.4). They share several features with the African baboons and mangabeys, including relatively long faces and a chromosome number of 44. In general,

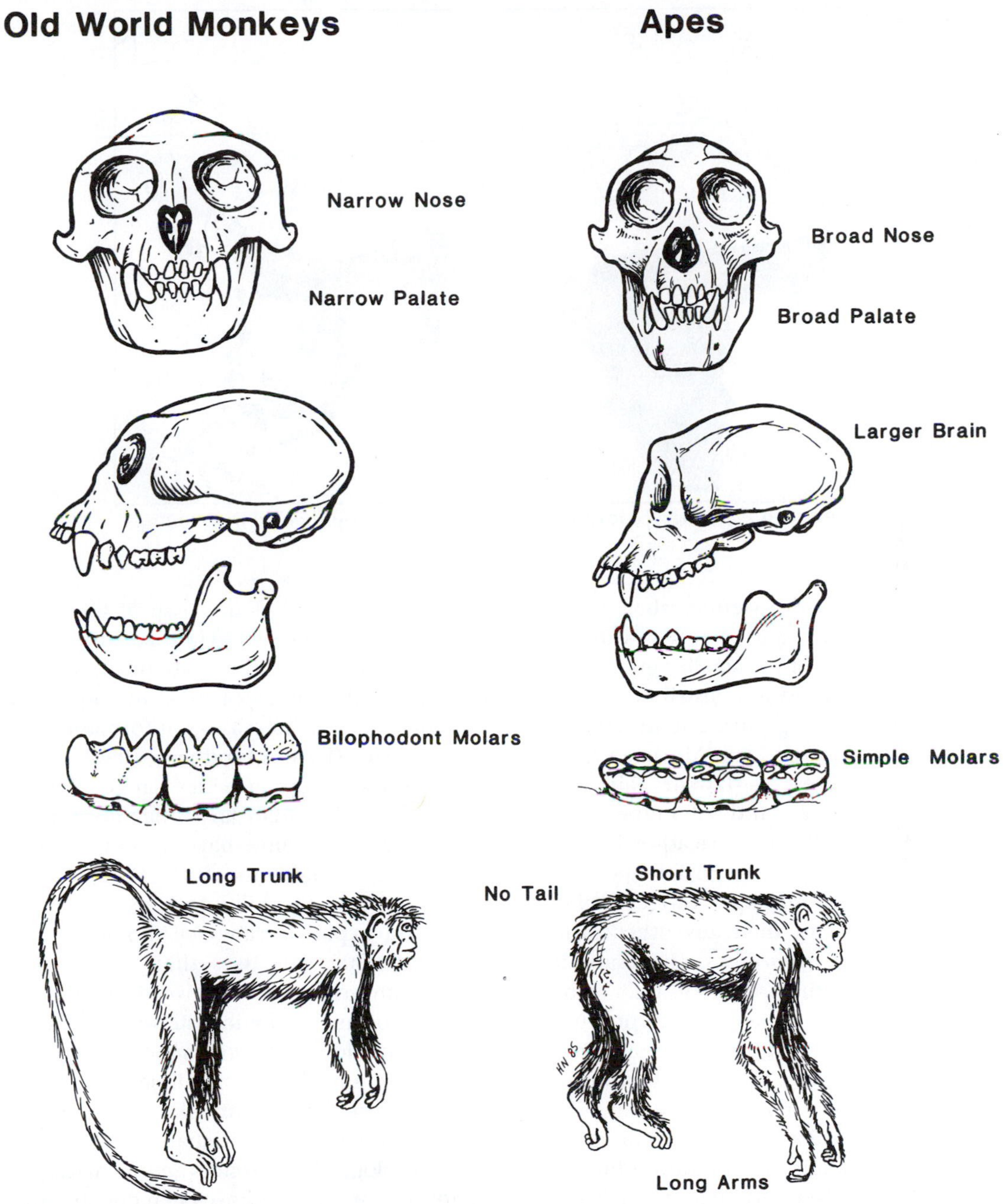

FIGURE 6.1 Characteristic features that distinguish the two groups of catarrhine primates, Old World monkeys and apes (Hominoidea).

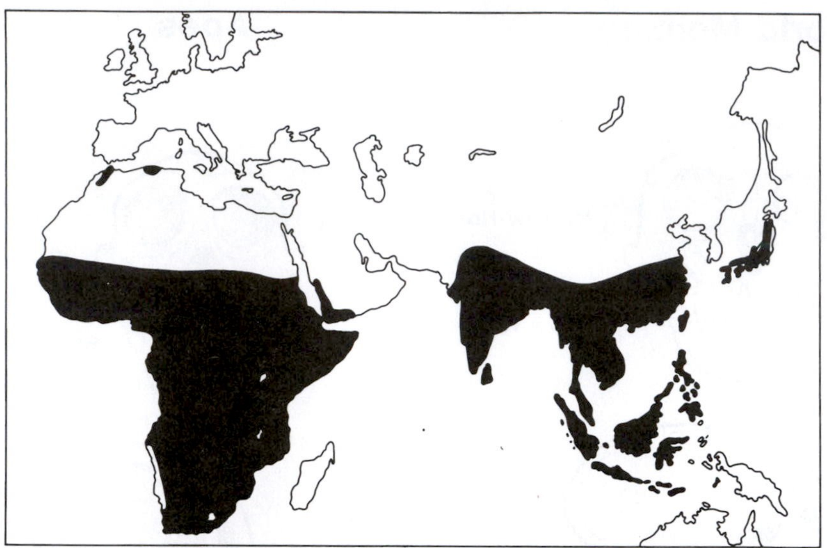

FIGURE 6.2 Geographic distribution of extant cercopithecoid monkeys.

their limbs are more slender than those of the African baboons and mangabeys and more robust than those of the smaller guenons.

Macaca has the widest distribution of any nonhuman primate genus. The nineteen species of *Macaca* range from Morocco and Gibraltar in the west to Japan, Taiwan, the Philippines, Sulawesi, and Bali in the east. *Macaca sylvanus,* the Barbary macaque, is the only living nonhuman primate in Europe; *M. fuscata,* the Japanese macaque, ranges farther to the north and east than any other primate species; and *M. fascicularis,* the crab-eating macaque, from the island of Bali, extends farthest to the southeast of any nonhuman primate species. Based on studies using a wide range of morphological and biomolecular data, the species of macaques are generally divided into five major groups, with broad areas of overlap in their distribution (Table 6:1; reviewed in Hoelzer and Melnick, 1996).

Because macaques are common laboratory primates, the anatomy, physiology, and captive behavior of this genus have been more thoroughly studied than those of any other nonhuman primate (e.g., Hartman and Straus, 1933). Much less is known about the natural behavior and ecology of most macaques, although there is increasing information on many species (Fa and Lindberg, 1996).

Macaques occupy a wider range of habitats and climates than any other nonhuman primate genus. Some have specific habitat preferences (Caldecott *et al.,* 1996). Other macaques have an ability to coexist with humans and exploit a range of modified environments that surpasses that of all other nonhuman primates. This "weed" adaptation is an important ecological strategy of several species, including *M. fascicularis, M. mulatta, M. radiata,* and *M. sinica.* All seem to reach their highest densities in places where they overlap with humans (Richard *et al.,* 1989).

The ecological differences among macaques in nonhuman settings have been documented for several species. The two Southeast Asian species, *M. nemestrina,* the pig-tailed macaque, and *M. fascicularis* (Figs. 6.5, 6.6) are among

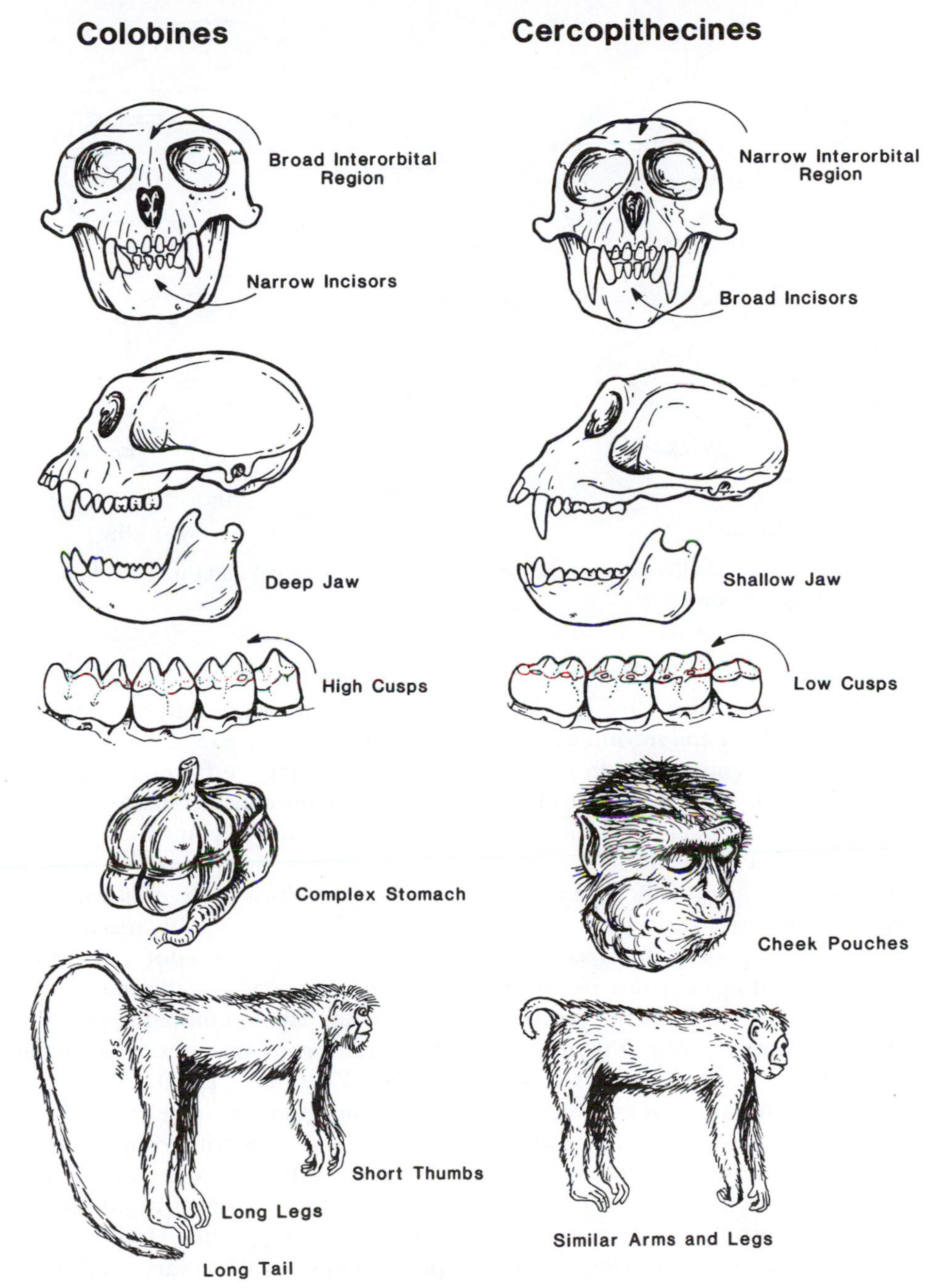

FIGURE 6.3 Characteristic features of the extant subfamilies of Old World monkeys, colobines and cercopithecines.

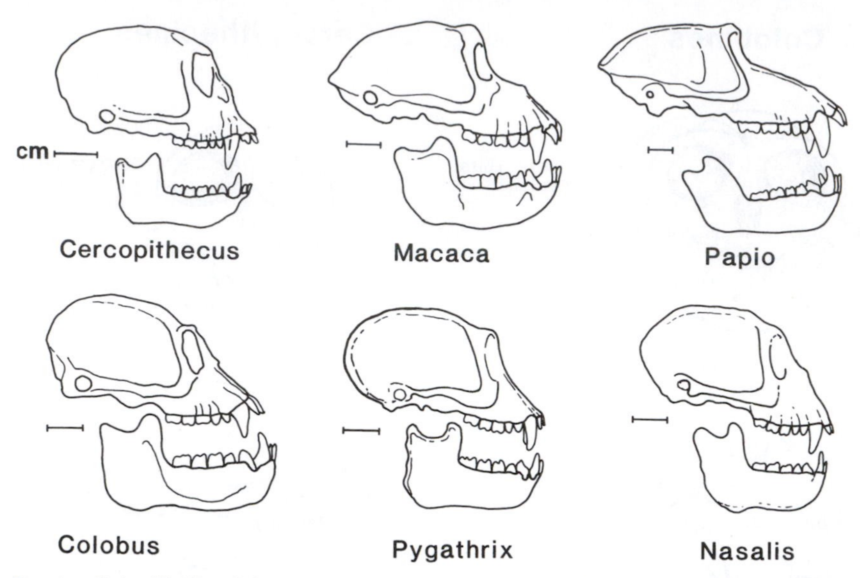

FIGURE 6.4 Skulls of three cercopithecine monkeys (above) and three colobine monkeys (below).

the best known. The smaller (3–5 kg) *M. fascicularis* prefers lowland and secondary forests with a denser, more continuous forest structure, often near rivers; the larger (6–10 kg) *M. nemestrina* prefers upland and more hilly environments with a less continuous canopy and less dense undergrowth (Rodman, 1991).

Macaque species vary considerably in the extent that they are arboreal or terrestrial. All species use both settings to some extent, but with different frequencies. *Macaca fascicularis* is primarily an arboreal species that normally feeds and travels in the trees. These macaques are most often found in the lower levels of the main canopy, but they utilize all levels, including the ground. *Macaca nemestrina* travels more on the ground, but it feeds frequently in the trees. The locomotion of macaques is almost totally quadrupedal walking and running, with very little leaping and no suspensory behavior aside from occasional hindlimb hanging during feeding. Macaques are extremely dexterous and have short fingers and an opposable thumb.

All macaques are frugivores (Fig. 6.6), but many consume considerable amounts of leaves, flowers, and other plant materials as well as various animal prey. Japanese macaques subsist on bark during the cold winters. *Macaca fascicularis,* the crab-eating macaque, eats a variety of invertebrates—not only crabs but also termites and small vertebrates. The very large (12–18 kg) Tibetan macaque (*M. thibetana*) is more folivorous than the smaller rain forest species (Zhao *et al.,* 1991).

All macaques live in relatively large, multimale social groups, with troops of some species containing fifty or more individuals. During the day these groups regularly split into smaller foraging parties. Home range size and patterns of habitat use vary considerably from species to species. Groups of about twenty *M. fascicularis* have home ranges of 40–100 ha and day ranges of less than a kilometer. Home

FIGURE 6.5 Two macaque species that are found sympatrically throughout Southeast Asia: upper right, the crab-eating or long-tailed macaque (*Macaca fascicularis*); below, the pig-tailed macaque (*Macaca nemestrina*).

TABLE 6.1
Infraorder Catarrhini
Family Cercopithecidae
Subfamily CERCOPITHECINAE, macaques

Common Name	Species	Intermembral Index	Body Mass (g)	
			F	M
Barbary macaque	*Macaca sylvanus*	—	11,000	16,000
Lion-tailed macaque	*M. silenus*	92	6,100	8,900
Pig-tailed macaque	*M. nemestrina*	98	6,500	11,200
Tonkean macaque	*M. tonkeana*	95	9,000	14,900
Moor macaque	*M. maura*	—	6,050	9,720
Ochre macaque	*M. ochreata*	100	2,600	5,300
Muna-Butung macaque	*M. brunnescens*	99		16,600
Heck's macaque	*M. hecki*	93	6,800	11,200
Gorontalo macaque	*M. nigriscens*	—	9,500	
Celebes black macaque	*M. nigra*	84	5,470	9,890
Toque macaque	*M. sinica*	—	3,200	5,680
Bonnet macaque	*M. radiata*	—	3,850	6,670
Assamese macaque	*M. assamensis*	96	6,900	11,300
Thibetan macaque	*M. thibetana*	95	9,500	12,200
Bear macaque	*M. arctoides*	98	8,400	12,200
Crab-eating macaque	*M. fascicularis*	93	3,590	5,360
Taiwan macaque	*M. cyclops*	—	4,940	6,000
Rhesus macaque	*M. mulatta*	93	5,370	7,710
Japanese macaque	*M. fuscata*	—	8,030	11,000

ranges and day ranges for the larger groups of *M. nemestrina* are considerably larger, because they use rapid terrestrial travel to exploit widespread, often erratically available food resources (Caldecott *et al.,* 1996). Social groups of *Macaca nemestrina* and also those of the Barbary macaque (*M. sylvanus*) have been reported to undergo fission and fusion (Menard *et al.,* 1990).

Social relations within macaque groups are complex. Female hierarchies and matrilines are particularly important in interindividual relations and troop politics. Males usually migrate from troop to troop many times during their lifetime. Most macaques have single births at yearly intervals.

Mangabeys

Mangabeys (Fig. 6.7; Table 6.2) are large, forest-living monkeys with long molars, very large incisors, relatively long snouts, and hollow cheeks. They have relatively long limbs and long tails. Formerly placed in a single genus, the living mangabeys are now widely recognized to be an unnatural group containing two distinct genera that differ in numerous aspects of their dental, cranial, and postcranial anatomy, biochemistry, and ecology (Groves, 1978; Disotell 1994, 1996; Nakatsukasa, 1996). Moreover, the two genera, *Cercocebus* and *Lophocebus,* have very different phylogenetic relationships within cercopithecines.

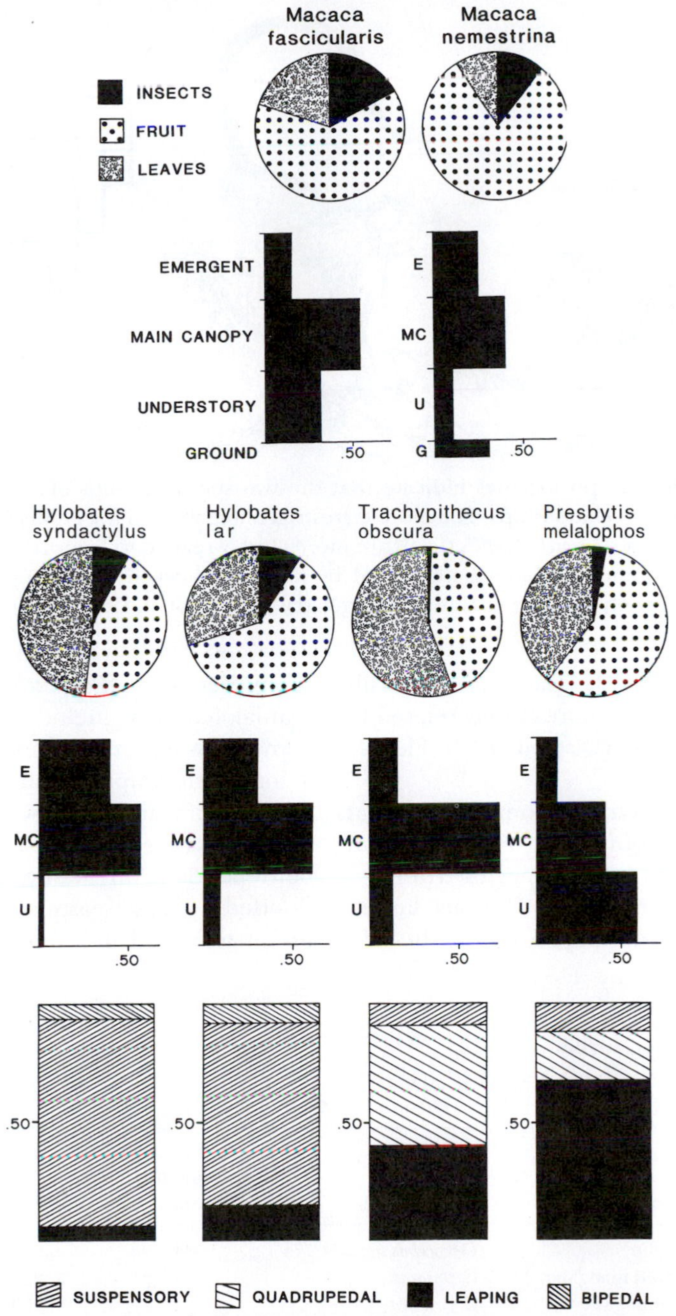

FIGURE 6.6 Diet, forest height preferences, and locomotor behavior of six sympatric catarrhines from Malaysia (data from Chivers, 1980).

FIGURE 6.7 Molecular phylogenies indicate that the two species groups of living mangabeys have separate phylogenetic relationships. The more terrestrial mangabeys of the genus *Cercocebus,* illustrated by the agile mangabey, *Cercocebus agilis* (left), are more closely related to mandrills. The more arboreal mangabeys of the genus *Lophocebus,* illustrated by the gray-cheeked mangabey *Lophocebus albigena* (right), are more closely related to baboons and geladas (see Disotell, 1994, 1996).

Cercocebus is most closely related to mandrills and drills; *Lophocebus* is more closely related to baboons and geladas (Disotell, 1996, Fleagle and McGraw, 1998).

Cercocebus is divided into four species that are found throughout much of western and central Africa, with a relic species from the Tana River in eastern Kenya. They are larger, more dimorphic monkeys and can be distinguished from *Lophocebus* by a number of cranial features, including very large teeth, convex nasal bones, a wide interorbital pillar, a long ectotympanic tube, and an upright mandibular ramus. These monkeys are found in a wide range of forest types, but seem to be dependent on swamps or forests that are flooded at least seasonally. They prefer the understory and are most commonly found

TABLE 6.2
Infraorder Catarrhini
Family Cercopithecidae
Subfamily CERCOPITHECINAE, mangabeys

Common Name	Species	Intermembral Index	Body Mass (g)	
			F	M
Agile mangabey	*Cercocebus agilis*	84	5,660	9,500
White-collared mangabey	*C. torquatus*	83	6,230	11,000
Tana River mangabey	*C. galeritus*	84	5,260	9,610
Sooty mangabey	*C. atys*	—	6,200	11,000
Gray-cheeked mangabey	*Lophocebus albigena*	78	6,020	8,250
Black mangabey	*L. aterrimus*	—	5,760	7,840

on the ground in both traveling and feeding (McGraw, 1996).

Cercocebus eat fruits, and especially hard nuts and seeds, which they find on the forest floor. They spend approximately 26–30 percent of feeding time foraging for invertebrates. *Cercocebus* live in groups that average 25 individuals, but these groups seem to divide regularly into subgroups and occasionally several come together in large supertroops of up to 100 individuals (e.g., Mitani, 1989, 1991).

Lophocebus contains two species (*L. albigena* and *L. aterrimus*) that are largely restricted to central Africa. They are smaller, less dimorphic monkeys that can be distinguished from *Cercocebus* by their smaller teeth, superiorly pinched interorbital pillar, an inferiorly curving zygomatic arch, and a generally elongated skull (Groves, 1978). These are strictly arboreal monkeys that prefer the main canopy levels in a variety of forests.

Lophocebus species also feed predominantly on fruits and invertebrates. They live in smaller groups averaging 15 individuals. These groups have occasionally been reported to form foraging subgroups, but this seems rare, and there is no evidence of supertrooping (N. Shah, Pers. comm.).

Baboons

Baboons (Fig. 6.8; Table 6.3) are the largest and among the best known of all cercopithecines. They were important figures in the mythology of ancient Egypt and were well known to Greek and Roman scholars. As savannah-dwelling primates, they have played an important role as models for various aspects of early human evolution (Washburn and DeVore, 1961; Strum and Mitchell, 1987).

Baboons are very large monkeys and are all sexually dimorphic in body size; in many species, females are only half the size of males. Baboons are characterized by long molars and broad incisors. Their canines are very sexually dimorphic, and the long anterior lower premolars form a sharpening blade for the daggerlike canines. Baboons have a long snout (Fig. 6.4), a long mandible, and pronounced brow ridges. Their limbs (Fig. 6.9) are nearly equal in length, and their forearm is much longer than their humerus; they have relatively short digits on their hands and feet. Compared with other cercopithecines, baboons have relatively short tails and large ischial callosities. Females have very pronounced sexual swellings during estrus.

Baboons are found throughout the forests and savannahs of sub-Saharan Africa. There are between seven and ten distinct populations of baboons, commonly, but not comfortably, placed in five species: *Papio papio, P. anubis, P. cynocephalus, P. ursinus,* and *P. hamadryas.* These five are all allopatric, with variable amounts of interbreeding at their boundaries, and there are several additional populations that are as distinct as the commonly recognized species. Because of the hybridization between populations, some authors recognize a single species, *P. hamadryas,* for the entire radiation. Indeed, the more that is known about baboon demography and genetics, the more obvious it becomes that identifying distinct species in a broad geographic radiation of allopatric populations involves very arbitrary boundaries (Jolly, 1993).

The ecology and behavior of **savannah baboons** has been the subject of many studies over the past four decades. These baboons live in woodland savannahs, grasslands, acacia scrubs, and other open areas, but also in gallery forests and some rain forest environments. They forage and travel primarily on the ground by quadrupedal walking and running, but they almost always climb trees or rocky cliffs for sleeping and often for resting. They are extremely eclectic feeders that subsist mainly on ripe fruits, roots, and tubers, as well as on grass seeds, gums, and leaves.

FIGURE 6.8 A group of savannah baboons (*Papio anubis*) in eastern Africa.

TABLE 6.3
Infraorder Catarrhini
Family Cercopithecidae
Subfamily CERCOPITHECINAE, baboons and geladas

Common Name	Species	Intermembral Index	Body Mass (g)	
			F	M
Hamadryas baboon	*Papio hamadryas*	95	9,900	16,900
Guinea baboon	*P. papio*	—	12,100	—
Olive baboon	*P. anubis*	97	13,300	25,100
Yellow baboon	*P. cynocephalus*	96	12,300	21,800
Chacma baboon	*P. ursinus*	96	14,800	29,800
Gelada	*Theropithecus gelada*	100	11,700	19,000

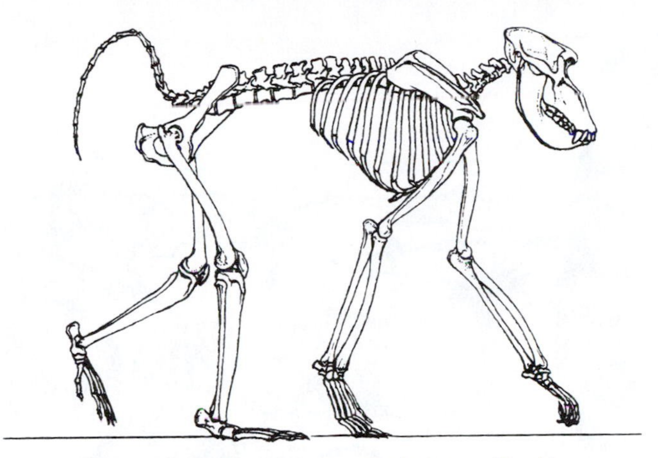

FIGURE 6.9 The skeleton of a baboon (*Papio*).

Most baboons are opportunistic faunivores (Strum, 1981) and have been reported to catch and eat numerous small mammals (hares, young gazelles, vervet monkeys) as well as many invertebrates and insects. They also eat bird eggs.

The four savannah baboons (*P. papio, P. anubis, P. cynocephalus,* and *P. ursinus*) normally live in large, socially complex multimale troops ranging from forty to eighty individuals, although some mountain populations of chacma baboons (*Papio ursinus*) are found in one-male groups (Byrne *et al.,* 1990). As in most Old World monkeys, baboon females generally remain in their natal troop and males usually emigrate to other troops. These female-bonded kin groups are generally considered to form the basic structure of a baboon troop (but see Berkovitch, 1990, and Byrne *et al.,* 1990). There is usually a pronounced dominance hierarchy among males and intense competition among the adult males for access to estrous females. This competition involves a whole repertoire of social maneuvers, not just simple physical prowess, but also coalitions and infant care(e.g., Smuts, 1985; Strum, 1987; Berkovitch, 1987; Noe and Sluijter, 1990, 1995). Females frequently mate with several males during the course of their cycle. Baboon troops occupy large (4000-ha) home ranges and travel long distances (over 5 km) every day, usually as a single group.

Social organization and foraging patterns in Hamadryas baboons (*P. hamadryas*) are quite different from those found among most other savannah species (Kummer, 1968; Nagel, 1973). These handsome silver baboons from the arid scrublands of Ethiopia live in groups of a single adult male with one to four females plus their offspring. Males guard the females in their harem jealously and actually herd them by chasing any straying females and biting them on the neck to keep the group together. Several one-male groups, probably led by related males, regularly associate to form clans. The individual harems forage separately during the day but congregate at night on rocky cliffs in troops of up to 150 animals.

The **gelada** (*Theropithecus gelada*) is a very distinctive baboon relative from the highlands of Ethiopia (Fig. 6.10). *Theropithecus gelada* is the only surviving species of a more successful and widespread radiation during the Pliocene and Pleistocene (see Chapter 16; also see Jablonski, 1993). Like other baboons, geladas are extremely sexually dimorphic in

FIGURE 6.10 Geladas (*Theropithecus gelada*) from the highlands of Ethiopia.

both size and appearance. Males have long, shaggy manes and pronounced facial whiskers, whereas the female pelage is much shorter. Both sexes have striking red hourglass patches of skin on their chests, and in females these are outlined with white vesicles. The distinctive molar teeth of *Theropithecus* are characterized by complex enamel foldings. Male canines are very large, even by baboon standards. The snout and mandible are relatively short and deep. The hands of geladas are characterized by a relatively long thumb compared with the length of the other digits, an adaptation for foraging for grass blades and seeds.

Geladas live in the treeless montane grasslands of the Ethiopian highlands, where they forage on the ground all day and sleep on rocky cliffs at night. They are the most terrestrial nonhuman primates and always move by quadrupedal walking and running. They are exclusively herbivorous, eating grass, seeds, and roots throughout the year, and occasionally, fruit (see Jwamoto *et al.*, 1996). They feed by sitting upright, plucking grass blades and seeds by hand.

Geladas live in one-male groups of three to twenty individuals, which may gather in bands. This relationship between bands is different from that found among the clans of Hamadryas baboons (e.g., Stammbach, 1986). Females stay together in matrilines, and unattached males converge into all-male groups. Groups occasionally join together in temporary herds of up to 400 individuals. Several groups usually share a single home range, and day ranges are relatively small for baboons

(1–2 km), reflecting both the small foraging units and sedentary feeding habits of geladas.

Mandrills and Drills

Mandrills (*Mandrillus sphinx*) and **drills** (*M. leucophaeus*) are large (male mean about 30 kg) forest monkeys from western Africa (Fig. 6.11; Table 6.4). Although they have been grouped with baboons traditionally, all molecular studies indicate that they are actually more closely related to terrestrial mangabeys of the genus *Cercocebus* (Figure 6.7; Disotell, 1994, 1996).

FIGURE 6.11 The mandrill (*Mandrillus sphinx*).

TABLE 6.4
Infraorder Catarrhini
Family Cercopithecidae
Subfamily CERCOPITHECINAE, mandrills and drills

Common Name	Species	Intermembral Index	Body Mass (g) F	Body Mass (g) M
Mandrill	*Mandrillus sphinx*	95	12,900	31,600
Drill	*M. leucophaeus*	—	12,500	20,000

They are extremely sexually dimorphic in both size and coloration. Males of both species are characterized by brightly colored faces and rumps. In both mandrills and drills, males have long muzzles with pronounced maxillary ridges and long tooth rows. Like baboons, mandrills and drills have forelimbs and hindlimbs of nearly equal length. Both have very short tails.

Drills and mandrills live in dense forest and tend to be shy as a result of human hunting. Within the forest they are primarily terrestrial, but females and juveniles regularly climb into trees. The diet of drills and mandrills has been described as fruit, leaves, pith, and insects. Like Cercocebus, they specialize on hard nuts and seeds when other foods are scarce (Hosino, 1985).

Mandrills and drills have been seen in one-male groups, multimale groups, and large congregations numbering up to 250 individuals. Although there are no detailed field studies so far, it seems most likely that mandrills and drills, like pig-tailed macaques and *Cercocebus,* normally live in multimale groups of twenty to forty individuals that break into smaller foraging parties and also congregate into supertroops in some seasons (Caldecott *et al.,* 1996). Mandrill groups have a ratio of adult females to adult males of approximately 9:1, and solitary males are common.

Guenons and Relatives

Guenons (*Cercopithecus*) are the small forest monkeys of sub-Saharan Africa. There are at least nineteen guenon species (Table 6.5), that are remarkably diverse in color and appearance (Fig. 6.12) but relatively uniform in size and body proportions (Schultz, 1970). Guenons range in size from 3 kg to 9 kg, with most species averaging about 4–5 kg. There is a moderate amount of sexual dimorphism in most species. All guenons have sexually dimorphic canines, relatively narrow molar teeth, and short third molars. They have relatively short snouts compared with larger cercopithecines (see Fig. 6.4). Guenons have longer hindlimbs than forelimbs and long tails. They are all forest dwellers and show considerable variation from species to species in their forest preference and use of canopy levels. All species are basically arboreal quadrupeds, but some frequently come to the ground, and some are quite good leapers (e.g., Gebo and Sargis, 1994; McGraw, 1996).

In general, guenons are predominantly frugivorous and insectivorous (Gautier-Hion, 1988); Some species eat leaves during some parts of the year. with larger species generally eating more leaves. In contrast with many primate females, female guenons actively defend group territories (Rowell, 1988). Most species live in single-male groups, and male guenons are generally antagonistic to one another. However, the relationship between male residence and reproductive success is not clear. In many species, large numbers of males join troops and mate with the females during the mating season, and resident males of one troop may also travel to mate with females in other troops (Cords, 1988). All guenons seem to have annual birth peaks associated with the time of greatest food abundance (Butyinski, 1988). Infanticide has been reported in some species.

Guenons often feed and travel in mixed-species groups, most probably as an antipredator strategy (Gautier-Hion *et al.,* 1983; Cords, 1990). Despite the overall structural uniformity of guenons, individual species have unique foraging strategies that distinguish them from sympatric species (Gautier-Hion, 1978). Three frequently associated guenon species that have been well studied in Gabon are *C. cephus,* the **moustached monkey;** *C. pogonias,* the **crowned monkey;** and *C. nictitans,* the **greater spot-nosed monkey** (Figs. 6.13, 6.14). *Cercopithecus cephus* is the smallest species, *C. pogonias* is slightly

TABLE 6.5
Infraorder Catarrhini
Family Cercopithecidae
Subfamily CERCOPITHECINAE, guenons

Common Name	Species	Intermembral Index	Body Mass (g)	
			F	M
Blue monkey	*Cercopithecus mitis*	82	4,250	7,930
Greater spot-nosed monkey	*C. nictitans*	82	4,260	6,670
Red-tailed monkey	*C. ascanius*	79	2,920	3,700
Lesser spot-nosed monkey	*C. petaurista*	—	2,900	4,400
Red-bellied monkey	*C. erythrogaster*	—	2,400	4,100
Moustached monkey	*C. cephus*	81	2,880	4,290
Red-eared monkey	*C. erythrotis*	—	2,900	3,600
Sclater's monkey	*C. sclateri*	—	2,500	4,000
Mona monkey	*C. mona*	86		5,100
Campbell's monkey	*C. campbelli*	—	2,700	4,500
Crowned monkey	*C. pogonias*	—	2,900	4,260
Wolf's monkey	*C. wolfi*	—	2,870	3,910
Diana monkey	*C. diana*	79	3,900	5,200
Zaire diana monkey	*C. salongo*	—		
Dryas monkey	*C. dryas*	—		3,000
Preuss's monkey	*C. preussi*	82		4,500
L'Hoest's monkey	*C. lhoesti*	80	3,450	5,970
Sun-tailed monkey	*C solatus*	—	3,920	6,890
De Brazza's monkey	*C. neglectus*	82	4,130	7,350
Hamlyn's monkey	*C. hamlyni*	—	3,360	5,490
Vervet monkey	*Chlorocebus aethiops*	83	2,980	4,260
Allen's swamp monkey	*Allenopithecus nigroviridis*	83	3,180	6,130
Talapoin monkey	*Miopithecus talapoin*	83	1,120	1,380
Patas monkey	*Erythrocebus patas*	92	5,770	10,600

larger, and *C. nictitans* is much larger and more sexually dimorphic. All are found in primary rain forests, but *C. cephus* is also very common in secondary forests. All are arboreal quadrupeds. *Cercopithecus pogonias* and *C. nictitans* are found primarily in the middle and upper levels of the main canopy, whereas *C. cephus* prefers the lower levels of the canopy and the understory and occasionally comes to the ground.

The diets of the three monkeys have been documented through examination of stomach contents (Fig. 6.14). *Cercopithecus pogonias* is the most frugivorous, the most insectivorous, and the least folivorous, *C. cephus* is intermediate, and *C. nictitans* is the most folivorous, with leaves, flowers, and other vegetable materials accounting for almost 30 percent of its diet. The two species that overlap most in canopy use overlap least in diet. The types of insects eaten by the species also differ considerably: *C. pogonias* appears to specialize on mobile prey, whereas *C. nictitans* eats cryptic

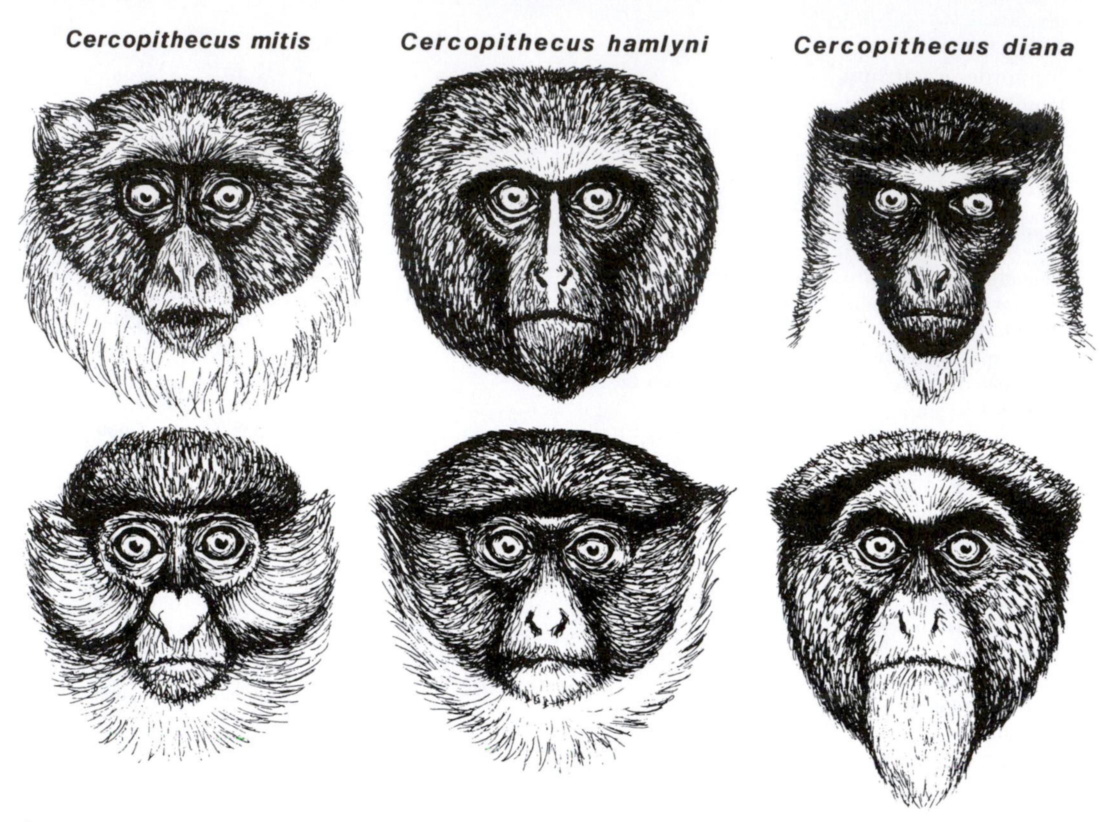

FIGURE 6.12 The faces of six guenons.

immobile prey almost exclusively. Again, *C. cephus* is intermediate, but it occupies a different level of the forest. All three species live in moderately small home ranges, with mean group size increasing from *C. cephus* (ten individuals) to *C. pogonias* to *C. nictitans*. In each species there is usually a single adult male.

Two other guenon species that have been particularly well studied are *C. mitis*, the **blue monkey,** and *C. ascanius*, the **red-tailed monkey;** both are found in the Kibale forest of Uganda. *Cercopithecus ascanius*, a close relative of *C. cephus*, is smaller and less dimorphic than *C. mitis*, a relative of *C. nictitans*.

One of the most handsome guenons is *Cercopithecus neglectus*, **De Brazza's monkey** (Fig. 6.15), from western and central Africa. *Cercopithecus neglectus* is one of the largest and most sexually dimorphic guenons, with males averaging over 7 kg and females less than 4 kg. De Brazza's monkeys prefer flooded forests, including islands in rivers, where they move primarily in the understory and on the ground. They are slow quadrupedal monkeys. Their diet is predominantly frugivorous and includes lesser amounts of leaves, animal matter, and mushrooms.

The most unusual feature of De Brazza's monkey is its social organization. In some parts of their range they live in polygynous groups of eight to twelve individuals (Wahome *et al.*, 1993), but in Gabon, where they have

FIGURE 6.13 Three guenon species from Gabon that frequently forage together in mixed species groups: above, the crowned guenon (*Cercopithecus pogonias*); middle, the greater spot-nosed guenon (*C. nictitans*); below, the moustached guenon (*C. cephus*).

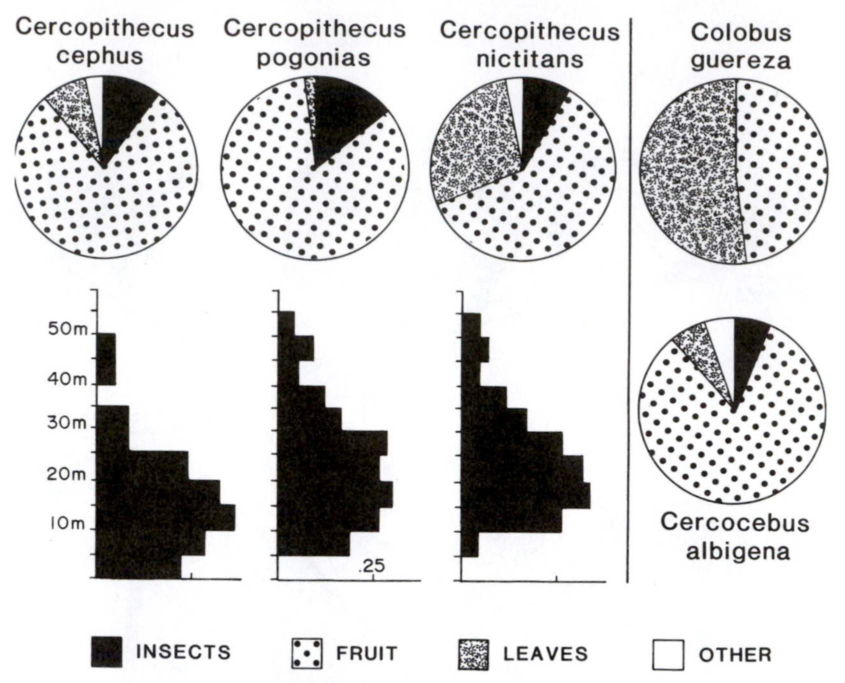

FIGURE 6.14 Diet and forest height preferences of five monkeys from Gabon.

been quite well studied (Gautier-Hion and Gautier, 1978), most live in monogamous families. The two sexes seem to forage somewhat independently and have different strategies for dealing with predators. The small females hide in the undergrowth, whereas the large males climb trees and call at the predator, apparently to distract it from the female and young. Both the day ranges and the home ranges of these family groups are very small.

Four other cercopithecines are closely related to guenons and are often placed in the genus *Cercopithecus*. However, the exact phylogenetic relationships of these monkeys to one another and to the genus *Cercopithecus* are difficult to resolve (Disotell, 1996). All are very distinct from guenons in both their morphology and their ecological adaptations and are best placed in separate genera.

Chlorocebus aethiops, the **vervet monkeys, grivets,** or **green monkeys** (Fig. 6.16), range throughout sub-Saharan Africa and are one of the most widespread cercopithecine species. They are often divided into several distinct species. The large range of this monkey is undoubtedly associated with its preference for woodland savannah and gallery forests. Vervet monkeys are more terrestrial than guenons and frequently cross open areas between feeding trees and forage on the ground. Terrestrial movements account for approximately 20 percent of their locomotion; the remainder is arboreal. Vervets are predominantly quadrupedal, both on the ground and in the trees; leaping accounts for only 10 percent of their locomotor activity. Their diverse diet consists of fruits, gums, shoots, and a variety of invertebrates. One population in Senegal catches crabs, a behavior reminiscent of the crab-eating macaque of Southeast Asia.

Vervet monkeys differ from most forest guenons in that they live in relatively large troops

FIGURE 6.15 A family of De Brazza's monkeys (*Cercopithecus neglectus*).

with several adult males. As with baboons, there is a clear dominance hierarchy among males, especially during the mating season.

The **talapoin monkey** (*Miopithecus talapoin*) is the smallest Old World monkey, weighing just over a kilogram, and also the smallest living catarrhine (Table 6.5). Many of the distinguishing features of talapoins—a relatively large head, large eyes, and a short snout—are characteristics of young animals, suggesting that talapoins are a neotenic guenon (Shea, 1992).

Talapoins live in the riverine forests of western and central Africa. They are found most frequently in the dense undergrowth, where they move by leaping and quadrupedal walking and running. They are among the best leapers of all cercopithecines, and they also seem to be good swimmers. Their diet contains large amounts of insects and fruit. They are among the most insectivorous of Old World monkeys (Gautier-Hion *et al.,* 1988).

Talapoins live in large social groups averaging between seventy and one hundred individuals (the larger groups studied were parasitic on manioc farms), with roughly one adult male for every two adult females. Talapoin troops are always on the move, flushing insects. They always sleep by the water, which they leap into when frightened.

Talapoins have an unusual social organization characterized by little interaction between males and females during most of the year (Rowell and Dixon, 1975). They usually forage in subgroups composed of individuals of the same sex. During the three-month

FIGURE 6.16 A troop of vervet monkeys (*Chlorocebus aethiops*) in a woodland savannah habitat.

breeding season, males join the female groups, and some females join all-male subgroups. After breeding, the individuals return to their single-sex groups.

The **swamp monkey** (*Allenopithecus nigroviridis*) is a medium-size monkey that lives in flooded forests in western and central Africa. Compared with other small cercopithecines, this species has broader macaque-like molar teeth suggestive of a frugivorous diet. On the basis of its dental and skeletal anatomy, several authorities have suggested that this species is the most primitive guenon and perhaps the most primitive cercopithecine. Virtually nothing is known of its behavior in a natural environment.

The **patas,** or **hussier, monkey** (*Erythrocebus patas*) is a close relative of the guenons, with extreme specializations for its life on the open grasslands. Patas are medium-size, very sexually dimorphic monkeys with slender bodies, long limbs, and long tails. They have narrow hands and feet with short digits and a reduced pollex and hallux.

Patas monkeys live on the grasslands and savannahs of western, central, and eastern Africa. Although they usually sleep in trees at the edge of the forest, most of their foraging takes place in the open grass, where they move by quadrupedal walking and running. They are extremely agile, fast runners (55 km/h, according to Kingdon, 1971), and they frequently stand bipedally to look over the tall grass for potential predators or conspecifics. The bulk of their diet seems to be grass seeds, new shoots, and acacia gums. They also eat the beans of tamarind trees and a variety of other tough savannah fruits, seeds, and berries. They supplement their herbivorous diet with insects and various other prey items. They seem normally to eat on the move, picking up bits of food items as they walk.

Groups of patas monkeys average about a dozen individuals, with usually a single adult male. Nongroup males often live together in all-male bands, but there seems to be considerable turnover of males in patas groups. Day ranges are extraordinarily variable, ranging from 700 to nearly 12,000 m, with a central tendency of between 2000 and 2500 m. Sometimes the group forages cohesively and other times members of a group are separated by as much as 800 m. The estimated home ranges are over 5000 ha, the largest known for any nonhuman primate species. It is not surprising that they can run so fast.

Colobines

The second subfamily of Old World monkeys are the colobines, or leaf-eating monkeys, of Africa and Asia. Colobines (Tables 6.6–6.8) are easily distinguished from cercopithecines by their sharp-cusped cheek teeth and relatively narrow incisors. Their skulls have relatively short snouts (see Fig. 6.4), narrow nasal openings, broad interorbital areas, and deep mandibles (see Fig. 6.3). They have complex, sacculated stomachs, similar to those of cattle, that enable them to maintain bacterial colonies for digesting cellulose (Bauchop, 1978; Chivas, 1995). Their skeletons are characterized by relatively long legs, long tails, and thumbs that are usually short or absent.

In general, the living colobine monkeys are more arboreal and folivorous than are cercopithecines, and they are also better leapers. Socially, most are characterized by single-male groups. **Aunting behavior,** in which infants are frequently passed around among females within a troop, is common in many colobines.

There are two major groups of colobine monkeys, the colobus monkeys of Africa and the langurs, or leaf monkeys, of Asia.

Colobus Monkeys

The African colobus monkeys come in three color schemes: black and white, red, and olive. The three groups are quite distinct behaviorally and ecologically (Table 6.6). The **black-and-white colobus monkey,** or **guereza** (Fig. 6.17), is the largest and most spectacular

TABLE 6.6
Infraorder Catarrhini
Family Cercopithecidae
Subfamily COLOBINAE, African colobus monkeys

Common Name	Species	Intermembral Index	Body Mass (g)	
			F	M
Guereza	*Colobus guereza*	79	9,200	13,500
King colobus	*C. polykomos*	78	8,300	9,900
Angolan colobus	*C. angolensis*	—	7,570	9,680
Black colobus	*C. satanas*	—	7,420	10,400
Red colobus	*Piliocolobus badius*	87	8,210	8,360
Olive colobus	*Procolobus verus*	80	4,200	4,700

FIGURE 6.17 Two sympatric colobines from Africa: above and below, the red colobus (*Piliocolobus badius*); center, the black-and-white colobus (*C. guereza*).

of the African colobine monkeys. It is a quite robust monkey, with considerable sexual dimorphism in size. There are three black-and white species: *Colobus guereza, C. polykomos,* and *C. angolensis.* Guerezas live in a wide range of forest types throughout sub-Saharan Africa, in both primary rain forests and patchy dry forests. They are extremely hardy and can survive in a variety of habitats. They are arboreal and prefer the main canopy levels. All colobines are good leapers, but guerezas move more frequently by quadrupedal walking and bounding. They virtually never engage in suspensory behavior. Guerezas usually feed by sitting on branches and pulling food toward themselves.

The diet of guerezas is predominantly leaves (see Fig. 6.14) and seeds, often of only a few tree species. They rely extensively on mature leaves for much of the year. Unripe fruits and young leaves are important only in some seasons.

Black-and-white colobus usually live in very small groups of a single adult male, one or two females, and offspring. They have extremely small home ranges, which they advertise with loud vocal calls, as well as very small day ranges. In some seasons they are virtually sedentary, and many spend days feeding in a single tree. The infants are white at birth and are commonly cared for by several females in the group.

The **black colobus** (*Colobus satanas*), from western and central Africa, is quite distinct from the guerezas, in both anatomy and ecology. This species has larger, flatter teeth and a more robust skull than guerezas. In contrast with the black-and-whites, black colobus feed predominantly on hard seeds; leaves and fruits are less important parts of their diet. As a result of this specialization on seeds as a protein source, *C. satanas* is able to thrive in areas that are uninhabitable for more folivorous colobines because the leaves have high levels of poisonous tannins.

Black colobus live in multimale groups that are larger than the single-male groups of guerezas, and they also have larger home ranges. Day ranges are, on average, quite small but vary dramatically from season to season, depending on the availability of food. When food is abundant, black colobus often feed in a few trees for several days; when food is scarce, they move long distances and feed briefly from many different trees.

Red colobus (*Piliocolobus badius*) are found sympatric with guerezas throughout most of sub-Saharan Africa. They are normally placed in a single species with many subspecies (Table 6.6). Red colobus are smaller than guerezas, and their relatively short pelage gives them a much more slender appearance (Fig. 6.17). They have broader incisors, more gracile skulls, longer legs, and less sexual size dimorphism than guerezas, and they are strikingly different from guerezas in many aspects of their behavior and ecology as well. They seem to reach higher densities in primary forests than black-and-whites but are not found in drier forests. They range through all levels of the main canopy and emergents. In comparison with guerezas, red colobus eat fewer mature leaves; they prefer fruit, young leaves, and shoots.

Red colobus live in large troops of forty to ninety animals, with numerous adults of both sexes. The troops have overlapping home ranges of 100 ha or more, and there seems to be direct competition and a dominance hierarchy between troops over access to specific feeding areas. Day ranges of red colobus are larger than those of black-and-white colobus. All of these ranging differences can be related to the distribution of the major food sources preferred by the two species. The guerezas' more spatially and temporally homogeneous diet of mature leaves can be exploited effectively in small groups. The fruit, shoots, and new leaves preferred by *Piliocolobus badius* are less evenly distributed, but when available they are quite abundant; thus the size of food resources probably does not limit their group size. Furthermore, larger groups are better able to obtain and defend large food sources.

Many aspects of the social organization of red colobus are unusual for colobines. Females have a large estrus swelling, like the cercopithecines that live in multimale groups, and it is females rather than males that generally emigrate from their natal group. Red colobus are commonly preyed on by chimpanzees, who can take as much as 10 percent of the red colobus population in a single year (Stanford *et al.,* 1994). The large groups of red colobus have probably evolved in part as an antipredator strategy (Noe and Bshary, 1997).

The **olive colobus** (*Procolobus verus*) is the smallest colobine and the most poorly known of the African species. It is sexually dimorphic in size and has the smallest thumb and the largest feet of all African colobines.

Olive colobus live in swamp forests and range almost exclusively in the understory. They are the most saltatory of the African colobines (McGraw, 1996). Olive colobus forage as a very dispersed group in thick vegetation and regularly associate with groups of guenons, especially *Cercopithecus diana.* They eat mainly new leaves.

The social organization of olive colobus seems to be similar to that of red colobus on a smaller scale. They live in small multimale groups of ten to fifteen animals. In contrast with red colobus, olive colobus seem to avoid predation through crypsis or associate with large groups of *Cercopithecus* species. Like red colobus, female olive colobus show sexual swellings. Females carry young infants in their mouth, like many prosimians. Little is known of their ranging behavior. Olive colobus have only soft vocalizations.

Langurs

It is in Asia that the colobine monkeys have reached their greatest diversity and abundance (Tables 6.7, 6.8). The diversity is almost certainly a result of Pleistocene fluctuations in climate and sea level in the Sunda Shelf (Brandon-Jones, 1996). Two or three sympatric species are a common pattern, and the density of leaf-eating monkeys in southern and eastern Asia exceeds that of any other forest vertebrates. The most common, most diverse, and most abundant are the langurs, formerly all placed in the genus *Presbytis,* but now divided into three or four separate genera. In addition, there are a number of most unusual monkeys commonly grouped as the "odd-nosed monkeys."

The systematics and phylogenetic relationships of Asian colobines are the subject of considerable debate, and numerous alternative arrangements have been proposed based on studies of skulls, teeth, pelage, and molecules (e.g., Brandon-Jones, 1984, 1996; Oates *et al.,* 1994; Collura and Stewart, 1996). In this chapter Asian langurs are divided into four distinct species groups, recognized here as separate genera: *Semnopithecus,* from India and Sri Lanka; *Kasi,* also from India and Sri Lanka; *Trachypithecus,* from Southeast Asia and parts of China; and *Presbytis,* restricted to Southeast Asia (Table 6.7; Figs. 6.18, 6.26).

The **sacred,** or **Hanuman, langur** (*Semnopithecus entellus*) of India (Fig. 6.19) is one of the most adaptable of all higher primates. This species is found from Sri Lanka in the south, to the Rajastan Desert in the west, and well into the Himalaya mountains of Nepal (Bishop, 1979). Over this broad geographic range the species shows considerable diversity in body weight and skeletal morphology. In general, they are long-limbed, gray to brown monkeys with a very long tail, short thumbs, and long feet. They thrive in virtually all

TABLE 6.7
Infraorder Catarrhini
Family Cercopithecidae
Subfamily COLOBINAE, langurs and leaf monkeys

Common Name	Species	Intermembral Index	Body Mass (g)	
			F	M
Hanuman langur	*Semnopithecus entellus*	83	9,890	13,000
Purple-faced langur	*Kasi vetulus*	—	5,900	8,170
John's or Nilgiri langur	*Kasi johnii*	80	11,200	12,000
Dusky leaf monkey	*Trachypithecus obscura*	83	6,260	7,900
Phayre's leaf monkey	*T. phayrei*	—	6,300	7,870
Silvered leaf monkey	*T. cristata*	—	5,760	6,610
Lutung	*T. auratus*	—		
Capped leaf monkey	*T. pileatus*	—	9,860	12,000
Golden leaf monkey	*T. geei*	—	9,500	10,800
Francois' leaf monkey	*T. francoisi*	—	7,350	7,700
Banded leaf monkey	*Presbytis melalophos*	78	6,470	6,590
Javan leaf monkey	*P. comata*	76	6,680	6,710
Maroon leaf monkey	*P. rubicunda*	76	6,170	6,290
White-fronted leaf monkey	*P. frontata*	76	5,670	5,560
Thomas's leaf monkey	*P. thomasi*	—	6,690	6,670
Hose's leaf monkey	*P. hosei*	75	5,630	6,180
Mentawai leaf monkey	*P. potenziani*	—	6,400	6,170

imaginable habitats found on the Indian subcontinent, including tropical rain forests, deciduous forests, temperate dry forests, and conifer forests, as well as deserts and cities. Hanuman langurs are the most terrestrial of the colobines. In the trees they use both quadrupedal gaits and leaps, but mainly the former. They feed primarily in seated postures.

Considering the diversity of habitats they occupy, it is not surprising that the diet of Hanuman langurs is quite eclectic. In general, however, they eat fruit, flowers, and new leaves and seem unable to subsist on a diet of mature leaves.

Social groups of Hanuman langurs vary considerably in size, from ten to one hundred individuals, with a mean of about twenty. There is typically one adult male per ten individuals. Home ranges vary, according to group size and habitat, from 24 to 200 ha.

Langur troops seem to be centered around groups of related adult females that aid one another and care for each other's offspring. By contrast, male residence in a troop is usually relatively short-lived. Most langurs live in one-male troops in which the single adult male fathers all of the offspring and drives out rival males. At fairly regular intervals, bands of roving males attack a group, drive out the resident male, and kill dependent infants. One of these intruders then establishes himself as the dominant male, drives away the others, and starts the cycle anew (Hrdy, 1977). Hanuman langurs have single births at sixteen- to eighteen-month intervals. Infants are regularly passed around among the adult females and

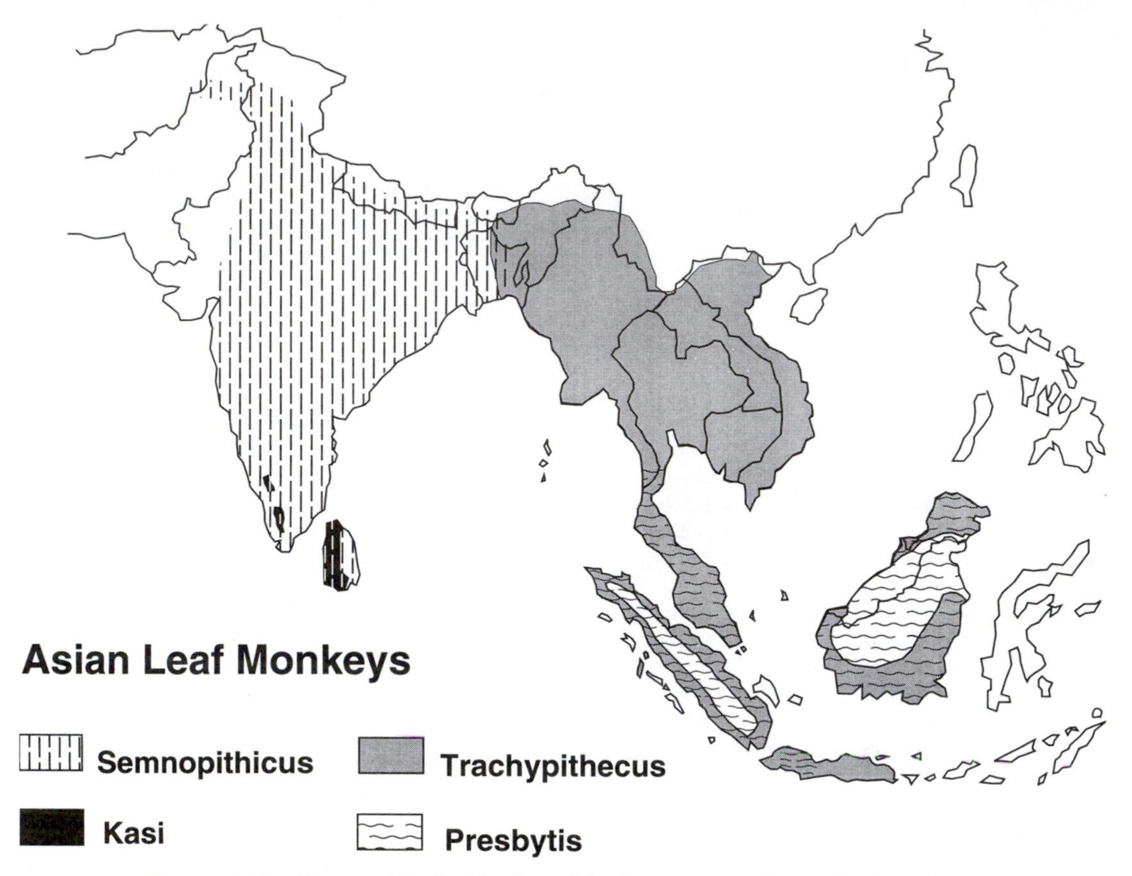

FIGURE 6.18 Geographic distribution of the four genera of Asian leaf-monkeys.

are often cared for by females other than their mother.

The other langurs from the Indian subcontinent are *Kasi vetulus* (= *senex*), the **purple-faced langur,** from Ceylon (Figs. 6.18, 6.19), and a closely related species, *Kasi johnii,* **John's langur,** from western India. Both species are characterized by purple faces, white sideburns, and relatively small size. The phylogenetic relationships of these monkeys is debated. Some authorities place them with the Hanuman langur in *Semnopithecus* (Fig. 6.18; Collura and Stewart, 1996); others put them with the *Trachypithecus* species (Oates *et al.,* 1994). In light of this debate, they are kept here as a separate genus, *Kasi* (Table 6.7; Pocock, 1935). Both are restricted to forested areas. They are almost totally arboreal and are excellent leapers. Both are more folivorous than *P. entellus* and can subsist almost exclusively on mature leaves.

Like the black-and-white colobus of Africa, purple-faced langurs live in very small, one-male groups or families, with home ranges of a hectare or less. Their day ranges are small and, like black-and-white colobus, they often

FIGURE 6.19 Two colobine species from India and Sri Lanka: above, the purple-faced monkeys (*Kasi vetulus* (= *senex*)); below, the Hanuman langur (*Semnopithecus entellus*).

spend an entire day in one tree. Male takeovers and infanticide have also been reported in purple-faced langurs (Rudran, 1973).

The langurs east of India are splendidly diverse in their coloration but can be divided into two distinct genera on the basis of body proportions, vocalizations, infant color patterns, and ecology (Figs. 6.6, 6.20). The most diverse of these is the genus **Presbytis,** which contains at least seven species from Thailand and west Malaysia (*P. melalophos*), Sumatra (*P. melalophos* and *P. thomasi*), Borneo (*P. aygula, P. frontata, P. hosei,* and *P. rubicunda*), the Mentawai Islands (*P. potenziani*), Java (*P. comata*), and various other islands of the Sunda Shelf (Table 6.7; Oates *et al.,* 1994). There are numerous subspecies with a dazzling variability in adult coat color. By contrast, all infants show a distinct banded or cruciform pattern with a dark head, dark body and tail, and dark arms.

The best-known species, *P. melalophos,* the banded leaf monkey, is a relatively small colobine (6 kg) that shows little sexual size dimorphism. Like other members of its species group, the banded leaf monkey has a relatively short face. Its postcranial skeleton is characterized by relatively long legs, a long trunk, and slender arms that are associated with its leaping locomotion (Fig. 6.21).

Leaf monkeys of this group seem to be found in a variety of inland forests, but not in swamps or montane forests. In west Malaysia the banded leaf monkey is more common in secondary forests than in upland primary forests and moves and feeds more frequently in the understory and lower canopy levels than do other sympatric langurs (see Fig. 6.6). Banded leaf monkeys are extraordinary leapers and move less frequently by arboreal quadrupedalism. They occasionally use forelimb suspension as well (Figs. 6.6, 6.20).

Banded leaf monkeys eat primarily young leaves, seeds, and fruit and rarely, if ever, partake of mature leaves. They live in small groups of a dozen or fewer animals, frequently with a single adult male. Home range sizes are quite variable from locality to locality and do not seem to be defended (Bennet 1986). Troops of banded leaf monkeys forage as a group over day ranges that average about 700 m. They are active throughout the day and often well into early evening. Males often show wounds, scars, and other evidence of conflicts, and infanticide seems likely in this species group (Sterck, 1995).

Presbytis potenziani, the **Mentawai Island leaf monkey,** is a colobine with an unusual social organization for an Old World monkey. Very little is known about the ecology and diet of these monkeys except that they live in monogamous family groups. Like many monogamous primates, the males and females have been reported to sing together in daily vocal duets (Tilson and Tenaza, 1976).

The other widespread group of Southeast Asian langurs is the genus ***Trachypithecus,*** with numerous, mostly allopatric species ranging from India in the West to China in the East (Table 6.7). These monkeys are similar in size to those of the *P. melalophos* group, but they are more sexually dimorphic. They are mostly silver, gray, or black monkeys as adults, but with bright yellow or orange infants. *Trachypithecus obscura,* the spectacled langur (Fig. 6.20), has striking white eye rings. This species group has a more robust skull, and arms and legs that are more similar in length. Some species have clear habitat preferences, but others do not. *Trachypithecus obscura* is more abundant in upland primary forests or secondary forests; *T. cristata* is commonly found in mangrove swamps and in flooded land along long rivers in west Malaysia and Borneo, but some populations are also found inland. In west Malaysia, where it has been most thoroughly studied, *P. obscura* prefers the main canopy levels and, to a lesser degree, the emergents

FIGURE 6.20 Two sympatric leaf monkey species from Malaysia: above, the spectacled langur, or dusky leaf monkey (*Trachypithecus obscura*); below, the banded leaf monkey (*Presbytis melalophos*).

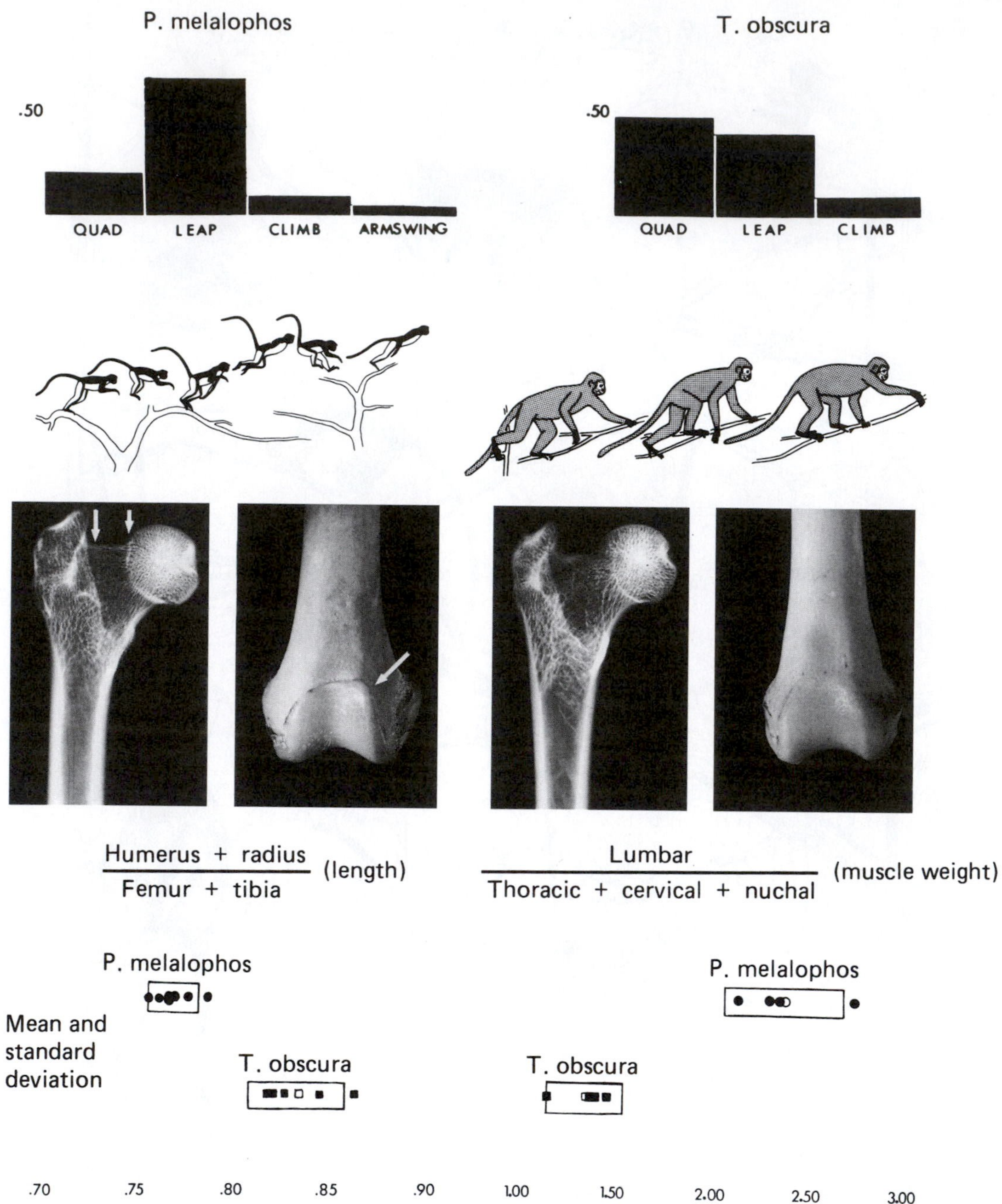

FIGURE 6.21 Locomotor and anatomical differences between *Presbytis melalophos* and *Trachypithecus obscura*. *Presbytis melalophos* leaps more frequently, and its femur is characterized by a short, straight neck and a prominent lateral ridge on the patellar groove (arrows). This species also has relatively longer hindlimbs and larger back muscles.

(see Fig. 6.6). These monkeys are primarily arboreal quadrupeds that leap much less frequently than *P. melalophos* (Fig. 6.21). Both *T. obscura* and *T. cristata* are primarily folivores, but they also eat unripe fruit.

The social organization and foraging patterns of *T. obscura* are quite different from those of *P. melalophos.* They live in slightly larger multimale groups and have larger home ranges. In addition they often do not forage as a group. The group, which may forage together in the early morning or late afternoon, splits up during the day into small foraging units, each of which travels a relatively short distance and may spend hours in one place. Like Hanuman langurs, these langurs pass their brightly colored infants around among the females. *Trachypithecus cristata* seem to live more frequently in single-male groups.

Odd-nosed Monkeys

The remaining colobine monkeys from Asia are all characterized by odd-shaped noses. The best known of these is *Nasalis larvatus,* the **proboscis monkey** of Borneo (Fig. 6.22, Table 6.8). This large, red monkey is the most sexually dimorphic of all colobines. Males are almost twice the size of females and have an enormous pendulous nose; females have a smaller, turned-up proboscis. Earlier suggestions that proboscis monkeys are basal members of the colobine radiation seem unlikely. Molecular studies group them clearly among the Asian langurs and separate from the African colobines (Collura and Stewart, 1966).

Proboscis monkeys are restricted to areas of riverine and coastal forest, where their main predators are the false gavials and clouded leopards. Proboscis monkeys practice a riverine refuging pattern, foraging widely during the day but returning to sleeping trees at the riverside each night. Groups regularly cross the river by swimming, usually at the narrowest spots. Their diet consists of about 50 percent (mostly new) leaves, 40 percent fruit and seeds, and 10 percent other plant parts but varies considerably throughout the year. For half of the year proboscis monkeys eat primarily fruit, and for half they eat predominantly leaves. Individual foraging groups, mostly polygynous one-male groups and nonreproductive groups of males and immatures, have large overlapping home ranges averaging 130 ha.

Proboscis monkey social organization seems to be organized in a two-tiered arrangement in which individual foraging groups (one-male groups and nonreproductive—mostly all-male—groups) regularly associate with adjacent groups to form a band. Individual groups spend about one-third of their time alone and two-thirds in association with other groups. Group interactions are not random; rather, each group seems to have preferred neighbors (Yaeger, 1992b). All males seem to disperse before adolescence, and many females disperse as well.

Simias concolor, the **Simakobu monkey** or **pigtailed langur,** is a close relative of *Nasalis* from the Mentawai Islands, off the west coast of Sumatra. Both males and females of this species have a short, turned-up nose, much like that of a female proboscis monkey. This species has very unusual body proportions for a colobine, with similar-size forelimbs and hindlimbs and a short tail. They are sexually dimorphic. Males are, on average, 30 percent larger than females in body size and have much larger canines. These monkeys' daily activities, including feeding and much of their travel, take place predominately in trees, but when frightened they often descend to the ground and flee terrestrially. They generally feed on leaves and also eat fruits, seeds, and berries.

Simakobu monkeys live in both monogamous family groups and polygynous groups of one male and two females. Nongroup males

FIGURE 6.22 The proboscis monkey (*Nasalis larvatus*) in a nipa mangrove swamp.

TABLE 6.8
Infraorder Catarrhini
Family Cercopithecidae
Subfamily COLOBINAE, odd-nosed monkeys

Common Name	Species	Intermembral Index	Body Mass (g)	
			F	M
Proboscis monkey	*Nasalis larvatus*	94	9,820	20,400
Simakobu or Pig-tailed langur	*Simias concolor*	—	6,800	9,150
Douc langur	*Pygathrix nemaeus*	94	8,440	11,000
Tonkin snub-nosed monkey	*Rhinopithecus avunculus*	—	8,500	14,000
Golden snub-nosed monkey	*R. roxellana*	—	11,600	17,900
Brelich's snub-nosed monkey	*R. brelichi*	—		15,800
Yunnan snub-nosed monkey	*R. bieti*	—	9,960	15,000

may be either solitary or in groups. Average group size on Pagai Island is 4.1 individuals (Tenaza and Fuentes, 1995). They have very few vocalizations. With increasing deforestation, these monkeys are rapidly becoming extinct.

Pygathrix nemaeus, the **Douc langurs** from Vietnam, Kampuchea, and Laos, are colorful monkeys with little or no sexual dimorphism. Little is known about their natural behavior. They live in mixed, partly deciduous forests and eat buds and leaves. They have been reported to live in small, single-male and multimale groups of three to eleven individuals that join to form larger bands. Food sharing among individuals is common.

Rhinopithecus, the **golden monkey** (Fig. 6.23), is probably the largest of all colobines, weighing 30 kg or more. One species is from the monsoon forests of Vietnam (*R. avunculus*); three others (*R. roxellana, R. brelichi,* and *R. bieti*) are from China (Table 6.8; Jablonski, 1995, 1997). All have a short, turned-up nose, a bluish face, and long, shaggy hair that forms a spectacular cape in males. They have a long tail, relatively short limbs, and short digits that are covered with long fur on the dorsal surface.

The behavior and ecology of *Rhinopithecus* is only just becoming known through the work of Chinese zoologists (see Jablonski, 1998). These monkeys are restricted to the mountainous conifer forests of southern China and some live in remarkably harsh environments. Little has been reported on their locomotion, but their limb anatomy, foot structure, and

FIGURE 6.23 The golden monkey (*Rhinopithecus roxellana*) from China.

some anecdotal field observations suggest that they are probably among the most terrestrial of all Asian colobines. They are largely folivorous and eat the foliage of the Chinese fir tree as well as lichens during the cold winter months.

Golden monkeys live in single-male and multimale groups of twenty to thirty animals in the winter. These groups form much larger congregations (up to 300 individuals) in the summer months. Their home ranges have been reported to be enormous: 30–40 km^2. Births seem to be strictly seasonal.

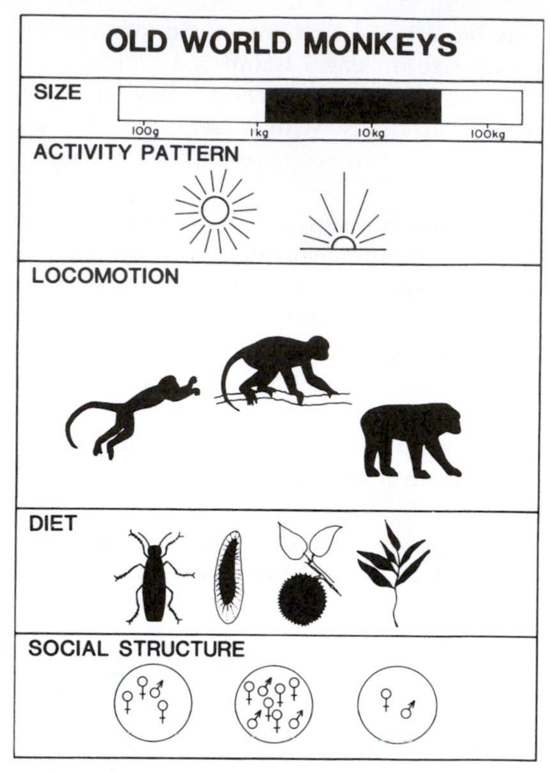

FIGURE 6.24 The adaptive radiation of Old World monkeys.

ADAPTIVE RADIATION OF OLD WORLD MONKEYS

Compared with either prosimians or neotropical anthropoids, Old World monkeys are a remarkably uniform group, in many aspects of morphology (Schultz, 1970) and behavior (Fig. 6.24). This uniformity may be due partly to the recency of their adaptive radiation. In size, they barely range over one order of magnitude, with no very small species and no extremely large species. Sexual dimorphism in canine teeth is the rule, and body size dimorphism is common. Although their bilophodont molars vary according to dietary habits in details of cusp height, length, and breadth and in the length of shearing crests, they are remarkably similar overall (Kay, 1978).

In their locomotor habits, cercopithecoid monkeys are all primarily quadrupedal walkers and runners, and some, especially colobines, are also good leapers (Gebo and Chapman, 1995; McGraw, 1996). When feeding in trees, they always sit above branches rather than use suspensory postures. No other primate radiation includes so many terrestrial species or is so limited in gait patterns and locomotor repertoire. Old World monkeys are primarily frugivorous or folivorous; a few are seed specialists. Although many species feed to varying degrees on insects (most cercopithecines) and gums (baboons, patas monkeys), there are no species that specialize on these foods to the degree found in many smaller prosimians and platyrrhines.

Old World monkeys generally live in either single-male or multimale polygynous groups; monogamy is rare. In social organization, they are a distinctive and remarkably uniform radiation compared with either platyrrhines or hominoids (Wrangham, 1980; Di Fiori and Rendall, 1994). In most species, females tend to spend their entire lives in their natal troop, and the troop's social organization centers around matrilines and female hierarchies, with a few notable exceptions, such as red colobus, hamadryas baboons, and some populations of chacma baboons. Males regularly emigrate from their natal group and may move through several troops during their lifetime. All have single births.

Despite, or perhaps because of, this relative uniformity, Old World monkeys have successfully colonized more vegetative zones and climates than has any other group of living primates. In numbers of individuals, numbers of species, and biomass density they are probably the most successful nonhuman primates. This success results from a variety of factors. For the colobines, it is almost certainly their ability to digest cellulose and to exploit folivorous food sources that are unavailable to other animals. For the cercopithecines, their terrestrial locomotor potential, manipulative abilities, and intelligence enable them to exploit a wide range of foods and environments that less flexible species cannot endure. Cercopithecines are also relatively prolific breeders for higher primates, with many species giving birth on an annual basis.

PHYLETIC RELATIONSHIPS OF OLD WORLD MONKEYS

There are relatively few major problems concerning the phyletic relationships among cercopithecoids. Above the genus level, most cercopithecoid phylogenies are in agreement. (Figs. 6.25, 6.26). However, there are numerous unresolved issues concerning the phyletic relationships among individual species and genera. Among cercopithecines, the most obvious unresolved problems concern the relationships of guenons, vervets, patas monkeys, swamp monkeys, and talapoins (Disotell, 1996).

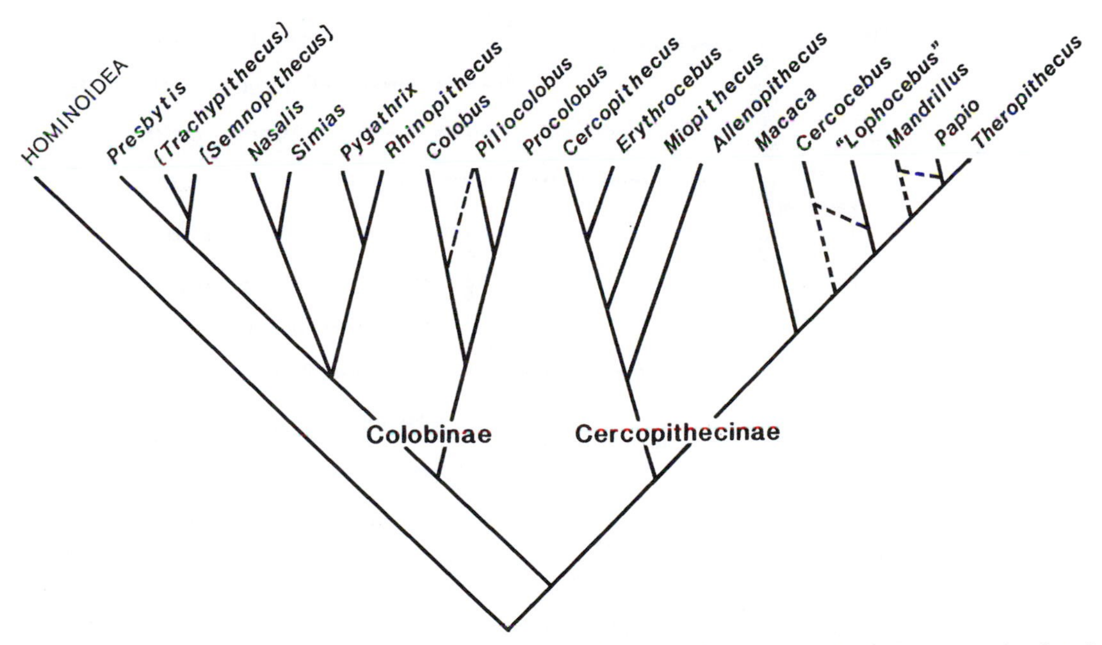

FIGURE 6.25 Traditional phylogeny of Old World monkeys based on morphological features (updated from Strasser and Delson, 1987).

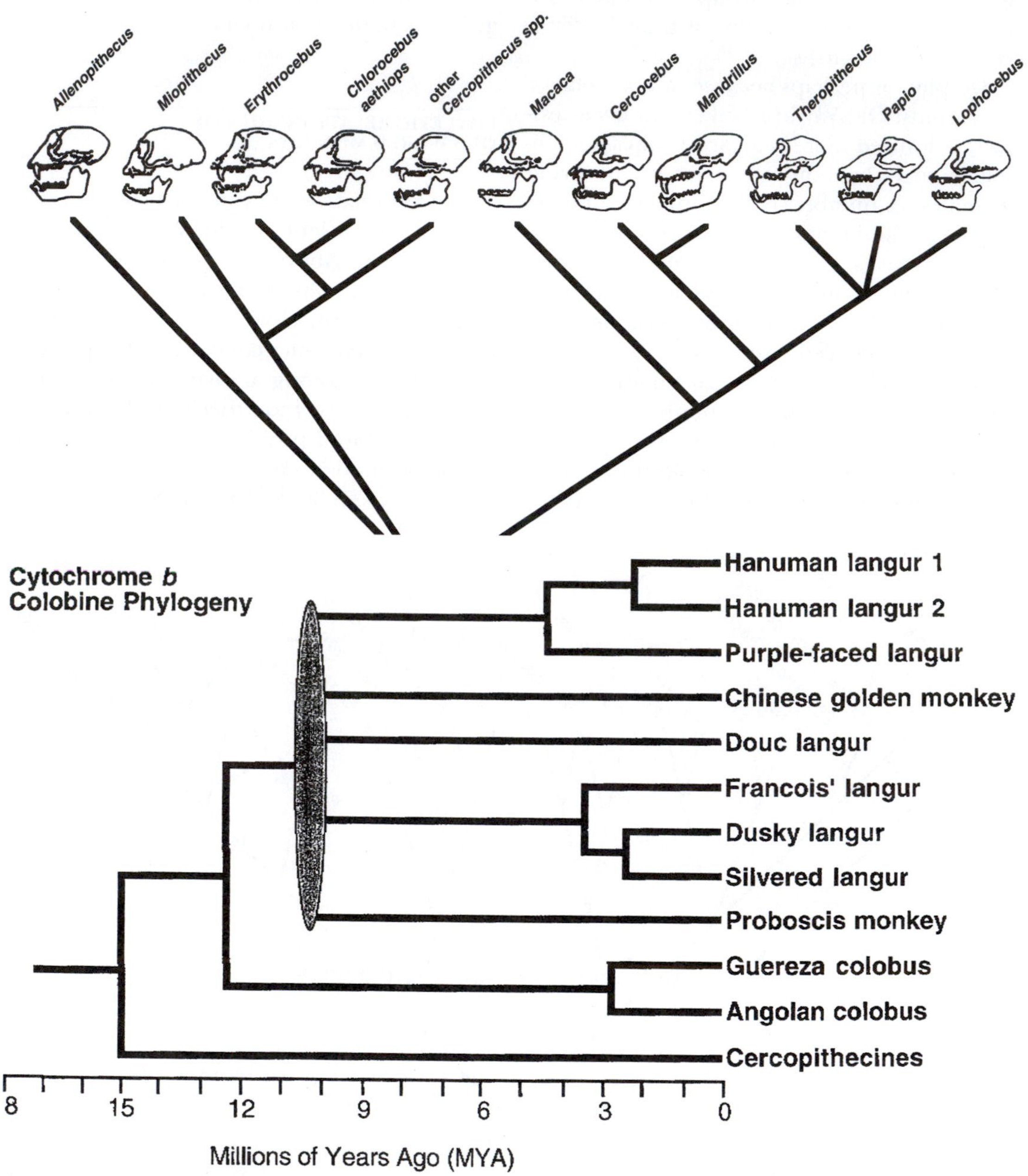

FIGURE 6.26 Recent molecular phylogenies for cercopithecines (above) and colobines (below). Courtesy of T. Disotell and C.-B. Stewart.

Among colobines there are many questions concerning the number of individual species and the generic allocation of different species of Asian colobines (Fig. 6.26; Oates *et al.*, 1994; Collura and Stewart, 1996; Brandon-Jones, 1996).

One major phylogenetic issue that appears to have been recently resolved is the phyletic relationships of mangabeys, baboons, geladas, and mandrills (Fig. 6.26). On morphological and behavioral grounds, mangabeys have generally been thought to be a coherent genus with two species-groups, and the larger papionins baboons, mandrills (and drills), and geladas were thought to be a closely related group. However, all molecular evidence as well as some morphological studies clearly indicate that one group of mangabey species—*Cercocebus*—is closely related to mandrills, while the other—*Lophocebus*—is more closely related to baboons and geladas (Fig. 6.7). A major challenge for primatology is to reconcile this phylogeny with the morphological development and evolution of this group (Fleagle and McGraw, 1998).

BIBLIOGRAPHY

GENERAL

Disotell, T. (1996). The phylogeny of Old World monkeys. *Evol. Anthropol.* **5**:18–24.

Gebo, D. L., and Chapman, C. A. (1995). Positional behavior in five sympatric Old World monkeys. *Am. J. Phys. Anthropol.* **97**:49–77.

McGraw, W. S. (1996). Cercopithecid locomotion, support use, and support availability in the Tai Forest, Ivory Coast. *Am. J. Phys. Anthropol.* **100**:507–522.

Napier, J. R., and Napier, P. H., eds. (1970). *Old World Monkeys.* New York: Academic Press.

Rose, M. D. (1974). Ischial tuberosities and ischial callosities. *Am. J. Phys. Anthropol.* **40**(3):375–384.

Strasser, E., and Delson, E. (1987). Cladistic analysis of cercopithecid relationships. *J. Human Evol.* **16**:81–100.

CERCOPITHECINES

Andelman, S. J. (1986). Ecological and social determinants of cercopithecine mating systems. In *Ecologcal Aspects of Social Evolution,* ed. D. I. Rubenstein and R. W. Wrangham, pp. 201–217. Princeton, N. J.: Princeton University Press.

Murray, P. (1975). The role of cheek pouches in cercopithecine monkey adaptive strategy. In *Primate Functional Morphology and Evolution,* ed. R. H. Tuttle, pp. 151–194. The Hague: Mouton.

Struhsaker, T. T. (1969). Correlates of ecology and social organization among African cercopithecines. *Folia Primatol.* **11**:80–118.

Struhsaker, T. T., and Leland, L. (1979). Socioecology of five sympatric monkey species in the Kibale Forest, Uganda. In *Advances in the Study of Behavior,* vol. 9, pp. 159–228. New York: Academic Press.

Macaques

Aldrich-Blake, F. P. G. (1980). Long-tailed macaques. In *Malayan Forest Primates: Ten Years' Study in the Tropical Forest,* ed. D. J. Chivers, pp. 147–167. New York: Plenum Press.

Caldecott, J. O. (1981). Findings on the behavioral ecology of the pig-tailed macaque. *Malays. Appl. Biol.* **10**(2): 213–220.

——— (1986). An ecological and behavioral study of the pig-tailed macaque. *Contrib. Primatol.* **21**:1–259.

Caldecott, J. O., Feistner, A. T. C., and Gadsby, E. L. (1996). A comparison of ecological strategies of pig-tailed macaques, mandrills and drills. In *Evolution and Ecology of Macaque Societies,* ed. J. E. Fa and D. G. Lindberg, pp.73–97. Cambridge: Cambridge University Press.

Chivers, D. J., ed. (1980). *Malayan Forest Primates: Ten Years' Study in the Tropical Forest.* New York: Plenum Press.

Crockett, C. M., and Wilson, M. L. (1980). The ecological separation of *Macaca nemestrina* and *Macaca fascicularis* in Sumatra. In *The Macaques. Studies in Ecology, Behavior and Evolution,* ed. D. G. Lindberg, pp. 148–181. New York: Van Nostrand–Reinhold.

Delson, E. (1980). Fossil macaques, phyletic relationships and a scenario of deployment. In *The Macaques: Studies in Ecology, Behavior and Evolution,* ed. D. G. Lindberg, pp. 10–30. New York: Van Nostrand–Reinhold.

Dittus, W. (1988). Group fission among wild toque macaques as a consequence of female resource competition and environmental stress. *Anim. Behav.* **36**: 1626–1647.

Fa, J. E., and Lindberg, D. G. (1996). *Evolution and Ecology of Macaque Societies.* Cambridge: Cambridge University Press.

Fittinghoff, N. A., Jr., and Lindberg, D. G. (1980). Riverine refuging in eastern Bornean *Macaca fascicularis.* In *The Macaques: Studies in Ecology, Behavior and Evolution,* ed. D. G. Lindberg, pp. 182–214. New York: Van Nostrand–Reinhold.

Fleagle, J. G. (1980). Locomotion and posture. In *Malayan Forest Primates: Ten Years' Study in the Tropical Forest,* ed. D. J. Chivers, pp. 191–207. New York: Plenum Press.

Fooden, J. (1980). Classification and distribution of living macaques (*Macaca* Lacepede, 1799). In *The Macaques: Studies in Ecology, Behavior and Evolution,* ed. D. G. Lindberg, pp. 1–9. New York: Van Nostrand–Reinhold.

Fooden, J., and Lanyon, S. M. (1989). Blood protein allele frequencies and phylogenetic relationships in *Macaca:* A review. *Am. J. Primatol.* **17**:209–241.

Furuya, Y. (1965). Social organization of the crab-eating monkey. *Primates* **6**:285–337.

Goldstein, S. J., and Richard, A. F. (1989). Ecology of rhesus macaques (*Macaca mulatta*) in northeast Pakistan. *Int. J. Primatol.* **10**:531–568.

Hartman, C. G., and Straus, W. L., eds. (1933). *The Anatomy of the Rhesus Monkey.* New York: Haffner.

Hoelzer, G. A., and Melnick, D. J. (1996). Evolutionary relationships of the macaques. In *Evolution and Ecology of Macaque Societies,* ed. J. E. Fa and D. G. Lindberg, pp. 3–19. Cambridge:Cambridge University Press.

Iwamoto, M. (1994). Morphology of the Japanese macaque. *Anthropological Science* **102**(suppl.):1–207.

Kawai, M., and Ohsawa, H. (1983). Ecology of Japanese monkeys, 1950–1982. In *Recent Progress of Natural Sciences in Japan,* vol. 8, ed. EDITOR(S), pp. 95–108. Tokyo: Sci. Council of Japan.

Kurland, J. A. (1973). A natural history of Kra macaques (*M. fascicularis* Raffies, 1821) at the Kutai Reserve, Kalimantan, Timur, Indonesia. *Primates* **14**:245–262.

Lindberg, D. G. (1971). The rhesus monkey in northern India: An ecological and behavioral study. In *Primate Behavior: Developments in Field and Lab Research,* vol. 2, ed. A. Rosenblum, pp. 2–107. New York. Academic Press.

———, ed. (1980). *The Macaques: Studies in Ecology, Behavior and Evolution.* New York: Van Nostrand–Reinhold.

MacKinnon, J. R., and MacKinnon, K. S. (1980). Niche differentiation in a primate community. In *Malayan Forest Primates: Ten Years' Study in the Tropical Forest,* ed. D. J. Chivers, pp. 167–190. New York: Plenum Press.

Melnick, D. J., and Hoelzer, G. A. (1992). Differences in male and female macaque dispersal lead to contrasting distributions of nuclear and mitochondrial DNA variation. *Int. J. Primatol.* **13**:379–393.

Menard, N., Hecham, R., Vallet, D., Chikki,H., and Gautier-Hion, A. (1990). Grouping patterns of a mountain population of *Macaca sylvanus* in Algeria—A fission–fusion system? *Folia Primatol.* **55**:166–177.

Menzel, C. (1996). Structure-guided foraging in long-tailed macaques. *Amr. J. Primatol.* **38**:117–132.

Richard, A. E, Goldstein, S. J., and Dewar, R. E. (1989). Weed macaques: The evolutionary implications of macaque feeding ecology. *Int. J. Primatol.* **10**: 569–594.

Rodman, P. S. (1991). Structural differentiation of microhabitats of sympatric *Macaca fascicularis* and *M. nemestrina* in East Kalimantan, Indonesia. *Int. J. Primatol.* **12**: 357–377.

Simonds, P. (1965). The bonnet macaques of south India. In *Primate Behavior,* ed. I. DeVore, pp. 175–197. New York: Holt, Rinehart and Winston.

Suzuki, A. (1965). An ecological study of wild Japanese monkeys in snowy areas: Focused on their food habits. *Primates* **6**:31–72.

Taub, D. M. (1977). Geographic distribution and habitat diversity of the Barbary macaque, *Macaca sylvanus* L. *Folia Primatol.* **27**:108–133.

van Noordwijk, M. A. (1985). *The socio-ecology of sumatran long-tailed macaques (Macaca fascicularis), II: The behavior of individuals.* Ph.D. Dissertation. Utrecht: Rijksuniversiteit.

van Noordwijk, M. A., and van Schaik, C. P. (1988). Male careers in Sumatran long-tailed macaques (*Macaca fascicularis*). *Behavior* **107**:24–43.

van Schaik, C. D. (1985). *The socio-ecology of Sumatran long-tailed macaques (Macaca fascicularis), I: Costs and benefits of group living.* Ph.D. Dissertation. Utrecht: Rijksuniversiteit.

Wheatley, B. P. (1980). Feeding and ranging of East Bornean *Macaca fascicularis.* In *The Macaques: Studies in Ecology, Behavior and Evolution,* ed. D. G. Lindberg, pp. 215–247. New York: Van Nostrand–Reinhold.

Yaeger, C. P. (1996). Feeding ecology of the long-tailed macaque (*Macaca fascicularis*) in Kalimantan, Tengah, Indonesia. *Int. J. Primatol.* **17**:51–62.

Zhao, Q., Deng, Z., and Xu, J. (1991). Natural foods and their ecological implications for *Macaca thibetana* at Mount Emei, China. *Folia Primatol.* **57**:1–17.

Mangabeys

Chalmers, N. R. (1968). Group composition, ecology and daily activity of free-living mangabeys in Uganda. *Folia Primatol.* **8**:247–262.

Cronin, J. E., and Sarich, V. M. (1976). Molecular evidence for dual origin of mangabeys among Old World monkeys. *Nature* **260**:700–702.

Disotell, T. R. (1994). Generic level relationships of the Papionini (Cercopithecoidea) *Am. J. Phys. Anthropol.* **94**:47–57.

______ (1996). The phylogeny of Old World monkeys. *Evol. Anthropol.* **5**:18–24.

Fleagle, J. G., and McGraw, W. S. (1998). Skeletal anatomy of African Papionins: Function, phylogeny or both? *Amer. J. Phys. Anthropol.* Suppl. **26**:82–83.

Freeland, W. J. (1979). Mangabey (*C. albigena*) social organization and population density in relation to food use and availability. *Folia Primatol.* **32**:108–124.

Groves, C. P. (1978). Phylogenetic and population systematics of the mangabey (Primates: Cercopithecoidea). *Primates* **19**:1–34.

Homewood, K. M. (1978). Feeding strategy of the Tana mangabey (*C. galeritus galeritus*). J. Zool (London) **186**: 375–392.

McGraw, W. S. (1996). Cercopithecid locomotion, support use, and support availability in the Tai Forest, Ivory Coast. *Am. J. Phys. Anthropol.* **100**:507–522.

Mitani, M. (1989). *Cercocebus torquatus:* Adaptive feeding and ranging behaviors related to seasonal fluctuations of food resources in the tropical rain forest of southwestern Cameroon. *Primates* **30**:307–323.

______ (1991). Niche overlap and polyspecific associations among sympatric Cercopithecids in the Campo Animal Reserve, southwestern Cameroon. *Primates* **32**: 137–157.

Nakatsukasa, M. (1994). Morphology of the humerus and femur in African mangabeys and guenons: Functional adaptation and implications for the evolution of positional behavior. *Afr. Study Monographs Suppl* **21**:1–61.

______ (1996). Locomotor differentiation and different skeletal morphologies in mangabeys (*Lophocebus and Cercocebus*). *Folia primatol.* **66**:15–24.

Olupot, W., Chapman, C., Brown, C. H., and Waser, P. (1994). Mangabey (*Cercocebus albigena*) population density, group size and ranging: a twenty-year comparison. *Amer. J. Primatol.* **32**:197–205.

Quiris, R. (1975). Ecologie et organisation sociale de *Cercocebus galeritus agilis* dans le Nord-Est du Gabon. *Terre Vie* **29**:337–398.

Waser, P. (1977). Feeding, ranging and group size in the mangabey *C. albigena*. In *Primate Ecology: Studies of Feeding and Ranging Behavior in Lemurs, Monkeys and Apes,* ed. T. H. Clutton-Brock, pp. 183–222. New York: Academic Press.

______ (1984). Ecological differences and behavioral contrasts between two mangabey species. In *Adaptations for Foraging in Non-human Primates,* ed. P. S. Rodman and J. G. H. Cant, pp. 195–217. New York: Columbia University Press.

Baboons

Alberts, S. C., and Altmann, J. (1995). Balancing costs and opportunities: Dispersal in male baboons. *Am. Nat.* **145**:279–307.

Altmann, J. (1980). *Baboon mothers and infants.* Cambridge, Mass.: Harvard University Press.

Altmann, S. A. (1974). Baboons, space, time and energy. *Am. Zool.* **14**:221–248.

Altmann, S. A., and Altmann, J. (1970). *Baboon Ethology.* Chicago: University of Chicago Press.

______ (1977). Life history of yellow baboons: Physical development, reproductive parameters and infant mortality. *Primates* **18**(2):315–330.

Barton, R. A., Whiten, A., Strum, S. C., Byrne, R. W., and Simpson, A. J. (1992). Habitat use and resource availability in baboons. *Anim. Behav.* **43**:831–844.

Barton, R. A., Byrne, R. W., and Whiten, A. (1996). Ecology, feeding competition, and social structure in baboons. *Behav. Ecol Sociobiol.* **38**:321–329.

Berkovitch, F. B. (1987). Reproductive success in male savanna baboons. *Behav. Ecolog. Sociobiol.* **21**:163–172.

______ (1990). Female choice, male reproductive success, and the origin of one male groups in baboons. In *Baboons: Behavior and Ecology, Use and Care,* ed. M. T. deMello, A. Whiten, and R. W. Byrne, pp.61–77. Brasilia: Sel. Proc. XII Cong. Int. Primat. Soc.

Byrne, R. W., Whiten, A., and Henzi, S. P. (1990). Social Relationships and affiliation in a non-female-bonded monkey. *Am. J. Primatol.* **20**:313–329.

deMello, M. T., Whiten A., and Byrne, R. W. (1990). *Baboons: Behavior and Ecology, Use and Care,* Brasilia: Sel. Proc. XII Cong. Int. Primat. Soc.

DeVore, I., and Hall, K. R. L. (1965). Baboon ecology. In *Primate Behavior,* ed. I. DeVore, pp. 20–52. New York: Holt, Rinehart and Winston.

DeVore, I., and Washburn, S. (1962). Baboon ecology and human evolution. In *African Ecology and Human Evolution.* Viking Fund Publ. in Anthropology, no. 36, pp. 335–367.

Hall, K. R. L. (1962). Numerical data, maintenance activities and locomotion of the wild Chacma baboon, *Papio ursinus. Proc. Zool. Soc. London* **139**:181–220.

Harding, R. S. O. (1975). Meat eating and hunting in baboons. In *Paleoanthropology, Morphology and Paleoecology,* ed. R. H. Tuttle, pp. 245–258. The Hague: Mouton.

Hausfater, G. (1975). Dominance and reproduction in baboons (*Papio cynocephalus*): A quantitative analysis.

In *Contributions to Primatology,* vol. 7. Basel, Switzerland: Karger.

Jablonski, N. G., ed.(1993). *Theropithecus: The Rise and Fall of a Primate Genus.* Cambridge: Cambridge University Press.

Jolly, C. J. (1970). The large African monkeys as an adaptive array. In *Old World Monkeys,* ed. J. R. Napier and P. H. Napier, pp. 139–174. New York: Academic Press.

——— (1993). Species, subspecies, and baboon systematics. In *Species, Species Concepts, and Primate Evolution,* ed. W. H. Kimbel and L. B. Martin, pp. 67–107. New York: Plenum Press.

Kummer, H. (1968). *Social Organization of Hamadryas Baboons.* Chicago: University of Chicago Press.

Maples, W. R. (1977). Differential habitat utilization of four Cercopithecidae in a Kenyan forest. *Folia Primatol.* **27**:85–107.

Nagel, U. (1973). A comparison of Anubis baboons, Hamadryas baboons and their hybrids at a species border in Ethiopia. *Folia Primatol.* **19**:104–167.

Noe, R., and Sluijter, A. A. (1990). Reproductive tactics of male savanna baboons. *Behavior* **113**:117–170.

——— (1995). Which adult male savanna baboons form coalitions? *Int. J. Primtol.* **16**:77–105

Popp, J. (1983). Ecological determinism in the life histories of baboons. *Primates* **24**(2):198–210.

Ransom, T. (1981). *Beach Troop of the Gombe.* London: Bucknell University Press.

Rose, M. D. (1976). Bipedal behavior of olive baboons (*Papio anubis*) and its relevance to an understanding of the evolution of human bipedalism. *Am J. Phys. Anthropol.* **44**:247–261.

Rowell, T. (1966). Forest living baboons in Uganda. *J. Zool. (London)* **149**:344–364.

Smuts, B. B. (1985). *Sex and Friendship in Baboons.* Hawthorne, N. Y.: Aldine.

Stammbach, E. (1986). Desert, forest, and montane baboons: Multilevel societies. In *Primate Societies,* ed. B. B. Smuts, D. L. Cheney, R. M. Seyfarth, R. W. Wrangham, and T. T. Struhsaker, pp. 112–120. Chicago: University of Chicago Press.

Strum, S. C. (1981). Processes and products of change: Baboon predatory behavior at Gilgil, Kenya. In *Omnivorous Primates: Gathering and Hunting in Human Evolution,* ed. R. S. O. Harding and G. Teleki, pp. 255–302. New York: Columbia University Press.

——— (1987). *Almost Human.* New York: Random House.

Strum, S. C., and Mitchell, W. (1987). Baboons: Baboon models and muddles. In *The Evolution of Human Behavior: Primate Models,* ed. W. G. Kinzey, pp. 87–104. Albany, N. Y.: SUNY Press.

Washburn, S. L., and DeVore, I. (1961). The social life of baboons. *Sci. Am.* **204**(6):62–71.

Geladas

Dunbar, R. I. M. (1980). Demographic and life history variables of a population of gelada baboons (*Theropithecus gelada*). *J. Anim. Ecol.* **49**:485–507.

——— (1984). *Reproductive Decisions: An Economic Analysis of Gelado Baboon Social Strategies.* Princeton, N. J.: Princeton University Press.

——— (1986). The social ecology of gelada baboons. In *Ecological Aspects of Social Evolution,* ed. D. Rubenstein and R. W. Wrangham, pp. 332–351. Princeton, N. J.: Princeton University Press.

Iwamoto, T., Mori, A., Kawai, M., and Bekele, A. (1996). Anti-predator behaviors of Gelada baboons. *Primates* **37**:389–397.

Jablonski, N. G. (1993) *Theropithecus: The Rise and Fall of a Primate Genus.* Cambridge: Cambridge University Press.

Kawai, M., Dunbar, R. I. M., Ohsawa, H., and Mori, U. (1983). Social organization of gelada baboons: Social units and definitions. *Primates* **24**(1):13–24.

Mori, A., Iwamoto, T., and Bekele, A. (1997). A case of infanticide in a recently found Gelada population in Arsi, Ethiopia. *Primates* **38**:79–88.

Stammbach, E. (1986). Desert, forest, and montane baboons: Multilevel societies. In *Primate Societies,* ed. B. B. Smuts, D. L. Cheney, R. M. Seyfarth, R. W. Wrangham, and T. T. Struhsaker, pp. 112–120. Chicago: University of Chicago Press.

Mandrills and Drills

Caldecott, J. O., Feistner, A. T. C., and Gadsby, E. L. (1996). A comparison of ecological strategies of pigtailed macaques, mandrills and drills. In *Evolution and Ecology of Macaque Societies,* ed. J. E. Fa and D. G. Lindberg, pp. 73–97. Cambridge:Cambridge University Press.

Disotell, T. R. (1994). Generic level relationships of the *Papionini* (Cercopithecoidea) *Am. J. Phys. Anthropol.* **94**: 47–57.

——— (1996). The phylogeny of Old World monkeys. *Evol. Anthropol.* **5**:18–24.

Gartlan, J. S. (1970). Preliminary notes on the ecology and behavior of the drill (*Mandrillus leucophaeus* Ritgen). In *Old World Monkeys,* ed. J. R. Napier and P. H. Napier, pp. 445–480. New York: Academic Press.

Gartlan, J. S., and Struhsaker, T. T. (1972). Polyspecific association and niche separation of rain forest anthropoids in Cameroon, West Africa. *J. Zool. (London)* **168**: 221–267.

Grubb, P. (1973). Distribution, divergence and speciation of the drill and mandrill. *Folia Primatol.* **20**:161–177.

Hosino, J. (1985). Feeding ecology of Mandrills (*Mandrillus sphinx*) in Campo Animal Reserve, Cameroon. *Primates* **26**:248–273.

Hosino, J., Mori, A., Kudo, H., and Kawai, M. (1984). Preliminary report on the grouping of mandrills (*Mandrillus sphinx*) in Cameroon. *Primates* **25**:295–307.

Jouventin, P. (1975). Observations sur la socio-ecologie du mandrill. *Terre Vie* **29**:493–532.

Lahm, S. (1986). Diet and habitat preference of *Mandrillus sphinx* in Gabon: Implications of foraging strategy. *Am. J. Primatol.* **11**:9–27.

Rogers, M. E., Abernathy, K. A., Fontaine, B., Wickings, E. J., White, L. J. T., and. Tutin, C. E. G. (1996). Ten days in the life of a mandrill horde in the Lope Reserve, Gabon. *Am. J. Primatol.* **40**:297–314.

Stammbach, E. (1986). Desert, forest, and montane baboons: Multilevel societies. In *Primate Societies,* ed. B. B. Smuts, D. L. Cheney, R. M. Seyfarth, R. W. Wrangham, and T. T. Struhsaker, pp. 112–120. Chicago: University of Chicago Press.

Guenons

Butyinski, T. M. (1988) Guenon birth seasons and correlates with rainfall and food. In *A Primate Radiation: Evolutionary Biology of the African Guenons,* ed. A. Gautier-Hion, F. Bourliere, J-P. Gautier, and J. Kingdon, pp. 284–322. Cambridge: Cambridge University Press.

Cords, M. (1986). Forest guenons and patas monkeys: Male–male competition in one-male groups. In *Primate Societies,* ed. B. B. Smuts, D. L. Cheney, R. M. Seyfarth, R. W. Wrangham, and T. T. Struhsaker, pp. 98–111. Chicago: University of Chicago Press.

______ (1988). Mating systems of forest guenons: A preliminary review. In *A primate radiation: Evolutionary biology of the African guenons,* ed. A. Gautier-Hion, F. Bourliere, J-P. Gautier, and J. Kingdon, pp. 323–339. Cambridge: Cambridge University Press.

______ (1990). Mixed-species association of east African guenons: General patterns or specific examples? *Am. J. Primatol.* **21**:101–114.

Disotell, T. (1996). The phylogeny of Old World Monkeys. *Evol. Anthropol.* **5**:18–24.

Galat, G., and Galat-Luong, A. (1985). La communaute de primates diurnis de la foret de Tai, Cote d'Ivoire. *Terre Vie* **40**:3–32.

Gautier, J., and Gautier-Hion, A. (1969). Associations polyspecifiques chez les Cercopitheques du Gabon. *Terre Vie* **23**:164–201.

Gautier-Hion, A. (1975). Dimorphisme sexual et organisation sociale chez les cercopithecines forestiers africains. *Mammalia* **39**(3):365–374.

Gautier-Hion, A., and Gautier, J. P. (1976). Croissance, maturite sexuelle et sociale, reproduction chez les cercopithecines forestiers africains. *Folia Primatol.* **26**: 165–184.

______ (1978). Le singe de Brazza: Une strategie originale. *Z. Tierpsychol.* **46**:84–104.

______ (1979). Niche ecologique et diversite des especes sympatriques dans le genre *Cercopithecus. Terre Vie* **33**: 493–507.

______ (1988) The diet and dietary habits of forest guenons In *A primate radiation: Evolutionary biology of the African guenons,* ed. A. Gautier-Hion, F. Bourliere, J-P. Gautier, and J. Kingdon, pp. 257–283. Cambridge: Cambridge University Press.

Gautier-Hion, A., Quris, R., and Gautier, J. P. (1978). Food niches and coexistence in sympatric primates in Gabon. In *Recent Advances in Primatology,* vol. 1, ed. D. J. Chivers and J. Herbert, pp. 269–287. London: Academic Press.

______ (1983). Monospecific vs. polyspecific life: A comparative study of foraging and antipredatory tactics in a community of *Cercopithecus* monkeys. *Behav. Ecol. Sociobiol.* **12**:325–337.

Gautier-Hion, A., Bourliere, F., Gautier, J-P. and Kingdon, J. (1988). *A Primate Radiation: Evolutionary Biology of the African Guenons.* Cambridge: Cambridge University Press.

Gebo, D. L., and Chapman, C. A. (1995). Habitat, annual, and seasonal effects on positional behavior in red colobus monkeys. *Am. J. Phys. Anthropol.* **96**:73–82.

Gebo, D.L., and Sargis, E. J. (1994). Terrestrial adaptations in the postcranial skeletons of guenons. *Am. J. Phys. Anthropol.* **93**:341–371.

Haddow, A. J. (1952). Field and laboratory studies of an African monkey, *Cercopithecus ascanius schmidti. Proc. Zool. Soc. London* **122**:297–394.

Hylander, W. L. (1975). Incisor size and diet in anthropoids with special reference to Cercopithecoidea. *Science* **189**:1095–1098.

Kay, R. F. (1978). Molar structure and diet in extant Cercopithecoidae. In *Development, Function and Evolution of Teeth,* ed. P. M. Butler and K. A. Joysey, pp. 309–339. New York: Academic Press.

Kay, R. F., and Hylander, W. F. (1978). The dental structure of mammalian folivores with special reference to primates and phalangeroids (Marsupialia). In *The*

Ecology of Arboreal Folivores, ed. G. G. Montgomery, pp. 173–192. Washington, D. C.: Smithsonian Institution Press.

Kingdon, J. (1971). *East African Mammals,* vol. 1. New York: Academic Press.

McGraw. W. S. (1996). Cercopithecid locomotion, Support Use, and Support Availability in the Tai Forest, Ivory Coast *Am. J. Phys. Anthropol.* **100**:507–522.

Oates, J. F., and Anadu, P. A. (1989). A field observation of Sclater's guenon (*Cercopithecus sclateri* Pocock, 1904). *Folia Primatol.* **52**:93–97.

Rowell, T. E. (1988). The social system of guenons, compared with baboons, macaques and mangabeys. In *A Primate Radiation: Evolutionary Biology of the African Guenons,* eds. A. Gautier-Hion, F. Bourliere, P.-P. Gauthier, and J. Kingdon, pp. 439–451. Cambridge: Cambridge University Press.

Rudran, R. (1978). Socioecology of the blue monkeys (*Cercopithecus mitis stuhlmanni*) of the Kibale Forest, Uganda. *Smithson. Contrib. Zool.* **249**:1–88.

Schultz, A. H. (1970). The comparative uniformity of the Cercopithecoidea. In *Old World Monkeys,* ed. J. R. Napier and P. H. Napier, pp. 39–52. New York: Academic Press.

Struhsaker, T. T. (1978). Food habits of five monkey species in the Kibale Forest, Uganda. In *Recent Advances in Primatology,* vol. 1, ed. D. J. Chivers and J. Herbert, pp. 225–248. London: Academic Press.

Wahome, J. M., Rowell, T. E., and Tsingalia, H. M. (1993). The natural history of deBrazza's monkey in Kenya. *Int. J. Primatol.* **14**:445–467.

Vervets

Cheney, D. L., and Setfarth, R. M. (1987). The influence of intergroup competition on the survival and reproductive success of female vervet monkeys. *Behav. Ecol. Sociobiol.* **21**:375–387.

Fedigan, L., and Fedigan, L. M. (1988). *Cercopithecus aethiops:* A review of field studies. In *A primate radiation: Evolutionary biology of the African guenons,* ed. A. Gautier-Hion, F. Bourliere, J-P. Gautier, and J. Kingdon, pp.383–411. Cambridge: Cambridge University Press.

Hall, K. R. L., and Gartlan, J. S. (1965). Ecology and behavior of the vervet monkeys, *Cercopithecus aethiops,* Lolui Island, Lake Victoria. *Proc. Zool. Soc. London* **145**: 37–57.

Isbell, L. A., Cheney, D. L., and Seyfarth, R. M. (1990). Costs and benefits of home range shifts among vervet monkeys (*Cercopithecus aethiops*) in Amboseli National Park, Kenya. *Behav. Ecol Sociobiol.* **27**:351–358.

Kavanagh, M. (1978). The diet and feeding behavior of *Cercopithecus aethiops tantalus. Folia Primatol.* **30**:30–63.

Struhsaker, T. T. (1967). Ecology of vervet monkeys (*Cercopithecus aethiops*) in the Masai-Amboseli Game Reserve, Kenya. *Ecology* **48**:891–904.

Talapoins

Gautier-Hion, A. (1966). L'ecologie et l'ethologie du talapoin (*Miopithecus talapoin talapoin*). *Revue Biol. Gabon* **2**:311-329.

——— (1970). L'organisation sociale d'une bande de talapoins dans le Nord-Est du Gabon. *Folia Primatol.* **12**: 116–141.

——— (1971). L'ecologie de talapoin du Gabon. *Terre Vie* **25**:427–490.

——— (1973). Social and ecological features of talapoin monkeys—comparisons with sympatric cercopithecines. In *Comparative Ecology and Behavior of Primates,* ed. R. P. Michael and J. H. Crook, pp. 147–170. New York: Academic Press.

Gautier-Hion, A., Bourliere, F., Gautier, J-P., and Kingdon, J. (1988). *A Primate Radiation: Evolutionary Biology of the African Guenons.* Cambridge: Cambridge University Press.

Rowell, T. E. (1972). Toward a natural history of the talapoin monkey in Cameroon. *Ann. Fac. Sci. Cameroun* **10**: 121–134.

——— (1973). Social organization of wild talapoin monkeys. *Am. J. Phys. Anthropol.* **38**(2):593–598.

Rowell, T. E., and Dixon, A. F. (1975). Changes in social organization during the breeding season of wild talapoin monkeys. *J. Reprod. Fert.* **43**:419–434.

Shea, B. (1992). Ontogenetic scaling of skeletal proportions in the talapoin monkey. *J. Human Evol.* **23**:283–307.

Swamp Monkeys

Zeeve, S. R. (1985). Swamp monkeys of the Lomako Forest, Central Zaire. *Primate Conservation* **5**:32–33.

Patas Monkeys

Chism, J., and Rowell, T. E. (1988). The natural history of patas monkeys. In *A Primate Radiation: Evolutionary Biology of the African Guenons,* ed. A. Gautier-Hion, F. Bourliere, J-P. Gautier, and J. Kingdon, pp. 412–438. Cambridge: Cambridge University Press.

Chism, J., Rowell, T. E., and Olson, D. (1984). Life history patterns of female patas monkeys. In *Female Primates: Studies by Women Primatologists,* ed. M. E Small, pp. 175–190. New York: Alan R. Liss.

Gartlan, S. J. (1974). Adaptive aspects of social structure in *Erythrocebus patas. Symp. Fifth Cong. Int. Primatol. Soc.,* pp. 161–171.

Hall, K. R. L. (1965). Behavior and ecology of the wild patas monkey, *Erythrocebus patas,* in Uganda. *J. Zool.* **148**:15–87.

Kingdon, J. (1971). *East African Mammals,* vol. 1. New York: Academic Press.

COLOBINES

Bauchop, T. (1978). Digestion of leaves in vertebrate arboreal folivores. In *The Ecology of Arboreal Folivores,* ed. G. G. Montgomery, pp. 193–204. Washington, D. C.: Smithsonian Institution Press.

Brandon-Jones, D. (1984). Colobus and leaf monkeys, In *The Encyclopedia of mammals,* ed. D. Macdonald, pp. 398–408. London:Allen and Unwin.

Davies, A. G., and Oates, J. F. (1994). *Colobine Monkeys: Their Ecology, Behavior and Evolution.* Cambridge: Cambridge University Press.

Delson, E. (1994). Evolutionary history of the colobine monkeys in paleoenvironmental perspective In *Colobine Monkeys: Their Ecology, Behavior and Evolution,* ed. A. G. Davies and J. F. Oates, pp. 11–44. Cambridge: Cambridge University Press.

Oates, J. F., and Davies, A. G. (1994). What are the colobines? In *Colobine Monkeys: Their Ecology, Behavior and Evolution Davies,* ed. A. G. Davies and J. F. Oates, pp. 1–10. Cambridge: Cambridge University Press.

Oates, J. F., Davies, A. G., and Delson, E. (1994). The Diversity of living colobines. In *Colobine Monkeys: Their Ecology, Behavior and Evolution,* ed. A. G. Davies and J. F. Oates, pp. 1–10. Cambridge: Cambridge University Press.

Struhsaker, T. T. (1986). Colobines: Infanticide by adult males. In *Primate Societies,* ed. B. B. Smuts, D. L. Cheney, R. M. Seyfarth, R. W. Wrangham, and T. T. Struhsaker, pp. 83–97. Chicago: University of Chicago Press.

Vogel, C. (1966). Morphologische Studien am Gesichtsschadel catarriner Primaten. *Bibl. Primatol. fasc.* **4,** Basel.

______ (1968). The phylogenetical evolution of some characteristics and some morphological trends in the evolution of the skull in catarrhine primates. In *Taxonomy and Phylogeny of Old World Primates with Reference to the Origin of Man,* ed. A. B. Chiarelli, pp. 21–57. Turin: Rosenberg and Sellier.

Guerezas

Clutton-Brock, T. H. (1975). Feeding behavior of red colobus and black and white colobus in East Africa. *Folia Primatol.* **23**:165–208.

Dasilva, G. (1994). Diet of Colobus polykomos on Tiwai island: Selection of food in relation to its seasonal abundance and nutritional quality. *Int. J. Primatol.* **15**: 655–680.

Hull, D. B. (1979). A craniometric study of the black-and-white colobus, *Colobus guereza* Illiger 1811 (Primates, Cercopithecoidea). *Am. J. Phys. Anthropol.* **51**:163–182.

Mittermeier, R. A., and Fleagle, J. G. (1976). The locomotor and postural repertoires of *Ateles geoffroy* and *Colobus guereza,* and a re-evaluation of the locomotor category of semibrachiation. *Am. J. Phys. Anthropol.* **45**: 235–257.

Morbeck, M. E. (1977). Positional behavior, selective use of habitat substrate and associated non-positional behavior in free-ranging *Colobus guereza* (Ruppel 1835). *Primates* **18**(1):35–58.

Oates, J. F. (1977a). The guereza and its food. In *Primate Ecology: Studies of Feeding and Ranging Behavior in Lemurs, Monkeys and Apes,* ed. C. H. Clutton-Brock, pp. 275–321. New York: Academic Press.

______ (1977b). The social life of a black-and-white colobus monkey, *C. guereza. Z. Tierpsychol.* **45**:1–60.

______ (1994). The natural history of African colobines. In *Colobine Monkeys: Their Ecology, Behavior and Evolution,* ed. A. G. Davies and J. F. Oates, pp. 75–128. Cambridge: Cambridge University Press.

Oates, J. F., and Trocco, T. F. (1983). Taxonomy and phylogeny of black-and-white colobus monkeys. *Folia Primatol.* **40**:83–113.

Rose, M. D. (1978). Feeding and associated positional behavior of black-and-white colobus monkeys (*Colobus guereza*). In *The Ecology of Arboreal Folivores,* ed. G. G. Montgomery, pp. 253–264. Washington, D. C.: Smithsonian Institution Press.

Struhsaker, T. T., and Oates, J. F. (1975). Comparison of the behavior and ecology of red colobus and black and-white colobus monkeys in Uganda: A summary. In *Socioecology and Psychology of Primates,* ed. R. H. Tuttle, pp. 103–124. The Hague: Mouton.

von Hippel, F. (1996). Interactions between overlapping multiple groups of black and white colobus monkeys (*Colobus guereza*) in the Kakamega Forest, Kenya. *Am. J. Primatol.* **38**:193–209.

Black Colobus

Harrison, M. J. S. (1986). Feeding ecology of black colobus (*Colobus satanas*) in central Gabon. In *Primate Ecology and Conservation,* ed. J. G. Else and P. C. Lee, pp. 31–38. Cambridge: Cambridge University Press.

McKey, D. B. (1978). Soils, vegetation and seed eating by black colobus monkeys. In *The Ecology of Arboreal*

Folivores, ed. G. G. Montgomery, pp. 423–437. Washington, D. C.: Smithsonian Institution Press.

McKey, D. B., and Gartlan, J. S. (1981). Food selection by black colobus monkeys (*C. satanas*) in relation to plant chemistry. *Biol. J. Linn. Soc.* **16**:115–147.

McKey, D. B., Waterman, P. G., Mbi, C. N., Gartlan, J. S., and Struhsaker, T. T. (1978). Phenolic content of vegetation in two African rain forests: Ecological implications. *Science* **202**:61–64.

Red Colobus

Clutton-Brock, T. H. (1973). Feeding levels and feeding sites of red colobus (*Colobus badius tephrosceles*) in the Gombe National Park. *Folia Primatol.* **219**:368–379.

______ (1975a). Feeding behavior of red colobus and black and white colobus in East Africa. *Folia Primatol.* **23**:165–208.

______ (1975b). Ranging behavior of red colobus monkeys. *Anim. Behav.* **23**:706–722.

Gebo, D. L., and Chapman, C. A. (1995). Habitat, annual, and seasonal effects on positional behavior in red colobus monkeys. *Am. J. Phys. Anthropol.* **96**:73–82.

Marsh, C. W. (1981). Time budget of Tana River red colobus. *Folia Primatol.* **35**:30–50.

Noe, R. and Bshary, R. (1997). The formation of red colobus-dinna monkey associations under pressure from chimpanzees. *Proc. Roy. Soc. B.* **264**:253–259.

Stanford, C. B., Wallis, J., Matama, H., and Goodall, J. (1994). Patterns of predation by chimpanzees on red colobus monkeys in Gombe National Park, 1982–1991. *Am. J. Phys. Anthro.* **94**:213–228.

Starin, E. D. (1981). Monkey moves. *Nat. Hist.* **90**(9):36–43.

Struhsaker, T. T. (1975). *The Red Colobus Monkey.* Chicago: University of Chicago Press.

______ (1978). Food habits of five monkey species in the Kibale Forest, Uganda. In *Recent Advances in Primatology,* vol. 1, ed. D. J. Chivers and J. Herbert, pp. 229–248. London: Academic Press.

Struhsaker, T. T., and Oates, J. F. (1975). Comparison of the behavior and ecology of red colobus and black-and-white colobus monkeys in Uganda: A summary. In *Socioecology and Psychology of Primates,* ed. R. H. Tuttle, pp. 103–124. The Hague: Mouton.

Struhsaker, T. T., and Pope, T. R. (1991). Mating syatem and reproductive success: A comparison of two African forest monkeys (*Colobus badius* and *Cercopithecus ascanius*). *Behaviour* **117**:182–204.

Olive Colobus

Booth, A. H. (1957). Observations on the natural history of the olive colobus monkey, *Procolobus verus* (Van Beneden). *Proc. Zool Soc. London* **129**:421–430.

Galat, G., and Galat-Luong, A. (1985). La communaute de primates diurnis de la foret de Tai, Cote d'Ivoire. *Terre Vie* **40**:3–32.

McGraw. W. S. (1996). Cercopithecid locomotion, support use, and support availability in the Tai Forest, Ivory Coast. *Am. J. Phys. Anthropol.* **100**:507–522.

Oates, J. F., and Whitesides, G. H. (1990). Assciation between olive colobus (*Procolobus verus*), Diana guenons (*Cercopithecus diana*) and other forest monkeys in Sierra Leone. *Am. J. Primatol.* **21**:129–147.

Asian Colobines

Bennet, E. L., and Davies, A. G. (1994) The ecology of Asian colobines In *Colobine Monkeys: Their Ecology, Behavior and Evolution Davies,* ed. A. G. Davies and J. F. Oates, pp. 129–174. Cambridge: Cambridge University Press.

Brandon-Jones, D. (1978). The evolution of the recent Asian Colobinae. In *Recent Advances in Primatology,* vol. 1, ed. D. J. Chivers and J. Herberi, pp. 323–327. London: Academic Press.

______ (1984) Colobus and leaf monkeys. In *The Encyclopedia of Mammals,* vol. 1, ed. D. MacDonald, pp. 398–408. London: Allen and Unwin. *All the World's Primates: Primates,* ed. D. McDonald, pp.102–112. New York: Torstar Books.

______ (1993). The taxonomic affinities of the Mentawi Islands Sureli, *Presbytis potenziani* (Bonaparte, 1856) (Mammalia: Primata: Cercopithecidae). *Raffles Bull. Zool.* **41**:331–357.

______ (1995). A revision of the Asian pied leaf monkeys (Mammalia: Cercopithecidae: superspecies *Semnopithecus auratus*), with a description of a new subspecies. *Raffles Bull. Zool.* **43**:3–43.

______ (1996). The Asian Colobinae (Mammalia: Cercopithecidae) as indicators of Quaternary climatic change. *Biological J. Linnaen Soc.* **59**:327–350.

Collura, R. V. and Stewart, C-B. (1996). Mitochondrial DNA phylogeny of the colobine Old World monkeys. *Abstracts XVIth Cong. Int. Primatol Soc.*

Pocock, R. I. (1935). On monkeys of the genera *Pithecus* (or *Presbytis*) and *Pygathrix* found to the east of Bengal. *Proc. Zool. Soc. London* (1935):895–961.

Hanuman Langurs

Bishop, N. H. (1979). Himalayan langurs: Temperate colobines: *J. Human Evol.* **8**:251–281.

Borries, C. (1993). Ecology of female social relationships: Hanuman langurs (*Presbytis entellus*) and the van Schaik model. *Folia Primatol.* **61**:21–30.

Borries, C., Sommer, V., and Srivastava, A. (1994). Weaving a tight social net: Allogrooming in free-ranging female langurs (*Presbytis entellus*). *Int. J. Primatol.* **15**: 421–443.

Hladik, C. M. (1977). A comparative study of the feeding strategies of two sympatric species of leaf monkeys: *Presbytis senex* and *Presbytis entellus*. In *Primate Ecology: Studies of Feeding and Ranging Behavior in Lemurs, Monkeys and Apes,* ed. C. H. Clutton-Brock, pp. 324–353. New York: Academic Press.

Hrdy, H. S. (1974). Male–male competition and infanticide among the langurs (*Presbytis entellus*) of Abu, Rajastan. *Folia Primatol.* **22**:19–58.

______ (1977). *The Langurs of Abu.* Cambridge, Mass.: Harvard University Press.

Jay, P. C. (1965). The common langur of northern India. In *Primate Behavior: Field Studies of Monkeys and Apes,* ed. I. DeVore, pp. 197–249. New York: Holt, Rinehart and Winston.

Newton, P. (1994). Feeding and ranging patterns of forest Hanuman langurs (*Presbytis entellus*). *Int. J. Primatol.* **13**:245–287.

Ripley, S. (1967). The leaping of langurs: A problem in the study of locomotor behavior. *Am. J. Phys. Anthropol.* **26**:149–170.

______ (1970). Leaves and leaf-monkeys: Social organization of foraging. In *Old World Monkeys,* ed. J. R. Napier and P. H. Napier, pp. 483–509. New York: Academic Press.

Sommer, V., and Mohnot, S. M. (1984). New observations on infanticide among Hanuman langurs (*Presbytis entellus*) near Jodhpur (Rajasthan, India). *Behav. Ecol. Sociobiol.* **16**:245–248.

Sommer, V., and Rajpurohit, L. S. (1989). Male reproductive success in harem troops of Hanuman langurs (*Presbytis entellus*) *Int. J. Primatol.* **10**:293–317.

Sugiyama, Y. (1964). Group composition, population density and some sociological observations of Hanuman langurs (*Presbytis entellus*). *Primates* **5**(3–4):7–48.

______ (1965). On the social change of Hanuman langurs (*Presbytis entellus*) in their natural condition. *Primates* **6**(3–4):381–418.

______ (1976). Characteristics of the ecology of the Himalayan langurs. *J. Human Evol.* **5**:249–277.

Yoshiba, K. (1968). Local and intertroop variability in ecology and social behavior of common Indian langurs. In *Primates: Studies in Adaptation and Variability,* ed. P. Jay, pp. 217–242. New York: Holt, Rinehart and Winston.

Purple-faced and John's Langurs

Collura, R. V., and Stewart, C.-B. (1996). Mitochondrial DNA phylogeny of the colobine Old World monkeys. *Abstracts XVIth Cong. Int. Primatol Soc.*:727.

Eisenberg, J. E, Muckenhirn, N. A., and Rudran, R. (1972). The relation between ecology and social structure in primates. *Science* **176**:863–874.

Hladik, C. M. (1977). A comparative study of the feeding strategies of two sympatric species of leaf monkeys: *Presbytis senex* and *Presbytis entellus*. In *Primate Ecology: Studies of Feeding and Ranging Behavior in Lemurs, Monkeys and Apes,* ed. C. H. Clutton-Brock, pp. 324–353. New York: Academic Press.

Hladik, C. M., and Hladik, A. (1972). Disponsibilites alimentaire et domaines vitaux des primates a Ceylan. *Terre Vie* **26**:149–217.

Poirier, F. E. (1970). The Nilgiri langur (*P. Johnii*) of South India. In *Primate Behavior: Developments in Field and Lab Research,* vol. 1, ed. L. A. Rosenblum, pp. 251–383. New York: Academic Press.

Rudran, R. (1973). Adult male replacement in one-male troops of purple-faced langurs (*Presbytis senex senex*) and its effect on population structure. *Folia Primatol.* **19**:166–192.

Leaf Monkeys

Bennet, E. L. (1986). Environmental correlates of ranging behavior in the banded langur, *Presbytis melalophos. Folia Primatol.* **47**:26–38.

Bernstein, L. S. (1968). The lutong of Kuala Selangor. *Behaviour* **32**:1–17.

Curtin, S. H. (1977). Niche separation in sympatic Malaysian leaf-monkeys (*Presbytis obscura* and *Presbytis melalophos*). *Yrbk. Phys. Anthropol.* **20**:421–439.

______ (1980). Dusky and banded leaf-monkeys. In *Malayan Forest Primates: Ten Years' Study in the Tropical Forest,* ed. D. J. Chivers, pp. 107–147. New York: Plenum Press.

Curtin, S. H., and Chivers, D. J. (1978). Leaf-eating primates of peninsula Malaysia: The siamang and the dusky leaf-monkey. In *The Ecology of Arboreal Folivores,* ed. G. G. Montgomery, pp. 441–464. Washington, D. C.: Smithsonian Institution Press.

Davies, A. G. (1987). Adult male replacement and group formation in *Presbytis rubicunda. Folia prilatol.* **49**:11–114.

Fleagle, J. G. (1977a). Locomotor behavior and muscular anatomy of sympatric Malaysian leaf-monkeys (*Presbytis obscura* and *Presbytis melalophos*). *Am. J. Phys. Anthropol.* **46**:297–308.

______ (1977b). Locomotor behavior and skeletal anatomy of sympatric Malaysian leaf-monkeys. *Yrbk. Phys. Anthropol.* **20**:440–453.

______ (1978). Locomotion, posture and habitat use of two sympatric leaf-monkeys in West Malaysia. In *Recent Advances in Primatology,* vol. 1, ed. D. J. Chivers and J. Herbert, pp. 331–337. London: Academic Press.

Fooden, J. (1971). Report on the primates collected in western Thailand, January–April, 1967. *Fieldiana Ser. Zool.* **59**:62.

——— (1976). Primates obtained in peninsular Thailand June–July, 1973, with notes on the distribution of continental Southeast Asian leaf-monkeys (*Presbytis*). *Primates* **17**:95–118.

Furuya, Y. (1962). The social life of silvered leaf monkeys (*Trachypithecus cristatus*). *Primates* **3**:41–60.

MacKinnon, J. R., and MacKinnon, K. S. (1978). Comparative feeding ecology of six sympatric primates in western Malaysia. In *Recent Advances in Primatology,* vol. 1, ed. D. J. Chivers and J. Herbert, pp. 309–321. London: Academic Press.

——— (1980). Niche differentiation in a primate community. In *Malayan Forest Primates,* ed. D. J. Chivers, pp. 167–190. New York: Plenum Press.

Mukherjee, R. P., and Saha, S. S. (1974). The golden langurs (*Presbytis geei* Khajuria 1956) of Assam. *Primates* **15**(4):327–340.

Sterck, E. H. M. (1998). Female dispersal, social organization, and infanticide in Langurs: Are they linked to human disturbance? *Amer. J. Primatol.* **44**:235–254.

Sterck, L. (1995). *Females, Foods and Fights.* Utrecht: Utrecht University.

Stott, K., Jr., and Selsor, C. J. (1961). Observations of the maroon leaf monkey in North Borneo. *Mammalia* **25**: 184–189.

Tilson, R. L. (1976). Infant coloration and taxonomic affinity of the Mentawai Islands leaf monkey, *Presbytis potenziani. J. Mammal.* **57**(4):766–769.

Tilson, R. L., and Tenaza, R. R. (1976). Monogamy and duetting in an Old World monkey. *Nature* **263**:230–231.

van Schaik, C. P., Assink, P. R., and Salafsky, N. (1992). Territorial behavior in Southeast Asian Langurs: resource defense or mate defense. *Amer. J. Primatol.* **26**: 233–242.

Washburn, S. L. (1944). The genera of Malaysian langurs. *J. Mammol.* **25**:289–294.

Wilson, C. C., and Wilson, W. L. (1977). Behavioral and morphological variation among primate populations in Sumatra. *Yrbk. Phys. Anthropol.* **20**:207–233.

Wilson, W. L., and Wlilson, C. C. (1975). Species-specific vocalizations and the determination of phylogenetic affinities of the *Preshytis aygula-melalophos* group in Sumatra. In *Contemporary Primatology,* ed. S. Kondo, M. Kawai, and A. Ehara, pp. 459–463. Basel, Switzerland: S. Karger.

Wolf, K. E., and Fleagle, J. G. (1977). Adult male replacement in a group of silvered leaf-monkeys (*Presbytis cristata*) at Kuala Selangor, Malaysia. *Primates* **18**: 949–957.

Proboscis Monkeys

Kawabe, M., and Mano, T. (1972). Ecology and behavior of the wild proboscis monkey (*Nasalis larvatus,* Wurmb) in Sabah, Malaysia. *Primates* **13**:213–228.

Yaeger, C. P. (1989). Feeding ecology of the proboscis monkey (*Nasalis larvatus*). *Int. J. Primatol.* **10**:497–530.

——— (1991). Proboscis monkey (*Nasalis larvatus*) social organization: Intergroup patterns of association. *Am. J. Primatol.* **23**:73–87.

——— (1992a). Changes in proboscis monkey (*Nasalis larvatus*) group size and density at Tanjung Puting National Park, Kalimantan Tenhgah, Indonesia. *Tropiocal Biodiversity* **1**:49–57.

——— (1992b). Proboscis monkey (*Nasalis larvatus*) social organization: Nature and possible functions of intergroup patterns of association. *Am. J. Primatol.* **26**: 133–137.

Simakobu Monkeys

Tenaza, R. R., and Fuentes, A. (1995). Monandrous social organization of pigtailed langurs (*Simias concolor*) in the Pagai Islands, Indonesia. *Int. J. Primatol.* **16**:295–310.

Tilson, R. L. (1977). Social organization of Simakobu monkeys (*Nasalis concolor*) in Siberut Island, Indonesia. *J. Mammal.* **58**(2):202–212.

——— (1979). Der uberkannte Affe: Die ersten Bilder der Pageh-Stumpfnasse. *Tier* **5**:20–23.

Watanabe, K. (1981). Variations in group composition and population density of the two sympatric Metawaian leaf monkeys. *Primates* **22**:145–160.

Douc Langurs

Jablonski, N. (1995). The phyletic position and systematics of the Douc Langur of Southeast Asia. *Am. J. Primatol.* **3S5**:185–207.

Kavanagh, M. (1972). Food sharing behavior within a group of Douc monkeys (*Pygathrix nemaeus*). *Nature* **239**:406–407.

Golden Monkeys

Davison, G. W. H. (1982). Convergence with terrestrial cercopithecines by the monkey *Rhinopithecus roxellanae. Folia Primatol.* **37**:209–217.

Jablonski, N., ed. (1998). *Natural History of the Doucs and Snub-nosed Monkeys.* Singapore: World Scientific.

Kirkpatrick, R. C., Long, Y. C., Zhong, T., and Xiao, L. (1998). Social organization and range use in the Yunnan Snub-nosed Monkey *Rhinopithecus bieti. Int. J. Primatol.* **19**:13–51.

Long, Y., Kirkpatrick, C. R., Zhongtai, and Xiaolin. (1994). Report on the distribution, population, and ecology of the Yunnan snub-nosed monkey (*Rhinopithecus bieti*). *Primates* **35**:241–250.

Schaller, G. B. (1982). Zhen-zhen, rare treasure of Sichuan. *Animal Kingdom* **85**(6):5–14.

——— (1985). First published photos of China's golden treasure. *Int. Wildlife* **15**(1):29–31.

Zhang, Y. Z., Wang, S., and Quan, C. Q. (1981). On the geographical distribution of primates in China. *J. Human Evol.* **10**:215–227.

Zhixiang, L., Shilai, M., Chenghui, H., and Yingxiang, W. (1982). The distribution and habitat of the Yunnan golden monkey, *Rhinopithecus bieti*. *J. Human Evol.* **11**: 633–638.

ADAPTIVE RADIATION OF OLD WORLD MONKEYS

Di Fiori, A., and Rendall, D. (1994). Evolution of social organization: A reappraisal for primates by using phylogenetic methods. *Proc. Natl. Acad. Sci. USA* **91**: 9941–9947.

Kay, R. F. (1978). Molar structure and diet in extant Cercopithecoidae. In *Development, Function and Evolution of Teeth,* ed. P. M. Butler and K. A. Joysey, pp. 309–339. New York: Academic Press.

Schultz, A. H. (1970). The comparative uniformity of the Cercopithecoidea. In *Old World Monkeys,* ed. J. R. Napier and P. H. Napier, pp. 39–52. New York: Academic Press.

Strum, S. C. (1987). *Almost Human.* New York: Random House.

van Noordwijk, M. A., and van Schaik, C. P. (1988) Male careers in Sumatran long-tailed macaques (*Macaca fascicularis*). *Behavior* **107**:24–43.

Wrangham, R. W. (1980). An ecological model of female-bonded primate groups. *Behaviour* **75**:262–300.

PHYLETIC RELATIONSHIPS OF OLD WORLD MONKEYS

Brandon-Jones, D. (1996). The Asian Colobinae (Mammalia: Cercopithecoidae) as indicators of quaternary climatic change. *Biol. J. Linnean Soc.* **43**:3–43.

Collura, R. V., and Stewart, C.-B. (1996). Mitochondrial DNA phylogeny of the colobine Old World monkeys. *Abstracts XVI Cong. Int. Primatol. Soc.:* 727.

Disotell, T. (1996). The phylogeny of Old World monkeys. *Evol. Anthropol.* **5**:18–24.

Fleagle, J. G., and McGraw, W. S. (1998). Skeletal anatomy of African Papionins: function, phylogeny or both? *Amer. J. Phys. Anthropol.* Suppl. **26**:82–83.

Jablonski, N. (1998). The evolution of the doucs and snub-nosed monkeys and the question of the phyletic unity of the odd-nosed colobines. In *The Natural History of the Doucs and Snub-nosed Monkeys,* ed. N. Jablonski, pp. 13–41. Singapore: World Scientific.

Oates, J. F., Davies, A. G., and Delson, E. (1994). The diversity of living colobines. In *Colobine Monkey: Their Ecology, Behavior and Evolution,* ed. A. G. Davies and J. F. Oates, pp. 1–10. Cambridge: Cambridge University Press.

Strasser, E. and Delson, E. (1987). Cladistic analysis of cercopithecid relationships. *J. Human Evol.* **16**:81–99.

SEVEN

Apes and Humans

SUPERFAMILY HOMINOIDEA

Hominoids are the less successful group of catarrhines. There are only five genera of living hominoids. They range in size from about 4 kg for the smallest gibbons to over 200 kg for male gorillas. With the notable exception of our own species, hominoids have a rather restricted distribution: the tropical forests of Africa and Southeast Asia (Fig. 7.1). As we see in later chapters, hominoids were much more diverse and abundant in Europe at earlier times. Humans are the only hominoids that occur naturally in the New World, despite persistent rumors of Sasquatch or other apelike animals from the western parts of North America. However, both humans and chimpanzees have been sighted from time to time in extraterrestrial environments.

Hominoids

Hominoids are distinguished from Old World monkeys by a variety of both primitive catarrhine features and unique specializations (see Fig. 6.1). Like cercopithecoids, all living hominoids have a tubular tympanic bone and a dental formula of 2.1.2.3. Compared to Old World monkeys, hominoids have relatively primitive molar teeth, with rounded cusps rather than the bilophodont crests of monkeys. The lower molars are characterized by an expanded talonid basin surrounded by five main cusps. The upper molars are quadrate and have a distinct trigon anteriorly and a large hypocone posteriorly. The anterior lower premolar varies in shape from an elongate shearing blade in gibbons to a bicuspid tooth in humans. Most hominoids have relatively broad incisors. Hominoid canines are much more variable than those of cercopithecoids in both shape and degree of sexual dimorphism.

Hominoids are characterized by relatively broad palates, broad nasal regions, and large brains. Hominoid skeletons show a variety of distinctive features (Fig. 7.2). The axial skeleton is characterized by a reduced lumbar region, an expanded sacrum, and the absence of a tail. All hominoids have a relatively broad thorax with a dorsally positioned scapula. Hominoids have relatively long upper limbs, and their elbow joint is characterized by a spool-shaped trochlea on the humerus and a short olecranon process on the ulna. Their wrist lacks an articulation between the ulna and the carpal bones; instead, a fibrous meniscus separates the two bones. The hindlimbs of hominoids are characterized by a broad ilium, broad femoral condyles, and usually a large, robust hallux.

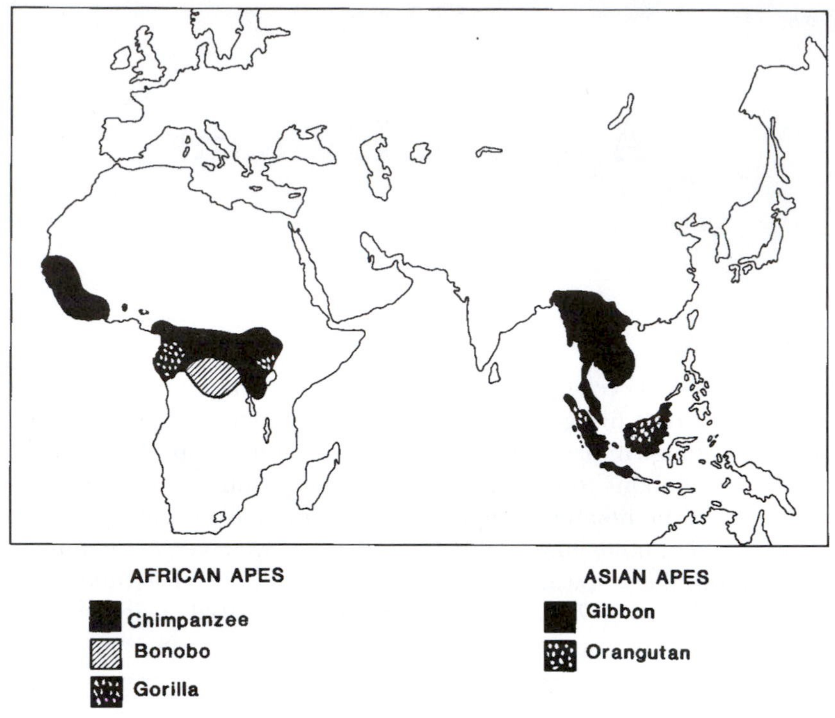

FIGURE 7.1 Geographic distribution of extant apes.

Hominoids are quite different from Old World monkeys in many aspects of their behavior and ecology as well as life history. However, it is unclear to what extent these are derived hominoid features or primitive anthropoid features. In general, where similar-sized animals can be compared, hominoids have longer gestation and longer time to first reproduction than cercopithecoids. Hominoid social organization is characterized by a general absence of female philopatry and their societies are not organized around female matrilines. Fission–fusion of foraging groups seem to be common in many species.

The five hominoid genera are traditionally placed in three separate families: hylobatids, pongids, and hominids. Although this chapter follows that classification, readers should appreciate that this is a gradistic rather than a phyletic arrangement. Humans and African apes are more closely related to each other than they are to orangutans. The phyletic relationships among hominoids are discussed later in the chapter (see also Chapter 1).

Hylobatids

The gibbons, *Hylobates,* from Southeast Asia (Fig. 7.3) are the smallest, the most specifically diverse, and the most numerically successful of living apes (Table 7.1). These lesser apes are anatomically the most primitive of living apes and retain many monkeylike features, but in some aspects, such as their limb proportions, they are the most specialized of the living hominoids. The numerous gibbon species are relatively uniform in morphology. All are relatively small (5–11 kg), with no sexual size dimorphism. They have simple molar teeth characterized by low, rounded cusps

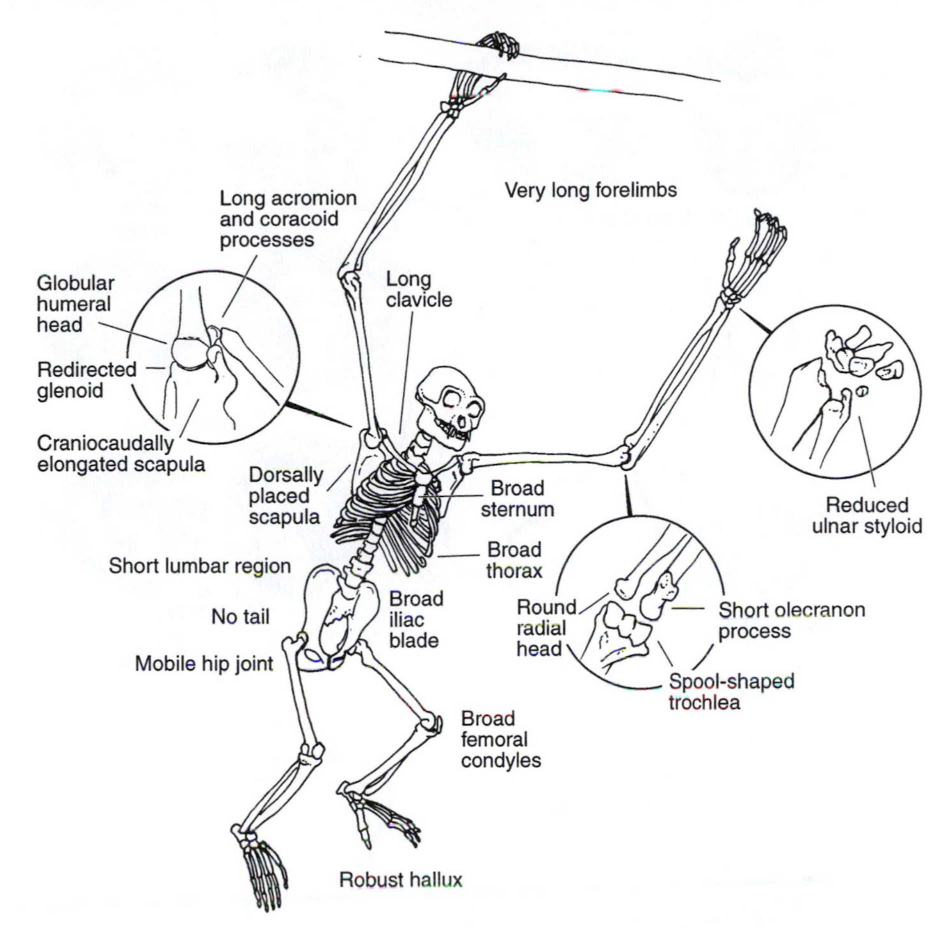

FIGURE 7.2 Characteristic skeletal features of extant apes, illustrated by a siamang.

and broad basins (Fig. 7.4). Their incisors are relatively short, but broad. Both sexes have long daggerlike canines and bladelike lower anterior premolars for sharpening the upper canine.

Gibbons have short snouts and shallow faces, large orbits with protruding rims, and a wide interorbital distance. Their braincase is globular and has no nuchal cresting, and only occasionally do they develop a sagittal crest. The mandible is shallow and has a broad ascending ramus.

Gibbons are outstanding among living primates in their limb proportions (Fig. 7.5). They have the longest forelimbs relative to body size of any living primates, and they also have very long legs. They have long, curved, slender digits on their hands and feet as well as a long muscular pollex and hallux. Gibbons are the only apes that consistently have ischial callosities, and females also show small sexual swellings that change shape and color during the estrus cycle (Dahl and Nadler, 1992) .

Gibbons are found throughout the evergreen forests of Southeast Asia, from eastern India to southern China on the mainland, as well as on Borneo, Java, Sumatra, and nearby islands of the Sunda Shelf. There is general

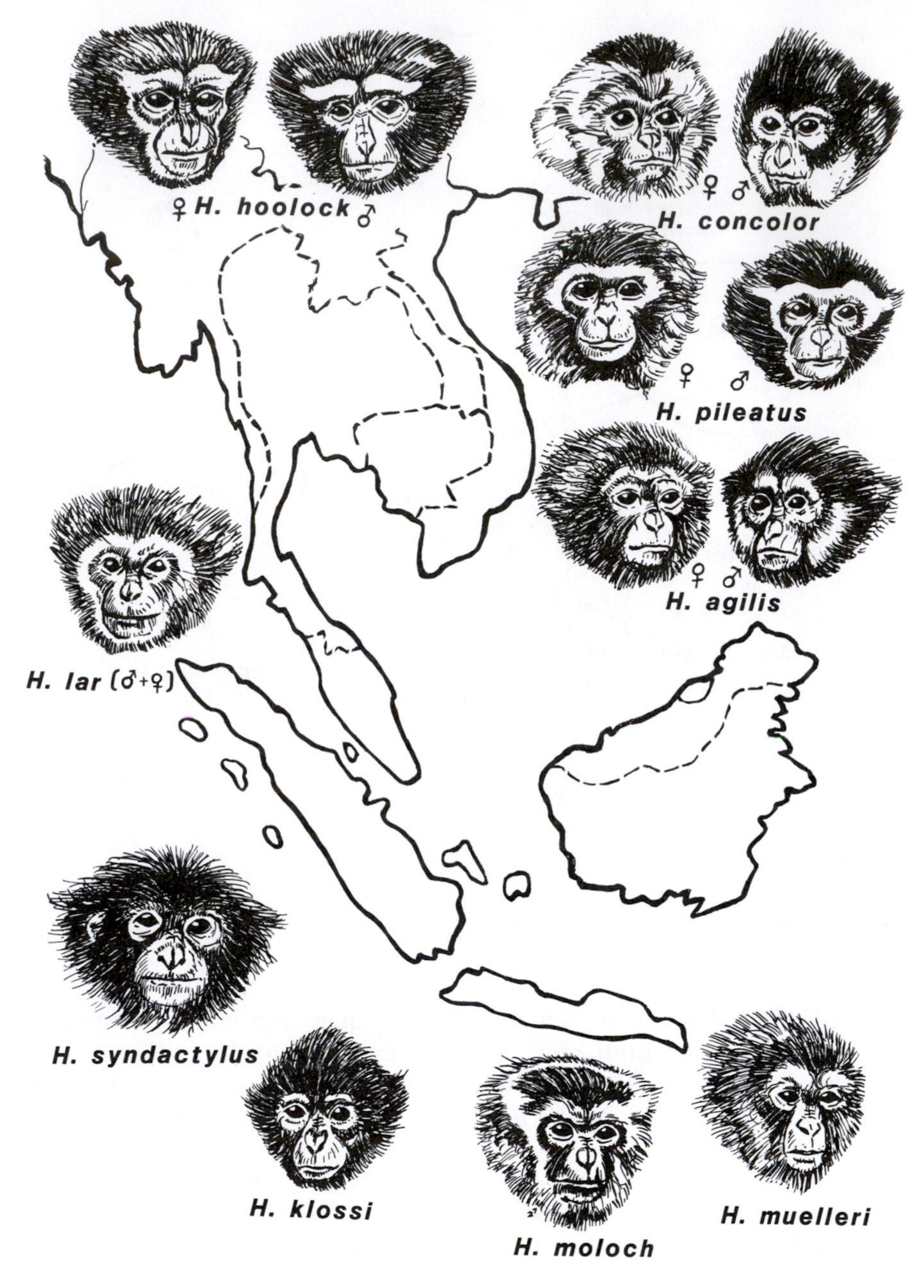

FIGURE 7.3 Geographic distribution and facial characteristics of extant gibbon populations.

TABLE 7.1
Infraorder Catarrhini
Family HYLOBATIDAE

Common Name	Species	Intermembral Index	Mass (g)	
			M	F
Siamang	*Hylobates syndactylus*	147	11,900	10,700
Crested gibbon	*H. concolor*	140	7,790	7,620
Hoolock gibbon	*H. hoolock*	129	6,870	6,880
Kloss's gibbon	*H. klossii*	126	5,670	5,920
White-handed gibbon	*H. lar*	130	5,900	5,340
Agile gibbon	*H. agilis*	129	5,880	5,820
Pileated gibbon	*H. pileatus*	—	5,500	5,440
Silvery gibbon	*H. moloch*	127	6,580	6,250
Mueller's gibbon	*H. muelleri*	129	5,710	5,350

agreement on the number of distinct morphological groups of gibbons, but debate on how many of these should be considered separate species. The four major groups of gibbons are the **siamang** (*H. syndactylus*), the **concolor gibbon** (*H. concolor*), the **hoolock gibbon** (*H. hoolock*), and the others (e.g., Groves, 1989). This last group contains numerous allopatric gibbon populations that some authorities recognize as distinct species but that others consider subspecies. The small Kloss's gibbon (*H. klossii*) is the most distinctive species in this group. The remaining populations, the white-handed, or lar, gibbon (*H. lar*), the agile gibbon (*H. agilis*), the pileated gibbon (*H. pileatus*), the silvery gibbon

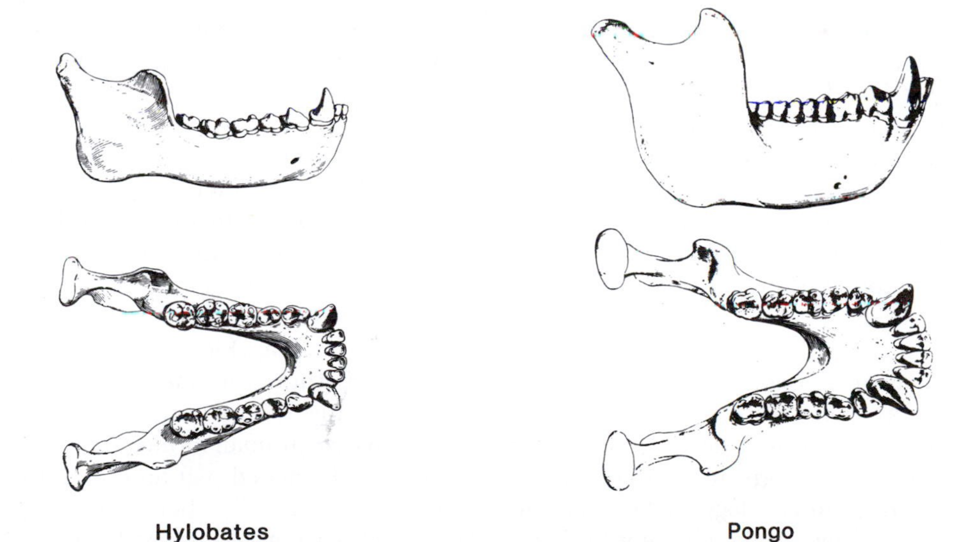

FIGURE 7.4 Lower jaws of a siamang (left) and an orangutan (right).

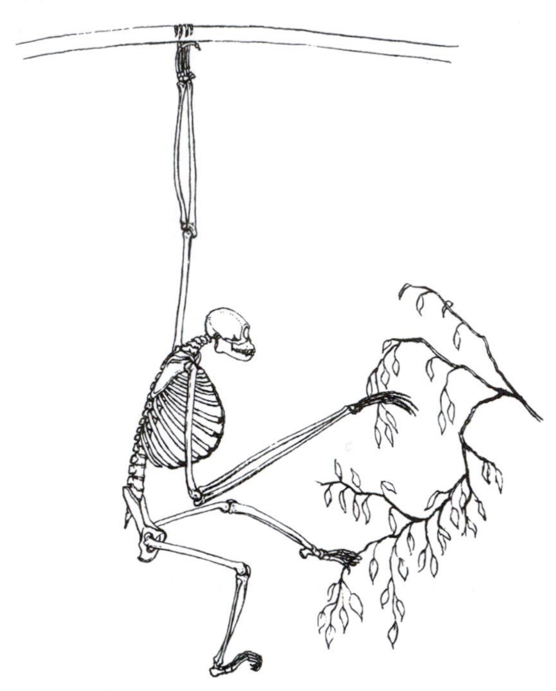

FIGURE 7.5 The skeleton of a gibbon (*Hylobates*).

(*H. moloch*), and Mueller's gibbon (*H. muelleri*), are considered by many authorities to be subspecies of *H. lar.* Like the savannah baboons of Africa, the "lar" gibbons form what is called a superspecies. The populations are all allopatric and have distinct vocalizations. There is, however, evidence of interbreeding at species borders, and the level of morphological differences in cranial morphology between the populations is less than that found between other so-called distinct species of primates.

The behavior and ecology of the different species of gibbons is remarkably uniform considering their broad geographic range (Chivers, 1984). The greatest differences in gibbon behavior and ecology are between the two sympatric species, the siamang and the white-handed gibbon, which are found together in the forests of west Malaysia (Fig. 7.6; see Fig. 6.6) and Sumatra (Palombit, 1997). All gibbon species show a preference for moist primary forests rather than secondary or riverine forests, but siamang are found at higher elevations and more commonly in mountain regions than are the sympatric white-handed gibbons. Gibbons move and feed mainly in the middle and upper levels of the canopy and virtually never descend to the ground. They are the most suspensory of all primates and are aerialists *par excellence,* moving almost exclusively by two-armed brachiation and by slower quadrumanous climbing (Fig. 7.7). The larger siamang travel mainly by slow, pendulum-like arm-over-arm brachiation, whereas the smaller gibbon species use more rapid ricocheting brachiation in which they throw themselves from one tree to the next over gaps of 10 m or more. During feeding, all gibbons use more deliberate quadrumanous climbing when moving among small terminal branches. They use a wide variety of both seated and suspensory feeding postures.

Gibbons specialize on a diet of ripe fruit, part of which is found in small, widely scattered clumps throughout the forest and part of which, such as many figs, occurs in large bonanzas. Gibbon species also eat varying amounts of new leaves and invertebrates such as termites and arachnids. The proportions of these foods in their diet vary from season to season, from species to species, and from locality to locality. In both Malaysia and Sumatra, the larger siamang rely more on new leaves than do the smaller lar gibbons, which eat more fruit. The Kloss's gibbon from the island of Siberut is unusual in that it does not eat leaves, only fruit and invertebrates.

Gibbons live in small monogamous families composed of a mated pair and up to four dependent offspring. The black-crested gibbon (*H. concolor*) from China has been reported to live in larger polygynous groups (Haimoff

FIGURE 7.6 Two sympatric gibbons from west Malaysia; upper left, the white-handed gibbon (*Hylobates lar*); below, the siamang (*Hylobates syndactylus*).

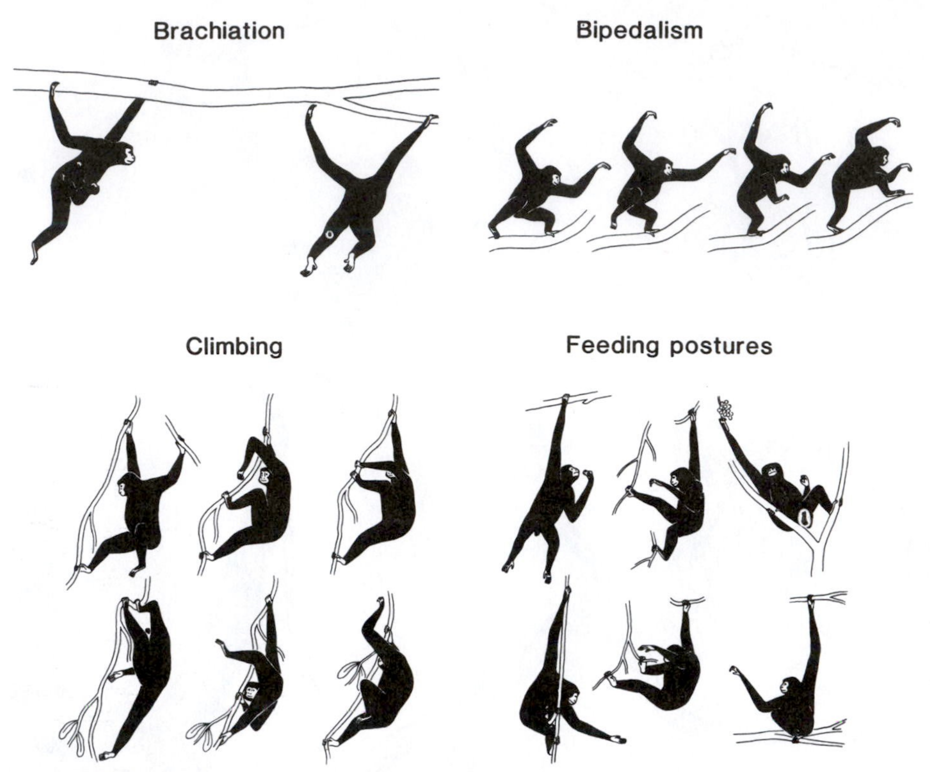

FIGURE 7.7 Locomotor behavior and feeding postures of the Malayan siamang (*Hylobates syndactylus*).

et al., 1986), but recent studies have failed to confirm this (Lan, 1993). Families of the more folivorous siamang usually forage as a unit and have smaller day ranges (1 km or less) and smaller home ranges (18–50 ha) than the smaller gibbons. The latter often forage individually, with the members of a single family separated by as much as several hundred meters, and they have larger day ranges (2 km) and home ranges. These differences in ranging pattern have been related to the different distribution of preferred foods: leaves for the siamang and scattered fruits for the gibbons. A clumped resource such as leaves can be readily exploited by a group, whereas widely scattered fruits are more easily sought out by individuals. Moreover, when feeding on clumped fruit resources, such as figs, siamang are dominant over the smaller lar gibbons and are able to exclude them from food trees until the siamang have had their fill.

All gibbons are fiercely territorial and defend their core areas with daily calling bouts and occasional intergroup conflict. Social interactions within a gibbon group are limited and consist primarily of occasional grooming bouts. Gibbons have single births every four or five years, with considerable individual variance in female reproductive success (Palombit, 1995). Siamang males carry the offspring during its second year of life. In other species, male investment is not so extensive. Young gibbons spend up to ten years in their family group before leaving, usually after harassment by their same-sex parents.

Pongids

There are four living species of pongids, or great apes (Table 7.2). The orangutan is the only great ape from Asia. Gorillas, chimpanzees, and bonobos, or pygmy chimpanzees, from Africa are a closely related group of apes that some authors regard as size variants of a single type. Great apes share many skeletal features with the lesser apes, such as relatively short trunks, the absence of a tail, a broad chest, long arms, and long hands and feet, but they are distinguished by their large size and many more detailed anatomical characteristics, such as more robust canine teeth, broader premolars, a very broad ilium, and a robust fibula (see Martin, 1986).

The great apes are primarily forest primates. All great apes are herbivorous and eat varying proportions of fruit and leaves. These large primates are less committed to a suspensory life than are the lesser apes, and all are to some extent terrestrial in their habits. They are, however, more suspensory than are living cercopithecoid monkeys, and all are characterized by some use of quadrumanous climbing. Even large male gorillas climb trees for feeding on particular foods in some seasons of the year.

All great apes build nests for sleeping and resting, but they share few unifying features in their social behavior. Each species seems to show a different pattern of social organization, sexual dimorphism, and individual grouping tendencies. In keeping with their large size, all have a relatively long life span and a long ontogeny. All have single births and gestation periods similar to those of humans.

Orangutans

The shaggy red orangutan (*Pongo pygmaeus*) is the largest living Asian ape (Fig. 7.8). There are two living subspecies, one on Borneo and one on Sumatra, but in prehistoric times orangutans had a much larger range, which included Java and parts of southern China. Orangutans are extremely sexually dimorphic in size, with females weighing approximately 60 kg and males roughly twice that. They are quite distinct morphologically from the African apes. Dentally, they are characterized by cheek teeth with thick enamel, low,

TABLE 7.2
Infraorder Catarrhini
Family PONGIDAE

Common Name	Species	Intermembral Index	Mass (g)	
			M	F
Orangutan	*Pongo pygmaeus pygmaeus*	139	77,900	35,600
	P. pygmaeus abelli	139	78,500	35,800
Chimpanzee	*Pan troglodytes schweinfurthii*	103	42,700	33,700
	Pan troglodytes troglodytes	106	59,700	45,800
	P. troglodytes verus	106	46,300	44,600
Pygmy chimpanzee, bonobo	*P. paniscus*	102	45,000	33,200
Western lowland gorilla	*Gorilla gorilla gorilla*	116	170,400	71,500
Mountain gorilla	*G. gorilla beringei*	116	162,500	97,500
Eastern lowland gorilla	*G. gorilla graueri*	—	175,200	71,000

FIGURE 7.8 The orangutan (*Pongo pygmaeus*).

flat cusps, and crenulated occlusal surfaces (Fig. 7.4). They have large upper central incisors and small, peglike upper laterals. The canines are large and sexually dimorphic. Orangutan crania are characterized by a relatively high, rounded braincase, poorly developed brow ridges, a deep face with small orbits set close together, and a uniquely prognathic snout with a large convex premaxilla (Fig. 15.18). The mandible is deep and has a high ascending ramus (Fig. 7.4).

The limbs of orangutans show extreme specializations for suspensory behavior. They have very long forelimbs and long, hooklike hands with long curved fingers and a short pollex. Their extremely mobile hindlimbs are relatively short, and they have handlike feet with long, curved digits and a reduced hallux.

Orangutans seem to prefer upland rather than lowland forest areas. Females and immature individuals are almost totally arboreal, whereas adult (especially old) males on Borneo frequently descend to the ground to travel. In the trees, orangutans move almost exclusively by slow quadrumanous climbing in which they use their hands and feet interchangeably as they move within tree crowns and transfer themselves from tree to tree. On the ground they move quadrupedally, with their hands held in a fist (Tuttle, 1969b). When feeding they use both seated and suspensory postures. They frequently use their strong arms to bend or break branches to bring food to their mouth rather than change positions. Orangutans eat primarily fruits (many of which contain hard seeds that they crush with their flat molars), and they also consume considerable amounts of new leaves, shoots, and bark. Choice of fruit trees in an orangutan's diet seems to be based on several variables, including the size of the available fruit crop and the amount of pulp in the fruit. They avoid tannins and seem to select fruit for its energy content rather than for its protein content (Leighton, 1993). There are differences in the diets of male and female orangutans on Borneo. Male orangutans eat more termites, in conjunction with their terrestrial forays.

Adult orangutans are usually solitary and have a noyau social organization similar to that found among nocturnal prosimians. The only consistent social groups among orangutans consist of females with their immature offspring. Individual females (with young) live in relatively small home ranges of approximately 70 ha and have day ranges of approximately 500 m. Mature adult males occupy much larger home ranges that overlap the ranges of several adult females. They move much farther each day, partly in search of more food to supply their greater bulk and also to monitor the whereabouts of their female consorts and male competitors. They also interact with other orangutans through a long call. Thus, even though, by other primate standards, orangutans appear to live solitary lives, it seems likely that they maintain extensive social networks of individual relationships. However, the details of the relationships among adult male orangutans in the larger "community" are not well understood (van Schaik and van Hoof, 1996)

Young adult males that have not acquired their own territory seem to have a very different ranging behavior and reproductive strategy (Galdikas, 1985; Mitani, 1985a,b). They forage with adult females for weeks or months at a time and forcibly mate with the usually uncooperative female. Interactions between adult male orangutans are usually aggressive, occasionally involving fierce battles but more often only vocal exchanges. Male–female sexual encounters vary dramatically from violent interactions between young males and adult females, which can best be described as rape, to occasional, long erotic treetop trysts, which usually occur between older adult males and females.

The care and upbringing of young orangutans is totally the responsibility of the females. Female orangutans become sexually mature at the age of about seven years. Sexual maturation in male orangutans is a more variable and interesting phenomenon. They become sexually competent subadults somewhere between the ages of eight and fifteen years, but they may not become fully mature with cheek pouches for many years after that. Final maturation seems to occur very rapidly and is influenced by social factors rather than by age alone.

Gorillas

The gorilla (*Gorilla gorilla*) is the largest living primate and shares with chimpanzees the distinction of being our closest primate relatives (Fig. 7.9). The single species of gorilla is normally divided into three geographically isolated subspecies, the western lowland gorilla, the eastern lowland gorilla, and the mountain gorilla (Table 7.2). However, there is extraordinary diversity both among and within gorilla "subspecies," suggesting that they may be genetically more comparable to distinct species in other primates (Ruvolo *et al.*, 1994).

Gorillas have extreme sexual size dimorphism; females weigh 70–90 kg and males up to 200 kg. This dimorphism is also evident in many aspects of their skeletal anatomy, where it manifests itself in the greater general robustness of the males. The molar teeth of gorillas have a greater development of crests than those of any other hominoid, a feature associated with their folivorous diet. They have large, tusklike canines and relatively small

FIGURE 7.9 The mountain gorilla (*Gorilla gorilla beringei*) of the Virunga volcanoes of Rwanda.

incisors. Gorillas have relatively long snouts, pronounced brow ridges, and, in males, well-developed sagittal and nuchal crests.

Gorillas have relatively long forelimbs (Fig. 7.10). Their hands are very broad and have a large pollex and (like all African apes) dermal ridges on the dorsal surface of their digits. Their trunk is relatively short and broad and has a wide thorax and a broad, basinlike pelvis. Their hindlimbs are relatively short, and their feet are broad. In mountain gorillas the hallux is somewhat adducted and connected to the other digits by webbing, giving them a very humanlike footprint.

Gorillas have a limited distribution in the tropical forests of sub-Saharan Africa. Mountain gorillas show a preference for secondary and herbaceous forests and are among the most terrestrial of all primates. Adult mountain gorillas rarely climb trees, and they nest on the ground. The lowland forms are found in a wide range of forests. They are much more arboreal, especially females and youngsters, and they normally feed, rest, and build their sleeping nests both in trees and on the ground. On the ground, all gorillas move by quadrupedal walking and running. Like chimpanzees, gorillas have an unusual hand posture for quadrupedal standing and moving, called knuckle-walking (Fig. 7.10). Rather than support their forelimb on the palm of their hand (like most primates) or on the palmar surface of their fingers (like many baboons), they support it on the dorsal surface of the third and fourth digits of their curled hands. In the trees they are relatively good climbers, but they rarely use suspensory feeding postures. Gorillas use their great strength to good advantage when foraging, often ripping apart branches or whole trees.

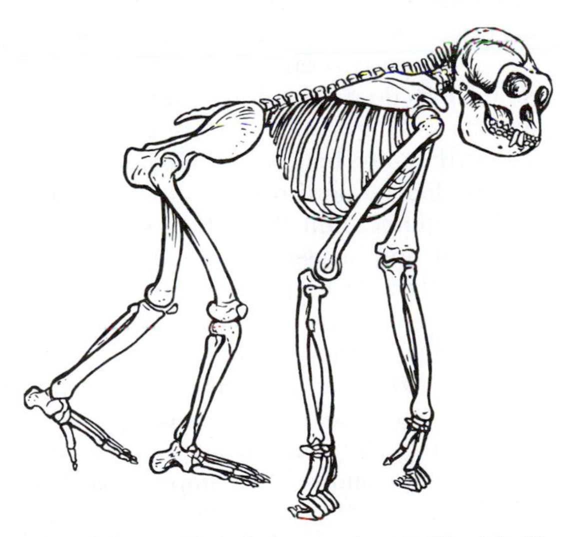

FIGURE 7.10 The skeleton of a gorilla (*Gorilla gorilla*).

Mountain gorillas have the most herbaceous diet of any living ape. They eat mainly leaves and pith in great quantities. In their feeding activities, gorillas are extremely destructive of the vegetation, and their ranging patterns seem to involve harvesting and destroying favorite patches of rapidly regenerating vegetation on a systematic and regular basis. Nevertheless, their foraging behavior often involves a complex series of detailed manipulations of individual food items. Lowland gorillas seem to eat more fruit than do the mountain subspecies—50 percent or more of their diet in some seasons. Reports of gorillas eating meat are extremely rare.

Mountain gorillas live in groups averaging between nine and ten individuals that usually contain one mature (silverback) adult male, one or more younger males, and several adult females with offspring. Mountain gorillas have relatively small, overlapping home ranges of approximately 100 ha and travel as a coherent group through day ranges of about 500 m. There are increasing numbers of reports about the foraging habits of lowland gorillas. Some suggest that are found in cohesive groups, like mountain gorillas (Tutin *et al.*, 1992); others report a more flexible, fission–fusion pattern, with small foraging groups regularly splitting and rejoining (Mitani, 1992; Remis, 1994). Day ranges of lowland gorillas are much larger than those of the mountain subspecies, averaging 1–2 km per day. Similarly, lowland gorillas groups live in very large home ranges of up to 500 ha and show seasonal changes in their ranging (Doran and McNeilage, 1998).

Although the composition of a gorilla group is similar to that of many primate groups, with a single adult male and several adult females, its formation and maintenance seem to be based on different demographic and social relationships. Most primate groups among Old World monkeys seem to be organized around groups of related females who grow up and remain in their natal group while males transfer from one troop to another. In gorilla society, both males and females migrate between groups. As a result, a normal gorilla group is composed of unrelated females who generally do not interact with each other very much. Most interactions are between the adult male and individual females rather than among group members of the same sex. Gorillas do, however, resemble other single-male groups of primates in the intense competition between males for control of a troop, and takeovers may be accompanied by infanticide.

Chimpanzees

Pan troglodytes, the chimpanzee (Fig. 7.11), has a broad distribution that extends in a broad belt across much of central Africa, from Senegal in the west to Tanzania in the east. Chimpanzee systematics is currently in flux as a result of many new molecular studies of ape phylogeny. There are currently three commonly recognized chimpanzee subspecies and the chimpanzee population from Nigeria and adjacent parts of Cameroon may be a fourth (Gonder *et al.,* 1997). The western subspecies, *P. troglodytes verus,* is quite distinct from the others and should probably be recognized as a separate species (Morin *et al.,* 1994). Chimpanzee subspecies differ considerably in body weight. All have moderate levels of sexual dimorphism (Table 7.2).

Compared with gorillas, chimpanzees have broader incisors and cheek teeth with broader basins and lower, more rounded cusps. Their skulls are very similar in overall shape to those of gorillas, but they have shallower faces and mandibles and do not show such extensive development of sagittal and nuchal crests. The limbs of chimpanzees are more similar in length than those of gorillas and are also less robust. They have narrower hands and feet, with more slender, curved digits.

Chimpanzees occupy a variety of habitats, from rain forests to dry savannah areas with very few trees. Most of our knowledge of chimpanzee behavior comes from woodland, open forested environments rather than jungles. It has become strikingly clear in recent decades that chimpanzee behavior is very different from site to site. Part of this can be attributed to differences in habitat and available resources, but part of it also sees to reflect local traditions. In all habitats chimpanzees generally feed in trees much of each day (depending on the season) and travel on the ground between feeding sites using a quadrupedal, knuckle-walking locomotion. In an arboreal setting they use both quadrupedal and suspensory locomotion to move about within a feeding source, and they also use a variety of seated and suspensory feeding postures (Doran, 1992). Chimpanzees are more suspensory than gorillas but considerably less suspensory than either gibbons or orangutans.

Chimpanzees eat primarily fruit and nuts (60 percent) as well as leaves (21 percent), withstrikingly different dietary habits among different populations. For example, nuts of the oil palm are a staple for chimpanzees at Gombe but are not eaten by the chimps at Mahale, less than 20 km away. Other populations may

FIGURE 7.11 The chimpanzee (*Pan troglodytes*).

eat different parts of the same fruit species or process the fruits in different ways (McGrew, 1992). Chimpanzees also seem to use a wide range of plants for medicinal purposes. Although primarily frugivores, chimpanzees also regularly eat social insects and various smaller mammals, including other primates (colobus monkeys and baboons). Predation and hunting patterns of chimpanzees has been well documented at a number of sites (Stanford, 1994; C. Boesch and H. Boesch, 1989; H. Boesch and C. Boesch, 1994). Hunting is predominantly an activity of adult males, who account for about 80 percent of all kills at both Gombe in Tanzania and Tai in Ivory Coast. Hunting also seems to involve considerable personal abilities (some individuals are much more interested in and adept at hunting than others), and it is commonly a group activity in which several individuals participate and cooperate. The hunting party subsequently shares the food with other members of the group. In addition to these general similarities, there are also striking differences in hunting habits of chimpanzees at different sites.

Chimpanzees utilize many types of tools and other aspects of material culture in the course of their daily activities, including using leaves as sponges, twigs as probes or digging sticks, and clubs and stones to smash hard nuts (McGrew, 1992a,b; H. Boesch and C. Boesch, 1994). As with many aspects of chimpanzee behavior, there are striking differences between populations in the amount and nature of tool use. Indeed, although many populations show tool use, no one behavior is universal. The most widespread behaviors are wielding hammers to crack nuts, using sticks as clubs, and throwing missiles. Certainly the most striking behavior for students of human evolution is the use of stone tools, by chimps at several sites in West Africa, to crack open nuts (C. Boesch and H. Boesch, 1990; H. Boesch and C. Boesch, 1994). Although no evidence exists that the stone tools are manufactured to any extent, particularly good ones may be carried around for some time or stored for future use. With long-term studies of chimpanzee behavior demonstrating more and more details of the patterns of their material culture, the distinction between chimps and humans has become increasingly hard to define.

Social groups of chimpanzees are more fluid than are those of many higher primates. They are best described as having a fission–fusion system like that of *Ateles*. Adults of both sexes spend large portions of their time foraging alone, but they join from time to time with other individuals in temporary associations or parties. The sociality of chimpanzees seems to vary considerably from population to population. In eastern Africa, females are more solitary than males. They spend more time alone, have shorter day ranges, and have smaller individual home ranges. Female–female social interactions are relatively rare except among close relatives such as mothers and daughters. In western Africa, chimpanzee females seem more social and frequently forage together. Adult male chimpanzees are more gregarious. They often groom each other, more frequently join in feeding parties, and also form patrol groups that monitor the boundaries of the community home range. Chimpanzee mating occurs in a number of contexts, including promiscuous mating within the group, possessive behavior on the part of an individual male within the group toward the fertile female, and consortship, in which an individual male and a receptive female travel together for several days at a time, apart from other individuals.

Foraging and feeding parties are drawn from a relatively closed social community of fifteen to eighty chimpanzees who share a common home range that is actively defended. Interactions between adults (except some estrous females) of neighboring communities are usually aggressive (Wrangham and Peterson, 1996). Among chimpanzees, young females regularly migrate between communities; males tend to remain in their natal groups. As a result, the adult males of a community are closely related, a fact that would account for their gregariousness and cooperation (Morin *et al.,* 1994). Long-term community structure in chimpanzees is based on continuity of the male members rather than of the more mobile females.

Bonobos

Pan paniscus, the pygmy chimpanzee, or bonobo (Fig. 7.12), is similar in adult body weight to the smaller subspecies of chimpanzee (*Pan troglodytes schweinfurthii*) but has a darker face, a more gracile skull, more slender limbs, and longer hands and feet. Although there seems to be sexual dimorphism in the body weight of bonobos, there is virtually no sexual dimorphism in either the dentition (only the canines) or the limb skeleton. Bonobos have a relatively restricted distribution in central Africa south of the Zaire River, where they live in a more forested environment than do most other chimpanzees.

Like chimpanzees, bonobos travel mainly on the ground by knuckle-walking and feed both on the ground and in trees. Their arboreal locomotion involves quadrupedal, suspensory, and bipedal activities. They also use a variety of seated, standing, and suspensory feeding postures. Bonobos are the most suspensory of the African apes (Susman, 1984; Doran and Hunt, 1994).

Like chimpanzees, bonobos eat primarily fruit, pith, and leaves, as well as occasional prey items, including small ungulates, insects, snakes, and fish. In contrast with other chimpanzees, they show little inclination to use tools. The limited studies on daily ranging patterns of bonobos indicate that day ranges average 2 km. A group of about fifty individuals ranges over an area of 2200 ha, and there seems to be overlap of group home ranges and peaceful relations between members of different communities. Like chimpanzees, bonobos are normally seen in small groups of four to five individuals, but their fission–fusion society differs from that of chimpanzees in the more frequent occurrence of feeding groups containing both males and females. Data from grooming partnerships, nearest neighbor choices, and part composition show that bonobos, unlike chimpanzees, exhibit a high degree of affiliation among females. However, in larger parties (associated with large food sources or provisioning), male–female affiliation predominates (White, 1992). Many friendly interactions between bonobos frequently involve sexual behavior. The unique genital–genital rubbing of female bonobos seems directly related to food competition. The greater cohesiveness of the sexes in bonobos compared with chimpanzees seems to be related to their consistent use of larger food patches throughout the year (White and Wrangham, 1988; Malenky and Wrangham, 1994). The role of terrestrial herbaceous vegetation in determining bonobo grouping behavior is an area of considerable debate (Chapman *et al.,* 1994; White, 1996).

Size and Evolution of the African Apes

Despite their differences in size, diet, and social organization, the three living African apes are quite similar in many aspects of cranial and skeletal anatomy and virtually identical in many biochemical assays. Many authors have suggested that most of their differences are a corollary of their differences in body size or simple differences in the timing of similar ontogenetic patterns. A simplistic summary this view is that gorillas are overgrown

FIGURE 7.12 The pygmy chimpanzee (*Pan paniscus*).

chimpanzees and that bonobos are morphologically similar to immature chimpanzees; that is, adult bonobos have a morphology that resembles the juvenile morphology of common chimpanzees, and adult chimpanzees have a morphology that corresponds to that of a juvenile gorilla. The adults of each species, in this view, represent different points on the same growth curve. For example, Doran (1992) has shown that adult bonobos resemble juvenile chimps in both limb proportions and locomotor behavior. In other aspects of morphology and behavior, the actual relationships between size and shape in the three African apes are somewhat more complex than this simple model suggests, but for many features of the skull and many limb proportions it accurately describes the species differences in shape (Fig. 7.13).

The relationships between relative growth, absolute growth rates, and sexual dimorphism among the African apes are a particularly intriguing and promising area of research (see, e.g., Shea, 1985). For example, despite their size difference, the temporal length of ontogeny in gorillas and chimps appears to be virtually the same. Gorillas attain their larger size by having a rate of absolute growth that is twice that of the common chimpanzee. Each species of African ape has a unique pattern of growth differences between the sexes, resulting in species-specific amounts of sexual dimorphism in the adults. Compared with common chimpanzees, gorillas are characterized by earlier female sexual maturity and thus greater differences in the timing of sexual maturity and in size dimorphism. In bonobos, however, both males and females retain more juvenile-like cranial features than common chimpanzees, and, like immature chimpanzees, they show less sexual dimorphism. These developmental and social differences

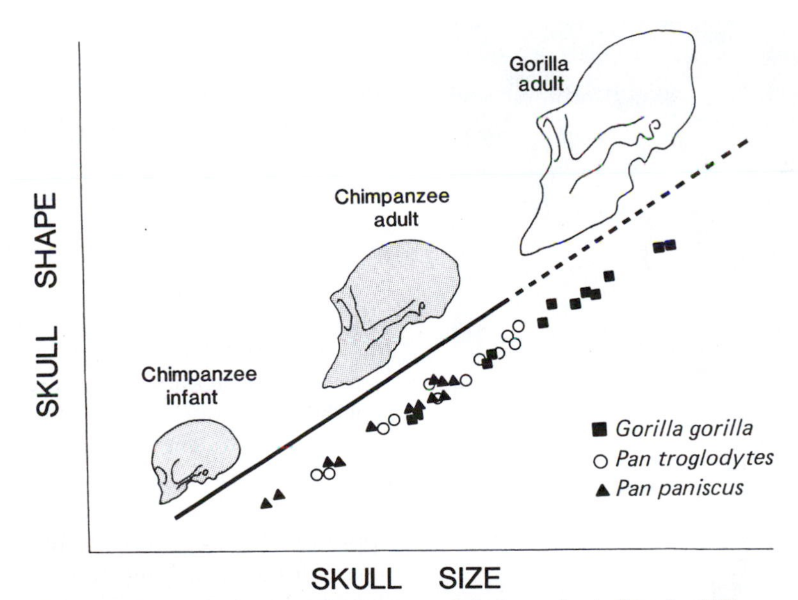

FIGURE 7.13 A schematic illustration of the hypothesis that the African apes occupy different positions on the same ontogenetic trajectory. The shape of a gorilla skull is just an extension of the chimpanzee growth curve to a larger size (courtesy of Brian Shea).

are probably related to the differing ecological adaptations of each ape.

Hominids

Homo sapiens, the only living hominid, is a very odd primate species—and a very unusual creature by any standard (Figs. 7.14, 7.15; Table 7.3). We are more similar to other hominoids in our dental and skeletal anatomy than our striking external and postural features would suggest, and we also have many outstanding specializations that set us apart. We are distinguished dentally by our relatively small canines, broad premolars, and reduced or absent third molars in many populations; otherwise our teeth are similar to those of many chimpanzees. Our mandible, with its protruding chin, is very unusual. Our skull has a small, short face and a large balloonlike cranium with poorly developed crests and a foramen magnum that lies well beneath the skull base. Relative to our body size, we have extremely large brains.

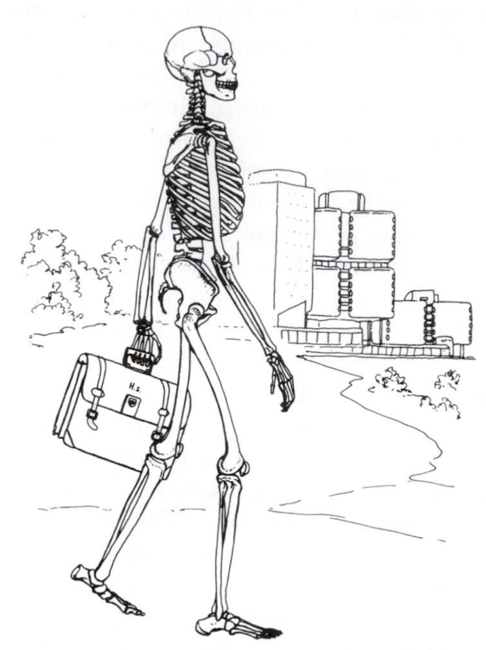

FIGURE 7.15 The skeleton of a human.

FIGURE 7.14 A member of the human race (*Homo sapiens*).

Like all hominoids, we have a relatively short trunk and long arms. Our hand is characterized by short, slender fingers and an extremely opposable thumb. The most distinctive features of our skeleton are those of our long lower extremities associated with our upright, bipedal locomotion. Our pelvic bone is extremely short and broad, and the femur, tibia, and fibula are extremely long. Unlike all other primates, we have a hallux that is not opposable but aligned with the other toes. We have a long heel, long metatarsals, and very short pedal phalanges. The bones of our foot form two arches, one longitudinal and one transverse. These give us our characteristic footprint, in which we land on the heel, pass

TABLE 7.3
Infraorder Catarrhini
Family HOMINIDAE

Common Name	Species	Intermembral Index	Mass (g) M	Mass (g) F
Human	*Homo sapiens*	72	47,000–78,000	42,000–73,000

our weight along the lateral border of the foot to the front, then push off with the ball of the foot and ultimately the great toe.

One of our most striking features is the apparent lack of hair over most of our body, which contrasts with the noticeable concentration of hair on our heads, under our arms, and in the genital region. The development of human facial hair is an extremely variable feature that differs not only between sexes but also among different human populations. In fact, the density of hairs on the human body is not that different from that in large apes such as chimpanzees and gorillas; human hairs are just very short and often lightly pigmented. Another striking feature that distinguishes humans from other primates is the distribution of subcutaneous body fat. Like body hair, this is not only quite different between sexes of many human populations—men tend to store fat in their abdomen, women store it in their breasts and on their hips and buttocks—but is also quite variable among major human population groups. Despite striking differences in body form and appearance among different human populations, we are remarkably uniform, genetically, compared with other hominoids; there is more genetic diversity in a single population of gorillas in west Africa than in all of humanity (Ruvolo *et al.*, 1994).

As primates, we humans are uniquely cosmopolitan. Only Antarctica has steadfastly resisted permanent colonization by our species. All other habitats, including tropical rain forests, woodlands, savannahs, plains, deserts, mountains, and arctic coastlines, have supported human populations for many thousands of years. We are also the most terrestrial of all primates. Only humans (and possibly gelada baboons) regularly live their entire lives without ever climbing a tree for food or sleep. Our bipedal gait is unique among mammals, but it does not seem to endow us with particularly striking speed or locomotor efficiency compared with other mammals, including nonhuman primates. There are indications, however, that it permits more endurance at slow speeds.

The "natural" human diet is probably something that exists only in television commercials and on billboards. Humans are opportunistic and probably omnivorous—we eat virtually anything. We lack the notable digestive specializations that characterize the more vegetarian primate species and show greater similarities to the faunivores. Even more than chimpanzees, humans regularly make use of tools of various kinds in their subsistence behavior, an aspect of behavior that has become extreme in recent centuries.

There is no single pattern of social organization among humans. There is more variability in the social organization of human societies than is found in any other primate species. Monogamous families and single-male groups of one male with several females are the most common arrangements throughout the world, but there are human populations in which polyandry (one female with several mates) or even more promiscuous multimale and multifemale groups are common. It is generally argued that culture complicates comparisons with nonhuman primate social organizations, since human social mores are often mandated by religion or law. Comparing human sexual dimorphism with that found in other primates does not offer any more convincing evidence of a natural social structure for our species. Our body size dimorphism allies us with polygynous mammal species (Alexander *et al.*, 1979). Our lack of canine dimorphism is more similar to that found among monogamous primates or among those living in fission–fusion societies. However, humans of both sexes have small canines, whereas in many monogamous primates (except *Callicebus*) both sexes have large canines. The morphological evidence thus suggests that human social structures are organized on a different morphological basis than those of other primates (e.g., Plavcan and van Schaik, 1994).

Several broad comparative studies have identified aspects of all human societies that are unusual compared with those of all other primates (Foley and Lee, 1989; Rodseth et al., 1990). Like many primates (e.g., chimpanzees, bonobos, spider monkeys), human societies are generally multilevel structures of small subsistence units grouped into larger communities. However, humans are unusual in that both males and females tend to maintain relationships with their kin regardless of normal dispersal (or residence) patterns. Intercommunity violence is generally the prerogative of males; human females do not form coalitions to fight coalitions of females from other communities. The evolutionary and adaptive origin of these unique human social features are subject of considerable debate.

Human communication is strikingly different from that of other primates in our extensive use of language, but there is increasing evidence of remarkable cognitive and communication abilities in chimpanzees and bonobos (e.g., deWaal, 1992; Savage-Rumbaugh and Lewin, 1994). Similarly, many aspects of nonverbal communication in humans can readily be recognized in chimpanzees (Goodall, 1986). Even human notions of morality have clear bases among other primates (deWaal, 1996).

Many of the details of human sexual behavior that have long been held to be unique to our species (Morris, 1967) seem to be found to various degrees and in various combinations among our primate relatives, from lovers' stares to female orgasms. From a primate perspective, the most striking reproductive features of humans are aspects of our life history (Charnov and Berrigan, 1993; Hill, 1993). Our newborn infants are extremely large compared to adult body size, despite being born at a relatively immature stage of development, and adult mortality among humans is strikingly low, leading to our long juvenile period and a very high life expectancy. More unusual than our long lifespan is the fact that a very large portion of the lifespan of human females is after menopause and seems nonreproductive (Fig. 7.16). The significance of menopause has long been a theoretical problem for evolutionary biologists (Williams, 1957). Most recently, Hawkes and colleagues (Hawkes *et al.*, 1997, 1998) have argued that the long postmenopausal lifespan of human females has been selected because in our species, older females or "grandmothers" make important contributions to the survival and reproductive success of their daughters and grandchildren.

ADAPTIVE RADIATION OF HOMINOIDS

Compared with other major radiations of primates, living hominoids (except humans) show a striking lack of taxonomic, morpho-

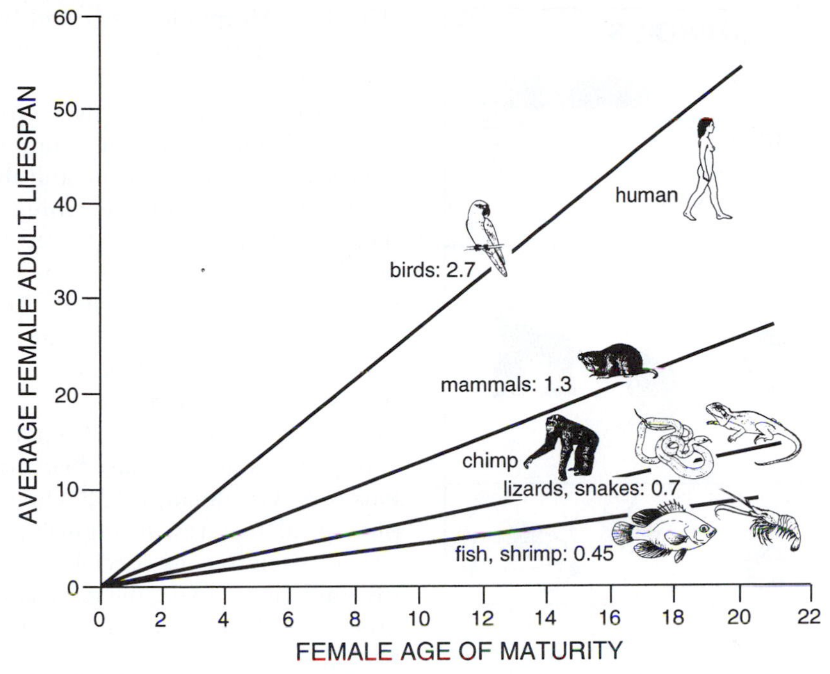

FIGURE 7.16 Humans are unusual among most other vertebrates in having a long female postreproductive lifespan (redrawn from Hill, 1993).

logical, and ecological diversity but a range in body size exceeding that found among any other extant primates (Fig. 7.17). Only the gibbons could be considered "normal-size" primates. Apes also show considerable interspecific differences in patterns of sexual dimorphism and in ontogeny (Leigh and Shea, 1995).

Ecologically, all apes are diurnal and all are more or less restricted to forested areas. Although all apes seem to utilize some suspensory locomotion, the African apes more commonly travel using arboreal and terrestrial quadrupedal (knuckle-walking) gaits. Human bipedalism is unique among primates. Living apes are all frugivorous and folivorous; there are no seed, insect, or gum specialists. Again, human dietary diversity and our regular use of agriculture and animal domestication are unique.

The most diverse aspect of the behavior of hominoids is their social organization. It is different for every genus, between chimpanzee species, and among populations of humans. Certainly this diversity belies most attempts to identify an ancestral social system for humans by simple extrapolation from our nearest relatives (Wrangham, 1987; but see Ghiglieri, 1987; Rodseth *et al.*, 1990). Several factors probably contribute to this diversity in hominoid social organization. One is the large size. Since hominoids are probably less subject to predation than are other primates, the antipredator advantages of group living are lessened (but see Boesch and Boesch, 1992) and we can afford a long juvenile period (Charnov and Berrigan, 1993; Hill, 1993). Associated with our large size is a considerable longevity compared with that of most primates. Our extended life span permits a

FIGURE 7.17 The adaptive radiation of living hominoids.

versatility in reproductive strategies at different life stages that is not available to animals with shorter life spans. Finally, the larger brain size and increased intelligence of apes probably permit a greater flexibility of social interactions based on memory and unique interindividual relationships (Dunbar, 1992; Whiten, 1996).

PHYLETIC RELATIONSHIPS OF HOMINOIDS

The proper systematic grouping of hominoids is a difficult and very subjective problem. As with the classification of prosimians and anthropoids, the issue is one of deciding between a gradistic and a phylogenetic classification (Fig. 1.3). Morphologically and behaviorally, there seem to be three groups of hominoids: hylobatids (the lesser apes), pongids (the great apes), and hominids (humans). Gibbons are undoubtedly the hominoids closest to cercopithecoid monkeys, and they seem to retain many primitive catarrhine features in their skeleton and visceral anatomy. For the great apes, I have adopted the gradistic approach in this book, grouping the great apes in Pongidae and separating out our own species (and our fossil relations) in Hominidae. But as noted in Chapter 1, there is no doubt that humans share a more recent heritage with chimpanzees and gorillas than with orangutans (e.g., Goodman, 1963). The unique morphological features distinguishing hominids are relatively recent specializations, whereas the similarities linking the great apes with each other are older, ancestral features that have been lost in the human lineage (Fig. 7.18).

The more difficult question concerns the phyletic relationships among humans and the African apes. Several types of biomolecular data indicate that humans and chimpanzees are more closely related to one another than either is to the gorilla (e.g., Bailey, 1993; Ruvolo, 1994). However this relationship seems in conflict with the fact that chimpanzees and

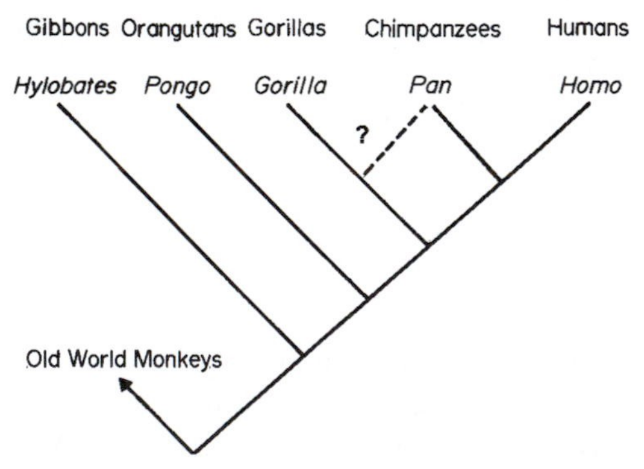

FIGURE 7.18 The phyletic relationships among hominoids.

gorillas share a number of unique dental and skeletal specializations, including thin dental enamel and the many musculoskeletal features of the forelimb associated with knuckle-walking (see, e.g., Andrews and Martin, 1987). Perhaps the exact relationships among chimpanzees, gorillas, and humans can never be clearly resolved, since there is already abundant evidence to support alternative views (Rogers, 1994), but the bulk of the evidence from studies of behavior and paleontology as well as genetics indicates that human ancestors were probably very much like chimpanzees.

BIBLIOGRAPHY

GENERAL

Doran, D. M. (1996). Comparative positional behavior of the African apes. In *Great Ape Societies,* ed. W. C. McGrew, L. F. Marchant, and T. Nishida, pp. 213–224. Cambridge: Cambridge Univ. Press.

Fleagle, J. G., and Jungers, W. L. (1982). Fifty years of higher primate phylogeny. In *A History of American Physical Anthropology (1930–1980),* ed. F. Spencer, pp. 187–230. New York: Academic Press.

Ghiglieri, M. D. (1987). Sociobiology of the great apes and the hominid ancestor. *J. Human Evol.* **16**:319–357.

Goodman, M. (1963). Man's place in the phylogeny of the primates as reflected in serum proteins. In *Classification and Human Evolution,* ed. S. L. Washburn, pp. 204–234. Chicago: Aldine.

Hamburg, D. A., and McCown, E. R., eds. (1979). *Perspectives on Human Evolution,* vol. 5: The Great Apes. Menlo Park, Calif.: Benjamin-Cummings.

McGrew, W. C., Marchant, L. F., and Nishida, T. (1996). *Great Ape Societies.* Cambridge: Cambridge University Press.

Preuschoft, H., Chivers, D. J., Brockelman, W. Y., and Creel, N., eds. (1984). *The Lesser Apes—Evolutionary and Behavioral Biology.* Edinburgh: Edinburgh University Press.

Reynolds, V. (1967). *The Apes.* New York: E. P. Dutton.

Schultz, A. H. (1968). The recent hominoid primates. In *Perspectives on Human Evolution,* ed. S. L. Washburn and P. C. Jay, pp. 122–195. New York: Holt, Rinehart and Winston.

Tuttle, R. H. (1987). *Apes of the World.* Park Ridge. N. J.: Noyes.

Wrangham, R. W. (1979) On the evolution of ape social systems. *Social Science Information* **18**:335–368.

______ (1987). The significance of African apes for reconstructing human social evolution. In *The Evolution of Human Behavior: Primate Models,* ed. W. G. Kinzey, pp. 51–71. Albany, N. Y.: SUNY Press.

HYLOBATIDS

Cannon, C. H., and Leighton, M. (1994). Comparative locomotor ecology of gibbons and macaques: Selection of canopy elements for crossing gaps. *Am. J. Phys. Anthropol.* **93**:505–524.

Carpenter, C. R. (1940). A field study in Siam of the behavior and social relations of the gibbon. *Comp. Psychol. Monogr.* **16**(5): 1–212.

Chivers, D. J. (1974). *The Siamang in Malaya. A Field Study of a Primate in a Tropical Rain Forest.* Contributions to Primatology, vol. 4. Basel, Switzerland: S. Karger.

______ (1977). The lesser apes. In *Primate Conservation,* ed. HSH Prince Ranier III and G. H. Bourne, pp. 539–598. New York: Academic Press.

______ (1984). Feeding and ranging in gibbons: A summary. In *The Lesser Apes—Evolutionary and Behavioral Biology,* ed. H. Preuschoft, D. J. Chivers, W. Y. Brockelman, and N. Creel, pp. 267–281. Edinburgh: Edinburgh University Press.

Creel, N., and Preuschoft, H. (1984). Systematics of the lesser apes: A quantitative taxonomic analysis of craniometric and other variables. In *The Lesser Apes—Evolutionary and Behavioral Biology,* ed. H. Preuschoft, D. J. Chivers, W. Y. Brockelman, and N. Creel, pp. 562–613. Edinburgh: Edinburgh University Press.

Dahl, J. F., and Nadler, R. D. (1992). Genital Swelling in Females of the Monogamous Gibbon, *Hylobates* (*H.*) *lar. Am. J. Phys. Anthrop.* **89**:101–108.

Fleagle, J. G. (1976). Locomotion and posture of the Malayan siamang and implications for hominoid evolution. *Folia Primatol.* **26**:245–269.

Gittens, G. P., and Raemaekers, J. J. (1980). Siamang, lar and agile gibbons. In *Maloyan Forest Primates,* ed. D. J. Chivers, pp. 63–105. New York: Plenum Press.

Groves, C. P. (1989). *A Theory of Human and Primate Evolution.* Oxford: Oxford University Press.

______ (1993). Speciation in living hominoid primates. . In *Species, Species Concepts, and Primate Evolution,* ed. W. H. Kimbel and L. B. Martin, pp. 109–121. New York: Plenum Press.

Haimoff, E. H., Yang, X. J., He, S.-J., and Chen, H. (1986). Census and survey of wild black-crested gibbons

(*Hylobates concolor concolor*) in Yunnan Province, People's Republic of China. *Folio Primatol.* **46**:205–214.

Lan, D-Y. (1993). Feeding and vocal behaviors of black gibbons (*Hylobates concolor*) in Yunnan: A preliminary study. *Folia primatol.* **60**:94–105.

Leighton, D. R. (1986). Gibbons: Territoriality and monogamy. In *Primate Societies,* ed. B. B. Smuts, D. L. Cheney, R. M. Seyfarth, R. W. Wrangham, and T. T. Struhsaker, pp. 135–145. Chicago: University of Chicago Press.

Marshall, J., and Sugardjito, J. (1986). Gibbon systematics. In *Comparative Primate Biology,* vol. 1: Systematics, Evolution, and Anatomy, ed. D. R. Swindler and J. Erwin, pp. 137–186. New York: Alan R. Liss.

Mitani, J. C. (1984). The behavioral regulation of monogamy in gibbons (*Hylobates muelleri*). *Behav. Ecol Sociobiol.* **15**:225–229.

______ (1990). Demography of the agile gibbons (*Hylobates agilis*). *Int. J. Primatol.* **11**:411–424.

Palombit, R. A. (1995). Longitudinal patterns of reproduction in wild female siamang (*Hylobates syndactylus*) and white-handed gibbons (*Hylobates lar*). *Int. J. Primatol.* **16**:739–760.

______ (1996). Pair bonds in monogamous apes: A comparison of the siamang Hylobates syndactylus and the white-handed gibbon *Hylobates lar. Behaviour* **133**:321–356.

______ (1997). Inter- and intra-specific variation in the diets of sympatric siamang (*Hylobates syndactylus*) and lar gibbons (*Hylobates lar*). *Folia. Primatol.* **68**:321–337.

Raemaekers, J. J. (1979). Ecology of sympatric gibbons. *Folia Primatol.* **31**:227–245.

Raemaekers, J. J., and Raemaekers, P. M. (1985). Field playback of loud calls to gibbons (*Hylobates lar*): Territorial, sex specific and species-specific responses. *Anim. Behav.* **33**:481–493.

Rumbaugh, D. M., ed. (1972–1976). *Gibbon and Siamang,* vols. 1–4. Basel, Switzerland: S. Karger.

Schultz, A. H. (1973). The skeleton of the Hylobatidae and other observations on their morphology. In *Gibbon and Siamang,* vol. 2, ed. D. M. Rumbaugh, pp. 1–54. Basel, Switzerland: S. Karger.

Tuttle, R. H. (1972). Functional and evolutionary biology of hylobatid hands and feet. In *Gibbon and Siamang,* vol. 1, ed. D. M. Rumbaugh, pp. 136–206. Basel, Switzerland: S. Karger.

PONGIDS

Orangutans

Cant, J. G. H. (1987). Positional behavior of female Bornean orangutans (*Pongo pygmaeus*). *Am. J. Primatol.* **12**: 71–90.

Dahl, J. F., Gould, K. G., and Nadler, R. D. (1993). Testicle size of orang-utans in relation to body size. *Am. J. Phys. Anthropol.* **90**:229–236.

Galdikas, B. M. E (1979). Orang-utan adaptations at Tanjung Puting Preserve: Mating and ecology. In *The Great Apes,* ed. D. A. Hamburg and E. R. McCown, pp. 195–233. Menlo Park, Calif.: BenjaminCummings.

______ (1985). Subadult male orangutan sociality and reproductive behavior at Tanjung Puting. *Am. J. Primatol.* **8**:87–99.

Galdikas, B. M. F., and Teleki, G. (1981). Variations in subsistence activities of female and male pongids: New perspectives on the origins of hominid labor division. *Curr. Anthropol.* **22**(3):241–256.

Leighton, M. (1993). Modeling dietary selectivity by Bornean orangutans: Evidence for integration of multiple criteria in fruit selection. *Int. J. Primatol.* **14**(2): 257–313.

MacKinnon, J. (1974a). The behavior and ecology of wild orang-utans (*Pongo pygmaeus*). *Anim. Behav.* **22**:3–74.

______ (1974b). *In Search of the Red Ape.* New York: Ballantine Books.

______ (1977). A comparative ecology of Asian apes. *Primates* **13**(4):747–772.

______ (1978). *The Ape within Us.* New York: Holt, Rinehart and Winston.

Maple, T. L. (1980). *Orang-utan Behavior.* New York: Van Nostrand Reinhold.

Markham, R., and Groves, C. P. (1990). Brief communication: Weights of wild orang-utans. *Am. J. Phys. Anthropol.* **81**:1–3.

Martin, L. (1986). Relationships among great apes and humans. In *Major Topics in Primate and Human Evolution,* ed. B. Wood, L. Martin, and P. Andrews, pp. 161–187. Cambridge: Cambridge University Press.

Mitani, J. (1985a). Mating behavior of male orangutans in the Kutai Game Reserve, Indonesia. *Anim. Behav.* **33**: 392–402.

______ (1985b). Sexual selection and adult male orangutan loud calls. *Anim. Behav.* **33**:272–283.

Mitani, J. C., Grether, G. F., Rodman, P. S., and Priatna, D. (1991). Associations among wild orang-utans: Sociality, passive aggregations or chance? *Anim. Behav.* **42**: 33–46.

Rijksen, H. D. (1978). A Field Study on Sumatran Orangutans (*Pongo pygmaeus abelii Lesson 1827*). Wageninger, The Netherlands: H. Veenman and Zonen.

Rodman, P. S. (1973). Population composition and adaptive organization among orang-utans of the Kutai Reserve. In *Comparative Ecology and Behavior of Primates,* ed. R. P. Michael and J. H. Crook, pp. 171–209. New York: Academic Press.

______ (1977). Feeding behavior of orang-utans of the Kutai Nature Reserve, East Kalimantan. In *Primate Ecology: Studies of Feeding and Ranging in Lemurs, Monkeys and Apes,* ed. T. H. Clutton-Brock, pp. 383–413. New York: Academic Press.

______ (1979). Individual activity patterns and the solitary nature of orangutans. In *The Great Apes,* ed. D. A. Hamburg and E. R. McCown, pp. 235–256. Menlo Park, Calif.: Benjamin-Cummings.

______ (1984). Foraging and social systems of orangutans and chimpanzees. In *Adaptations for Foraging in Nonhuman Primates,* ed. P. S. Rodman and J. G. H. Cant, pp. 134–160. New York: Columbia University Press.

______ (1988). Diversity and consistency in ecology and behavior. In *Orang-utan Biology,* ed. J. H. Schwartz, pp. 31–51, New York: Oxford University Press.

Rodman, P. S., and Mitani, J. (1986). Orangutans: Sexual dimorphism in a solitary species. In *Primate Societies,* ed. B. B. Smuts, D. L. Cheney, R. M. Seyfarth, R. W. Wrangham, and T. T. Struhsaker, pp. 146–154. Chicago: University of Chicago Press.

Schultz, A. H. (1941). Growth and development of the orang-utan. In *Contributions to Embryology.* Carnegie Institute Washington Pub. no. 525, vol. 29, pp. 57–110.

Schwartz, J. H., ed. (1988). *Orang-utan Biology.* New York: Oxford University Press.

Shea, B. T. (1988). Phylogeny and skull form in the hominoid primates. In *Orang-utan Biology,* ed. J. H. Schwartz, pp. 233–245, New York: Oxford University Press.

Sugardjito, J. (1982). Locomotor behavior of the Sumatran orang-utan (*Pongo pygmaeus abelii*) at Kekambe, Gunung Leuser National Park. *Malay Nat. J.* **35**:57–64.

Sugardjito, J., te Boekhorst, I. J. A., and van Hooff, J. A. R. A. M. (1987). Ecological constraints on the grouping of wild orang-utans (*Pongo pygmaeus*) in the Gunung Leuser National Park, Sumatra, Indonesia. *Int. J. Primatol.* **8**: 17–42.

Tuttle, R. H. (1969a). Knuckle-walking and the problem of human origins. *Science* **166**:953–961.

______ (1969b). Quantitative and functional studies on the hands of the Anthropoidea—I. The Hominoidea. *J. Morphol.* **128**(3):309–364.

van Schaik, C. P., and van Hoof, J. A. R. A. M. (1996). Toward an understanding of the orangutan's social system. In *Great Ape Societies,* ed. W. C. McGrew, L. F. Marchant, and T. Nishida, pp. 3–15. Cambridge: Cambridge University Press.

Gorillas

Byrne, R. W., and Byrne, J. M. E. (1993), Complex leaf-gathering skills of mountain gorillas (*Gorilla g. beringei*) variability and stabilization. *Am. J. Primatol.* **31**: 241–261.

Casimer, M. J. (1975). Feeding ecology and nutrition of an eastern gorilla group in the Mt. Kuhuzi region (Republic de Zaire). *Folia Primatol.* **24**:1–136.

Dixon, E. (1981). *The Natural History of the Gorilla.* New York: Columbia University Press.

Doran, D. and McNeilage, A. (1998). Gorilla behavior and ecology. *Evol. Anthrop.* **6**:120–131.

Fay, J. M., Agnagna, M., Moore, J., and Oko, R. (1989). Gorillas (*Gorilla gorilla gorilla*) in the Likouala swamp forests of north central Congo: Preliminary data on populations and ecology. *Int. J. Primatol.* **10**(5): 477–486.

Fossey, D. (1983). *Gorillas in the Mist.* Boston: Houghton Mifflin.

Harcourt, A. H. (1978). Strategies of emigration and transfer by primates, with particular reference to gorillas. *Z. Tierpsychol.* **48**:401–420.

______ (1979). The social relations and group structure of wild mountain gorillas. In *The Great Apes,* ed. D. A. Hamburg and E. R. McCown, pp. 187–192. Menlo Park, Calif.: Benjamin-Cummings.

Harcourt, A. H., and Stewart, K. J. (1989). Functions of alliances in contests within wild gorilla groups. *Behaviour* **109**:176–190.

Mace, G. M. (1990). Birth sex ratio and infant mortality rates in captive Western Lowland gorillas. *Folia Primatol.* **55**:156–165.

Maple, T. L., and Hoff, M. P. (1982). *Gorilla Behavior.* New York: Van Nostrand Reinhold.

Merfield, E. G., and Miller, H. (1956). *Gorilla Hunter: The African Adventures of a Hunter Extraordinary.* New York: Farrar, Straus and Cadahy.

Mitani, M. (1992). Preliminary results of the studies on wild Western Lowland gorillas and other sympatric diurnal primates in the Ndoki Forest, northern Congo. In *Topics in Primatology,* vol. 2, *Behavior, Ecology and Conservation,* ed. N. Itoigawa, Y. Sugiyama, G. P. Sackett, and R. K. R. Thompson, pp. 215–224. Tokoyo: University of Tokyo Press.

Mwanza, N., Yamagiwa, J., Yumoto, T. and Maruhashi, T. (1992). Distribution and range utilization of Eastern Lowland gorillas. In *Topics in Primatology,* vol. 2, *Behavior, Ecology and Conservation,* ed. N. Itoigawa, Y. Sugiyama, G. P. Sackett, and R. K. R. Thompson, pp.283–300 Tokyo: University of Tokyo Press.

Nishihara, T. (1992). A preliminary report on the feeding habits of Western Lowland gorillas (*Gorilla gorilla gorilla*) in the Ndoki Forest, northern Congo. In *Topics in Primatology,* vol. 2, *Behavior, Ecology and Conservation,* ed. N. Itoigawa, Y. Sugiyama, G. P. Sackett, and

R. K. R. Thompson, pp. 225–240. Tokyo: University of Tokyo Press.

Remis, M. J. (1994). *Feeding Ecology and Positional Behavior of Western Lowland Gorillas (Gorilla gorilla gorilla) in Central African Republic.* Ph.D. Dissertation, Yale University.

Rogers, M. E., and Williamson, E. A. (1987). Density of herbaceous plants eaten by gorillas in Gabon: Some preliminary data. *Biotropica* **19**(3):278–281.

Rogers, M. E., Maisels, F., Williamson, E. A., Tutin, C. E. G., Fernandez, M. (1992). Nutritional aspects of gorilla food choice in the lop reserve, Gabon. In *Topics in Primatology,* vol. 2, *Behavior, Ecology and Conservation,* ed. N. Itoigawa, Y. Sugiyama, G. P. Sackett, and R. K. R. Thompson, pp. 255–266. Tokoyo: University of Tokyo Press.

Ruvolo, M., Pan, D., Zehr, S., Goldberg, T., Disotell, T. R., and von Dornum, M. (1994). Gene trees and hominoid phylogeny. *Proc. Natl. Acad. Sci.* **91**:8900–8904.

Sarmiento, E. E., Butynski, T. M., and Kalina, J. (1996). Gorillas of Bwindi–Impenetrable forest and the Virunga volcanoes: taxonomic implications of morphological and ecological differences. *Amer. J. Primatol.* **40**:1–22.

Schaller, G. B. (1963). *The Mountain Gorilla.* Chicago: University of Chicago Press.

Stewart, K. J., and Harcourt, A. H. (1986). Gorillas: Variation in female relationships. In *Primate Societies,* ed. B. B. Smuts, D. L. Cheney, R. M. Seyfarth, R. W. Wrangham, and T. T. Struhsaker, pp. 155–164. Chicago: University of Chicago Press.

Tutin, C. E. G., and Fernandez, M. (1985). Foods consumed by sympatric populations of *Gorilla g. gorilla* and *Pan t. troglodytes* in Gabon: Some preliminary data. *Int. J. Primatol.* **6**:27–43.

Tutin, C. E. G., Fernandez, M., Rogers, M. E., and Williamson, E. A. (1992). A preliminary analysis of the social structure of Lowland gorillas in the Lope Reserve, Gabon. In *Topics in Primatology,* vol. 2, *Behavior, Ecology and Conservation,* ed. N. Itoigawa, Y. Sugiyama, G. P. Sackett, and R. K. R. Thompson, pp. 245–253. Tokyo: University of Tokyo Press.

Tuttle, R. H., and Watts, D. P. (1985). The positional behavior and adaptive complexes of Pan gorilla. In *Primate Morphophysiology, Locomotor Analyses and Human Bipedalism,* ed. S. Kondo, pp. 261–288. Tokyo: University of Tokyo Press.

Vedder, Amy L. (1984). Movement patterns of a group of free-ranging mountain gorillas (*Gorilla gorilla beringei*) and their relation to food availability. *Am. J. Primatol.* **7**(2):73–88.

Watts, D. (1984). Composition and variability of mountain gorilla diets in the central Virungas. *Am. J. Primatol.* **7**:323–356.

______ (1985). Relations between group size and composition and feeding competition in mountain gorilla groups. *Anim. Behav.* **33**:72–85.

______ (1991). Strategies of habitat use by mountain gorillas. *Folia Primatol.* **56**:1–16.

______ (1994). Social relationships of immigrant and resident female mountain gorillas, II: Relatedness, residence, and relationships between females. *Am. J. Primatol.* **32**:13–30.

______ (1996). Comparative socio-ecology of gorillas in great ape societies. In *Great Ape Societies,* ed. W. C. McGrew, L. F. Marchant, and T. Nishida, pp. 16–28. Cambridge:Cambridge University Press.

Yamagiwa, J., and Goodall, A. G. (1992). Comparative socio-ecology and conservation of gorillas. In *Topics in Primatology,* vol. 2, *Behavior, Ecology and Conservation,* ed. N. Itoigawa, Y. Sugiyama, G. P. Sackett, and R. K. R. Thompson, pp. 209–213. Tokoyo: University of Tokyo Press.

Yamagiwa, J., Mwanza, N., Yumoto, T., and Maruhashi, T. (1992). Travel distances and food habits of Eastern Lowland gorillas: A comparative analysis. In *Topics in Primatology,* vol. 2, *Behavior, Ecology and Conservation,* ed. N. Itoigawa, Y. Sugiyama, G. P. Sackett, and R. K. R. Thompson, pp. 267–281. Tokoyo: University of Tokyo Press.

Chimpanzees

Baldwin, P. J., McGrew, W. C., and Tutin, C. E. G. (1982). Wide-ranging chimpanzees at Mt. Asserik, Senegal. *Int. J. Primatol.* **3**(4):367–385.

Boesch, C. (1991a). The effects of leopard predation on grouping patterns in forest chimpanzees. *Behaviour* **117**:220–241.

______ (1991b). Teaching among wild chimpanzees. *Anim. Behav.* **41**:530–532.

Boesch, C., and Boesch, H. (1981). Sex differences in the use of natural hammers by wild chimpanzees: A preliminary report. *J. Human Evol.* **10**:265–286.

______ (1989). Hunting behavior of wild chimpanzees in the Taï National Park. *Am. J. Phys. Anthropol.* **78**:547–573.

______ (1990). Tool use and tool making in wild chimpanzees. *Folia Primatol.* **54**:86–99

Boesch, C., Marchesi, P., Marchesi, N., Fruth, B., and Joulian, F. (1994). Is nut cracking in wild chimpanzees a cultural behavior? *J. Human Evol.* **26**:325–338.

Boesch, H., and Boesch, C. (1994). Hominization in the rainforest: The chimpanzee's piece of the puzzle. *Evol. Anthropol.* **3**:171–178

Bourne, G. H., ed. (1969–1970). The *Chimpanzee,* vols. 1–2. Basel, Switzerland: S. Karger.

Byrne, R. W., and Byrne, J. M. E. (1988). Leopard killers of Mahale. *Nat. Hist.* **97**(3):22–24.

______ (1993). Complex leaf-gathering skills of mountain gorillas (*Gorilla g. beringei*): Variability and standardization. *Am. J. Primatol.* **31**(4):241–262.

Chapman, C. A., and Wrangham, R. W. (1993). Range use of the forest chimpanzees of Kibale: Implications for the understanding of chimpanzee social organization. *Am. J. Primatol.* **31**(4):263–274.

Collins, D. A., and McGrew, W. C. (1988). Habitats of three groups of chimpanzees (*Pan troglodytes*) in western Tanzania compared. *J. Human Evol.* **17**:553–574.

deWaal, E. (1982). *Chimpanzee Politics: Power and Sex among Apes.* New York: Harper and Row.

______ (1988). The communicative repertoire of captive bonobos (*Pan paniscus*), compared to that of chimpanzees. *Behavior* **106**:183–251.

Doran, D. M. (1992). Comparison of instantaneous and locomotor bout sampling methods: A case study of adult male chimpanzee locomotor behavior and substrate use. *Am. J. Phys. Anthropol.* **89**:85–99.

Doran, D. M. (1992). The ontogeny of chimpanzee and pygmy chimpanzee locomotor behavior: a case study of paedomorphism and its behavioral corelates. *J. Human Evol.* **23**:139–157.

Doran, D. M. (1993). The comparative locomotor behavior of chimpanzees and bonobos: the influence of morphology on locomotion. *Amer. J. Phys. Anthropol.* **91**: 83–98.

Doran, D. M., and Hunt, K. D. (1994). The comparative locomotor behavior of chimpanzees and bonobos: species and habitat differences. In *Chimpanzee Cultures,* ed. R. W. Wrangham, W. C. McGrew, F. B. M. deWaal, and P. G. Heltne, pp. 93–108. Cambridge, MA: Harvard University Press.

Gagneux, P., Woodruff, D. S., and Boesch, C. (1997). Furtive mating in female chimpanzees. *Nature* **387**:358–359.

Ghiglieri, M. P. (1984). *The Chimpanzees of Kibale Forest.* New York: Columbia University Press.

______ (1985). The social ecology of chimpanzees. *Sci. Am.* **252**:36–40.

______ (1988). *East of the Mountains of the Moon.* New York: The Free Press.

Gonder, M. K., Oates, J. F., Disotell, T. R., Forster, A. R. J., Morales, J. C., and Melnick, D. J. (1997). A new west African chimpanzee subspecies? *Nature* **388**:337.

Goodall, J. van L. (1965). Chimpanzees of the Gombe Stream Reserve. In *Primate Behavior,* ed. I. Devore, pp. 425–473. New York: Holt, Rinehart and Winston.

______ (1968). The behavior of free-living chimpanzees of the Gombe Stream Reserve. *Anim. Behav. Monographs* **1**:161–311.

______ (1971). *In the Shadow of Man.* New York: Dell.

______ (1983). Population dynamics during a fifteen-year period in one community of free-living chimpanzees in the Gombe National Park, Tanzania. *Z. Tierpsychol.* **64**:1–60.

______ (1986). *Chimpanzees of Gombe.* Cambridge, Mass.: Harvard University Press.

Heltne, P. G., and Marquardt, L. A., eds. (1989). *Understanding Chimpanzees.* Cambridge, Mass.: Harvard University Press, 407 pp.

Hunt, K. D. (1992). Positional behavior of *Pan troglodytes* in the Mahale Mountains and Gombe Stream National Parks, Tanzania. *Am. J. Phys. Anthropol.* **87**:83–105.

McGrew, W. C. (1983). Animal foods in the diets of wild chimpanzees (*Pan troglodytes*): Why cross-cultural variation? *J. Ethol.* **1**:46–61.

______ (1992a). *Chimpanzee Material Culture: Implications for Human Evolution.* Cambridge: Cambridge University Press, 277 pp.

______ (1992b). Tool-use by free-ranging chimpanzees: The extent of diversity. *J. Zool., Lond.* **228**:689–594.

McGrew, W. C., and Marchant, L. F. (1992). Chimpanzees, tools, and termites. Hand preference or handedness? *Current Anthrop.* **33**(1):114–119.

McGrew, W. C., Baldwin, C. J., and Tutin, C. E. G. (1981). Chimpanzees in a hot, dry and open habitai: Mt. Asserik, Senegal, West Africa. *J. Human Evol.* **10**: 227–244.

McGrew, W. C., Marchant, L. F., and Nishida, T. (1996). *Great Ape Societies.* Cambridge: Cambridge University Press

Moore, J. (1992). "Savanna" chimpanzees." In *Topics in Primatology,* vol. 1, *Human Origins,* ed. T. Nishida, W. C. McGrew, P. Marler, M. Pickford, and F. B. M. de Waal, pp. 99–118. Tokoyo: University of Tokyo Press.

Morin, P. A., Moore, J. J., Chakraborty, R., Jin, L., Goodall, J., and Woodruff, D. S. (1994). Kin selection, social structure, gene flow, and the evolution of chimpanzees. *Science* **265**:1193–1201.

Nishida, T. (1979). The social structure of chimpanzees of the Mahale Mountains. In *The Great Apes,* ed. D. A. Hamburg and E. R. McCown, pp. 73–121. Menlo Park, Calif.: Benjamin-Cummings.

Nishida, T., ed. (1990). *The Chimpanzees of the Mahale Mountains.* Tokyo: University of Tokyo Press.

Nishida, T., and Hiraiwa-Hasegawa, M. (1986). Chimpanzees and bonobos: Cooperative relationships among males. In *Primate Societies,* ed. B. B. Smuts, D. L. Cheney, R. M. Seyfarth, R. W. Wrangham, and T. T. Struhsaker, pp. 165–177. Chicago: University of Chicago Press.

Pusey, A. E. (1990). Behavioral changes at adolescence in chimpanzees. *Behavior* **115**(3–4):203–246.

Reynolds, V. (1965). *Budongo: An African Forest and Its Chimpanzees.* New York: Natural History Press.

Rohles, F. H., ed. (1972). *The Chimpanzee: A Topical Bibliography.* Seattle, Wash.: Primate Information Center, Regional Primate Research Center.

Shea, B. T., and Coolidge, H. J. (1988). Craniometric differentiation and systematics in the genus *Pan. J. Human Evol.* **17**:671–685.

Shea, B. T. , Leigh, S. R., and Groves, C. P. (1993). Multivariate craniometric variation in chimpanzees. In *Species, Species Concepts, and Primate Evolution.* ed. W. H. Kimbel and L. B. Martin, pp. 265–296. New York: Plenum Press.

Stanford, C. B., Wallis, J., Matama, H., and Goodall, J. (1994). Patterns of predation by chimpanzees on red colobus monkeys in Gombe National Park, 1982–1991. *Am. J. Phys. Anthropol.* **94**:213–228.

Sugiyama, Y. (1984). Population dynamics of wild chimpanzees at Bossou, Guinea, between 1976 and 1983. *Primates* **25**:391–400.

——— (1994). Age-specific birth rate and lifetime reproductive success of chimpanzees at Bossou, Guinea. *Am. J. Primatol.* **32**(4):311–318.

Teleki, G. (1981). The omnivorous diet and eclectic feeding habits of chimpanzees in Gombe National Park, Tanzania. In *Omnivorous Primates: Gathering and Hunting in Human Evolution,* ed. R. S. O. Harding and G. Teleki, pp. 303–343. New York: Columbia University Press.

Tsukaharam, T. (1993). Lions eat chimpanzees: The first evidence of predation by lions on wild chimpanzees. *Am. J. Primatol.* **29**:1–11.

Tutin, C. E. G. (1980). Reproductive behavior of wild chimpanzees in the Gombe National Park, Tanzania. *J. Reprod. Fert., Suppl.* **28**:43–57.

Tutin, C. E. G., and McGinnis, P. R. (1981). Chimpanzee reproduction in the wild. In *Reproductive Biology of the Great Apes,* ed. C. E. Graham, pp. 239–264. New York: Academic Press.

Wrangham, R. W. (1979). On the evolution of ape social systems. *Social Science Information* **18**(3):335–368.

Wrangham, R. W., and Peterson, D. (1996). *Demonic Males.* Boston: Houghton Mifflin.

Wrangham, R. W., and Smuts, B. B. (1980). Sexual differences in the behavioral ecology of chimpanzees in the Gombe National Park, Tanzania. *J. Reprod. Fert., Suppl.* **28**:13–31.

Wrangham, R. W., Clark, A. P., and Isabirye-Basuta, G. (1992). Female social relationships and social organization of Kibale Forest chimpanzees. In *Topics in Primatology,* vol. 2, *Behavior, Ecology and Conservation,* ed. N. Itoigawa, Y. Sugiyama, G. P. Sackett, and R. K. R. Thompson, pp. 81–98. Tokoyo: University of Tokyo Press.

Wrangham, R. W., Conklin, N. L., Etot, G., Obua, J., Hunt, K. D., Hauser, M. D., and Clark, A. P. (1993). The value of figs to chimpanzees. *Int. J. Primatol.* **14**(2): 243–256.

Wrangham, R. W., McGrew, W. C., deWaal, F. B. M., and Heltne, P. G. (1994). *Chimpanzee Cultures.* Cambridge, Mass.: Harvard University Press.

Bonobos

Badrian, A., and Badrian, N. (1980). The other chimpanzee. *Animal Kingdom* (Aug.–Sept. 1980):173–181.

Badrian, N., Badrian, A., and Susman, R. L. (1981). Preliminary observations on the feeding behavior of *Pan paniscus* in the Lomako Forest of Central Zaire. *Primates* **22**(2): 173–181.

Chapman, C. A., White, F. J., and Wrangham, R. W. (1994). Party size in chimpanzees and bonobos: A reevaluation of theory based on two similarly forested sites. In: *Chimpanzee Cultures,* ed. R. W. Wrangham, W. C. McGrew, F. M. B. de Waal, and P. G. Heltne, pp. 41–57. Cambridge, MA.: Harvard University Press.

de Waal, F. B. M. (1992). Appeasement, celebration, and food sharing in the two Pan species. In *Topics in Primatology,* vol. 1, *Human Origins,* ed. T. Nishida, W. C. McGrew, P. Marler,M. Pickford, and F. B. M. Waal, pp. 37–51. Tokoyo: University of Tokyo Press.

Doran, D. M., and Hunt, K. D. (1994). The comparative locomotor behavior of chimpanzees and bonobos: species and habitat differences. In *Chimpanzee Cultures,* ed. R. W. Wrangham, W. C. McGrew, F. B. M., de Waal, and P. G. Heltne, pp. 93–108. Cambridge, MA: Harvard University Press.

Furuichi, T. (1989). Social interactions and the life history of female *Pan paniscus* in Wamba, Zaire. *Int. J. Primatol.* **10**:173–197.

Kano, T. (1979). A pilot study on the ecology of pygmy chimpanzees, *Pan paniscus.* In *The Great Apes,* ed. D. A. Hamburg and E. R. McCown, pp. 123–136. Menlo Park, Calif.: Benjamin-Cummings.

——— (1992). *The Last Ape: Pygmy Chimpanzee Behavior and Ecology.* Stanford, Calif.: Stanford University Press.

Malenky, R. K., and Wrangham, R. W. (1994). A quantitative comparison of terrestrial herbaceous food consumption by *Pan paniscus* in the Lamako Forest, Zaire, and *Pan troglodytes* in the Kibale Forest, Uganda. *Am. J. Primatol.* **32**:1–12.

Savage-Rumbaugh, S., and Lewin, R. (1994). *Kanzi: The Ape at the Brink of the Human Mind.* New York: Wiley.

Susman, R. L., ed. (1984). *The Pygmy Chimpanzee: Evolutionary Biology and Behavior.* New York: Plenum Press.

Susman, R. L., and Jungers, W. L. (1981). Comments on: Bonobos—General hominid prototype or special insular dwarfs? *Curr. Anthropol.* **22**(4):369–370.

Susman, R. L., Badrian, N. L., and Badrian, A. J. (1980). Locomotor behavior of *Pan paniscus* in Zaire. *Am. J. Phys. Anthropol.* **53**:69–80.

White, F. J. (1992a). Pygmy chimpanzee social organization: Variation with party size and between study sites. *Am. J. Primatol.* **26**(3):203–214.

______ (1992b). Activities, budgets, feeding behavior, and habitat use of pygmy chimpanzees at Lomako, Zaire. *Am. J. Primatol.* **26**(3):215–224.

______ (1996a). Comparative socio-ecology of *Pan paniscus*. In *Great Ape Societies,* ed. W. C. McGrew, L. F. Marchant, and T. Nishida, pp. 29–41. Cambridge: Cambridge University Press.

______ (1996b). *Pan paniscus* 1973 to 1996: Twenty-three years of field research. *Evol. Anthropol.* **5**:11–17.

White, F. J., and Lanjouw, A. (1993). Feeding competition in Lomako bonobos: Variation in social cohesion. In *Topics in Primatology,* vol. 1, *Human Origins,* ed. T. Nishida, W. C. McGrew, P. Marler, M. Pickford, and F. B. M. de Waal, pp. 67–79. Tokoyo: University of Tokyo Press.

White, F. J., and Wrangham, R. W. (1988). Feeding competition and patch size in the chimpanzee species *Pan paniscus* and *Pan troglodytes. Behavior* **105**(2):148–164.

Wrangham, R. W. (1986). Ecology and social relationships in two species of chimpanzee. In *Ecological Aspects of Social Evolution,* ed. D. I. Rubenstein and R. W. Wrangham, pp. 352–378. Princeton, N. J.: Princeton University Press.

SIZE AND EVOLUTION OF AFRICAN APES

Doran, D. M. (1992). The ontogeny of chimpanzee and pygmy chimpanzee locomotor behavior: A case study of paedomorphism and its behavioral correlates. *J. Human Evol.* **23**:139–157.

Shea, B. T. (1985). Ontogenetic allometry and scaling: A discussion based on the growth and form of the skull in African apes. In *Size and Scaling in Primate Biology,* ed. W. L. Jungers. New York: Plenum Press

HOMINIDS

Alexander, R. D., Hoogland, J. L., Howard, R. D., Noonan, K. M., and Sherman, P. W. (1979). Sexual dimorphisms and breeding systems in pinnipeds, ungulates, primates and humans. In *Evolutionary Biology and Human Social Organization,* ed. N. A. Chagnon and W. Irons, pp. 402–435. Boston, Mass.: Duxbury Press.

Charnov, E. L., and Berrrigan, D. (1993). Why do female primates have such long lifespans and so few babies? or Life in the slow lane. *Evol. Anthrop.* **1**(6):191–194.

deWaal, F. (1996). *Good Natured: The Origins of Right and Wrong in Humans and Other Animals.* Cambridge, Mass.: Harvard University Press

Dunbar, R. I. M. (1992). Neocortex size as a constraint on group size in primates. *J. Human Evol.* **20**:469–493.

Foley, R. A., and Lee, P. C. (1989). Finite social space, evolutionary pathways, and reconstructing hominid behavior. *Science* **243**:901–906.

Harrison, G. A., Tanner, J. M., Pilbeam, D. R., and Baker, P. T. (1988). *Human Biology.* Oxford: Oxford University Press.

Hawkes, K., O'Connell, J. F., and Blurton-Jones, N. (1997). Hadza women's time allocation, offspring provisioning, and the evolution of post-menopausal lifespans. *Curr. Anthrop.* **38**:551–577.

Hawkes, K., O'Connell, J. F., Blurton-Jones, N., Alvarez, N., and Charnov, E. L. (1998). Grandmothering and the evolution of human life histories. *Proc. Natl. Acad. Sci. USA* **95**:1336–1339.

Hill, K. (1993). Life history theory and evolutionary anthropology. *Evol. Anthrop.* **2**(3):78–88.

Hrdy, S. B. (1981). *The Woman That Never Evolved.* Cambridge, MA: Harvard University Press.

Manson, J. H., and Wrangham, R. W. (1991). Integroup aggression in chimpanzees and humans. *Curr. Anthrop.* **32**(4):369–390.

Morris, D. (1967). *The Naked Ape.* New York: McGrawHill.

Plavcan, M. J., and van Schaik, C. P. (1994). Canine dimorphism. *Evol. Anthrop.* **2**(6):208–214.

Rodseth, L., Wrangham, R. W., Harrigan, A. M., and Smuts, B. B. (1990). The human community as a primate society. *Curr. Anthrop.* **32**(3):221–254.

Williams, C. C. (1957). Pleiotropy, natural selection, and the evolution of senescence. *Evolution* **11**:398–411.

ADAPTIVE RADIATION OF HOMINOIDS

Fleagle, J. G. (1976). Locomotion and posture of the Malayan siamang and implications for hominoid evolution. *Folia Primatol.* **26**:245–269.

Ghiglieri, M. P. (1987). Sociobiology of the great apes and the hominid ancestor. *J. Human Evol.* **16**:319–357.

Leigh, S. R., and Shea, B. T. (1995). Ontogeny and the evolution of adult body size dimorphism in apes. *Am. J. Phys. Anthropol.* **36**:37–60.

MacKinnon, J. (1978). *The Ape within Us.* New York: Holt, Rinehart and Winston.

McGrew, W. C., Marchant, L. F., and Nishida, T. (1996). *Great Ape Societies.* Cambridge: Cambridge University Press

Schultz, A. H. (1968). The recent hominoid primates. In *Perspectives on Human Evolution,* ed. S. L. Washburn and P. C. Jay, pp. 122–195. New York: Holt, Rinehart, and Winston.

Whiten, A. (1996). Ape mind, monkey mind. *Evol. Anthrop.* **5**:3–4.

Wrangham, R. W. (1987). The significance of African apes for reconstructing human social evolution. In *The Evolution of Human Behavior: Primate Models,* ed. W. G. Kinzey. Albany, N. Y.: SUNY Press.

PHYLETIC RELATIONSHIPS OF HOMINOIDS

Andrews, P., and Martin, L. (1987). Cladistic relationships of extant and fossil hominoids. *J. Human Evol.* **16**: 101–118.

Bailey, W. J. (1993). Hominoid trichotomy: A molecular overview. *Evol. Anthrop.* **2**(3):100–108.

Cronin, J. E. (1983). Apes, humans and molecular clocks: A reappraisal. In *New Interpretations in Ape and Human Ancestry,* ed. R. L. Ciochon and R. S. Corruccini, pp. 115–150. New York: Plenum Press.

Fleagle, J. G., and Jungers, W. L. (1982). Fifty years of higher primate phylogeny. In *A History of American Physical Anthropology (1930–1980),* ed. F. Spencer, pp. 187–230. New York: Academic Press.

Goodman, M. (1963). Man's place in the phylogeny of the primates as reflected in serum proteins. In *Classification and Human Evolution,* ed. S. L. Washburn, pp. 204–234. Chicago: Aldine.

Miyamoto, M. M., Slightom, J. L., and Goodman, M. (1987). Phylogenetic relations of humans and African apes from DNA sequences in the ψ- & η-globin region. *Science* **238**:369–373.

Rogers, J. (1994). Levels of the geneological hierarchy and the problem of hominoid phylogeny. *Am. J. Phys. Anthropol.* **94**:81–88.

Ruvolu, M., Pan, D., Zehr, S., Goldberg, T., Disotell, T. R., and von Dornum, M. (1994). Gene trees and hominoid phylogeny. *Proc. Natl. Acad. Sci. USA* **91**:8900–8904.

Simpson, G. G. (1966). The biological nature of man. *Science* **152**(3721):472–478.

EIGHT

Primate Communities

A BIOGEOGRAPHICAL PERSPECTIVE

In the preceding chapters, we have discussed the behavior and ecology of living primates as individual species or as phyletic radiations of related taxa. This is not the way that primates are encountered in their natural habitats. Rather, if one walks through a tropical forest in South America, Africa, Madagascar, or Asia, one normally encounters a collection of distantly related, sympatric species sharing various parts of a common habitat. In this chapter we move from the earlier chapters' focus on systematic organization to a consideration of living primates from a biogeographical perspective. We will compare and contrast the numbers and the ecological adaptations of primates that are found on different continents in order to examine patterns of primate diversity on a global scale.

As the previous chapters document, most living species of primates have been studied under naturalistic conditions to some degree. Virtually all have been the subject of short-term observations or surveys, and many have been the subject of longer studies lasting a year or more. In some cases these long-term studies have been part of extensive research projects that compare and contrast a number of species in a single community. Much of our understanding about primate behavior and ecology has come from comparative studies of sympatric species at sites such as Manu and Raleighvallen-Voltzberg in South America; Kibale, Tai, and Makokou in Africa; Ranomafana, Marosalaza, Kirindi, and Beza Mahafaly in Madagascar; and Kuala Lompat, Kutai, and Ketembe in Asia. Despite the large number of comparative studies of the similarities and difference among species within communities, there have been relatively few attempts to compare patterns of species diversity and ecology of primates among communities and on different continents (e.g., Bourliere, 1985; Terborgh and van Schaik, 1987; Fleagle and Reed, 1996). What are the broad biogeographical patterns that describe the distribution of primate species on different continents (Fig. 8.1)? How similar or different are the primates and communities of primates that we find in South America, Africa, Madagascar, and Asia? This chapter describes an area of primatology that is very much in progress.

Primate Biogeography

There are approximately fifty genera and 250 species of living primates. What general factors determine the distribution of these species on a global basis? It has been well established for many types of organisms that the number of species in any given region is a function of

FIGURE 8.1 Living primates are found in four major biogeographical regions: the neotropics of Central and South America, Africa, Madagascar, and Southern Asia.

the geographical area of the region. This is known as the **species–area relationship** (e.g., Rosenzweig, 1995). The causal mechanisms underlying this pattern are not clearly understood, but it seems reasonable that larger areas provide more potential geographical barriers and more habitat diversity, thus giving rise to or maintaining greater numbers of species.

As discussed in Chapter 4, primates are primarily animals of tropical forests, with only a few hardy species extending to temperate areas or nonforested habitats. The number of primate species found in major continental areas, and large islands is largely a function of the amount of tropical forest in those regions (Fig. 8.2; Reed and Fleagle, 1995). Thus the islands of Java, Sumatra, and Borneo have increasing numbers of species, in accordance with the increasing amount of tropical forest found on those islands, and South America and Africa have the largest forest blocks and the largest numbers of species. Madagascar has many more species than predicted by the size of the island. This may partly reflect the fact that the present forest area is greatly reduced from the time when the species evolved, but it also probably results from the absence of many other groups of mammals and birds on Madagascar to compete with primates for resources. Thus primates make up a larger proportion of the species on Madagascar than they do in areas with a more diverse mammalian fauna.

In addition to showing a species–area relationship, primates are similar to many other organisms in showing greater diversity near the equator than at higher latitudes. This latitudinal gradient has been demonstrated for South America (Ruggiero, 1994) and for Africa (Eeley and Lawes, 1998). Like species–area curves, increased diversity at low latitudes is a widespread, but poorly understood,

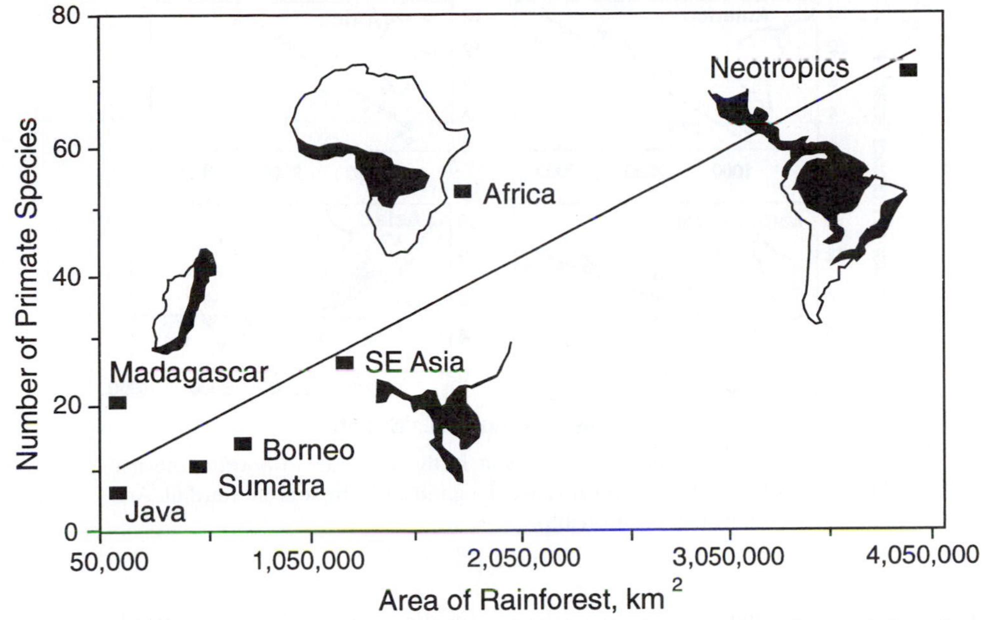

FIGURE 8.2 The number of primate species on large islands or continental areas is largely a function of the area of tropical rain forest.

phenomenon. Common explanations include higher productivity near the equator associated with greater solar radiation and less climatic seasonality, as well as more long-term patterns of species evolution (e.g., Gaston and Williams, 1996). However, others have argued that the apparent latitudinal gradients of many taxa may be heavily driven by species–area effects, since for South America and Africa, the greatest area of these continents surrounds the equator (Terborgh, 1973; Rosenzweig, 1995).

In addition to the increased density of primate species at low latitudes, primate species found near the equator tend to have smaller geographic ranges than those found at higher latitudes (Ruggiero, 1994; Eeley and Lawes, 1998). This pattern is called **Rapoport's Rule** (Stevens, 1989). As with other biogeographical rules, the underlying causality of this pattern is the subject of considerable debate. Most authorities feel that it reflects increasing climatic harshness with distance from the equator. Species that can live in more temperate habitats are likely to be generalists with wider habitat tolerances than more equatorial species.

Although these general patterns provide a broad guide to the number of species one is likely to find on islands and continents, in many parts of the world the number of species at any individual site seems to be closely correlated with rainfall (Fig. 8.3). The neotropics, Africa, and Madagascar exhibit a positive relationship between rainfall and the number of primate species found at individual sites, up to rainfall levels of about 2000 mm per year (Reed and Fleagle, 1995). There is some evidence that at higher rainfall levels, the number of species declines, perhaps due to leaching of soils (Kay *et al.*, 1997). In Asia, the relationship between rainfall and number of species does not seem to follow any obvious pattern.

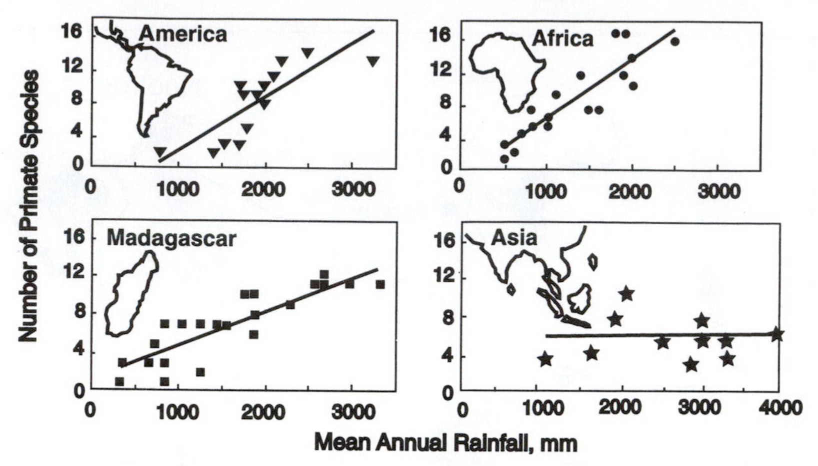

FIGURE 8.3 The number of species at individual sites in South America, Africa, and Madagascar is positively correlated with the annual rainfall. Asia exhibits no such clear relationship.

This lack of a pattern may be due in part to species–area effects on the different islands of the Sunda Shelf and in part to the fact that the diversity–rainfall relationship doesn't seem to hold at high levels of rain.

A number of hypotheses have been put forward to explain why primate species diversity is positively related to rainfall. Several authors have found that species diversity of plants is also positively associated with rainfall, and primates species diversity may reflect the diversity of plant foods available (e.g., Ganzhorn *et al.*, 1997). Others have argued, more generally, that primate diversity reflects overall productivity of the forests, in leaves, fruits, and flowers that primates eat. Indeed, Kay *et al.* (1997) have recently argued that productivity is the factor driving primate species diversity and that species diversity and productivity show the same curvilinear relationship with rainfall.

In contrast with the relationships between primate species diversity and rainfall or productivity, attempts to explain diversity in primate biomass have not yielded any simple results (e.g., Oates *et al.*, 1990). In some regions, primate biomass seems positively correlated with rainfall; other areas exhibit a negative correlation.

Ecology and Biogeography

The preceding paragraphs discussed general patterns that describe the distribution of numbers of primate species as a function of continental area, latitude, and climate. However, no general prediction was made about the behavior and ecology of the primate species found in different parts of the world. How similar or different are the primate faunas found in different continental areas?

Primates today are largely restricted to the tropical forested regions of Central and South America, sub-Saharan Africa, Madagascar, and southern Asia (Fig. 8.1). Each of these four major biogeographical regions has a distinctive fauna, with no common species between them and, in the case of the neotropics and Madagascar, no common families. Indeed the only genus shared by any two regions is *Macaca,* which has numerous species in Asia and a single species in north Africa. Africa and Asia also share several common families.

Although these different biogeographical regions inhabited by primates are similar in supporting tropical forests of various types, there are many differences in the forest habitats, both within and between regions. These include differences in climate and seasonality, in the nature of the underlying geology and soils, in the taxonomic and structural composition of the flora, and in the other animals that live in association with the primates. Several recent studies have compared aspects of life history, ecology, and behavior of the primates found on different continents (Reed and Fleagle, 1993; Kappeler and Heymann, 1996; Fleagle and Reed, 1996).

In body size distributions, neotropical primates are significantly smaller than those of other continents, if the subfossil species are included for Madagascar (Kappeler and Heymann, 1996). If the subfossil Malagasy species are not included, Madagascar shows a similar size range as the neotropics, but the distribution of sizes is significantly different among all major faunas (Fig. 8.4). Although there are both nocturnal and diurnal species in all major primate faunas, Madagascar is significantly different from other regions in the high proportion of nocturnal and cathemeral species (Fig. 8.5; Kappeler and Heymann, 1996).

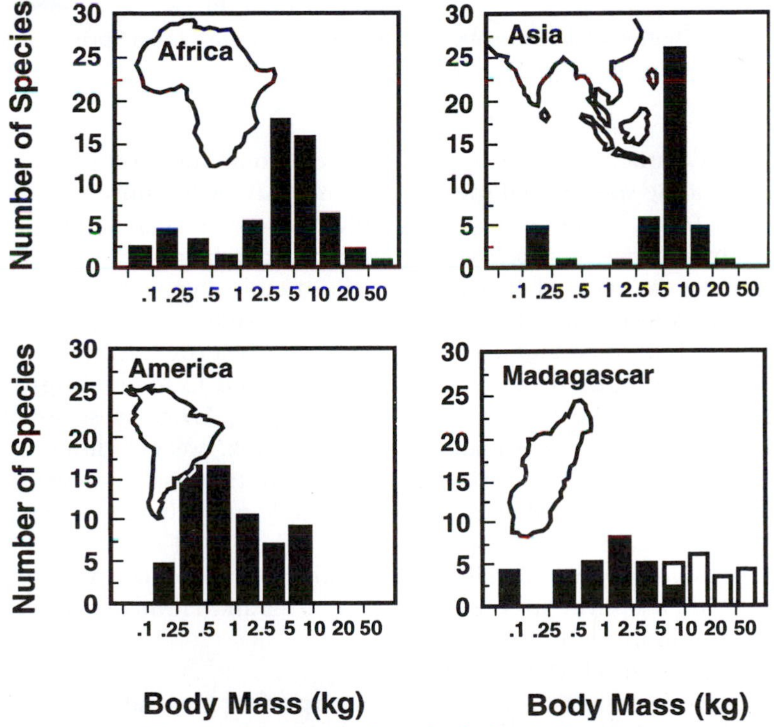

FIGURE 8.4 The distribution of primate body sizes in four biogeographical areas. White columns indicate recently extinct Malagasy species (courtesy of Peter Kappeler).

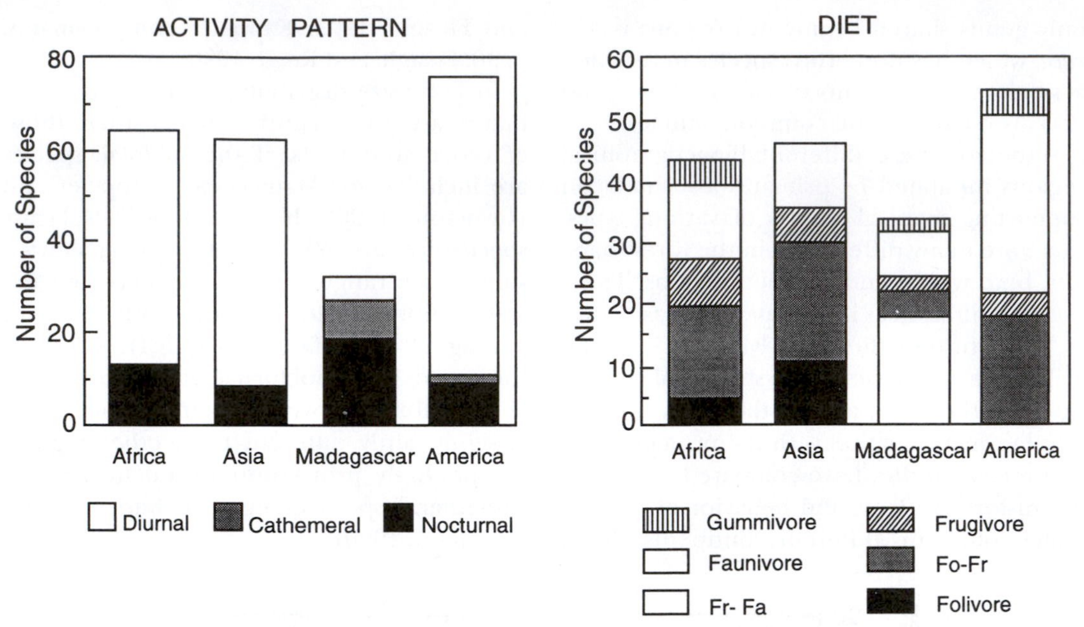

FIGURE 8.5 The distribution of activity pattern (left) and diet (right) among primates species of four major geographical areas (courtesy of Peter Kappeler).

All primate fauna contain primates with a variety of diets, including species that specialize on fauna, gums, fruits, leaves, and various combinations of these foods. As discussed in Chapter 3, classifying the diets of individual species is a difficult undertaking that ignores many important details of inter- and intraspecific behavior. Nevertheless, there are a number of striking patterns in the frequency of different dietary habits among the primates of major continental areas (Fig. 8.5; Terborgh and van Schaik, 1987; Fleagle and Reed, 1996; Kappeler and Heymann, 1996). Compared with other regions, Madagascar is dominated by folivorous species. This is true for both the living fauna and even more so if the subfossil species are included (Tattersall, 1982; Godfrey *et al.*, 1997). In contrast, folivory is relatively rare in the neotropics, which are dominated by frugivorous and insectivorous species. In Asia it is the frugivorous-folivorous species that are dominant. Compared with other continents, Africa is unusual in the balance of different dietary habits and the lack of any dominant dietary type.

Even more than diet, grouping behavior is difficult to reduce to simple, quantitative comparisons. Nevertheless, we can say that most primates tend to forage in groups of a characteristic size. When these are plotted by continental area, we get different distributions for each of the major regions (Fig. 8.6). Most noticeable is the high frequency among Malagasy primates of species that forage solitarily and their absence in neotropical species (Kappeler and Heymann, 1996).

Comparing Primate Communities

In addition to examining the frequency of different ecological and behavioral features in

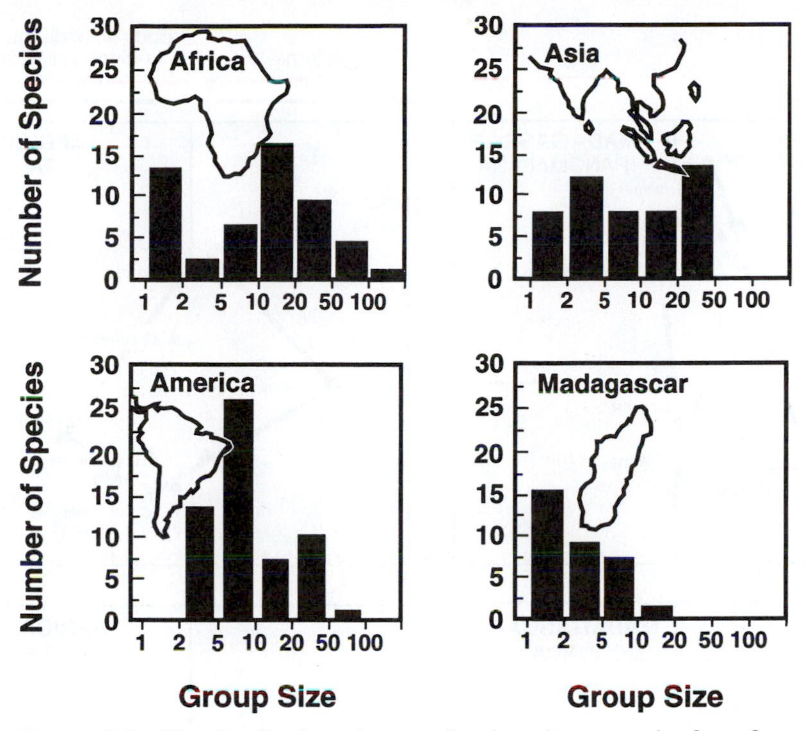

FIGURE 8.6 The distribution of group sizes in primate species from four major geographical areas (courtesy of Peter Kappeler).

the primates of different continents, we can also look specifically at the ecological composition of individual communities. Despite the different phylogenetic histories of the primates of South America, Africa, Madagascar, and Asia, do they occupy roughly the same type of "ecological space?" Or are primates on different continents doing very different things? Are the ecological differences between communities of primates within a single biogeographical area greater or less than those between regions?

To address these questions, Fleagle and Reed (1996) analyzed ecological data on body size, activity pattern, diet, and locomotion for seventy-one species of primates from eight communities—two each from South America, Africa, Madagascar, and Southeast Asia. Using a principal coordinates analysis, they mapped each of the species in a common multivariate ecological space defined by these variables and then plotted the position of the species of individual communities in this overall space (Figs. 8.7, 8.8).

Primate communities within biogeographical regions are remarkably similar, despite differences in species number and local ecology. For example, the rain forest of Ranomafana in Madagascar (Fig. 8.9) has three times the rainfall and nearly twice as many species as the dry forest of Marazolaza, near Morondava in western Madagascar (Fig 8.10; also Fig. 4.9), but the two communities are strikingly similar in the overall ecology of the component species (Fig. 8.7). Both contain numerous small species, nocturnal species, and many folivorous

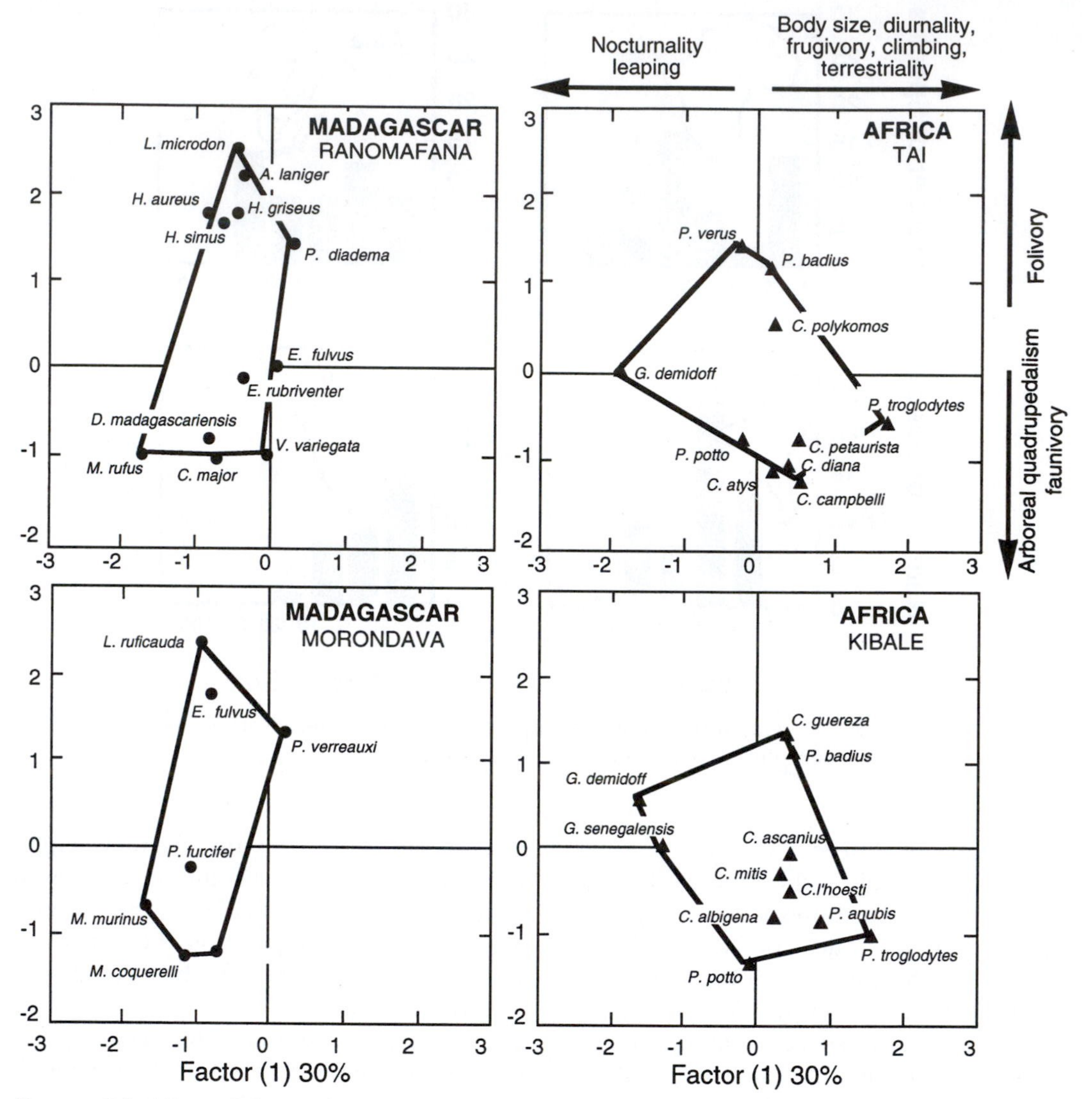

FIGURE 8.7 Plots of the ecological distribution of individual species of two primate communities from Madagascar—Ranomafana and Marazolaza (near Morondava)—and two communities from Africa—Tai Forest, Ivory Coast, and Kibale Forest, Uganda. Note the overall similarity of the communities from the same biogeographical area, despite differences in the individual species at each site.

species compared with primate communities in other parts of the world. The main differences are that Ranomafana has more medium-size, quadrupedal frugivorous species (*Eulemur rubriventor, E. fulvus,* and *Varecia variegata*) as well as three species that specialize on bamboo (*Hapalemur*).

Likewise, even though they are separated by 5000 km and have fewer than half of their primate species in common, the communities of

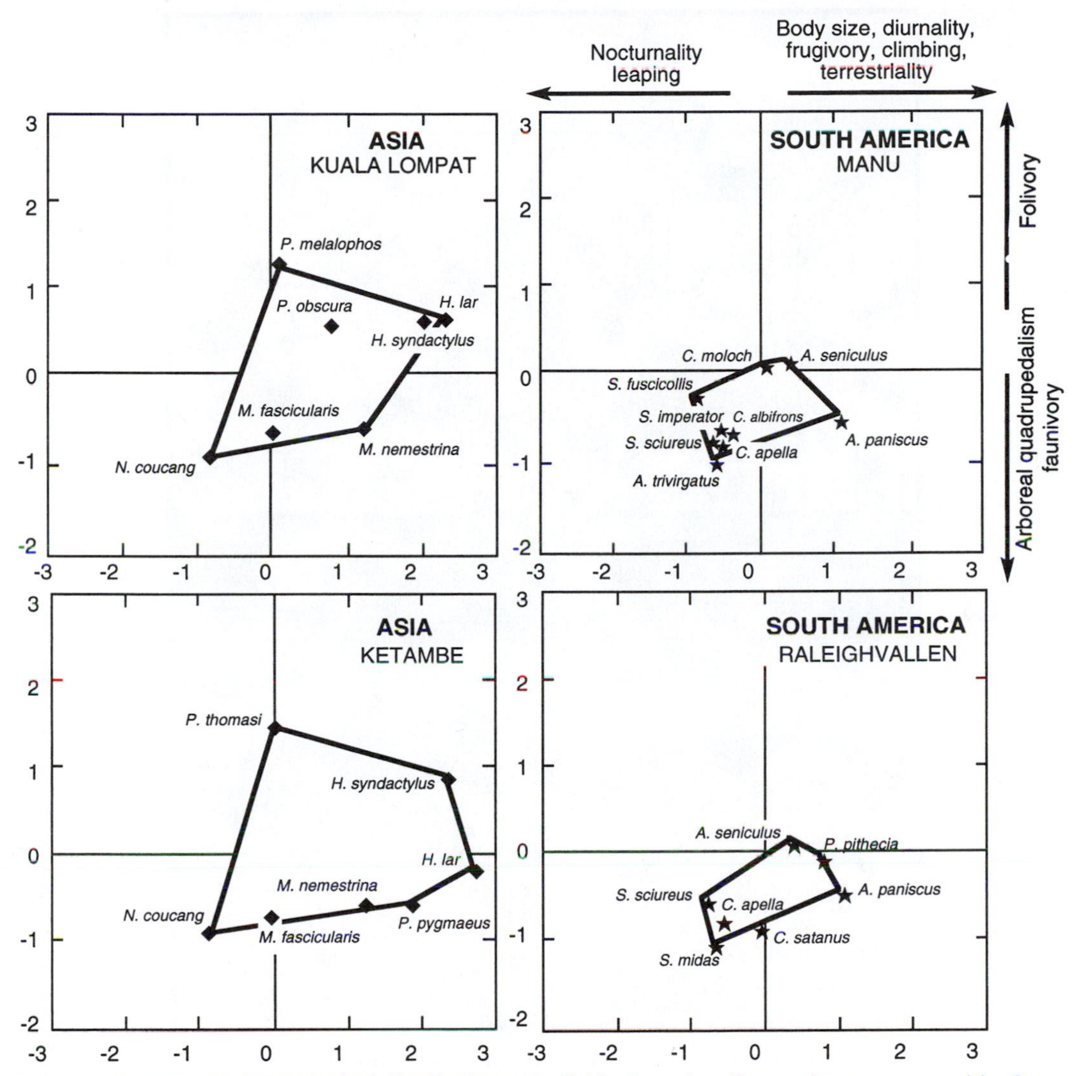

FIGURE 8.8 Plots of the ecological distribution of individual species of two primate communities from Asia—Kuala Lompat, Malaysia, and Ketambe, Sumatra—and two from South America—Raleighvallen-Voltsberg, Suriname, and Manu, Peru. Note the overall similarity of the communities from the same biogeographical area, despite differences in the individual species found at each site.

Tai Forest in Ivory Coast (Fig. 8.11) and Kibale Forest in Uganda are very similar. The distinctive shape of the African communities (Fig. 8.7) compared with those of other continents is defined by the small, nocturnal, leaping, insectivorous galagos and the numerous larger, frugivorous, quadrupedal species (*Pan, Papio, Cercopithecus*).

The two Asian communities—Kuala Lompat, Malaysia (Fig. 8.12), and Ketambe, in Indonesia—are distinctive in the paucity of small, nocturnal, and insectivorous species

FIGURE 8.9 The diurnal primate communities (above) and the nocturnal primate communities (below) of the eastern rain forest of Ranomafana, Madagascar.

and in having the suspensory gibbons. Compared with communities from other parts of the world, Asian primate communities generally have very few species (Fig. 8.8).

The two communities from South America are virtually identical to one another, despite coming from different biogeographical provinces in South America and having very different numbers of species (Fig. 8.8). Raleighvallen-Voltzberg (Fig. 8.13) is in Suriname on the Guiana shield, whereas Manu is in the upper reaches of the Amazon drainage in Peru.

FIGURE 8.10 The diurnal primate communities (above) and the nocturnal primate communities (below) from the western dry forest of Marazolaza, near Morondava. Note the dramatic differences in the typical vegetation and the smaller number of primate species compared with Ranomafana.

Nevertheless, the primate communities occupy the same ecological space, despite the greater number of species at Manu. The most striking feature of the South American communities is the remarkable uniformity of the different species and the limited amount of ecological space that they define. Compared with the primates in other parts of the world, neotropical monkeys are mostly diurnal, medium-size, frugivorous quadrupeds. They lack the extreme specialists of other continents. There is only one nocturnal species (*Aotus*) and one somewhat folivorous species (*Alouatta seniculus*), and even the most saltatory or suspensory species (such as *Ateles*) are mostly quadrupedal.

These community comparisons accord well with other broad comparisons of primate communities on a global scale (Bourliere, 1985;

FIGURE 8.11 The diurnal primate communities (above) and the nocturnal primate communities (below) of the Tai Forest, Ivory Coast.

Terborgh and van Schaik, 1987; Kappeler and Heymann, 1996). Malagasy primates are characterized by an abundance of nocturnal and folivorous species; South American communities lack folivores; Africa has an abundance of large frugivores; and Asian communities are relatively low in species diversity, especially in small species, but are unusual in the number of suspensory taxa. The communities reflect the overall characteristics of the faunas of these

FIGURE 8.12 The diurnal primates of Kuala Lompat, Malaysia.

different regions, and individual communities within regions are extremely similar to one another. Why are the primates on different continents so distinctive? At present there is no satisfactory answer to this question, just a series of hypotheses and interesting observations.

Three types of arguments are commonly offered to explain the major differences among primate faunas throughout the world. One view is that the ecological differences among

FIGURE 8.13 The primate community of the Raleighvallen-Voltsberg Nature Reserve in Suriname.

the primate faunas of South America, Africa, Madagascar, and Asia reflect major differences in the nature of the forests and the available resources on these continents, perhaps associated with broad geographical patterns of soils or climate. For example, Terborgh and van Schaik (1987) have argued that the lack of primate folivores in South America reflects the synchronous productivity of leaves and fruits due to the extreme seasonality of rainfall in much of South America. Thus, when fruits are not available, a species cannot switch to a diet of leaves, and vice versa. Likewise it has been argued that Asian forests have relatively few understory insects compared with other continents, and Madagascar has very few plants that produce fleshy fruits (Goodman and Ganzhorn, 1998). The low numbers of primate species in Asia has been related to a generally low level of productivity in Asian forests and to the abundance of a group of trees, dipterocarps, that are not utilized extensively by primates.

Another possible explanation is that the differences in primate faunas are the result of competitive interactions with other groups of vertebrates. The availability of primate niches may be constrained in many areas by the presence of other mammals or birds. Thus, the lack of primate folivores in South America has been related to an earlier presence of sloths on that continent and the lack of nocturnal insectivorous species to the diversity of small nocturnal marsupials. These resource availability and competition explanations are to some degree complementary and provide tests of one another.

For example, the presence of sloths, often in great numbers, indicates that South American forests can certainly support mammalian folivores, albeit very specialized ones. Likewise, the presence and diversity in the neotropics of many nocturnal, frugivorous opossums, many of which are very primatelike, demonstrates that this niche (or guild) is both available and well occupied. In contrast, many aspects of the vertebrate fauna of Madagascar and Asia parallel the ecological patterns seen in primates, suggesting differences in resource availability. For example, Madagascar has both a low number of primate frugivores and very few frugivorous birds compared with other parts of the world (Wright, 1996). The dearth of small insectivorous primates in much of Asia is paralleled by a low diversity of insectivorous frogs and lizards in these forests as well (Duellman and Pianka, 1990).

Finally, some of the continental differences in primate ecology may be the result of historical accidents of biogeography, of differences in the adaptive potential of the radiations that initially colonized the regions, or of recent extinctions. For example, it is clear that the lack of large species among the living primates of South America and Madagascar is a recent phenomenon. Much larger primates inhabited both regions in the relatively recent past. Likewise, the extinct Malagasy species include numerous suspensory and more quadrupedal species than the living fauna. However, in diet they only further emphasize the distinctiveness of that fauna, for they include a preponderance of folivores (e.g., Godfrey *et al.*, 1997). Other patterns in the biogeography of primate adaptations are difficult to explain. Why are there no galagos in Asia, despite the presence of lorises? Is this a historical accident? Does it reflect a scarcity of resources? Or is it due to competition from the very specialized and geographically restricted tarsiers? Why is the radiation of anthropoids in South America so narrow ecologically?

Despite the clear pattern of biogeographical differences in primate ecological adaptations among the primate faunas of the world, there are at present few general explanations. In most cases the present-day patterns seem to reflect a combination of ecological and historical factors rather than a single unitary cause. Further comparisons of many communities from each area will likely complicate this picture, but may provide greater insight into the ways in which community composition is determined (e.g., Ganzhorn, 1996). In addition to being an area of major interest for understanding the origins of the patterns of biodiversity we see around us, this area is critical for conservation planning as primate habitats become increasingly limited.

BIBLIOGRAPHY

Bourliere, F. (1985). Primate communities: Their structure and role in tropical ecosystems. *Int. J. Primatol.* **6**:1–26.

Charles-Dominique, P., Cooper, H. M., Hladik, A., Hladik, C. M., Pages, E., Pariente G. F., Petter-Rousseaux, A., Petter, J. J., and Schilling, A. (1980). *Nocturnal Malagasy Primates: Ecology, Physiology, and Behavior.* New York: Academic Press.

Chivers, D. J. (1980) *Malayan Forest Primates: Ten Years' Study in Tropical Rain Forest.* New York: Plenum Press.

Duellman, W. E., and Pianka, E. R. (1990). Biogeography of nocturnal insectivores: Historical events and ecological filters. *Annu. Rev. Ecol. Syst.* **21**:57–68.

Eeley, H., and Lawes, M. (1998) Large-scale patterns of species richness and species range size in African and South American primates. In *Primate Communities,* ed. J. G. Fleagle, C. H. Janson, and K. E. Reed (in press). Cambridge: Cambridge University Press.

Emmons, L. H., Gautier-Hion, A., and Dubost, G. (1983), Community structure of the frugivorous-folivorous forest mammals of Gabon. *J. Zool. London* **199**:209–222.

Fleagle, J. G., and Reed, K. E. (1996). Comparing primate communities: A multivariate approach. *J. Human Evol.* **30**:489–510.

Ganzhorn, J. U. (1996). Test of Fox's assembly rule for functional groups in lemur communities in Madagascar. *J. Zool.* **241**:533–542.

Ganzhorn, J. U., and Sorg, J.-P., eds. (1997). *Ecology and Economy of a Tropical Dry Forest in Madagascar,* Göttingen: Primate Report, 46-1.

Ganzhorn, J. U., Malcomber, S., Andrianantoanina, O., and Goodman, S. M. (1997). Habitat characteristics and lemur species richness in Madagascar. *Biotropica* (in press).

Gaston, K. J., Blackburn, T. M., and Spicer, J. J. (1998). Rapoport's rule: time for an epitaph? *TREE* **13**:70–74.

Godfrey, L., Jungers, W., Reed, K. E., Simons, E. L., and Chatrath, P. S. (1997). Primate subfossils inferences about past and present primate community structure. In *Natural and Human-Induced Change in Madagascar,* ed. S. Goodman and B. Patterson, pp. 218–256. Washington, D. C.: Smithsonian Institution Press.

Goodman, S. M., and Ganzhorn, J. U. (1998). Rarity of figs (*Ficus*) on Madagascar and its relationship to a depauperate frugivore community. *Rev. Ecol.* **52**:321–323.

Kappeler, P. M., and Heymann E. W. (1996). Nonconvergence in the evolution of primate life history and socio-ecology. *Biol. J. Linn. Soc.* **59**:297–326.

Kay, R. F., Madden, R. H., van Schaik C., and Higdon, D. (1997). Primate species richness is determined by plant productivity: Implications for conservation. *Proc. Natl. Acad. Sci. USA* **94**:13023–13027.

Oates, J. F., Whitesides, G. H., Davies, A. G., Waterman, P. G., Green S. M., Dasilva, G. L., and Mole S. (1990). Determinants of variation in tropical forest biomass: New evidence from West Africa. *Ecology* **71**:328–343.

Reed, K. E., and Fleagle J. G. (1993). Comparing primate communities. *Am. J. Phys. Anthropol. Suppl.* **16**:163.

______ (1995). Geographic and climatic control of primate diversity. *Proc. Natl. Acad. Sci. USA* **92**:7874–7876.

Rosenzweig, M. L. (1995). *Species Diversity in Space and Time.* Cambridge: Cambridge University Press.

Ruggiero, A. (1994). Latitudinal correlates of the sizes of mammalian geographical ranges in South America. *J. Biogeog.* **21**:545–559.

Stevens, G. C. (1989). The latitudinal gradient in geographical range: How so many species coexist in the tropics. *Am. Nat.* **133**:240–256.

Tattersall, I. (1982). *The Primates of Madagascar.* New York: Columbia University Press.

Terborgh, J. (1973). On the notion of favorableness in plant ecology. *Am. Nat.* **107**:481–501.

______ (1983). *Five New World Primates: A Study in Comparative Ecology.* Princeton, N. J.: Princeton University Press.

______ (1992). *Diversity and the Tropical Rain Forest.* New York: Scientific American Library.

Terborgh, J., and van Schaik, C. P. (1987). Convergence vs. nonconvergence in primate communities. In *Organization of Communities,* ed. J. H. R. Gee, and P. S. Giller, pp.205–226. Oxford: Blackwell Scientific Publications.

Whitmore, T. C. (1990). *An Introduction to Tropical Rain Forests.* New York: Oxford University Press.

Wright, P. C. (1996). The future of biodiversity in Madagascar: A view from Ranomafana National Park. In *Natural and Human-Induced Change in Madagascar,* ed. S. Goodman and B. Patterson, pp. 381–405. Washington, D. C.: Smithsonian Institution Press.

NINE

Primate Adaptations

ANATOMY AND BEHAVIOR

In the preceding chapters we have discussed the anatomy, behavior, and comparative ecology of the major radiations of living primates. In this chapter we examine size, diet, locomotor behavior, and other aspects of comparative ecology for consistent associations between anatomical and behavioral characteristics as well as for correlations among the different aspects of behavior and ecology. By investigating the functional relationships between morphological features such as size, tooth shape, bone shape, and behavioral habits, we can understand why primate species have evolved many of their anatomical differences. We can also use this information to reconstruct some aspects of the behavior and ecology of extinct species.

Effects of Size

Body size is a basic aspect of the adaptive strategy of any primate species. An animal's size is associated with both opportunities and restrictions on its ecological options, and many of the differences between species in structure, behavior, and ecology are correlated with absolute body size. Much of this size-dependent variation in morphology, physiology, and ecology can be explained by the impact of simple mathematical considerations on basic physiological and mechanical phenomena.

As linear dimensions of any object—including an animal—increase, so too do its areal (e.g., cross-sectional) dimensions and its volume. But linear dimensions, area, and volume do not increase at the same rates. Area increases as a function of the square of linear dimensions (L^2) and volume increases as a function of the cube of linear dimensions (L^3) (Fig. 9.1). If an animal were to double in length, breadth, and width, for example, its cross-sectional dimensions would increase fourfold, and its volume would be eight times as great. When these simple mathematical considerations are applied to animal bodies, the consequences are great. An animal's weight is a function of its volume; the strength of any of its bones is a function of the cross-sectional area of the bone. Thus an animal whose linear dimensions double would weigh eight times as much, but its structural supports would be only four times as strong. We expect, then, that animals of greatly different size will not be similarly proportioned—and this is usually what we find. Figure 9.2 shows the femur of a pygmy marmoset expanded to the same length as a gorilla femur. The gorilla femur is thicker—an adaptation to support the gorilla's much greater volume and weight.

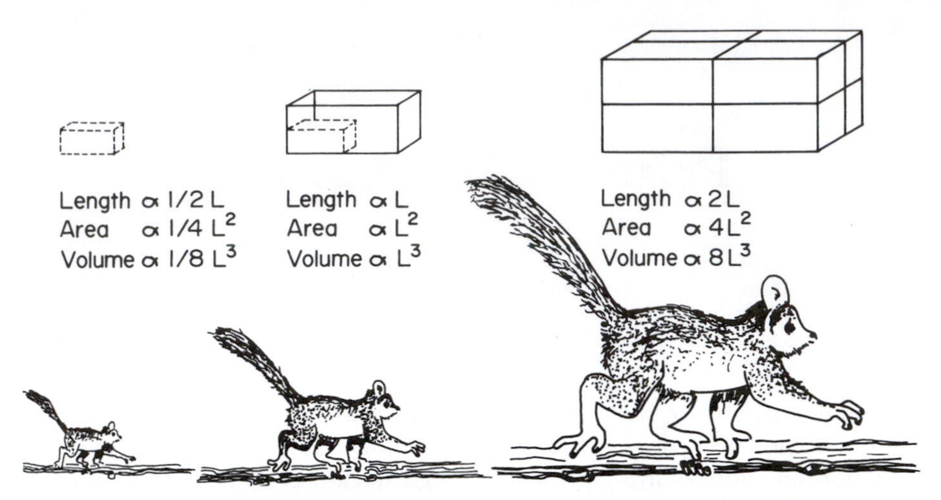

FIGURE 9.1 In a simple geometrical model, the halving or doubling of the length of an animal involves much greater changes in area or volume, which affect many physiological and structural aspects of an animal's life.

Such size-related scaling, first discussed by Galileo, is apparent throughout the animal kingdom. It explains why there are no 1-ton flying birds and why humans cannot jump like grasshoppers or carry weight like ants do. It also explains why cinematic fantasies of incredible shrinking or growing men and women are indeed fantasies. Similar considerations affect the scaling of other physiological functions, such as absorption in the digestive system. If the surface area of the digestive system increased in proportion to L^2 while the mass that must be fed increased in proportion to L^3, larger animals would have a relatively small digestive tract with which to process foods for much larger bodies (see Chivers and Hladik, 1980). If primate brains were all the same shape, larger species would have a relatively smaller brain surface area for any particular brain weight. Some primates have relatively longer intestines; some have different kinds of digestive organs; some have different diets. Larger species tend to have more convoluted brains, so the ratio of surface area to brain weight remains approximately the same from species to species.

Metabolism is another physiological function that does not scale linearly with changes in body size, and thus it is an important consideration for understanding size-related differences in ecology and behavior. A primate's metabolic rate, or the amount of energy the individual requires for either basic body functions (basal metabolism) or daily activities (daily metabolic rate), scales in proportion to body mass raised to the power 0.75, not simply in direct proportion to body mass—or to surface area of the body, as had long been believed. Thus larger animals expend relatively less energy and consequently need proportionately less food than smaller animals need. Put more simply, two 5-kg monkeys require more food than a single 10-kg monkey.

Primates have evolved in two ways to accommodate the constraints of scaling: by evolving different physical proportions (as in increasing brain convolutions) and by adapting lifestyles that capitalize on the scaling con-

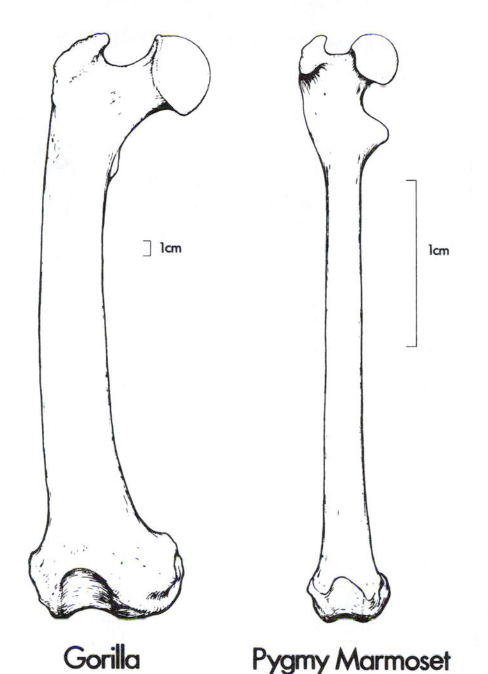

FIGURE 9.2 The right femur of a gorilla (*Gorilla gorilla*) and a pygmy marmoset (*Cebuella pygmaea*) drawn to the same length. The bone of the gorilla is relatively much thicker than that of the small marmoset because it must support a relatively greater body mass for its length.

sequences. For example, whereas mechanical scaling would require that bone diameters scale in proportion to body mass$^{0.375}$ and simple geometric scaling would yield a scaling of diameters proportional to mass$^{0.333}$, most studies find that the scaling of bone diameter with body size is slightly greater than geometric scaling but not as great as a mechanical scaling would require (e.g., Alexander, 1985; Demes and Jungers, 1993). However, this slight positive scaling of bone thickness is often associated with postural and behavioral changes that reduce stress on bones with increasing body size (e.g., Biewener, 1990). This combination of mechanical and behavioral change thus generates a pattern of similar bone strain that has been called "dynamic similarity" (e.g., Demes and Jungers, 1993).

There is an increasingly large and sophisticated literature on size-related differences in many aspects of primate biology, including limb length, brain size, reproductive physiology, tooth size, and locomotor behavior (see Jungers, 1985). Altogether these studies are termed **allometry,** of which there are three general types (Fig. 9.3): (a) **growth allometry** is the study of shape changes associated with size changes in ontogeny; (b) **intraspecific allometry** is the study of size-related differences in adults of the same species; and (c) **interspecific allometry** examines size-related differences across a wide range of different species for broader principles of scaling.

Considerations of size are critical to our understanding of both primate adaptation and evolution, but a detailed discussion of this topic is beyond the scope of this book. In this discussion we concentrate instead on the role of size in ecological adaptation—the way primates of different size tend to have different ways of life. In considering these adaptive differences, it is often impossible to determine whether size-related differences in behavior and ecology are behavioral adaptations a species has adopted to "accommodate its size" or whether size changes themselves are better viewed as gross morphological adaptations that enable a species to exploit a particular ecological niche better. Size and adaptation are so intertwined that determining which precedes the other in evolution is akin to determining whether the chicken comes before the egg. Furthermore, any particular body size comes with both advantages and disadvantages. The best approach is to look for consistent associations between size and behavioral ecology, associations that may provide us with insight into the structure of both living and fossil primate communities.

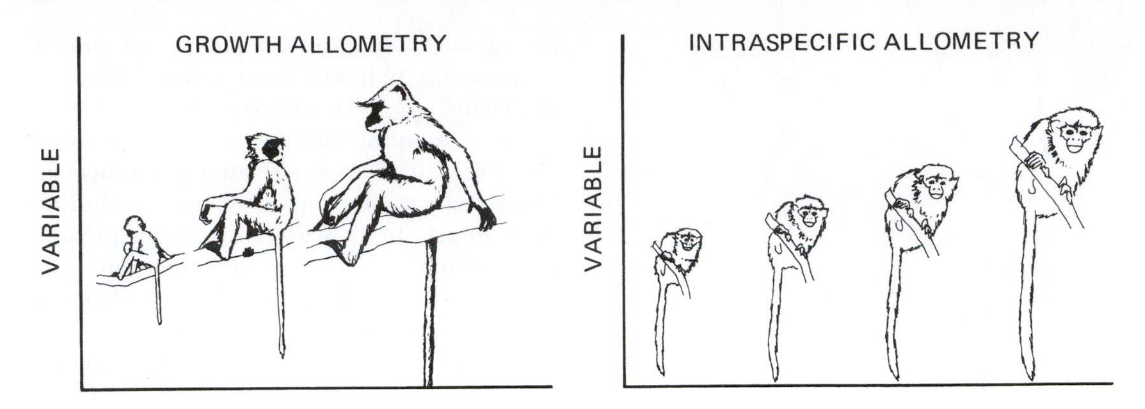

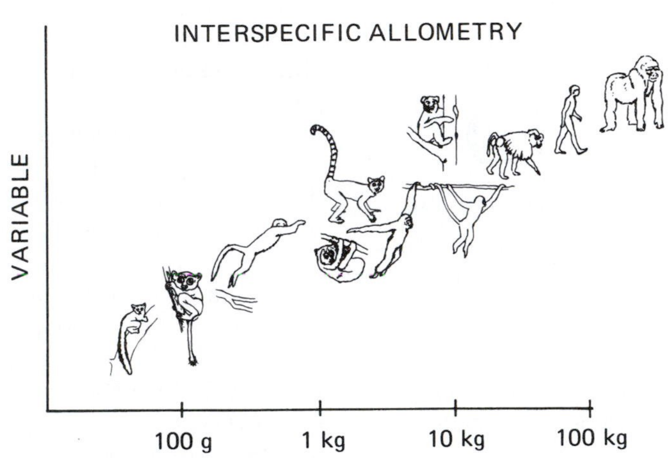

FIGURE 9.3 Graphs representative of three ways of examining the association of shape changes with size changes: growth allometry examines the shape changes associated with ontogenetic size increase; intraspecific allometry examines the shape changes associated with size differences among adults of a single species; interspecific allometry examines shape changes associated with size differences across a wide sample of different species. Allometric change are usually plotted on a logarithmic scale so that exponential relationships appear linear.

Size and Diet

Primate diets are closely linked with body size (Fig. 9.4). Species that eat insects tend to be relatively small, whereas those that eat leaves tend to be relatively large. Fruit eaters tend to supplement their diets with either insects or leaves, depending on their size. These patterns result from the interaction of several independent, size-related phenomena. First, primates need a balanced diet that meets not only their caloric (energy) needs but also their other nutritional requirements, such as for protein and a variety of trace elements and vitamins. Although fruits are high in calories, they are very low in protein content; most pri-

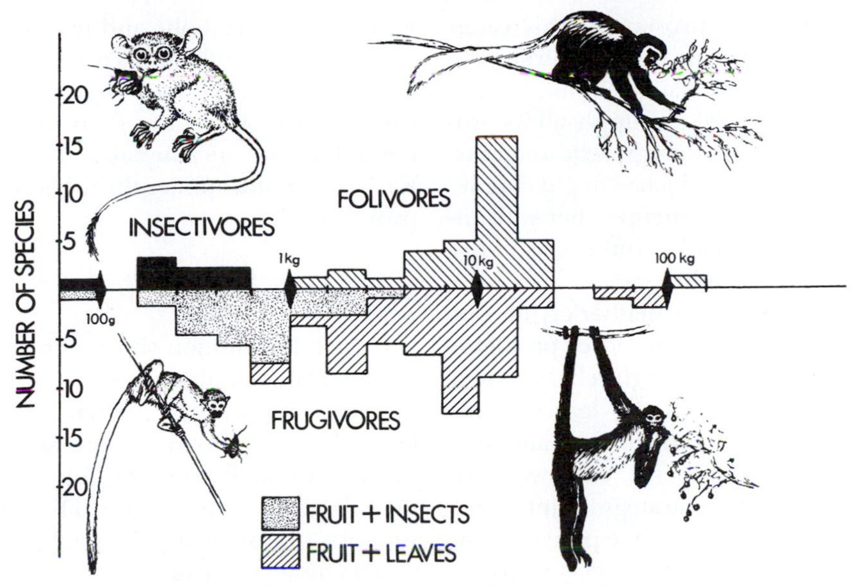

FIGURE 9.4 Primate dietary habits are correlated with body size. Insectivorous primates are relatively smaller than folivorous species. Smaller frugivorous species tend to supplement their diet with insects, and larger frugivorous species supplement their diet with leaves (redrawn from Kay, 1984).

mates must therefore turn to other sources for their protein. The two most abundant sources of dietary protein for primates are other animals (such as insects) and folivorous materials, such as leaves, shoots, and buds. Why, then, do small primates tend to eat insects and large ones folivorous material? Although these two protein strategies are in a nutritional sense complementary, the physiological and behavioral problems faced by a primate that feeds on these two dietary items are quite different.

Insects, and animal material in general, are an excellent source of nutrients (except calcium), fulfilling nearly all of a primate's requirements. Furthermore, insects are relatively high in calories per unit weight. This is particularly important for small animals, which have relatively higher energy requirements than large ones (the shrew must eat several times its body weight in food every day). Insects are so good a food source that the real question is not why small primates eat them but why large ones do not. The answer seems to lie in the time normally involved in catching and handling insects. No primates have evolved the specialized abilities of anteaters to prey on large colonies of social insects; rather, they depend largely on locating and catching isolated individuals. It has been suggested, and it seems quite reasonable, that the number of insects that a primate can find and catch in a given day (or night) is likely to be relatively similar from species to species—regardless of size—assuming, of course, that they look in the appropriate places and have appropriate adaptations for insect catching, such as good grasping abilities and keen eyesight. In an 8-hour active period, any two primates might be able to ingest forty insects of one type or another. For a small prosimian, this catch could supply all the energy requirements needed for the day; for a larger monkey, however, this much food might supply all its protein needs but not its energy requirements. Thus, although larger primates might supplement their fruity (high-energy) diet with insects, they cannot rely solely on insects in the way a small primate might.

Unlike insects, leaves are neither cryptic nor hard to catch, but they pose other problems for foraging primates. Although relatively high in protein (particularly young leaves, buds, and shoots), leaves also contain large amounts of less palatable components, such as cellulose or even toxins, a strategy plants have evolved to prevent or discourage predation on their leaves. Compared with insects or fruits, leaves are generally low in energy yield for their weight. Large body size helps a primate overcome some of these problems inherent in a leafy diet. First, large animals need less energy per kilogram of mass than do small animals. Thus they can more easily afford to have a diet that is relatively low in energy sources. Second, although primates do not have the enzymes needed to break down the cellulose in leaves, many are able to maintain colonies of microorganisms in part of their digestive tract to perform this task for them. This kind of digestion takes time, but the time it takes food to travel through an animal's digestive tract is roughly proportional to the length of the gut and thus to the animal's size. For this reason, a small primate with a short gut has less opportunity to digest plant fibers than does a larger animal with a longer gut. Furthermore, these longer, slower guts with special chambers for fermenting cellulose also seem to help detoxify some of the poisons. Thus, whereas the upper size limit of insect eaters seems to be imposed by the time required to locate and catch their prey, the lower size limit of folivores seems to be determined by metabolic and digestive parameters. In general, folivorous primates have body weights of no less than 500 g, whereas insectivores tend to weigh less than this limit. This natural physiological break at 500 g, known as **Kay's threshold,** applies throughout the order primates.

Size and Locomotion

Like diet, locomotion shows general patterns of size-related scaling in primates. Terrestrial primates are usually larger than arboreal ones, both within taxonomic groups and for the order as a whole. Presumably this difference reflects both the limited capability of arboreal supports to sustain large animals and perhaps also some amount of selection for large size among terrestrial species as a means of deterring potential predators.

Within arboreal primates, there are size-related trends in the use of different types of locomotion. Although we lack the extensive quantitative data on primate locomotion that we have for diet, the allometry of locomotor behavior has been quantitatively assessed for South American monkeys, and similar patterns seem to hold for the rest of the order (with some notable exceptions). In general, we find that leaping is more common among small primates (Fig. 9.5a), whereas suspensory behavior is more common in larger species (Fig. 9.5b). Like fruit eating, quadrupedal walking and running does not seem to show any pattern with respect to body size; there are small, medium, and large quadrupeds.

The trends we find in leaping and suspensory behavior seem to be the result primarily of simple mechanical phenomena (Fig. 9.6). Two primates, one small and one large, traveling through the forest canopy will each encounter gaps between trees that they somehow must cross to continue their journey. In the same forest, the smaller one will more

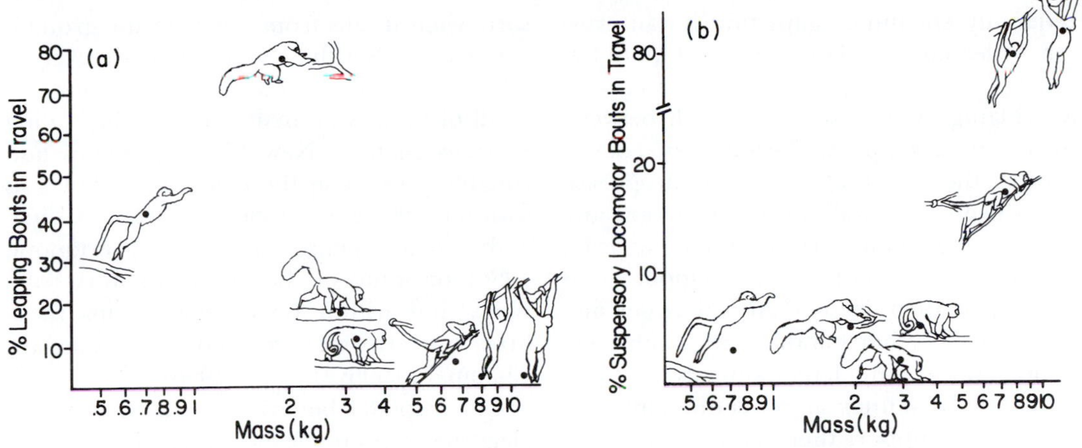

FIGURE 9.5 Primate locomotor behavior is correlated with body size. Among platyrrhine monkeys, (a) leaping is more common for smaller than for larger species and (b) suspensory behavior is more common for larger than for smaller species.

FIGURE 9.6 A small primate and a large primate traveling through the same forest are confronted with different locomotor problems because of the difference in their size. The small primate encounters relatively more gaps that can be crossed only by leaping, while the larger species encounters relatively more gaps that can be crossed by suspensory behavior or bridging.

frequently encounter gaps that it can cross only by leaping; the larger one will more frequently encounter gaps that can be crossed by bridging or by suspending itself between the terminal supports. Leaping, of course, involves the generation of high propulsive forces from the hindlimbs—and larger animals must generate greater forces to leap. Smaller animals will find more supports that can sustain their leaps than will larger animals. On the other hand, during both locomotion and feeding, larger animals will more frequently encounter supports too narrow or too weak to support their larger bodies and will more often need to suspend themselves below multiple branches for both support and balance (Fig. 9.7). Another relevant factor is the amount of energy a tree climber must absorb when it falls from a tree to the ground. Those animals with greater weight are likely to adopt the more cautious form of locomotion.

FIGURE 9.7 During feeding, small primates encounter more supports that can easily support their weight, while larger primates have to spread their weight over a large number of supports to feed at the same place.

All of these arguments support the scaling patterns seen in New World primates and roughly present in the order as a whole. As with diet, there are notable exceptions, such as the small suspensory lorises (e.g., Terranova, 1996) or some of the larger saltatory colobines, but within taxonomic groups these broad patterns seem to hold (e.g., Gebo and Chapman, 1995; McGraw, 1996).

Quadrupedal behavior seems to show no clear size restrictions; there are both large and small quadrupeds. Larger quadrupeds tend to move on larger supports, however, and the largest support is the ground. The interesting exceptions to this pattern are animals that show other special adaptations. such as marmosets, which have claws for clinging to large tree trunks, or very suspensory animals, such as spider monkeys, which spread their weight over several relatively small branches.

Size and Life History

In addition to diet and locomotion, primate reproduction seems particularly closely linked to size (see Chapter 3). Although life history variables such as gestation period and life span increase with body size (larger primates generally take longer to grow up and live longer), they have a negative allometric scaling, so a doubling of body size does not correspond to a doubling of gestation period or life span. Similarly, litter weight, or the size of the newborns, scales negatively with body size; that is, smaller primates have relatively larger babies. Obviously, the energetics of reproduction and growth are a major determinant of an individual's metabolic costs, especially for females. Although the causal relationships among size, life history variables, and ecology remain poorly understood, they are all closely integrated (e.g., Janson and van Schaik, 1993; Ross, 1998).

Size and Ecology

Various other aspects of primate ecology show important relationships with size. Within any habitat, smaller primates are certainly more susceptible to predation than are larger species, because they can be hunted by a larger range of predators (Terborgh, 1986). In some cases, size-related features of ecology may be just alternate expressions of the factors just discussed. For example, home range size for primate species increases with body size. This presumably reflects the need for larger animals to cover a wider area to support themselves, since the amount of food a primate ingests also increases with body size (Figure 9.8; Barton, 1992). It has also been shown that primate group size increases with body size—larger species live in larger groups—but this relationship is more suspect and difficult to explain. Terrestrial species frequently aggregate into large groups, especially for sleeping and resting, as a strategy to avoid predation, but they may still forage in smaller units.

Adaptations to Diet

Diet is generally recognized as the single most important parameter underlying the behavioral and ecological differences among living primates, and primate diets have been more thoroughly documented than any other aspect of behavior. Food provides the energy that primates need for reproduction and seems to be the main objective of most of their daily activities. The use of hands to obtain and prepare

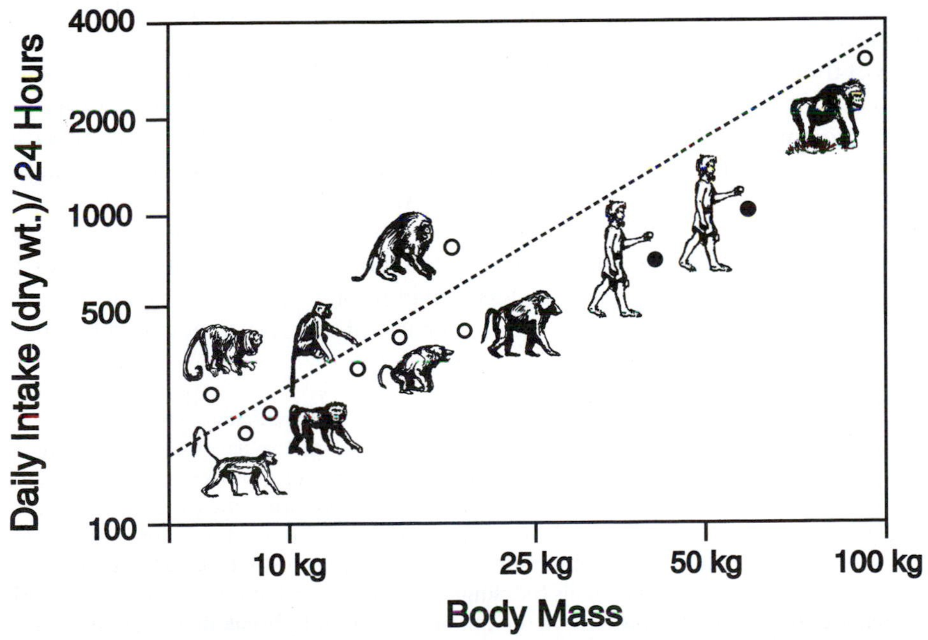

FIGURE 9.8 Total daily food intake increases with body size in living primates (from Barton, 1992).

food is a distinctive feature unifying the feeding habits of all primates, but, as the previous chapters emphasize, primate species show a wide range of behavioral and morphological adaptations for obtaining and processing different types of food (Fig. 9.9).

Dental Adaptations

The best-documented morphological adaptations to diet are those found in primate teeth, the organs primarily responsible for initial processing of food once it has been located. Fortunately, since teeth are the parts most commonly preserved in the fossil record, they also provide us with an opportunity for reconstructing the diets of extinct species.

The primary function of the anterior part of the tooth row, the incisors and canines, is ingestion (e.g., Ungar, 1994), but these teeth also serve a wide range of nondietary functions, such as grooming and fighting. The role of canines and incisors in procuring or ingesting food is often not as food-specific as that of other parts of the tooth row. For example, primates that eat bark, insects, wood, or exudates may all have procumbent incisors for removing bark from trees to obtain food. Nevertheless, there are some general patterns linking incisor form with diet. Relative to the size of their molars, folivores tend to have smaller incisors than do frugivores, because leaves require less incisive preparation. Primates that feed extensively on exudates frequently have large procumbent incisors for digging holes in the bark of trees to elicit the flow of these fluids.

The cheek teeth—the premolars and particularly the molars—break up food mechanically and prepare it for additional chemical processing further along the digestive system. Thus the particular adaptations we see in molar teeth are generally not for specific foods but for food items with particular structural properties or consistencies (e.g., Strait, 1997). There are major functional differences among primate molar teeth in the development of cusps and shearing crests or dental blades for reducing food items into small particles. Physiological experiments have demonstrated that the digestion of both insect skeletons and leaves is enhanced by reducing these food items into small pieces and thereby increasing the surface area Thus we find that insect eaters and folivores are characterized by molars with extensive development of these shearing crests. In folivores, this development of shearing crests is also associated with thin enamel on the tooth crown, an adaptation that creates even more shearing edges on the border between the superficial enamel and the underlying dentine once the teeth are slightly worn.

Although both insect eaters and folivores are characterized by well-developed shearing crests or blades, the optimal shapes of dental blades vary according to the physical properties of the food being cut (Strait, 1997;

FIGURE 9.9 Morphological adaptations to diet among living primates. Fruit eaters tend to have relatively large incisors for ingesting fruits, simple molar teeth with low cusps for crushing and pulping soft fruits, and relatively simple digestive tracts without any elaboration of either the stomach or the large intestine. Leaf eaters have relatively small incisors, molar teeth with well-developed shearing crests, and an enlargement of part of the digestive tract for the housing of bacteria for the breakdown of cellulose. Gum (exudate) eaters usually have specialized incisor teeth for digging holes in bark and scraping exudates our of the holes, and claws or clawlike nails for clinging to the vertical trunks of trees. Many also have an enlarged caecum, suggesting that bacteria in the gut may function to break down the structural carbohydrates in gums or resins. Insect eaters are characterized by molar and premolar teeth with sharp cusps and well-developed shearing crests and a digestive tract with a simple stomach and a short, large intestine.

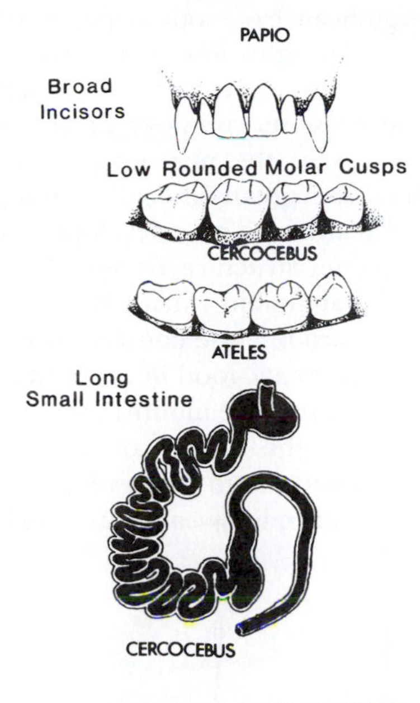

FRUIT EATERS

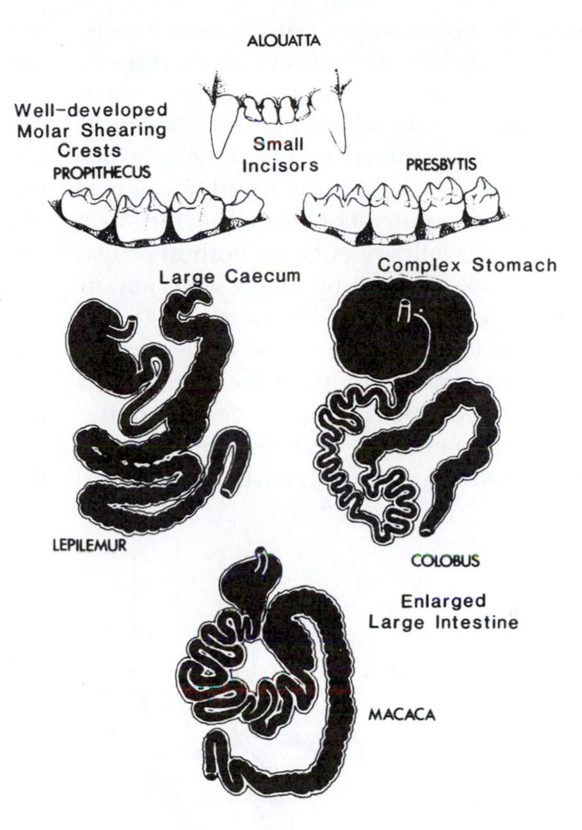

LEAF EATERS

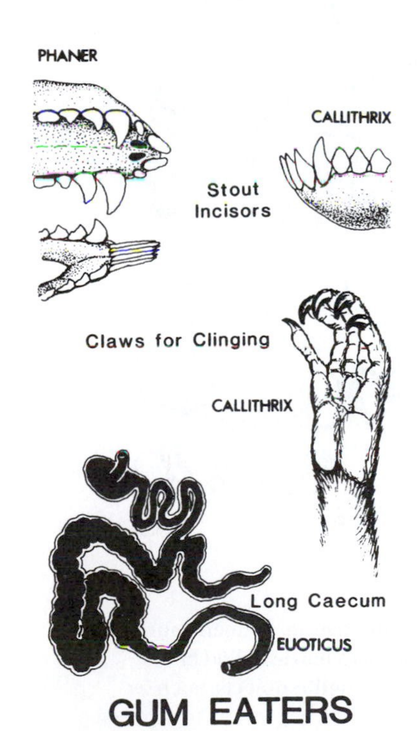

GUM EATERS

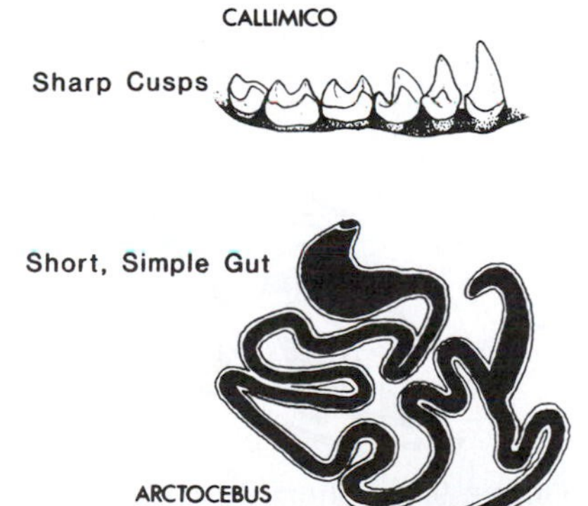

INSECT EATERS

Fig. 9.10). For example, hard-bodied insects, which will shatter, are best cut with short shearing blades that concentrate force in a small area and generate self-propagating cracks. In contrast, softer insects require longer blades to separate the cut parts (Strait, 1997). Similar mechanical considerations show that the lophs of colobine teeth are effective both in propagating cracks to open up tough seeds and in cutting leaves (Lucas and Teaford, 1994). For other soft foods, such as fruits, teeth seem to be designed mainly to maximize surface area between teeth and food items. Studies of the physical properties of primate foods, in conjunction with a better understanding of the mechanical significance of tooth shape, are providing many new insights into primate dental function. In addition to the many adaptations to diet in the gross morphology of primate teeth, differences in the microscopic structure of tooth enamel greatly affect the strength of its occlusal features. Thus, developmental patterns of enamel structure frequently can be related to dietary habits (Maas, 1993)

Because teeth are in close contact with the food that a primate eats, food items (and any other material entering the mouth) frequently leave scratches on the surface of teeth. Although not adaptations to processing food, these scratches, or **microwear,** can provide

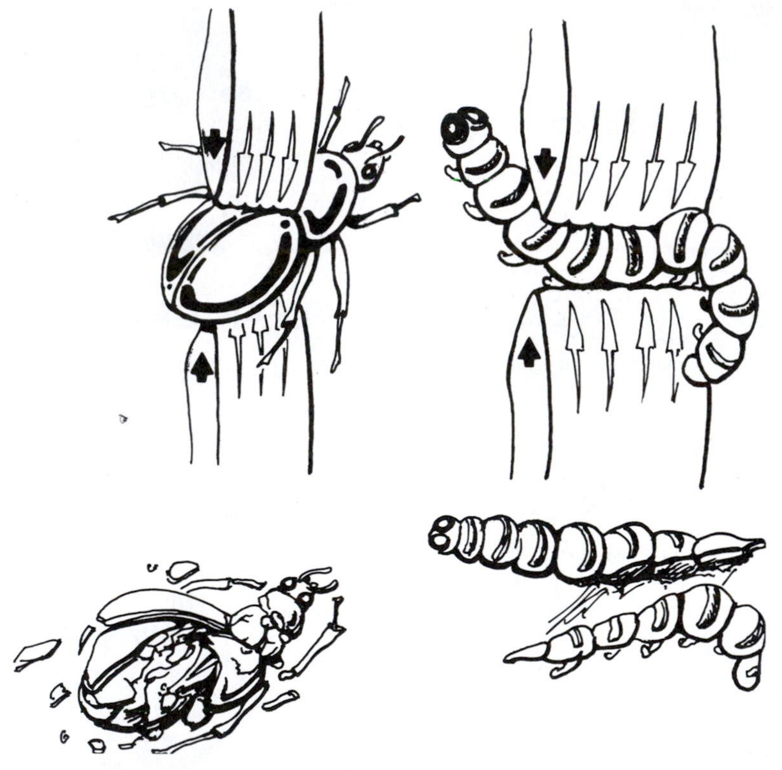

FIGURE 9.10 Short dental blades are more effective in concentrating forces to induce shattering of brittle objects, such as a hard-shelled beetle, whereas long blades are more effective in slicing ductile objects, such as a caterpillar (courtesy of S. Strait).

clues to the diet of living and fossil animals (Teaford, 1994; Ungar, 1994). For example, primates eating hard insects show more pits on their teeth than do those eating soft insects (Strait, 1993a), and primate folivores show fewer pits than do most frugivores, although their teeth lack the etching caused by acid fruits (Teaford, 1994).

In contrast with the considerable success primatologists have had in relating dental anatomy to dietary habits, attempts to link differences in mandible shape and skull form with dietary differences have been considerably less successful, probably because cranial morphology serves so many diverse and often conflicting functions (e.g., Hylander and Johnson, 1997).

Other Oral Adaptations

Although teeth and jaws are the most well-studied aspects of the primate mouth, other aspects of the oral cavity play important roles in the ingestion and initial preparation of foods (Chapter 2). Some primates, especially great apes, use their lips extensively in grasping and processing food items. Chimpanzees, for example, often use their lips to extract the fluids from foods and then spit out the fibrous remains. Cercopithecine monkeys have cheek pouches that enable them to ingest large numbers of fruits rapidly without processing them immediately. Primates differ dramatically in the size and structure of the salivary glands, which play an important role in the initial digestion of foods as well as in helping control the pH balance of both the oral cavity and the initial part of the gut (e.g., Dumont *et al.*, 1997).

Digestive Tract Adaptations

Although of little use to the paleontologist, the soft anatomy of the primate digestive system shows dietary adaptations as distinctive as those seen in the dentition (Fig. 9.9; Chivers and Langer, 1994). Whereas the dentition shows adaptations to relatively gross features, such as the size and mechanical characteristics of particular foods, the remainder of the digestive system shows adaptations to the chemical or nutritive properties of dietary items. Leaves and gums, for instance, which are very different in their consistency, require different dental adaptations but present similar problems for the remainder of the digestive system. Both are composed of long chains of structural carbohydrates and require extra processing chambers and the action of microorganisms.

In general, primate digestive systems show three different patterns of dietary adaptation. Faunivorous primates (mainly insect eaters but also some omnivorous species) have a relatively short, simple digestive system with a small, simple stomach, usually a small caecum, and a very small colon relative to the size of the small intestine. In essence, the digestive system of a faunivore is devoted to absorption, the function of the small intestine. Frugivores also have relatively simple digestive systems, although large frugivores tend to have relatively large stomachs.

Folivores show the most elaborate adaptations in the visceral part of their digestive system because they must process foods containing large amounts of structural carbohydrates and also must overcome various toxins. Because primates lack the natural capability to digest the cellulose contained in the cell walls of plants, these elaborations of the visceral digestive system involve forming an enlarged pouch somewhere in the digestive tract to maintain a colony of microorganisms that can digest cellulose or other structural carbohydrates. The host primates then digest both the products of the bacterial action and the bacteria themselves. There are several possible solutions to this ranching situation, and different primate folivores grow their bacteria and break down cellulose in at least three different places.

Some folivorous prosimians, such as *Lepilemur* and *Hapalemur*, have an enlarged caecum, a feature also seen in rabbits and horses. Colobine monkeys have an enlarged stomach with numerous sections, similar to but much less elaborate than that of cows. Most other partly folivorous species, including indriids, apes (siamang and gorillas), New World monkeys (*Alouatta*), and some cercopithecine monkeys (*Macaca sylvanus*), accommodate the leafy portion of their diet by means of an enlarged colon. In addition to their role in breaking down the cellulose, it seems likely that the "fermenting" areas in the digestive systems of primate folivores help them overcome the various toxins found in many plant parts. This detoxification seems to be facilitated both directly, through actual chemical breakdown, and indirectly, by slowing down the rate at which food is processed to allow the liver more time to detoxify.

Although the visceral modifications for digestion of plant materials have been well studied in primates, there is less evidence about how and where primates break down other structural carbohydrates, such as those in gums (see Nash, 1986; Power and Oftedal, 1996) and the chitinous exoskeleton of invertebrates. There are anatomical indications, and a few physiological studies, suggesting that the processes used to digest these substances may be similar to those involved in cellulose digestion, since primates with specialized diets of gums (*Galago*), insects (*Tarsius*), or both (*Cebuella*) are also characterized by a large caecum.

Diet and Ranging

Adaptations to diet extend well beyond the digestive system. The many foods primates eat are found in various places, and many of the differences we see in ranging patterns seem to be adaptations for harvesting foods with unique distributions in both time and space. Primate ranging behavior is clearly correlated with diet (see Oates, 1986; Janson, 1992; Janson and Goldsmith, 1995). Folivores tend to have relatively smaller home ranges for their size than do frugivores, reflecting the fact that foliage is more uniformly distributed and more common than fruits. Folivores tend to have shorter day ranges for the same reason. Because of their smaller ranges, folivores also are found in higher population densities and biomass densities than are frugivores.

In conjunction with the different distributions of primate foods, it has been argued that fruit-eating primates have relatively larger brains for their body weight than do leaf-eating primates (but see Dunbar, 1992), and several authors have suggested that the need to remember the location and fruiting cycles of trees may have been the most important factor leading to the relatively large brain size and intelligence that characterize higher primates.

Diet and Social Groups

Broad correlations between simple characterizations of diet and social structure have proven to be more difficult to identify than the anatomical associations just discussed. Although the social organization of any single species or pair of species can readily be explained in terms of dietary differences and the distribution of preferred foods, broader predictive patterns are more elusive. Monogamous species include frugivores like gibbons, folivores like *Indri*, and faunivores like *Tarsius*. For any dietary group such as folivores, we can find species that live in monogamous families (indriids or siamang), single-male groups (*Colobus guereza, Alouatta, Gorilla*), or large, multimale groups (*Piliocolobus badiius, Theropithecus*), not to mention *Lepilemur,* which lives in a noyau arrangement.

Why does group organization seem to have so little to do with the single activity that occupies most of a primate's time? There are several reasons. One is that a broad categorization of diets into insects, fruits, and leaves does not reflect the patterns of food distribution that are likely to be important for determining foraging group size. For example, although mature leaves may be ubiquitous and easy to harvest, new leaves and shoots are more like fruits in their seasonal abundance and restricted availability. Thus folivores specializing on these two types of foliage may show dramatic differences in both ranging patterns and social organization. Among frugivores, some primates specialize on fruits that are found in large numbers at a given time but may be widely distributed in time and space. Such foods may be exploited best by a large, wide-ranging group. Other primates specialize on fruits that are found in small numbers but on a more regular temporal basis. Unfortunately, such details about spatial distribution of primate food items have not been as well documented as other aspects of diet (Oates, 1986).

In addition, although food is certainly the major determinant of molar shape, it is only one of many factors likely to influence the grouping behavior of primates. Other factors, such as activity pattern, predation, and reproductive considerations (access to mates, parental care), are discussed in Chapter 3 (see Janson, 1992; van Schaik, 1996). There is also a large phylogenetic component in the distribution of primate social organization (e.g., DiFiori and Rendall, 1994). Closely related species with different diets often seem to show subtle modifications of a basically similar social system rather than a dramatically different arrangement. Nevertheless, increasingly sophisticated field studies and comparative analyses are succeeding in identifying the most important aspects of diet, such as patch size and density, that seem to have the greatest influence on group structure (e.g., Janson and Goldsmith, 1995).

Locomotor Adaptations

Primate locomotor adaptations are found in many parts of the body. Most of the differences we see in the anatomy of the limbs and trunk of living primates are clearly related to differences in their locomotor and postural abilities—the way they move, hang, and sit. Locomotion and posture also affect the orientation of the head on the trunk, the shape of the thorax, and the positioning of abdominal viscera.

Like many other adaptations, the modifications of the musculoskeletal system related to locomotor differences are influenced by the ancestry of the group being considered, and primates often have evolved different solutions to the same problem. Evolution by natural selection has worked with the available material. Thus quadrupedal lemurs, quadrupedal monkeys, and quadrupedal apes all show similarities related to their quadrupedal habits, but they show affinities to other lemurs, monkeys, and apes as well. For the paleontologist, this is a real advantage; it means that bones can provide information about both phylogeny and adaptation—if the two can be accurately distinguished.

Because locomotor adaptations may have different expressions in different species, our best approach is to examine the mechanical problems that different types of locomotion present. Then we can consider how living primate species have evolved musculoskeletal differences to meet these mechanical demands. We will concentrate on features of the skeleton that can be related to different postures and methods of progression (Figs. 9.11–9.15), because these are the best documented aspects of primate locomotor anatomy and those that are most useful in reconstructing the locomotor habits of fossils. It is important to realize that such correlations between bony morphology and locomotor behavior are constantly being tested and refined by experimental studies that permit a clearer understanding of the biomechanical and physiological mechanisms of primate locomotion.

Arboreal Quadrupeds

Arboreal quadrupedalism is the most common locomotor behavior among primates, and most radiations of primates include arboreal quadrupeds. In many respects, arboreal quadrupeds show a generalized skeletal morphology that can easily be modified into any of the more specialized locomotor types, and it is likely that this type of locomotor behavior characterized both the earliest mammals and the earliest primates (Fig. 9.11).

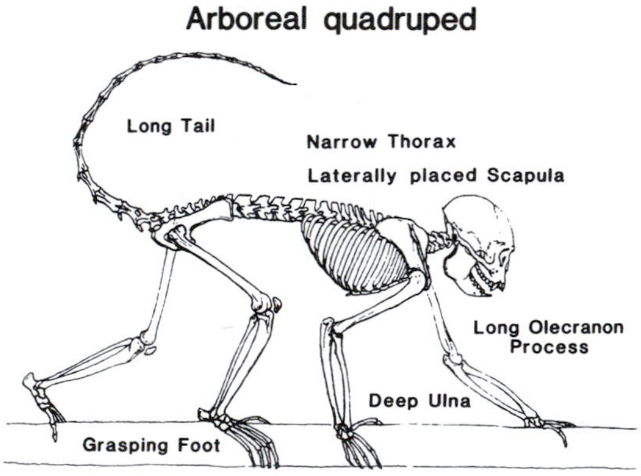

FIGURE 9.11 The skeleton of a primate arboreal quadruped, illustrating some of the distinctive anatomical features associated with that type of locomotion.

Quadrupeds, by definition, use four limbs in locomotion. Experimental evidence suggests that in primates, as in most mammals, the hindlimbs play a greater role than the forelimbs in propulsion (Demes et al., 1994). The major problem arboreal quadrupeds face in their locomotion is providing propulsion on an inherently unstable, uneven support that is usually very small compared with the size of the animal. Thus, stability and balance are their major concerns.

The overall body proportions of arboreal quadrupeds are adapted in several ways to meeting these problems of balance and stability. These primates have forelimbs and hindlimbs that are more similar in length than are those of either leapers, which have relatively long hindlimbs, or climbers, which have relatively long forelimbs. In addition, arboreal quadrupeds' forelimbs and hindlimbs are usually short, to bring the center of gravity closer to the arboreal support. Most arboreal primates may also bring the center of gravity closer to the support by using more flexed limbs when they walk on arboreal supports than on terrestrial supports (Schmitt, 1994). In addition, they tend to move more slowly. Finally, many have a long tail, which aids in balancing. The grasping hands and feet of most arboreal quadrupedal primates provide both a firm base for propulsion and a guard against falling.

The forelimbs of arboreal quadrupeds show a number of distinctive osteological features related to their typical postures and method of progression. The shoulder joint is characterized by an elliptically shaped glenoid fossa on the scapula and a broad humeral head surrounded by relatively large tubercles for the attachment of the scapular muscles that control the position of the head of the humerus. The humeral shaft is usually moderately robust, since the forelimb plays a major role in both support and propulsion.

The elbow region of an arboreal quadruped is particularly diagnostic. On the distal end of the humerus, the medial epicondyle is large and directed medially. This process, where the major flexors of the wrist and some of the finger flexors originate, provides leverage for these muscles when the hand and wrist are in different degrees of pronation and supination. The olecranon process of the ulna is long, to provide leverage for the triceps muscles when the elbow is in the flexed position characteristic of arboreal quadrupeds. Because the elbow rarely reaches full extension, the olecranon fossa of the humerus is shallow. The ulna shaft is relatively robust and often more bowed and deep in arboreal quadrupeds than in many other locomotor types. At the wrist, arboreal quadrupeds are characterized by a relatively broad hamate, presumably for weight bearing, and a midcarpal joint that seems to permit extensive pronation.

As a group, primates are characterized by relatively long digits and grasping hands. Among primates, however, arboreal quadrupeds usually have digits of moderate length—longer than those of terrestrial quadrupeds but shorter than those of suspensory species. They show a wide range of grasps.

The most distinctive features of the hindlimb joints of an arboreal quadruped reflect the characteristic abducted posture of that limb. The femoral neck is set at a moderately high angle relative to the shaft, enhancing abduction at the hip. At the knee, the abduction of the hindlimb is expressed in the asymmetrical size of the femoral condyles and their articulating facets on the top of the tibia. At the ankle, the tibio–talar joint is also asymmetrical. The lateral margin of the proximal talar surface is higher than the medial margin, reflecting the normally inverted posture of the grasping foot. Arboreal quadrupeds all have a large hallux and moderately long digits.

Terrestrial Quadrupeds

Compared with their abundance among other orders of mammals, terrestrial quadrupeds are relatively rare among primates, and none show the striking morphological adaptations found in such runners as cheetahs and antelopes. The main group of primate terrestrial quadrupeds are the larger Old World monkeys: -baboons, some macaques, and the patas monkey. These species show a number of distinctive anatomical features that separate them from more arboreal species. Most of these features relate to the use of more extended, adducted limb postures on a broad flat surface. Since balance is not a problem, these primates have a narrow, deep trunk and relatively long limbs, designed for long strides and speed, and their tails are often short or absent (Fig. 9.12).

The limbs of terrestrial quadrupeds seem designed for speed and simple fore–aft motions rather than for power and more complex rotational movements at the joints. At the shoulder joint, the articulating surfaces of the scapula and head of the humerus provide only a limited anterior–posterior motion, and the greater tuberosity of the humerus is high and positioned in front of the shoulder joint to stabilize that joint during the support phase of locomotion (Larson and Stern, 1989).

Terrestrial quadrupeds have an elbow joint that reflects their more extended limb postures. Instead of being long and extending proximally, as in arboreal quadrupeds, the olecranon process extends dorsally to the long axis of the ulna, an orientation that maximizes the leverage of the elbow-extending muscles when the elbow is nearly straight rather than flexed. A related feature is that the olecranon fossa on the posterior surface of the humerus is deep, with an expanded articular surface on the lateral aspect of the fossa for articulation with the ulna. The

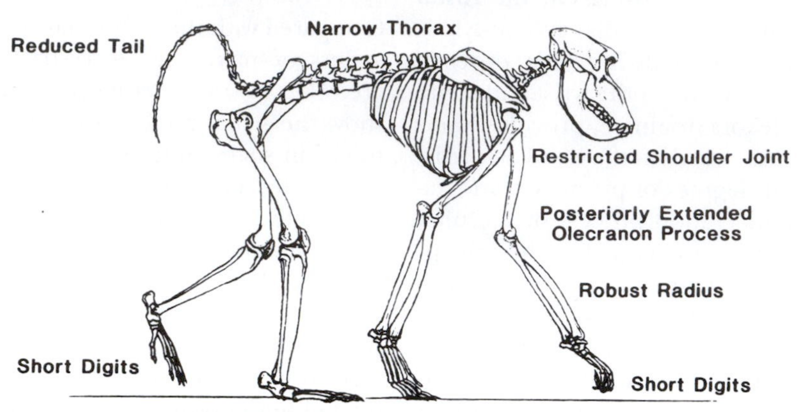

FIGURE 9.12 The skeleton of a primate terrestrial quadruped, illustrating some of the distinctive anatomical features associated with that type of locomotion.

articulation of the ulna with the humerus is relatively narrow, whereas the head of the radius is relatively large, oval in shape, and flattened proximally (Rose, 1988), suggesting that the latter bone plays a more important role in transmitting weight from the elbow to the wrist in terrestrial quadrupeds than in other primates. The medial epicondyle on the humerus is short and directed posteriorly, an orientation that facilitates the use of the wrist and hand flexors when the forearm is pronated, the normal position for terrestrial species.

The carpal bones of terrestrial quadrupeds are relatively short and broad, more suitable for weight bearing and less adapted for rotational movements. Their hands are characterized by robust metacarpals and short, straight phalanges.

The hindlimbs of terrestrial quadrupeds, like their forelimbs, are long. Their feet have robust tarsal elements, robust metatarsals, and short phalanges.

Leapers

Many primates are excellent leapers, and leaping adaptations have almost certainly evolved independently in many primate groups. Although there are many differences among primate leapers, there are also a number of similarities resulting from the mechanical demands of such movement (Fig. 9.13). In leaping, most of the propulsive force comes from a single rapid extension of the hindlimbs, with little or no contribution from the forelimbs. The leaper's takeoff speed, and hence the distance the animal can travel during a leap, is proportional to the distance over which the propulsive force is applied—the length of its hindlimbs. Longer legs thus enable a longer leap from the same locomotor force. Relative to the length of the hindlimb, leapers have relatively short, slender forelimbs. Although the forelimbs are certainly used in landing after leaps, in clinging between leaps, and for various other tasks, including feeding, their

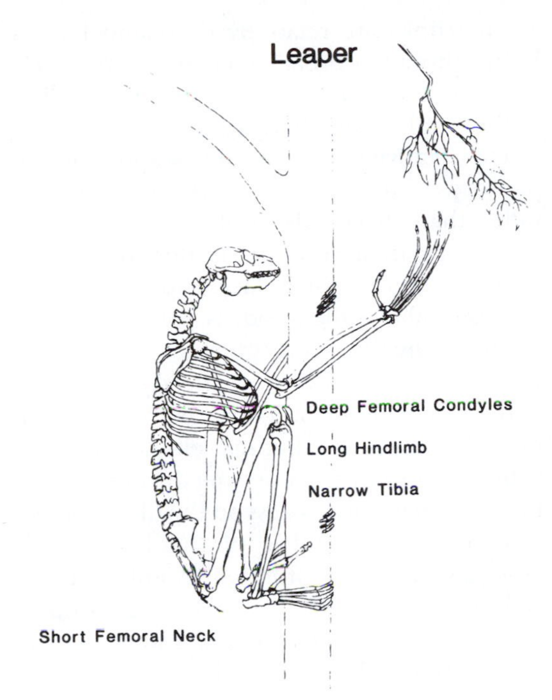

FIGURE 9.13 The skeleton of a primate leaper, illustrating some of the distinctive anatomical features associated with that type of locomotion.

most important role during leaping is probably for control of rotational forces during flight (Demes et al., 1996). The anatomy of the vertebral column varies considerably among leapers, in association with trunk posture. More quadrupedal species seem to have relatively long and flexible lumbar regions; however, indriids, which leap from vertical postures, have a relatively short and stiff lumbar region (Shapiro, 1995).

There are many skeletal adaptations for leaping to be found in the hindlimb. Because hip extension is a major source of propulsive force in leaping, primate leapers usually have a long ischium, which increases the leverage of the hamstring muscles. The direction in which the ischium is extended depends on the postural habits of the species. In primates that leap from a quadrupedal position, the ischium extends distally in line with the blade of the ilium, enhancing hip extension when the hindlimb is at a right angle to the trunk. In prosimians that normally leap from a vertical clinging posture, the ischium is usually extended posteriorly rather than distally, increasing the moment arm of the hamstrings when the limb is near full extension, a common situation for vertical clingers (Fleagle and Anapol, 1992).

Whereas arboreal quadrupeds use abducted limbs for balancing on small supports, leapers restrict their limb excursions to simple, hinge-like flexion and extension movements, both for greater mechanical efficiency and to avoid twisting and damaging joints during the powerful takeoff. In this regard, leapers resemble swift quadrupedal mammals. Many features of the hindlimbs of leapers seem related to this alignment of movement and to increasing the range of flexion and extension. For example, the neck of the femur is very short and thick in leapers, and in many species the head of the femur has a cylindrical shape for simple flexion-extension movements rather than the ball-and-socket joint found at the hips of most primates. At the knee joint, the femoral condyles are very deep, to permit an extensive range of flexion and extension, and they are symmetrical because of the adducted limb postures. The patellar groove has a pronounced lateral lip to prevent displacement of the patella during powerful knee extension. The tibia is usually very long and laterally compressed, reflecting the emphasis on movement in an anterior–posterior plane, and the attachments for the hamstring muscles on the tibial shaft are relatively near the proximal end so that when these muscles extend the hip they do not flex the knee as well. In many leapers, the fibula is very slender and bound to the tibia distally so that the ankle joint becomes a simple hinge joint for flexion and extension. The morphology of the tarsal region varies considerably among leapers. In many small leapers, the calcaneus and navicular are extremely long, providing a long load arm for rapid leaping. The digits of leapers seem to reflect postural habits rather than adaptations directly related to leaping.

Suspensory Primates

Many living primates hang below arboreal supports by various combinations of arms and legs. Because of the acrobatic nature of such behavior, the skeletons of suspensory primates show features that enhance their abilities to reach supports in many directions (Fig. 9.14). In their body proportions, suspensory primates have long limbs, especially forelimbs.

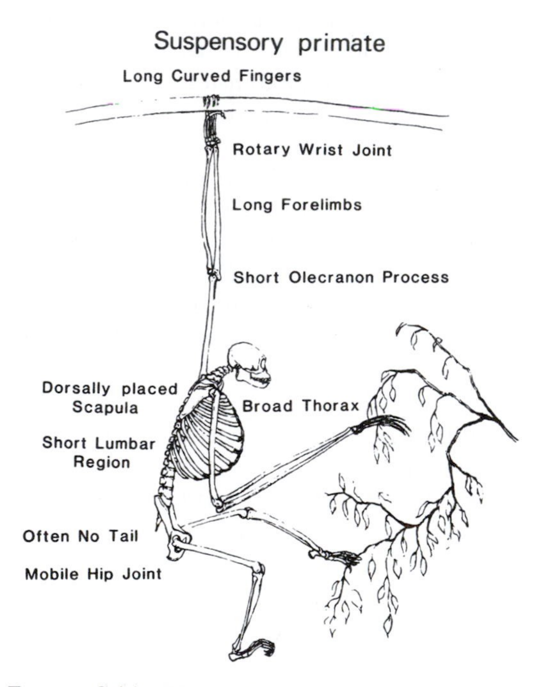

FIGURE 9.14 The skeleton of a suspensory primate, illustrating some of the distinctive anatomical features associated with that type of locomotion.

Their trunks are relatively short and have a broad thorax, a broad fused sternum, and a very short lumbar region to reduce bending of the trunk during hanging and reaching.

The relatively deep, narrow scapula of suspensory primates is positioned on the dorsal rather than the lateral side of the broad thorax, enhancing their reach in all directions. The shoulder joint, which faces upward to aid reaching above the head, is composed of a relatively small, round glenoid fossa and a very large, globular humeral head with low tubercles, a combination that permits a wide range of movement. Because elbow extension is important but does not need to be powerful, the olecranon process on the ulna is short. The medial epicondyle of the humerus is large and medially oriented to enhance the action of the wrist flexors at all ranges of pronation and supination. Both the ulna and the radius are usually relatively long and slender.

Suspensory primates show numerous features of the wrist that seem to increase the mobility of that joint. In many species the ulna does not articulate with the carpals, and the distal and proximal rows of carpal bones form a ball-and-socket joint with increased rotational ability. Suspensory species have long fingers with curved phalanges for grasping a wide range of arboreal supports. Like the forelimb, the hindlimb of suspensory primates is characterized by very mobile joints. Mobility at the hip joint is increased by a spherical head of the femur set on a highly angled femoral neck to permit extreme degrees of abduction. The knee joint is characterized by broad, shallow femoral condyles and a shallow patellar groove. There is very little bony relief on the talus at the ankle joint, a condition that allows movement in many directions rather than restricting it in one direction. In most species, the calcaneus has a short lever arm for the calf muscles that extend the ankle; there is, however, an additional process for the origin of the short flexor muscle of the toes to enhance grasping. The feet of suspensory primates, like their hands, have long, curved phalanges for grasping branches.

Bipeds

One of the most distinctive types of primate locomotion is the bipedalism that characterizes humans. The mechanics and dynamics of human locomotion have been more thoroughly studied than those of any other type of animal movement, but many aspects of human locomotion are still poorly understood. Compared with other types of primate locomotion, bipedalism is unusual in that only one living species habitually moves in this way.

The major bony features associated with bipedalism are found in the trunk and lower extremity. The upper extremity of humans, like that of leapers, does not normally play a role in locomotion and is adapted for other functions. The major mechanical problems faced by a bipedal primate are balance, particularly from side to side, and the difficulty of supporting all of the body weight on a single pair of limbs. One of the most striking correlates of our upright posture is the dual curvature of our spine, with a dorsal convexity (kyphosis) in the thoracic region and a ventral convexity (lordosis) in the lumbar region. In most other primates the kyphosis extends the entire length of the spine; the unique human lumbar curvature moves the center of mass of the trunk forward and also brings the center of mass closer to the hip joint. In keeping with our vertical posture, the size of each vertebra increases dramatically from the cervical region to the lumbar region, for each successive vertebra must support a greater part of the body mass (Fig. 9.15).

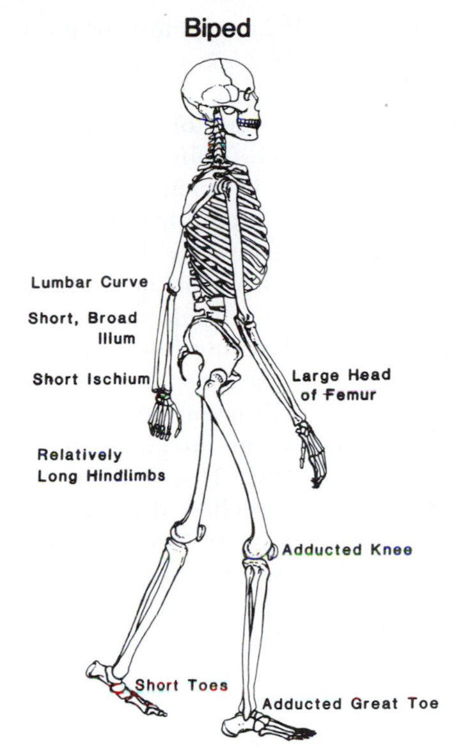

FIGURE 9.15 The skeleton of a bipedal primate, illustrating some of the distinctive anatomical features associated with that type of locomotion.

The human pelvis is the most unusual in the entire primate order. It has a very short, broad iliac blade that serves to lower the center of gravity and to provide better balance and stability. This arrangement also places many of the large hip muscles on the side of the lower limb rather than behind it; in this position, they can act to balance the trunk over the lower limbs during walking and running. The human ischium, where the hip extensors originate, is extended posteriorly (as in vertical leapers) rather than inferiorly (as in most other primates). This position provides greater leverage for the major hip extensors to move the lower limbs behind the trunk.

The human femur is characterized by a very large head, which must support the weight of the entire body during much of the locomotor cycle. Unlike most other primates, humans are naturally knock-kneed; our femur is normally aligned obliquely, with the proximal ends much further apart than the distal ends. This alignment of the femur (called a valgus position) has the effect of placing the knees—as well as the legs, ankles, and feet—directly beneath the body rather than at the sides. As a result, successive footsteps involve less lurching from side to side, and, during those parts of the walking cycle when only one limb is on the ground, that limb is always near the midline of the body (its center of gravity). This oblique orientation of the femur is reflected in many of its bony details, such as the long oblique neck and the angle between the distal condyles and the shaft. A disadvantage of this oblique alignment of our femur is that it predisposes us to a dislocation of the patella, because the muscles extending the knee are now located lateral to the knee itself. To keep the small patella in place we have developed a very large bony lip on the lateral side of the patellar groove.

In contrast with the grasping, handlike foot of most primates, our foot has been transformed into a rather rigid lever for propulsion. The long tuberosity on the calcaneus forms the lever arm, while the stout metatarsals and the large hallux aligned with the other digits provide a firm load arm. The phalanges on our toes are extremely small, because they are not used for grasping, only for pushing off. The strong ligaments on the sole of the foot bind the tarsals and metatarsals together to form two bony arches that act to some degree as springlike shock absorbers. In addition, they direct the body weight through the outside of the foot during each stride, providing us with our characteristic human footprint.

Locomotor Compromises

In the previous sections we have portrayed primates that are somewhat hypothetical and idealistic—primates adapted for a single type of locomotion. But, as we discussed in earlier chapters, most primates habitually use many types of locomotion, just as they eat many types of food. For example, many arboreal quadrupeds often leap, and some leapers are also suspensory. Nevertheless, it is reassuring that the features discussed earlier seem to distinguish not only primates that always leap from primates that always move quadrupedally, but also those species that leap more and are less quadrupedal from those that leap less and are more quadrupedal (see Fig. 6.21). We can therefore have confidence that these features are likely to be useful in reconstructing the habits of extinct primates known only from bones.

There are other factors to consider in trying to understand how primate skeletons are related to locomotor habits. The same parts of the body that are used in locomotion play other roles in the animal's life. Hands are used both in locomotion and in obtaining food, perhaps catching insects, picking leaves, or opening seed pods. The bony pelvis is an anchor for the hindlimb and a site for the origin of many hip muscles, but it also supports the abdominal viscera and serves as the birth canal in females. Such multiple, often-conflicting functional demands are present throughout the body, and often complicate attempts to identify features that are uniquely related to one type of movement or to reconstruct the locomotor abilities of an extinct primate from bits of the skeleton. Still, many of the bony features discussed earlier, as well as numerous others (which can be found in more technical treatises), have proved to be generally characteristic of animals with particular locomotor habits and should provide useful evidence for reconstructing fossils.

Locomotion, Posture, and Ecology

Why do primates show such diverse locomotor and postural abilities and all the morphological specializations that accompany them? One factor is certainly size. As we discussed earlier in this chapter, within the same habitat large and small primates are likely to face very different problems in terms of balance and the availability of strong-enough supports. Thus larger species are more likely to be suspensory or terrestrial.

Apart from size, the major adaptive significance of different locomotor habits seems to be the access they provide to different parts of a forest habitat. In different types of forests and at different vertical levels within a forest, the density and the arrangement of available supports for a primate to move on are often quite different. Primates that live in openareas are best adapted to terrestrial walking and running. Even within a tropical rain forest, the available supports in the understory are different from those higher in the canopy, and species that travel and feed in different levels have different methods of moving (e.g., Cannon and Leighton, 1994). The lowest levels of most forests are characterized by many vertical supports, such as tree trunks and lianas, but there are few pathways that are continuous in a horizontal direction (see Fig. 3.3). Primates that feed and travel in the understory are often leapers that can move best between discontinuous vertical supports. Higher, in the main canopy levels, the forest is usually more continuous horizontally and suitable for other methods of progression, such as quadrupedal walking and running or suspensory behavior.

We know remarkably little about the relationship between postural activities and aspects of primate ecology. Although most postural behavior elicits lower levels of bone strain and muscle activity than does locomotor behavior, it certainly occupies a major portion of an individual's activity budget. Clearly, there are particular types of posture that seem related to specific habitats or food sources. Primates that regularly eat gums or other tree exudates often have claws or clawlike nails so that they can cling to large tree trunks (Fig. 9.9). Among Old World monkeys, colobines tend to sit while feeding, whereas more insectivorous cercopithecines tend to stand (McGraw, 1998). It is likely that different postural abilities enable species to exploit different parts of the same resource.

However, except for special cases such as gum eaters, there are very few general associations between the patterns of locomotion and posture used by primates and their dietary habits. Primates that live in bamboo forests (*Hapalemur, Callimico*) are almost always leapers because of the predominance of vertical supports. However, it is more frequently the case that, among sympatric species, those with the most similar diets show the greatest locomotor differences; at the same time, those with the most similar locomotion show the greatest dietary differences. This suggests that primates have often evolved locomotor differences for exploiting similar foods in different parts of their environment, and vice versa (Fleagle and Mittermeier, 1980; Walker, 1996).

It is also likely that many of the ways locomotion contributes to a species' foraging habits have not been properly studied. As noted in Chapter 3, we normally categorize foods into fruits, leaves, and insects, a classification that accords well with the mechanical and nutritional properties of dietary items. But for understanding locomotion, we should perhaps classify foods according to their distribution in the forest, the shapes of the trees in which they are found, or the size of the branches from which they can best be harvested. Locomotor habits are certainly an integral part of primate feeding strategies, and the subtle nature of this relationship deserves more study (e.g., Grand, 1972).

Anatomical Correlates of Social Organization

As we have discussed in previous chapters, primates live in many different types of social groups, and the reproductive strategies of individuals of different ages and sexes vary dramatically from species to species. There are a number of general anatomical and physiological features that seem to characterize species that live in particular types of social groups. Among higher primates, the degree of canine dimorphism is closely associated with the type and amount of intrasexual competition normally found among the males and females of that species (Plavcan and van Schaik, 1997). Those species characterized by intense male–male competition, compared with the amount of female–female competition, have greater canine dimorphism than those in which intrasexual competition is less, or equal, between the sexes (Figs. 9.16, 9.17, 9.18). Body size dimorphism shows a strong correlation with patterns of intrasexual competition, but other variables, such as predation and phylogeny, seem to be important as well (Plavcan and van Schaik, 1992, 1997).

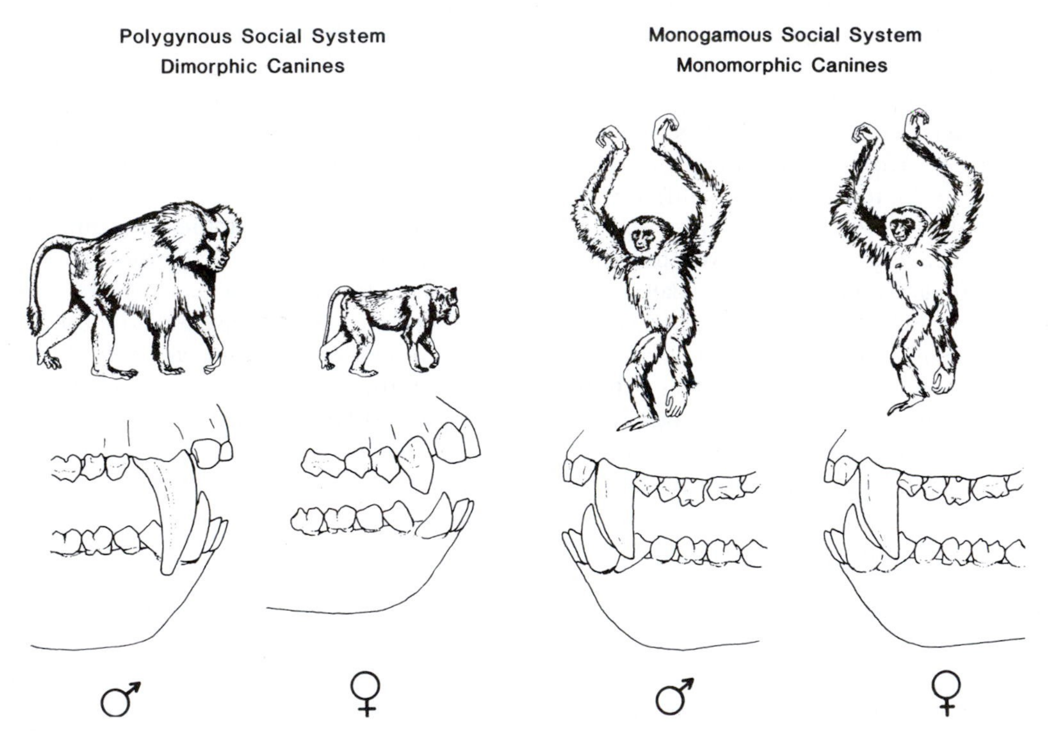

FIGURE 9.16 Canine differences between monogamous gibbons (*Hylobates*) and polygynous baboons (*Papio*)

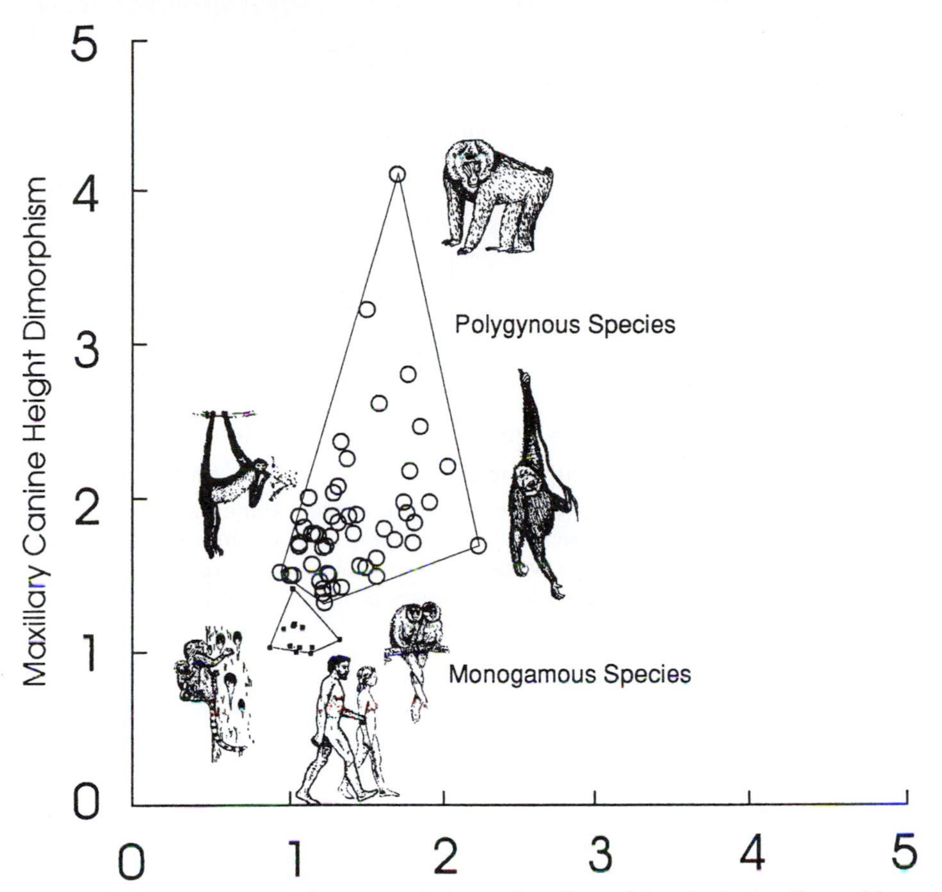

FIGURE 9.17 Relationship between relative canine dimorphism, body size dimorphism, and mating system (courtesy of M. Plavcan).

Not surprisingly, soft tissues of the reproductive system also show strong associations with patterns of social behavior. Testes size shows a close association with the amount of intrasexual competition among males (Fig. 9.18). Monogamous and single-male species show little mating competition within a group, so testes are relatively small. On the other hand, multimale and polyandrous groups exhibit considerable male–male competition for mating success, and males have relatively large testes (Harcourt, 1995). Concomitantly, female catarrhines that live in multimale groups usually have sexual swellings that advertise their reproductive status throughout the menstrual cycle. There are, however, several very different hypotheses as to why these swellings evolved (Hrdy and Whitten, 1986; Sillen-Tullberg and Moller, 1993).

Probably the most intriguing relationship between anatomy and social behavior is Dunbar's (e.g., 1992, 1997) "gossip hypothesis" linking primate brain size to groups. Primate brain size, and especially neocortex size, is closely correlated with the average social group size for a species (Fig. 9.19). Thus, according to

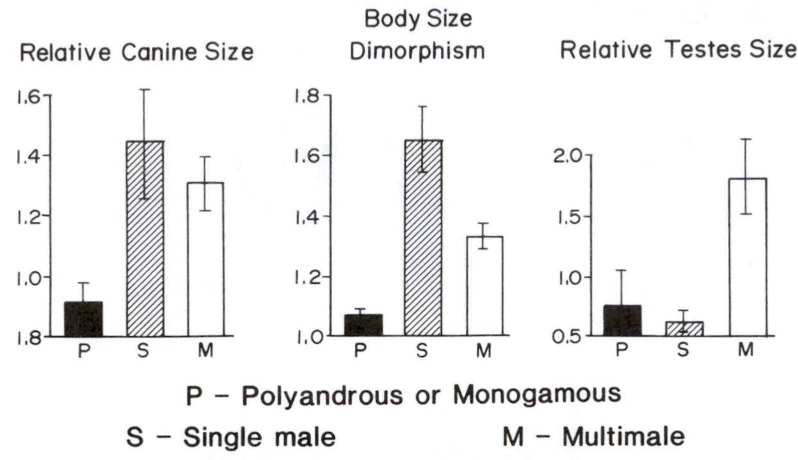

FIGURE 9.18 Morphology and social organization. Canine dimorphism and body size dimorphism separate monogamous and polyandrous species from polygynous species; relative testes size separates primate species living in multimale groups (adapted from Harvey and Harcourt, 1987).

Dunbar, primate neocortex size reflects primarily the number of other individuals one has to keep track of socially. In essence, brain size is driven by social interactions rather than by the requirements of remembering food trees or of monitoring some aspect of the physical environment.

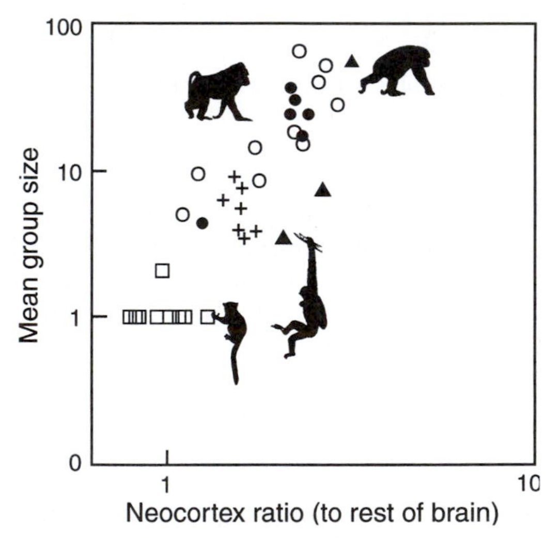

FIGURE 9.19 Relative neocortex size is correlated with the size of primate social groups. Redrawn from Dunbar (1992).

ADAPTATION AND PHYLOGENY

Although primates show tremendous diversity in morphology and behavior related to diet, locomotion, and social organization, it is largely true in any overall assessment of either morphology or behavior that closely related species are more similar to one another than to distantly related species. The reason for this is that evolution by natural selection does not begin with a blank slate for every species; rather, evolution, by definition, is modification of ancestral forms through descent. Thus, despite showing many anatomical differences related to their locomotor adaptations, leaping and quadrupedal langurs are, in general, more similar to one another than they are to leaping and quadrupedal platyrrhines, respectively. Likewise, the teeth of frugivorous and folivorous Old World monkeys can be readily distinguished from one another, but they are more similar to each other in many respects that they are to the teeth of frugivorous and folivorous lemurs.

Not only does natural selection work with the raw material available locally, but it also does not necessarily generate identical optimal solutions to similar ecological problems. In many cases the adaptations found in a species or group of species may be constrained or guided by the ancestral condition from which the species or group has evolved as well as by other aspects of the species' anatomy or behavior. Thus indriids show very different types of adaptations to both the mastication and the digestion of leaves than do colobine monkeys. Similar patterns of phylogenetic similarity is evident in many aspects of social behavior. Despite their tremendous ecological success in many different habitats, all Old World monkeys share many features of their social behavior that distinguish them from other primates (DiFiori and Rendall, 1994; Rendall and DiFiori, 1995).

The fact that closely related species may be more similar to one another for historical reasons rather than strictly adaptive ones is a potentially very serious complication for identifying adaptations, and it is essential that comparative studies make some effort to take phylogeny into account (Harvey and Pagel, 1991; Larson and Losos, 1996). However, adaptation and phylogeny should not be viewed as mutually exclusive, alternative explanations for primate diversity. We must make sure not to ascribe all morphological change to just one or the other process, since both play a role in evolution. Phylogenetic evolution is largely the result of adaptation, and all adaptations have evolved in a phylogenetic context. Indeed, the relationship of adaptation and phylogeny is at the heart of evolutionary biology, and we can never properly study one without the other (Rose and Lauder, 1996; Harvey *et al.*, 1997).

BIBLIOGRAPHY

GENERAL

Bock, W., and von Wahlert, G. (1965). Adaptation and the form–function complex. *Evolution* **19**:269–299.

Gebo, D. L. (1993). *Postcranial Adaptation in Nonhuman Primates.* DeKalb, Ill.: Northern Illinois University Press

Hildebrand, M., Bramble, D. M., Liem, K. F., and Wake, D. B., eds. (1985). *Functional Vertebrate Morphology.* Cambridge, Mass.: Harvard University Press.

Kay, R. E, and Cartmill, M. (1977). Cranial morphology and adaptations of *Palaechthon nacimienti* and other Paromomyidae (Plesiadapoidea, ?Primates), with a description of a new genus and species. *J. Human Evol.* **6**:19–35.

Morbeck, M. E., Preuschoft, H., and Gomberg, N. (1979). *Environment, Behavior, and Morphology: Dynamic Interactions in Primates.* New York: Gustav Fischer.

Rose, M. R., and Lauder, G. V. (1996). *Adaptation.* San Diego: Academic Press.

Strasser, E., and Dagosto, M., eds. (1988). *The Primate Postcranial Skeleton: Studies in Adaptation and Evolution.* London: Academic Press

EFFECTS OF SIZE

Alexander, R. M. (1985). Body support, scaling, and allometry. In *Functional Vertebrate Morphology,* ed. M. Hildebrand, D. M. Bramble, K. F. Liem, and D. B. Wake, pp. 26–37. Cambridge, Mass.: Belknap Press.

Barton, R. A. (1992). Allometry of food intake in free-ranging anthropoid primates. *Folia Primatol.* **58**:56–59.

Biewener, A. A. (1990). Biomechanics of mammalian terrestrial locomotion. *Science* **250**:1097–1103.

Chivers, D. J., and Hladik, C. M. (1980). Morphology of the gastrointestinal tract in primates: Comparisons with other mammals in relation to diet. *J. Morphol.* **166**: 337–386.

Damuth, J., and MacFadden, B. J. (1990). *Body Size in Mammalian Paleobiology: Estimation and Biological Implications.* Cambridge: Cambridge University Press

Demes, B., and Jungers, W. L. (1993). Long bone cross-sectional dimensions, locomotor adaptations and body size in prosimian primates. *J. Human Evol.* **25**:57–74.

Fleagle, J. G., and Mittermeier, R. A. (1980). Locomotor behavior and comparative ecology of seven Surinam monkeys. *Am. J. Phys. Anthrop.* **52**:301–322.

Gebo, D. L., and Chapman, C. A. (1995). Positional behavior in five sympatric Old World monkeys. *Am. J. Phys. Anthropol.* **97**:49–76.

Janson. C. H., and van Schaik, C. P. (1993). Ecological risk aversion in juvenile primates: Slow and steady wins the race. In *Juvenile Primates,* ed. M. E. Pereira and L. A. Fairbanks, pp.57–74. New York: Oxford University Press.

Jungers, W. L. (1985). *Size and Scaling in Primate Biology.* New York: Plenum Press.

Martin, R. D. (1989). Size, shape and evolution. In *Evolutionary Studies—A Centenary Celebration of the Life of Julian Huxley,* ed. M. Keynes, pp. 96–141. Eugenics Society: London.

McGraw, W. S. (1996). Cercopithecoid locomotion, support use, and support availability in the Tai Forest, Ivory Coast. *Am. J. Phys. Anthropol.* **100**:507–522.

Preuschoft, H., Witte, H., Christian, A., and Fischer, M. (1996). Size influences on primate locomotion and body shape, with special emphasis on the locomotion of "small mammals." *Folia Primatol.* **66**:93–112.

Ross, C. (1992). Basal metabolic rate, body weight and diet in primates: An evaluation of the evidence. *Folia Primatol.* **58**:7–23.

______ (1997). Primate life histories. *Evol. Anthropol.* **6**: 54–63.

Smith, R. J. (1993). Categories of allometry: Body size versus biomechanics. *J. Human Evol.* **24**:173–182.

Terborgh, J. (1986). The social system of New World primates: An adaptationist view. In *Primate Ecology and Conservation,* ed. J. G. Else and P. C. Lee, pp. 199–211. Cambridge: Cambridge University Press.

Terranova, C. J. (1996). Variation in the leaping of lemurs. *Am. J. Phys. Anthropol.* **40**:145–166.

ADAPTATIONS TO DIET

Chivers, D. J., and Hladik, C. M. (1980). Morphology of the gastrointestinal tract in primates: Comparisons with other mammals in relation to diet. *J. Morphol.* **166**: 337–386.

Chivers, D. J., and Langer, P. (1994). *The Digestive System in Mammals: Food, Form and Function.* Cambridge: Cambridge University Press.

Chivers, D. J., Wood, B. A., and Bilsborough, A. (1984). *Food Acquisition and Processing in Primates.* New York: Plenum Press.

Clutton-Brock, T. H. (1974). Primate social organization and ecology. *Nature* **250**:539–542.

Clutton-Brock, T. H., and Harvey, P. H. (1977). Primate ecology and social organization. *J. Zool.* **183**:1–39.

Di Fiori, A., and Rendall, D. (1994). Evolution of social organization: A reappraisal for primates by using phylogenetic methods. *Proc. Natl. Acad. Sci USA* **91**:9941–9945.

Dumont, E. R., Strait, S. G., and Overdorff, D. J. (1997). Oral pH in fruit-, leaf-, and insect-feeding primates and bats. *Am. J. Phys. Anthropol. Suppl* **24**:103.

Dunbar, R. I. M. (1992). Neocortex size as a constraint on group size in primates. *J. Human Evol.* **20**:469–493.

Glander, K. E. (1982). The impact of plant secondary compounds on primate feeding behavior. *Yrbk. Phys. Anthropol.* **25**:1–18.

Hylander, W. L. (1979). The functional significance of primate mandibular form. *J. Morphol.* **160**:223–240.

Hylander, W. L., and Johnson, K. R. (1997). *In vivo* bone strain patterns in the zygomatic arch of macaques and the significance of these patterns for functional interpretations for interpreting the fossil record. *Am. J. Phys. Anthropol.* **102**:203–232.

Janson, C. J. (1992). Evolutionary ecology of primate social structure. In *Evolutionary Ecology and Human Behavior,* ed. E. A. Smith and B. Winterhalder, pp. 95–130. Hawthorne, N.Y.: Aldine de Gruyter.

Janson, C. H., and Goldsmith, M. L. (1995). Predicting group size in primates: Foraging costs and predation risks. *Behav. Ecol.* **6**:326–336.

Kay, R. F. (1984). On the use of anatomical features to infer foraging behavior in extinct primates. In *Adaptations for Foraging in Nonhumon Primates: Contributions to an Organismal Biology of Prosimians, Monkeys and Apes,* ed. P. S. Rodman and J. G. H. Cant, pp. 21–53. New York: Columbia University Press.

Kay, R. E, and Hylander, W. L. (1978). The dental structure of mammalian folivores with special reference to primates and phalangeroidea (Marsupialia). In *The Ecology of Arboreal Folivores,* ed. G. G. Montgomery, pp. 173–191. Washington, D. C.: Smithsonian Institution. Press.

Lucas, P. (1991). Fundamental physical properties of fruits and seeds in primate diets. In *Primatology Today, Proceedings of the XIIIth Congress of the International Primatological Society,* ed. A. Ehara, T. Kimura, O. Takenaka, and M. Iwamoto, pp. 235–238. Amsterdam: Elsevier.

Lucas, P. W., and Teaford, M. F. (1994). Functional morphology of colobine teeth. In *Colobine Monkeys: Their Ecology, Behavior, and Evolution,* ed. A. G. Davies and J. F. Oates, pp.173–204. Cambridge: Cambridge University Press.

Lucas, P. W., Corlett, R. T., and Luke, D. A. (1986). A new approach to postcanine tooth size applied to Plio-Pleistocene hominids. In *Primate Evolution,* ed. J. G. Else and P. C. Lee, pp. 191–201. Cambridge: Cambridge University Press.

Maas, M. C. (1993). Enamel microstructure and molar wear in the greater galago, *Otolemur crassicaudatus* (Mammalia, Primates). *Am. J. Phys. Anthropol.* **92**: 217–234.

Milton, K. (1978). The quality of diet as a possible limiting factor on the Barro Colorado Island howler monkey population. In *Recent Advances in Primatology,* vol. 1, *Behaviour,* ed. D. J. Chivers and K. A. Joysey, pp. 387–389. London: Academic Press.

Milton, K., and May, M. L. (1976). Body weight, diet, and home range size in primates. *Nature* **259**:459–462.

Nash, L. T. (1986). Dietary, behavioral, and morphological aspects of gumnivory in primates. *Yrbk. Phys. Anthropol.* **29**:113–138.

Oates, J. E. (1986). Food distribution and foraging behavior. In *Primate Societies,* ed. B. B. Smuts, D. L. Cheney, R. M. Seyfarth, R. W. Wrangham, and T. T. Struhsaker, pp. 197–209. Chicago: University of Chicago Press.

Parra, R. (1978). Comparison of foregut and hindgut fermentation in herbivores. In *The Ecology of Arboreal Folivores,* ed. G. G. Montgomery, pp. 205–230. Washington, D. C.: Smithsonian Institution Press.

Power, M. L., and Oftedal, O. T. (1996). Differences among captive callitrichids in the digestive responses to dietary gum. *Am. J. Primatol.* **40**:131–144.

Rodman, P. S., and Cant, J. G. H., eds. (1984). *Adaptations for Foraging in Nonhuman Primates: Contributions to an Organismal Biology of Prosimians, Monkeys and Apes.* New York: Columbia University Press.

Rubenstein, D. I., and Wrangham, R. W. (1986). *Ecological Aspects of Social Evolution.* Princeton, N. J.: Princeton University Press.

Smith, B. H. (1992). Life history and the evolution of human maturation. *Evol. Anthropol.* **1**:134–142.

Strait, S. G. (1993a). Differences in occlusal morphology and molar size in frugivores and faunivores. *J. Human Evol.* **25**:471–484.

______ (1993b). Molar microwear in extant small-bodied faunivorous mammals: An analysis of feature density and pit frequency. *Am. J. Phys. Anthropol.* **92**:63–79.

______ (1997). Tooth use and the physical properties of food. *Evol. Anthropol.* **5**:199–211.

Teaford, M. F. (1994). Dental microwear and dental function. *Evol. Anthropol.* **3**:17–30.

Teaford, M. F., and Runestad, J. A. (1992). Dental microwear and diet in Venzuelan primates. *Am. J. Phys. Anthropol.* **88**:347–364.

Ungar, P. S. (1994). Incisor microwear of Sumatran anthropoid primates. *Am. J. Phys. Anthropol.* **94**:339–363.

van Schaik, C. P. (1996). Social evolution in Primates: The role of ecological factors and male behavior. In *Evolution of Social Behavior Patterns in Primates and Man,* ed. W. G. Runciman, J. Maynard Smith, and R. J. M. Dunbar, Proceedings of the British Academy **88**:9–31.

LOCOMOTOR ADAPTATIONS

Cannon, C. B., and Leighton, M. (1994). Comparative locomotor ecology of gibbons and macaques: Selection of canopy elements for crossing gaps. *Am. J. Phys. Anthropol.* **93**:505–524.

Cant, J. G. H., and Temerin, L. A. (1984). A conceptual approach to foraging adaptations of primates. In *Adaptations for Foraging in Nonhuman Primates: Contributions to an Organismal Biology of Prosimians, Monkeys and Apes,* ed. P. C. Rodman and J. G. H. Cant, pp. 304–342. New York: Columbia University Press.

Demes, B., Larson, S. G., Stern, J. T., Jr., Jungers, W. L., Biknevicius, A. R., and Schmitt, D. (1994). The kinetics of primate quadrupedalism: "Hindlimb drive" reconsidered. *J. Human Evol.* **26**:353–374.

Demes, B., Jungers, W. L., Gross, T. S., and Fleagle, J. G. (1995). Kinetics of leaping primates: Influence of substrate orientation and compliance. *Am. J. Phys. Anthropol.* **96**:419–429.

Demes, A. B., Jungers, W. L., Fleagle, J. G., Wunderlich, R. E., Richmond, B. G., and Lemelin, P. (1996). Body size and leaping kinematics in Malagasy vertical clingers and leapers. *J. Human Evol.* **31**:367–388.

Fleagle, J. G. (1977a). Locomotor behavior and muscular anatomy of sympatric Malaysian leaf monkeys (*Presbytis obscura* and *melalophos*). *Am. J. Phys. Anthropol.* **46**:297–308.

______ (1977b). Locomotor behavior and skeletal anatomy of sympatric Malaysian leaf monkeys (*Presbytis obscura* and *melalophos*). *Yrbk. Phys. Anthropol.* **20**:440–453.

______ (1979). Primate positional behavior and anatomy: Naturalistic and experimental approaches. In *Environment, Behavior and Morphology: Dynamic Interactions in Primates,* ed. M. E. Morbeck, H. Preuschoft, and N. Gomberg, pp. 313–325. New York: Gustav Fischer.

______ (1984). Primate locomotion and diet. In *Food Acquisition and Processing in Primates,* ed. D. J. Chivers, B. A. Wood, and A. L. Bilsborough, pp. 105–117. New York: Plenum Press.

Fleagle, J. G., and Anapol, F. C. (1992). The indriid ischium and the hominid hip. *J. Human Evol.* **22**:285–305.

Fleagle, J. G., and Meldrum, D. J. (1988). Locomotor behavior and skeletal anatomy of two sympatric pitheciine monkeys, *Pithecia pithecia* and *Chiropotes satanas. Am. J. Primatol* **16**:249–277.

Fleagle, J. G., and Mittermeier, R. A. (1980). Locomotor behavior, body size and comparative ecology of seven Suriname monkeys. *Am. J. Phys. Anthropol.* **52**:301–322.

Gebo, D. L., and Sargis, E. J. (1994). Terrestrial adaptations in the postcranial skeletons of guenons. *Am. J. Phys. Anthropol.* **93**:341–371.

Grand, T. I. (1972). A mechanical interpretation of terminal branch feeding. *J. Mammal.* **53**:198–201.

Hamrick, M. W. (1996a). Articular size and curvature as determinants of carpal joint mobility and stability in strepsirhine primates. *J. Morphol.* **230**:113–127.

——— (1996b). Functional morphology of the lemuriform wrist joints and the relationship between wrist morphology and positional behavior in arboreal primates. *Am. J. Phys. Anthropol.* **99**:319–344.

Jenkins, F. A., Jr. (1974). *Primate Locomotion.* London: Academic Press.

Jolly, C. J. (1966). The evolution of the baboon. In *The Baboon in Medical Research,* vol. 2, ed. H. Vogtborg, pp. 23–50. Austin: University of Texas Press.

Kay, R. E, and Covert, H. H. (1984). Anatomy and behavior of extinct primates. In *Food Acquisition and Processing in Primates,* ed. D. J. Chivers, B. A. Wood, and A. Bilsborough, pp. 467–508. New York: Plenum Press.

Larson, S. G. (1993). Functional morphology of the shoulder in primates. In *Postcranial Adaptation in Nonhuman Primates,* ed. D. L. Gebo, pp. 45–69. DeKalb: Northern Illinois University Press.

——— (1995). New characters for the functional interpretation of primate scapulae and proximal humeri. *Am. J. Phys. Anthropol.* **96**:13–36.

Larson, S. G., and Stern, J. T., Jr., (1989). The role of supraspinatus in the quadrupedal locomotion of vervets (*Cercopithecus aethiops*): Implications for interpretation of humeral morphology. *Am. J. Phys. Anthropol.* **79**:369–377.

McGraw, W. S. (1996). Cercopithecoid locomotion, support use, and support availability in the Tai Forest, Ivory Coast. *Am. J. Phys. Anthropol.* **100**:507–522.

——— (1988). Posture and support use of Old World monkeys (Cercopithecidae): The influence of foraging strategies, activity patterns and the spatial distribution of preferred food items. *Am. J. Primatol.* **44**.

Morbeck, M. E., Preuschoft, H., and Gomberg, N. (1979). *Environment, Behavior and Morphology: Dynamic Interactions in Primates.* New York: Gustav Fischer.

Rose, M. D. (1993). Functional anatomy of the elbow and forearm in primates. In *Postcranial Adaptation in Nonhuman Primates,* ed. D. L. Gebo, pp. 70–95. DeKalb: Northern Illinois University Press.

Schmitt, D. (1994). Forelimb mechanics as a function of substrate type during quadrupedalism in two anthropoid primates. *J. Human Evol.* **26**:441–457.

Shapiro, L. (1993). Functional morphology of the vertebral column in primates. In *Postcranial Adaptation in Nonhuman Primates,* ed. D. L. Gebo, pp. 121–149. DeKalb: Northern Illinois University Press.

——— (1995). Functional morphology of indrid lumbar vertebrae. *Am. J. Phys. Anthropol.* **98**:323–342.

Walker, S. E. (1996). The evolution of positional behavior in the saki-uakaris (*Pithecia, Chiropotes,* and *Cacajao*). In *Adaptive Radiations of Neotropical Primates,* ed. M. A. Norconk, A. L. Rosenberger, and P. A. Garber, pp. 335–367. New York: Plenum.

SOCIAL ORGANIZATION

Dunbar, R. I. M. (1992). Neocortex size as a constraint on group size in primates *J. Human Evol.* **20**:469–493.

——— (1995). Neocortex sixe and group size in primates: A test of the hypothesis *J. Human Evol.* **28**: 287–296

——— (1997). *Grooming, Gossip, and the Evolution of Language.* Cambridge, Mass.: Harvard University Press

——— (1998). The social brain hypothesis. *Evol. Anthropol.* **6**:178–190.

Harcourt, A. H. (1995). Sexual selection and sperm competition in primates: What are male genitalia good for? *Evol. Anthropol.* **4**:121–129.

Harvey, P. H., and Harcourt, A. H. (1984). Sperm competition, testes size and breeding systems in primates. In *Sperm Competition and the Evolution of Animal Mating Systems,* ed. R. L. Smith, pp. 589–600. London: Academic Press.

Hrdy, S. B., and Whitten, P. L. (1986). Patterning of sexual activity. In *Primate Societies,* ed. B. B. Smuts, D. L. Cheney, R. M. Seyfarth, R. W. Wrangham, and T. T. Struhsaker, pp. 370–384. Chicago: University of Chicago Press.

Kay, R. F., Plavcan, M., Wright, P. C., Glander, K., and Albrecht, G. H. (1988). Behavioral and size correlates of canine dimorphism in platyrrhine primates. *Am. J. Phys. Anthropol.* **77**:385–397.

Plavcan, J. M., and van Schaik, C. P. (1992). Intrasexual competition and canine dimorphism in anthropoid primates *Am. J. Phys, Anthropol.* **87**:461–477.

——— (1994). Canine dimorphism. *Evol. Anthropol.* **2**: 208–214.

Plavcan, J. M., and van Schaik, C. P. (1997). Intrasexual competition and body weight dimorphism in anthropoid primates. *Am. J. Phys. Anthropol.* **103**:37–68.

Plavcan, J. M., van Schaik, C. P., and Kappeler, P. M. (1995). Competition, coalitions and canine dimorphism in anthropoid primates. *J. Human Evol.* **28**:245–276.

Sillén-Tullberg, B., and Mller, A. P. (1993). The relationship between concealed ovulation and mating systems in anthropoid primates: A phylogenetic analysis. *Am. Nat.* **141**:1–25.

Adaptation and Phylogeny

Larson, A., and Losos, J. B. (1996). Phylogenetic systematics of adaptation. In *Adaptation,* ed., M. R. Rose and G. V. Lauder, pp. 187–220. San Diego: Academic Press.

Di Fiori, A., and Rendall, D. (1994). Evolution of social organization: A reappraisal for primates by using phylogenetic methods. *Proc. Natl. Acad. Sci USA* **91**:9941–9945.

Harvey, P. H., and Pagel, M. D. (1991). *The Comparative Method in Evolutionary Biology.* Oxford: Oxford University Press.

Harvey, P. H., Leigh Brown, A. J., Maynard Smith, J., and Nee, S. (1997). *New Uses for New Phylogenies.* Oxford: Oxford University Press.

Rendall, D., and Di Fiori, A. (1995). The road less traveled: Phylogenetic perspectives in primatology. *Evol. Anthropol.* **4**:43–52.

Rose, M. R., and Lauder, G. V., eds. (1996). *Adaptation.* San Diego: Academic Press.

TEN

The Fossil Record

PALEONTOLOGICAL RESEARCH

In the previous chapters we have discussed the anatomy, behavior, and ecology of extant primates, with only a passing mention of their evolutionary history. In the following chapters we discuss primate adaptation and evolution from a paleontological perspective. Although most of our understanding of the relationships among living organisms is based on the study of living species themselves, the fossil record provides us with many types of information about the biology of primates that we could never know from the living species alone.

The unique aspect of the fossil record is that it establishes a temporal framework for evolution. It provides a crude dating for individual events, such as the first appearance of particular taxonomic groups or particular anatomical features. It provides evidence for the patterns and rates of evolutionary change—whether it was gradual or occurred in fits and starts.

We can extract several kinds of information valuable for understanding the phylogeny of living primate species from the fossil record. It often shows us intermediate or primitive forms that link more distinct living groups, and it demonstrates how the living species came to be the way they are by documenting the sequence of evolutionary changes that led to their present differences. Phylogenies based on living organisms are best considered as hypotheses of evolutionary changes that can be tested by the fossil record.

The fossil record also enables us to examine adaptive changes through time. Knowledge of past adaptations can help us understand how the adaptive characteristics of extant radiations came to be the way they are and can also suggest tests for examining causal changes between morphology and environment.

Most important, the fossil record provides us with a record of life in the past. It is our only evidence of extinct primates—in most cases, animals whose existence we could never have predicted or even imagined had we not been confronted with their bones. As we shall see, some groups of primates were far more diverse in morphology, ecology, and biogeography during the very recent past than they are today, and other successful radiations from previous epochs have no living representatives.

The information available from the fossil record is quite different from what we can obtain about living species, most noticeably in its incompleteness. Time extracts its price, and our insights into the past are, alas, more often glimpses rather than panoramas. If we hurry, we can still observe living primates in the forest as they go about their daily activities and we can record their behavior in scientific papers, books, photographs, and films. We can examine their pelage, measure and dissect their bodies, and study their physiology, communication, and learning abilities, in addition to measuring their bones and teeth and sequencing their DNA. For fossils we have only bones and teeth, mainly the latter. The occasional impression of the bushy tail of an archaic primate or the footprint of an early hominid is, unfortunately, a rare and remarkable occurrence. As a result, our discussions of the behavior of extinct primates, and even the identification of different sexes and species, require a much larger dose of guesswork than our descriptions of living species.

Our greatest tool is, of course, our ability to extrapolate from the consistent patterns we see among living species to these more poorly known animals in the fossil record. We must keep an open mind, however; the fossil record is likely to be full of unique events. Thus, before we discuss primate evolution, we must consider briefly the special attributes of the fossil record and the types of information that are available for understanding primate history.

Geological Time

The evolution of primates has taken place on a time scale that is virtually impossible to comprehend in anything but a comparative sense (Fig. 10.1). As individuals, 100 years is the most we are likely to ever experience, yet few events in primate evolution can be dated to

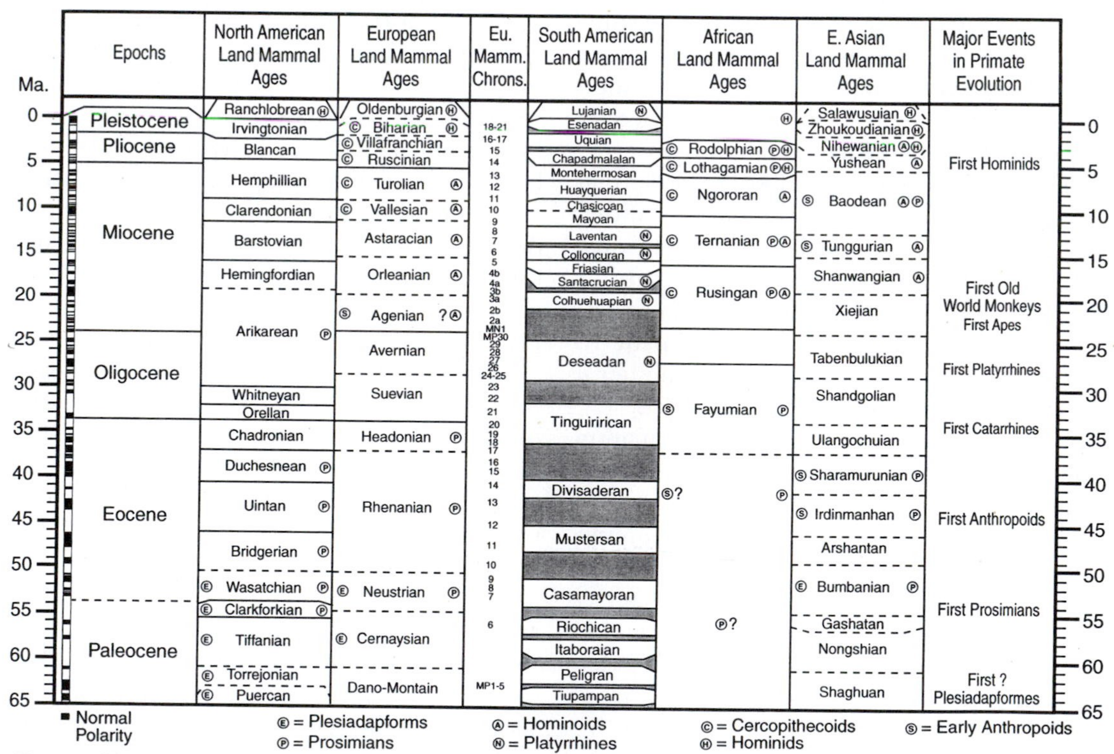

FIGURE 10.1 A geological time scale for the Cenozoic era, showing the epoch series, major land mammal ages, paleomagnetic changes, distribution of fossil primates on different continents, and first appearance of major phyletic groups.

within 1 million or even 5 million years. The scale of events is more commonly on the order of tens of millions of years.

The evolution of primates, like that of most other groups of modern mammals, has occurred almost totally within the Cenozoic era—the Age of Mammals—roughly the last 65 million years. Paleontologists have traditionally divided this period into smaller units (epochs and land mammal ages) on the basis of animals commonly contained in the sediments. Through **faunal correlation,** sediments from different places and the fossils in them can be placed in a relative time scale. However, faunal correlation is a method of relative dating. It can enable a researcher to compare different sequences of rocks and faunas, but provides no absolute age for fossils and geological deposits. These relative scales, such as the epoch series, must be calibrated by some method of absolute dating that can assign a specific age to either geological deposits containing fossils or the fossils themselves

An increasing array of techniques has become available for calibrating the fossil record (Fig. 10.2). These techniques vary considerably in the time range over which they are applicable, depending on the half-life of the elements being used, and in the materials that can be dated.

In general, **radiometric dating** techniques for absolute dating of geological deposits rely on the normal decay patterns of radioactive elements. Many elements on earth are naturally unstable and change to more stable elements at a characteristic rate. By examining the percentages of the parent isotope and of the product, it is possible to calculate how long ago the rock was formed. Thus **carbon-14 dating** is based on the rate of radioactive decay of carbon isotopes found in many organic materials, such as wood and bone; unfortunately, it

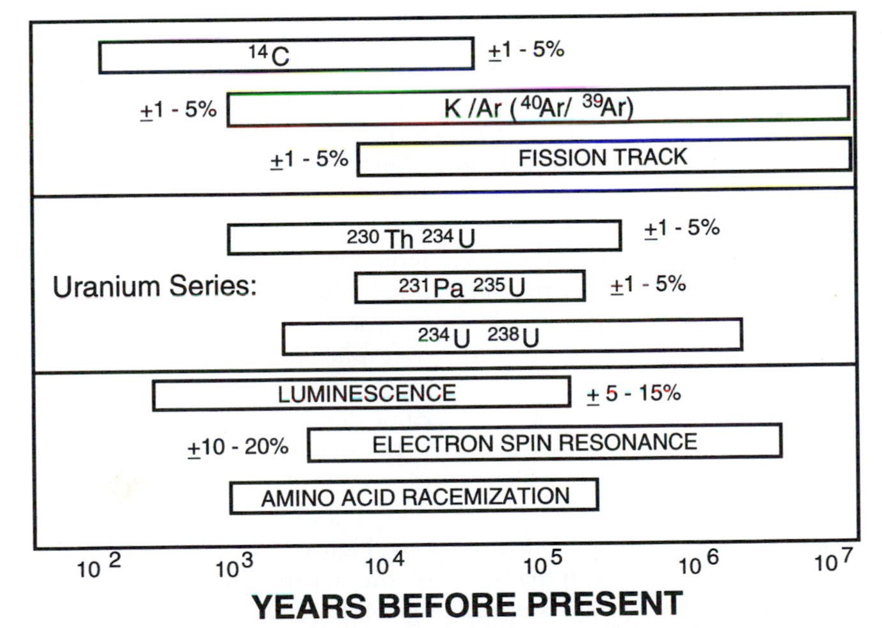

FIGURE 10.2 Time ranges for common methods of geological dating used in paleontology and anthropology (from Schwarcz, 1992).

can only be used for dating events that occurred during the last 70,000–90,000 years (Taylor, 1996). **Potassium/argon dating** and **argon/argon dating** can be applied to a much longer range of ages, but these dating techniques are limited to material containing these elements, usually volcanic rocks (Deino *et al.*, 1998). However, these represent the most precise method currently available, and are responsible for most of the absolute dates used to calibrate primate and early human evolution.

Uranium series dating can be applied to other types of rocks over a much younger time range (Schwarcz, 1992). In **fission track dating,** the investigator measures the number of scars left in a crystal by nuclear particles given off during radioactive decay (Wagner, 1996). **Electron spin resonance** (Grun, 1994), **thermoluminescence** (Feathers, 1996), and **amino acid racemization** are valuable techniques when one has some estimates of the environmental history of the materials being dated.

Radiometric dating of geological sediments in absolute numbers of years is generally possible only for certain types of rocks, for exfjample, relatively pure volcanic ashes or lava flows that were clearly formed at a particular point in time. However, some techniques can measure time elapsed since a particular "resetting event." For example, thermoluminescence or fission tracks may be reset by high temperatures and can be used to date hearths or fired ceramics in the archeological record (Wagner, 1996).

Paleomagnetism

One of the many startling geological discoveries of the last few decades has been that the earth's magnetic field has reversed its north and south poles many times during the past—approximately once every 700,000 years, but not at regular intervals. How and why these changes have occurred is not well understood, but geologists are continually compiling a history of magnetic reversals over the past 500 million years through combined studies of paleomagnetism and radiometric dating techniques. Paleomagnetism provides another method of relative dating of sediments and fossils (Kappelman, 1993). Thus, at many fossil sites that may lack rocks suitable for absolute dating, geologists can use the sequence of magnetic reversals in conjunction with faunal correlation to determine the position of the rocks in the geological timetable and to estimate the absolute age.

Tephrostratigraphy is another way of correlating sediments in different areas. Each volcanic eruption produces ashes with a characteristic chemical signature. By careful examination of the pyroclastic deposits, it is often possible to identify a unique chemical signature for volcanic ashes from a single eruption event. Thus geologists can identify ashes from the identical eruption in several different places and correlate geographically separated stratigraphic sections. For example, it has been possible to identify and correlate individual ash falls across many parts of East Africa and even into ocean cores from the Red Sea (Sarna-Wojcicki *et al.*, 1985). Although tephracorrelation is a method of relative dating, it is a very precise one because individual eruptions usually take place within a very limited time span. Moreover, it is often possible to obtain absolute dates from these same volcanic deposits.

Determining the age of particular events in primate evolution usually requires a combination of both relative and absolute dating methods. Figure 10.1 summarizes determinations of the age of geological epochs and faunal ages relevant to primate evolution, together with major events in the history of primates.

Plate Tectonics and Continental Drift

Far from being a stable, unchanging sphere, earth is a very dynamic body. Its surface is made up of a patchwork of individual plates that form different continents and ocean floors. These plates are constantly in motion with respect to one another. Thus the sizes, orientations, and connections of the continents and the positions of the oceans surrounding them have changed considerably in the past, just as they are changing today (Fig. 10.3). As you might guess, these geographic rearrangements have greatly influenced the routes of migration and dispersal available to plants and animals. In addition, the relative positions of land masses have had major effects on ocean currents and climate—effects with global consequences for primates and all other living things. Many of the most dramatic changes in the earth's surface took place well before the first appearance of primates and so have little bearing on the subject of this book. Nevertheless, during the past 65 million years a number of changes in continental positions and connections have influenced primate evolution.

Paleoclimate

Through studies of fossil land plants and mammals as well as marine organisms, geologists and paleontologists have been able to reconstruct the major changes in the earth's climate during the Cenozoic era. These studies show several general trends over the past 65 million years (Fig. 10.4) that have undoubtedly been important in primate evolution. It is important to remember, however, that climatic changes in a restricted area are quite likely to show different patterns than

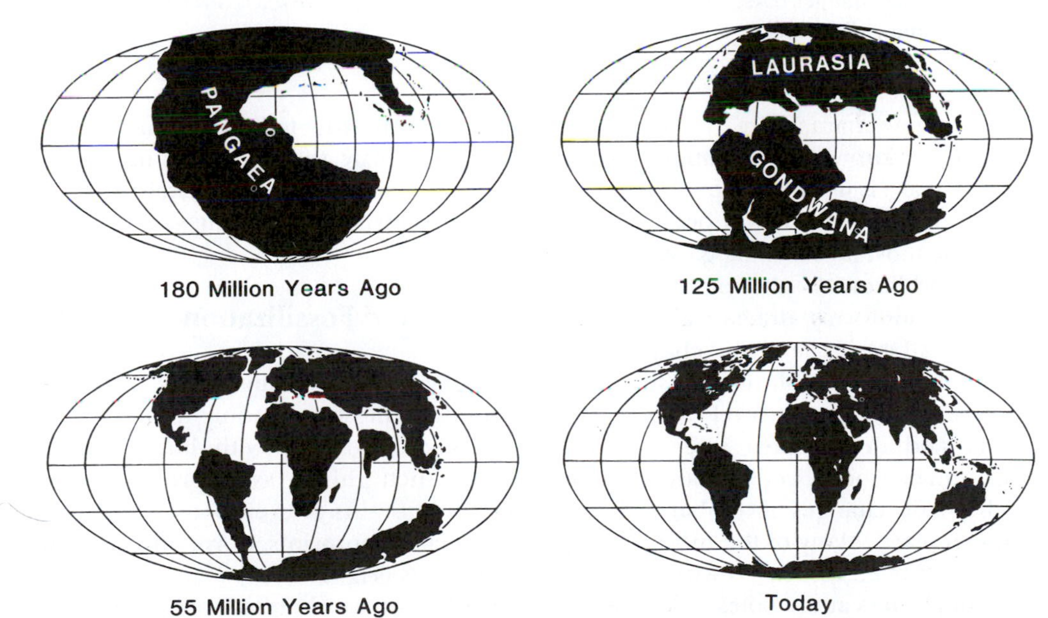

FIGURE 10.3 Positions of the continents at various times during the past 180 million years.

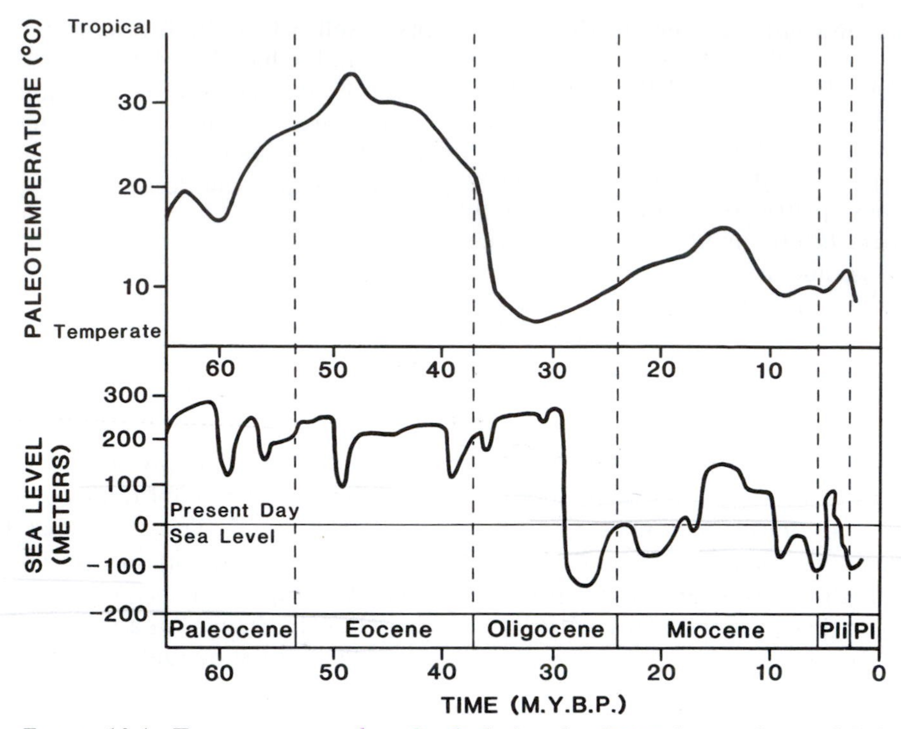

FIGURE 10.4 Temperatures and sea levels during the Cenozoic era; above, global climate changes based on North Sea foraminifera; below, relative sea levels based on seismic reflections.

those that may characterize the earth as a whole, and our knowledge of climatic changes in any one place is usually quite crude.

The formation of glaciers at polar latitudes is one of the most far-reaching global climatic events. In addition to altering regional climates and landforms dramatically, glaciers profoundly affect sea levels by changing the distribution of water on the earth's surface. In turn, these changes in sea level can affect the erosional and depositional rates of rivers, streams, and beaches. Over the past 65 million years dramatic changes in global sea levels have taken place. Many of the most dramatic of these have been associated with the development of glaciers at the poles. Like the positions of continents, changes in sea level can have important effects on plant and animal dispersal. In the following chapters we attempt to relate these changes in continental position, climate, and sea level to the major events in primate evolution.

Fossils and Fossilization

Fossils are any remains of life preserved in rocks. We most commonly think of fossils as petrified bones and teeth, but fossils also include such things as impressions, natural molds of brains or even bodies, and traces of life such as footprints, worm burrows, and termite nests (Fig. 10.5).

Although fossils often preserve shapes of bones or teeth very accurately, most fossils are usually formed by replacement of the original

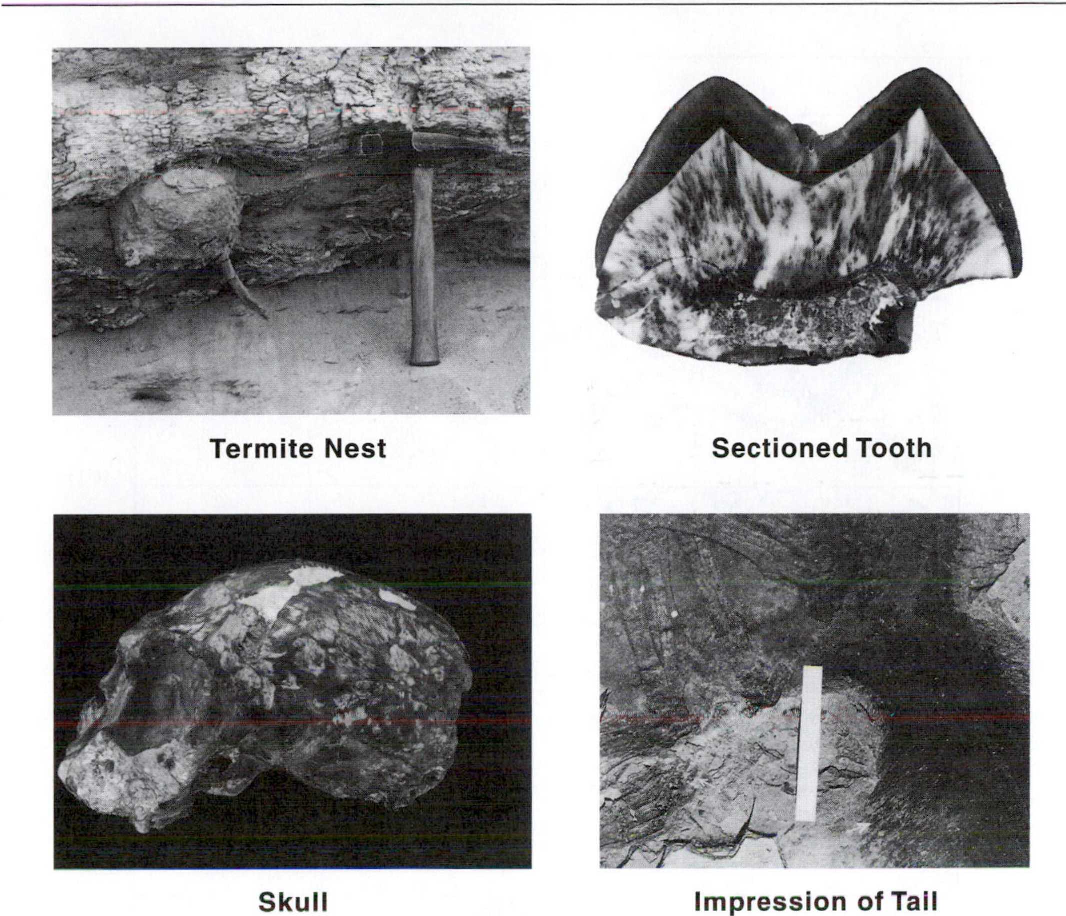

FIGURE 10.5 Different kinds of fossils.

biological materials with minerals derived from the sedimentary environments in which they are buried. In many cases, however, this replacement takes place at a molecular level, so even microscopic details of morphology, such as muscle attachments, fine tooth-wear scratches, dental enamel prisms, or delicate bone structures are preserved and can be analyzed with many of the same tools we apply to the study of living primate skeletons.

The type of remains available to a scientist today from an animal that lived some time in the geological past is determined by many events and processes. The study of the factors that determine which animals become fossils, what parts of their bodies are preserved, how they are preserved, and how they appear to scientists many millions of years later is **taphonomy.** Taphonomists seek to reconstruct as well as possible everything that has happened to a bone between the time it was, say, climbing a tree 35 million years ago in the body of an early fossil monkey until the time it was discovered along with other fossils in Egyptian sandstone (Fig. 10.6). They want to know such things as why teeth and ankle bones are commonly found as fossils but other parts may not be, or why some fossils are

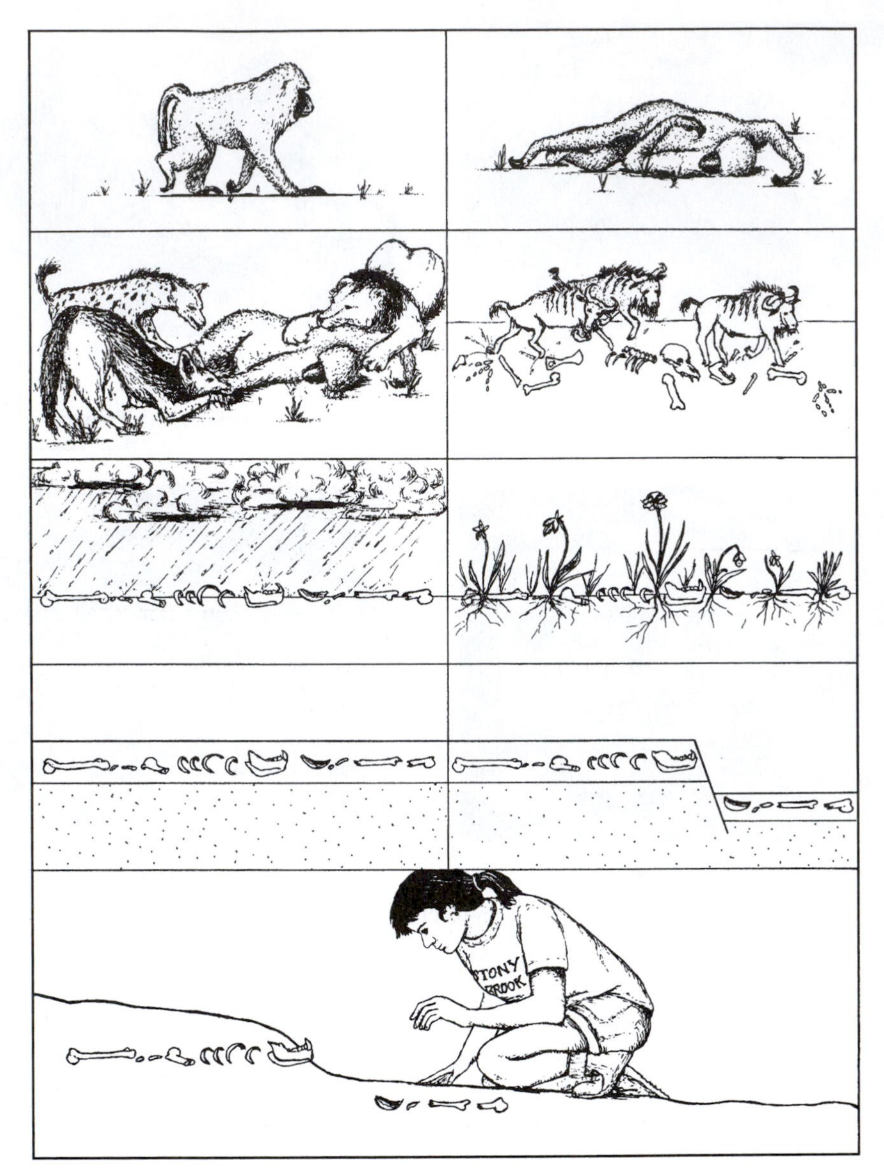

FIGURE 10.6 The taphonomic history of a fossil (adapted from Shipman, 1981).

found as whole bodies and others as fragments. In pursuit of answers to such questions, taphonomists engage in many unusual activities, such as staking out dead antelopes on the Serengeti Plains to see what happens to them, or placing bones in cement mixers to simulate the effects of rolling down a rocky stream. Taphonomy is a young science, barely in its infancy, but as it progresses it is providing many new insights into primate evolution.

Taphonomic studies enable paleontologists to determine if the remains of animals found at a particular locality or site have been transported to the site by the action of streams or perhaps predatory birds, or whether the animals are more likely to have lived and died where their bones are recovered. Studies of the proportions of different skeletal elements recovered, the absence of abrasion, and the abundance of bite marks on bones of fossil primates and other mammals from the early Eocene of Wyoming, for example, indicate that the fossils are the result of long-term accumulations on the surface and were not transported long distances and concentrated by stream action. Thus the proportions of species in the fossil record at this site probably represent a relatively accurate estimate of the proportions of different species living in this area 50 million years ago.

Studies of the proportions of bony elements found in archeological sites have been widely used as evidence for different patterns of carcass transport associated with different strategies of procurement, such as hunting or scavenging. While such inferences are intriguing, they are not without potential pitfalls arising from differential destruction of different body parts by carnivores as well as simple errors due to collecting bias (e.g., Lyman, 1994; Marean, 1995; Marean and Kim, 1998).

Paleoenvironments

A primate fossil is usually found along with other fossils, both plant and animal, and within a particular geological setting—all of which can yield useful information about the environment in which the animal lived and died. Whether a fossil primate is found associated with forest rodents or savannah rodents, for example, can provide clues to its habitat preferences. Land snails seem to have narrow habitat preferences, and fossil snails have proved very useful in determining the extent to which a particular fossil locality represents a forested or an open habitat. Similarly, fossil plants can yield information about both local habitat and climate.

The sediments containing fossils can provide many kinds of information about the fossils' origin. They can tell us if a fossil deposit was preserved on a floodplain, on a river delta, in a stream channel, or on the shores of a lake. This information about where an animal's bones were preserved provides clues to where it lived or died. In addition, sediments can provide detailed information about the climatic regime during which they were formed. Was it hot, cold, wet, or dry? Was the weather relatively uniform or was it seasonal? In addition to the information they convey about a particular fossil site, sedimentary deposits can tell us about climatic trends in a particular region. Many fossil primates are found in ancient soils, and these soils usually contain considerable information about the conditions under which they were formed (Bown, *et al.,* 1992; Bown and Krause, 1993).

In reconstructing fossil environments, there are obvious limits to the amount of detailed information we can infer about the life of an extinct primate. There are also potential pitfalls in extrapolating from the events surrounding fossilization to the habits of an animal. For example, most fossil primates are found in sediments that were originally deposited by streams, rivers, or lakes, often channels within a stream or along floodplains of rivers that overflowed their banks during floods. One early worker, finding fossil lemur bones mixed with the bones of turtles and crocodiles. argued that the lemurs must have been aquatic. Obviously, finding lemur bones in deposits formed by water need not imply aquatic lemurs; more probably, the bones of many different animals were just buried together in stream or lake deposits. While the crocodiles may have lived in the river, the lemurs probably lived in trees overhanging the water, or perhaps their bodies were washed into the river during a rainstorm.

However, by careful comparisons of the ecological characteristics of animals found in different habitats today, we can often reconstruct the nature of ancient habitats by the ecological characteristics of the animals preserved as fossils (Andrews, 1996). For example, comparison of the percentage of arboreal to terrestrial mammals in a fauna usually provides a good estimate of the amount of tree cover in a habitat and can distinguish rain forests from woodlands and savannahs (Fig. 10.7). This is true for several continents and regardless of the particular species of mammals being compared (e.g., Reed, 1997; Kay and Madden, 1997). Thus fossil faunas containing high numbers of species with arboreal adaptations most likely represent forested environments.

Reconstructing Behavior

Generally, the best and most reliable information about the habits of an extinct primate is obtained by comparing details of its dental and skeletal anatomy with those of living primates. Sediments may tell us where it died, and taphonomy may tell us how and why it was preserved, but its teeth and bones can tell us how it lived—what it ate, how it moved, and

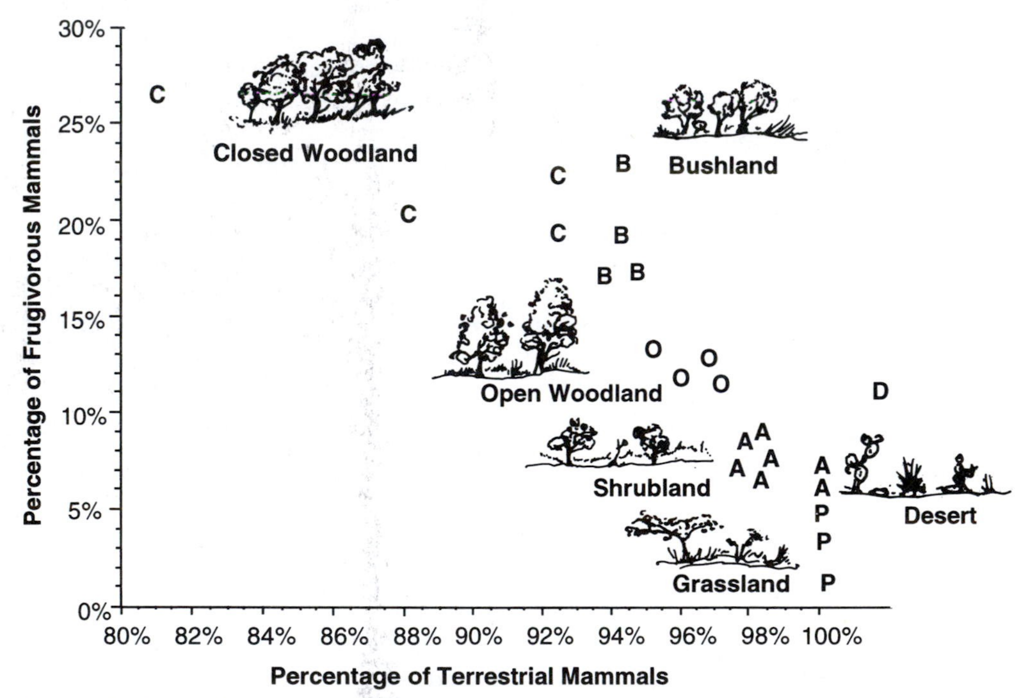

FIGURE 10.7 Modern African habitats can be distinguished by the relative proportions of terrestrial and frugivorous mammals. The same criteria can be used to interpret the habitats represented by assemblages of fossil bones in paleontological sites (from Reed, 1997).

possibly in what kind of social group it lived. In the previous chapter we discussed many of the associations between behavior and anatomy among living primates that form the basis for our interpretations of fossil behavior. Our ability to reconstruct the habits of an extinct primate from its bones is intimately linked to our understanding of how the shape of bones in living primates varies with their behavior. Associations between bony morphology and behavior that are true only "some of the time" among living primates cannot be expected to yield reliable reconstructions when applied to fossils (see Kay and Covert, 1984).

Furthermore, we have to remain always aware that uniformitarianism has its limits; the present is our best key to the past, but the past was not necessarily just like the present. We know, for example, that tooth size and many aspects of behavior are highly correlated with body size among living primates, but we cannot necessarily extrapolate these relationships based on a finite sample of living species to a fossil primate whose teeth are considerably larger or smaller than those of any living species. Likewise, many fossil primates had anatomical features that were quite different from anything we find among living species. We are sure to have problems interpreting such structures and may need to compare the fossil primates with another type of mammal for an analogy.

We commonly find that fossil primates differ from living species in the combinations of anatomical features they exhibit. A fossil ape may have a humerus that resembles that of a howling monkey in some features, that of a variegated lemur in others, and that of a macaque in still others. In such a case we must examine closely the mechanical implications of the individual features rather than simply look for a living species that matches the fossil in all respects. Our reconstructions of the behavior of extinct primates from their bones and teeth must be based not on simple analogy but on an understanding of the physiological and mechanical principles underlying the associations between bony structure and behavior (Fleagle, 1979).

A number of very promising approaches use the chemical composition of plant and animal remains to reconstruct aspects of the behavior and ecology of extinct species (e.g., Schoeninger, 1996). For example, the ratio of carbon isotopes in the teeth of both fossil and living mammals seems to be correlated with the chemical composition of their diet. Thus it is possible to distinguish grazing ungulates from browsing ungulates (MacFadden and Cerling, 1994). Other isotopes seem to distinguish species on the basis of their height preferences (Schoeninger *et al.*, 1997).

Paleobiogeography

It is a common tale that primate fossils are rare because primates typically live in jungles, which have acid soils that destroy their bones before they can be preserved, whereas animals such as horses live on savannahs, where bones are more easily saved for posterity. Although different soils may well affect the chances of fossilization in different environments, there are many examples of tropical environments in the Cenozoic fossil record indicating that the tree-dwelling habits of primates are not primarily responsible for the gaps in our knowledge of primate evolution. In fact, the primate fossil record is, overall, more complete than that of most other groups of mammals.

The large gaps in the primate fossil record are more directly the result of a remarkably meager geological record from those parts of the world in which primates have almost certainly been most successful for tens of millions of years: the Amazon Basin in South America, the Zaire Basin in central Africa, and the tropical forests of Southeast Asia. For huge amounts of time and space we lack not just fossils but even rocks from critical places and ages. Thus the seemingly poor fossil record of primates compared with, for example, that of horses is likely due to the fact that primates have evolved in places with virtually no fossil record or one that is still covered with forests and recent sediments, whereas horses were evolving in temperate areas of Europe and western North America, which have an excellent fossil record and miles of well-exposed sediments resulting from recent climatic events. For the most part, our knowledge of extinct primates comes from places that today are too dry and poorly vegetated to support living primates: Wyoming, Egypt, and northern Kenya. This terrain is excellent for geological and paleontological research; however, all of the paleoenvironmental evidence tells us that, during the earlier epochs when primates were abundant, these places were lush forests.

Because so much of our understanding of major events in primate evolution is based a limited sampling of past life in both time and space, no aspect of primate evolution is open to more surprises than biogeography. As new fossils are discovered from parts of the world that were previously poorly known, such as Africa and Asia, many of our notions about the evolution, diversity, and biogeography of primates will be dramatically revised. For example, it now seems most likely that platyrrhines arrived in South America about 30 million years ago, because we have no record of earlier primates on that continent. But an unsuspected discovery of fossil prosimians from Brazil would dramatically change our view of the evolutionary history and biogeography of higher primates. Similarly, our current view that hominids originated in Africa is based on a lack of early hominids from other continents. We must keep in mind that our current understanding of primate evolution will continue to change with new finds and new interpretations. In the following chapters we try to evaluate the nature of the evidence for our present understanding of primate evolution, with an eye toward particular issues that are presently unresolved.

BIBLIOGRAPHY

GEOLOGICAL TIME

Berggren, W. A., Kent, D. V., Flynn, J. J., and van Couvering, J. A. (1985). Cenozoic geochronology. *G. S. A. Bulletin* **96**:1407–1418.

Bown, T. M., and Krause, M. J. (1993). Soils, time and primate paleoenvironments. *Evol. Anthropol.* **2**:11–21.

Brown, F. H. (1992). Methods of dating. In *The Cambridge Encyclopedia of Human Evolution,* ed. S. Jones, R. Martin, and D. Pilbeam, pp. 179–186. Cambridge: Cambridge University Press.

Cande, S. C., and Kent, D. V. (1992). A new geomagnetic polarity time scale for the Late Cretaceous and Cenozoic. *J. Geophysical Res.* **97**:13,917–13,951.

——— (1995). Revised calibration of the geomagnetic polarity time scale for the Late Cretaceous and Cenozoic. *J. Geophysical Res.* **100**:6093–6095.

Deino, A. L., Renne, P. R., and Swisher III, C. C. (1998). ^{40}Ar/^{39}Ar dating in paleoanthropology and archeology. *Evol. Anthropol.* **6**:63–75.

Delcourt, H. R., and Delcourt, P. A. (1991). *Quaternary Ecology, A Paleoecological Perspective.* New York: Chapman and Hall

Feathers, J. M. (1996). Luminescence dating and modern human origins. *Evol. Anthropol.* **5**:25–36.

Grun, R. (1993). Electron spin resonance dating in paleoanthropology. *Evol. Anthropol.* b:172–181.

Haq, B. U., Hardenbol, J., and Vail, P. R. (1987). Chronology of Fluctuating sea levels since the Triassic. *Science* **235**:1156–1167.

Kappelman, J. (1993). The Attraction of Paleomagnetism. *Evol. Anthropol.* **2**:89–99.

MacFadden, B. J. (1990). Chronology of Cenozoic primate localities in South America. *J. Human Evol.* **19**:7–21.

McKenna, M. C. and Bell, S. K. (1997). *Classification of Mammals above the species level.* New York: Columbia Univ. Press.

Pickford, M. (1986). The geochronology of Miocene higher primate faunas of East Africa. In *Primate Evolu-*

tion, ed. J. G. Else and P. C. Lee, pp. 19–33. Cambridge: Cambridge University Press.

Sarna-Wojcicki, A. M., Mayer, C. E., Roth, P. H., and Brown, F. H. (1985). Ages of tuff beds at East African early hominid sites and sediments in the Gulf of Aden. *Nature* **313**:306–308.

Savage, D. E., and Russell, D. E. (1983). *Mammalion Paleofaunas of the World*. Reading, Mass.: Addison-Wesley.

Schwarcz, H. P. (1992). Uranium series dating in paleoanthropology. *Evol. Anthropol.* **1**:56–62.

Taylor, R. E. (1996). Radiocarbon dating: The continuing revolution. *Evol. Anthropol.* b:169–181.

Wagner, G. A. (1996). Fission-track dating in paleoanthropology. *Evol. Anthropol.* **5**:164–171

Wolfe, J. A. (1978). A paleobotanical interpretation of Tertiary climates in the northern hemisphere. *Am. Sci.* **66**:694–703.

FOSSILS, ENVIRONMENT, BEHAVIOR

Andrews, P. (1996). Palaeoecology and hominoid palaeoenvironments. *Biol. Rev.* **71**:257–300.

Behrensmeyer, A. K., and Hill, A., eds. (1980). *Fossils in the Making*. Chicago: University of Chicago Press.

Behrensmeyer, A. K., and Kidwell, S. M. (1985). Taphonomy's contributions to paleobiology. *Paleobiology* **11**(1): 105–119.

Behrensmeyer, A. K., Damuth, J. D., Di Michele, W. A., Potts, R., Sues, H.-D., and Wing, S. L. (1992). *Terrestrial Ecosystems through Time*. Chicago: University of Chicago Press.

Bown, T. M., and Krause, M. J. (1993). Soils, time and primate paleoenvironments. *Evol. Anthropol.* **2**:11–21.

Bown, T. M., Kraus, M. J., Wing, S. L., Fleagle, J. G., Tiffany, B., Simons, E. L., and Vondra, C. F. (1982). The Fayum forest revisited. *J. Human Evol.* **11**(7):603–632.

Fleagle, J. G. (1979). Primate postural behavior and anatomy: Naturalistic and experimental approaches. In *Environment, Behavior, and Morphology*, ed. M. E. Morbeck, H. Preuschoft, and N. Gomberg, pp. 313–325. New York: Gustav Fischer.

Jablonski, D., Erwin, D. H., and Lipps, J. H. (1996). *Evolutionary Paleoecology*. Chicago: University of Chicago Press.

Kay, R. E, and Covert, H. H. (1984). Anatomy and behavior of extinct primates. In *Food Acquisition and Processing in Primates*, ed. D. J. Chivers, B. A. Wood, and A. Bilsborough, pp. 467–508. New York: Plenum Press.

Kay, R. F., and Madden, R. H. (1997). Mammals and rainfall: Paleoecology of the Middle Miocene of La Venta (Colombia, South America). *J. Human Evol.* **32**:161–200.

Lyman, R. L. (1994). *Vertebrate Taphonomy*. Cambridge: Cambridge University Press.

MacFadden, B. J., and Cerling, T. E. (1994). Fossil horses, carbon isotopes and global change. *Tree* **9**:481–486

Marean, C. W. (1995). Of taphonomy and zooarcheology. *Evol. Anthropol.* **4**:64–72.

Marean, C. W., and Kim, S. Y. (1997). Moustarian large mammal remains from Kobeh Cave (Zagros Mountains, Iran): Behavioral implications for neanderthals and early modern humans. *Curr. Anthropol.* **39**:579.

Reed, K. E. (1997). Early hominid evolution and ecological change through the African Plio-Pleistocene. *J. Human Evol.* **32**:289–322.

Schoeninger, M. J. (1995). Stable isotope studies in human evolution. *Evol. Anthropol.* **4**:83–98.

Schoeninger, M. J., Iwaniec, U. T., Nash, L. T., and Glander, K. (1997). Hair $\delta^{13}C$ values reflect aspects of primate ecology. *Am. J. Phys. Anthropol. Suppl.* **24**: 205

Shipman, P. (1981). *Life History of a Fossil*. Cambridge, Mass.: Harvard University Press.

ELEVEN

Primate Origins

PALEOCENE EPOCH

The Paleocene, the first epoch in the Age of Mammals, is a poorly known part of earth history, but it documents the first major radiation of placental mammals. At the end of the Cretaceous, the dinosaurs that had dominated terrestrial faunas for the past 160 million years had all disappeared. No one is quite sure why they disappeared or whether their departure was abrupt or gradual. All we know is that, beginning about 65 million years ago, the fossil record contains no more dinosaurs; instead, the most abundant vertebrates are mammals of various sorts.

Geologically, the late Cretaceous and Paleocene were relatively active times in earth history and were marked by the rise of several major mountain groups, including the American Rockies. Geographically, the world looked somewhat different than it does today (Fig. 11.1). The North Atlantic was considerably

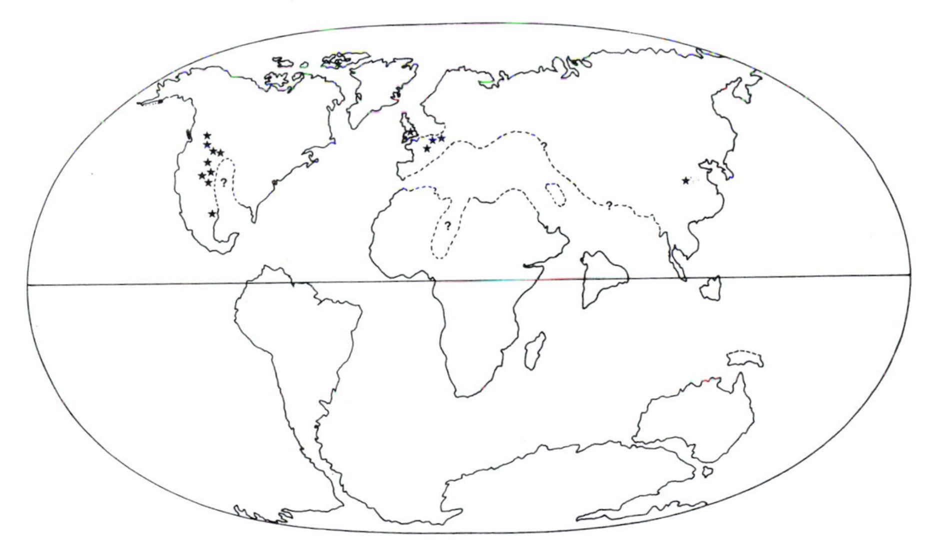

Plesiadapiform Localities

FIGURE 11.1 Map of the world during the middle Paleocene (60 million years ago), with locations (*) of plesiadapiform fossil sites.

narrower than it is today, particularly in the vicinity of Greenland. The intermittent occurrence of land connections between North America and Europe is indicated by the similarity of the Paleocene faunas of the two continents. There is also faunal evidence of occasional connections between North America and Asia, presumably across the Bering Strait.

South America, Africa, and India were all island continents, except that South America and Antarctica were apparently connected until the Oligocene. The South Atlantic was an open ocean, although somewhat narrower than it is today, and perhaps there were land surfaces lying between South America and Africa. The Panama land bridge, which currently connects North and South America, would not come into being for another 60 million years. Africa was separated from Europe by the great Tethys Seaway extending from China on the east to southern France on the west. India was adrift in the Indian Ocean and had not yet collided with the Asian mainland.

Paleocene climates were relatively cooler than those of either the preceding late Cretaceous or the succeeding Eocene, but temperatures fluctuated throughout the epoch (see Fig. 10.4). The flora of western North America, which has been carefully studied, was characterized by deciduous trees and conifers rather than the more tropical plants characteristic of immediately earlier and later epochs.

Archontans—Primates and Other Mammals

The earliest primates evolved from an insectivorous mammal some time in the latest part of the Cretaceous period or the early Paleocene. It is impossible to determine exactly how primates are related to other orders of mammals, but there are indications from paleontology, comparative anatomy, and biomolecular studies that, among living mammals, primates, tree shrews, flying lemurs, and bats are more closely related to one another than to other mammals (Wible and Covert, 1987; MacPhee, 1993). These groups are frequently placed in the superorder Archonta. Although the fossil plesiadapiforms have traditionally played a prominent role in discussions of primate origins, most authorities no longer place them in the order Primates, but rather in a separate order, Plesiadapiformes, that lies near the origin of primates along with the mammalian orders listed above.

Tree Shrews

Tree shrews, order Scandentia, are small squirrel-like mammals from Asia. The five diurnal, partly terrestrial genera—*Tupaia* (Fig. 11.2), *Anathana, Dendrogale, Lyonogale,* and *Urogale*—are placed in the subfamily Tupaiinae; and the nocturnal, arboreal *Ptilocercus* is placed in its own subfamily, the Ptilocercinae. Tree shrews are insectivorous and frugivorous. For many decades, following the early studies of LeGros Clark (e.g., 1926, 1934), tree shrews were considered primitive primates.

FIGURE 11.2 The superorder Archonta includes Primates and three other orders of mammals that are thought to be closely related to primates: below, Scandentia (tree shrews); middle, Dermoptera (flying lemurs, or colugos); above, Chiroptera (bats) (drawing by William Yee).

William Yee ©95

In the late 1960s and early 1970s, however, numerous studies demonstrated that many of the similarities between tree shrews and primates in brain morphology and reproductive anatomy were not homologous features, and tree shrews were no longer classified with primates (Martin, 1968). Recent studies of cranial anatomy have pointed to more robust features (including a postorbital bar, the position of the maxillary artery, the bony canals for the arteries of the middle ear, and the dual jugular foramen) that link tree shrews and primates more closely with each other than with any other mammals (Wible and Covert, 1987; Kay et al., 1992; Wible and Zeller, 1994).

Flying Lemurs

The flying lemurs, or Dermoptera, are among the most interesting living mammals, even though they are not lemurs and do not fly. Rather, they are gliding mammals from Southeast Asia that resemble lemurs only in possessing a tooth comb. However, in Dermoptera, each of the lower incisors contains numerous tines arranged like a comb (Fig. 11.3). There is a single genus, *Cynocephalus* (Fig. 11.2), with two species. The limbs of Dermoptera show numerous specializations associated with their gliding habits, including unusual phalangeal proportions related to the webbing between their fingers (Beard, 1994). Some authorities have argued in favor of a close phylogenetic grouping between Dermoptera and bats (Chiroptera), based in part on neurophysiological wiring of the visual system, the membranous connections between their limbs associated with aerial locomotion, and a digital locking mechanism in their feet. In contrast, Beard (e.g., 1993, 1994) has argued that Primates and Dermoptera are sister taxa (see the later discussion).

Bats

Living bats are divided into two suborders: the large, fruit-eating megabats, or Megachiroptera, and the smaller, more frequently insectivorous microchiroptera. The idea that bats and primates are closely related is the result of neurophysiological studies by Pettigrew and colleagues (1989) showing that megabats (but not microbats) share features of the visual system found otherwise only in primates. Thus, they argue that megabats are more closely related to primates than to other bats. These results have been countered by many studies demonstrating that the similarities shared by the two bat groups are far more numerous (and convincing) than the features linking megabats with primates (e.g., Simmons, 1993). However, the neuroanatomical similarities remain unexplained.

Plesiadapiforms

Traditionally considered a separate suborder of Primates, the plesiadapiforms were an extremely successful group of primatelike mammals that flourished in the Paleocene and early Eocene of North America and Europe (Fig. 11.1) and have recently been reported from Asia (Beard and Wang, 1995). They are the most common mammals in many Paleocene faunas. Their known taxonomic diversity (more than thirty-five genera and seventy-five species) is greater than that of living prosimians, and their diversity in size is comparable to that of either living prosimians or New World anthropoids. However, the primate status of plesiadapiforms has been questioned for decades (Martin, 1968; Cartmill, 1972). Although dental similarities seem to link plesiadapiforms with primates (Van Valen, 1994), in recent years it has become widely recognized that the evidence linking plesiadapiforms with primates is no stronger than that linking primates with either tree shrews or Dermoptera. Thus Beard has argued that Primates, Dermoptera, and Plesiadapiformes form a supraordinal group that he calls Primatomorpha. In our discussion, Plesiadapiformes are ranked as a distinct order of mammals, and they will be discussed in some detail because they are the best known of the early Cenozoic mammals that may be related to primate origins.

Plesiadapiforms have long been known mainly from fragmentary jaws and teeth, and many families show quite distinctive dental specializations (Figs. 11.3, 11.4). Several dental features of plesiadapiforms seem to link them with later primates. These include molar teeth with relatively low cusps (compared with contemporary or extant insectivores), lower molars with low trigonids and basin-shaped talonids, and unreduced lower third molars with an extended talonid. Their upper molars have prominent conules, a poorly developed or absent stylar shelf, and a well-developed postprotocingulum (nannopithex fold) or comparable wear facet distal to the protocone. The primitive dental formula for plesiadapiforms is 3.1.3.3. Most later members of all lineages show reduction and loss of teeth, especially incisors and the anterior premolar. Since all plesiadapiforms have a dental formula

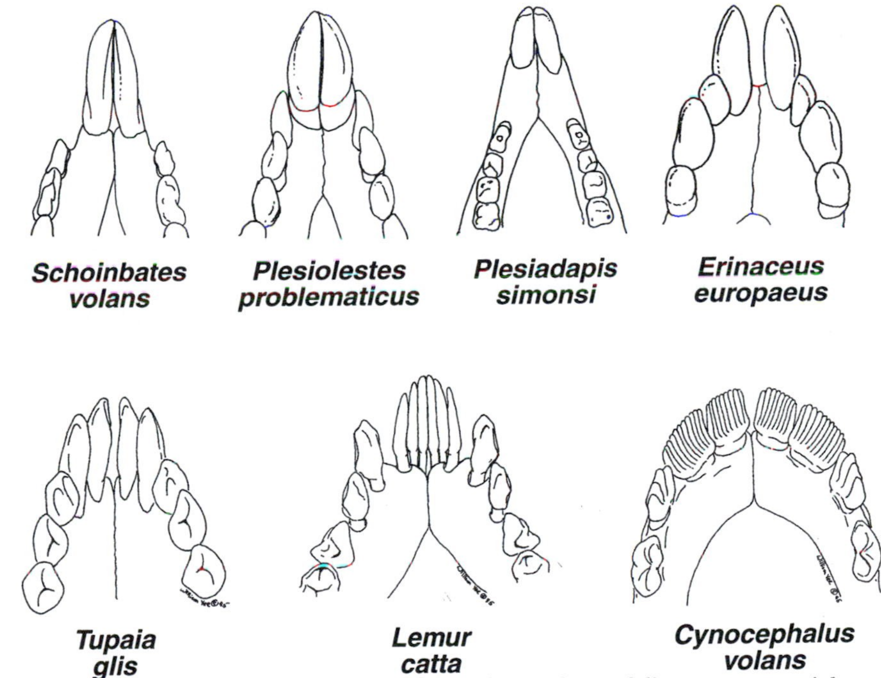

FIGURE 11.3 The anterior dentitions of *Schoinobates volans,* a folivorous marsupial; *Plesiolestes problematicus* and *Plesiadapis simonsi,* two plesiadapiforms; *Erinaceus europaeus,* a hedgehog; *Tupaia glis,* a tree shrew; *Lemur catta,* a ring-tailed lemur; and *Cynocephalus volans,* a colugo, or flying lemur.

with three or fewer premolars, they are too specialized to have given rise to the earliest prosimians, many of which have four premolars. In addition, most plesiadapiforms have extremely large and procumbent upper and lower central incisors, which are more derived than those teeth in the earliest primates (Figs. 11.3, 11.4; see Fig. 12.2).

The sharp cusps on the teeth of many species, as well as their small size, suggest that many of the plesiadapiform species were largely insectivorous. Nevertheless, many have

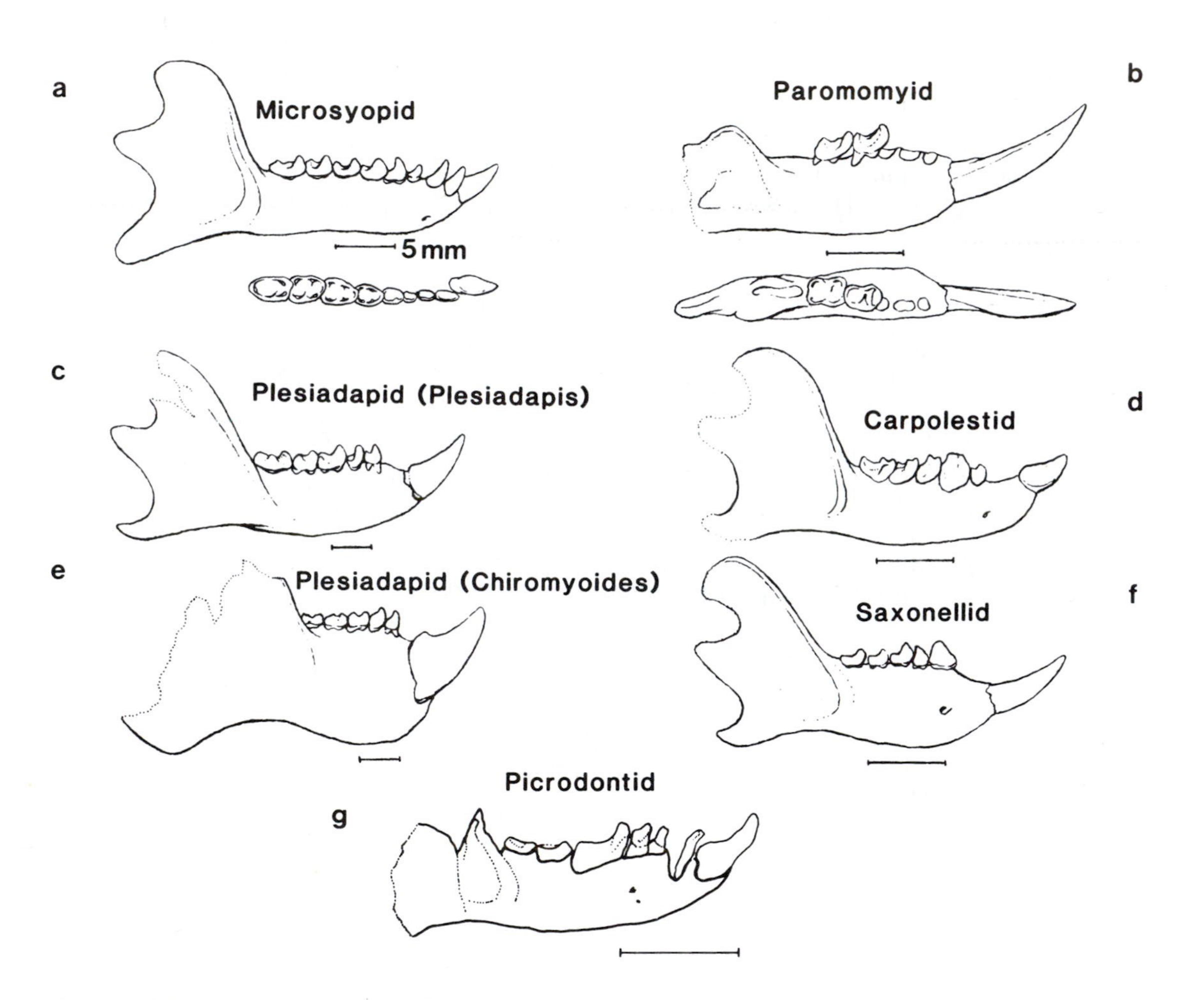

FIGURE 11.4 Mandibles of several Plesiadapiformes, showing the diversity in shape and size of the dentition. a. *Plesiolestes problematicus,* lateral and occlusal views; b. *Elwynella oreas,* lateral and occlusal views; c. *Plesiadapis rex,* lateral view; d. *Elphidotarsius florencae,* lateral view; e. *Chiromyoides campanicus,* lateral view; f. *Saxonella crepaturae,* lateral view; g. *Picrodus silberlingi,* lateral view.

teeth that indicate they were capable of more crushing and thus more omnivory and herbivory compared with contemporary insectivores, and some show adaptations for harvesting exudates (Beard, 1991).

Plesiadapiforms have a low, flat skull with a long snout, a small brain, large zygomatic arches, and no bony ring surrounding the orbits (Fig. 11.5). In these features they are more primitive than all later primates. There has been considerable debate about the likely arterial circulation to the braincase and whether Plesiadapiformes had an auditory bulla made up of the petrosal bone (e.g., MacPhee *et al.*, 1983; Kay *et al.*, 1992). New evidence seems to indicate that the auditory bulla of Plesiadapiformes was probably not formed by the petrosal bone, and thus there are no definite cranial features linking plesiadapiforms with extant primates.

Analyses of the limb and trunk skeleton, particularly the foot, the elbow, and the wrist, have indicated several features that link plesiadapiforms with primates and others that link them with Dermoptera (Beard, 1994). Such analyses are, however, severely limited by the paucity of material. One genus for which ample skeletal remains are known, *Plesiadapis,* has relatively short, robust limbs, a nonopposable hallux, and clawed digits, features that are clearly more primitive than those found in the limbs of later primates (see Fig. 12.2). The paromomyid *Phenacolemur* has digital proportions (Beard, 1993), but not long bone proportions (Runestad and Ruff, 1995), that suggest gliding habits.

There is considerable diversity of opinion among current authorities regarding the composition of the order Plesiadapiformes. The taxonomic scheme adopted here (Tables 11.1–11.4) is a composite arrangement based on the results of numerous studies (e.g., Gunnell, 1989; Kay *et al.*, 1992; Beard, 1993, 1994; Rose

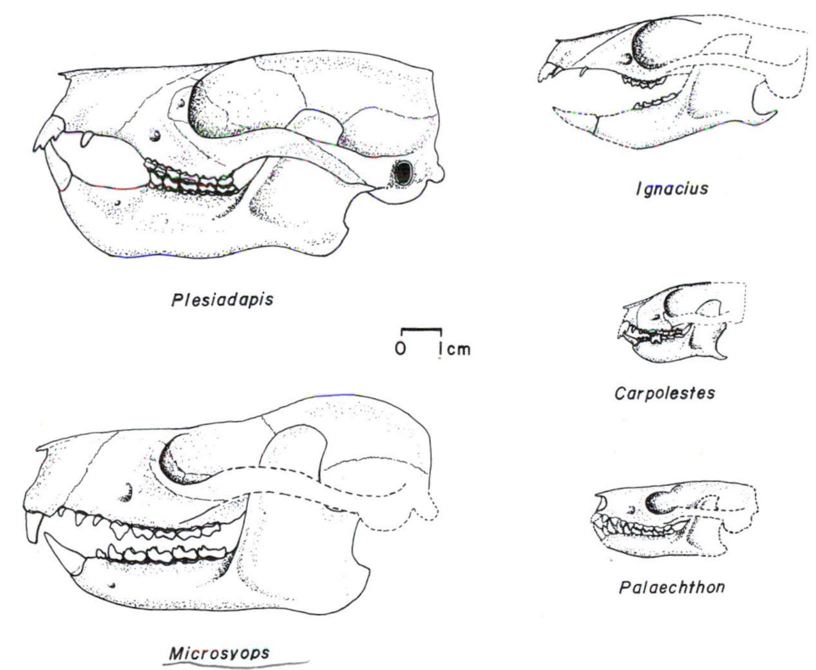

FIGURE 11.5 Skulls of five plesiadapiforms.

et al., 1993) and is a conservative arrangement. The order Plesiadapiformes contains ten families placed in two superfamilies. Five families—plesiadapids, carpolestids, saxonellids, paromomyids, and micromomyids—are placed in the superfamily Plesiadapoidea. Two other families—microsyopids and palaechthonids—are placed in the a separate superfamily, Microsyopoidea. Three families—purgatoriids, picromomyids, and picrodontids—are not placed in any superfamily. Purgatoriids are probably near the ancestry of all plesiadapiforms, picromomyids are very poorly known, and picrodontids are so strangely derived that their phyletic relationships are obscure.

Purgatorius

Purgatorius, a tiny mammal from the earliest part of the Paleocene of Montana, is often regarded as the most primitive plesiadapiform (Table 11.1). With a dental formula of 3.1.4.3 and very primitive molars, *Purgatorius* is generalized enough to be ancestral to both Plesiadapiformes and later primates. Although Van Valen (1994) has identified intermediate forms linking *Purgatorius* with other plesiadapiforms, and there are dental features, including relatively low molar trigonids, broad second lower molars, and elongate third molars, that link *Purgatorius*, plesiadapiforms, and later primates, most authorities feel it cannot confidently be placed in any order (Buckley, 1997).

TABLE 11.1
Order *INCERTAE SEDIS*

Species	Estimated Mass (g)
Family PURGATORIIDAE	
Purgatorius (early Paleocene North America)	
P. unio	92
P. titusi	62
P. janisae	—

Plesiadapids

The best known of the Plesiadapiformes, plesiadapids, were very diverse and abundant in the Paleocene and early Eocene of both North America and Europe (Table 11.2). There are

TABLE 11.2
Order Plesiadapiformes
Superfamily PLESIADAPOIDEA

Species	Estimated Mass (g)
Family PLESIADAPIDAE	
Pandemonium (early Paleocene, North America)	
P. dis	220
Pronothodectes (middle–late Paleocene North America)	
P. matthewi	153
P. jepi	208
P. gaoi	153
Nannodectes (late Paleocene, North America)	
N. intermedius	221
N. gazini	188
N. simpsoni	329
N. gidleyi	396
Plesiadapis (late Paleocene–early Eocene, North America, Europe)	
P. praecursor	316
P. anceps	427
P. rex	506
P. gingerichi	1752
P. churchilli	728
P. fodinatus	540
P. dubius	376
P. simonsi	1217
P. cookei	3055
P. walbeckensis	389
P. remensis	759
P. tricuspidens	1278
P. russelli	—
Chiromyoides (late Paleocene–early Eocene, North America, Europe)	
C. campanicus	314
C. caesor	204
C. minor	150
C. potior	—
C. major	—

TABLE 11.2 *(continued)*
Order Plesiadapiformes
Superfamily PLESIADAPOIDEA

Species	Estimated Mass (g)
Platychoerops (early Eocene, Europe)	
P. daubrei	1887
P. richardsoni	—
Family CARPOLESTIDAE	
Chronolestes (early Eocene, Asia)	
C. simul	35
Elphidotarsius (middle–late Paleocene North American)	
E. florencae	30
E. shotgunensis	29
E. russelli	38
E. wightoni	26
Carpodaptes (late Paleocene, North America)	
C. aulacodon	53
C. hazelae	53
C. jepseni	96
Carpolestes (late Paleocene–early Eocene, North America)	
C. nigridens	87
C. dubius	146
Carpocristes (early Eocene, Late Paleocene, N. America, Asia)	
C. orlens	18
C. hobackensis	33
C. cygneus	53
Family SAXONELLIDAE	
Saxonella (late Paleocene, North America, Europe)	
S. crepaturae	88
S. naylori	59
Family PAROMOMYIDAE	
Paromomys (middle Paleocene, North America)	
P. maturus	312
P. depressidens	94
Ignacius (middle Paleocene–late Eocene, North America)	
I. graybullianus	152
I. frugivorus	96
I. fremontensis	46
I. mcgrewi	115
Phenacolemur (late Paleocene–middle Eocene, North America, Europe)	
P. praecox	414
P. simonsi	48
P. pagei	192
P. jepseni	121
Elwynella (middle Eocene, North America)	
E. oreas	184
Simpsonlemur (early Eocene, North America)	
S. citatus	171
Dillerlemur (early Eocene, North America)	
D. robinettei	419
Pulverflumen (early Eocene, North America)	
P. magnificum	619
Arcius (early Eocene, Europe)	
A. rougieri	76
A. fuscus	—
A. lapparenti	—
Family MICROMOMYIDAE	
Micromomys (late Paleocene–early Eocene, North America)	
M. silvercouleei	—
M. willwoodensis	—
M. vossae	7
M. fremdi	13
Tinimomys (early Eocene, North America)	
T. graybulliensis	17
Chalicomomys (early Eocene, North America)	
C. antelucanus	12
Myrmekomomys (early Eocene, North America)	
M. loomisi	17

six genera. The smallest species were comparable in size to a marmoset (150 g); the largest were the size of a guenon (3 kg).

Compared with many other plesiadapiforms, plesiadapids generally have molar and premolar teeth that often resemble those of primates. (Figs. 11.4). *Pronothodectes* has a dental formula of 2.1.3.3, but later genera show considerable reduction and loss of incisors, canines, and premolars. All plesiadapids have relatively broad, procumbent lower incisors that occlude in a pincerlike fashion with the mitten-shaped upper central incisors (Figs. 11.4, 11.5). Like primates, plesiadapids have

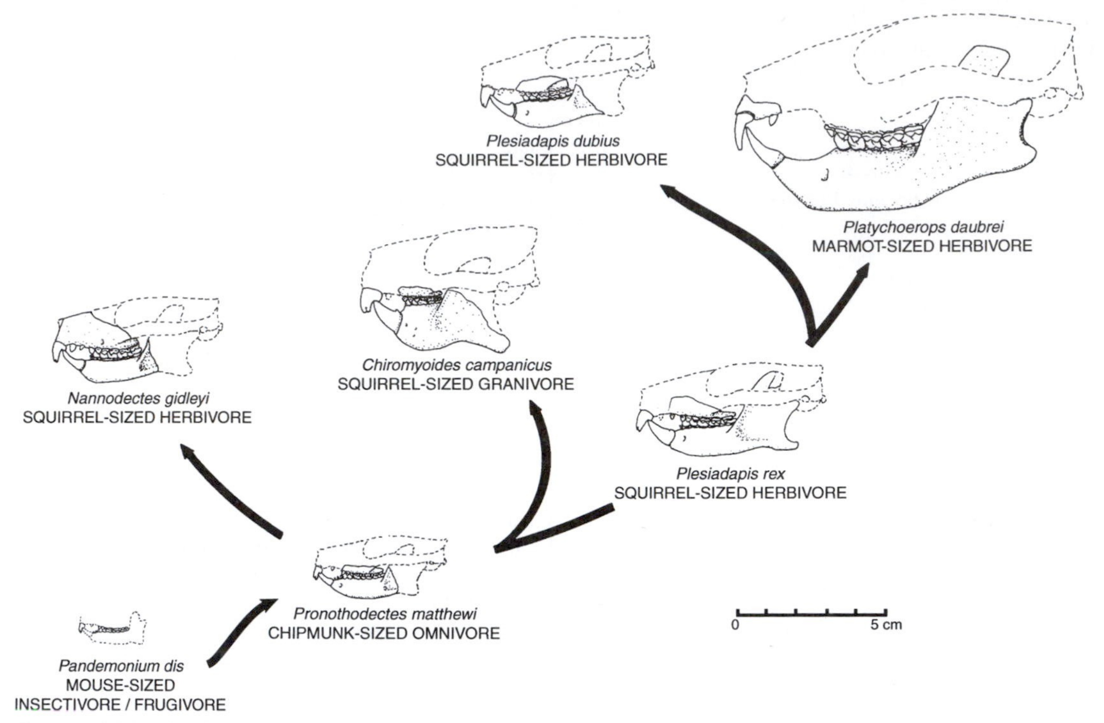

FIGURE 11.6 A phylogeny of plesadapids, showing diversity in dental, mandibular, and cranial form probably associated with dietary diversity (from Gingerich, 1976).

premolars and molars with low, bulbous cusps. The posterior two upper premolars are short and broad. The lower molars have a relatively low trigonid and broad talonid. These low-crowned cheek teeth, together with the relatively large size of most species, suggest that plesiadapids were predominantly herbivorous with a diversity of adaptations including some more insectivorous species and the odd *chiromyoides* with a deep aye-aye like jaw (Fig. 11.6).

There are several skulls of *Plesiadapis*. The cranium of *Plesiadapis* (Fig. 11.5) has a long snout with a large premaxillary bone and a diastema between the large incisors and the cheek teeth in both the upper and the lower jaws. The auditory bulla in adult individuals is continuous with the petrosal bone, but there is debate over whether it is actually a petrosal bulla as in primates or an entotympanic bone whose suture with the petrosal has been obliterated (e.g., Kay *et al.,* 1992). The tympanic ring (tympanic bone) is fused to the bulla and extends laterally to form a bony tube (Szalay, 1975).

Considerable skeletal material is known for *Plesiadapis*. The short, robust limbs, the long, laterally compressed claws, and the long, bushy tail (known from a delicate limestone impression; see Fig. 10.5) indicate that it was an arboreal quadruped.

The phylogenetic relationships among plesiadapids have been thoroughly studied, and there is excellent documentation of the patterns of evolutionary change and adaptive diversity in this family (Gingerich, 1976; Fig. 11.6). The recently named *Pandemonium* seems to be intermediate between the basal *Purgatorius* and later plesiadapids, thus providing insight into the history of the family as well as helping to unify Plesiadapiformes as an order (Van Valen, 1994). Unfortunately, it is known only from isolated teeth.

Carpolestids

The carpolestids (Table 11.2) are a North American and Asian family of small, mouse-size primates (20–150 g) characterized by the enlargement of their last lower premolar and last two upper premolars. The family has traditionally comprised three North American genera that more or less follow one another in time: *Elphidotarsius,* from the middle and late Paleocene; *Carpodaptes,* from the late Paleocene; and *Carpolestes,* from the latest Paleocene and earliest Eocene (Fig. 11.7). Because of their distinctive morphological specializations and short species durations, these North American genera have been useful as biostratigraphic indicators in early Tertiary sediments. However, recent discoveries, from the early Eocene of China, of the most primitive (*Chronolestes*) and the most derived (*Carpocristes*) of all carpolestid genera indicate a much more complex phylogenetic and biogeographical history for this family (Beard and Wang, 1995).

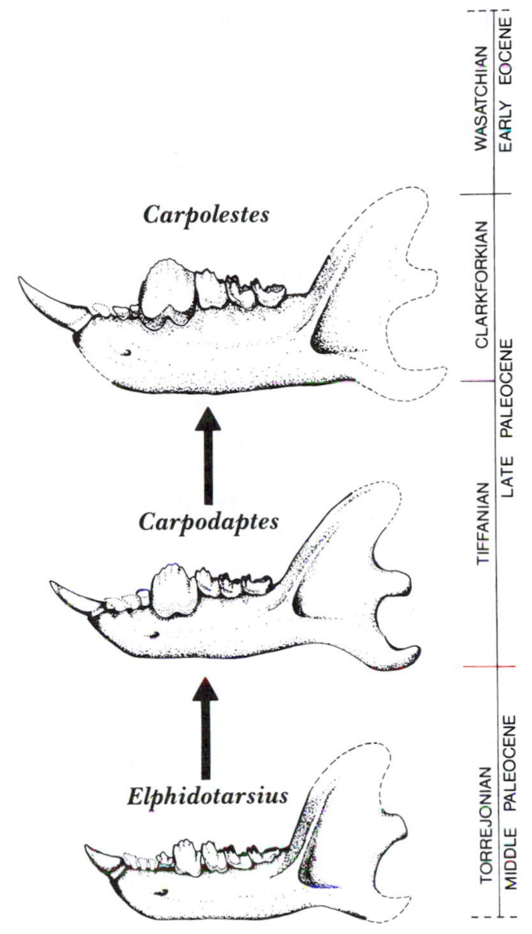

FIGURE 11.7 Differences in premolar shape among three genera of North American carpolestids. Note the increasing size of the last premolar and first molar (from Rose, 1975).

Saxonellids

Saxonella (Fig. 11.4, Table 11.2) is a relative of the plesiadapids from the late Paleocene of North America and Germany. Like carpolestids, *Saxonella* seems to be a derivative of the plesiadapids that evolved a very large lower premolar. In contrast with carpolestids, however, which enlarged the last premolar, *Saxonella* enlarged P_3. Thus, although probably adaptively similar, the two groups are not closely related.

Paromomyids

The paromomyids (Table 11.2) are a group of plesiadapiforms with teeth suggesting affinities to plesiadapids (e.g., Rose *et al.,* 1993). They are among the most long-lived and geographically widespread families. They ranged in time from the middle Paleocene through the latest Eocene and have been found in North America as far north as the Arctic Circle, as well as in Europe. Paromomyids were small to medium-size plesiadapiforms with long, slender lower central incisors (Fig. 11.4). The posterior lower premolar is usually tall and pointed. Paromomyids have relatively flat, low-crowned lower molars with short, squared trigonids and broad, shallow talonid basins; the upper molars are square with expanded basins. Both upper and lower posterior molars are conspicuously elongated.

The function of the large lower incisor, which occludes with the lobate upper incisors, is uncertain. Some authors note similarities to the incisors of shrews and suggest that they functioned in procuring insects; others suggest that it was used to gnaw holes in trees to elicit the flow of exudates (Fig. 11.8). The pointed P_4 seems to be adapted for puncturing food during initial preparation, and the broad, flat, lower molars suggest a herbivorous rather than an insectivorous diet, probably including gum or nectar, for most paromomyids.

Partial skulls are known for both *Ignacius* (Fig. 11.5) and *Phenacolemur.* In both genera, the face is long and narrow and has a large infraorbital foramen, suggesting a richly innervated snout with tactile vibrissae. In *Ignacius* the bony auditory bulla is made up of the entotympanic bone. The cerebral blood supply appears to have been predominantly via the vertebral arteries, for there is no enlarged foramen for the internal carotid.

Numerous skeletal remains of *Phenacolemur* were the subject of a detailed study by Beard (1988, 1993). On the basis of unique similarities to flying lemurs in having relatively long, straight intermediate phalanges as well as many other osteological similarities in the remainder of the skeleton, including the wrist, the femur, and the pelvis, Beard has argued that *Phenacolemur* was a glider. However, Runstead and Ruff (1995) have demonstrated that the limbs of *Phenacolemur* are far more robust than those of any extant gliding or flying mammals and have questioned the likelihood that paromomyids were capable of any type of flight.

Paromomyids are the only nonhuman primates with a geographic range that extends above the Arctic Circle. During the early Eocene, a paromomyid similar to *Ignacius* (but much larger) thrived on Ellesmere Island, which has been located near the Arctic Circle at 78° north latitude during most of the Cenozoic era. Because there are several months of total darkness at that latitude today, it seems likely that the fauna was composed of cathemeral or crepuscular mammals.

Micromomyids

This family with four genera (Table 11.2), contains the smallest of the plesiadapiforms; they probably weighed no more than 30 g. Their molars resemble those of many microsyopids, but they have an upper central incisor similar to that of plesiadapids and carpolestids and a large posterior lower premolar. From its small size and its molars with very acute cusps, *Micromomys* appears to have been almost totally insectivorous. *Tinimomys* has more rounded molar cusps.

Beard (1993, 1994) has described several limb bones of micromomyids. They show many similarities to paromomyids in femoral anatomy, and the very long, straight, isolated phalanges, attributed to *Tinimomys,* suggest possible gliding habits.

Microsyopids

The Microsyopidae (Table 11.3) are a family of primitive placental mammals usually linked with plesiadapiforms. It was a very successful family, with species in both North America and Europe, ranging through much of the Eocene. Microsyopids are relatively diverse in ap-

TABLE 11.3
Superfamily MICROSYOPOIDEA

Species	Estimated Mass (g)
Family PALAECHTHONIDAE	
Palaechthon (middle–late Paleocene, North America)	
P. alticuspis	93
P. nacimienti	166
P. woodi	57
Plesiolestes (middle–late Paleocene, North America)	
P. problematicus	147
P. sirokyi	563
Talpohenach (middle Paleocene, North America)	
T. torrejonia	300
Torejonia (middle Paleocene, North America)	
T. wilsoni	575
Palenochtha (middle–late Paleocene, North America)	
P. minor	22
Premnoides (middle Paleocene, North America)	
P. douglassi	201
Family MICROSYOPIDAE	
Navajovius (late Paleocene–early Eocene, North America)	
N. kohlhaasae	32
Berruvius (late Paleocene–early Eocene, Europe)	
B. lesseroni	20
B. gingerichi	31
Niptomomys (early Eocene, North America)	
N. doreenae	29
N. thelmae	41
Uintasorex (middle–late Eocene North America)	
U. parvulus	45
Avenius (early Eocene, Europe)	
A. amatorum	8
Microsyops (Eocene, North America)	
M. angustidens	433
M. latidens	525
M. scottianus	790
M. elegans	622
M. annectens	1355
M. kratos	1817
Arctodontomys (early Eocene, North America)	
A. wilsoni	248
A. simplicidens	335
A. nuptus	476
Craseops (late Eocene, North America)	
C. sylvestris	2015
Megadelphus (early Eocene N. America)	
M. lundeliusi	2721

pearance and include both very small species (e.g., *Vintasorex*), no larger than a shrew, and large species (*Microsyops*), the size of a raccoon. Microsyopids share many features of the molar and premolar teeth with plesiadapiforms, but have a distinctive caniniform upper incisor (Rose *et al.*, 1993). All microsyopids have a narrow, lanceolate (spearhead-shaped), specialized lower central incisor (Fig. 11.4).

Cranially, microsyopids have most of the primitive mammalian features that we have already described for plesiadapiforms (Fig. 11.5). The best-known genus, *Microsyops,* has a cranial blood supply that more closely resembles the cranial arterial pattern of living primates than does that of any plesiadapiform, but it also has an auditory structure that is more primitive than that of any plesiadapiforms in that it lacks a bony bulla (MacPhee *et al.*, 1983).

Palaechthonids

Allied with the microsyopids, but perhaps more closely related to plesiadapids (e.g., Rose *et al.*, 1993) are the palaechthonids, a diverse family from the middle Paleocene of North America with six genera: *Talpohenach, Palenochtha, Palaechthon, Plesiolestes, Premnoides* and *Torrejonia*. In size, they are comparable to the smallest living primates (20–200 g). Most species have a dental formula of 2.1.3.3, but the canine and anteriormost premolar are very small in many species and probably absent in some. The enlarged, lanceolate lower first incisors form a scooplike apparatus for cutting, an adaptation that suggests a partly herbivorous diet (Figs. 11.3, 11.4; Szalay, 1981). Yet the molars have relatively acute cusps compared with those of many living primate species, suggesting that insects were a major part of the diet as well.

For one of these small middle Paleocene species, *Palaechthon nacimienti*, there is a relatively complete but crushed skull (Fig. 11.5) with relatively small, laterally directed orbits, suggesting limited stereoscopic abilities; a broad interorbital region, suggesting a large olfactory fossa and greater reliance on a sense of smell; and a large infraorbital foramen, suggesting the presence of a richly innervated snout bearing sensitive facial vibrissae.

Kay and Cartmill (1977) suggest that the small size and the cranial features of *Palaechthon* indicate that it was probably a terrestrial forager that hunted for concealed insects and other animal prey by "nosing around the ground," guided more by hearing, smell, and its sensitive snout than by vision (Fig. 11.8), and that it was probably nocturnal. Unfortunately, there are no associated skeletal elements for palaechthonids.

Picrodontids

The family Picrodontidae (Table 11.4) is known only by dental remains from the middle and late Paleocene of western North America. Their relationship to other plesiadapiforms in very unclear. They resemble plesiadapiforms in their incisor morphology (Fig. 11.3), but their cheek teeth are quite unusual. The first upper and lower molars are enlarged and oddly shaped. The lower molars have very small trigonids and large, shallow talonids with crenulated enamel. Because of notable similarities between the molars of picrodontids and those of bats, Szalay (1972) has suggested a diet of fruit and nectar (Fig. 11.8).

TABLE 11.4
Order Plesiadapiformes
Superfamily INCERTAE SEDIS

Species	Estimated Mass (g)
Family PICRODONTIDAE	
Picrodus (middle–late Paleocene, North America)	
P. silberlingi	40
Zanycteris (late Paleocene, North America)	
Z. paleocenus	—
Draconodus (middle Paleocene, North America)	
D. apertus	200
Family PICROMOMYIDAE	
Picromomys (early Eocene, North America)	
P. petersonorum	10
Alveojunctus (middle Eocene, North America)	
A. minutus	—

Picromomyids

Rose and Bown (1996) have recently described a new genus *Picromomys* that is smaller than any other plesiadapiform. This tiny (10 g) insectivorious species is placed in its own family that is probably most closely related to the micromomyids.

ADAPTIVE RADIATION OF PLESIADAPIFORMS

The plesiadapiforms were a very successful group of early primatelike mammals that evolved a wide range of body sizes as well as

FIGURE 11.8 Reconstruction of a scene from the late Paleocene of North America showing several plesiadapiforms. A small group of *Plesiadapis rex* feeds in a tree, and *Ignacius frugivorous* feeds on exudates from the trunk. A small *Picrodus silberlingi* feeds on nectar in a bush. On the ground, *Chiromyoides minor* chews on a seed, and a small microsyopid grasps its prey.

dental and postcranial adaptations (Fig. 11.8). They include several species that were almost as large as the largest living prosimians or New World monkeys, as well as several species that were much smaller than any living primate. Their cranial structure is so different from that of any living primate that we have no real evidence of whether they were diurnal or nocturnal. Their great diversity in dental morphology suggests considerable diversity in dietary adaptations. It seems likely that many species specialized on insects, and many others relied on fruit, leaves, seeds, or other plant parts. In addition, it seems likely that some relied on nectar or gums. The size and shape differences in their incisor and mandible structure indicate that plesiadapiform feeding habits were probably quite different from anything found among living primates. This is especially true for picrodontids and carpolestids, which had very odd dental specializations by any standards.

The limb skeletons of plesiadapiforms are so poorly known in most cases or so different from living species that it is difficult to reach any firm conclusions regarding their locomotor adaptations. Their size range suggests considerable locomotor diversity. *Plesiadapis* was quadrupedal and partly arboreal, but some of the smallest species may well have been terrestrial. Paromomyids were probably largely arboreal quadrupeds but may have had some gliding abilities, like some extant marsupials. The claws suggest likely climbing abilities for many genera.

The social habits of these archaic primates are certainly beyond our ken, but we can speculate. If they were nocturnal, they probably lived in a noyau arrangement, like many primitive mammals. Diurnal species may have lived in larger groups.

The radiation of plesiadapiforms was largely during the Paleocene. Only a few carpolestids (in Asia), microsyopids, and paromomyids survived past the early Eocene. There are several explanations commonly offered for the rapid decline and disappearance, in the beginning of the Eocene, of this once very successful group. The most common view has been that plesiadapiform decline and extinction resulted from competition with rodents (Van Valen and Sloan, 1966). Others have suggested that early prosimians (Szalay, 1972) and possibly bats (Sussman and Raven, 1978) also played a role in their decline. In addition, Gingerich (e.g., 1976) has suggested that the diversity of some plesiadapiforms is closely linked with climatic changes (see Fig. 18.11) and that their decline and extinction at the beginning of the Eocene related to the more tropical environments of that epoch (see Fig. 10.4) as well as to competition from new groups of mammals. In a review of this problem, Maas *et al.* (1987, 1988) found that the changes in climate during the late Paleocene and early Eocene do not correlate well with changes in the diversity of plesiadapiforms, and that the radiation of early prosimians (see Chapter 12) came after the extinction of most plesiadapiforms. The increasing diversity of early rodents is, however, inversely correlated with the decline of the plesiadapiforms (see Fig. 18.12). Moreover, functional comparisons show that plesiadapiforms and rodents were likely to have been similar in many aspects of their ecological adaptations, so it is not unlikely that they were competitors to some extent.

PLESIADAPIFORMS AND PRIMATES

Although plesiadapiforms are the most primatelike mammals from the Paleocene, all have unique specializations that preclude them from the ancestry of the early prosimians that immediately succeeded them in the beginning of the Eocene epoch, as well as from any relationship with other later primates (Fig. 11.9). Only *Purgatorius* is generalized enough in its dental formula to be a suitable ancestor for all later primates, but

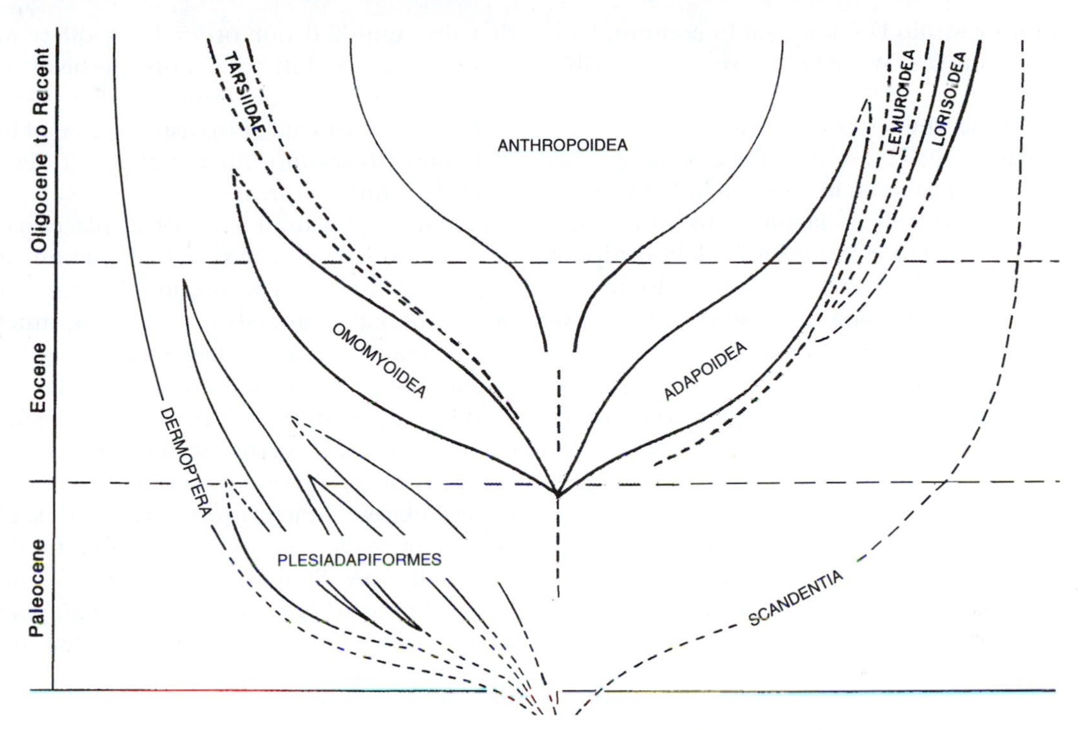

FIGURE 11.9 Phylogenetic relationships of plesiadapiforms, primates (and other archontans).

it is so generalized and poorly known that its primate status is marginal. Given the features that have traditionally been used to link plesiadapiforms with primates (petrosal bulla, shape of the elbow, overall molar shape), in most comparative studies these seem less convincing than features linking other orders with primates. Moreover, Beard (1993) has argued that plesiadapiforms are actually more closely related to Dermoptera than to primates and places them in that order (which is in turn linked with primates).

THE PHYLOGENETIC ORIGINS OF PRIMATES AMONG THE ARCHONTA

As already noted, primates are usually grouped with Plesiadapiformes, Scandentia (tree shrews), Dermoptera, and Chiroptera (bats) in a single superorder, the Archonta. However, this superordinal classification implies more agreement about the phylogenetic relationships among these four orders than actually exists. It is difficult to identify more than a few common features linking any of these different orders amid their many specializations. Moreover, although most authorities accept the concept of a superorder Archonta, every analysis and every anatomical system (e.g., globin genes, basicranial anatomy, postcranial anatomy) seems to indicate a different composition for the group and a different phylogeny among the included taxa (see MacPhee, 1993). Thus, on the basis of many aspects of postcranial anatomy and a few features of the skull and dentition, Beard (1993) considers Primates and Dermoptera as sister taxa, with plesiadapiform families placed variously within Dermoptera. In contrast, basicranial studies (Wible and Covert, 1987; Wible and Martin, 1993) find tree shrews (Scandentia) as the sister taxon to Primates. Molecular studies often find these same groups related to primates, but also include elephant shrews, rodents, or lagomorphs within the same clade. The supraordinal relationships of Primates are far from resolved. Undoubtedly, this reflects the long time between the separation of Primates from other orders of living mammals over 65 million years ago and the numerous evolutionary specializations all groups have accumulated in the meantime that may mask true relationships and/or suggest false ones.

THE ADAPTIVE ORIGIN OF PRIMATES

In contrast to the lack of consensus over the phylogenetic origin of primates, there is increasing consensus on the adaptive nature of the common ancestor of all living primates and the ecological factors that were probably most important in primate origins. Primates are distinguished from other mammals, including plesiadapiforms, by numerous anatomical specializations, including more convergent orbits with postorbital bars, and grasping extremities with nails rather than claws on most digits, as well as details of their basicranial anatomy. For most of this century, theories of primate origins have emphasized the role of arboreality as the driving selective force that has given rise to the special visual and grasping abilities of primates; however, there are a number of competing views concerning the specific details of primate arboreal foraging that were most critical in primate origins (Cartmill, 1992).

In the view of F. S. Szalay (1973), the arboreal habits of primates represent a primitive archontan feature, and the specific behaviors that distinguished our order from other archontans was a shift to a more herbivorous diet (to account for the rounded molar cusps) and a more acrobatic grasp-leaping type of locomotion (to account for the grasping digits and other limb features).

In contrast, Cartmill (e.g., 1992) places particular emphasis on the visual adaptations in accounting for primate origins. Noting that most arboreal mammals do not look much like primates, because they possess neither orbital convergence nor grasping digits, he argues that we must look beyond arboreality per se to account for the distinctive features of our order. Stereoscopic vision, he notes, is particularly characteristic of predators, such as cats or owls, that rely on vision to detect their prey. Similarly, he noted that the hands of most arboreal mammals are equipped with claws rather than nails, and that the nailed, grasping hands of primates are probably adaptations for grasping prey rather than arboreal supports. Thus, in Cartmill's hypothesis, the ancestral primate was specifically a visual predator that stalked and grasped its prey in either the canopy or the forest undergrowth.

Bob Sussman (1991) has a third view of primate origins. In his hypothesis, primates evolved in conjunction with the radiation of angiosperm plants to exploit the products of flowering plants (fruit, flowers, and nectar) in a small branch setting. While he admits that his theory does not specifically account for primate visual adaptation, he argues that visual predation is a rare behavior among primates and is unlikely to have been important in the earliest primates. It is also important to note that although the convergent orbits of primates are similar to those of visual predators such as cats and owls, the mammals that seem to show the greatest resemblance to primates in their visual system are fruit bats, a totally herbivorous group.

Finally, Crompton (1995) has emphasized the role of locomotion, especially leaping behavior in a nocturnal animal, as an important selective factor for the evolution of primate stereoscopic vision.

As many authors have emphasized, the key to reconstructing the history of any seemingly integrated suite of evolutionary adaptations lies in having a record of the sequence in which the features were acquired. Did the primate visual adaptations precede or follow the grasping abilities and the reduced molar cusps? In the absence of a clear sequence of intermediate forms in early primate evolution, several authors have looked to other groups of mammals, specifically marsupials, to provide evolutionary models of early primate adaptations. Thus, Rasmussen (1990) conducted a field study of the neotropical marsupial *Caluromys derbianus,* a didelphid that shows striking anatomical similarities to primates in having a relatively large brain, a nearly complete postorbital bar, a relatively short snout, and primatelike proportions in its (clawed) digits. Similarly, in a study of the hand morphology and grasping abilities of marsupials and prosimian primates, Lemelin (1995) found that *Caluromys* was strikingly similar to small prosimians such as *Cheirogaleus* and *Microcebus* in details of digital proportions and grasping and manipulation. Since Rasmussen (1990) found that *Caluromys* was more arboreal than any of its relatives and foraged for both fruits and insects among terminal branches, he was unable to falsify any of the competing theories. Nevertheless, he speculated that primate visual adaptations evolved for visual predation in an animal that was already adapted to foraging for fruit in terminal branches, a scenario that supports each of the different adaptive explanations for primate origins, but at different times. Unfortunately, until we have a better fossil record of intermediate forms preceding the first appearance of "real primates" at the beginning of the Eocene, the details of primate origins will remain shaded in the mists of time and phylogeny.

BIBLIOGRAPHY

PALEOCENE EPOCH

Adams, C. G. (1981). An outline of Tertiary paleogeography. In *The Evolving Earth,* ed. L. R. M. Cocks, pp. 221–235. Cambridge, Mass.: Cambridge University Press.

Hickey, L. J. (1980). Paleocene stratigraphy and flora of the Clark's Fork Basin. In *Early Cenozoic Paleontology and Stratigraphy of the Bighorn Basin, Wyoming 1880–1980,* ed. P. D. Gingerich, pp. 33–50. University of Michigan, Museum of Paleontology, Papers on Paleontology no. 24.

Krause, D. (1984). Mammalian evolution in the Paleocene: The beginning of an era. In *Mammals: Notes for a Short Course,* ed. T. D. Broadhead, pp. 87–109. Knoxville: University of Tennessee, Dept. of Geological Sciences.

ARCHONTANS

Bailey, W. J., Slighton, J. L., and Goodman, M. (1992). Rejection of the "Flying Primate" hypothesis by phylogenetic evidence from the ε-globin gene. *Science* **256**: 86–89.

Beard, K. C. (1990). Do we need the newly proposed order Proprimates? *J. Human Evol.* **19**:817–820.

——— (1993). Phylogenetic systematics of the Primatomorpha, with special reference to Dermoptera. In: *Mammal Phylogeny: Placentals,* ed. F. S. Szalay, M. J. Novacek, and M. C. McKenna, pp. 129–150. New York: Springer-Verlag.

MacPhee, R. D. E., ed. (1993). *Primates and Their Relatives in Phylogenetic Perspective.* New York: Plenum Press.

MacPhee, R. D. E., Novacek, M. J., and Storch, G. (1988). Basicranial morphology of early Tertiary erinaceomorphs and the origin of primates. *Am. Mus. Novitates 2921.*

Martin, R. D. (1993). Primate origins: Plugging the gaps. *Nature* **363**:223–234.

Simmons, N. B. (1993). The importance of methods, Archontan phylogeny and cladistic analysis of morphological data. In: *Primates and Their Relatives in Phylogenetic Perspective,* ed. R. D. MacPhee, pp. 1–61. New York: Plenum Press.

——— (1994). The case for Chiropteran monophyly. *Am. Mus. Novitates 31103.*

Wible, J. R. and Covert, H. H. (1987). Primates: Cladistic diagnosis and relationships. *J. Human Evol.* **16**:1–22.

Tree Shrews

Clark, W. E. Le Gros (1934). *Early Forerunners of Man.* London: Bailleire, Tindall and Cox.

Kay, R. F., Thewissen, J. G. M., and Yoder, A. (1992). Cranial anatomy of *Ignacius graybullianus* and the affinities of Plesiadapiformes. *Am. J. Phys Anthropol.* **89**:477–498.

Luckett, W. P., ed. (1980). *Comparative Biology and Evolutionary Relationships of Tree Shrews.* New York: Plenum Press.

Martin, R. D. (1968). Towards a new definition of primates. *Man* **3**:377–401.

Wible, J. R., and Covert, H. H. (1987). Primates: Cladistic diagnosis and relationships. *J. Human Evol.* **16**:1–22.

Wible, J. R., and Martin, J. (1993). Ontogeny of the tympanic floor and roof in *Archontans.* In *Primates and Their Relatives in Phylogenetic Perspective.* ed. R. D. E. MacPhee, pp. 111–148. New York: Plenum Press.

Wible, J. R., and Zeller, U. (1994). Cranial circulation of the pen-tailed tree shrew *Ptilocercus lowii* and relationships of Scandentia. *J. Mammal. Evol.* **2**(4):209–230.

Dermoptera

Beard, K. C. (1993). Phylogenetic systematics of the Primatomorpha, with special reference to Dermoptera. In *Mammal Phylogeny: Placentals,* ed. F. S. Szalay, M. J. Novacek, and M. C. McKenna, pp. 129–150. New York: Springer-Verlag.

Ducrocq, S., Buffetaut, E., Buffetaut-Tong, Jaeger, J.-J., Jongkanjanasoontorn, Y., and Suteethorn, V. (1992). First fossil flying lemur: A dermopteran from the Late Eocene of Thailand. *Palaeontology* **35**(2):373–380.

Kay, R. F., Thewissen, J. G. M., and Yoder, A. (1992). Cranial anatomy of *Ignacius graybullianus* and the affinities of Plesiadapiformes. *Am. J. Phys. Anthropol.* **89**:477–498.

Thewissen, J. G. M., and Babcock, S. K. (1993). The implications of the propatagial muscles of flying and gliding mammals for Archontan systematics. In *Primates and Their Relatives in Phylogenetic Perspective,* ed. R. D. E. MacPhee, pp. 91–109. New York: Plenum Press.

Simmons, N. B., and Quinn, T. H. (1994). Evolution of the digital tendon locking mechanism in bats and dermopterans: A phylogenetic perspective. *J. Mammal. Evol.* **2**(4):231–254.

Chiroptera

Bailey, W. J., Slightom, J. L., and Goodman, M. (1992). Rejection of the "Flying Primate" hypothesis by phylogenetic evidence from the ϵ-globin gene. *Science* **256**: 86–89.

Pettigrew, J. D., Jamieson, B. G. M., Robson, S. K., Hall, L. S., McAnally, K. I., and Cooper, H. M. (1989). Phylogenetic relations between microbats, megabats and primates (Mammalia: Chiroptera and Primates). *Philos. Trans. R. Soc. London b* **325**:489–559.

Simmons, N. B. (1994). The case for Chiropteran monophyly. *Am. Mus. Novitates 31103.*

PLESIADAPIFORMS

Beard, K. C. (1993). Phylogenetic systematics of the Primatomorpha, with special reference to Dermoptera. In *Mammal Phylogeny: Placentals,* ed. F. S. Szalay, M. J. Novacek, and M. C. McKenna, pp. 129–150. New York: Springer-Verlag.

Covert, H. H. (1986). Biology of early Cenozoic primates. In *Comparative Primate Biology,* vol. 1: *Systematics, Evolution, and Anatomy,* ed. D. R. Swindler and J. Erwin, pp. 335–359. New York: Alan R. Liss.

Gidley, J. W. (1923). Paleocene primates of the Fort Union, with discussion of relationships of Eocene primates. *Proc. U. S. Nat. Mus.* **63**:1–38.

Gingerich, P. D. (1986). Pleisadapis and the delineation of the order Primates. In *Major Topics in Primate and Human Evolution,* ed. B. Wood, L. Martin, and P. Andrews, pp. 32–46. Cambridge: Cambridge University Press.

Krause, D. W. (1978). Paleocene primates frorn Western Canada. *Canadian J. Earth Sci.* **15**:1250–1271.

MacPhee, R. D. E., Cartmill, M., and Gingerich, P. D. (1983). New Palaeogene primate basicrania and the definition of the order Primates. *Nature* **301**:509–511.

Simons, E. L. (1967). Fossil primates and the evolution of some primate locomotor systems. *Am. J. Phys. Anthropol.* **26**:241–253.

Simpson, G. G. (1937). The Fort Union of the Crazy Mountain Field, Montana and its mammalian faunas. *Bull. U. S. Nat. Mus.* **169:**1–287.

Szalay, F. S. (1972). Paleobiology of the earliest primates. In *The Functional and Evolutionary Biology of Primates,* ed. R. Tuttle, pp. 3–35. Chicago: Aldine-Atherton.

Szalay, F. S., and Dagosto, M. (1980). Locomotor adaptations as reflected in the humerus of Paleogene primates. *Folia Primatol.* **34**:1–45.

Szalay, F. S., Tattersall, I., and Decker, R. (1975). Phylogenetic relationships of *Plesiadapis*—Postcranial evidence. *Contrib. Primatol.* **5**:136–166.

Teilhard de Chardin, P. (1922). Les mammiferes de l'Eocene inferieur Francais et leurs gisements. *Ann. Paleontol.* **11**:9–116.

Van Valen, L. M. (1994). The origin of the plesiadapid primates and the nature of *Purgatorius. Evolutionary Monographs* **15**:1–79.

Purgatorius

Buckley, G. A. (1997). A new species of *Purgatorius* (Mammalia; Primatomorpha) from the Lower Paleocene Bear Formation, Crazy Mountain Basin, South-Central Montana. *J. Paleont.* **71**:149–155.
Clemens, W. A. (1974). *Purgatorius,* an early paromomyid primate (Mammalia). *Science* **184**:903–906.
Van Valen, L. M. (1994). The origin of the plesiadapid primates and the nature of *Purgatorius. Evolutionary Monographs* **15**:1–79.
Van Valen, L., and Sloan, R. E. (1965). The earliest primates. *Science* **150**:743–745.

Plesiadapids

Gingerich, P. D. (1973). First record of the Paleocene primate *Chiromyoides* from North America. *Nature* **244**: 517–518.
——— (1974). Dental function in the Paleocene primate *Plesiadapis*. In *Prosimian-Biology,* ed. R. D. Martin, G. A. Doyle, and A. C. Walker, pp. 531–541. London: Duckworth.
——— (1975). New North American Plesiadapidae (Mammalia, Primates) and a biostratigraphic zonation of the middle and upper Paleocene. *Contr. Mus. Paleontol. Univ. Michigan* **24**:135–148.
——— (1976). Cranial anatomy and evolution of early Tertiary Plesiadapidae (Mammalia, Primates). University of Michigan, Museum of Paleontology, Papers on Paleontology no. 15.
Russell, D. (1964). Les mammiferes paleocenes d'Europe. *Mem. Mus. Nat. d'Hist. Natur., ser. C.* **13**:1–324.
Szalay, F. S. (1975). Phylogeny of primate higher taxa: The basicranial evidence. In *Phylogeny of the Primates: A Multidisciplinary Approach,* ed. W. P. Luckett and F. S. Szalay, pp. 91–125. New York: Plenum Press.
Szalay, F. S., Tattersall, I., and Decker, R. (1975). Phylogenetic relationships of *Plesiadapis*—Postcranial evidence. *Contrib. Primatol.* **5**:136–166.

Carpolestids

Beard, K. C., and Wang, J. (1995). The first Asian Plesiadapoids (Mammalia: Primatomorpha). *Ann. Carnegie Mus.* **64**(1):1–33.
Biknevicius, A. (1986). Dental function and diet in the Carpolestidae (Primates: Plesiadapiformes) *Am. J. Phys. Anthropol.* **71**:157–172.
Fox, R. C. (1984). A new species of the Paleocene primate *Elphidotarsius:* Its stratigraphic position and evolutionary relationships. *Can. J. Earth Sci.* **21**(11):1268 1277.
Fox, R. C. (1993). The primitive dental formula of the Carpolestidae (Plesiadapiformes, mammalia) and its phylogenetic implications. *J. Vert. Paleo.* **13**:516–524.
Rose, K. D. (1975). The Carpolestidae, early Tertiary primates from North America. *Bull. Mus. Comp. Zool.* **147**: 1–74.
——— (1977). Evolution of carpolestid primates and chronology of the North American middle and late Paleocene. *J. Paleontol.* **51**(3):536–542.

Saxonellids

Fox, R. C. (1984). First North American record of the Paleocene primate *Saxonella. J. Paleontol.* **58**(3):892–894.
Fox, R. C. (1991). *Saxonella* (Plesiadapiformes: ? Primates) in North America: *S. naylori,* sp. nov., from the late paleocene of Alberta, Canada. *J. Vert. Paleont.* **65**:700–701.
Russell, D. (1964). Les mammiferes paleocenes d' Europe. *Mem. Mus. Nat. d'Hist. Natur., ser. C.* **13**:1–321.

Paromomyids

Beard, K. C. (1990). Gliding behavior and palaeoecology of the alleged primate family Paromomyidae (Mammalia, Dermoptera). *Nature* **345**:340–341.
——— (1993). Origin and evolution of gliding in early Cenozoic Dermoptera (Mammalia, Primatomorpha). In *Primates and Their Relatives in Phylogenetic Perspective,* ed: R. D. E. MacPhee, pp. 63–90. New York: Plenum Press.
Gingerich, P. D. (1974). Function of pointed premolars in *Phenacolemur* and other mammals. *J. Dent. Res.* **53**:497.
Godinot, M. (1984). Un noveau genre de Paromomyidae (Primates) de l'Eocene Inferieur d'Europe. *Folia Primatol.* **43**:84–96.
Hickey, L. J., West, R. M., Dawson, M. R., and Choi, D. K. (1983). Arctic terrestrial biota: Paleomagnetic with mid-northern latitudes during the late Cretaceous and early Tertiary. *Science* **221**:1153–1156.
Kay, R. F., Thewissen, J. G. M., and Yoder, A. D. (1992). Cranial anatomy of *Ignacius graybullianus* and the affinities of the Plesiadapiformes. *Am. J. Phys. Anthropol.* **89**:477–498.
Krause, D. W. (1991). Were paromomyids gliders? Maybe, maybe not. *J. Human Evol.* **21**:177–188.
MacPhee, R. D. E., Cartmill, M., and Gingerich, P. D. (1983). New Palaeogene primate basicrania and the definition of the order Primates. *Nature* **301**:509–511.
McKenna, M. C. (1980). Eocene paleolatitude, climate and mammals of Ellesmere Island. *Palaeogeogr., Palaeoclimatol., Palaeoecol.* **30**:349–362.

Robinson, P., and Ivy, L. D. (1994). Paromomyidae (?Dermoptera) from the Powder River Basin, Wyoming, and a discussion of microevolution in closely related species. *Contrib. Geol.* **30**(1):91–116.

Rose, K. D., and Bown, T. M. (1982). New plesiadapiform primates from the Eocene of Wyoming and Montana. *J. Vert. Paleontol.* **2**(1):63–69.

Rose, K. D., and Gingerich, P. D. (1976). Partial skull of the plesiadapiform primate *Ignacius* from the early Eocene of Wyoming. *Contrib. Mus. Paleontol., Univ. Michigon* **24**:181–189.

Runestad, J. A., and Ruff, C. B. (1995). Structural adaptations for gliding in mammals with implications for locomotor behavior in paromomyids. *Am. J. Phys. Anthropol.* **98**:101–119.

Russell, D. E., Louis, P., and Savage, D. E. (1967). Primates of the French early Eocene. *Univ. California Geol. Sci.* **73**:1–46.

Simpson, G. G. (1955). The Phenacolemuridae, a new family of early primates. *Bull. Am. Mus. Nat. Hist.* **105**: 415–441.

Szalay, F. S. (1972). Cranial morphology of the early Tertiary *Phenacolemur* and its bearing on primate phylogeny. *Am. J. Phys. Anthropol.* **36**:59–76.

Micromomyids

Beard, K. C., and Houde, P. (1989). An unusual assemblage of diminutive Plesiadapiforms (Mamalia, ?Primates) from the early Eocene of the Clark's Fork Basin, Wyoming. *J. Vert. Paleontol.* **9**(4):388–399.

Fox, R. C. (1984). The dentition and relationships of the Paleocene primate *Micromomys* Szalay with description of a new species. *Can. J. Earth Sci.* **21**(11):1262–1267.

Robinson, P. (1994). *Myrmekomomys,* a new genus of Micromomyine (Mammalia, Microsyopidae) from the lower Eocene rocks of the Powder River Basin, Wyoming. *Contrib. Geology, Univ. of Wyoming* **30**(1):85–90.

Microsyopids

Bown, T. M., and Gingerich, P. D. (1973). The Paleocene primate *Plesiolestes* and the origin of Microsyopidae. *Folia Primatol.* **19**:1–18.

Bown, T. M., and Rose, K. D. (1976). New early Tertiary primates and a reappraisal of some Plesiadapiformes. *Folia Primatol.* **26**:109–138.

Gunnell, G. F. (1989). Evolutionary history of *Microsyopoidea* (Mammalia, Primates) and the relationship between Plesiadapiformes and Primates. *Univ. Michigan Papers Paleont. 27.*

Hoffstetter, R. (1986). Paleontologie. Limite entre primates et non-primates; position des Plesiadapiformes et des Microsyopidae. *C. R. Acad. Sci. (Paris),* t. 302, serie 11, no. 1, pp. 43–45.

Rose, K. D., and Bown, T. M. (1982). New plesiadapiform primates from the Eocene of Wyoming and Montana. *J. Vert. Paleontol.* **2**(1):63–69.

Szalay, F. S. (1968). The beginnings of primates. *Evolution* **22**:19–36.

——— (1973). New Paleocene primates and a diagnosis of the new suborder Paramomyiformes. *Folia Primatol.* **19**:73–87.

——— (1981). Phylogeny and the problems of adaptive significance: The case of the earliest primates. *Folia Primatol.* **36**:157–182.

Palaechthonids

Kay, R. F., and Cartmill, M. (1977). Cranial morphology and adaptations of *Palaechthon nacimienti* and other Paromomyidae (Plesiadapoidea, Primates), with a description of a new genus and species. *J. Human Evol.* **6**: 19–53.

Rose, K. D., Beard, K. C., and Houde, P. (1993). Exceptional new dentitions of the diminutive plesiadapiforms *Tinimomys* and *Niptomomys* (Mammalia), with comments on the upper incisors of Plesiadapiformes. *Ann. Carnegie Mus.* **62**(4):351–361.

Szalay, F. S. (1969). Mixodectidae, Microsyopidae, and the insectivore–primate transition. *Bull. Am. Mus. Nat. Hist.* **140**:195–330.

Picrodontids

Gingerich, P. D., Houde, P., and Krause, D. W. (1983). A new earliest Tiffanian (Late Paleocene) mammalian fauna from Bangtail Plateau, Western Crazy Mountain Basin, Montana. *J. Paleontol.* **57**:957–970.

Szalay, F. S. (1968). The Picrodontidae, a family of early primates. *Am. Mus. Nov.* no. 2329, pp. 1–55.

——— (1972). Paleobiology of the earliest primates. In *The Functional and Evolutionary Biology of Primates,* ed. R. Tuttle, pp. 3–35. Chicago: Aldine-Atherton.

Tomida, Y. (1982). A new genus of picrodontid primate from the Paleocene of Utah. *Folia Primatol.* **37**:37–43.

Picromomyids

Rose, K. D. and Bown, T. M. (1996). A new plesiadapiform (Mammalia: Plesiadaformes) from the early Eocene of the Bighorn Basin, Wyoming. *Ann. Carnegie Museum* **65**:305–321.

ADAPTIVE RADIATION OF PLESIADAPIFORMS

Gingerich, P. D. (1976). Cranial anatomy and evolution of early Tertiary Plesiadapidae (Mammalia, Primates).

University of Michigan, Papers on Paleontology no. 15, pp. 1–40.

Maas, M. C., Krause, D. W., and Strait, S. G. (1988). Decline and extinction of plesiadapiforms in North America: Displacement or replacement? *Paleobiology* **14**:410–431.

Sussman, R. W., and Raven, P. H. (1978). Pollination by lemurs and marsupials: An archaic coevolutionary system. *Science* **200**:731–736.

Szalay, F. S. (1972). Paleobiology of the earliest primates. In *The Functional and Evolutionary Biology of Primates,* ed. R. L. Tuttle, pp. 3–35. Chicago: Aldine-Atherton.

Van Valen, L., and Sloan, R. E. (1966).The extinction of the multituberculates. *Syst. Zool.* **15**:261–278.

PLESIADAPIFORMS AND LATER PRIMATES

Cartmill, M. (1972). Arboreal adaptations and the origin of the order Primates. In *The Functional and Evolutionary, Biology of Primates,* ed. R. Tuttle, pp. 97–122. Chicago: Aldine-Atherton.

______ (1974). Rethinking primate origins. *Science* **184**: 436–443.

Gingerich, P. D. (1981). Why study fossils? *Am. J. Primatol.* **1**:293–295.

______ (1986). *Plesiadapis* and the delineation of the order Primates. In *Major Topics in Primate and Human Evolution,* ed. B. Wood, L. Martin, and P. Andrews, pp. 32–46. Cambridge: Cambridge University Press.

Luckett, W. P., ed. (1980). *Comparative Biology and Evolutionary Relationships of Tree Shrews.* New York: Plenum Press.

MacPhee, R. D. E., Cartmill, M., and Gingerich, P. D. (1983). New Palaeogene primate basicrania and the definition of the order Primates. *Nature* **301**:509–511.

Martin, R. D. (1968). Towards a new definition of primates. *Man* **3**(3):377–401.

______ (1986). Primates: A definition. In *Major Topics in Primate and Human Evolution,* ed. B. Wood, L. Martin, and P. Andrews, pp. 1–31. Cambridge: Cambridge University Press.

Simpson, G. G. (1940). Studies on the earliest primates. *Bull. Am. Mus. Nat. Hist.* **77**:185–212.

Szalay, F. S. (1975a). Phylogeny of primate higher taxa: The basicranial evidence. In *Phylogeny of the Primates: A Multidisciplinary Approach,* ed. W. P. Luckett and F. S. Szalay, pp. 91–125. New York: Plenum Press.

______ (1975b). Where to draw the nonprimate–primate taxonomic boundary. *Folia Primatol.* **23**:158–163.

Szalay, F. S., and Decker, R. L. (1974). Origins, evolution and function of the tarsus in late Cretaceous eutherians and Paleocene primates. In *Primate Locomotion,* ed. F. A. Jenkins, pp. 239–259. New York: Academic Press.

Szalay, F. S., and Delson, E. (1979). *Evolutionary History of the Primates.* New York: Academic Press.

Szalay, F. S., Rosenberger, A. L., and Dagosto, M. (1987). Diagnosis and differentiation of the order Primates. *Yrbk. Phys. Anthropol.* **30**:75–105.

Wible, J. R., and Covert, H. H. (1987). Primates: Cladistic diagnosis and relationships. *J. Human Evol.* **16**:1–20.

PRIMATE ORIGINS

Archibald, J. D. (1977). Ectotympanic bone and internal carotid circulation of eutherians in reference to anthropoid origins. *J. Human Evol.* **6**:609–622.

Cartmill, M. (1992). New views on primate origins. *Evol. Anthropol.* 105–111.

Crompton, R. H. (1995). Visual Predation, Habitat Structure, and the Ancestral Primate niche. In *Creatures of the Dark: The Nocturnal Prosimians,* ed. L. Alterman, G. A. Doyle, and M. Kay Izard, pp. 11–30. New York: Plenum Press.

Gingerich, P. D. (1986). *Plesiadapis* and the delineation of the order Primates. In *Major Topics in Primate and Human Evolution,* ed. B. Wood, L. Martin, and P. Andrews, pp. 32–46. Cambridge: Cambridge University Press.

Rasmussen, D. T. (1990). Primate origins: Lessons from a neotropical marsupial. *Am. J. Primatol.* **22**:263–277.

Rose, K. D., and Fleagle, J. G. (1981). The fossil history of nonhuman primates in the Americas. In *Ecology and Behavior of Neotropical Primates,* vol. 1, ed. A. E. Coimbra-Filho and R. A. Mittermeier, pp. 111–167. Rio de Janeiro: Academeia Brasileria de Ciencias.

Sussman, R. W. (1991). Primate origins and the evolution of angiosperms. *Am. J. Primatol.* **23**:209–223.

Szalay, F. S. (1972). Paleobiology of the earliest primates. In *The Functional and Evolutionary Biology of Primates,* ed. R. Tuttle, pp. 3–35. Chicago: Aldine-Atherton.

______ (1975). Where to draw the nonprimate–primate taxonomic boundary. *Folia Primatol.* **23**:158–163.

Szalay, F. S., and Drawhorn, J. (1980). Evolution and diversification of the Archonta in an arboreal milieu. In *Comparative Biology and Evolutionary Relationships of Tree Shrews,* ed. W. P. Luckett, pp. 133–169. New York: Plenum Press.

Van Valen, L., and Sloan, R. E. (1965). The earliest primates. *Science* **150**:743–745.

TWELVE

Fossil Prosimians

EOCENE EPOCH

In North America and Europe, the Eocene epoch (54–34 million years ago) was marked by a major change in faunas. Many modern types of mammals, including the earliest artiodactyls, perissodactyls, and rodents, replaced more archaic types of mammals. In primate evolution, the beginning of this epoch is marked by the first appearance of primates that resemble living prosimians (Fig. 12.1). These faunal changes took place in a series of waves rather than in a single broad sweep and seem to be the result of both climatic changes and new connections between continents or major continental areas.

Eocene paleogeography was not strikingly different from that at the beginning of the Paleocene (see Fig. 11.1). North America and Europe were connected at the beginning of the Eocene but became increasingly separated and distinct throughout the epoch, resulting in an increasing distinctiveness in their mammalian faunas. Additionally, there is faunal evidence for intermittent connections between North America and Asia. Whereas the Tethys

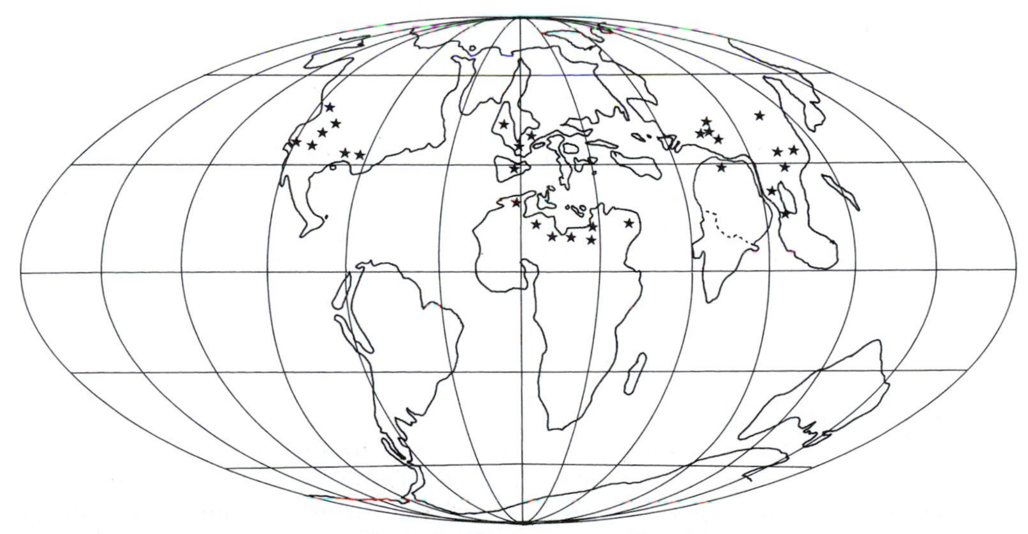

FIGURE 12.1 Geographic distribution of Eocene fossil prosimian sites, shown on an early Eocene paleogeography.

Seaway remained open across most of the Mediterranean region and western Asia, India was coming in contact with the Asian mainland. South America remained isolated from all other continents except Antarctica. Little is known of the paleogeography of Africa during this time, but there are indications that a large seaway separated the northwest corner from the rest of the continent.

Eocene climates in Europe and North America were warmer and more equable than those of the preceding epoch (see Fig. 10.4). Both the sediments and the flora indicate tropical climates for North America. Climates were so warm during the early-middle part of the Eocene that there was a relatively diverse fauna of mammals, including a paromomyid, living well within the Arctic Circle. It has been suggested that changes in Eocene faunas of northwestern North America reflect migrations of new taxa and lineages from farther south, along with the increasingly warmer climates.

The First Modern Primates

The primates that made their debut in the early Eocene were quite different from the plesiadapiforms of the preceding Paleocene epoch. They have all the anatomical features characteristic of living primates. They had shorter snouts, smaller infraorbital foramina, and a postorbital bar completing the bony ring around their orbits (Fig. 12.2). They had larger, more rounded braincases, and their auditory regions and cerebral blood supplies were like those of living prosimians. Their skeletons had more slender limbs with divergent, grasping halluces, and they possessed nails rather than claws on most digits (Dagosto, 1993).

All of these morphological differences indicate that the Eocene primates practiced a very different way of life from the plesiadapiforms they succeeded. Many of the cranial differences indicate an increased reliance on vision rather than smell and tactile vibrissae. The postcranial changes suggest an increased importance of manipulative abilities, with the replacement of claws by nails, and the locomotor skeletons of many species suggest leaping abilities and more primatelike, acrobatic locomotion. In several species there are indications of canine sexual dimorphism, as in later anthropoids. When plesiadapiforms were generally regarded as early primates, Simons (1972) aptly dubbed the Eocene prosimians "the first primates of modern aspect"; now they are generally regarded as the first primates.

Like plesiadapiforms, the early prosimians were among the most abundant mammals of their day, but they are not equally well documented on all continents. They are common in mammalian faunas of North America and Europe and are becoming better known from Asia and Africa, but are unknown from South America and Antarctica (Fig. 12.1). From their first appearance in the early Eocene, the prosimians of North America and Europe can be readily divided into two distinct groups: the lemurlike adapoids and the tarsierlike or galagolike omomyoids. The earliest members of the two superfamilies (*Donrussellia, Cantius,* and *Teilhardina*) are very similar (Fig. 12.15), suggesting a divergence just prior to the earliest Eocene. Both families subsequently produced adaptive radiations of species that flourished throughout the epoch, and their collateral relatives are thriving today in the forests of Africa, Madagascar, and Asia.

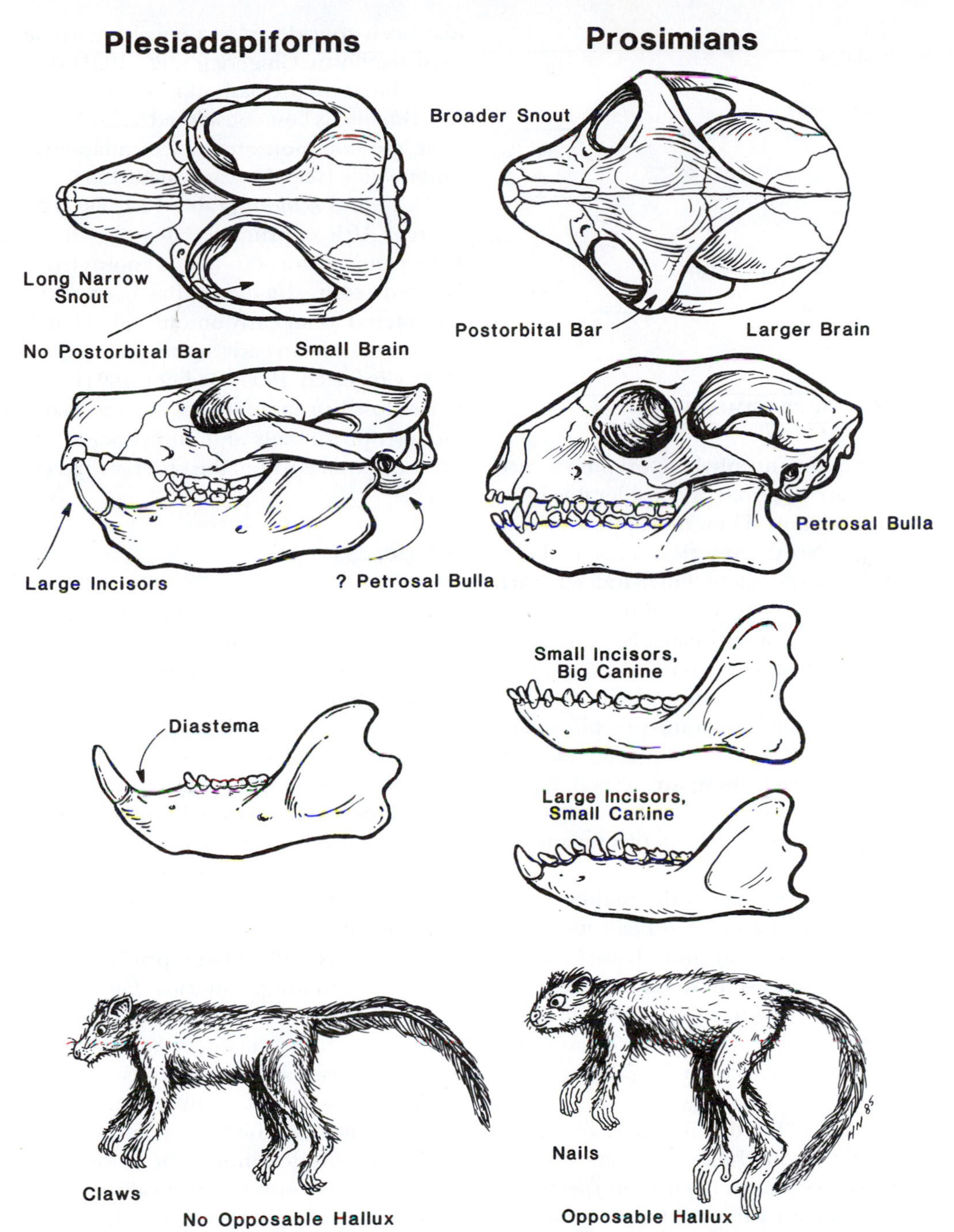

FIGURE 12.2 Comparison of fossil prosimians and more archaic plesiadapiforms, showing major anatomical contrasts.

TABLE 12.1
Order Primates
Suborder Prosimii
Family *Incertae sedis*

Species	Estimated Mass (g)
Altanius (early Eocene, Asia)	
A. orlovi	10
Altiatlasius (late Paleocene, Africa)	
A. koulchii	50–100

THE ORIGIN OF PROSIMIANS

The geographic and phyletic origins of early prosimians are a source of considerable interest and speculation. Their first appearances in Europe and North America seem to be at approximately the same time; indeed, early Eocene faunas of those two continents are virtually identical, and primates appear at the beginning of the epoch. As we noted in Chapter 11, there are no good phyletic ancestors for early prosimians among the plesiadapiforms or among any other group of early mammals. However, there are several, poorly known species from Africa and Asia that may lie close to the origins of fossil prosimians and all later primates.

Decoredon elongatus, from the middle Paleocene of southern China, has been identified by some authors (Szalay and Li, 1986) as the earliest primate, possibly an omomyoid. However, it is unclear from the available material whether *Decoredon* is actually a primate or belongs to some other order (e.g., Rose *et al.,* 1994).

Altanius orlovi (Table 12.1; Fig. 12.3) is a tiny primate (10 g) from the early Eocene of Mongolia that is now known from many specimens preserving much of the dentition. It has elevated trigonids and tall premolars. Originally described as an early omomyoid, *Altanius* has also been considered as a possible carpolestid plesiadapiform. Gingerich *et al.* (1991) demonstrated that on the basis of its dental morphology, *Altanius* is best considered a basal primate, near the common ancestry of adapoids and omomyoids (see also Van Valen, 1994).

Altiatlasius koulchii (Table 12.1; Fig. 12.3) is a larger African primate (50–100 g) from the Paleocene of Morocco that is known from ten isolated teeth. *Altiatlasius* has been variously considered as an early omomyoid, a basal primate, and even an early anthropoid (Sige *et al.,* 1990; Gingerich, 1990; Godinot, 1994). It is the oldest fossil primate that is clearly related to modern prosimians and anthropoids, but its precise affinities within the order are uncertain.

Adapoids

In many aspects of their dental anatomy, adapoids are the most primitive of all known primates, fossil or living. Most of the dental specializations found among later primates could easily be derived from an early adapoid morphology. As we discuss later, such a basically primitive morphology poses interesting difficulties and virtually unlimited possibilities in ascertaining the phyletic relationships of adapoids with later primate groups.

Compared with the earlier plesiadapiforms and the contemporaneous omomyoids, most adapoids were rather large primates, comparable in size to living lemurids (Fig. 12.4). The primitive adapoid dental formula (Fig. 12.5), retained by many relatively late members of the family, is 2.1.4.3. Adapoids differ from plesiadapiforms and many omomyoids, and superficially resemble living anthropoids in their anterior dentition. The lower incisors are small and positioned vertically in the mandible, and the uppers are relatively broad, but short, and are separated by a median gap. Both upper and lower canines are larger than the incisors and, in some taxa, are sexually dimorphic. The anterior premolars are often caniniform, and the posterior ones are often molariform. The upper molars are broad, and the two major lineages evolved a hypocone independently. Lower molars are relatively long and narrow in most taxa. The numerous shearing crests, presumably an adaptation to folivory, appear to have evolved independently in many adapoid lineages, along with fusion of the two halves of the mandible.

Adapoids have relatively long but broad snouts with a small infraorbital foramen (Figs. 12.4, 12.6). As in living prosimians, each orbit is encircled by a complete bony ring. They have a large ethmoid recess with numerous ethmoturbinates, as in lemurs and primitive mammals generally. The braincase is larger than that of the archaic primates but smaller than in extant lemurs or anthropoids. The tympanic ring is suspended within the inflated bony bulla, much as in extant lemurs. The bony canals for stapedial and promontory branches of the internal carotid artery are apparently quite variable. Even within a single species, some individuals apparently have a larger canal for the stapedial, some have a larger promontory canal, and still others have similar-size canals for the two (Gunnell, 1995).

The skeletal anatomy, which is well known for several North American and European genera (Figs. 12.4, 12.7), shows that adapoid limbs are similar to those of living strepsirhines but more robust. These Eocene prosimians have relatively long legs, a long trunk, and a long tail. Their hands and feet have nails rather than claws, and they have a divergent pollex and a grasping foot (Dagosto, 1993).

The systematics of adapoids has been studied by many workers—and not without disagreement. The biostratigraphy of species from the western United States is particularly well documented. Adapoids are divided into four subfamilies that are largely, but not com-

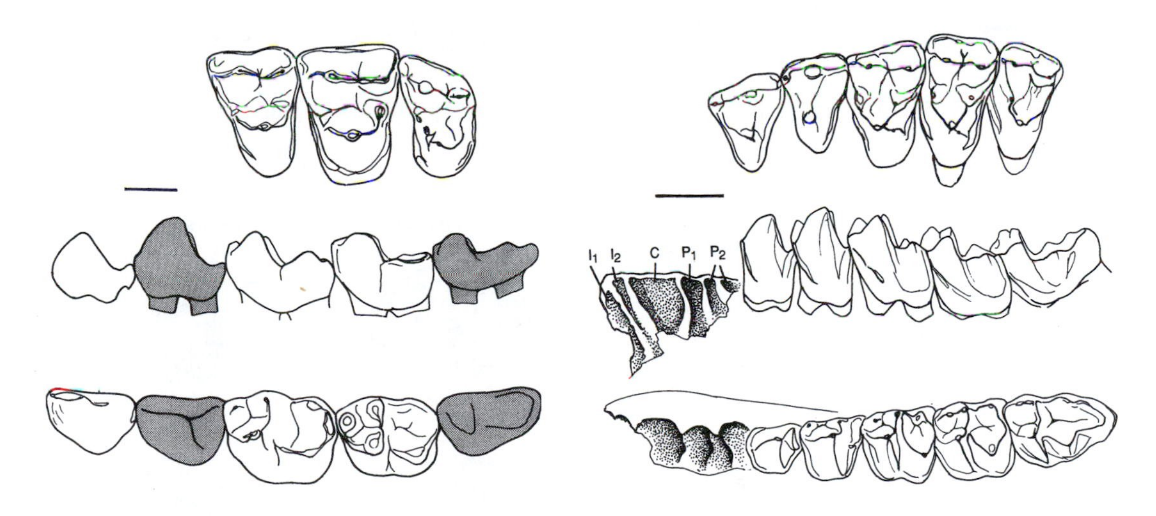

FIGURE 12.3 *Altiatlasius koulchii* (left), from the Paleocene of Morocco, and *Altanius orlovi* (right), from the early Eocene of Mongolia, are generally considered the earliest fossil primates (courtesy of K. D. Rose).

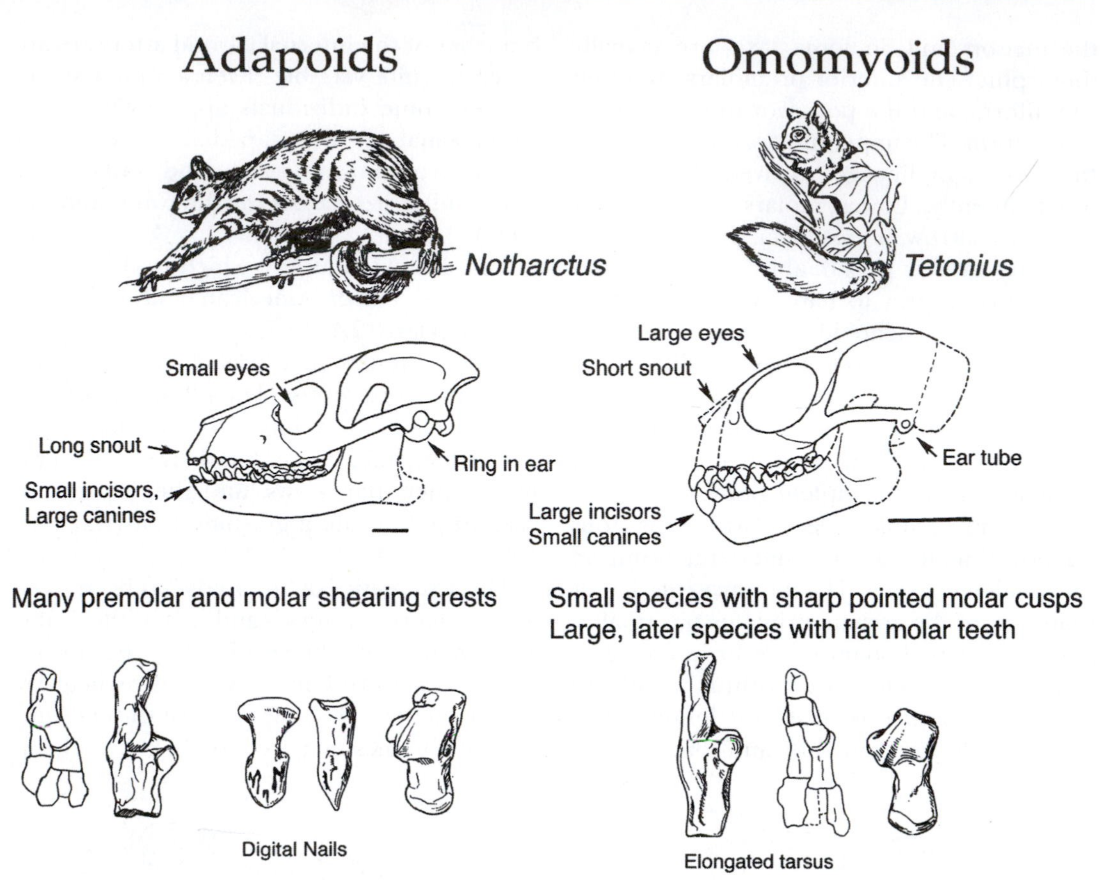

FIGURE 12.4 Comparison of adapoids and omomyoids (courtesy of K. D. Rose).

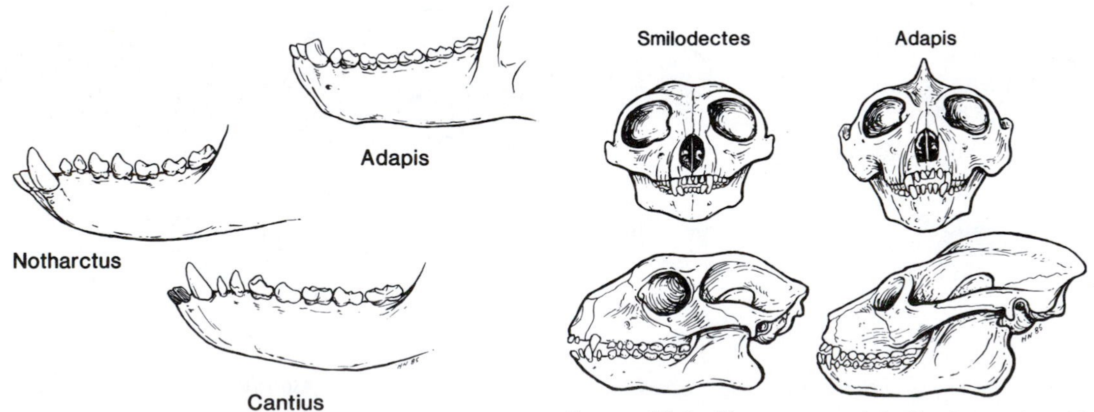

FIGURE 12.5 Mandibles of several adapoid primates.

FIGURE 12.6 Reconstructed skulls of two adapoid primates (approximately one-half natural size).

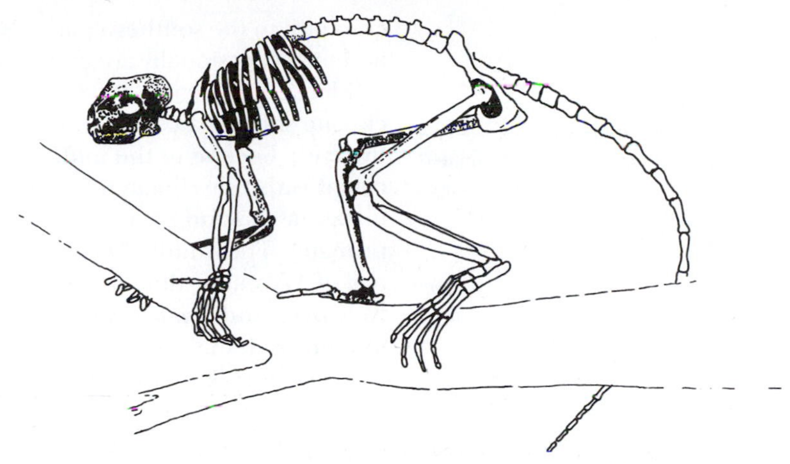

FIGURE 12.7 Reconstructed skeleton of *Smilodectes gracilis* (redrawn from Simons, 1964).

pletely, distinct biogeographically. The notharctines are a predominantly early-middle Eocene group from North America. Cercamoniines (also called protoadapines) are a predominantly early-middle Eocene European radiation closely related to notharctines, with a few genera from North America, Africa, and Asia. The adapines are a predominantly European group that appeared abruptly in the middle Eocene and became extinct by the early Oligocene. The fourth subfamily of adapoids, the sivaladapines, is from the late Miocene of Asia. In addition, there are various African and Asian fossil primates with uncertain adapoid affinities.

Notharctines

The notharctines (Table 12.2) were among the most common mammals in the early and middle Eocene faunas of western North America; there are numerous time-successive species, but they have limited diversity in size or adaptations (Fig. 12.8). There were never more than two or three synchronic species and only a total of six genera from the early and middle Eocene. The earliest notharctine, and one of the earliest adapids, is ***Cantius,*** with numerous species from North America (Gingerich, 1986) and two from Europe. *Cantius* was a small to medium-size prosimian ranging from about 1 kg in the earliest and smallest species to over 3 kg in the latest.

Cantius has a dental formula of 2.1.4.3 (Fig. 12.5). The lower molars have a simple trigonid with three cusps and a broad-basined talonid; the upper molars are simple tritubercular teeth in the early species, but later species (in North America) developed a hypocone from the postprotocingulum (or nannopithex fold). All species have four premolars, prominent canines, and two small vertical incisors. The mandibular symphysis is unfused in this early genus. *Cantius* was probably largely frugivorous.

The partial skulls and few skeletal remains of *Cantius* resemble those of the better-known, later genera, *Notharctus* and *Smilodectes,* in most aspects. They indicate a diurnal species that moved primarily by arboreal quadrupedal running and leaping (Rose and Walker, 1985). Some species of *Cantius* were sexually

TABLE 12.2
Superfamily Adapoidea
Family Notharctidae
Subfamily NOTHARCTINAE

Species	Estimated Mass (g)
Cantius (early Eocene, North America, Europe)	
C. torresi	1100
C. ralstoni	1300
C. mckennai	1600
C. trigonodus	2000
C. abditus	3000
C. angulatus	
C. frugivorus	2800
C. venticolis	3000
C. eppsi	1000
C. savagei	1760
Copelemur (early Eocene, North America)	
C. australotutus	
C. tutus	3600
C. feretutus	2000
C. consortutus	1600
C. praetutus	1300
Notharctus (middle Eocene, North America)	
N. robinsoni	4700
N. tenebrosus	4200
N. pugnax	5500
N. robustior	6900
Smilodectes (early–middle Eocene, North America)	
S. gingerichi	
S. mcgrewi	3000
S. gracilis	2100
Pelycodus (early Eocene, North America)	
P. jarrovii	4500
P. danielsae	6300
Hesperolemur (middle Eocene, North America)	
H. actius	4000

dimorphic in canine size, similar to many extant higher primates (Gingerich, 1995).

Cantius is the most common early Eocene taxon in the northern parts of the American West (Wyoming). In contrast, two related genera, ***Pelycodus*** (with large, broad teeth) and ***Copelemur*** (with small, narrow teeth), are more common in the southern parts (New Mexico), and only occasionally are they found in northern localities (Beard, 1988). North American climate showed a considerable warming from the early Eocene to the middle Eocene. Associated with this climate change was a change in faunas, including the appearance of new primates. The middle Eocene Bridgerean faunas of Wyoming document two new genera, ***Notharctus*** and ***Smilodectes,*** both of which show numerous dental specializations for folivory.

Notharctus is larger (up to 7 kg) than *Cantius* and has larger hypocones and mesostyles on the upper molars, reduced paraconids on the lower molars, and a fused mandibular symphysis. Because the transition from *Cantius* to *Notharctus* was a gradual and essentially continuous one, this last feature is arbitrarily used to delineate the two genera (Fig. 12.8). The cheek teeth of *Notharctus* have well-developed shearing crests, and the genus was certainly folivorous (Covert, 1986, 1995). *Notharctus* also had sexually dimorphic canines (Krishtalka *et al.*, 1990; Alexander, 1994).

Notharctus is similar to *Lemur* in both overall cranial proportions and in details of its basicranial anatomy (Fig. 12.4). The Eocene genus is more robustly built and has a smaller braincase and more pronounced sagittal and nuchal crests. There is a moderately long snout with a large premaxillary bone. The lacrimal bone is at the edge of the orbit rather than anterior to it, as in extant lemurs. The auditory region has a free tympanic ring lying within the bulla. The canal for the stapedial artery was generally smaller than the canal for the promontory artery. Although the size and position of these canals are widely used to reconstruct patterns of cranial circulation in fossil mammals, it is important to keep in mind that there is not necessarily a one-to-one correspondence between bony canals and arteries in living primates (see Conroy and Wible, 1978).

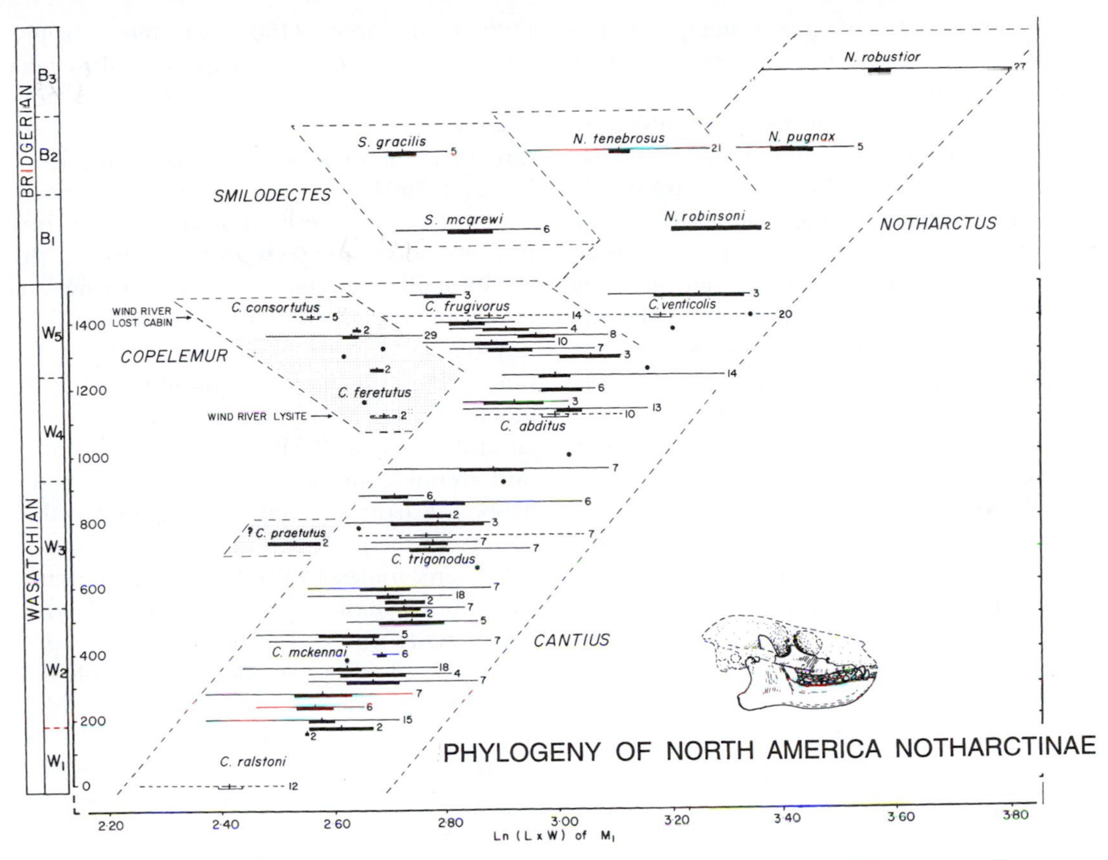

FIGURE 12.8 A phylogeny of notharctines from northern Wyoming (from Gingerich, 1984).

Several virtually complete skeletons are known for *Notharctus*. Gregory (1920) found that the Eocene genus is most similar (but not identical) in skeletal proportions and details of limb architecture to the extant genera *Lemur, Varecia, Lepilemur,* and *Propithecus* but has relatively more robust bones. *Notharctus* has extremely long hindlimbs (intermembral index = 60), a long, flexible trunk, and a long tail. The ilium is sickle-shaped, as in extant lemurs, and the ischium is rather long. The pollex and hallux are large and opposable; the digits are long and tipped with nails. In most but not all details of muscle attachment that could be reconstructed, it is similar to living prosimian leapers. The calcaneus is rather short, as in *Varecia*. There is little doubt that *Notharctus* was an adept leaper and quadrupedal runner, but it probably was not so restricted to vertical supports as are living indriids (Dagosto, 1993; Alexander, 1994).

Smilodectes (Figs. 12.6, 12.7), a smaller (2 kg) middle Eocene contemporary of *Notharctus*, was characterized by narrower teeth, a shorter snout, and a more rounded frontal bone. Like *Notharctus*, it was diurnal and folivorous. Its external brain morphology is known from several endocasts. Compared with other Eocene mammals, *Smilodectes* had an expanded visual cortex and reduced olfactory bulbs; its brain was larger than that of most contemporaneous mammals but smaller than that of extant prosimians (Radinsky, 1975, 1977). The phylogenetic origin of *Smilodectes* is the subject of some debate. Traditionally, it has been accepted that the two middle Eocene taxa shared a common ancestry (e.g., Gingerich, 1984; Covert, 1990); however, Beard (1988) has argued that whereas *Notharctus* is derived from *Cantius* in the north, *Smilodectes* is a descendant of the more southern *Copelemur* lineage that appeared in the northern part of the continent in conjunction with the climatic warming. Both *Notharctus* and *Smilodectes* apparently became extinct early in the middle Eocene (Fig. 12.8).

Hesperolemur is a newly described notharctine from the slightly younger Uintan Land Mammal Age of southern California (Gunnell, 1995). It is unusual in that the tympanic ring is partly fused to the inside of the auditory bulla.

Cercamoniines (Protoadapines)

Until most recently, the adapoid primates of Europe (Fig. 12.9) were placed in a single subfamily, the Adapinae. However, this geographic grouping includes two very different phyletic radiations that are more appropriately placed in separate subfamilies and families (e.g., Franzen, 1987; Thalmann *et al.*, 1989).

Cercamoniines are an Old World radiation of early adapoids closely related to notharctines and placed in a common family, the Notharctidae. Although cercamoniines are best known from western Europe, related taxa have been described from North America and increasingly from Asia and Africa (Table 12.3). Cercamoniines had a much more diverse evolutionary radiation than the relatively uniform notharctines (Fig. 12.9). They ranged in size from tiny, presumably insectivorous species the size of a pygmy marmoset (100 g) to larger (1500+ g), more frugivorous or partly folivorous species. Most taxa are known primarily from their dentition. Cranial and associated skeletal remains are rare (Thalmann *et al.*, 1989; Dagosto, 1993; Franzen, 1994).

Donrussellia is the earliest and most primitive cercamoniine and is probably close to the origin of all adapoids as well as all prosimians (Godinot, 1992; Bown and Rose, 1991). This tiny genus has a full dental formula of 2.1.4.3, simple tritubercular upper molars, and lower molars with a simple trigonid and a broad talonid. In contrast with the North American notharctines, the cercamoniines developed a hypocone from the lingual cingulum rather than from the protocone.

Cercamoniines evolved numerous dental adaptations, indicative of considerable dietary diversity within the subfamily. The small (110 g) *Anchomomys gaillardi,* for example, has extremely simple upper molars, not unlike those of marmosets (Gingerich, 1977a). Judging from its sharp molar cusps and tiny size, this species was almost certainly insectivorous. The larger *Pronycticebus gaudryi* has relatively simple molar teeth with sharp cusps, a robust, tusklike upper canine, and a long row of sharp premolars, suggesting a carnivorous diet (Szalay and Delson, 1979). *Periconodon* has molars with broader, more bulbous cusps, suggestive of fruit eating.

The most complete skeletal remains of a cercamoniine are those of *Pronycticebus neglectus* from Germany (Thalmann *et al.,* 1989). This species is known from a nearly complete skeleton that has been crushed flat. The bones resemble those of notharctines rather than those of adapines in many details. In its locomotor behavior it was probably a quadrupedal, leaping, climbing form; there is less evidence of leaping from vertical supports than with the notharctines and evidence of less extensive slow quadrupedal adaptations than with the adapines *Adapis* and *Leptadapis* (see later discussion). A particularly unusual cercamoniine fossil is a half-skeleton of a small primate from the oil shales of Messel, Germany. Because only the lower half of the skeleton has been found, it cannot be confidently assigned to any genus or species. The hindlimb suggests a leaper. Like living strepsirhines, this species has a "grooming claw" on the second digit of its foot. It also has a very large baculum (penis bone) for an animal of its size.

Mahgarita stevensi, from the late Eocene of Texas, is the only North American cercamoniine and occurs after the apparent extinction of the notharctines at the end of the middle Eocene. *Mahgarita* has relatively small premolars, and, as with the European cercamoniines, the hypocone on the upper molars is derived from the lingual cingulum. The mandibular symphysis is fused. The strong development of crests on the molar teeth, as well as its moderate size (1200 g), suggests that it was probably folivorous. Although *Mahgarita* frequently has been mentioned as a possible anthropoid ancestor because of its deep snout, fused mandibular symphysis, and enlarged canal for the promontory artery (Rasmussen, 1990; but see Ross, 1994), it shows no evidence of postorbital closure.

Many new adapoids similar to the European cercamoniines have been described from the Eocene of Africa in recent years (Rasmussen, 1994; Simons, 1997). *Azibius trerki* (Sudre,

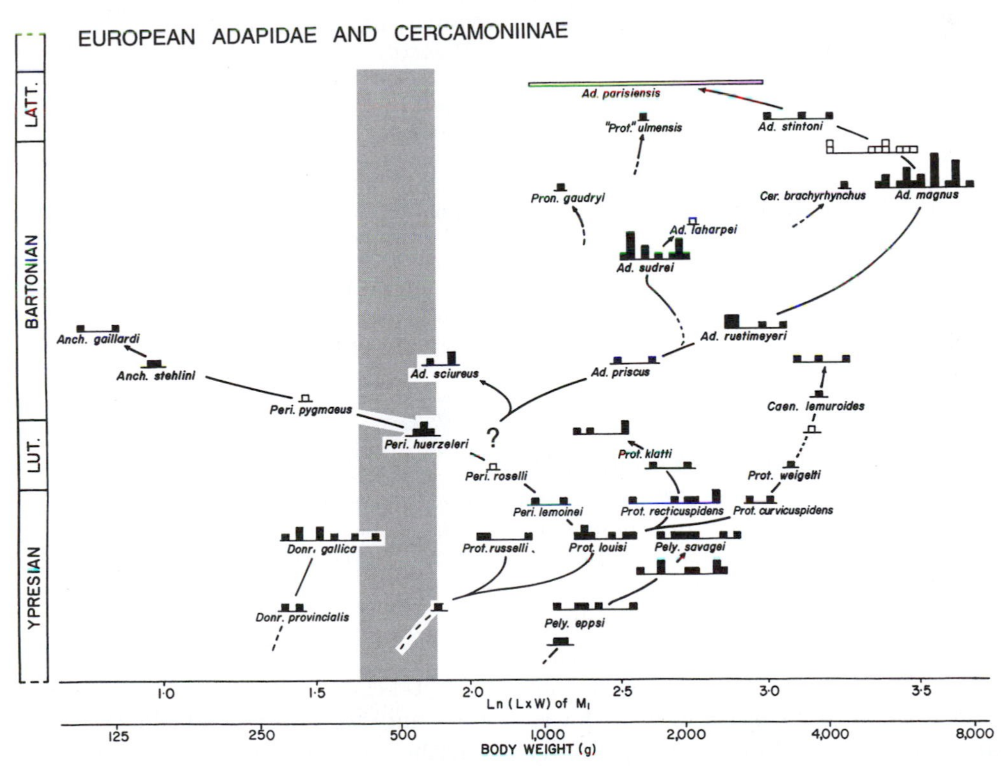

FIGURE 12.9 A phylogeny of European adapoid primates; cross-hatching indicates Kay's threshold (from Gingerich, 1977).

TABLE 12.3
Superfamily Adapoidea
Family Notharctidae
Subfamily CERCAMONIINAE

Species	Estimated Mass (g)
Donrussellia (early Eocene, Europe)	
D. gallica	210
D. provincialis	270
D. magna	730
Protoadapis (early–middle Eocene, Europe)	
P. curvicuspidens	2500
P. filholi	870
P. lemoinei	—
P. recticuspidens	1600
P. russelli	700
P. louisi	1100
P. weigelti	3000
P. ulmensis	1400
P. ignoratus	1860
P. muechelnensis	1800
Europolemur (middle to late Eocene, Europe)	
E. klatti	1700
E. koenigswaldi	—
E. dunaefi	1360
E. collinsonae	1500
Periconodon (middle Eocene, Europe)	
P. helveticus	250
P. huerzleri	570
P. roselli	650
P. lemoinei	—
P. jaegeri	920
Caenopithecus (late Eocene, Europe)	
C. lemuroides	3500
Pronycticebus (middle to late Eocene, Europe)	
P. gaudryi	1100
P. neglectus	825
Cercamonius (late Eocene, Europe)	
C. brachyrhynchus	4000
Anchomomys (middle to late Eocene, Europe, Africa)	
A. gaillardi	110
A. pygmaea	250
A. crocheti	160
A. milleri	100
Huerzeleria (late Eocene, Europe)	
H. quercyi	190
Buxella (middle Eocene, Europe)	
B. prisca	550
B. magna	620
Agerinia (middle Eocene, Europe, Asia)	
A. roselli	—
A. sp.	—
Panobius (?early–middle Eocene, Asia)	
P. afridi	130
Mahgarita (late Eocene, North America)	
M. stevensi	700
Djebelemur (early Eocene, Africa)	
D. martinezi	100
Aframonius (late Eocene, Africa)	
A. dieides	1600
Omanodon (early Oligocene, Arabia)	
O. minor	100
Shizarodon (early Oligocene, Arabia)	
S. dhofarensis	200
Wadilemur (late Eocene, Africa)	
W. elegans	

1975), a tiny mammal from the Eocene of Algeria, is known from a single jaw with three teeth. The present material is insufficient either to deny or to confirm adapoid affinities (Table 12.4). *Djebelemur* (Fig. 12.10) is a tiny (100 g) adapoid from the early Eocene site of Chambi in Tunisia that was described by Hartenberger and Marandat (1992) as very similar to *Protoadapis* or *Cercamonius* in its lower dentition. *Aframonius* (Table 12.3; Fig. 12.11) is a large cercamoniine from the late Eocene of the Fayum, Egypt (Simons *et al.,* 1995).

Most recently, two tiny cercamoniines, *Omanodon* and *Shizarodon* (Table 12.3), have been described from the early Oligocene of Oman in association with early anthropoids (Gheerbrant *et al.,* 1993). Unfortunately, they are

TABLE 12.4
Superfamily Adapoidea
Family *Incertae sedis*

Species	Estimated Mass (g)
Azibius (Eocene, Africa)	
A. trerki	120
Hoanghonius (Eocene, Asia)	
H. stehlini	700
Lushius (late Eocene, Asia)	
L. qinlinensis	2900
Rencunius (middle Eocene, Asia)	
R. zhoui	700
Wailekia (Eocene, Asia)	
W. orientale	2000

known only from limited dental remains. A similar tiny cercamoniine, *Wadilemur,* and a new species of *Anchomomys* (Table 12.3) have been described from the late Eocene of Egypt (Simons, 1997). The tiny Fayum primates are strikingly similar in molar morphology to the living mouse lemur, but there is no evidence of a tooth comb to suggest a special relationship with living strepsirhines.

There are several primates from the Eocene of Asia that have been allied with cercamoniines (Table 12.3). *Panobius afridi* is a tiny primate from the late early Eocene of Pakistan that is very similar to *Donrussellia* (Russell and Gingerich, 1987). The same deposits also include specimens referred to the European genus *Agerina. Lushius, Hoanghonius,* and *Rencunius* are slightly younger Chinese adapoids that have less clear-cut affinities (Fig. 12.4). Most recently, Beard *et al.* (1994) have described a cercamoniine similar to *Europolemur* from the Eocene Shanghuang deposits of southern China. More enigmatic are *Amphipithecus* and *Pondaungia,* from the late Eocene of Burma, which are often discussed as possible early anthropoids but which also show many similarities to notharctines or cercamoniines (Ciochon and Holroyd, 1994). *Wailekia,* originally described by Ducrocq *et al.,* (1955) as an early anthropoid is probably a cercamoniine.

Adapines

The genus *Adapis* was the first fossil nonhuman primate named. However, it was so un-

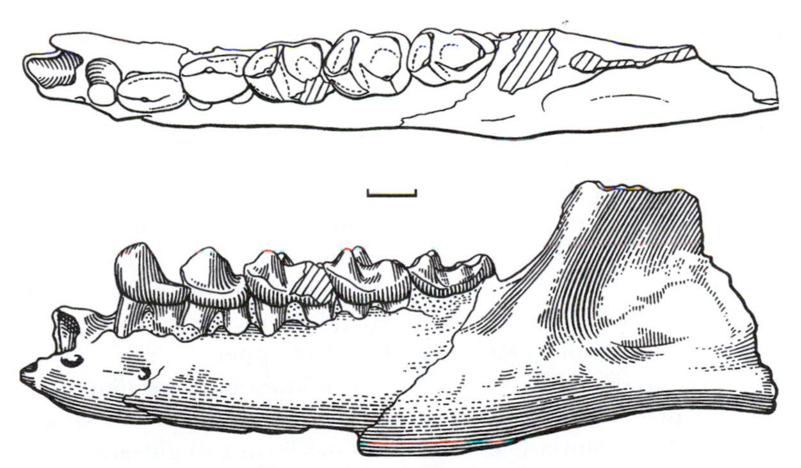

FIGURE 12.10 *Djebelemur martinezi,* a tiny cercamoniine from the Eocene of Tunisia. (Courtesy of J. L. Hartenberger)

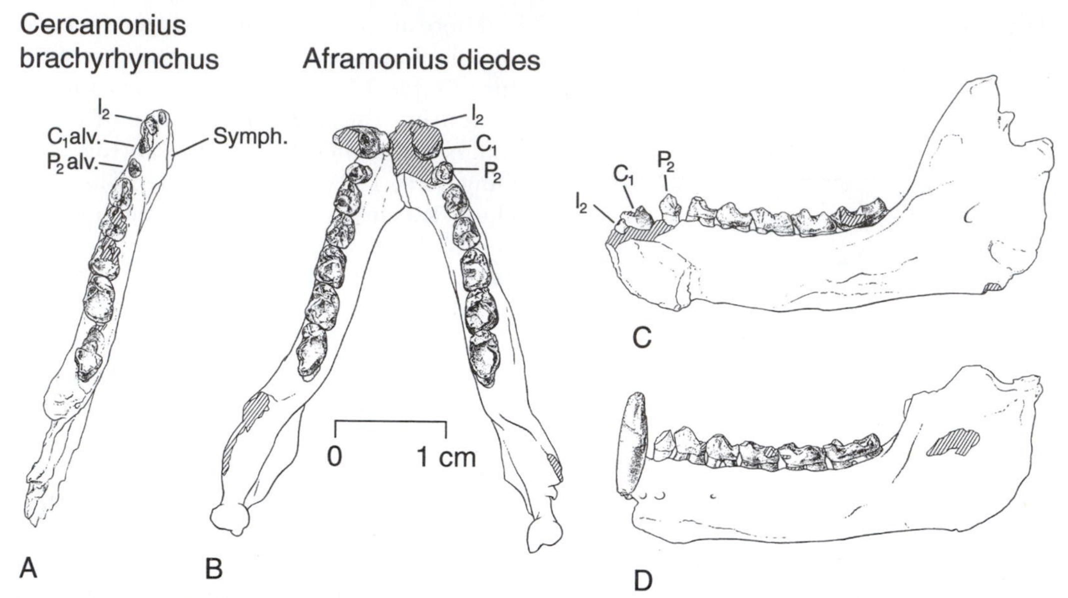

FIGURE 12.11 *Aframonius dieides* (B and D), a cercamoniine from the late Eocene of Egypt, compared with *Cercamonius brachyrhynchus* (A and C) from the Eocene of Europe.

usual that its describer, Cuvier, did not recognize it as a primate. Although *Adapis* and its close relatives *Leptadapis, Microadapis, Cryptadapis,* and possibly *Caenopithecus* have traditionally been grouped with the cercamoniines in a single subfamily, it is now generally recognized that they cannot be derived from any of the earlier cercamoniines and that the cercamoniines are actually more closely related to notharctines. As a result, the subfamily Adapinae is now restricted to a few genera (Table 12.5), all characterized by molarized last premolars, well-developed shearing crests, and many postcranial peculiarities (Franzen, 1987; Thalmann *et al.,* 1989; Dagosto, 1993). They are extremely common mammals in the late Eocene of France; *Adapis* and *Leptadapis* are known from numerous skulls.

The best-known adapine is ***Adapis parisiensis*** (Figs. 12.5, 12.6), a medium-size species from several late Eocene deposits in France. This species was first described by Cuvier in 1822, well before the discovery of any other fossil primates, but its primate affinities were not recognized until some fifty years later. *Adapis parisiensis,* the latest adapine in Europe, disappeared during the major European faunal turnover known as the Grande Coupure, which coincided with a major drop in temperature near the Eocene–Oligocene boundary.

Adapis parisiensis has a full primate dental formula of 2.1.4.3 (Fig. 12.5). Like most adapoids and living lemurs, it has upper central incisors that are small, relatively broad, and spatulate with a gap between their bases, presumably for an organ of Jacobson. The upper lateral incisors are smaller and positioned behind the upper centrals.

The lower anterior dentition of *A. parisiensis* is unusual in that the lower incisors and canines form a single cutting edge. Gingerich has suggested that this morphology represents incipient development of a tooth comb

TABLE 12.5
Superfamily Adapoidea
Family Adapidae
Subfamily ADAPINAE

Species	Estimated Mass (g)
Adapis (late Eocene–early Oligocene, Europe)	
A. betillei	—
A. parisiensis	1300
A. sudrei	1400
A. laharpei	1700
Cryptadapis (late Eocene, Europe)	
C. tertius	2500
Microadapis (late Eocene, Europe)	
M. sciureus	600
Leptadapis (late Eocene–early Oligocene, Europe)	
L. magnus	4000
L. assolicus	3000
L. capellae	—
L. priscus	1300
L. ruetimeyeri	2500
Adapoides (Eocene, Asia)	
A. troglodytes	500

as seen in extant strepsirhines. However, microwear analysis of the occlusal surfaces showed no evidence of the characteristic hair scratches found on other mammalian tooth combs (Asher, pers. comm.). *Adapis* has long, narrow molars and premolars with well-developed shearing crests. They are strikingly similar to the molars of *Hapalemur,* suggesting a folivorous diet for *Adapis.*

Adapis has a very low, broad skull (Fig. 12.6) with flaring zygomatic arches and a small braincase. Prominent sagittal and nuchal crests are found on the larger individuals, suggesting they are probably males. The orbits are relatively small, suggesting diurnal habits, and are oriented slightly upward rather than directly forward. The snout is moderately short. From the robust zygomatic arches and the extremely large temporal fossa, it is clear that *Adapis* had extremely large chewing muscles, concordant with the extensive shearing abilities seen it its dentition.

The auditory region of the *Adapis* skull has an inflated bulla with a free tympanic, as in extant strepsirhines. There is always a canal for the stapedial artery and a groove for the promontory artery, but the relative sizes of these conduits vary from one specimen to another. The brain is relatively small compared with that of extant prosimians and has a large olfactory bulb.

There are several relatively complete limb bones of *A. parisiensis.* Initial analyses of these bones suggested that *Adapis* was most similar to the living lorises *Nycticebus* and *Perodicticus*—slow arboreal quadrupeds (Fig. 12.12; Dagosto, 1983; Spoor, 1998). More recent studies indicate a greater diversity of species and locomotor adaptations among the fossils attributed to *Adapis* (Bacon and Godinot, 1998). In addition, the joint between the ulna and the wrist in *Adapis* shows features linking it with extant lemurs and lorises (Beard *et al.,* 1988).

Leptadapis magnus was a large (4+ kg), earlier relative of *Adapis* that is often placed in the same genus. Like *Adapis,* it was probably a diurnal folivore that moved by quadrupedal climbing. *Leptadapis* shows evidence of sexual dimorphism in both cranial size and canine size, and it seems likely that this large Eocene adapine lived in polygynous social groups.

The phyletic and geographic origins of the adapines are unknown; they cannot be readily derived from any of the earlier European cercamoniines or the notharctines of North America. Their appearance in the late Eocene of Europe was just as abrupt as their extinction shortly thereafter, suggesting an immigration from some other continental region. The recent discovery of an adapine, ***Adapoides troglodytes,*** from the Eocene of southern China suggests an Asian origin for the group (Beard *et al.,* 1994).

FIGURE 12.12 Scene from the late Eocene of the Paris basin. Above, the diurnal *Adapis parisiensis* feed on leaves. Below are several nocturnal microchoerines: the tiny *Pseudoloris* attempts to catch an insect while *Necrolemur* (left) and *Microchoerus* (right) cling to branches.

TABLE 12.6
Superfamily Adapoidea
Family Sivaladapidac
Subfamily SIVALADAPINAE

Species	Estimated Mass (g)
Indraloris (late Miocene, Asia)	
I. himalayensis	2500
Sivaladapis (late Miocene, Asia)	
S. nagrii	2700
S. palaeindicus	4450
Sinoadapis (late Miocene, Asia)	
S. carnosus	—

Sivaladapines

Well after the notharctines, cercamoniines, and adapines disappeared from North America and Europe, there were a number of adapoid primates thriving alongside fossil apes in the late Miocene of India, Pakistan, and China (Table 12.6). The best known of these, ***Sivaladapis nagrii***, from the late Miocene of India, was fairly large (3 kg) with a dental formula of 2.1.3.3 (Fig. 12.13). The sharp crests on its molars and premolars suggest a folivorous diet. Unlike the latest members of either the European adapines or the North American notharctines, *Sivaladapis* has simple upper molars with no hypocone. There is a similar genus, *Sinoadapis,* from the latest Miocene site of Lufeng in China known from dental remains (Wu and Pan, 1985). Most recently, Beard (1998) has argued that sivaladapines are derived from the Eocene *Hoanghonius.*

ARE ADAPOIDS STREPSIRHINES?

Since adapoids were first identified as primates, virtually all authors have noted their many anatomical similarities to living strepsirhines and particularly to lemurs. Adapoids are lemurlike in their cheek teeth, in the overall configuration of their skull with its simple postorbital bar and moderately long snout, and in the morphology of the nasal region. The auditory region is also lemurlike, with an inflated bulla and a free ectotympanic ring. The carotid circulation is more similar to that of lemurs than to that of either haplorhines or lorises, in that most individuals have a stapedial canal of moderate size. However, in virtually all of these features, adapoids and strepsirhines retain the primitive primate condition found in many other mammals rather than sharing unique specializations. Furthermore, adapoids lack a tooth comb, the derived feature that most clearly distinguishes living strepsirhines from other primates, and they also seem to have retained more primitive hands and feet than many Malagasy species.

Adapoids and living strepsirhines share only a few anatomical features that may be unique specializations linking the two and also precluding ancestral relations to other primates. One is the grooming claw of the second toe, which is present in the Messel cercamoniine and in all extant strepsirhines. Eocene adapoids and strepsirhines also share two unusual features of the ankle, a flaring fibular surface on the talus and the arrangement of the cuneiform facets of the navicular (Dagosto, 1988). More specifically, adapines (but not notharctines) are linked with extant strepsirhines by a unique articulation between the ulna and the carpus (Beard *et al.,* 1988). It has also been argued that adapoids share with extant strepsirhines reduced upper incisors with a large median gap for Jacobson's organ (Rosenberger *et al.,* 1985).

The overall anatomical similarity between adapoids and strepsirhines clearly demonstrates that living strepsirhines have retained many aspects of an adapoid-like morphology for nearly 60 million years, but at present there is very little evidence demonstrating a

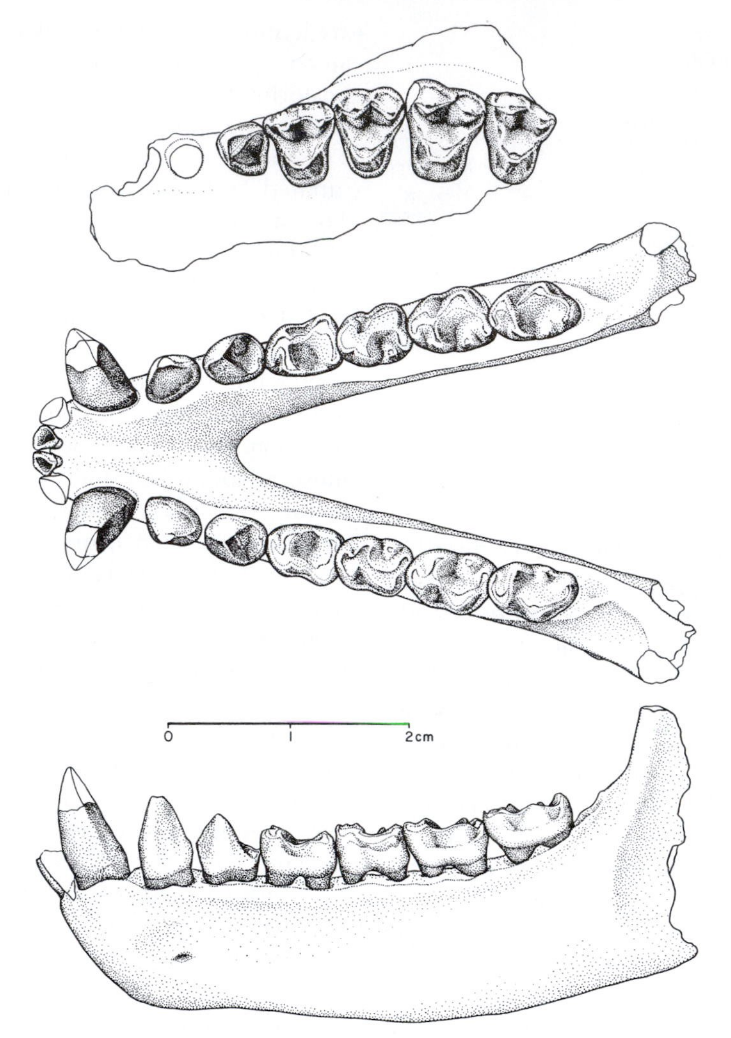

FIGURE 12.13 Upper and lower dentition of *Sivaladapis nagrii* (courtesy of P. Gingerich).

unique phyletic relationship between tooth-combed strepsirhines and the Eocene adapoids. The many new discoveries from Africa and Asia may help clarify these relationships. In particular, several new tiny primates from North Africa and the Arabian peninsula are strikingly similar to both cercamoniines and living cheirogaleids, suggesting an African adapoid origin for strepsirhines (e.g., Gheerbrant *et al.*, 1993; Simons, 1997).

In addition to their traditional link with strepsirhines, adapoids (especially cercamoniines) have frequently been proposed as the ancestors of higher primates (Rasmussen, 1994; Gingerich *et al.*, 1994; Simons, 1997; but see Kay *et al.*, 1997). This suggestion has been based on their anthropoid-like anterior dentition, fused mandibular symphysis, and common size range as well as on dental similarities between early anthropoids and European and African cercamoniines. However, we must defer consideration of the relationship between adapoids and anthropoids until the next chapter.

Fossil Lorises and Galagos

In addition to the recently extinct Malagasy species (see Chapter 4), one group of fossil prosimians that can be linked clearly to living strepsirhines are the fossil lorisoids from the Miocene, Pliocene, and Pleistocene of Africa and possibly the Eocene and Oligocene. There are also fossil lorises from the Miocene of Asia (Table 12.7). The earliest possible record of this group is a single upper molar from the late Eocene/early Oligocene of Egypt that Simons (Simons *et al.*, 1987) has identified as that of a loris. More recently, the late Eocene *Plesiopithecus* (Simons, 1992; Simons and Rasmussen, 1994) has been identified as an aberrant lorisoid based on its molar morphology. However, it lacks a tooth comb; instead it has a large procumbent anterior tooth. Clarification of the affinities of these fossils must obviously await additional material (Rasmussen and Nekaris, 1998).

More clearly related to modern galagos and lorises are several genera and species from the early Miocene of Kenya and Uganda. One genus, ***Mioeuoticus*** (Fig. 12.14), seems to be related to the lorises, and two others, ***Komba*** and ***Progalago,*** seem to be closer to living galagos (Walker, 1978; but see McCrossin, 1992).

These Miocene prosimians are very similar to living African genera in their dental and cranial anatomy, and the shape of the incisor roots indicates that they had tooth combs. Although they can generally be identified (not without debate) as lorises or galagos, none can be positively linked to any living genus or species. The dental remains indicate a size range comparable to that of modern lorises and galagos (100–1000g) as well as considerable dietary diversity, including frugivores and faunivores. The skulls and facial fragments indicate large orbits, which are suggestive of nocturnal habits. The galagos have elongated limbs, but their tarsals are not as elongated as those of living galagos. They are more similar to the tarsals of cheirogaleids (Gebo, 1989). Thus far no postcranial remains of the African Miocene lorisoids indicate slow climbing habits.

Younger fossil galagos, from 2–4 million years ago in Ethiopia and Tanzania, are similar to the living *Galago* and *Otolemur.*

The earliest fossil record of Asian lorises comes from the middle to late Miocene of

TABLE 12.7
Infraorder Lemuriformes
Superfamily LORISOIDEA

Species	Estimated Mass (g)
Family GALAGIDAE	
Progalago (early Miocene, Africa)	
P. dorae	1200
P. songhorensis	800
Komba (early-m. Miocene, Africa)	
K. robusta	300
K. minor	125
K. winamensis	1000
Galago (Pliocene-Recent, Africa)	
G. howelli	700
G. sadimanensis	200
Family LORISIDAE	
Mioeuoticus (early Miocene, Africa)	
M. bishopi	?300
M. spp.	—
Nycticeboides (late Miocene, Asia)	
N. simpsoni	500
Family PLESIOPITHECIDAE	
Plesiopithecus (late Eocene, Africa)	
P. teras	

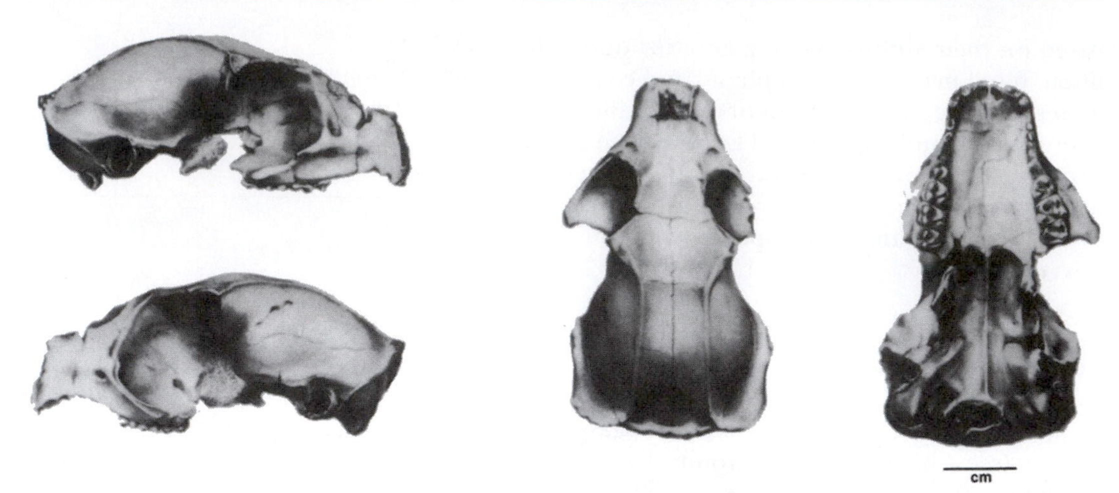

FIGURE 12.14 The skull of a fossil loris, *Mioeuoticus,* from the Miocene of eastern Africa (from LeGros Clark, 1950).

Pakistan. Numerous isolated teeth from 7–14 million years ago and one relatively complete skeleton have been attributed to a single species, ***Nycticeboides simpsoni.*** This species seems closely related to the living slow loris, *Nycticebus,* in both cranial and postcranial anatomy (MacPhee and Jacobs, 1986).

Omomyoids

Like the lemurlike adapoids, the tarsierlike omomyoids first appeared in the earliest Eocene of North America, Europe, and possibly Asia (Fig. 12.1; Gingerich *et al.,* 1992). Omomyoids, like adapoids, had a very different evolutionary history in North America and in Europe, where their fossil record has been well documented for over a century. They are just becoming known from Asia and possibly Africa. In North America, they were very diverse taxonomically throughout the Eocene. In Europe, omomyoids were less diverse, with only a single, poorly known genus, *Teilhardina,* from the early Eocene, and four genera from the middle and late Eocene. There are three widely recognized subfamilies of omomyoids: Anaptomorphinae and Omomyinae, both predominantly North American, and the European Microchoerinae. In recent years the phyletic distinctiveness of the two North American subfamilies has become increasingly blurred, whereas the European microchoerines remain more clearly a distinct radiation (Rose and Bown, 1991; Rose *et al.,* 1994; Williams, 1994). Thus, it seems best to recognize two separate families: Omomyidae, for the diverse, predominantly North American anaptomorphines and omomyines (each divided into many tribes), and Microchoeridae, for the smaller, but distinct, radiation in the European middle-late Eocene.

The most primitive omomyoids, *Teilhardina* and *Steinius* (and *Altanius*), are very similar to early adapoids such as *Donrussellia* and *Cantius* in their dental morphology and in retaining a primitive dental formula of 2.1.4.3. (Bown and Rose, 1991). However, all subsequent taxa are characterized by reduction and reorganization of the antemolar dentition, changes that occurred independently in several lineages (Figs. 12.15–12.18). Many omomyoids have a relatively large, procumbent lower central incisor and a smaller lateral one, and the canines are usually small—never large as in adapids or absent as in some plesiadapiforms (Figs. 12.4; 12.17). The premolars are reduced to three or fewer in all but a few species, and these teeth vary considerably in shape among subfamilies and tribes. In some, they are tall and pointed; in others, they are broad and molariform. The lower molars usually have relatively small, low, mesiodistally compressed trigonids and broad-basined talonids. The upper molars are usually broad. Many early species have a prominent postprotocingulum (nannopithex fold) joining the protocone distally, and later species developed a hypocone from the lingual cingulum. The mandibular symphysis of omomyoids is always unfused.

The skulls of most omomyoids resemble those of extant tarsiers and galagos in their relatively short, narrow snout, posteriorly broadening palate, and large eyes (Figs. 12.4, 12.18). The auditory region of some species has an inflated auditory bulla and a tympanic ring that is fused to the bullar wall and extends laterally to form a bony tube. The internal carotid circulation is known in only a few genera. In *Tetonius, Shoshonius, Necrolemur,* and *Rooneyia,* both the stapedial and promontory canals are present. In the former two they are similar in size, but in the latter two the promontory is larger (Ross, 1994).

There are only a few partial skeletons known for omomyoids (Dagosto, 1993). In at least four genera, the calcaneus is moderately elongated, as in extant cheirogaleids (Fig. 12.4). In both North American omomyids and European microchoerids the distal tibia and fibula either are joined by an extensive fibrous joint or show evidence of some fusion as in extant *Tarsius.* Most known skeletal elements indicate habits of leaping, but not clinging, for these early prosimians, and they show greater overall similarities to the skeletons of cheirogaleids than to those of the extant *Tarsius* (Covert, 1995).

Omomyids

Although North American omomyids have traditionally been placed in two separate subfamilies, anaptomorphines and omomyines, the postulated composition of the subfamilies is very unstable, with many taxa being regularly shuffled from one to another by different authorities (Rose, 1995). The hypothesized distribution of individual genera into distinct tribes is no more stable (e.g., Honey, 1990; Beard *et al.,* 1992; Williams, 1994; Gunnell, 1995). The recent discovery in Asia of omomyids with North American affinities further emphasizes the breadth and complexity of this radiation. Moreover, several tribes of omomyines appear to have separate anaptomorphine ancestors, making that subfamily polyphyletic as normally constituted (Williams, 1994; Rose, 1995).

Anaptomorphines

Anaptomorphines (Table 12.8) are the most primitive omomyoids and are probably ancestral to both omomyines and microchoerids. The earliest and most primitive genus, ***Teilhardina*** (Figs. 12.15, 12.16), from the early Eocene of both Europe and North America, is near the base of the entire radiation. The remaining members of the subfamily (over a dozen genera) are from the early and middle Eocene of North America. Anaptomorphines are very common and speciose in the early Eocene, but are relatively rare in the middle Eocene and later. The early evolution of anaptomorphines in Wyoming is one of the most detailed records of population and species

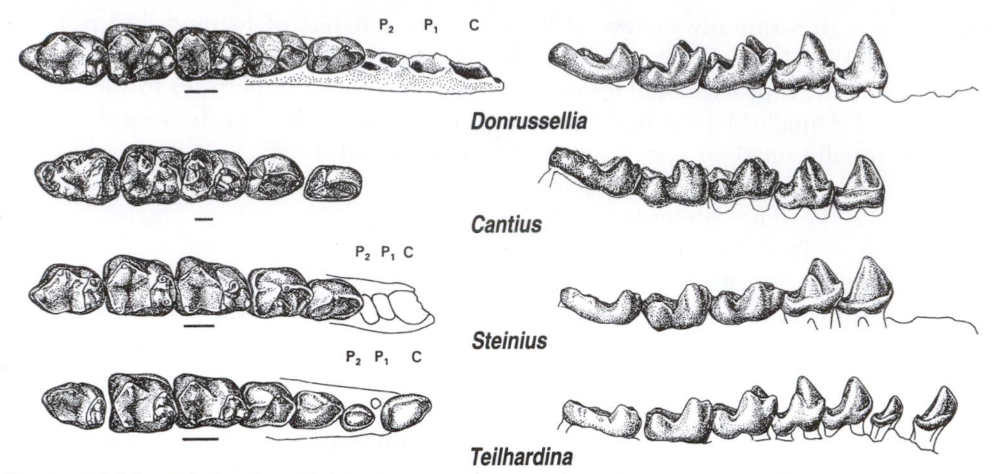

FIGURE 12.15 Occlusal and lingual views of the lower left dentitions of early omomyoids and adapoids. Scale = 1 mm (courtesy of K. D. Rose).

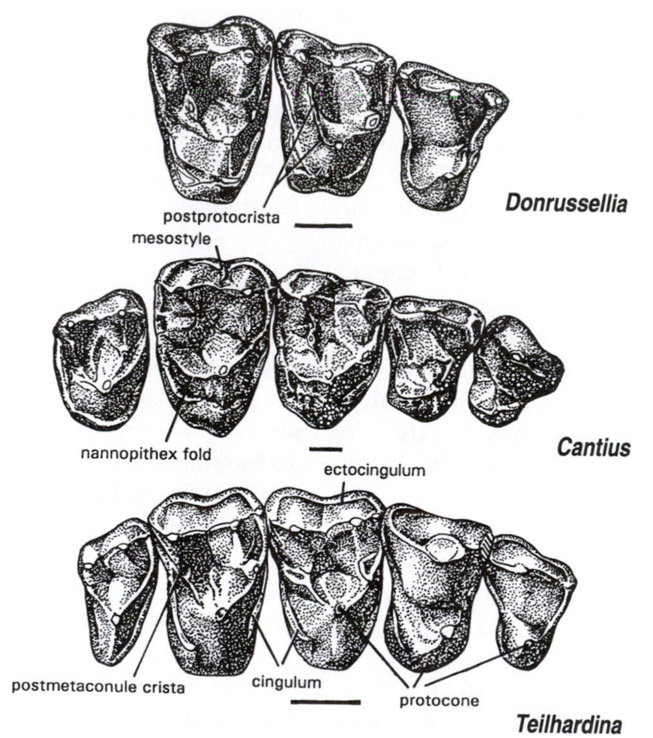

FIGURE 12.16 Upper right dentitions of primitive adapoids and omomyoids. Scale = 1 mm (courtesy of K. D. Rose).

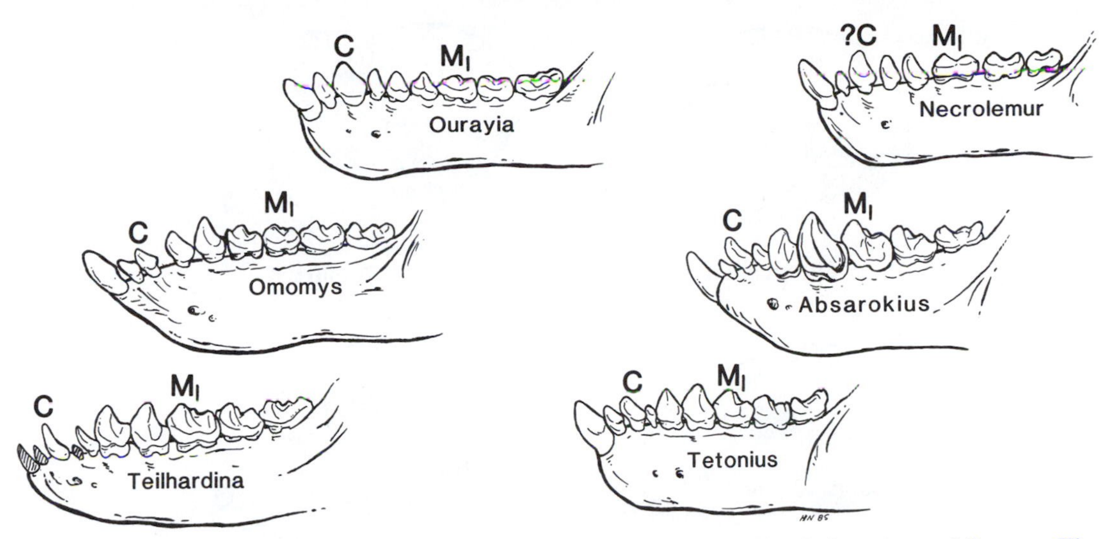

FIGURE 12.17 Mandibles of representative omomyoid primates from North America and Europe. The positions of the canine (C) and the first molar (M_1) are indicated.

changes through time in the entire fossil record (Bown and Rose, 1986). Evolutionary analyses of this radiation have provided remarkable documentation of transitions between paleospecies (Rose and Bown, 1993) and evidence of considerable parallelism in omomyid phylogeny. There is considerable uncertainty regarding the taxonomic and phyletic relationships of and the Washakiine tribe, including *Shosharius* and *Dyseolemur.* Many authorities place them in the omomyines. Here they are considered anaptomorphines (see Williams, 1994).

Despite their systematic diversity, anaptomorphines are all relatively similar in many aspects of their morphology. All are very small, probably ranging from about 50 to 500 g. Later members of the subfamily are usually characterized by a tall, pointed P_4 and a reduced M_3. Many species have only two premolars. Their lower molars have relatively low trigonids with bulbous cusps, and shallow talonids. The lower incisors are small relative to the canines in the earliest taxa (based on alveolus size), but incisor enlargement and canine reduction is common in many later genera. ***Dyseolemur*** (Fig. 12.20), from the late Eocene of California, has anterior teeth that are strikingly similar to those of *Tarsius* (Rasmussen *et al.,* 1995). Studies of the dental morphology of the early anaptomorphines indicate that most were frugivorous, with a few species showing adaptations for processing invertebrates with hard shells (Strait, 1991).

Two anaptomorphines are known from relatively complete cranial remains. The skull of ***Tetonius homunculus*** (Figs. 12.4, 12.18), from the early Eocene of Wyoming, was recovered over 100 years ago. It has a short snout, large eyes, and a relatively globular braincase (Fig. 12.18). Unfortunately, the auditory region is extremely damaged. The teeth of *Tetonius* suggest that it was probably largely insectivorous (Fig. 12.19). Its orbits are similar in size to those of a living cheirogaleid or a small galago, suggesting that it was nocturnal. Because the orbits are relatively smaller than those of *Tarsius,* it seems likely that it had a tapetum

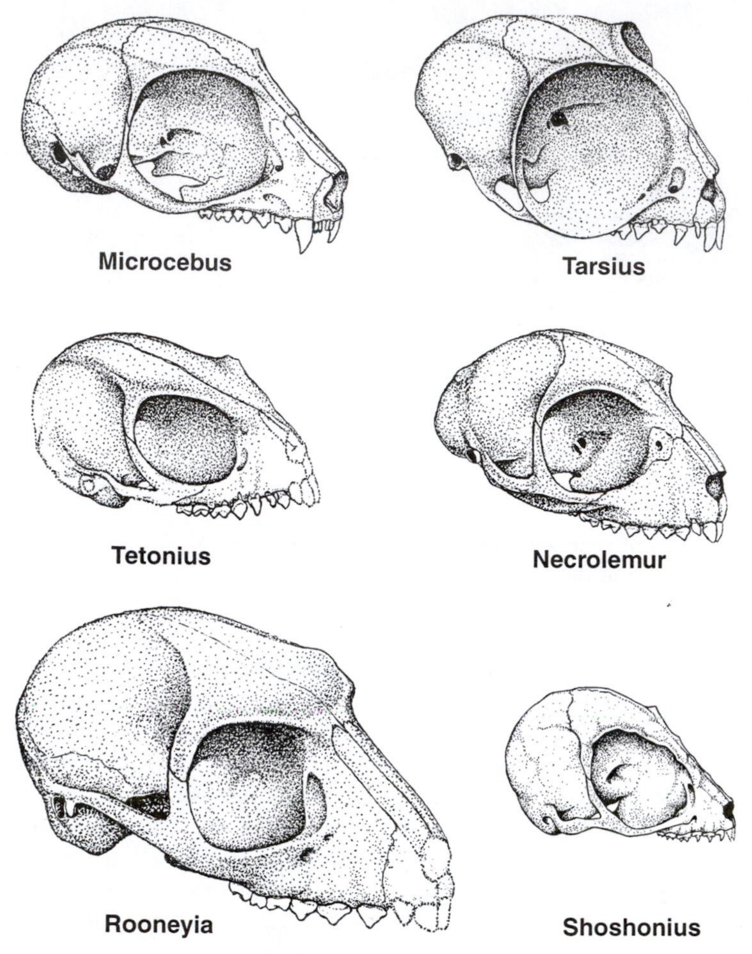

FIGURE 12.18 Skulls of two small nocturnal living primates, *Microcebus murinus* and *Tarsius syrichta,* compared with reconstructed skulls of several omomyoids, *Tetonius homunculus, Necrolemur antiquus, Rooneyia viejensis,* and *Shoshonius cooperi.* Note that *Tarsius* and *Shoshonius* have relatively larger eyes than the other primates. 1.5 × natural size

lucidum like living strepsirhines do. In recent years, ***Shoshonius cooperi*** (Fig. 12.18) has become one of the best known of all omomyids. There are now several skulls and many parts of the postcranial skeleton (Beard *et al.,* 1991; Beard and MacPhee, 1994). *Shoshonius* shows many striking cranial similarities to the living *Tarsius,* including very large orbits, overlap of basicranial structures on the enlarged auditory bullae, and possibly some features of cranial circulation.

Very little is known of the postcranial anatomy of anaptomorphines. One of the best known taxa, ***Absarokius,*** shows adaptations for quadrupedal leaping, including an extensive fibrous connection between the tibia and the

TABLE 12.8
Suborder Prosimii
Superfamily Omomyoidea
Family Omomyidae
Subfamily ANAPTOMORPHINAE

Species	Estimated Mass (g)	Species	Estimated Mass (g)
Tribe ANAPTOMORPHINI		Tribe TROGOLEMURINI	
Teilhardina (early Eocene, North America, Europe)		*Trogolemur* (middle–l. Eocene, North America)	
T. belgica	90	*T. myodes*	75
T. brandti	—	*T. amplior*	
T. americana	120	*Sphacorhysis* (middle Eocene, North America)	
T. crassidens	90		
T. tenuicula	135	*S. burntforkensis*	140
T. demissa	—	*Anemorhysis* (early Eocene, North America)	
Anaptomorphus (middle Eocene, North America)		*A. sublettensis*	70
		A. natronensis	93
A. aemulus	275	*A. wortmani*	130
A. wortmani	160	*A. pattersoni*	144
A. westi	465	*A. savagei*	100
Gazinius (middle Eocene, North America)		*Tetonoides* (early Eocene, North America)	
G. amplus	875	*T. pearcie*	91
G. bowni	600	*T. coverti*	70
Tetonius (early Eocene, North America)		*Arapahovius* (early Eocene, North America)	
T. homunculus	290	*A. gazini*	290
T. mckennai	100	*A. advena*	130
T. matthewi	180	*Chlororhysis* (early Eocene, North America)	
Pseudotetonius (early Eocene, North America)		*C. knightensis*	165
		C. incomptus	—
P. ambiguus	170	Tribe WASHAKIINI	
Absarokius (early–m. Eocene, North America)		*Washakius* (middle–late Eocene, North America)	
A. abbotti	200	*W. insignis*	165
A. noctivagus	200	*W. woodringi*	130
A. witteri	500	*W. izetti*	170
Tatmanius (early Eocene, North America)		*W. laurae*	
T. szalayi	160	*Shoshonius* (early–middle Eocene, North America)	
Strigorhysis (middle Eocene, North America)		*S. cooperi*	155
S. bridgeriensis	500	*S. bowni*	160
S. rugosus	320	*Dyseolemur* (late Eocene, North America)	
S. huerfanensis	?600	*D. pacificus*	165
Acrossia (middle Eocene, North America)		*Loveina* (early Eocene, North America)	
A. lovei	275	*L. zephyri*	170
		L. minuta	95
		L. wapitiensis	

fibula, but no indications of vertical clinging (Covert and Hamrick, 1993).

Omomyines

The omomyines (Table 12.9), a predominantly North American group with at least one Asian representative, were almost certainly derived from an anaptomorphine-like ancestor similar to *Steinius*. The composition of the subfamily is under considerable flux; several genera (*Stockia, Uintanius, Utahia*) are frequently placed in different tribes by various researchers, and others, such as *Rooneyia* (Fig. 12.18) and *Ekgmowechashala*, are often placed in other families or orders. The major adaptive radiation of omomyines was later than that of the more primitive anaptomorphines; omomyines were most abundant from the middle to late Eocene. They ranged in size from about 100 g to over 2 kg. Many authorities have suggested that the replacement of anaptomorphines by omomyines in the middle Eocene is associated with the general climatic warming and a northward movement of southern faunas (Beard *et al.*, 1992; Gunnell, 1995).

Despite their lower diversity, omomyines show a far greater range of dental adaptations than the anaptomorphines. Their molars often have lower cusps, and the trigonid cusps are less inflated; the last molar is usually elongated. Other, later members of the family developed very flat molars with accessory cusps and crenulated enamel. Omomyines probably occupied a variety of dietary niches (Fig. 12.19). Like anaptomorphines, the earlier, smaller species included both frugivores and species with adaptations for processing hard invertebrates such as beetles. Later, larger species with broad, flat molars and rounded cusps, such as *Rooneyia* and *Ekgmowechashala*, were almost certainly frugivorous. ***Macrotarsius***, the largest omomyine, has well-developed shearing crests and large stylar cusps, indicative of folivory.

The skull is well known only for one possible omomyoid, ***Rooneyia viejaensis*** (Fig. 12.18), a relatively late genus from the late Eocene of Texas that may not be at all representative of the subfamily. *Rooneyia* has a relatively broad, short snout and moderately large orbits surrounded by a complete postorbital bar. On the basis of orbit size, it seems most likely that *Rooneyia* was diurnal. The braincase is relatively large, in the range of that of extant prosimians. The auditory region has an uninflated bulla with a tubular bony ectotympanic partly enclosed by the bulla. The details of its carotid circulation are unknown. The omomyine affinities of *Rooneyia* have been contested by several workers. Some have suggested microchoerid affinities, others place it in its own family.

The limb skeleton is best known for ***Hemiacodon gracilis*** and ***Omomys carteri*** (Covert, 1995). All of the bones that have been reported indicate adaptations for both leaping and quadrupedal behavior. They suggest that omomyines had greater overall similarities to cheirogaleids or galagos than to the extant *Tarsius* in proportions and details of individual elements. The distal parts of the tibia and fibula are not fused, as in *Tarsius* or some European omomyids, but seem to have been firmly conjoined by connective tissues.

Microchoerids

The microchoerids (Table 12.10) were a small but diverse group of omomyoids from the middle Eocene through the latest Eocene of western Europe. They were probably derived from an early anaptomorphine such as *Teilhardina* or *Tetonius* (Rose, 1995). The four genera vary in size from tiny ***Pseudoloris*** (50–120 g) to the medium-size ***Microchoerus*** (500–1800 g), and all are relatively abundant in the fossil record.

The dental formula for microchoerines has never been satisfactorily resolved, which complicates attempts to understand the relationships between this group of fossils and later primates (Fig. 12.21). The upper dentition has a formula of 2.1.3.3, like that of *Tarsius,*

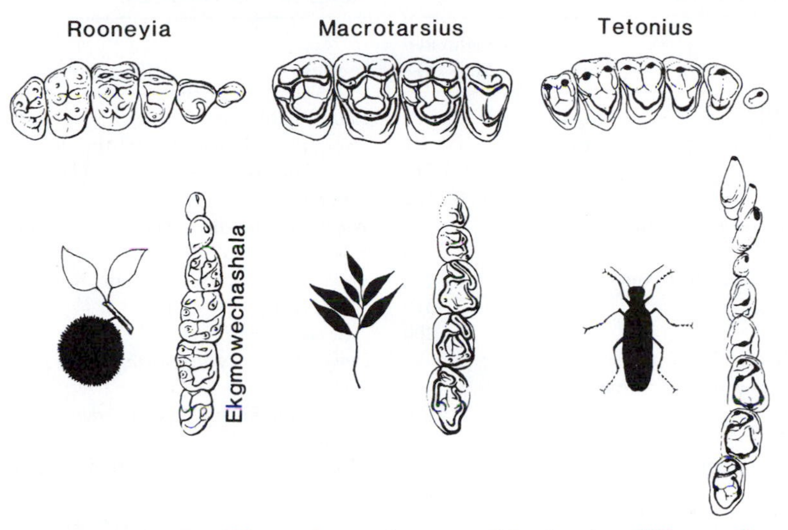

FIGURE 12.19 Dentitions of several omomyoids, showing different dietary adaptations.

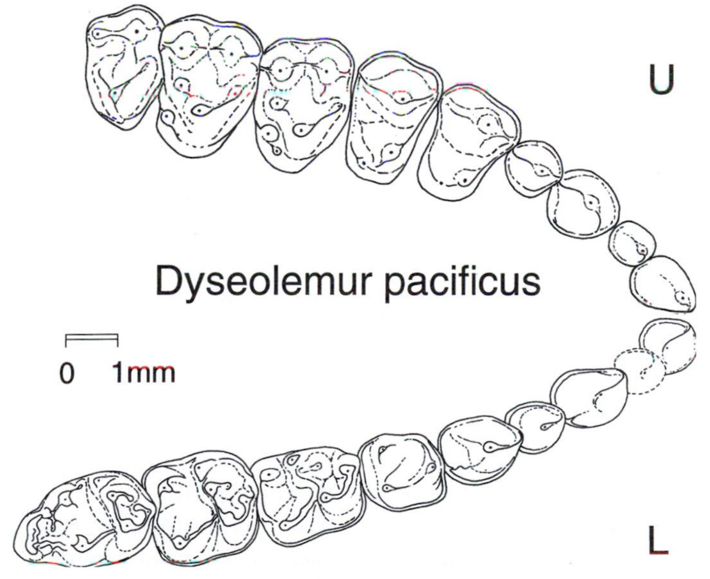

FIGURE 12.20 Upper (U) and lower (L) right dentitions of *Dyseolemur* (courtesy of D. T. Rasmussen).

TABLE 12.9
Suborder Prosimii
Superfamily Omomyoidea
Family Omomyidae
Subfamily OMOMYINAE

Species	Estimated Mass (g)
Tribe OMOMYINI	
Omomys (middle Eocene, North America)	
O. carteri	310
O. lloydi	180
Chumashius (late Eocene, North America)	
C. balchi	295
Steinius (early Eocene, North America)	
S. vespertinus	310
S. annectens	400
Tribe UINTANIINI	
Uintanius (?early Eocene, North America)	
U. ameghini	150
U. rutherfurdi	170
Jemezius (early Eocene, North America)	
J. szalayi	155
Tribe MACROTARSIINI	
Macrotarsius (late Eocene–early Oligocene, North America)	
M. seigerti	1635
M. montanus	2520
M. jepseni	1620
M. roederi	1640
M. macrorhysis (middle Eocene, China)	—
Hemiacodon (middle Eocene, North America)	
H. gracilis	1005
Yaquius (middle Eocene, North America)	
Y. travisi	2160

TABLE 12.9 *(continued)*
Suborder Prosimii
Superfamily Omomyoidea
Family Omomyidae
Subfamily OMOMYINAE

Species	Estimated Mass (g)
Tribe OURAYINI	
Ourayia (late Eocen, North America)	
O. uintensis	2170
O. hopsoni	1150
Wyomomys (middle Eocene, North America)	
W. bridgeri	320
Ageitodendron (middle Eocene, North America)	
A. matthewi	840
Utahia (late Eocene, North America)	
U. kayi	95
Stockia (late Eocene, North America)	
S. powayensis	475
Chipetaia (middle Eocene, N. America)	
C. lamporea	1015
Asiomomys (?early Eocene, China)	
A. changbaicus	475
Tribe *Incertae sedis*	
Ekgmowechashala (late Oligocene, North America)	
E. philotau	1870

with a large upper central incisor followed by a small lateral incisor, a moderate-size canine, three relatively simple premolars, and three molars. The most primitive genus, ***Nannopithex,*** has no hypocone on the upper molars but has a long postprotocingulum, or nannopithex fold. In the three other genera there is a hypocone derived from the lingual cingulum. The lower dentition has one less tooth than the upper tooth row, but one of the teeth is so small that it does not occlude with anything, so the occlusal relationships cannot be used to interpret the homologies of the teeth (Fig. 12.21; Schmid, 1983). Thus, microchoerine lower dentitions contain a large procumbent tooth, probably I_1, followed by two small teeth (either I_2, C; C, P_2; or P_1, P_2). The large tooth is roughly similar in shape to the canine in *Tarsius,* but it clearly functioned differently from that tooth in the living tarsier, which is used primarily to kill animal prey. The large tooth in both *Necrolemur* and *Microchoerus* developed heavy wear on the tips (from scraping and gouging) as well as fine parallel striations on its mesial surface, indicating that it also functioned as a grooming tooth (Schmid, 1983).

The cheek teeth of microchoerines vary considerably among the genera. The tiny *Nannopithex* has an enlarged, pointed premolar and anaptomorphine-like molars with a high trigonid and deep, narrow talonid; it is most comparable to living frugivorous primates. The tiny *Pseudoloris,* however, probably had a faunivorous diet. The larger genera, *Necrolemur* and *Microchoerus,* have molars with low, rounded cusps and elaborate crenulations of the enamel; these suggest a more frugivorous diet or, considering their anterior dentition, perhaps a diet supplemented by gums.

There are many complete, usually crushed, skulls of *Necrolemur* (Fig. 12.18) and cranial fragments of *Microchoerus, Pseudoloris,* and *Nannopithex.* All have a relatively short, narrow snout with a bell-shaped palate, a gap between the upper central incisors, large eyes, and a moderately large infraorbital foramen. The olfactory bulb apparently passes above the orbits as in all extant haplorhines, but the back of the orbit is not walled off from the temporal fossa as in *Tarsius* and anthropoids. In the ear region, the ectotympanic forms a ring within the bulla but extends laterally to form a bony tube. This unique condition makes them resemble strepsirhines in the position of the ring and *Tarsius* in the tube. The canal for the stapedial artery and the groove for the promontory artery are similar in size. There is extensive inflation of the mastoid region behind the middle and inner ear. The large eyes of microchoerines suggest that they were all nocturnal animals, but the orbits are more like those of strepsirhines in relative size rather than like those of *Tarsius* or *Aotus,* suggesting that they probably had a tapetum lucidum.

Although there are no complete skeletons for microchoerids, numerous isolated hindlimb elements have been attributed to species of this subfamily; these include a nearly complete femur, a partly fused tibia-fibula, a talus and a calcaneus for *Necrolemur,* and isolated tarsal bones probably attributable to *Microchoerus.* All of these postcranial elements indicate leaping abilities. In their elongation, however, the calcanei of microchoerids are more like those of cheirogaleids than those of *Tarsius* (Dagosto, 1993). In several aspects of femoral morphology, microchoerines show striking similarities to anthropoids, especially parapithecids (Dagosto and Schmid, 1996).

TABLE 12.10
Suborder Prosimii
Superfamily Omomyoidea
Family MICROCHOERIDAE

Species	Estimated Mass (g)
Nannopithex (early-middle Eocene, Europe)	
N. pollicaris	125
N. raabi	170
N. filholi	155
N. quaylei	—
N. abderhalderi	170
N. barnesi	200
N. humilidens	140
N. zuccolae	180
Pseudoloris (middle-late Eocene, Europe)	
P. parvulus	45
P. isabenae	50
P. crusafonti	75
P. requanti	120
P. saalae	60
Necrolemur (late Eocene, Europe)	
N. zitteli	290
N. antiquus	320
Microchoerus (late Eocene-early Oligcocne Europe)	
M. erinaceus	1775
M. edwardsi	930
M. ornatus	915
M. wardi	560
M. creechbarrowensis	900

Asian Omomyoids

The fossil evidence of omomyoids in Asia (Table 12.11) has expanded in recent years. In

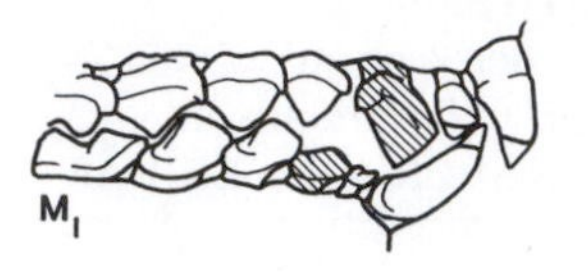

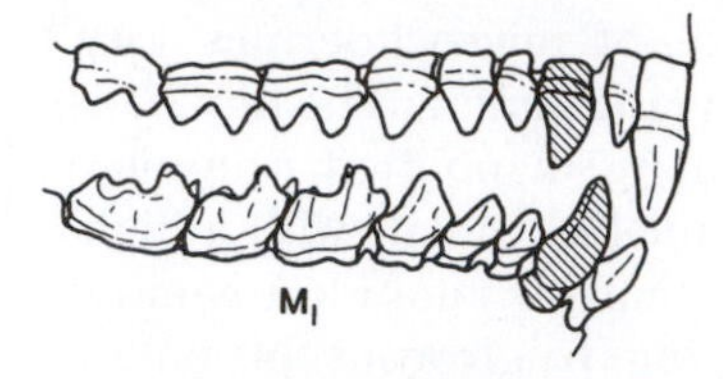

FIGURE 12.21 A lateral view of the anterior dentition of *Necrolemur antiquus* and *Tarsius syrichta,* showing the dental proportions. Various authorities have identified each of the first three teeth as the canine. It seems most likely that the shaded tooth is the canine and the teeth anterior to it are incisors. Note that regardless of how the dental formula of *Necrolemur* is interpreted, the dental proportions are very different from those of *Tarsius* (adapted from Schmid, 1983).

TABLE 12.11
Suborder Prosimii
Superfamily Omomyoidea
Subfamily *Incertae sedis*

Species	Estimated Mass (g)
Rooneyia (late Eocene, North America)	
R. viejaensis	1475
Kohatius (early–middle Eocene, Asia)	
K. coppensi	190

TABLE 12.12
Suborder Prosimii
Family TARSIIDAE

Species	Estimated Mass (g)
Afrotarsius (early Oligocene, Africa)	
A. chatrathi	100
Tarsius (Eocene-Recent, Asia)	
T. eocaenus	50
T. thailandica	—
Xanthorhysis (Eocene, Asia)	
X. tabrumi	75

addition to *Altanius,* once considered an early omomyoid but now regarded as a basal primate, there are fragmentary remains from Pakistan and two well-documented genera from China with very close affinities to Eocene taxa from North America. ***Kohatius*** is a moderate-size (200 g) species from the early to middle Eocene of Pakistan known from only a few teeth. *Asiomomys* is strikingly similar to the late Eocene *Stockia* from California (Beard and Wang, 1991), a genus whose subfamily affinities within omomyids are far from clear. More recently, Beard and colleagues (1994) have described an omomyid very similar to the North American *Macrotarsius.* These new discoveries clearly document continuity between North American and Asian primate communities during the Eocene.

African Omomyoids

Several isolated omomyoid-like teeth have been recovered from the Eocene-Oligocene deposits of the Fayum and from similar-age deposits in Oman, which was connected to Africa in the Oligocene. However, there are few

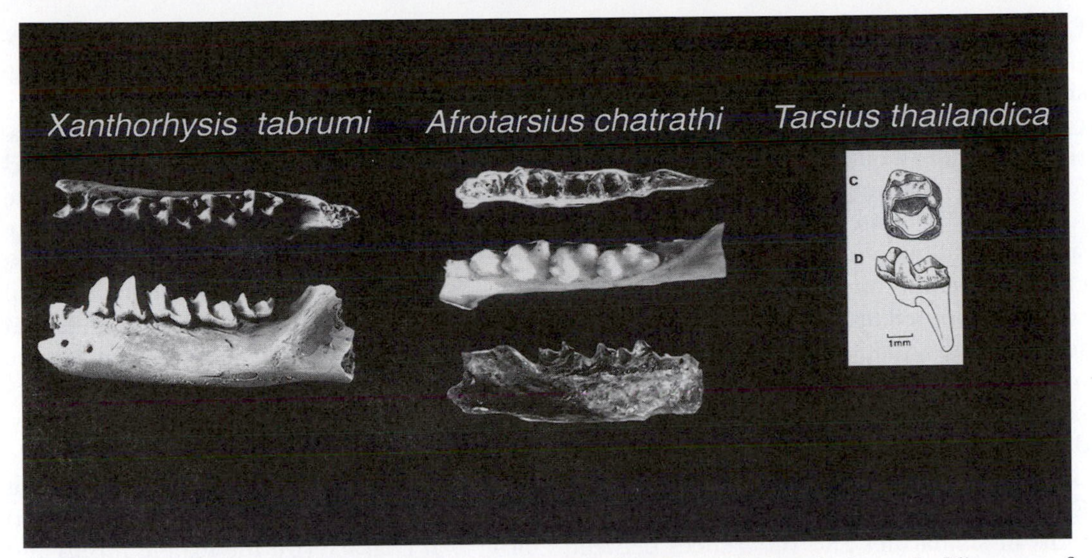

FIGURE 12.22 Fossil tarsiids from the Eocene of China (*Xanthorhysis tabrumi*), the early Oligocene of Africa (*Afrotarsius chatrathi*) and the Miocene of Thailand (*Tarsius thailandica*). Scale = 2.5 ×. (Courtesy of K. C. Beard, E. L. Simons, and L. Ginsburg).

detailed discussions, and no species have been named as yet.

Tarsiids

Numerous fossil primates have been described in recent years that are attributable to the same family or even the same genus as the living *Tarsius* (Table 12.12). ***Afrotarsius chatrathi*** (Fig. 12.22) is from the Oligocene Fayum of Egypt. Because the anterior dentition is not known, it has been debated whether *Afrotarsius* is more closely related to the living *Tarsius,* to the European microchoerines, or even to early anthropoids, all of which it resembles to some degree. However a newly discovered tibia-fibula is virtually identical to that of *Tarsius* (Rasmussen, pers. comm).

In the Eocene of China, Beard and colleagues (1994, 1998) have described both dental (Fig. 12.22) and postcranial remains (Dagosto *et al.,* 1996; Gebo *et al.,* 1996) that are virtually identical to the extant tarsier including *Xanthorhysis* tabrumi. These remains appear to document the presence of tarsiids in Asia for at least 50 million years, an observation that fits well with their morphological and biomolecular distinctiveness from other primates (Beard, 1998). Unfortunately, no cranial remains have been described so far. From Miocene deposits in Thailand, *Tarsius thailandica* (Ginsburg and Mein, 1986) is known from a single lower molar.

OMOMYOIDS, TARSIERS, AND HAPLORHINES

As small prosimians with large eyes, elongate calcanei, and in some species a fused tibia-fibula, omomyoids have been traditionally linked with the extant *Tarsius,* just as their contemporaries, the adapoids, have been allied with extant strepsirhines. Several authorities (e.g., Simons, 1972) have even placed the

European microchoerines into the family Tarsiidae. As with the adapoid–lemur relationship discussed earlier, the omomyoid–*Tarsius* connection has been debated extensively in recent decades in light of new anatomical information, new analytical techniques, and new fossils. Despite their small size, short snouts, and large eyes, omomyids and microchoerids are not simply Eocene tarsiers. Many of their supposed tarsierlike resemblances are superficial similarities or features common to other Eocene prosimians as well; moreover, all known omomyoids clearly lacked many of the distinguishing features of the ear, orbit, and skeleton that characterize the living *Tarsius*. Even more confusing are the different patterns of similarities and differences between Eocene taxa and the extant genus, indicating that any phylogenetic scenario involved considerable mosaic evolution and parallelism.

Until most recently, the microchoerids were considered the most tarsierlike of the omomyoids (e.g., Simons, 1972; Rosenberger and Dagosto, 1992). The genus *Pseudoloris* has been identified as an ancestor of *Tarsius*. However, although the molar teeth of *Pseudoloris* are strikingly like those of *Tarsius*, the teeth of other microchoerids are less obviously indicative of this relationship. Furthermore, the anterior dentition of all microchoerids is clearly different from that of *Tarsius* in both number and relative proportions of the upper and lower teeth (Fig. 12.21).

More recently, there have been more extensive arguments for an omomyoid–tarsier relationship based on the cranial morphology of both the microchoerid *Necrolemur* (Rosenberger, 1985) and the washakiine anaptomorphine *Shoshonius* (e.g., Beard and MacPhee, 1994). However, the pattern of similarities is a complex one. Much of the tarsierlike appearance of omomyids derives from their large orbits. Although large, the orbits of all omomyoids are structurally more similar to those of strepsirhines in lacking any postorbital closure. The relative size of omomyid orbits suggests that most of these early prosimians were like strepsirhines in having an eye with a tapetum lucidum rather than lacking that structure as do *Tarsius* and anthropoids. Because all living haplorhines lack a tapetum, the light-catching efficiency of their eyes is less than that of strepsirhines, and both nocturnal haplorhines (*Tarsius* and *Aotus*) have eyes that are much larger than those of similar-size nocturnal strepsirhines. Only *Shoshonius* approaches tarsiers in relative orbit size. Thus, although omomyids had large orbits and were almost certainly nocturnal, it is unlikely that they had the derived features of the eye and orbit that characterize the extant *Tarsius* and other haplorhines (see Gonzalez *et al.*, 1998). However, there are a number features that link *Shoshonius, Necrolemur,* and possibly *Tetonius* to *Tarsius* in the structure of the basicranium, where bony sheets from the pterygoid bones and the sphenoid bone overlap with the enlarged auditory bulla, and the jaw articulation has a narrow, gutterlike shape. Phylogenetic analysis of the basicranial anatomy in a wide range of fossil and living prosimians groups *Shoshonius, Tarsius,* and *Necrolemur* as a clade (Beard and MacPhee, 1994; Kay *et al.*, 1997).

The postcranial similarities linking omomyoids and tarsiers are largely limited to the tendency toward fusion of the fibrous joint between the tibia and the fibula. Otherwise they are more similar to cheirogaleid strepsirhines. All omomyoid limbs lack the extremely elongated calcaneus found in tarsiers, and their elbow region resembles that of small anthropoids more than it does that of the genus *Tarsius*.

Analysis of new omomyid and microchoerid cranial and skeletal remains have greatly confused the relationship among the omomyoids, the extant tarsiers, and anthropoids. It appears that in all fossil omomyoids, as in *Tarsius* and anthropoids, the olfactory bulb lies above the interorbital septum and there is no extensive sphenoethmoid recess. This organization of the anterior part of the cranium is unquestionably a derived condition, and, assuming this feature did not evolve in parallel in the various lineages, it supports a broad haplorhine grouping. However, although a number of derived features of the cranium have been described that link *Shoshonius* and *Necrolemur* with *Tarsius,* these are not the same cranial features that link *Tarsius* with anthropoids, specifically, postorbital closure and an anterior accessory cavity of the middle ear. Thus, if tarsiers are uniquely derived from an omomyid or microchoerid ancestry, then these unique cranial features linking tarsiers and anthropoids must have evolved independently. Alternatively, if these cranial features are really derived features linking tarsiers and anthropoids (e.g., Ross, 1994), the omomyid and microchoerid similarities to tarsiers must be primitive haplorhine features or similarities acquired in parallel. Either solution implies considerable parallel evolution within these three groups. Cranial remains of some of the Eocene tarsiids from China (Beard *et al.,* 1997) could help resolve these alternatives. The presence of dental tarsiers in Asia contemporary with or older than omomyids and microchoerids in North America, Europe, and Asia indicates that any phylogenies linking one or two Eocene omomyoids with either *Tarsius* or anthropoids, beyond indicating a broad haplorhine cranial organization, is undoubtedly overly simplistic.

ADAPTIVE RADIATIONS OF EOCENE PROSIMIANS

Our understanding of the early evolution of prosimians is currently in an exciting, but very awkward state of complexity. These groups have been well known in North America and Europe for over a century, and the broad patterns of their taxonomic and adaptive diversity on those continents are well established. At the same time, our understanding of the biogeography of these early primates is minimal. We have only the first hints of their presence in Africa and Asia, with no way to estimate their likely diversity in either taxonomic or adaptive realms. Thus, for the present we must concentrate on the better-known continents.

In North America and Europe, the adapoids and omomyoids were a diverse group of primates that occupied a wide range of ecological niches. There seem to be clear temporal trends in the adaptive radiations of these early prosimians on both continents. The adapoids started out at a relatively large size compared to the omomyids. Throughout the Eocene and early Oligocene, adapoids seem to have occupied adaptive niches that characterize extant higher primates (large size, diurnality, frugivory, and folivory), whereas omomyoids were perhaps more comparable to galagos. Only in the later part of the Eocene and the Oligocene do the omomyids appear to have expanded into the adaptive zones of large size and folivory. Equally striking are the phyletic and adaptive differences between the Eocene prosimian faunas on the two continents from which they are well known.

In North America (Fig. 12.23), the omomyoids of the early and middle Eocene were taxonomically diverse, but all were relatively small (less than 500 g). Their teeth suggest diets that were predominantly frugivorous, with some specializing in hard insects. The two skulls indicate nocturnal habits. In contrast, the North American notharctines from the early and middle Eocene were much less taxonomically diverse, with only five or six genera, and all were considerably larger (1.5–7 kg), frugivorous or folivorous, and probably diurnal. Only after the disappearance of notharctine adapoids at the end of the middle

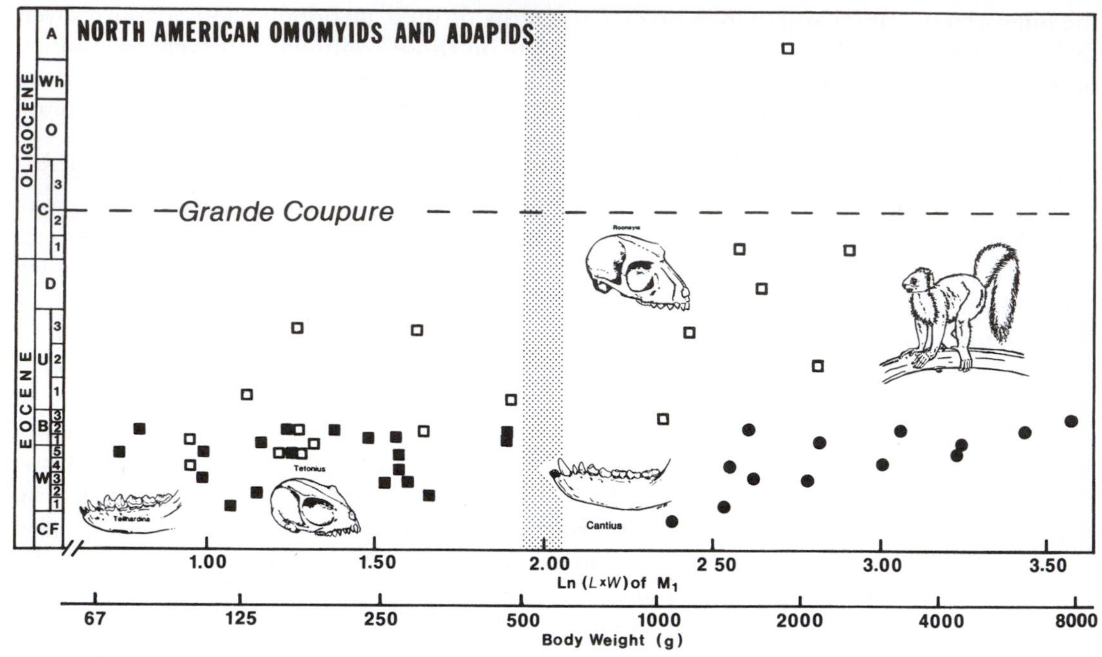

FIGURE 12.23 Size of North American adapoids, anaptomorphines, and omomyines through time. Note that adapoids (circles) are larger than contemporary anaptomorphines (filled squares) and that the radiation of larger omomyines (open squares) takes place after the extinction of most of the adapoids. Cross-hatching indicates Kay's threshold.

Eocene do we find larger, probably frugivorous and folivorous omomyids in North America. The locomotor adaptations of Eocene prosimians are poorly known, but most remains indicate quadrupedal and leaping abilities for both omomyoids and adapoids.

In Europe (Fig. 12.24), the notharctid cercamoniines were more diverse, and the microchoerids were limited to only four genera after the basal omomyid *Teilhardina*. In the latest Eocene there were numerous medium to large adapoids. Although the European adapoids were generally larger than synchronic microchoerids, the size range of the two groups overlapped somewhat in the late Eocene and early Oligocene with the evolution of very small cercamoniines such as *Anchomomys gaillardi* and large microchoerines such as *Microchoerus*. Associated with their size diversity was considerable dietary diversity among the cercamoniines and adapines. There seem to have been insectivorous, frugivorous, and possibly carnivorous (*Pronycticebus*) species among the cercamoniines as well as many folivorous adapids. The microchoerids, although less diverse, included small insectivorous species and other species that probably specialized on fruits or gums. One ecological parameter that seems to have separated the two radiations was their activity cycle. Most microchoerines seem to have been nocturnal, and the cercamoniines and adapines were probably mostly diurnal, judging from orbit size. Furthermore, the microchoerines seem to have been leapers

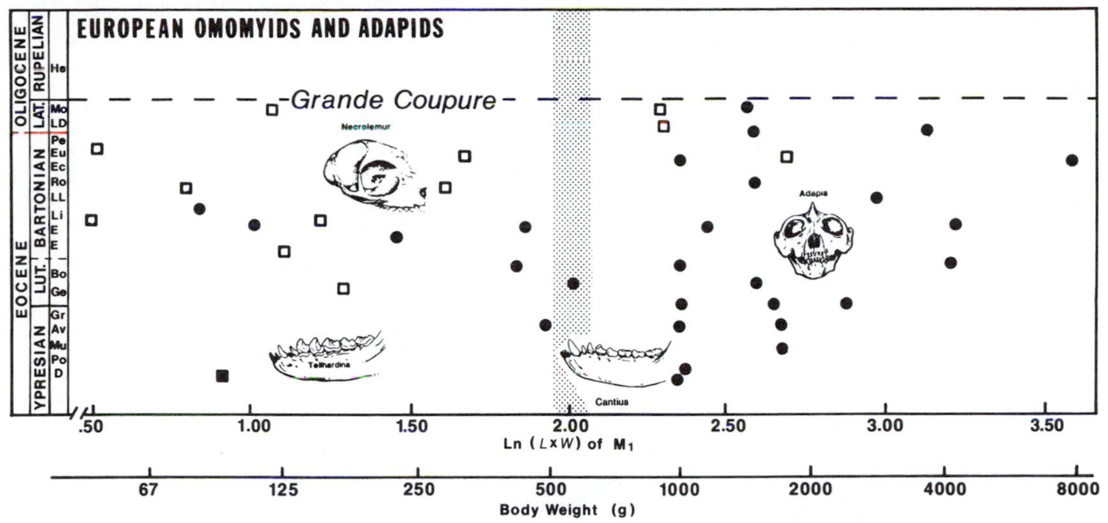

FIGURE 12.24 Size of European adapoids (circles) and microchoerine omomyoids (squares) through time. Note that there is considerable overlap in body size in the two radiations and that the adapoids are more diverse. Compare with 12.23.

or cheirogalid-like arboreal runners, whereas the skeletal remains from adapoids suggest slower quadrupedal climbing for some and leaping for others.

Our only information about the social organization of Eocene prosimians is the sexual dimorphism in canine size that seems to characterize many adapoids. This suggests that some type of polygynous social system was common in this radiation (Gingerich, 1995).

PHYLETIC RELATIONSHIPS OF ADAPOIDS AND OMOMYOIDS

Although adapoids and omomyoids have traditionally been identified as Eocene lemurs and tarsiers, respectively, both Eocene families are decidedly more primitive in some respects than the recent prosimians. Moreover, it seems quite clear from the paleontological record that the earliest adapids and omomyids were extremely similar. Indeed, *Donrussellia* has been allocated by some authorities to the omomyids and by others to the adapines. As discussed earlier in this chapter, it is more appropriate to consider these Eocene taxa as basal "primates of modern aspect" from which the modern prosimians evolved. Both are "missing links" that have phyletic affinities with the modern taxa and preserve information about more primitive morphological stages in primate evolution (Fig. 12.25). Compared with later primate taxa, the adapoids are clearly very primitive in virtually all aspects of their anatomy, but they may show a few derived features that link them with later strepsirhines. Although clearly distinct from the adapoid radiation, omomyoids are nevertheless very similar in retaining a more primitive morphology with respect to most later primate groups, and there are indications of a few features in the structure of the orbit, leg, and foot that link them with tarsiers and an-

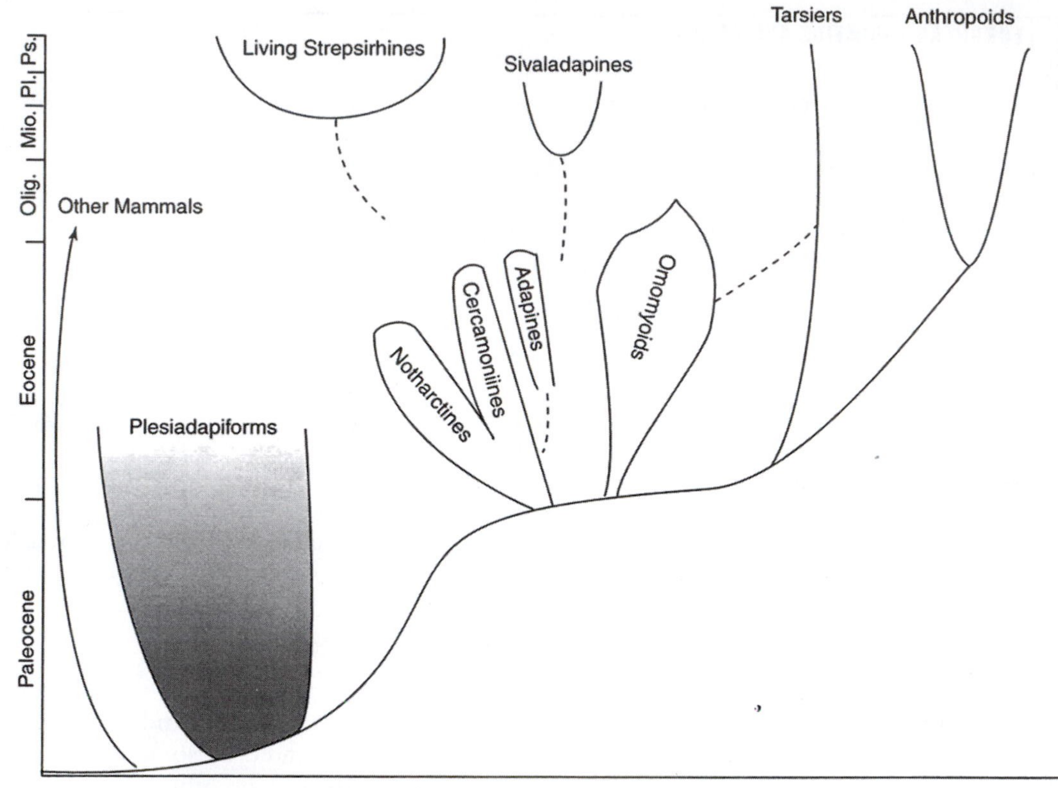

FIGURE 12.25 The phyletic relationships of adapoids and omomyoids.

thropoids and place them at the base of the haplorhine radiation. Thus it seems likely that the divergence between modern haplorhine and strepsirhine primates corresponds to the initial divergence of omomyoids from adapoids (Kay *et al.,* 1997). Although the Eocene adapoids and omomyoids can be placed in this general phyletic position, more specific details concerning the divergence and radiation of modern prosimians are more difficult to reconstruct. The presence of fossil strepsirhines in the Eocene through Miocene of Africa and tarsiers in the Eocene through Miocene of Asia suggests that many of the details of prosimian phylogeny lie on those continents.

BIBLIOGRAPHY

EOCENE EPOCH

Adams, C. G. (1981). An outline of Tertiary paleogeography. In *The Evolving Earth,* ed. L. R. M. Cocks, pp. 221–235. Cambridge: Cambridge University Press.

Rose, K. D. (1981) The Clarkforkian land-mammal age and mammalian faunal composition across the Paleocene–Eocene boundary. University of Michigan, Museum of Paleontology, Papers on Paleontology, no. 26.

——— (1984). Evolution and radiation of mammals in the Eocene, and the diversification of modern orders. In *Mammals: Notes for a Short Course,* ed. T. D. Broadhead, pp. 110–127. Knoxville: University of Tennessee, Dept. of Geological Sciences.

Wolfe, J. A. (1978). A paleobotanical interpretation of Tertiary climates in the Northern Hemisphere. *Am. Sci.* **66**:694–703

THE FIRST MODERN PRIMATES

Beard, K. C., Dagosto, M.., Gebo, D. L., and Godinot, M. (1988). Interrelationships among primate higher taxa. *Nature* **331**:712–714.

Covert, H. H. (1986). Biology of early Cenozoic primates. In *Comparative Primate Biology,* vol. 1: *Systematics, Evolution, and Anatomy,* ed. D. R. Swindler and J. Erwin, pp. 335–359. New York: Alan R. Liss.

______ (1995). Locomotor adaptations of Eocene primates: Adaptive diversity among the earliest prosimians. In *Creatures of the Dark,* ed. L. Alterman, G. A. Doyle, and M. K. Izard, pp. 495–509. New York: Plenum Press

Covert, H. H., and Williams, B. A. (1994). Recently recovered specimens of North American Eocene omomyids and adapids and their bearing on debates about anthropoid origins In *Anthropoid Origins,* ed. J. G. Fleagle and R. F. Kay, pp. 29–54. New York: Plenum Press.

Dagosto, M. (1988). Implications of postcranial evidence for the origin of euprimates. *J. Hum. Evol.* **17**:35–56.

______ (1993). Postcranial anatomy and locomotor behavior in Eocene primates. In *Postcranial Adaptation in Nonhuman Primates,* ed. D. L. Gebo, pp. 199–219. DeKalb: Northern Illinois University Press.

Dagosto, M.,. and Terranova, C. J. (1992). Estimating the body size of Eocene primates: A comparison of results from dental and postcranial variables. *Int. J. Primatol.* **13**:307–344.

Gebo, D. L. (1988). Foot morphology and locomotor adaptation in Eocene primates. *Folia primatol.* **50**:3–41.

Gingerich, P. D. (1990). African dawn for primates. *Nature* **346**:411

Gingerich, P. D., Dashzeveg, D., and Russell, D. E. (1991). Dentition and systematic relationships of *Altanius orlovi* (Mammalia, Primates) from the early Eocene of Mongolia. *Geobios* **24,** fasc. 5, pp. 637–646.

Godinot, M. (1992). Early euprimate hands in evolutionary perspective. *J. Human Evol.* **22**:267–283.

Rose, K. D. (1995). The earliest primates. *Evol. Anthropol.* **3**:159–173.

Rose, K. D., and Bown, T. M. (1991). Additional fossil evidence on the differentiation of the oldest Euprimates. *Proc. Natl. Acad. Sci. USA* **88**:98–101

Rose, K. D.,. Godinot, M., and Bown, T. M. (1994). The early radiation of Euprimates and the initial divirsification of omomyidae. In *Anthropoid Origins,* ed. J. G. Fleagle and R. F. Kay, pp. 1–28. New York: Plenum Press.

Sige, B., Jaeger, J. J., Sudre, J., and Vianey-Liaud, M. (1990). *Altiatlasius kuolchii* n.gen. et sp. Primate omomyide du paleocene superieur du Maroc, et les origines des Euprimates. *Palaeontographica A* **214**:31–56.

Simons, E. L. (1972). *Primate Evolution: An Introduction to Man's Place in Nature.* New York: Macmillan.

Szalay, F. S., and Li, C.-K. (1986). Middle Paleocene Euprimate from southern China and the distribution of primates in the Paleogene *J. Human Evol.* **15**: 387–397

van Valen, L. M. (1994). The origin of the plesiadapid primates and the nature of *Purgatorius. Evolutionary Monographs* **15**:1–79.

ADAPOIDS

Beard, K. C. and Godinot, M. (1988). Carpal anatomy of *Smilodectes gracilis* (Adapiformes, Northactinae) and its implications. *J. Human Evol.* **17**:71–92

Dagosto, M. (1993). Postcranial anatomy and locomotor behavior in Eocene primates. In *Postcranial Adaptation in Nonhuman Primates,* ed. D. L. Gebo, pp. 199–219. DeKalb:Northern Illinois University Press.

Franzen, J. L. (1994). The Messell primates and anthropoid origins. In *Anthropoid Origins,* ed. J. G. Fleagle and R. F. Kay, pp. 99–122. New York: Plenum Press.

Gingerich, P. D. (1980). Eocene Adapidae, paleobiogeography and the origin of the South American Platyrrhini. In *Evolutionary Biology of the New World Monkeys and Continental Drift,* ed. R. L. Ciochon and A. B. Chiarelli, pp. 123–138. New York: Plenum Press.

Gregory, W. K. (1920). On the structure and relation of *Notharctus,* an American Eocene primate. *Mem. Am. Mut Nat. Hist. n.s.* **351**:243.

Gunnell, G. F. (1995). New notharctine (Primates, Adapiformes) skull from the Uintan (middle Eocene) of San Diego County, California. *Am. J. Phys. Anthropol.* **98**:447–470.

Stehlin, H. G. (1916). Die Saugetiere des schweizerischen Eocaens. Siebenter Tel, zweite Halfte. *Caenopithecus—Necrolemur—Microchoerus—Nannopithex —Anchomomis—Periconodon—Heterochiromys*-Nachtrade zu Adapis Schlussbetrachtugen zu den Primaten. *Abh. schweiz. Palaeontol. Gessellsch.* **41**:1299–1552.

Notharctines

Alexander, J. P. (1992). Alas, poor *Notharctus. Nat. Hist.* 8/92:55–58.

——— (1994). Sexual Dimorphism in Notharctid Primates. *Folia Primatol.* **63**:59–62.

Beard, K. C. (1988). New notharctine primate fossils from the early Eocene of New Mexico and southern Wyoming and the phylogeny of Notharctinae. *Am. J. Phys. Anthropol.* **75**:439–469.

Conroy, G. C., and Wible, J. R. (1978). Middle ear morphology of *Lemur variegatus. Folia Primatol.* **29:** 81–85.

Covert, H. H. (1986). Biology of early Cenozoic primates. In *Comparative Primate Biology,* vol. 1: *Systematics, Evolution, and Anatomy,* ed. D. R. Swindler and J. Erwin, pp. 335–359. New York: Academic Press.

——— (1990). Phylogenetic relationships among Notharctinae of North America. *Am. J. Phys. Anthropol.* **81**:381–398.

Dagosto, M. (1993). Postcranial anatomy and locomotor behavior in Eocene primates. In *Postcranial Adaptation in Nonhuman Primates,* ed. D. L. Gebo, pp. 199–219. DeKalb:Northern Illinois University Press.

Gazin, C. L. (1958). A review of the middle and upper Eocene primates of North America. *Smithson. Misc. Coll.* **136**:1–112.

Gebo, D. L., Dagosto, M., and Rose, K. D. (1991). Foot morphology and evolution in early Eocene *Cantius. Am. J. Phys. Anthropol.* **86**:51–73.

Gingerich, P. D. (1979). Phylogeny of middle Eocene Adapidae (Mammalia, Primates) in North America: *Smilodectes and Notharctus. J. Paleontol.* **53**(1):153–163.

——— (1980). Dental and cranial adaptation in Eocene Adapidae. *Z. Morphol. Anthropol.* **71**(2):135–142.

——— (1984). Primate evolution. In *Mammals: Notes for a Short Course,* ed. T. D. Broadhead, pp. 167–181. Knoxville: University of Tennessee, Dept. of Geological Sciences.

——— (1986). Early Eocene *Cantius torresi*—Oldest primate of modern aspect from North America. *Nature (London)* **319**:319–321.

——— (1995). Sexual dimorphism in earliest Eocene *Cantius torresi* (Mammalia, Primates, Adapoidea). *Contrib. Mus. Paleontol., Univ. Michigan* **29**(8):185–199.

Gingerich, P. D., and Simons, E. L. (1977). Systematics, phylogeny and evolution of early, Eocene Adapidae (Mammalia, Primates) in North America. *Mus. Paleontol.* **24**(22):245–279.

Gregory, W. K. (1920). On the structure and relation of *Notharctus,* an American Eocene primate. *Mem. Am. Mus. Nat. Hist. n.s.* **351**:243.

Gunnell, G. F. (1995). New notharctine (Primates, Adapiformes) skull from the Uintan (middle Eocene) of San Diego County, California. *Am. J. Phys. Anthropol.* **98**:447–470.

Hamrick, M. W. (1996). Locomotor adaptations reflected in the wrist joints of early Tertiary primates (Adapiformes). *Am. J. Phys. Anthropol.* **100**:585–604.

Hamrick, M. W. and Alexander, J. P. (1996). The hand skeleton of *Notharctus tenebrosus* (Primates, Notharctidae) and its signiificance for the origin of the primate hand. *Novitates* No. 3182, pp. 1–20.

Krishtalka, L., Stucky, R. K., and Beard, K. C. (1990). The earliest fossil evidence for sexual dimorphism in primates. *Proc. Natl. Acad. Sci.* **87**:5223–5226.

Radinsky, L. (1975). Primate brain evolution. *Am. Sci.* **63**(6):656–663.

——— (1977). Early primate brains: Facts and fiction. *J. Human Evol.* **6**:79–86.

Rose, K. D., and Walker, A.. (1985). The skeleton of early Eocene *Cantius,* oldest lemuriform primate. *Am. J. Phys. Anthropol.* **66**:73–89.

Rosenberger, A. L., Strasser, E., and Delson, E. (1985). Anterior dentition of *Notharctus* and the adapid-anthropoid hypothesis. *Folia Primatol.* **44**:15–39.

Simons, E. L. (1962). A new Eocene primate genus, *Cantius,* and a revision of some allied European lemuroids. *Bull. Brit. Mus. (Nat. Hist.) Geol.* **7**:1–30.

Cercamoniines and Adapines

Bacon, A.-M., and Godinot, M. (1998). Analyse morphofonctionnele des femurs et des tibias des "*Adapis*" du Quercy: mise en evidence de cinq types morphologiques. *Folia Primatol.* **69**:1–24.

Beard, K. C., Dagosto, M., Gebo, D. L., and Godinot, M. (1988). Interrelationships among primate higher taxa. *Nature* **331**:712–714.

Beard, K. C., Tao, Q., Dawson, M. R., Wang, B., and Chuanhuei, L., (1994). A diverse new primate fauna from middle Eocene fissure-fillings in southeastern China. *Nature* **368**:604–609.

Ciochon, R. L., and Holroyd, P. A. (1994). The Asian origin of Anthropoidea revisited. In *Anthropoid Origins,* ed. J. G. Fleagle and R. F. Kay, pp.143–162. New York: Plenum Press.

Dagosto, M. (1983). Postcranium of *Adapis parisiensis* and *Leptadapis magnus* (Adapiformes): Adaptational and phylogenetic significance. *Folia Primatol.* **41**:49–101.

Ducrocq, S., Jaeger, J.-J., Chaimanee, Y., and Suteethorn, Y. (1995). New primate from the Palaeogene of Thailand, and the biogeographical origin of anthropoids. *J. Human Evol.* **28**:477–485.

Filhol, H. (1883). Observations relatives au Memoire de M. Cope initule: Relation des horizons renfermant des debris d'animaux vertebres fossiles en Europe et en Amerique. *Ann. Sci. Geol., Paris* **14**:1–51.

Franzen, J. L. (1987). Ein neuer Primate aus dem Mitteleozan der Grube messel (deutschland, S-Hessen). *Cour. Forsch.-Inst. Senckenberg* **91**:151–187.

______ (1994). The Messel primates and anthropoid origins. In *Anthropoid Origins,* ed. J. G. Fleagle and R. F. Kay, pp. 99–122. New York: Plenum Press.

Gheerbrant, E., Thomas, H., Roger, J., Sen, S., and Al-Sulaimani, Z. (1993). Deux nouveaux primates dans L'Oligocene inferieur de Taqah (Sultanat d'Pman): premiers adapiformed (?Anchomomyini) de la Peninsula Arabique. *Palaeovertebrata, Montpelier* **22**:141–196.

Gingerich, P. D. (1975). Dentition of *Adapis parisiensis* and the evolution of lemuriform primates. In *Lemur Biology,* ed. I. Tattersall and R. W. Sussman, pp. 65–80. New York: Plenum Press.

______ (1977a). New species of Eocene primates and the phylogeny of European Adapidae. *Folia Primatol.* **28**:60–80.

______ (1977b). Radiation of Eocene Adapidae in Europe. *Geobios, Mem. Spec.* **1**:165–182.

______ (1981). Cranial morphology and adaptations in Eocene Adapidae, I: Sexual dimorphism in *Adapis magnus* and *Adapis parisiensis. Am. J. Phys. Anthropol.* **56**: 217–234.

Gingerich, P. D., and Martin, R. D. (1981). Cranial morphology and adaptations in Eocene Adapidae, II: The Cambridge skull of *Adapis parisiensis. Am. J. Phys. Anthropol.* **56**:235–257.

Gingerich, P. D., Holroyd, P. A., and Ciochon, R. L. (1994). *Rencunius zhoui,* new primate from the late middle Eocene of Hunan, China, and a comparison with some early Anthropoidea. In *Anthropoid Origins,* ed. J. G. Fleagle and R. F. Kay, pp. 163–177. New York: Plenum Press.

Godinot, M. (1984). Un nouveau genre temoignant de la diversite des Adapines (Primates, Adapidae) a l'Eocene terminal. *C. R. Acad. Sci. (Paris), ser. II* **299**(18):1291–1296.

______ (1991). Toward the locomotion of two contemporaneous *Adapis* species. *Z. Morph. Anthropol.* **78**: 387–405.

______ (1992). Apport a la systematique de quatre genres d'Adapiformes (Primates, Eocene). *C. R. Acad. Sci.* **314**:237–242.

Godinot, M., and Jouffroy, E. K. (1984). La main d'Adapis (Primates, Adapidae). In *Actes du Symposium Paleontologique* G. Cuvier, ed. E. Buffetaut, J. M. Mazin, and E. Salmion, pp. 221–242. Paris: Montbeliard.

Hartenberger, J.-L., and Marandat, B. (1992). A new genus and species of an early Eocene primate from North Africa. *Human Evol.* **7**:9–16.

Rasmussen, D. T. (1990). The phylogenetic position of *Mahgarita stevensi:* Protoanthropoid or lemuroid? *Int. J. Primatol.* 1B1:439–460.

______ (1994). The different meanings of a tarsioid-anthropoid clade and a new model of anthropoid origins. In *Anthropoid Origins,* ed. J. G. Fleagle and R. F. Kay, pp. 335–360. New York: Plenum Press.

Ross, C. (1994). The craniofacial evidence for anthropoid and tarsier relationships. In *Anthropoid Origins,* ed. J. G. Fleagle and R. F. Kay, pp. 469–547. New York: Plenum Press.

Russell, D. E., and Gingerich, P. D. (1987). Nouveaux primates de l'Eocene du Pakistan. *C. R. Acad. Sci. t. 304, ser. II(5),* pp. 209–214.

Simons, E. L. (1997). Discovery of the smallest Fayum Egyptian primates (Anchomomyini, Adapidae). *Proc. Natl. Acad. Sci. USA* **94**:180–184.

Simons, E. L., Rasmussen, D. T., and Gingerich, P. D. (1995). New cercamoniine adapid from Fayum, Egypt. *J. Human Evol.* **29**:577–589.

Sudre, J. (1975). Un Prosimien du Paleogene ancien du Sahara nord-occidental: *Azibius trerki* n.g., n. sp. *C. R. Acad. Sci. (Paris)* **280**:1539–1542.

Thalmann, U., Haubold, H., and Martin, R. D. (1989). *Pronycticebus neglectus*—An almost complete Adapid primate specimen from the Geiseltal (GDR). *Palaeovertebrata* **19**:115–130.

Wilson, J. A., and Szalay, E. S. (1976). New adapid primate of European affinities from Texas. *Folia Primatol.* **25**:294–312.

______ (1977). *Mahgarita,* a new name for *Margarita* Wilson and Szalay, 1976 non Leach 1814. *J. Paleontol.* **51**: 643.

Sivaladapines

Beard, K. C. (1998). Unmasking an Eocene primate enigma: the true identity of *Hoanghonius stehlinii. Am. J. Phys. Anthropol. Suppl.* **26**:69.

Chopra, S. R. K., and Vasishat, R. N. (1979). A new Mio-Pliocene *Indraloris* (Primate) material with comments on the taxonomic status of Sivanasuo (Carnivora) from the Siwaliks of the Indian subcontinent. *J. Human Evol.* **9**:129–132.

______ (1980). Premiere indication de la presence dans le Mio-Pliocene des Siwaliks de l'Inde d'un Primate Adapidae, *Indoadapis shivaii,* nov. gen., nov. sp. *C. R. Acad. Sci. (Paris), ser. D* **290**:511–513.

Gingerich, P. D. (1979). *Indraloris* and *Sivaladapis:* Miocene adapid primates from the Siwaliks of India and Pakistan. *Nature (London)* **279**(5712): 415–416.

Gingerich, P. D., and Sahni, (1984). Dentition of *Sivaladapis nagrii* (Adapidae) from the late Miocene of India. *Int. J. Primatol.* **5**:63–69.

Pan, Y., and Wu, R. (1986). A new species of *Sinoadapis* from the hominoid site, Lufeng. *Acta Anthropol. Sinica* **5**:39–50.

Wu, R., and Pan, Y. (1985). A new adapid primate from the Lufeng Miocene, Yunnan. *Acta Anthropol. Sinica* **4**(1):1–6.

ADAPOIDS AND STREPSIRHINES

Beard, K. C., Dagosto, M.., Gebo, D. L., and Godinot, M. (1988). Interrelationships among primate higher taxa. *Nature* **331**:712–714.

Dagosto, M. (1988). Implications of postcranial evidence for the origin of Euprimates. *J. Human Evol.* **17**:35–56.

Gheerbrant, E., Thomas, H., Roger, J., Sen, S., and Al-Sulaimani, Z. (1993). Deux nouveaux primates dans L'Oligocene inferieur de Taqah (Sultanat d'Pman): premiers adapiformed (?Anchomomyini) de la Peninsula Arabique. *Palaeovertebrata, Montpelier* **22**:141–196.

Gingerich, P. D. (1980). Eocene Adapidae, paleobiogeography and the origin of the South American Platyrrhini. In *Evolutionary Biology of the New World Monkeys and Continental Dhift,* ed. R. L. Ciochon and A. B. Chiarelli, pp. 123–138. New York: Plenum Press.

Gingerich, P. D., Holroyd, P. A., and Ciochon, R. L. (1994). *Rencunius zhoui,* new primate from the late middle Eocene of Hunan, China, and a comparison with some early Anthropoidea. In *Anthropoid Origins,* ed. J. G. Fleagle and R. F. Kay, pp. 163–177. New York: Plenum Press.

Kay, R. F., Ross, C., and Williams, B. A. (1997). Anthropoid origins. *Science* **275**:797–803.

Rasmussen, D. T. (1986). Anthropoid origins: A possible solution to the Adapidae–Omomyidae Paradox. *J. Human Evol.* **15**:1–12.

——— (1994). The different meanings of a tarsiod-anthropoid clade and a new model of anthropoid origins. In *Anthropoid Origins,* ed. J. G. Fleagle and R. F. Kay, pp. 335–360. New York: Plenum Press.

Rosenberger, A. L., Strasser, E., and Delson, E. (1985). Anterior dentition of *Notharctus* and the adapid anthropoid hypothesis. *Folia Primatol.* **44**:15–39.

Simons, E. L. (1997). Discovery of the smallest Fayum Egyptian primates (Anchomomyini, Adapidae). *Proc. Natl. Acad. Sci. USA* **94**:180–184.

Szalay, F. S., Rosenberger, A. L., and Dagosto, M. (1987). Diagnosis and differentiation of the order Primates *Yrbk. Phys. Anthropol.* **30**:75–105.

FOSSIL LORISES AND GALAGOS

Gebo, D. L. (1986). Miocene lorisids—The foot evidence. *Folia Primatol.* **47**:217–225.

Jacobs, L. L. (1981). Miocene lorisid from the Pakistan Siwaliks. *Nature (London)* **189**:585–587.

Le Gros Clark, W. E. (1956). A Miocene lemuroid skull from East Africa. Fossil mammals of Africa, no. 9. *Brit. Mus. (Nat. Hist.) London,* pp. 1–6.

Le Gros Clark, W. E., and Thomas D. P. (1952). The Miocene lemuroids of East Africa. Fossil mammals of Africa, no. 5. *Brit. Mus. (Nat. Hist.) London,* pp. 1–20.

MacPhee, R. D. E., and Jacobs, L. L. (1986). *Nycticeboides simpsoni* and the morphology, adaptations, and relationships of Miocene Siwalik Lorisidae. In *Vertebrates, Phylogeny, and Philosophy,* ed. K. M. Flanagen and J. A. Lillegraven. *Contrib. Geol. Univ. Wyoming, Special Papers* **3**:131–162.

McCrossin, M. (1992). New species of bushbaby from the middle Miocene of Maboko Island, Kenya. *Am. J. Phys. Anthropol.* **89**:215–234.

Simons, E. L., Bown, T. M., and Rasmussen, D. T. (1987). Discovery of two additional prosimian primate families (Omomyidae, Lorisidae) in the African Oligocene. *J. Human Evol.* **15**:431–437

Simpson, G. G. (1967). The Tertiary lorisiform primates of Africa. *Bull. Mus. Comp. Zool.* **136**:39–62.

Walker, A. C. (1970). Postcranial remains of the Miocene Lorisidae of East Africa. *Am. J. Phys. Anthropol.* **33**:249–262.

——— (1974). A review of the Miocene Lorisidae of East Africa. In *Prosimian Biology,* ed. R. D. Martin, G. A. Doyle, and A. C. Walker, pp. 435–447. London: Duckworth.

——— (1978). Prosimian primates. In *Evolution of African Mammals,* ed. V. J. Maglio and H. B. S. Cooke, pp. 90–99. Cambridge, MA: Harvard University Press.

——— (1987). Fossil galagines from Laetoli. In *Laetoli: A Pliocene Site in Northern Tanzania,* ed. M. D. Leakey and J. M. Harris, pp. 88–90. Oxford: Clarendon Press.

Wesselman, H. B. (1984). The Omo micromammals. *Contrib. Vert. Evol.* **7**:1–22.

OMOMYOIDS

Beard, K. C. and MacPhee, R. D. E. (1994). Cranial anatomy of *Shoshonius* and the antiquity of the Anthropoidea. In *Anthropoid Origins,* ed. J. G. Fleagle and R. F. Kay, pp. 55–97. New York: Plenum Press.

Beard, K. C., Krishtalka, L., and Stucky, R. (1992). Revision of the Wind River faunas, early Eocene of central Wyoming. Part 12. New species of omomyid primates (Mammalia: Primates: Omomyidae) and the omomyid taxonomic composition across the early-middle Eocene boundary. *Ann. Carnegie Mus.* **61**:39–62.

Covert, H. H. (1995). Locomotor adaptations of Eocene primates: Adaptive diversity among the earliest prosimians. In *Creatures of the Dark,* ed. L. Alterman, G. A. Doyle, and M. K. Izard, pp. 495–509. New York: Plenum Press

Dagosto, M. (1985). The distal tibia of primates with special reference to the Omomyidae. *Int. J. Primatol.* **6**:45–75.

Gingerich, P. D. (1981). Early Cenozoic Omomyidae and the evolutionary history of tarsiiform primates. *J. Human Evol.* **10**:345–374.

Gunnell, G. F. (1995). Omomyid primates (Tarsiiformes) from the Bridger Formation, middle Eocene, Southern Green River Basin, Wyoming. *J. Human Evol.* **28**: 147–187.

Honey, J. (1990). New Washakiin primates (Omomyidae) from the Eocene of Wyoming and Colorado, and comments on the evolution of the Washakiini. *J. Vert. Paleontol.* **10**:206–221.

Rose, K. D. (1995). The earliest primates. *Evol. Anthropol.* **3**:159–173.

Rose, K. D., and Bown, T. M. (1991). Additional fossil evidence on the differentiation of the oldest Euprimates. *Proc. Natl. Acad. Sci. USA* **88**:98–101.

Rose, K. D., Godinot, M., and Bown, T. M. (1994). The early radiation of Euprimates and the initial diversification of Omomyidae. In *Anthropoid Origins,* ed. J. G. Fleagle and R. F. Kay, pp. 1–28. New York: Plenum Press.

Rosenberger, A. L., and Dagosto, M. (1992). New craniodental and postcranial evidence of fossil tarsiiforms, In *Topics in Primatology,* ed. S. Matano, R. H. Tuttle, H. Ishida, and M. Goodman, Vol. 3, pp. 37–51. Kyoto: University of Kyoto Press.

Ross, C. (1994). The craniofacial evidence for anthropoid and tarsier relationships. In *Anthropoid Origins,* ed. J. G. Fleagle and R. F. Kay, pp. 469–547. New York: Plenum Press.

Szalay, F. S. (1976). Systematics of the Omomyidae (Tarsiiformes, Primates): Taxonomy, phylogeny and adaptations. *Bull. Am. Mus. Nat. Hist.* **156**(3):157–450.

Williams, B. A. (1994). Phylogeny of the Omomyidae and Implications for Anthropoid Origins. Ph.D. Dissertation, Univeristy of Colorado.

Anaptomorphines and Omomyines

Beard, K. C. (1987). *Jemezius:* A new omomyid primate from the early Eocene of northwestern New Mexico. *J. Human Evol.* **16**:457–468.

Beard, K. C., and MacPhee, R. D. E. (1994). Cranial anatomy of *Shoshonius* and the antiquity of the Anthropoidea. In *Anthropoid Origins,* ed. J. G. Fleagle and R. F. Kay, pp. 55–97. New York: Plenum Press.

Beard, K. C., and Wang, B. (1990). Phylogenetic and biogeographic significance of the tarsiiform primate *Asiomomys changbaicus* from the Eocene of Jilin Province, People's Republic of China. *Am. J. Phys. Anthropol.* **85**:159–166.

Beard, K. C., Krishtalka, L., and Stucky, R. K. (1991). First skulls of the early Eocene primate *Shoshonius cooperi* and the anthropoid–tarsier dichotomy. *Nature* **349**:64–67

______ (1992). Revision of the Wind River faunas, early Eocene of central Wyoming. Part 12. New species of omomyid primates (Mammalia: Primates: Omomyidae) and the omomyid taxonomic composition across the early-middle Eocene boundary. *Ann. Carnegie Mus.* **61**:39–62.

Beard, K. C., Tao, Q., Dawson, M. R., Wang, B., and Chuanhuei, L., (1994). A diverse new primate fauna from middle Eocene fissure-fillings in southeastern China. *Nature* **368**:604–609.

Bown, T. M. (1976). Affinities of *Teilhardina* (Primates, Omomyidae) with description of a new species from North America. *Folia Primatol.* **25**:62–72.

Bown, T. M., and Rose, K. D. (1987). Patterns of dental evolution in early Eocene Anaptomorphine primates (Omomyidae) from the Bighorn Basin, Wyoming. *Paleont. Soc. Mem.* **23**:1–162

______ (1991). Evolutionary relationships of a new genus and three new species of omomyid primates (Willwood Formation, lower Eocene, Bighorn Basin, Wyoming). *J. Human Evol.* **20**:465–480.

Covert, H. H., and Hamrick, M.. (1993). Description of new skeletal remains of early Eocene anaptomorphine primate *Absarokius* (Omomyidae) and a discussion about its adaptive profile. *J. Human Evol.* **25**:351–362.

Covert, H. H., and Williams, B. A. (1991). The anterior lower dentition of *Washakius insignus* and adapid-anthropoidean affinities. *J. Human Evol.* **21**:463–467.

Dagosto, M. (1985). The distal tibia of primates with special reference to the Omomyidae. *Int. J. Primotol.* **6**:45–75.

Gingerich, P. D. (1981). Early Cenozoic Omomyidae and the evolutionary history of tarsiiform primates. *J. Human Evol.* **10**:345–374.

Gunnell, G. F. (1995). Omomyid primates (Tarsiiformes) from the Bridger Formation, middle Eocene, southern Green River Basin, Wyoming. *J. Human Evol.* **28**: 147–187.

Honey, J. (1990). New Washakiin primates (Omomyidae) from the Eocene of Wyoming and Colorado, and comments on the evolution of the Washakiini. *J. Vert. Paleontol.* **10**:206–221.

Rasmussen, D. T. (1996). A new middle Eocene omomyine primate from the Uinta Basin, Utah. *J. Human Evol.* **31**:75–87.

Rasmussen, D. T., Shekele, M., Walsh, S. L., and Riney, B. O. (1995). The dentition of *Dyseolemur* and comments on the use of the anterior teeth in primate systematics. *J. Human Evol.* **29**:301–320.

Rose, K. D. (1995). Anterior dentition and relationships of the early Eocene omomyids *Arapahovius advena* and *Teilhardina demissa*, sp. nov. *J. Human Evol.* **28**:231–244.

Rose, K. D., and Bown, T. M.. (1991). Additional fossil evidence on the differentiation of the oldest Euprimates. *Proc. Natl. Acad. Sci. USA* **88**:98–101

——— (1993). Species concepts and species recognition in Eocene primates. In *Species, Species Concepts and Primate Evolution,* ed. W. H. Kimbel and L. B. Martin, pp. 299–330. New York: Plenum Press.

Rose, K. D., and Rensberger, J. M. (1983). Upper dentition of *Ekgmowechashala* (Omomyid, Primate) from the John Day Formation, Oligo-Miocene of Oregon. *Folia Primatol.* **41**:102–113.

Simpson, G. G. (1940). Studies on the earliest primates. *Bull. Am. Mus. Nat. Hist.* **77**:185–212.

Strait, S. (1991). Dietary Reconstruction on Small-Bodied Fossil Primates. Ph.D. Disssertation, SUNY Stony Brook.

Williams, B. A., and Covert, H. H. (1994). New early Eocene anaptomorphine primate (Omomyidae) from the Washakie Basin, with comments on the phylogeny and paleobiology of anaptomorphines. *Am. J. Phys. Anthropol.* **93**:323–340.

Wilson, J. A. (1966). A new primate from the earliest Oligocene, west Texas, preliminary report. *Folia Primatol.* **4**.227–248.

Microchoerids

Dagosto, M. (1993). Postcranial anatomy and locomotor behavior in Eocene primates. In *Postcranial Adaptation in Nonhuman Primates,* ed. D. L. Gebo, pp. 199–219. DeKalb:Northern Illinois University Press.

Dagosto, M., and Schmid, P. (1996). Proximal femoral anatomy of omomyiform primates. *J. Human Evol.* **30**: 29–56.

Godinot. M. (1985). Evolutionary implications of morphological changes in Paleogene primates. *Special papers in Palaentology* **33**:39–47.

Godinot, M., and Dagosto, M. (1983). The astragalus of *Necrolemur* (Primates, Microchoerinae). *J. Paleontol.* **57**: 1321–1324.

Godinot, M., Russell, D. E., and Louis, P. (1992). Oldest known *Nannopithex* (Primates, Omomyiformes) from the early Eocene of France. *Folia Primatol.* **58**:32–40.

Krishtalka, L., and Schwartz, J. H. (1978). Phylogenetic relationships of plesiadapiform-tarsiiform primates. *Ann. Carnegie Mus.* **47**:515–540.

Rose, K. D. (1995). The earliest primates. *Evol. Anthropol.* **3**:159–173.

Rosenberger, A. L. (1985). In favor of the *Necrolemur*–Tarsier hypothesis. *Folia Primatol.* **45**:179–194.

Schmid, P. (1979). Evidence of microchoerine evolution from Dielsdorf (Zurich region, Switzerland)—A preliminary report. *Folia Primatol.* **31**:301–313.

——— (1982). Comparison of Eocene nonadapids and *Tarsius.* In *Primate Evolutionary Biology,* ed. A. B. Chiarelli and R. E. Corruccini, pp. 6–13. Berlin: Springer-Verlag.

——— (1983). Front dentition of the Omomyiformes (Primates). *Folia Primatol.* **40**:1–10.

Simons, E. L. (1961). Notes on Eocene tarsioids and a revision of some Necrolemurinae. *Bull. Brit. Mus. (Nat. Hist.) Geol.* **5**:43–49.

Simons, E. L., and Russell, D. E. (1960). The cranial anatomy of *Necrolemur. Breviora* **127**:1–14.

Szalay, F. S., and Dagosto, M. (1980). Locomotor adaptations as reflected on the humerus of Paleogene primates. *Folia Primatol.* **34**:1–45.

Asian Omomyids

Beard, K. C., and Wang, B. (1990). Phylogenetic and biogeographic significance of the tarsiiform primate *Asiomomys changbaicus* from the Eocene of Jilin Province, People's Republic of China. *Am. J. Phys. Anthropol.* **85**:159–166.

Beard, K. C., Tao, Q., Dawson, M. R., Wang, B., and Chuanhuei, L., (1994). A diverse new primate fauna from middle Eocene fissure-fillings in southeastern China. *Nature* **368**:604–609.

Dagosto, M., Gebo, D. L., Beard, C., and Qi, T. (1996). New primate postcranial remains from the middle Eocene Shanghuang fissures, southeastern China. *Am. J. Phys. Anthropol. Suppl.* **22**:92–93.

Gebo, D. L., Dagosto, M., Beard, C., and Qi, T. (1996). New Primate tarsal remains from the middle Eocene Shanghuang fissures, southeastern China. *Am. J. Phys. Anthropol. Suppl.* **22**:113.

Szalay, F. S., and Li, C. K. (1986). Middle Paleocene Euprimate from southern China and the distribution of primates in the Paleogene. *J. Human Evol.* **15**:387–398.

Tarsiids

Beard, K. C. (1998). A new genus of Tarsiidae (Mammalia: Primates) from the middle Eocene of Shanxi Province, China, with notes on the historical biogeography of tarsiers. *Bull. Carnegie Mus. Nat. Hist.* **34**:260–277.

Beard, K. C., Tao, Q., Dawson, M. R., Wang, B., and Chuanhuei, L., (1994). A diverse new primate fauna from middle Eocene fissure-fillings in southeastern China. *Nature* **368**:604–609.

Dagosto, M., Gebo, D. L., Beard, C., and Qi, T. (1996). New Primate postcranial remains from the middle Eocene Shanghuang fissures, southeastern China. *Am. J. Phys. Anthropol. Suppl.* **22**:92–93.

Gebo, D. L., Dagosto, M., Beard, C., and Qi, T. (1996). New Primate tarsal remains from the middle Eocene Shanghuang fissures, southeastern China. *Am. J. Phys. Anthropol. Suppl.* **22**:113.

Ginsburg, L., and Mein, P. (1986). *Tarsius thailandica* nov. sp., Tarsiidae (Primates, Mammalia) fossile d'Asie. *C. R. Acad. Sci. (Paris), t.304, ser. II, no. 19,* pp. 1213–1215.

Simons, E. L., and Bown, T. M. (1985). *Afrotarsius chatrathi,* first tarsiiform primate (?Tarsiidae) from Africa. *Nature (London)* **313**:475–477.

OMOMYOIDS, TARSIERS, AND HAPLORHINES

Beard, K. C., and MacPhee, R. D. E. (1994). Cranial anatomy of *Shoshonius* and the antiquity of the Anthropoidea. In *Anthropoid Origins,* ed. J. G. Fleagle and R. F. Kay, pp. 55–97. New York: Plenum Press.

Cartmill, M. (1980). Morphology, function and evolution of the anthropoid postorbital septum. In *Evolutionary Biology of New World Monkeys and Continental Drift,* ed. R. L. Ciochon and A. B. Chiarelli, pp. 243–274. New York: Plenum Press.

Cartmill, M., and Kay, R. F. (1978). Cranio-dental morphology, tarsier affinities, and primate suborders. In *Recent Advances in Primatology,* vol. 3, ed. D. J. Chivers and K. A. Joysey, pp. 205–213. London: Academic Press.

Dagosto, M. (1988). Implications of postcranial evidence for the origin of Euprimates. *J. Human Evol.* **17**:35–56.

Fleagle, J. G., and Simons, E. L. (1983). The tibio-fibular articulation in *Apidium phiomense,* an Oligocene anthropoid. *Nature (London)* **301**(5897):238–239.

Gingerich, P. D. (1981). Early Cenozoic Omomyidae and the evolutionary history of tarsiiform primates. *J. Human Evol.* **10**:345–374.

Gonzalez, W. G., Kay, R. F., and Kirk, E. C. (1998). Optic canal and orbit size implications for the origins of diurnality and visual acuity in primates. *Am. J. Phys. Anthropol. Suppl.* **26**:86.

Kay, R. F., Ross, C., and Williams, B. A. (1997). Anthropoid origins. *Science* **275**:797–803.

Rosenberger, A. L., and Szalay, F. S. (1981). On the tarsiiform origins of Anthropoidea. In *Evolutionary Biology of New World Monkeys and Continental Drift,* ed. R. L. Ciochon and A. B. Chiarelli, pp. 139–157 New York: Plenum Press.

Ross, C. (1994). The craniofacial evidence for anthropoid and tarsier relationships. In *Anthropoid Origins,* ed. J. G. Fleagle and R. F. Kay, pp. 469–547. New York: Plenum Press.

Schmid, P. (1982). Comparison of Eocene nonadapids and *Tarsius.* In *Primate Evolutionary Biology,* ed. A. B. Chiarelli and R. L. Corruccini, pp. 6–13. Berlin: Springer-Verlag.

______ (1983). Front dentition of the Omomyiformes (Primates). *Folia Primatol.* **40**:1–10.

Simons, E. L. (1961). The dentition of *Ourayia*—Its bearing on relationships of omomyid prosimians. *Postilla* **54**:1–20.

______ (1972). *Primate Evolution.* New York: Macmillan.

Szalay, F. S., Rosenberger, A. L., and Dagosto, M. (1987). Diagnosis and differentiation of the order Primates. *Yrbk. Phys. Anthropol.* **30**:75–105.

Wortman, J. L. (1903). Classification of the primates. *Am. J. Sci.* **15**:399–414.

THIRTEEN

Early Anthropoids

EOCENE–OLIGOCENE TRANSITION

Although the Oligocene epoch proper is currently dated to between approximately 34 million and 23 million years ago, the Eocene-to-Oligocene transition involved a long series of geological, climatic, and paleontological changes that took place over about 10 million years (Prothero and Berggren, 1992). Some of these changes took place in the late Eocene, others in the early part of the Oligocene, and many in a series of steps. By the Oligocene, the continents were beginning to look as they do today except for the lack of a connection between North America and South America. India was colliding with the Asian mainland, closing off the Tethys Seaway on the east, and both South America and Australia were separated from Antarctica. These last events made possible the first deep water currents around Antarctica. As a result, the Eocene–Oligocene transition was marked by a major drop in global temperatures from the more tropical climates of the early and middle Eocene. The early Oligocene also saw a dramatic lowering of sea level (see Fig. 10.4), probably as a result of glaciations at the poles. These climatic changes led to equally dramatic changes in the primate fossil record. In the northern hemisphere, the prosimians that had been abundant in the Eocene disappeared by the beginning of the Oligocene in Europe, and they became increasingly rare in North America, so primates are virtually unknown from northern continents during the Oligocene. However, this same time period has yielded increasingly abundant remains of early anthropoids in Africa and Asia.

Eocene Anthropoids from Asia

One of the most exciting developments in recent years has been the discovery of a new possible early anthropoid, ***Eosimias,*** from the middle Eocene of southern China (Beard *et al.,* 1994, 1996; Table 13.1; Figs. 13.1, 13.2). This tiny primate has been identified as a very primitive anthropoid on the basis of its small, spatulate incisors, enlarged canines, broad premolars with oblique roots, and several molar features, including relatively broad trigonids as well as an anthropoid-like mandible. There are two species, *E. sinensis* from Shanghuang and *E. centennicus* from the Yuangu Basin. The Chinese fossils are much more primitive than any other taxa allocated to Anthropoidea (Kay *et al.,* 1997) and are very different from any of the early anthropoids from the Eocene and Oligocene of Egypt except possibly *Afrotarsius.* A petrosal bone attributed to *Eosimias* (MacPhee *et al.,* 1995) is omomyid-

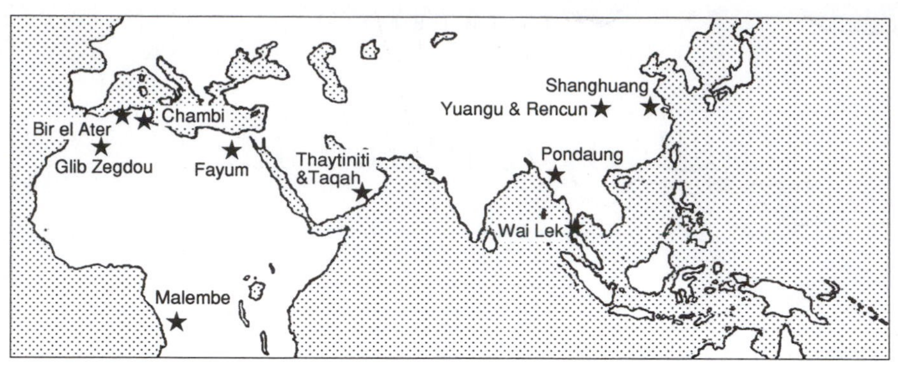

FIGURE 13.1 Geographic distribution of early fossil anthropoids.

like with no definitively anthropoid features. However, there are anthropoid-like limb bones from the same fissure-fills as *Eosimias* (Gebo *et al.*, 1996; Dagosto *et al.*, 1996) The anthropoid affinities of *Eosimias* are likely to be debated until additional cranial material becomes available. If the new Chinese fossils are indeed basal anthropoids as proposed, it is particularly notable that they show none of the dental similarities to adapids found in the oligopithecines from Egypt (see upcoming discussion). Indeed, among potential anthropoid sister taxa, they show the greatest similarities to tarsiers, a group that is also common in the Chinese Eocene faunas (see Chapter 12).

TABLE 13.1
Suborder cf. Anthropoidea
Superfamily *Incertae sedis*

Species	Estimated Mass (g)
Family EOSIMIIDAE	
Eosimias (middle Eocene, China)	
E. sinensis	100
E. centennicus	
Family *incertae sedis*	
Amphipithecus (middle to late Eocene, Myanmar)	
A. mogaungensis	8600
Pondaungia (middle to late Eocene, Myanmar)	
P. cotteri	7000
Siamopithecus (late Eocene, Thailand)	
S. eocaenus	6800

Two other Eocene possible anthropoids from Asia are ***Amphipithecus mogaungensis*** and ***Pondaungia cotteri*** (Fig. 13.3, Table 13.1) from the late Eocene of Myanmar (Burma). A few tantalizing fossils of each were recovered earlier this century, and additional material has come to light in recent years. In both species, the broad, low-crowned molars and deep mandibles suggest higher primate rather than either adapid or omomyid affinities, but the material presently available is insufficient to confirm this suggestion. Authorities continue to debate whether either is an anthropoid (e.g., Ciochon and Holroyd, 1994). Both were moderate-size primates (6–10 kg) with molars that suggest a frugivorous diet.

Siamopithecus eocaenus (Fig. 13.3) is another potential early anthropoid, similar to *Pondaungia,* from the late Eocene Krabi Basin (Wailek) in Thailand (Chaimanee *et al.,* 1997). It was a large primate with an estimated body mass of between 6.5 and 7 kg. As in anthropoids, the mandibular and maxillary molars are broad, bunodont teeth, and the mandibular corpus is very deep.

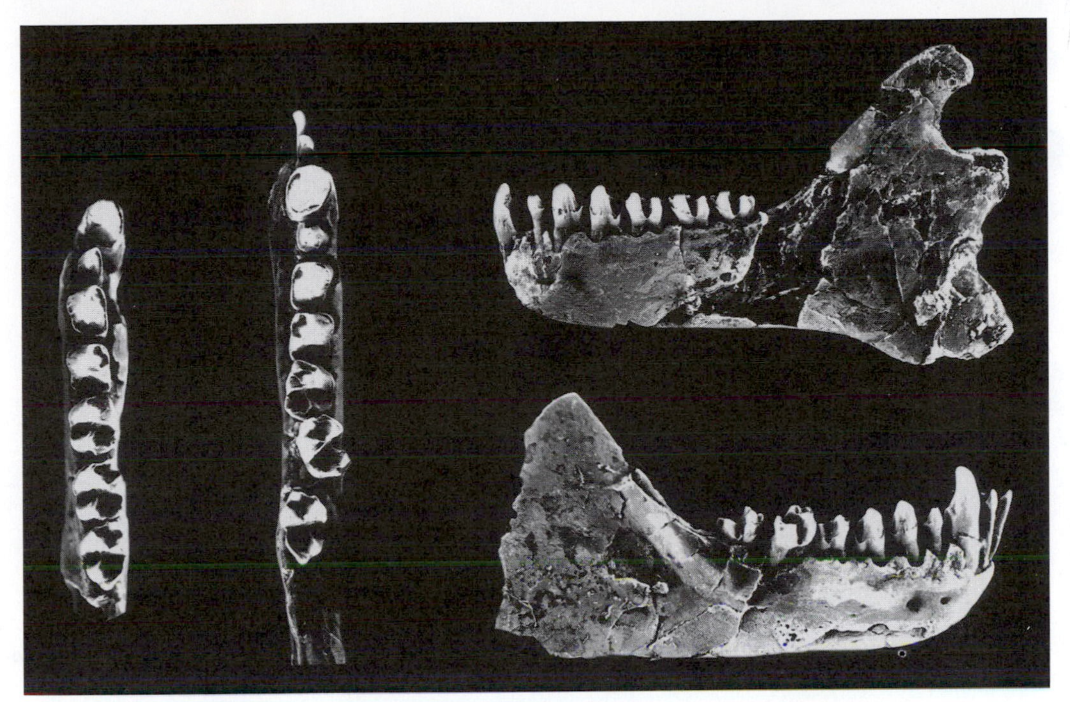

FIGURE 13.2 *Eosimias centennicus* from the middle Eocene of China (courtesy of K. C. Beard).

Eocene and Oligocene Anthropoids from Africa and Arabia

In recent years there has also been a virtual explosion of fossil anthropoids from Eocene deposits in North Africa and Arabia that pushes the record of early anthropoids back into the late or even middle Eocene (Simons and Rasmussen, 1994). However, many of these new Eocene anthropoids are still very poorly known, and even their Eocene dates are subject to debate. In discussing these new early anthropoids we will begin with the best-known record from Egypt and then consider the more fragmentary remains from other sites.

Fossil Primates from Fayum, Egypt

Since the beginning of this century, most of our knowledge of early higher primate evolution in the Old World has come from an area in Egypt known as the Fayum Depression. Here, in an expanse of eroded badlands on the eastern edge of the Sahara Desert (Fig. 13.4), is a sequence of highly fossiliferous sedimentary deposits, the Jebel Qatrani Formation (Fig. 13.5), that preserves the best record of early mammalian evolution in Africa. There is considerable debate over the exact age of this formation and of the primate fossils in these deposits. Some have argued that the entire formation is of Eocene age; others have argued that the entire sequence is Oligocene in age. This formation lies very near the Eocene/Oligocene boundary, about 34 million years ago (Prothero and Berggren, 1992). It seems most likely that the Jebel Qatrani Formation straddles the Eocene/Oligocene boundary, with the lower part of the formation dating from the latest Eocene and the upper part

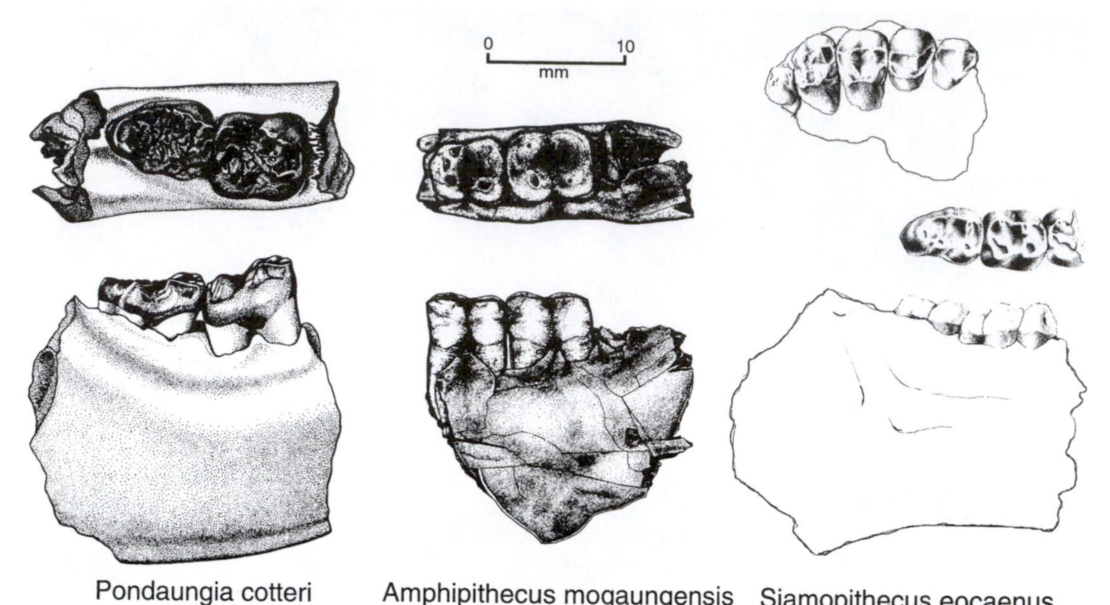

FIGURE 13.3 Potential early anthropoids from the Eocene of Asia. *Pondaungia, Amphipithecus,* and *Siamopithecus* (courtesy of R. L. Ciochon and S. Ducrocq).

FIGURE 13.4 The desert landscape of the Fayum Depression of Egypt.

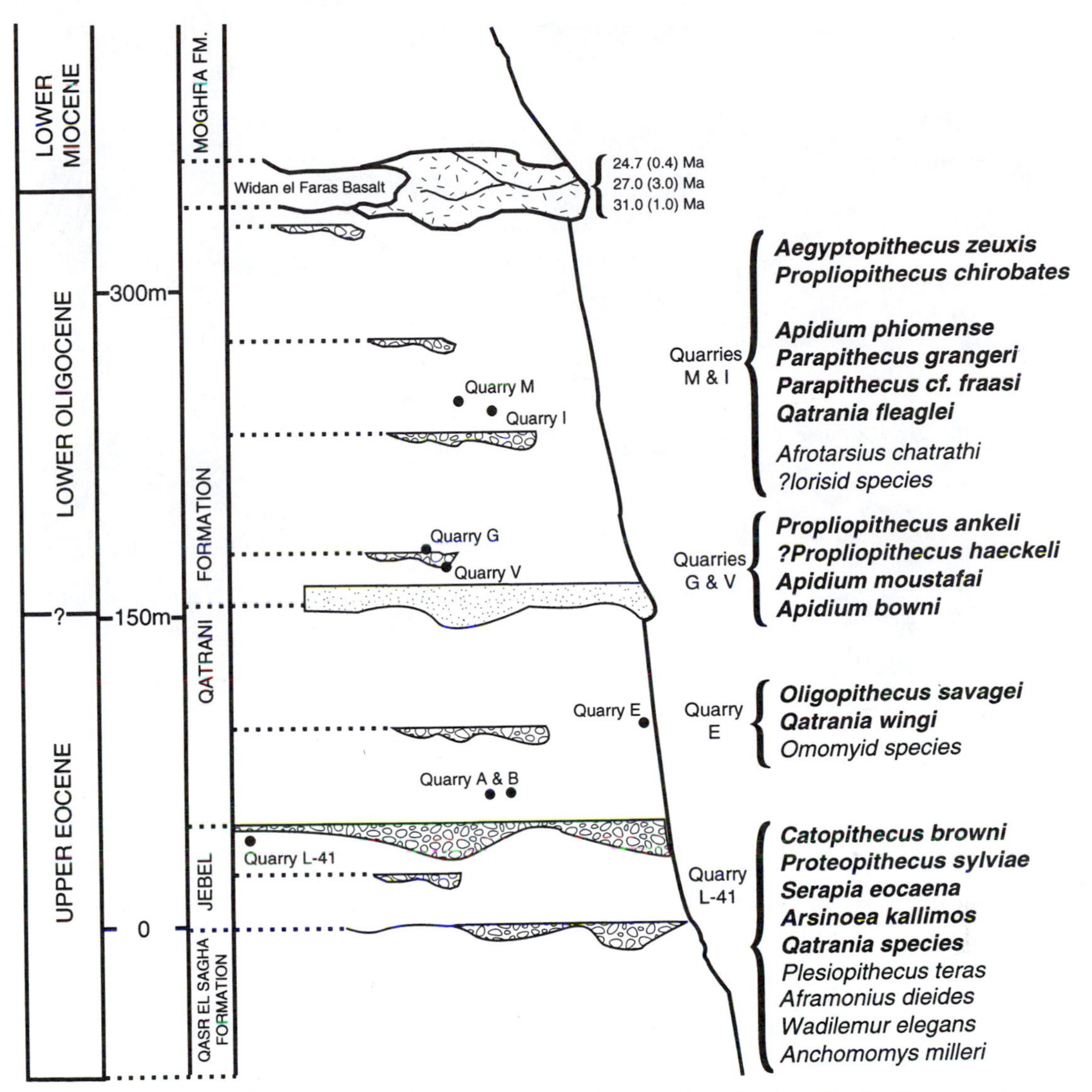

FIGURE 13.5 Stratigraphic section of the Jebel Qatrani Formation showing the fossil taxa and major quarries. Anthropoids are in bold type.

from the earliest Oligocene (Simons and Rasmussen, 1994; Kappelman *et al.*, 1992).

Despite the uncertainty over the exact age of the Fayum deposits, we know a great deal about the environment under which the sediments and the fossil primates within them were deposited (Fig. 13.6). From the sediments, we know that the climate was warm, wet, and somewhat seasonal. The fossil plants are most similar to species currently found in

FIGURE 13.6 A reconstruction of the early Oligocene environment and fauna from the Fayum.

the tropical forests of Southeast Asia. The quarries yielding the primate fossils are very diverse. Some were laid down as sandbars in river channels and show repeated sequences of standing water (probably oxbow lakes) as well as roots of mangrovelike plants. This evidence, together with abundant fossil remains of water birds, indicates a swampy environment at the time of deposition (Figs. 13.6, 13.7); others were probably quicksands that were dry in some seasons.

Most of the mammals that are found with the primates are different from Eocene or Oligocene mammals in other parts of the world and from younger African faunas from Kenya and Uganda. The rodents are the earliest members of the porcupine suborder (hystricomorphs), which includes the guinea pig–like rodents of South America as well as several species from Africa. These Oligocene rodents were more arboreal than their living African relatives. There were opossums as well as insectivores, bats, primitive carnivores, and anthracotheres, an archaic group of artiodactyls related to the hippopotamus. In addition, the Fayum provides the first substantial record of several African groups of mammals, such as hyraces, elephants, and elephant shrews, as well as numerous fossil primates.

As a result of ongoing research by Elwyn Simons and his colleagues over the past three decades, the Fayum has yielded one of the

FIGURE 13.7 Three Fayum anthropoids: above, the propliopithecids *Aegyptopithecus zeuxis* (left) and *Propliopithecus chirobates* (right); below, the parapithecid *Apidium phiomense*.

TABLE 13.2
Suborder Anthropoidea
Superfamily Parapithecoidea

Species	Estimated Mass (g)
Family PARAPITHECIDAE	
Serapia (late Eocene, Egypt)	
S. eocaena	?1000
Qatrania (late Eocene to early Oligocene, Egypt)	
Q. wingi	300
Q. fleaglei	600
Apidium (early Oligocene, Egypt)	
A. phiomense	1600
A. moustafai	850
A. bowni	750
Parapithecus (early Oligocene, Egypt)	
P. fraasi	1700
P. grangeri	3000
Biretia (Eocene, Algeria)	
B. piveteaui	?300
Superfamily *Incertae sedis*	
Proteopithecus (late Eocene, Egypt)	
P. sylviae	500
Arsinoea (late Eocene, Egypt)	
A. kallimos	?350
Algeripithecus (Eocene, Algeria)	
A. minutus	150–300
Tabelia (Eocene, Algeria)	
T. hammadae	?450

most diverse primate faunas from anywhere in the world (Fig. 13.5). There are at least five groups of prosimians (Chapter 12), including cercamoniine adapids, the tarsierlike *Afrotarsius,* and the odd *Plesiopithecus* with a lorislike cranium and an enlarged anterior tooth like that of plesiadapiforms. In addition there is a single tooth attributed to an omomyid and another tooth attributed to a loris. There are three well-known groups of higher primates: parapithecids, propliopithecids, and oligopithecids (Fig. 13.7). Finally, there are two taxa, *Proteopithecus* and *Arsinoea,* that cannot readily be placed in any of the preceding groups (Simons and Rasmussen, 1994). This diverse array provides many insights into the initial diversification of higher primates.

The fossil primates have been recovered primarily from three levels within the overall sequence of sediments of the Jebel Qatrani Formation (Fig. 13.5). In the uppermost level (Quarries M, I), primates are the most common mammals, and one species, *Apidium phiomense,* is known from hundreds of fossils. The primates from the uppermost levels of the Fayum have been well documented in the past few decades (e.g., Simons, 1967; Fleagle and Kay, 1983, 1985, 1987; Simons and Rasmussen, 1991). The middle levels (Quarries G, V) have yielded relatively few primates, but these may contain some crucial intermediate species. It is the lowest, late Eocene levels (E and especially L-41) that have yielded the most exciting new discoveries of primitive anthropoids (Simons and Rasmussen, 1994). Many of the new fossils have not yet been thoroughly described, but they have already restructured our understanding of the initial radiation of anthropoids from views that were current less than ten years ago.

Parapithecids

Although the first parapithecid was discovered near the turn of the century, an appreciation of the diversity of this group has come only in recent years. There are eight species and four genera placed in the family Parapithecidae. Parapithecid species ranged in size from tiny, marmoset-size species such as *Serapia* and *Qatrania,* which are among the smallest Old World higher primates, to the guenon-size *Parapithecus grangeri* (Table 13.2). These late Eocene and early Oligocene anthropoids are the most primitive of all known higher primates, and they have a number of anatomical features that distinguish them from all other Old World anthropoids.

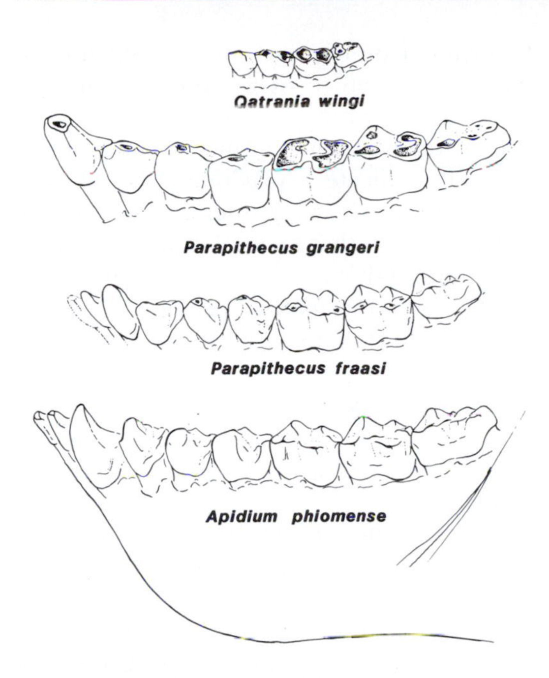

FIGURE 13.8 Lower dentitions of parapithecids (courtesy of Richard Kay).

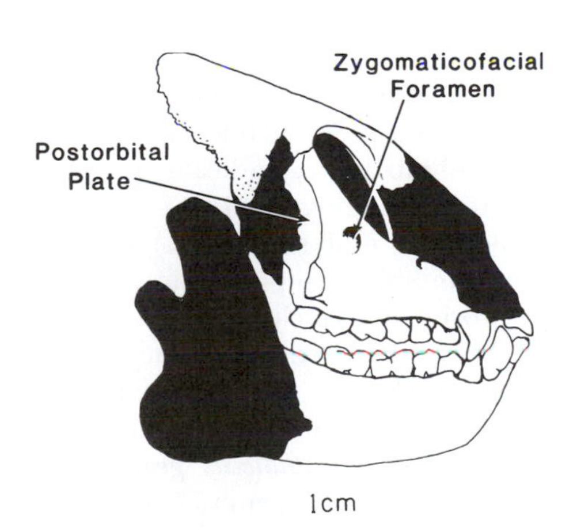

FIGURE 13.9 A reconstructed facial skeleton of *Apidium phiomense* (after Simons, 1971).

Parapithecids have a primitive dental formula of 2.1.3.3, as in New World monkeys. This is probably the primitive dental formula for all higher primates, but it is not found in any living catarrhines. In the best-known species, *Apidium phiomense,* the lower incisors are small and spatulate, but one species, *Parapithecus grangeri,* lost its permanent lower incisors altogether (Fig. 13.8). Upper incisors are poorly known. The canines in *Apidium* are similar to those of most platyrrhines, but in *P. grangeri* they are large and tusklike (Simons, 1986). The three lower premolars increase in size and complexity from front to back, but in all species the last premolar resembles the premolars of earlier prosimians rather than of later anthropoids in having a metaconid that is smaller and distally positioned relative to the protoconid. The upper premolars of parapithecids are broad, with three cusps rather than two as in other higher primates.

Parapithecid molars are characterized by low, rounded cusps. The upper molars are quadrate with well-developed conules and a large hypocone. The lower molars have a small trigonid (often with paraconid) and a broad talonid basin. In some species accessory cusps are common, and often there is a buccolingual alignment of the molar cusps and a narrowing in the center of the tooth, giving parapithecid molars a "waisted" shape superficially similar to that seen in cercopithecoid monkeys. The mandible is fused at the symphysis.

The skull of parapithecids is known from only a few relatively complete remains (Fig. 13.9), but these clearly show higher primate features, such as fused frontal sutures and postorbital closure (Simons, 1995). The arrangement of the cranial sutures in the pterion region of the skull seems to be similar to that of platyrrhines in having a zygomatic–parietal contact exposed on the skull wall. Several frontal fragments that preserve an endocast of the anterior part of the brain show a relatively large olfactory bulb. The auditory region in parapithecids is poorly known but seems to be characterized by a large promontory artery and anterior accessory cavity as in anthropoids (and *Tarsius*) and the lack of a tubular ectotympanic.

Dozens of parts of the limb skeleton have been recovered for one species, *Apidium phiomense* (Fig. 13.10). In many features of their limbs, parapithecids are more primitive than any later Old World higher primates and resemble platyrrhines and omomyid prosimians or are unique among primates (Fleagle and Simons, 1995). In *Apidium,* the tibia and fibula are joined for approximately 40 percent of their length, a similarity to some microchoerines, some platyrrhines, and *Tarsius.*

Serapia eocaena, from the late Eocene Quarry, L-41, is the earliest and most primitive parapithecid. Only known from lower jaws, the tiny *Serapia* has a dental formula of ?.1.3.3 and relatively high molar trigonids for a parapithecid.

Qatrania wingi, from Quarry E in the lower part of the Jebel Qatrani Formation, is another very early and primitive parapithecid (Fig. 13.8). This tiny primate (less than 300 g), known from only two lower jaws and a few isolated teeth, has remarkably low and bulbous molar cusps. The absence of shearing crests on the teeth indicates that its diet was probably fruits or gums rather than insects. A second species of *Qatrania, Q. fleaglei,* is from the upper part of the formation.

There are three species of *Apidium,* the best known parapithecid. Two smaller species, ***A. bowni*** (Quarry V) and ***A. moustafai*** (Quarry G), are from the intermediate levels of the Jebel Qatrani Formation; the larger ***A. phiomense*** is from the upper part of the formation. *Apidium phiomense* is known from hundreds of specimens, including jaws, limb bones, and a few recently described cranial remains. *Apidium* has tiny incisors, moderate-size, sexually dimorphic canines, and molars with numerous low, rounded cusps and very few shearing crests (Fig. 13.8). All species of *Apidium* have a fused mandibular symphysis. Functionally, the teeth indicate a predominantly frugivorous diet, but the very thick enamel on the molars suggests that seeds also may have been an important dietary component. The canine dimorphism, unusual in a primate this small, suggests that *Apidium* lived in polygynous social groups.

The few cranial remains of *Apidium* show a short snout, a small infraorbital foramen, and relatively small eyes (Fig. 13.9). It was a diurnal monkey.

The many postcranial bones attributed to *Apidium* show it was an excellent leaper (Fig. 13.10). The hindlimb is relatively long compared with the forelimb (intermembral index = 70), the ischium is extremely long, the femoral neck is oriented at a right angle to the shaft, and the distal femoral condyles are very deep, more so than in any other higher primate. The tibia is extremely long and laterally compressed, and the fibula is tightly attached to it distally. The ankle joint is hinged for rapid flexion and extension. *Apidium* probably had a divergent hallux. The scapula is similar to that seen in many living anthropoid quadrupedal leapers such as *Saimiri,* and the short forelimb bones indicate quadrupedal rather than clinging habits (Fig. 13.7). In many details of limb structure, *Apidium* shows greatest similarities to platyrrhines and to Eocene prosimians (including omomyids and microchoerids) rather than to later Old World anthropoids (Fleagle and Simons, 1995).

The most unusual parapithecid, and also the largest, is ***Parapithecus grangeri,*** often placed in a separate genus, *Simonsius.* Like *Apidium,* this species has three premolars and three molars. The cusps on the lower molars are arranged in two lophs, superficially similar to the condition found in cercopithecoid

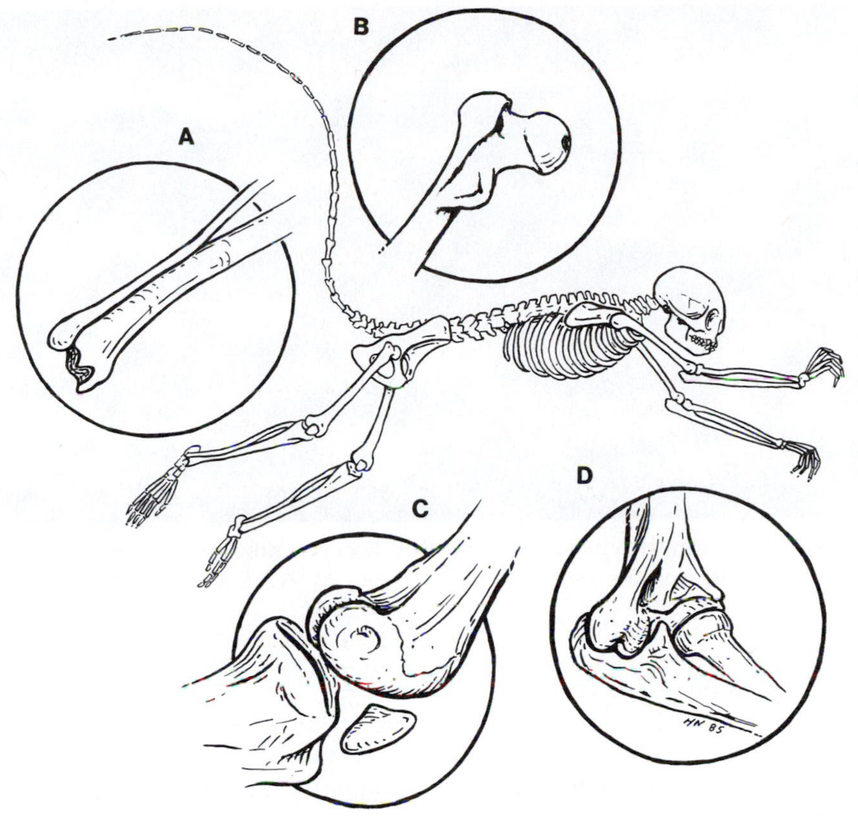

FIGURE 13.10 A restored skeleton of *Apidium phiomense* showing many of the distinctive features of this species: A, the tibia and fibula nearly fused for the distalmost 40 percent of their length; B, the large lesser trocanter of the femur; C, the deep distal condyles of the femur; D, the entepicondylar foramen and elongate capitulum of the humerus. In these features, *Apidium* is more like omomyoid prosimians and small platyrrhines than modern catarrhines.

monkeys. The lower premolars are short with bulbous cusps, and the upper premolars have three major prominent cusps. A most unusual feature of is the anterior dentition: large, tusklike canines and no permanent incisors (Kay and Simons, 1983; Simons, 1986). The function of this tusklike arrangement is unclear, but it seems likely that *P. grangeri* was partly folivorous or perhaps a seed predator. The few facial parts known indicate a short, pointed snout (Simons, 1995).

Although it was the first parapithecid described, ***Parapithecus fraasi*** remains a poorly understood species. The type specimen was described earlier this century from an unknown site in the Fayum, and this medium-size species is still known from only a few jaws, none of which preserve the incisors intact. The dental formula of *P. fraasi* has been debated since its initial discovery. Because *Apidium* has a dental formula of 2.1.3.3, it was assumed that *P. fraasi* is similar and that the

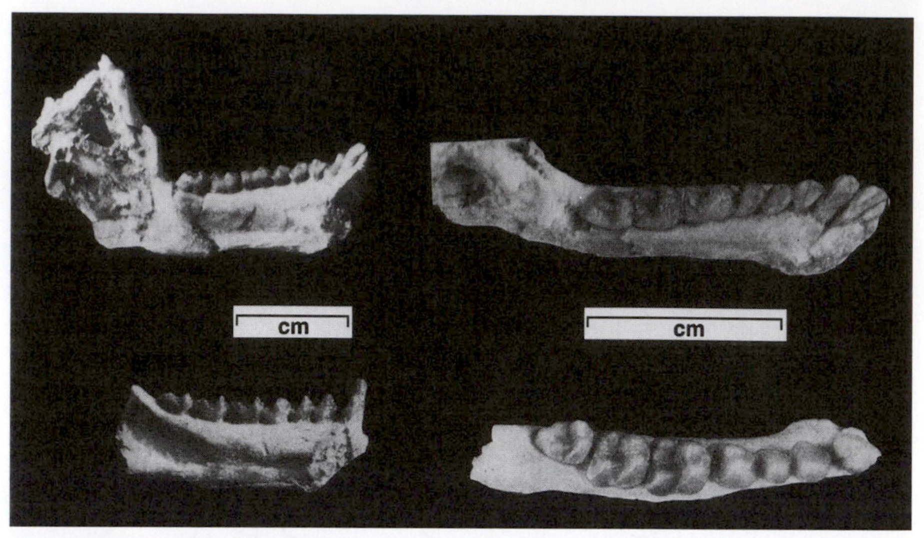

FIGURE 13.11 Dental remains from two late Eocene anthropoids from the Fayum: above, *Arsinoea;* below, *Serapia.*

lateral incisors of the type specimen were lost during collecting. But it is now known that *P. grangeri* lacks permanent incisors altogether, so it is quite possible that *P. fraasi* also lacks permanent incisors and that the tiny anterior teeth preserved in the type specimen are deciduous incisors. More complete fossils are needed to resolve this question. *Parapithecus fraasi* has distinct, rounded cusps on its molars, suggesting a frugivorous diet, relatively simple premolars, and a reduced third molar.

In addition to the three genera definitely placed in the Parapithecidae, three other poorly known genera from the Eocene of North Africa may be loosely allied with the parapithecids or may belong to totally different groups of early anthropoids. ***Biretia*** is a small, parapithecid-like primate from the late Eocene of eastern Algeria (DeBonis *et al.*, 1988). It is known from a single lower molar and has been likened to *Qatrania.*

Proteopithecus and ***Arsinoea*** (Fig. 13.11) are two late Eocene genera from quarry L-41 in the lowest level of the Fayum Jebel Qatrani Formation; each has a dental formula of 2.1.3.3. Both are primitive anthropoids (Simons, 1992; Simons and Rasmussen, 1994) that lack derived features found in *Apidium* and *Parapithecus.* It has been suggested that the premolar morphology of *Arsinoea* is intermediate between the primitive parapithecid morphology and the derived condition found in other anthropoids, including platyrrhines, modern catarrhines, oligopithecines, and propliopithecines. The cheek teeth of *Arsinoea* are extremely flat and crenulated, suggesting a diet of nuts or seeds. *Proteopithecus sylviae* is another late Eocene species with three premolars whose affinities are uncertain. However, recently described dental, cranial, and postcranial remains of *Proteopithecus* are very platyrrhine-like, and this genus shows no specializations that would preclude it from platyrrhine ancestry (Miller and Simons, 1997; Simons, 1997).

PHYLETIC RELATIONSHIPS The phyletic position of parapithecids in anthropoid evolution has long been debated, but new fossils and comparative analyses have greatly expanded our understanding of this group. Parapithecids are among the earliest and most primitive fossil higher primates. They have many primitive features in their dentition, including three simple lower premolars and occasional paraconids on their molars. Many skeletal features of *Apidium,* such as lack of expanded ischial tuberosities, a large greater trochanter, deep condyles on the femur, and retention of an entepicondylar foramen on the humerus, are also primitive features not found in most later Old World higher primates. The arrangement of the cranial bones on the skull wall and the morphology of the ear region seem to be similar to those in platyrrhines. Although some authorities (see, e.g., Hoffstetter, 1977) have advocated linking parapithecids directly with platyrrhines, most of the similarities are likely to be primitive anthropoid features retained in the two groups.

In earlier decades, many authors considered parapithecids, and especially *P. grangeri,* to be directly ancestral to Old World monkeys (Simons, 1970, 1972; Kay, 1977; Gingerich, 1978). Although some parapithecids, particularly *P. grangeri,* have lower molars and canines that are superficially similar to those of cercopithecoid monkeys, all parapithecoids lack many anatomical features characteristic of catarrhines, such as the presence of two rather than three premolars, broad ischial tuberosities, and a tubular tympanic. If parapithecids are uniquely ancestral to cercopithecoids, then many of the bony features that living apes and monkeys have in common must have evolved independently. In addition, the species that shows the greatest similarity to cercopithecoids in its molar morphology, *P. grangeri,* is the species with the most aberrant anterior dentition. It seems more likely that the bilophodont appearance of the parapithecid molars is an evolutionary convergence with later monkeys rather than an indication of a phyletic relationship (Delson, 1975).

The more difficult question is whether parapithecids preceded or followed the divergence of platyrrhines and catarrhines (Fig. 13.12). In contrast to the large number of primitive prosimian and platyrrhine features in parapithecids, the presence of a large hypoconulid on the lower molars links parapithecids with extant catarrhines to the exclusion of other anthropoids. There is, however, reason to suspect that this feature may well be a primitive anthropoid feature lost in platyrrhines. Moreover, platyrrhines and later catarrhines share many features lacking in parapithecids, including shallow femoral condyles and broad lower fourth premolars with a crest joining the protoconid and metaconid. Thus it seems more likely that parapithecids preceded the divergence of platyrrhines and lie near the origin of anthropoids (Fleagle and Kay, 1987; Harrison, 1987; Kay *et al.,* 1997). However, new fossils of early parapithecids and other primitive anthropoids from the late Eocene of Africa will undoubtedly increase and complicate our understanding of the radiation of these basal anthropoids.

Propliopithecids

The best-known group of early anthropoids from the Fayum, the propliopithecids, were as large as or larger than the largest parapithecid. They have a dental formula of 2.1.2.3. and a dental morphology more like that of later apes than of cercopithecoid monkeys in that they lack bilophodont molars. However, in details of their dental, cranial, and postcranial anatomy they are much more primitive than any living catarrhines. There are two genera (Table 13.3).

The first fossil "ape" described from the Fayum is ***Propliopithecus.*** There are four species: *P. haeckeli,* the type species, described early in the century; *P. chirobates,* from the uppermost levels of the Fayum; *P. ankeli,* a species with large premolars from the middle levels; and *P. markgrafi,* another species first described early in the century and sometimes

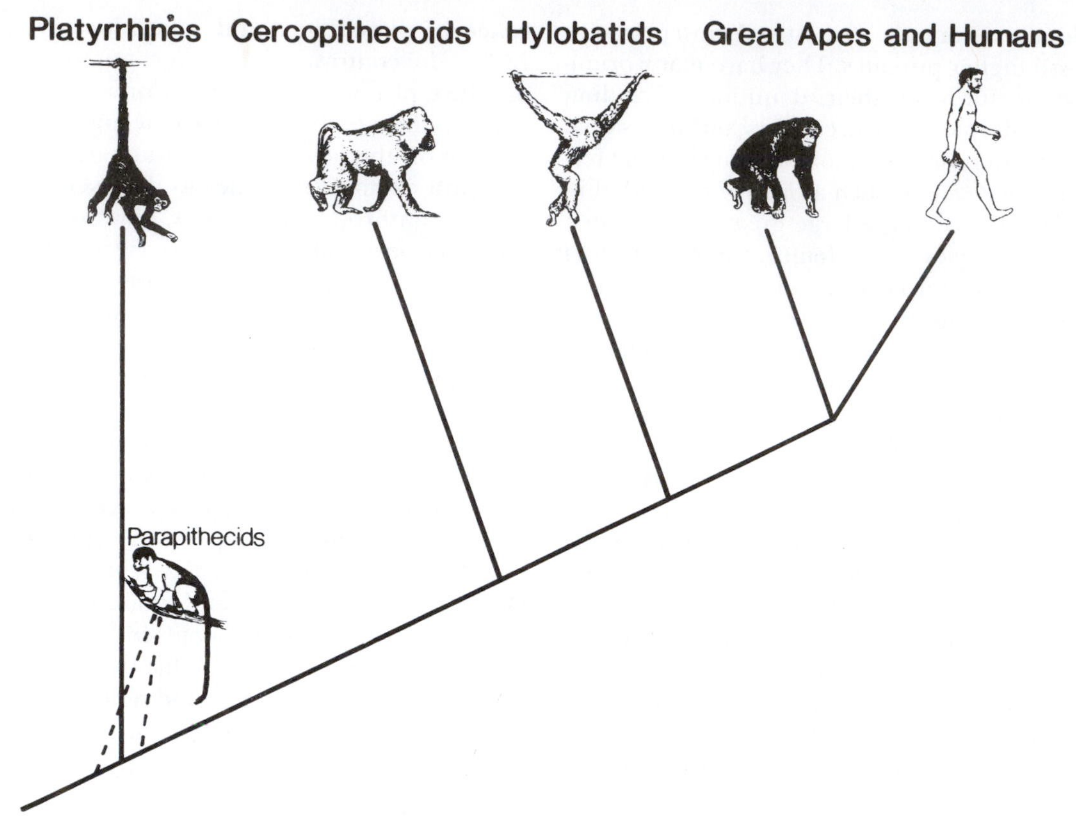

FIGURE 13.12 The phyletic position of parapithecids.

often placed in a separate genus, *Moeripithecus*. *Propliopithecus chirobates,* the best-known species, is a medium-size (4 kg) anthropoid. *Propliopithecus* has relatively broad, spatulate lower incisors and large, sexually dimorphic canines (Figs. 13.13). As in most living anthropoids, the anterior lower premolar shears against the posterior surface of the upper canine to sharpen it; the posterior premolar is semimolariform, with protoconids and metaconids of equal size. The lower molars resemble those of later apes in that they are formed by a broad talonid basin surrounded by five rounded cusps. There is no paraconid, and the trigonid is small. The three lower molars are similar in size. The upper premolars are bicuspid, and the upper molars are broad and quadrate, with a small hypocone connected to a pronounced lingual cingulum. There are no conules or stylar cusps on the upper molars. The simple molars with low, rounded cusps and the broad incisors suggest that *Propliopithecus* was frugivorous.

There are no described cranial remains of *Propliopithecus.* Several isolated limb elements indicate that *Propliopithecus* was an arboreal quadruped (Fig. 13.7) with a strong grasping foot and was probably capable of hindlimb suspension.

Aegyptopithecus zeuxis, from the same quarries in the Fayum as *P. chirobates,* was a much larger animal (6–8 kg) and is one of the best

TABLE 13.3
Infraorder Catarrhini
Superfamily Propliopithecoidea

Species	Estimated Mass (g)
Family PROPLIOPITHECIDAE	
Propliopithecus (early Oligocene, Egypt, Oman)	
P. haeckeli	4000
P. chirobates	4200
P. markgrafi	4000
P. ankeli	5700
Aegyptopithecus (early Oligocene, Egypt)	
A. zeuxis	6700
Family OLIGOPITHECIDAE	
Oligopithecus (late Eocene, Egypt, Oman)	
O. savagei	1000
O. rogeri	1500
Catopithecus (late Eocene, Egypt)	
C. browni	900

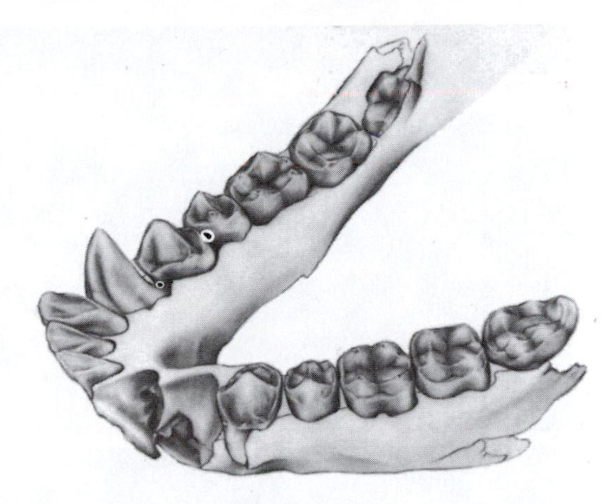

FIGURE 13.13 A mandible of *Propliopithecus chirobates.* Note the dental formula of 2.1.2.3, the apelike arrangement of cusps on the lower molars, the elongate anterior premolar that shears against the upper canine, and the fused mandibular symphysis. Scale: approximately 2× (courtesy of R. F. Kay).

known of all fossil anthropoids (Figs. 13.7, 13.14, 13.15). Dentally, *Aegyptopithecus* differs from *Propliopithecus* in having narrower incisors, lower molars with larger cusps and a more restricted talonid basin, and upper molars with better-developed conules and stylar cusps. In contrast to *Propliopithecus,* in which the three lower molars are similar in length, the molars of *Aegyptopithecus* increase in size posteriorly. Overall, the dental differences suggest that *A. zeuxis* was largely frugivorous but probably more folivorous than *Propliopithecus.* Like *Propliopithecus, Aegyptopithecus* has sexually dimorphic canines; it probably lived in polygynous social groups.

The cranial anatomy of *Aegyptopithecus* is more primitive than that of any living Old World anthropoid but more advanced than that of any prosimian (Fig. 13.14). The skull resembles other anthropoids in that the lacrimal bone lies within the orbit, and the relatively small orbits (indicating diurnal habits) are completely walled off posteriorly with a bony configuration similar to that found in extant catarrhines rather than in platyrrhines (Fleagle and Rosenberger, 1983). *Aegyptopithecus* has a premaxillary bone that is very large for an anthropoid, and the superficial cranial morphology changes dramatically with age (Simons, 1987). Older individuals develop a pronounced sagittal crest that divides anteriorly and extends over the brow ridges. There is also a large nuchal crest along the posterior border of the occiput. The auditory region is like that of other anthropoids in having an anterior accessory cavity and in lacking a stapedial artery. Among anthropoids, the auditory region is most similar to that in platyrrhines; the tympanic is a bony ring fused to the lateral surface of the bulla, with no bony tube.

The brain of *Aegyptopithecus* was relatively small compared with the brains of living anthropoids, and more like a prosimian brain. However, compared with contemporaneous Oligocene mammals or Eocene prosimians, it

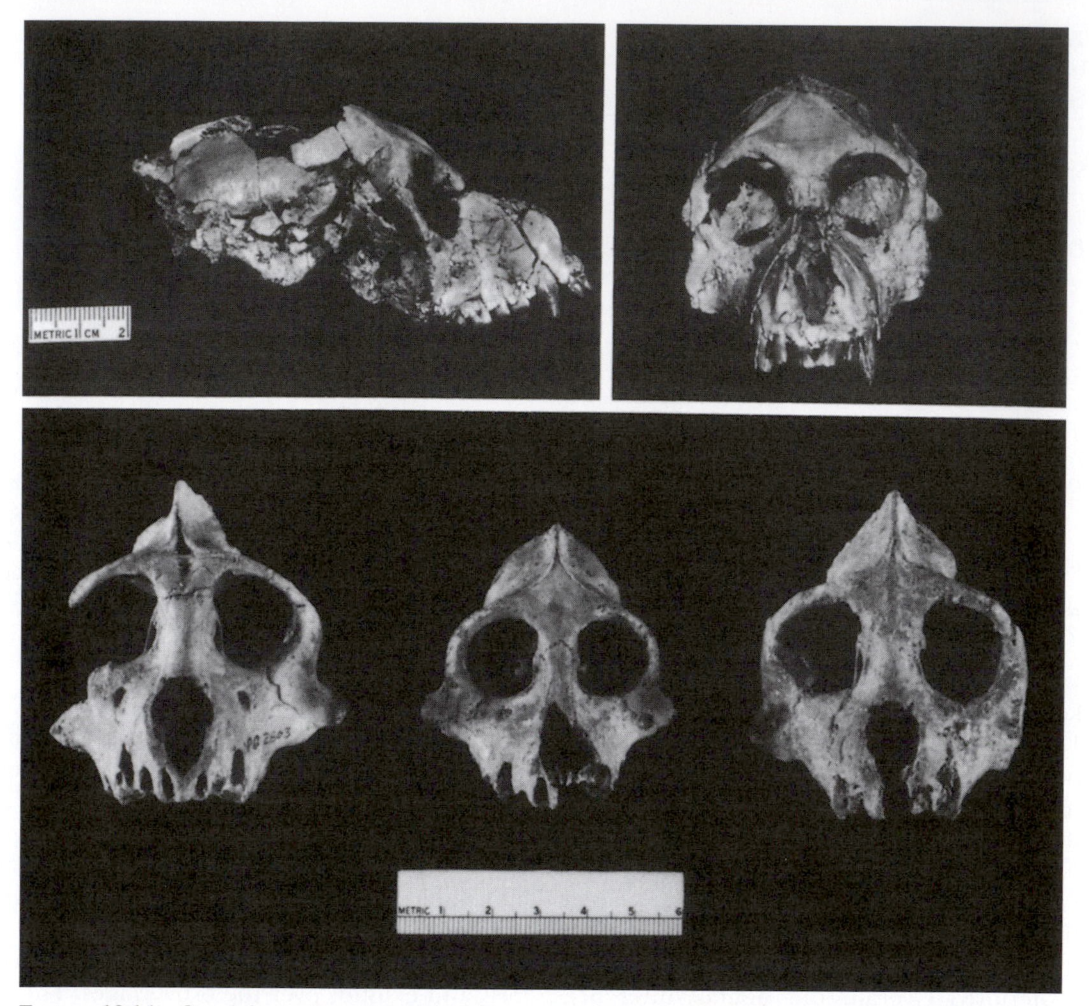

FIGURE 13.14 Cranial remains of *Aegyptopithecus zeuxis*. Note the long snout, small orbits, sagittal and nuchal crests, and converging temporal lines in older individuals (courtesy of E. L. Simons).

was relatively large, with an expanded parietal region (Simons, 1993).

The forelimb of *Aegyptopithecus* is known from the humerus and ulna, and the hindlimb is known from the femur, talus, calcaneus, and first metatarsal (Fig. 13.15). All of these elements indicate that *Aegyptopithecus* was a robust arboreal quadruped (see Fig. 13.7). The foot bones indicate that it had a grasping hallux and was capable of considerable inversion of the foot. In many anatomical details, the limb elements of *Aegyptopithecus* are more similar to those of platyrrhines and prosimians than to those of either living apes or cercopithecoid monkeys. This early ape retained many primitive features lost in later catarrhines.

PHYLETIC RELATIONSHIPS Since their initial discovery, *Propliopithecus* and *Aegyptopithecus*

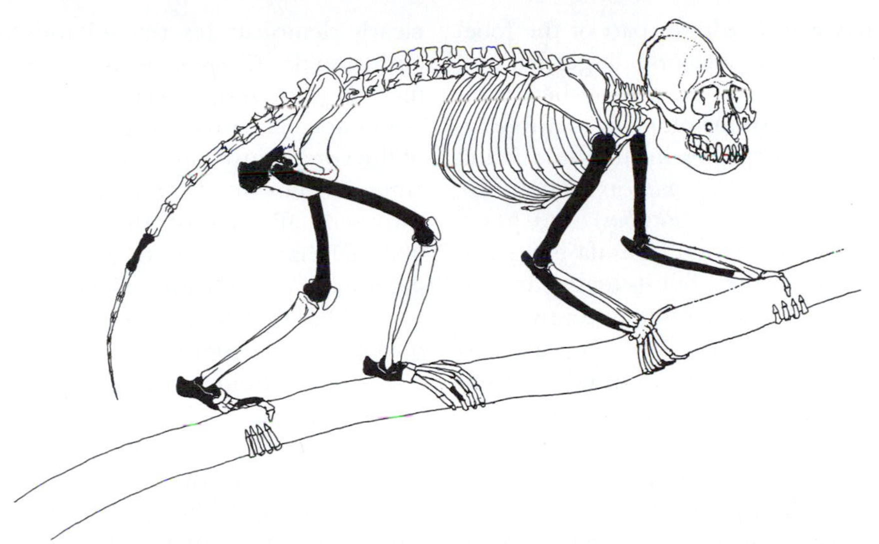

FIGURE 13.15 A reconstructed skeleton of *Aegyptopithecus zeuxis* showing, in black, the bones that have been recovered for this species.

have been identified as early apes on the basis of their dental similarities to living hominoids and to later fossil apes from Europe and Africa (Schlosser, 1911; Simons, 1972; Szalay and Delson, 1979). The similarities to living apes are, however, primitive anthropoid features rather than specializations, and increasing knowledge of their cranial and postcranial anatomy has shown that these early anthropoids were more like platyrrhines than catarrhines in many aspects of their anatomy. They have all of the characteristic features of anthropoids (fused mandibular symphysis, postorbital closure, lacrimal bone within the orbit) but are linked with living catarrhines only by their dental formula of 2.1.2.3. In the anatomy of their auditory region and limbs, they lack common specializations found both in living apes and in living Old World monkeys, and they have the more primitive platyrrhine morphology. Thus the Fayum "apes" are neither Old World monkeys nor apes but a primitive group of catarrhines that preceded the evolutionary divergence and subsequent radiations of both living groups (Fig. 13.16). They are usually placed in a primitive family of catarrhines, the Proliopithecidae.

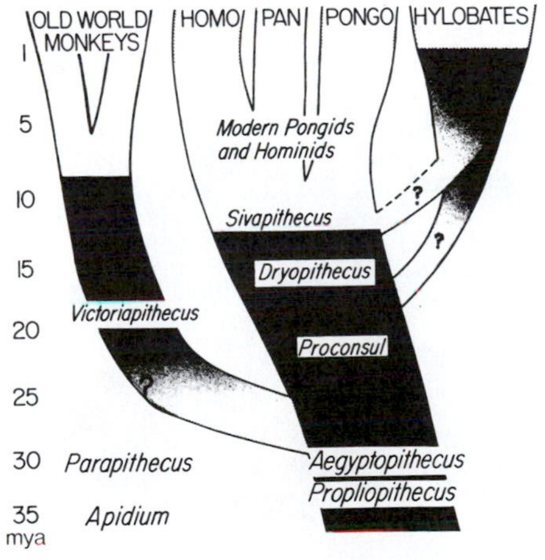

FIGURE 13.16 The phyletic position of propliopithecids.

Oligopithecids

One of the earliest and the most enigmatic of the Fayum primates is ***Oligopithecus savagei***, from Quarry E in the lower part of the Jebel Qatrani Formation. A second, larger species, ***O. rogeri,*** has recently been described from Oman, with suggestions of a considerable radiation of related species in the same fauna (Gheerbrant, 1995). *Oligopithecus* is about the size of a titi monkey (*Callicebus*) and has a deep mandible. Its dental formula is 2.1.2.3, as in propliopithecines, but its teeth show an odd mixture of features quite different from those of other Fayum primates. The canine is small and mesiodistally compressed, and the simple P_3 is narrow with a honing blade on its mesial edge. The last premolar is strikingly similar to the same tooth in propliopithecids. The molars are very primitive compared with those of other anthropoids in having a relatively high trigonid and a small paraconid on the first molar. On both the first and second lower molars there is a large hypoconulid near the entoconid and no posterior fovea.

For many years, *Oligopithecus* was known from a single jaw, and paleoanthropologists debated its affinities. Many features of the molars, such as the twinned hypoconulid and entoconid, suggest adapoid affinities, the premolar morphology and the dental formula, suggest a relationship with propliopithecids and later catarrhines.

Catopithecus browni (Fig. 13.17), from Quarry L-41, is a close relative of *Oligopithecus* that clearly demonstrates that oligopithecids are anthropoids. *Catopithecus* is known from numerous jaws, several skulls, and some limb bones and has provided an excellent picture of the comparative anatomy of the oligopithecines. While clearly demonstrating the anthropoid affinities of this group, this new material has confirmed the morphological paradox that had been indicated by *Oligopithecus* and has fueled ongoing debates about the phylogenetic origins of anthropoids. *Catopithecus* is a medium-size primate with an estimated body mass of between 600 and 900 g. It has a dental formula of 2.1.2.3 with relatively broad upper and lower incisors, like anthropoids (Simons, 1995). Likewise, the lower premolars resemble those of propliopithecids and other catarrhines, with a honing blade on P_3 and a P_4 with similar-size protocone and metacone. The molar teeth show a mixture of anthropoid features and others that are reminiscent of cercamoniine adapids. The upper molars have a very tiny hypocone like adapoids, and the lower molars have a small paraconid on the trigonid and a small hypoconulid adjacent to the entoconid. The dental morphology suggests a frugivorous/insectivorous diet (Rasmussen and Simons, 1992).

The cranial anatomy of *Catopithecus* is well known from several skulls (Simons and Rasmussen, 1996). Overall, the skull is clearly an-

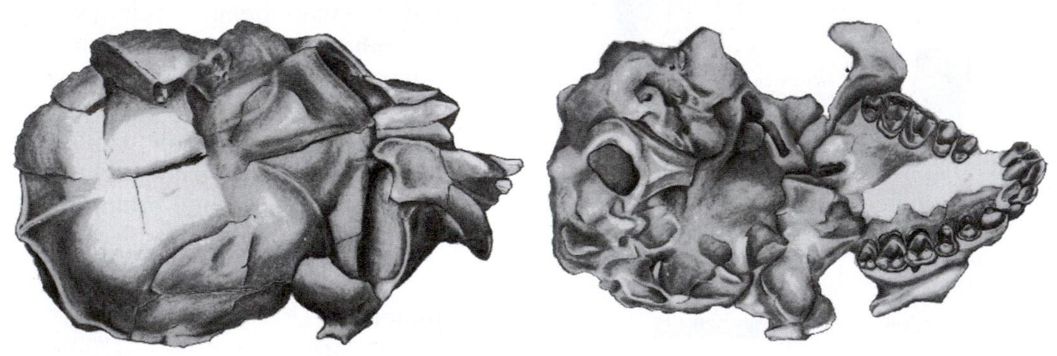

FIGURE 13.17 The cranium and dentition of *Catopithecus browni,* an oligopithecid from the late Eocene of Egypt (drawing by Jennifer Reig).

thropoid and very much like that of a small platyrrhine, with a fused frontal bone, complete postorbital closure, a large promontory artery and anterior accessory cavity of the ear, and a platyrrhine-like ectotympanic ring fused to the bulla wall (Simons, 1995). The orbit size indicates diurnal habits. Like *Aegyptopithecus, Catopithecus* has a relatively large premaxillary bone.

A few limb bones have been attributed to *Catopithecus* (Gebo *et al,* 1994). These suggest arboreal quadrupedal habits, with no indication of either leaping or clinging behavior.

In addition to the material of *Oligopithecus* and *Catopithecus* that has already been described from Egypt and Oman, there are indications that oligopithecids have a much greater diversity and a broader distribution than currently recognized. Gheerbrandt and colleagues (1995) suggest as many as four oligopithecine species in Oman (Fig. 13.1; Table 13.3). Ducrocq *et al.* (1995) have described a new oligopithecid, ***Wailekia orientale*** from the upper Eocene of Thailand but *Wailekia* is more likely an adapoid prosimian.

PHYLETIC RELATIONSHIPS The phyletic affinities of oligopithecids have been much debated since the discovery of *Oligopithecus* in the early 1960s. Simons originally described one new species of *Oligopithecus,* and considered it a primitive propliopithecid related to *Propliopithecus* and *Aegyptopithecus* because of the similar dental formula. Others argued that the dentition is more suggestive of adapid affinities. The material of *Catopithecus* has confirmed the morphological similarities to both adapoids and anthropoids but indicates that oligopithecids have undoubted anthropoid features of postorbital closure and a fused frontal as well as an anthropoid anterior dentition. However, the relationship of this group to other anthropoids and the implications for the prosimian origin of anthropoids are issues of current debate. On the basis of the unusual combination of dental features, some authorities (e.g., Kay *et al.,* 1997) place oligopithecids in a distinct family of basal anthropoids that precedes the platyrrhine-catarrhine divergence, implying that the premolar loss in catarrhines and oligopithecids was independent. The alternative view, adopted here (Table 13.3), is that oligopithecids are related to propliopithecids. Indeed, Rasmussen and Simons (1992) suggest that *Propliopithecus* (*Moeripithecus*) *markgrafi* is intermediate in molar morphology between oligopithecids and *Propliopithecus* or *Aegyptopithecus.* However, this placement of the oligopithecids, with their simple tritubercular upper molars and primitive lower molars, as basal catarrhines seem inconsistent with some anatomical features (particularly in the upper dentition) shared by platyrrhines and catarrhines and implies that these dental features of living anthropoids must have evolved in parallel. An alternative scenario, recently advocated by Gheerbrant and colleagues (1995), is that the oligopithecids are a group of early catarrhines whose molar teeth secondarily became simplified for an insectivorous diet. In this view, the similarities to adapines are parallel developments and not indicative of an evolutionary relationship.

Other North African and Arabian Early Anthropoids

In the past few years, early anthropoids have been found in many other areas of North Africa and the Arabian peninsula that are either comparable in age or older than those of the Fayum (Fig. 13.1).

Oman

Taqah and Thaytiniti are two localities in Oman that seem to be near the Eocene/Oligocene boundary Taqah is the younger of the two. The exact correlation between these localities and the Fayum sequence are still being debated, but faunal and paleomagnetic comparisons suggest that they are most comparable to the middle and lower parts of the Jebel Qatrani Formation (Quarries G and E). Taqah has yielded abundant remains of a propliopithecid that seems most comparable to *Propliopithecus (Moeripithecus) markgrafi* or possibly *P. ankeli,* (Thomas *et al.,* 1991) as well as several new species of oligopithecines.

Algeria

Two small early anthropoids have been described from Eocene deposits at Glib Zegdou in Algeria (Godinot and Mahboubi, 1992, 1994; Godinot, 1994); both are known only from isolated teeth. ***Algeripithecus minutus*** is a tiny species with broad, bunodont upper molars that seems most closely related to parapithecids (Godinot, 1994). ***Tabelia hammadae*** is a slightly larger species with more crested molars that is perhaps related to propliopithecids. As already mentioned, the parapithecid *Biretia* is from Bir el Ater in Algeria.

Tunisia

The early to middle Eocene locality of Chambi has yielded a relatively complete jaw of a primate called ***Djebelemur martinezi*** (Fig. 12.10) and a second unnamed species (Hartenberger and Marandat, 1992). Although originally described as a cercamoniine adapoid, Godinot (1994) has argued that it is possibly an early anthropoid.

EARLY ANTHROPOID ADAPTATIONS

For many decades, the fossil primates from the early Oligocene of Egypt provided our only record of Old World higher primate evolution. These Oligocene anthropoids were all small to medium in size, comparable to extant platyrrhines. Their dentitions indicate that they ate fruits, seeds, and perhaps gums, but there is no evidence of predominantly folivorous species. From the available limb bones, they seem to have been arboreal quadrupeds and leapers; there is no evidence of either terrestrial quadrupeds or suspensory species. Overall, the adaptive breadth of the early Oligocene primates from the Fayum is more like that of extant platyrrhines than that found among later catarrhine primates of the Old World. It thus seemed likely that these platyrrhine-like morphologies represented the primitive anthropoid adaptations.

However, the new discoveries from the older Eocene deposits from the Fayum, as well as the new fossils from other parts of North Africa and Asia, have greatly expanded our knowledge of early anthropoids and also enlarged our view of the adaptive diversity of this group. With all of the diversity among early anthropoids, it is not easy to characterize the basal anthropoid adaptations, but the "large platyrrhine" morphology of the Oligocene anthropoids seems to have been a later phenomenon that greatly postdated the origin of higher primates as a group. Rather, the new, Eocene anthropoids are generally very small (100–300 g), with more primitive, "insectivorous" teeth. Overall, they are more like prosimians in their size and likely dietary adaptations, but there are also adaptive oddities, such as the tiny, marmoset-size *Algeripithecus* with broad, quadrate, frugivorous molars like the later parapithecids and propliopithecids. All early anthropoids that are known from skulls (*Catopithecus, Apidium,* and *Aegyptopithecus*) were probably diurnal.

Several features of the new, Eocene anthropoids have important implications for reconstructing adaptive scenarios to account for anthropoid origins (Kay *et al.,* 1997). Ross

(1996) has hypothesized that the distinctive anthropoid feature of postorbital closure evolved in conjunction with a shift to diurnality in a small primate that was probably a visual predator. As small primates, often less than 500 g in mass, the earliest anthropoids were probably partly insectivorous. Moreover, all anthropoids have a retinal fovea, a pit in the retina that provides greater visual acuity in part of the field of vision; in other vertebrates, this feature is usually associated with diurnal visual predation. The shift to a diurnal activity pattern in the ancestors of anthropoids would have resulted in increased convergence of the orbits. In addition, anthropoid skulls have increased frontation because of enlarged frontal lobes of the brain. This increase in the size of the frontal lobes may have been associated with living in larger social groups, a characteristic feature of diurnal primates. All of these changes in orbit shape would lead to a need for a bony septum between the orbit and the temporal fossa so that contraction of the temporal muscles during chewing would not move the eyeball and disturb an animal's vision. This scenario is very compatible with our current understanding of both the size and the habits of early anthropoids.

Although little is known of the postcranial skeleton of the very early anthropoids, their small size suggests that they may well have included more leaping in their locomotor repertoire than larger species.

PHYLETIC RELATIONSHIPS OF EARLY ANTHROPOIDS

The early anthropoids from the Fayum and elsewhere in Africa, Arabia, and Asia are all more primitive than later Old World higher primates, but how they relate to specific extant lineages and to one another is far from being resolved (Kay *et al.*, 1997). In general, the major groups of Fayum primates seemed to be intermediate forms that fill in many of the morphological gaps between the major radiations of extant anthropoids (Fig. 13.18). In Africa, the parapithecids are the most primitive and the closest to the origin of anthropoids, and they may run the risk of becoming a wastebasket group of primitive anthropoids. They share postcranial features with omomyoid prosimians as well as with platyrrhines, suggesting that the extant radiation of higher primates probably originated in Africa. *Aegyptopithecus* and *Propliopithecus* are more advanced than platyrrhines but more primitive than later catarrhines. Although oligopithecids clearly are anthropoids and more primitive than propliopithecids, their phyletic position relative to parapithecids and platyrrhines is not resolved.

The positions of *Siamopithecus, Pondaungia,* and *Eosimias* are much more difficult to resolve on present evidence. All show anthropoid-like features in some aspects of their anatomy, but other features suggest that they may not be anthropoids at all. Since there are no skulls for any of these taxa, inclusion of them in Anthropoidea requires a definition of anthropoids based solely on dental features (Kay *et al.*, 1997).

The many new Eocene taxa and new morphological information about the oligopithecids have provided evidence of a much bushier phylogeny at the base of the anthropoid radiation than anyone would have imagined just a few years ago. Many of these early anthropoids may well be members of separate clades that are collateral to the evolution of later higher primates. Likewise, there has undoubtedly been considerable parallelism in the evolution of key anthropoid features in the dentition and the postcranial skeleton. What we understand now that was less evident only a few years ago is that anthropoids were a diverse and successful group well before the

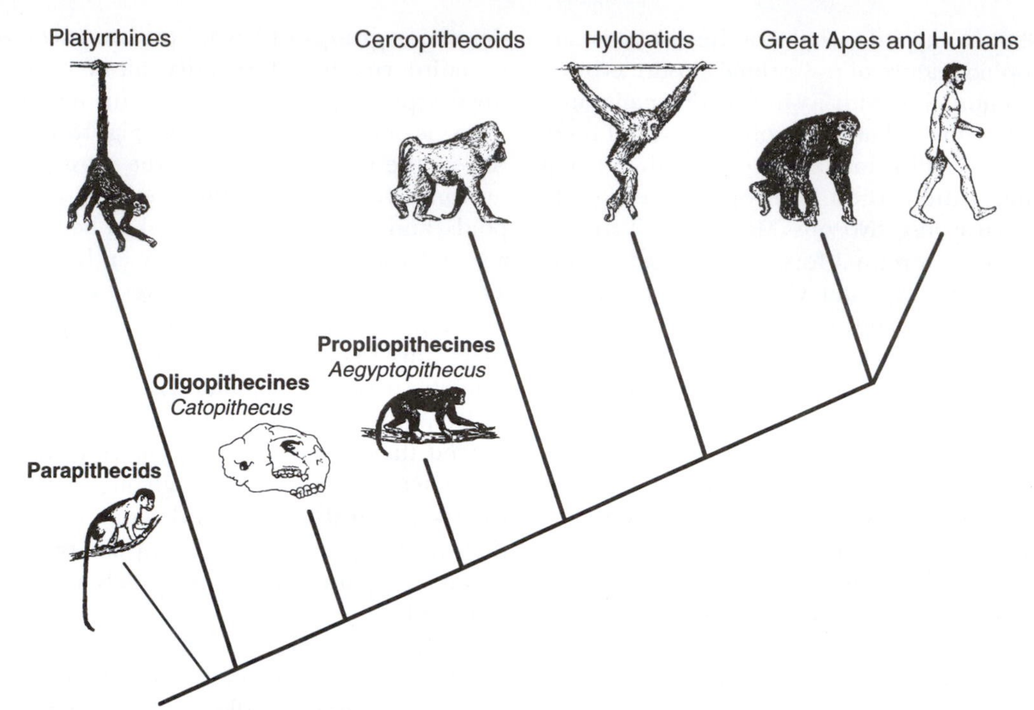

FIGURE 13.18 Phyletic relationships of early anthropoids.

early Oligocene, with an inordinate number of tiny species the likes of which are not found in the suborder today. However, as the geographical and morphological scope of this early diversity becomes clearer, our views of early anthropoid evolution will continue to evolve.

PROSIMIAN ORIGINS OF ANTHROPOIDS

The enlarging fossil record of early anthropoids has both expanded and intensified earlier debates over the broader question of their phyletic origins, or, more specifically, which group of prosimians gave rise to higher primates (Fleagle and Kay, 1994; Kay *et al.*, 1997). There is general agreement among primatologists and mammalogists that among living primates, anthropoids seem to be more closely related to the living *Tarsius* than to living lemurs and lorises. When we consider fossil prosimians, however, the possible phyletic relationships between anthropoids and various group of living and fossil prosimians are much more complicated. Many of these complications arise from the conflicting views regarding the phyletic relationships among Eocene prosimians and living lemurs, lorises, and tarsiers as well as different interpretations about what features characterized the earliest anthropoids. Analyses of the new fossil anthropoids from the Eocene of Africa, Arabia and Asia indicate that few if any of the dental and gnathic features traditionally used to identify anthropoids unequivocally distinguish these new fossil anthropoids from other primates (Kay *et al.*, 1997). Many of the early anthropoids retain primitive features, such as unenlarged upper molar hypocones, lower molar paraconids, simple premolars, and an unfused mandibular symphysis; likewise many Eocene prosimians have characteristic anthropoid-like features, including a fused mandibular symphysis, large upper molar hypocones, reduced molar paraconids, and canine sexual dimorphism. Unfortunately, many of the earliest anthropoids and many fossil prosimians are known primarily from dental remains. Cranial and postcranial anatomy is either unknown or known from material not clearly associated with the dental remains. As a result, there are currently four distinct hypotheses concerning the sister taxon of anthropoids, each with its own strengths and weaknesses: tarsier origin, omomyid origin, adapid origin, and separate, ancient origin (Fig. 13.19).

TARSIER ORIGIN As noted in Chapter 4, tarsiers share with anthropoids many similarities in reproductive anatomy, eye structure, and cranial anatomy, as well as biochemical similarities not found in other living primates. Moreover, the features of postorbital closure and development of an anterior accessory chamber of the middle ear that unite tarsiers and anthropoids are unique among primates or even among mammals rather than being similar features that appear to have evolved in numerous groups (homoplasies) (Ross, 1994). Although the cranial similarities linking tarsiers with anthropoids have been questioned (Simons and Rasmussen 1989; Beard and MacPhee, 1994), there seems little reason to do so on morphological grounds (Cartmill, 1994; Ross, 1994). Indeed, perhaps the most striking result from all the new remains of fossil primates that have been recovered in recent decades has been the demonstration that postorbital closure and middle ear morphology are the main features distinguishing anthropoids from prosimians. Only tarsiers among all known fossil and extant prosimians share any evidence of either of these traits with anthropoids. The extensive phylogenetic analysis of anthropoid origins by Kay, Ross and Williams (1997) supports a tarsier origin for anthropoids as the most parsimonious.

Nevertheless, the argument that one or more Eocene omomyoids without postorbital closure shares unique cranial features with tarsiers suggests that some tarsier–anthropoid similarities may indeed be homoplasies (see previous paragraph; Beard *et al.*, 1991; Beard and MacPhee, 1994). Likewise, although the dentition of fossils such as *Eosimias* could be interpreted as supporting tarsier–anthropoid similarities, the absence of common tarsier/anthropoid features in the ear region of *Eosimias* suggests that these features may have evolved in parallel (Kay *et al.*, 1997).

OMOMYOID ORIGIN Many authorities have argued that the sister taxon of anthropoids is not the genus *Tarsius* but a more generalized omomyoid. In their view, both tarsiers and anthropoids are descended from some common tarsiiform or omomyoid ancestry (e.g., Rosenberger and Szalay, 1980; Rosenberger, 1986; Rosenberger and Dagosto, 1992). And the haplorhine features uniting *Tarsius* and anthropoids were presumably present in the omomyid ancestor of anthropoids, but not so the unique features that are characteristic of *Tarsius*. This hypothesis has the advantage of being in accord with notions of a haplorhine–strepsirhine dichotomy among primates, with molecular studies linking *Tarsius* with anthropoids, and with studies that argue for the origin of *Tarsius* from particular omomyid taxa.

As many have pointed out, even though it would be tidy to be able to apply a haplorhine–strepsirhine division to fossil prosimians, such a division of fossil prosimians is not as clear as among extant primates. It may well be that both omomyoids and adapoids, as well as tarsiers, would cluster with anthropoids, making the living strepsirhines the odd group (Rasmussen, 1986). Moreover, the big-

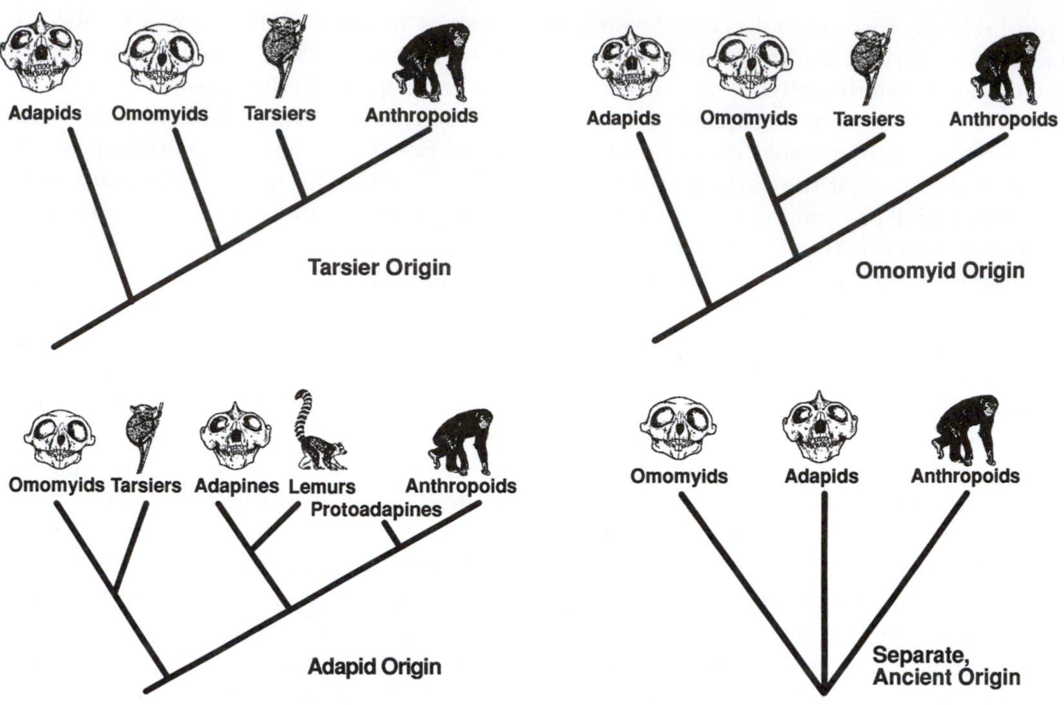

FIGURE 13.19 Hypotheses of anthropoid origins.

gest weakness of this scenario is that the cranial features that most clearly link tarsiers with anthropoids (postorbital closure and anterior accessory chamber of the middle ear) are not present in known omomyoids, including the taxa that share various other cranial features with *Tarsius.* Thus, there are very few derived features (possibly an apical interorbital septum and reduced nasal region, and a few postcranial characters) that link generalized omomyoids with anthropoids. And in addition, if tarsiers are indeed derived from any of the taxa most frequently proposed (*Tetonius, Shoshonius,* and *Necrolemur*), then the unique tarsier–anthropoid cranial features have evolved independently in tarsiers and anthropoids, and the morphological evidence linking these two groups as haplorhines becomes substantially weaker and very difficult to identify among fossils.

Another stumbling block to any of these arguments is the extremely poor fossil record of omomyoid cranial remains. There are only a few skulls that document either the orbital anatomy or the ear region, and practically nothing is known about the postcranial anatomy of most taxa (Covert and Williams, 1994; Dagosto and Gebo, 1994).

ADAPOID (CERCAMONIINE) ORIGIN Largely on the basis of similarities in the anterior dentition (small incisors, large canines, and, frequently, fused mandibular symphysis), many workers have long supported an adapid–anthropoid relationship (e.g., Simons, 1995). Supporters of such a relationship have argued that omomyids tend to be too specialized dentally (but see Covert and Williams, 1994) and that many of the haplorhine features do not necessarily exclude adapoids, because they are not clearly strepsirhine (Rasmussen 1986). The postcanine similarities between oligopithecids and cercamoniines have strengthened the notion of an adapoid–anthropoid link and have led Rasmussen (1994) to predict that some of the taxa currently recognized as cercamoniines on the basis of dental remains may prove to have anthropoid crania.

One major weakness of the adapoid–anthropoid hypothesis is that adapoids share with strepsirhines a number of possibly derived features (Beard *et al.*, 1988; Covert and Williams, 1994; Kay *et al.*, 1997). In addition, some of the features that adapoids share with anthropoids are almost certainly primitive primate features, whereas the anthropoid features shared by tarsiers (postorbital closure) and perhaps omomyoids (apical interorbital septum, reduced nasal region) are almost certainly derived features. Finally, as with omomyoids, there is no indication among any of the known fossils that any adapoid ever approximated the diagnostic anthropoid features of the orbit and ear.

ANCIENT OR OTHER ORIGIN The seemingly endless frustration and difficulty of showing that anthropoids are derived from any known group of Eocene prosimians from either Europe or North America has led many experts to argue that the divergence of anthropoids from any known group of prosimians was an ancient event and that the anthropoid ancestor was an unknown Paleocene group inhabiting some relatively unknown continent, more probably Africa or Asia (Sige *et al.*, 1990; Beard and MacPhee, 1994). It is certainly not unreasonable to expect that the ancestors of this group are not among currently known primate taxa, given both how little we know about the fossil record from many parts of the world and the increasing evidence for numerous Eocene anthropoids, but it is not very satisfying. Arguing for a nonadapoid, nonomomyoid, nontarsier origin for anthropoids is little more than a claim for ignorance regarding the origin of the group. A relatively ancient split among adapoids, omomyoids, tarsiers, and anthropoids does not preclude one of these three from being more closely related to anthropoids than the others, but it does suggest that we might expect identification of their correct phylogenetic relationships to be more difficult to reconstruct, because each would have had even more time to develop parallelisms and autapomorphies.

SOLVING ANTHROPOID ORIGINS

Anthropoids certainly had nonanthropoid ancestors. The origin of this group, just like the evolution of any other group of organisms, is a biological phenomenon that took place. It is also certain that most of the hypotheses regarding anthropoid origins from tarsiers, omomyoids, adapoids, or some unknown group are not correct. Anthropoids most probably had a single origin from a single "prosimian" ancestor. How will we get past the current impasse and resolve the apparently contradictory and mutually exclusive scenarios implied by the current evidence? The answers will most likely come through discovery of additional new fossils with more intermediate morphologies spanning the current gaps between prosimians and anthropoids in cranial and postcranial anatomy.

BIBLIOGRAPHY

GENERAL

Fleagle, J. G. (1994). Anthropoid origins. In: *Integrative pathways to the Past: Paleoanthropological Advances in Honor of F. Clark Howell.* eds. R. S. Corruccini and R. L. Ciochon, pp. 17–35. New York: Prentice Hall.

Fleagle, J. G., and Kay, R. F. (1994). *Anthropoid Origins.* New York: Plenum Press.

Kay, R. F., Ross, C., and Williams, B. A. (1997). Anthropoid origins. *Science* **275**:797–803.

Prothero, D. R., and Berggren, W. A., eds. (1992). *Eocene–Oligocene Climatic and Biotic Evolution.* Princeton, N. J.: Princeton University Press.

EOCENE ANTHROPOIDS FROM ASIA

Beard, K. C., Qi, T., Dawson, M. R., Wang, B., and Li, C. (1994). A diverse new primate fauna from middle Eocene fissure-fillings in southeastern China. *Nature* **368**:604–609.

Beard, K. C., Tong, Y., Dawson, M., Wang, J., and Huang, X. (1996). Earliest complete dentition of an anthropoid primate from the late Middle eocene of Shanxi Province, China. *Science* **272**:82–85.

Chaimanee, Y., Suteethorn, V., Jaeger, J. J., and Ducrocq, S. (1997). A new late Eocene anthropoid from Thailand. *Nature* **385**:429–431.

Ciochon, R. L., and Holroyd, P. A. (1994). The asian origin of Anthropoidea revisited. In *Anthropoid Origins,* ed. J. G. Fleagle and R. F. Kay, pp. 143–162. New York: Plenum Press..

Dagosto, M., Gebo, D. L., Beard, K. C., and Qi, T. (1996). New primate postcranial remains from the middle Eocene Shanghuang fissures, southeastern China. *Am. J. Phys. Anthropol. Suppl.* **22**:92–93.

Gebo, D. L., Dagosto, M., Beard, C., and Qi, T. (1996). New primate tarsal remains form the middle Eocene Shanghuang fissures, southeastern China. *Am. J. Phys. Anthropol. Suppl.* **22**:113.

Kay, R. F., Ross, C., and Williams, B. A. (1997). Anthropoid origins. *Science* **275**:797–803.

MacPhee, R. D. E., Beard, K. C., and Qi, T. (1995) Significance of primate petrosal from middle Eocene fissure-fillings at Sahnghuang, Jiangsu Province, People's Republic of China. *J. Human Evol.* **29**:501–514.

FAYUM FOSSIL ANTHROPOIDS

Bown, T. M., and Kraus, M. J. (1988). *Geology and Paleoenvironment of the Oligocene Jebel Qatrani Formation and Adjacent Rocks, Fayum Depression, Egypt.* U. S. Geological Survey Professional Paper 1452. Washington, D. C.: U.S. Government Printing Office.

Bown, T. M., Kraus, M. J., Wing, S. L., Fleagle, J. G., Tiffany, B., Simons, E. L., and Vondra, C. F. (1982). The Fayum forest revisited. *J. Human Evol.* **11**(7):603–632.

Fleagle, J. G., and Kay, R. E (1983). New interpretations of the phyletic position of Oligocene hominoids. In *New Interpretations of Ape and Human Ancestry,* ed. R. L. Ciochon and R. Corruccini, pp. 181–210. New York: Plenum Press.

——— (1985). The paleobiology of catarrhines. In Ancestors: *The Hard Evidence,* ed. E. Delson, pp. 23–36. New York: Alan R. Liss.

——— (1987). The phyletic position of the Parapithecidae. *J. Human Evol.* **16**:483–531.

Fleagle, J. G., Bown, T. M., Obradovich, J. O., and Simons, E. L. (1986). How old are the Fayum primates? In *Primate Evolution,* ed. J. G. Else and P. C. Lee, pp. 133–142. Cambridge: Cambridge University Press.

Kappelman, J., Simons, E. L., and Swisher, C. C. III (1992). New Age determinations for the Eocene–Oligocene boundary sediments in the Fayum depression, northern Egypt. *J. Geol.* **100**:647–667.

Kay, R. E, and Simons, E. L. (1980). The ecology of Oligocene African Anthropoidea. *Int. J. Primatol.* **1**:21–37.

Rasmussen, D. T., Bown, T. M., and Simons, E. L. (1992). The Eocene–Oligocene transition in continental Africa. In *Eocene–Oligocene Climatic and Biotic Evolution,* ed. D. R. Prothero and W. A. Berggren, pp. 548–568. Princeton, N. J.: Princeton University Press.

Schlosser, M. (1910). Uber einige fossile Saugertiere aus dem Oligocan von Agypten. *Zool. Anz.* **34**:500–508.

——— (1911). Beitrage zur Kenntnis der Oligozanen Lansaugetiere aus dem Fayum, Aegypten. *Beitr. Palaeontol. Oesterreich-Ungarns Chients* **6**:1–227.

Simons, E. L. (1967). Review of the phyletic interrelationships of Oligocene and Miocene Old World Anthropoidea. In *Evolution des vertebres: Problemes actuels de Palaeontologie. Actes CNRS Coll. Int.* **163:** 597–602.

——— (1995). Egyptian Oligocene primates: A review. *Yrbk. Phys. Anthropol.* **38**:199–238.

Simons, E. L., and Rasmussen, D. T. (1991). The generic classification of Fayum Anthropoidea. *Int. J. Primatol.* **12**:163–178.

——— (1994). A whole new world of ancestors: Eocene anthropoideans from Africa. *Evol. Anthropol.* **3**:128–138.

Simons, E. L., Rasmussen, D. T., Bown, T. M., and Chatrath, P. J. (1994). The Eocene origin of anthropoid primates. In *Anthropoid Origins,* ed. J. G. Fleagle and R. F. Kay, pp. 179–202. New York: Plenum Press.

Parapithecids, *Arsinoea, Proteopithecus*

Conroy, G. C. (1976). Primate postcranial remains from the Oligocene of Egypt. *Contrib. Primatol.* **8**:1–134.

deBonis, L., Jaeger, J.-J., Coiffait, P.-E. (1988). Decouverte du plus ancien primate Catarrhinien connu dans l'Eocene superieur d'Afrique du Nord. *C. R. Acad. Sci. Paris* **306**:(II):929–934.

Delson, E., and Andrews, P. (1975). Evolution and interrelationships of the catarrhine primates. In *Phylogeny of the Primates: A Multidisciplinary Approach,* ed. W. C. Luckett and E. S. Szalay, pp. 405–446. New York: Plenum Press.

Fleagle, J. G. (1986). The fossil record of early catarrhine evolution. In *Major Topics in Primate and Human Evolution,* ed. B. Wood, L. Martin, and P. Andrews, pp. 130–149. Cambridge: Cambridge University Press.

Fleagle, J. G., and Kay, R. E. (1987). The phyletic position of the Parapithecidae. *J. Human Evol.* **16**:483–531.

Fleagle, J. G., and Simons, E. L. (1983). The tibio-fibular articulation in *Apidium phiomense,* an Oligocene anthropoid. *Nature (London)* **301**(5897):238–239.

——— (1995). Skeletal Anatomy of *Apidium phiomense,* an early anthropoid from Egypt. *Am. J. Phys. Anthropol.* **97**:235–289.

Fleagle, J. G., Kay, R. E, and Simons, E. L. (1980). Sexual dimorphism in early anthropoids. *Nature (London)* **287**: 328–330.

Hoffstetter, R. (1982). Les primates simiiformes (Anthropoidea) (Comprehension, Phylogenie, Histoire Biogeographique). *Ann. Paleontol. (Vert. Invert.)* **68**(3):241–290.

Harrison, T. (1987). The phyletic relationships of the early catarrhine primates: A review of the current evidence. *J. Human Evol.* **16**:41–80.

Hoffstetter, R. (1977). Primates: Filogenia e historia biogeographica. *Studia Geol.* **13**:211–253.

Kay, R. F. (1977). The evolution of molar occlusion in the Cercopithecoidea and early catarrhines. *Am. J. Phys. Anthropol.* **46**:327–352.

Kay, R. F., and Williams, B. A. (1994). Dental evidence for anthropoid origins. In *Anthropoid Origins,* ed. J. G. Fleagle and R. F. Kay, pp. 361–445. New York: Plenum Press.

Kay, R. F., and Simons, E. L. (1983). Dental formulae and dental eruption patterns in Parapithecidae (Primates, Anthropoidea). *Am. J. Phys. Anthropol.* **62**:363–375.

Kay, R. F., Ross, C., and Williams, B. A. (1997). Anthropoid origins. *Science* **275**:797–803.

Miller, E. R., and Simons, E. L. (1997). Dentition of *Proteopithecus sylviae,* an archaic anthropoid from the Fayum, Egypt. *Proc. Nat. Acad. Sci. USA* **94**:13760–13764.

Osborn, H. F. (1908). New fossil mammals from the Fayum Oligocene, Egypt. *Bull. Am. Mus. Nat. Hist.* **24**:265–272.

Simons, E. L. (1970). The deployment and history of Old World monkeys (Cercopithecoidea, Primates). In *Old World Monkeys,* ed. J. R. Napier and P. H. Napier, pp. 92–147. New York: Academic Press.

——— (1972). *Primate Evolution: An Introduction to Man's Place in Nature.* New York: Macmillan.

Simons, E. L. (1974). *Parapithecus grangeri* (Parapithecidae, Old World Higher Primates): New species from the Oligocene of Egypt and the initial differentiation of Cercopithecoidea. *Postilla* **166**:1–12.

——— (1986). *Parapithecus grangeri* of the African Oligocene: An archaic catarrhine without lower incisors. *J. Human Evol.* **15**:205–213.

——— (1992). Diversity in early Tertiary anthropoidean radiation in Africa. *Proc. Natl. Acad. Sci. USA,* **89**: 10743–10747.

——— (1997). Preliminary description of the cranium of *Proteopithecus sylviae,* an Egyptian late Eocene anthropoidean primate. *Proc. Natl. Acad. Sci. USA* **94**: 14970–14975.

Simons, E. J., and Kay, R. F. (1983). *Qatrania,* new basal anthropoid primate from the Fayum, Oligocene of Egypt. *Nature (London)* **304**:624–626.

——— (1988). New material of *Qatrania* from Egypt with comments on the phylogenetic position of Parapithecidae (Primates, Anthropoidea). *Am. J. Primatol.* **15**: 337–347.

Simons, E. L., and Rasmussen, D. T. (1994). A whole new world of ancestors: Eocene anthropoideans from Africa. *Evol. Anthropol.* **3**:128–138.

Propliopithecids

Andrews, P. (1985). Family group systematics and evolution among catarrhine primates. In *Ancestors: The Hard Evidence,* ed. E. Delson, pp. 14–22. New York: Alan R. Liss.

Delson, E., and Andrews, P. (1975). Evolution and interrelationships of the catarrhine primates. In *Phylogeny of the Primates: A Multidisciplinary Approach,* ed. W. C. Luckett and F. S. Szalay, pp. 405–446. New York: Plenum Press.

Fleagle, J. G. (1983). Locomotor adaptations of Oligocene and Miocene hominoids and their phyletic implications. In *New Interpretations of Ape and Human Ancestry,* ed. R. L. Ciochon and R. Corruccini, pp. 301–324. New York: Plenum Press.

Fleagle, J. G. (1986). The fossil record of early catarrhine evolution. In *Major Topics in Primate and Human Evolution,* ed. B. Wood, L. Martin, and P. Andrews, pp. 130–149. Cambridge: Cambridge University Press.

Fleagle, J. G., and Kay, R. F. (1983). New interpretations of the phyletic position of Oligocene hominoids. In *New Interpretations of Ape and Human Ancestry,* ed. R. L. Ciochon and R. Corruccini, pp. 181–210. New York: Plenum Press.

——— (1985). The paleobiology of catarrhines. In *Ancestors: The Hard Evidence,* ed. E. Delson, pp. 23–36. New York: Alan R. Liss.

Fleagle, J. G., and Rosenberger, A. L. (1983). Cranial morphology of the earliest anthropoids. In *Morphologie Evolutive, Morphogenese du Crane et Origine de l'Homme.,* ed. M. Sakka, pp.141–153. Paris: Centre National del Recherche Scientifique.

Fleagle, J. G., and Simons, E. L. (1978). Humeral morphology of the earliest apes. *Nature (London)* **276**: 705–707.

——— (1982a). Skeletal remains of *Propliopithecus chirobates* from the Egyptian Oligocene. *Folia Primatol.* **39**: 161–177.

——— (1982b). The humerus of *Aegyptopithecus zeuxis,* a primitive anthropoid. *Am. J. Phys. Anthropol.* **59**:175–193.

Fleagle, J. G., Kay, R. F., and Simons, E. L. (1980). Sexual dimorphism in early anthropoids. *Nature (London)* **287**: 328–330.

Harrison, T. (1987). The phyletic relationships of the early catarrhine primates: A review of the current evidence. *J. Human Evol.* **16**:41–80.

Kay, R. F., Fleagle, J. G., and Simons, E. L. (1981). A revision of the Oligocene apes from the Fayum Province, Egypt. *Am. J. Phys. Anthropol.* **55**:293–322.

Radinsky, L. (1974). The fossil evidence of anthropoid brain evolution. *Am. J. Phys. Anthropol. n.s.* **41**:15–27.

Schlosser, M. (1910). Uber einige fossile Saugertiere aus dem Oligocan von Agypten. *Zool. Anz.* **34**:500–508.

——— (1911). Beitrage zur Kenntnis der Oligozanen Lansaugetiere aus dem Fayum, Aegypten. *Beitr. Palaeontol. Oesterreich-Ungarns Orients* **6**:1–227.

Simons, E. L. (1965). New fossil apes from Egypt and the initial differentiation of Hominoidea. *Nature (London)* **205**:135–139.

——— (1967). The earliest apes. *Sci. Am.* **217**:28–35.

——— (1987). New faces of *Aegyptopithecus* from the Oligocene of Egypt. *J. Human Evol.* **16**:273–289.

——— (1993). New endocasts of *Aegyptopithecus:* Oldest well-preserved record of the brain in Anthropoidea. *Am. J. Sci.* **293** A:383–390.

——— (1995). Crania of *Apidium:* Primitive Anthropoidea (Primates, Parapithecidae) from the Egyptian Oligocene. *Am. Mus. Novitates* **3124**:1–10.

Simons, E. L., and Pilbeam, D. R. (1972). Hominoid paleoprimatology. In *The Functional and Evolutionary Biology of Primates,* ed. R. H. Tuttle, pp. 36–62. Chicago: Aldine-Atherton.

Simons, E. L., and Rasmussen, D. T. (1991). The generic classification of Fayum Anthropoidea. *Int. J. Primatol.* **12**:163–178.

Thomas, H., Sen, S., Roger, J., and Al-Sulaimani, Z. (1991). The discovery of *Moeripithecus markgrafi* Schlosser (Propliopithecidae, Anthropoidea, Primates), in the Ashawq Formation (Early Oligocene of Dhofar Province, Sultanate of Oman). *J. Human Evol.* **20**: 33–49.

Oligopithecids

Ducrocq, S., Jaeger, J. J., Chaiminee, Y., and Suteethorn, V. (1995). New primate from the Paleogene of Thailand, and the biogeographical origin of anthropoids. *J. Human Evol.* **28**:477–485.

Gheerbrandt, E., Thomas, H., Sen, S., and Al-Sulaimani, Z. (1995). Noveau primate Oligopithecinae (Simiiformes) de l'Oligocene inferiur de Taqah, Sultinat d'Oman. *C. R. Acad. Sci. Paris* **321**:425–432.

Gebo, D. L., Simons, E. L., Rasmussen, D. T., and Dagosto, M. (1994). Eocene anthropoid postcrania from the Fayum, Egypt. In *Anthropoid Origins,* ed. J. G. Fleagle and R. F. Kay, pp. 203–233. New York: Plenum Press.

Kay, R. F., Ross, C., and Williams, B. A. (1997). Anthropoid Origins. *Science* **275**:797–804.

Rasmussen, D. T., and Simons, E. L. (1992). Paleobiology of the oligopithecines, the earliest known anthropoid primates. *Int. J. Primatol.* **13**:477–508.

Simons, E. L. (1989). Description of two genera and species of late Eocene Anthropoidea from Egypt. *Proc. Natl. Acad. Sci. U.S.A.* **86**:9956–9960.

Simons, E. L. (1990). Discovery of the oldest known anthropoidean skull from the Paleogene of Egypt. *Science* **247**:1507–1509.

Simons, E. L. (1995). Skulls and anterior teeth of *Catopithecus* (Primates: Anthropoidea) from the Eocene shed light on anthropoidean origins. *Science* **268**: 1885–1888.

Simons, E. L., and Rasmussen, D. T. (1996). Skull of *Catopithecus browni,* an early Tertiary catarrhine. *Amer. J. Phys. Anthropol.* **100**:261–292.

Other Early Anthropoids

Godinot, M. (1994). Early North African primates and their significance for the origins of Simiformes (= Anthropoidea). In *Anthropoid Origins,* ed. J. G. Fleagle and R. F. Kay, pp. 235–296. New York: Plenum Press.

Godinot, M., and Mahboubi, M. (1992). Earliest simian primate found in Algeria. *Nature* **357**:324–326.

——— (1994). Les petits primates simiiformes de Glib Zegdou (Eocene, Algeria). *C. R. Acad. Sci. Paris* **319**: 357–364.

Hartenberger, J.-L., and Marandat, B. (1992). A new genus and species of an early Eocene primate from North Africa. *Human Evol.* **7**:9–16.

Kay, R. F., Ross, C., and Williams, B. A. (1997). Anthropoid origins. *Science* **275**:797–803.

Thomas, H., Sen, S., Roger, J., and Al-Sulaimani, Z. (1991). The discovery of *Moeripithecus markgrafi* Schlosser (Propliopithecidae, Anthropoidea, Primates), in the Ashawq Formation (Early Oligocene of Dhofar Province, Sultanate of Oman). *J. Human Evol.* **20**: 33–49.

EARLY ANTHROPOID EVOLUTION

Fleagle, J. G., and Kay, R. F. (1985). The paleobiology of catarrhines. In *Ancestors: The Hard Evidence,* ed. E. Delson, pp. 23–36. New York: Alan R. Liss.

Gebo, D. L., Simons, E. L., Rasmussen, D. T., and Dagosto, M. (1994). Eocene anthropoid postcranial from the Fayum, Egypt. In *Anthropoid Origins,* ed. J. E. Fleagle and R. F. Kay, pp. 203–233. New York: Plenum Press.

Kay, R. F., Ross, C., and Williams, B. A. (1997). Anthropoid origins. *Science* **275**:797–803.

Ross, C. F. (1995). Allometric and functional influences on primate orbit orientation and the origins of the Anthropoidea. *J. Human Evol.* **29**:201–227.

——— (1996). Adaptive explanation for the origins of the Anthropoidea (Primates). *Am. J. Primatol.* **40**: 205–230.

PROSIMIAN ORIGINS OF ANTHROPOIDS

Beard, K. C., and McPhee, R. D. E. (1994). Cranial anatomy of *Shoshonius* and the antiquity of the Anthropoidea. In *Anthropoid Origins,* ed. J. G. Fleagle and R. F. Kay, pp.55–97. New York: Plenum Press.

Beard, K. C., Dagasto, M., Gebo, D. L., and Godinot, M. (1988). Interrelationships among primate higher taxa. *Nature* **331**:712–714.

Beard, K. C., Krishtalka, L., and Stucky, R. K. (1991). First skulls of the early Eocene primate *Shoshonius cooperi* and the anthropoid–tarsier dichotomy. *Nature* **349**: 64–67.

Beard, K. C., Qi, T., Dawson, M. R., Wang, B., and Li, C. (1994). A diverse new primate fauna from middle Eocene fissure-fillings in southeastern China. *Nature* **368**:604–609.

Cartmill, M. (1980). Morphology, function and evolution of the anthropoid postorbital septum. In *Evolutionory Biology of the New World Monkeys and Continental Drift,* ed. R. L. Ciochon and A. B. Chiarelli, pp. 243–274. New York: Plenum Press.

——— (1994). Anatomy, antinomies and the problem of anthropoid origins. In *Anthropoid Origins,* ed. J. G. Fleagle and R. F. Kay, pp. 549–566. New York: Plenum Press.

Cartmill, M., and Kay, R. F. (1978). Cranio-dental morphology, tarsier affinities, and primate suborders. In *Recent Advances in Primatology,* vol. 3, ed. D. J. Chivers and K. A. Joysey, pp. 205–213. London: Academic Press.

Cartmill, M., MacPhee, R., and Simons, E. L. (1981). Anatomy of the temporal bone in early anthropoids, with remarks on the problem of anthropoid origins. *Am. J. Phys. Anthropol.* **56**:3–22.

Conroy, G. C. (1978). Candidates for anthropoid ancestry: Some morphological and paleozoogeographical considerations. In *Recent Advances in Primatology,* vol. 3, ed. D. J. Chivers and K. A. Joysey, pp. 27–41. London: Academic Press.

——— (1981). Review of *Evolutionary Biology of the New World Monkeys and Continental Drift,* edited by R. L. Ciochon and A. B. Chiarelli. *Folia Primatol.* **36**:155–156.

Covert, H. H, and Williams, B. A. (1994). Recently discovered specimens of North American Eocene omomyids and adapids and their bearing on debates about anthropoid origins. In *Anthropoid Origins,* ed. J. G. Fleagle and R. F. Kay, pp 29–54. New York: Plenum Press.

Dagasto, M., and Gebo, D. L. (1994). Postcranial anatomy and the origin of the Anthropoidea. In *Anthropoid Origins,* ed. J. G. Fleagle and R. F. Kay, pp. 567–595. New York: Plenum Press

Fleagle, J. G., and Kay, R. F., eds. (1994). *Anthropoid Origins.* New York: Plenum Press.

Gingerich, P. D. (1980). Eocene Adapidae: Paleobiogeography and the origin of South American Platyrrhini. In *Evolutionary Biology of the New World Monkeys and Continental Drift,* ed. R. L. Ciochon and A. B. Chiarelli, pp. 123–138. New York: Plenum Press.

Godinot, M. (1994). Early North African primates and their significance for the origins of Simiformes (= Anthropoidea). In *Anthropoid Origins,* ed. J. G. Fleagle and R. F. Kay, pp. 235–296. New York: Plenum Press.

Kay, R. F. (1980). Platyrrhine origins: A reappraisal of the dental evidence. In *Evolutionary Biology of the New World Monkeys and Continental Drift,* ed. R. L. Ciochon and A. B. Chiarelli, pp. 159–188. New York: Plenum Press.

Kay, R. F., Ross, C., and Williams, B. A. (1997). Anthropoid origins. *Science* **275**:797–803.

MacPhee, R. D. E., and Cartmill, M. (1986). Basicranial structures and primate systematics. In *Comparative Primate Biology,* vol. 1: *Systematics, Evolution and Anatomy,* ed. D. R. Swindler and J. Erwin, pp. 219–275. New York: Alan R. Liss.

Rasmussen, D. T. (1986). Anthropoid origins: A possible solution to the Adapidae-Omomyidae paradox. *J. Human Evol.* **15**:1–12.

——— (1994). The different meanings of the tarsioid-anthropoid clade and a new model of anthropoid origins. In *Anthropoid Origins,* ed. J. G. Fleagle and R. F. Kay, pp. 335–360. New York: Plenum Press

Rosenberger, A. L. (1986). Platyrrhines, catarrhines, and the anthropoid transition. In *Major Topics in Primate and Human Evolution,* ed. B. Wood, L. Martin, and P. Andrews, pp. 66–88. Cambridge: Cambridge University Press.

Rosenberger, A. L., and Dagasto, M. (1992). New craniodental and postcranial evidence of fossil tarsiiforms. In *Topics in Primatology,* vol. 3, ed. S. Matano, R. H. Tuttle, H. Ishida, and M. Goodman, pp. 37–51. Kyoto: University of Kyoto Press.

Rosenberger, A. L., and Szalay, F. S. (1980). The tarsiiform origins of Anthropoidea. In *Evolutionary Biology of the New World Monkeys and Continental Drift,* ed. R. L. Ciochon and A. B. Chiarelli, pp. 139–157. New York: Plenum Press.

Ross, C. F. (1994). The craniofacial evidence for anthropoid tariser relationships. In *Anthropoid Origins,* ed. J. G. Fleagle and R. F. Kay, pp. 469–548. New York: Plenum Press.

Sigé, B., Jaeger, J. J., Sudre, J., and Vianey-Liaud, M. (1990). *Altiatlasius kuolchii* n. gen. et sp. Primate omomyide du Paleocene superieur du Maroc, et les origines des Euprimates. *Paleontographica A* **214**:31–56.

Simons, E. L. (1995). Skulls and anterior teeth of *Catopithecus* (Primates: Anthropoidea) from the Eocene and anthropoid origins. *Science* **268**:1885–1888.

Simons, E. L., and Rasmussen, D. T. (1989). Cranial morphology of *Aegyptopithecus* and *Tarsius* and the question of the Tarsier-Anthropoidean clade. *Am. J. Phys. Anthropol.* **79**:1–23.

Szalay, F. S., Rosenberger, A. L., and Dagosto, M. (1987). Diagnosis and differentiation of the order Primates. *Yrbk. Phys. Anthropol.* **30**:75–105.

FOURTEEN

Fossil Platyrrhines

MAMMAL EVOLUTION IN SOUTH AMERICA

The earliest occurrence of platyrrhines in the fossil record of South America comes from the late Oligocene, 5–10 million years later than the Fayum primates. Although South America has an extensive record of Paleocene and Eocene deposits, mostly in Argentina, these ancient deposits have not yet yielded any primate fossils. For most of the Cenozoic Era, South America was an island with connections to no continent other than Antarctica. Much of Central America, including what is now Panama, was much farther west and joined North and South America only in the end of the Miocene. The biogeography of the Caribbean during most of the Cenozoic remains a mystery.

The early mammalian fossil record of South America reflects its isolation. It contains many unusual mammals unique to that continent, such as armadillos, sloths, many types of marsupials, and a large radiation of endemic ungulates, rather than the artiodactyls, perissodactyls, rodents, and prosimian primates common to the Eocene of North America and Europe (Simpson, 1980). The first appearance of primates and rodents marks novel additions to the South American fauna. There are no similar appearances of exogenous mammals until the joining of the northern and southern hemispheres in the latest Miocene or early Pliocene. Until recently, the earliest appearance of both primates and rodents was in Oligocene deposits of Bolivia and Argentina. However, rodents have now been recovered from latest-Eocene deposits in Chile, suggesting that monkeys may soon be found at a similarly early date. Where New World monkeys came from and how they got there are two of the most fascinating and difficult questions in the study of primate evolution (Hartwig, 1994). Before we tackle these questions, we examine the fossil record.

The Platyrrhine Fossil Record

Fossil New World monkeys are relatively scarce (Table 14.1), considering the extensive radiation of living primates found in the neotropics today and the relatively good fossil record for other South American mammals. Until recently, a large shoe box would have been sufficient to contain the primate fossils of South America and the Caribbean from the last 30 million years. The paucity of primates among the well-documented mammalian faunas of South America presumably indicates that much of the evolution of this group took place in areas from which there are very few fossil mammals at all, such as the vast Amazo-

TABLE 14.1
Suborder Anthropoidea
Infraorder PLATYRRHINI

Species	Estimated Mass (g)	Species	Estimated Mass (g)
Subfamily PITHECIINAE		Subfamily ATELINAE	
Soriacebus (early to middle Miocene, Argentina)		*Stirtonia* (middle to late Miocene, Colombia, Brazil)	
S. ameghinorum	2,000	*S. tatacoensis*	5,800
S. adrianae	600	*S. victoriae*	10,000
Carlocebus (early to middle Miocene, Argentina)		*S.* sp.	
C. carmenensis	3,500	*Protopithecus* (Pleistocene, Brazil)	
C. intermedius	2,000	*P. brasiliensis*	23,500
Homunculus (early to middle Miocene, Argentina)		*Caipora* (Pleistocene, Brazil)	
H. patagonicus	2,700	*C. bambuiorum*	24,000
Cebupithecia (middle to late Miocene, Colombia)			
C. sarmientoi	2,200	Subfamily CALLITRICHINAE	
Nuciruptor (middle to late Miocene, Colombia)		*Micodon* (middle to late Miocene, Colombia)	
N. rubricae	2,000	*M. kiotensis*	—
Propithecia (middle Miocene, Argentina)		*Patasola* (middle to late Miocene, Colombia)	
P. neuquenensis	1,600	*P. magdalena*	1,000
		Lagonimico (middle to late Miocene, Colombia)	
Subfamily AOTINAE		*L. conclutatus*	1,300
Tremacebus (early Miocene, Argentina)			
T. harringtoni	1,800	Subfamily *Incertae sedis*	
Aotus (middle Miocene, Colombia, to Recent)		*Branisella* (late Oligocene, Bolivia)	
A. dindensis	1,000	*B. boliviana*	1,000
		Szalatavus (late Oligocene, Bolivia)	
Subfamily CEBINAE		*S. attricuspis*	550
Dolichocebus (early Miocene, Argentina)		*Mohanamico* (middle to late Miocene, Colombia)	
D. gaimanensis	2,700	*M. hershkovitzi*	1,000
Chilecebus (early Miocene, Chile)		*Paralouatta* (Pleistocene, Cuba)	
C. carrascoensis	1,000	*P. varonai*	
Neosaimiri (middle to late Miocene, Colombia)		*Xenothrix* (Recent, Jamaica)	
N. fieldsi	840	*X. mcgregori*	—
Laventiana (middle to late Miocene, Colombia)		*Antillothrix* (Recent, Dominican Republic)	
L. annectens	800	*A. bernensis*	—

nian Basin. Although it is not extensive, the platyrrhine fossil record is expanding rapidly and provides us a broad overview and many tantalizing hints about the evolutionary history of the group.

On the basis of geography (Fig. 14.1) and age (Fig. 14.2), fossil platyrrhines can conveniently be divided into four groups: (a) the earliest platyrrhine fossils from a single late Oligocene locality in Bolivia; (b) several difficult-to-

FIGURE 14.1 A map of the Caribbean and Central and South America showing fossil platyrrhine localities.

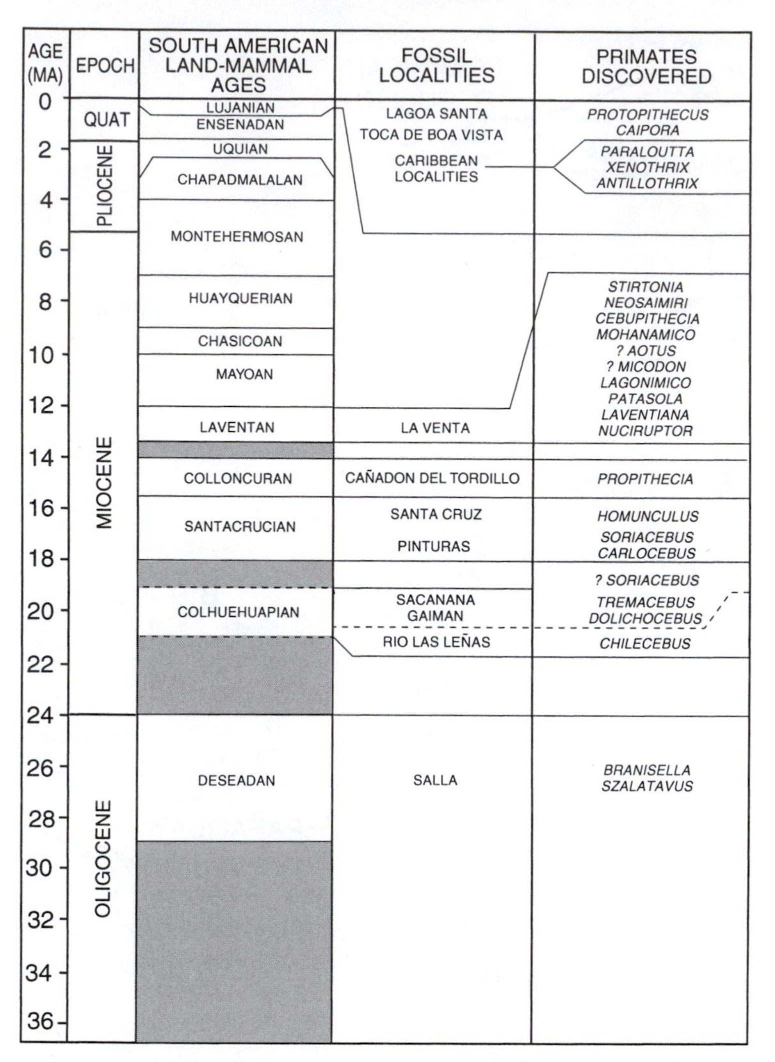

FIGURE 14.2 Geological time scale for South American land-mammal ages and primate-bearing fossil sites (redrawn and updated from MacFadden, 1990).

interpret genera from the early Miocene and middle Miocene of southern Argentina and Chile; (c) a diversity of relatively modern genera from the late Miocene of Colombia; and (d) several unusual species from Pleistocene or Recent deposits in the Caribbean and Brazil.

The Earliest Platyrrhines

The earliest platyrrhine fossils come from the late Oligocene (Deseadan) locality of Salla in Bolivia. Two genera have been described: ***Branisella*** and ***Szalatavus*** (Fig. 14.3). Both are small monkeys the size of an owl monkey, with

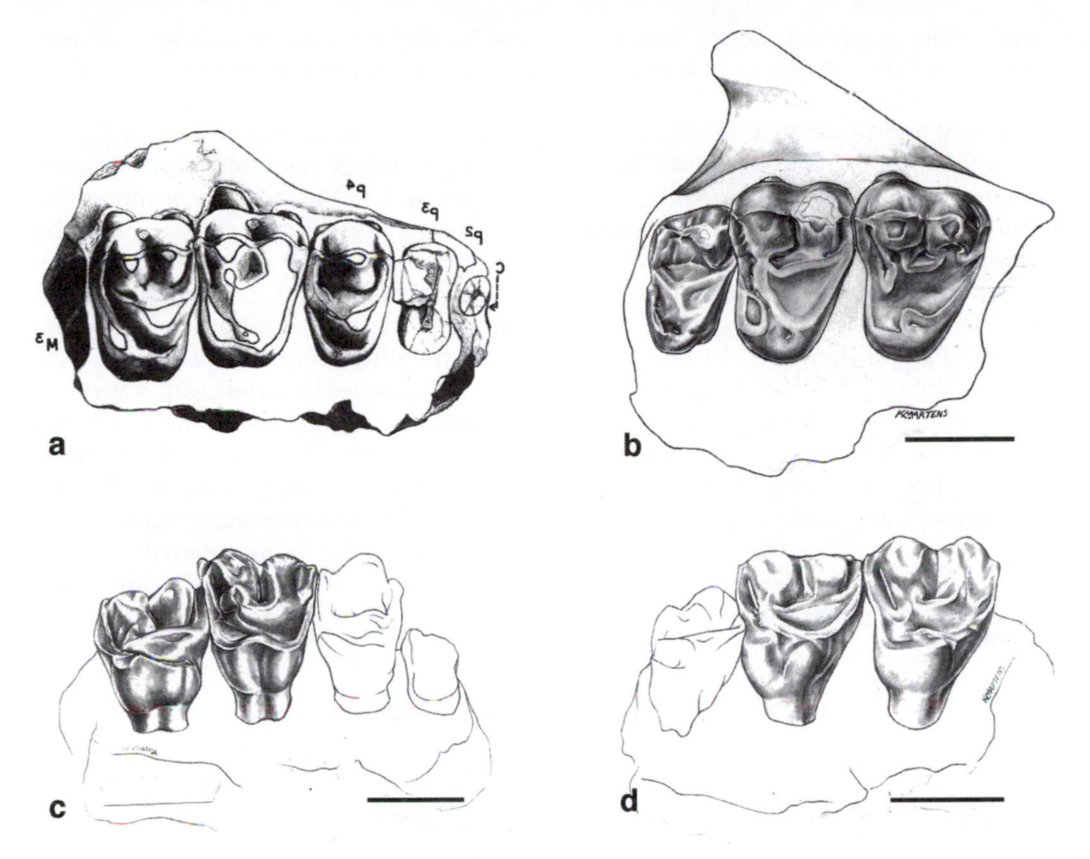

FIGURE 14.3 Maxillary remains attributed to *Branisella boliviana* (a, c) and *Szalatavus attricuspis* (b, d) (courtesy of A. L. Rosenberger).

three premolars and three molars. Both have four-cusped upper molars with a small hypocone and a well-developed lingual cingulum. They differ only in the shape of the molars: *Szalatavus* has narrower, more triangular molars than *Branisella*. The small P_2 and the shape of the mandible suggest short-faced monkeys. The low, rounded cusps of both taxa suggest frugivorous diets, and *Branisella* has very high-crowned lower molars, suggesting possible semiterrestrial habits (Kay and Williams, 1995). Although there have been few indications of any special phylogenetic relationship between *Branisella* and any modern platyrrhine subfamily, the triangular molars of *Szalatavus* have been suggested to indicate possible tamarin and marmoset affinities.

The Patagonian Platyrrhines

The southern parts of Argentina and Chile, a region informally known as Patagonia, have yielded over a half dozen genera and species of platyrrhines from early and middle Miocene deposits (Table 14.1; Figs. 14.1, 14.2). The primates were part of a rich fauna dominated by rodents, endemic ungulates, sloths, and marsupials. After many decades of fossil collecting, these monkeys are now known from hundreds of fossils—a few skulls, but mostly isolated dental and postcranial remains. Like the earlier fossil monkeys from Bolivia, these Patagonian fossils are clearly related to New World monkeys as a group, but any relationships to particular living subfamilies are difficult to identify and are the subject of considerable debate.

Dolichocebus gaimanensis (Fig. 14.4) is from sediments of the Colhuehuapian Land Mammal Age (earliest Miocene) near Gaiman, in Chubut Province, Argentina. It is known from a nearly complete but damaged skull, numerous isolated teeth, and a talus. *Dolichocebus* was a medium-size platyrrhine; the dentition suggests a body weight of between 2 and 3 kg, the size of a capuchin monkey, although the cranium is relatively smaller. *Dolichocebus* has dimorphic canines, three premolars, and three broad upper molars with a moderate-size hypocone and a broad lingual cingulum. The molar morphology resembles that of *Saimiri, Callicebus,* or *Aotus,* but is more primitive than these genera in such features as broad molars and a paraconule. The molar morphology of *Dolichocebus* suggests a frugivorous diet.

The skull of *Dolichocebus* has a narrow, posteriorly widening snout, complete postorbital closure, moderate-size orbits with a very narrow interorbital dimension, and relatively large tooth roots. The relative brain size is similar to that of extant platyrrhines. The distortion of the cranium suggests that in *Dolichocebus,* as in many living platyrrhines, the cranial sutures fused late in adulthood. The cranial morphology of *Dolichocebus* has been

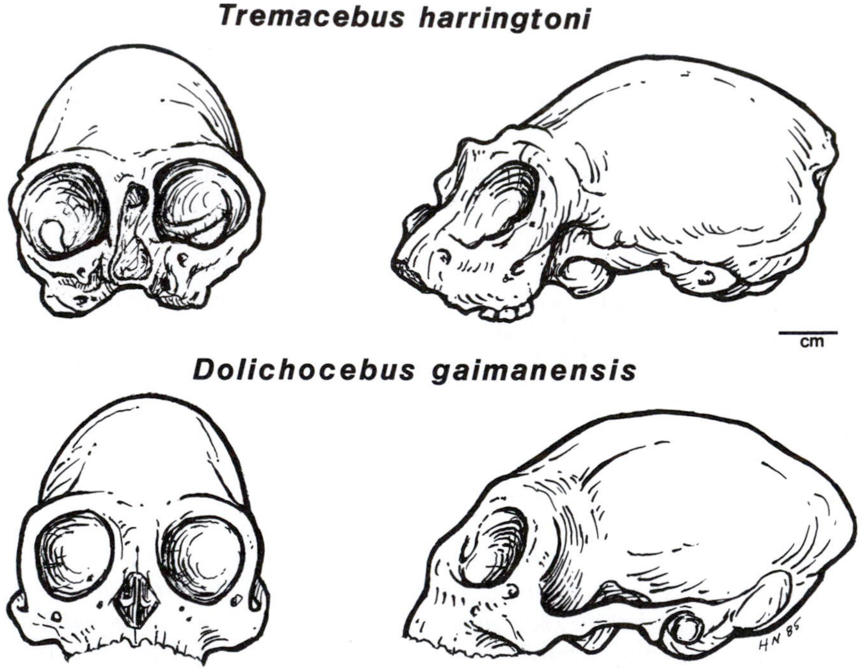

FIGURE 14.4 Reconstructed skulls of *Dolichocebus gaimanensis* and *Tremacebus harringtoni.*

the subject of considerable debate. Rosenberger has argued that *Dolichocebus* had an interorbital foramen linking the right and left orbits, an unusual cranial feature found only in *Saimiri* among living primates. Hershkovitz has argued that the supposed interorbital fenestra is an artifact of breakage. The talus of *Dolichocebus* is most similar to that of *Cebus* or *Saimiri,* suggesting it was either a rapid arboreal quadruped or a leaper.

On the basis of the presumed interorbital foramen and several other aspects of the cranial morphology of *Dolichocebus,* Rosenberger (1979) has argued that this genus is uniquely related to the living squirrel monkey. Hershkovitz has argued that the Oligocene monkey is too distinctive in other cranial features, such as the palate shape and molar root morphology, to bear any close relationship to living platyrrhines. Analysis of the isolated teeth found in association with *Dolichocebus* yields a similarly dichotomous picture of the relationships of this genus. In a phylogenetic analysis it is equally parsimonious to place *Dolichocebus* as either a close relative of the squirrel monkey or the sister group of all living platyrrhines (Fleagle and Kay, 1989).

Tremacebus harringtoni (Fig. 14.4) is from the Colhuehuapian (early Miocene) locality of Sacanana, also in Chubut, Argentina. It was a smaller (1–2 kg) monkey than *Dolichocebus.* The type specimen and only fossil clearly attributable to this species is a nearly complete but broken skull. *Tremacebus* has relatively small canines, three premolars, and three molars. The broken upper molars on the skull are quadrate with a large hypocone and a broad lingual cingulum. They are most similar to the teeth of *Callicebus* or *Aotus.* In addition to a relatively short, broad snout, *Tremacebus* has larger orbits than most diurnal platyrrhines but smaller ones than the nocturnal *Aotus,* suggesting to Hershkovitz (1974) that the species was possibly crepuscular. The posterior wall of the orbit is not completely walled off in the type specimen. Hershkovitz (1974) has argued from this evidence that *Tremacebus* was more primitive than any known anthropoids, for which full postorbital closure is a defining feature, but there are other indications that the large opening in the back of the orbit is due to breakage of the fossil and that *Tremacebus* is similar to living platyrrhines in its postorbital wall.

Tremacebus shows greatest dental and cranial similarities to the extant platyrrhines *Callicebus* and *Aotus.* Rosenberger (1984) has suggested that it is an ancestor of the living owl monkey, and other authorities have noted similarities to *Callicebus.*

Chilecebus carrascoensis is a newly described platyrrhine from the early Miocene (20 million years ago) of Chile (Flynn *et al.,* 1995). Its dentition resembles that of *Saimiri* in the relatively large upper premolars. The cranium is similar to that of *Tremacebus* in the posteriorly directed occipital region.

There are several genera of fossil platyrrhines from slightly younger deposits of the Santacrucian Land Mammal Age (early-middle Miocene) in Santa Cruz Province, Argentina (Fig. 14.1). These fossil monkeys come from two main geographical and geological areas: the slightly older Pinturas Formation in the west, and the younger Santa Cruz Formation on the east.

The Pinturas Formation, preserved in the foothills of the Andes of southern Argentina, is approximately 17.5–16.5 million years old (Fleagle *et al.,* 1995) and has yielded an abundant fauna of fossil birds, reptiles, and mammals, including four primate species (Fleagle, 1990). Evidence from the sediments themselves, fossil pollen, fossil birds, and abundant nests of fossil insects indicate that the Pinturas primates lived in a forested habitat in what must have been a time of climatic fluctuations, with periods of relative wetness separated by periods of desiccation (Bown and Larriestra, 1990). The fossil monkeys, which are known from over 300 specimens, including several facial skeletons, mandibles, and postcranial elements as well as isolated teeth, are found at several separate levels within the formation.

The best-known and most unusual of the Pinturas primates is ***Soriacebus*** (Fig. 14.5). There are two species: the saki-size *S. ameghinorum* from the lowest levels of the formation, and the tamarin-size *S. adrianae* from younger levels. *Soriacebus* has a dental formula of 2.1.3.3, with large procumbent lower incisors that form a continuous arcade with the large canine, a tall P_2, tiny posterior premolars, and three narrow, marmoset-like molars. The jaw deepens posteriorly as in extant pitheciines. The upper teeth have large, daggerlike canines, broad premolars with a hypocone on P^3 and P^4, and small, triangular molars. The facial skeleton is very deep. *Soriacebus* was probably frugivorous and used its large front teeth for some type of gnawing. The few postcranial elements of *Soriacebus* suggest quadrupedal running and leaping habits, with some clinging. The affinities of *Soriacebus* have been debated since its initial discovery, but it is most probably a primitive pitheciine on the basis of the large incisors, broad upper premolars, and the deep mandible.

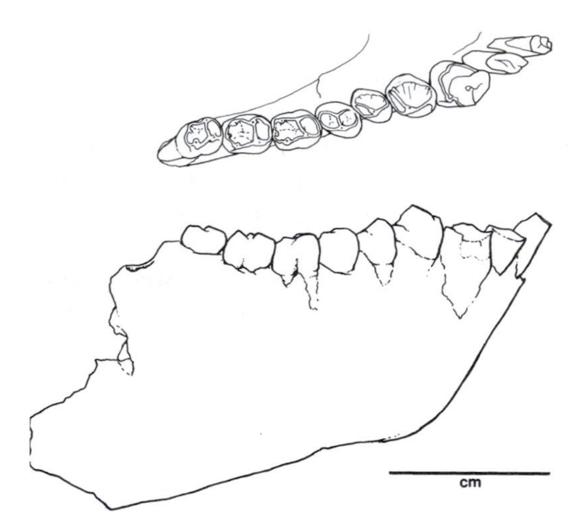

Soriacebus ameghinorum

FIGURE 14.5 Mandibular and maxillary remains of *Soriacebus ameghinorum* from the Pinturas Formation (early Miocene) of southern Argentina.

The other platyrrhine genus from Pinturas is ***Carlocebus,*** also with two species: the saki-size *C. intermedius* from the lower levels, and the larger *C. carmenensis* from throughout the formation. The lower dentition of *Carlocebus* is more generalized than that of *Soriacebus,* with small vertical incisors, a small canine, a moderate-size anteriormost premolar, and relatively larger posterior premolars and molars. As in *Soriacebus,* the mandible is relatively deep. The upper dentition is characterized by very broad premolars and molars with large hypocones. Dentally, *Carlocebus* appears to have been frugivorous and folivorous. Its dentition is most comparable to that of *Callicebus,* the titi monkey, only it is much larger. Postcranial remains of *Carlocebus* suggest arboreal quadrupedal habits.

Homunculus patagonicus (Fig. 14.6), from the early-middle Miocene (16.5 million years ago) Santa Cruz Formation on the Atlantic coast of southern Argentina, was one of the earliest fossil platyrrhines discovered (Ameghino, 1891), and for many years all fossil platyrrhines were placed in this genus. It was a medium-size monkey, with the largest individuals probably weighing nearly 3 kg. The dental formula is 2.1.3.3. The lower incisors are narrow and spatulate; the canines are probably sexually dimorphic. The molars are characterized by relatively small cusps connected by long shearing crests; they have a small, square trigonid and a broader talonid with a prominent cristid obliquid. The mandible is relatively shallow compared with that of the Pinturas primates. *Homunculus* was probably

FIGURE 14.6 A cranium of *Homunculus patagonicus* (courtesy of Adan Tauber).

frugivorous and folivorous. The upper dentition is less well know but seems most comparable to that of *Carlocebus* with smaller teeth.

The facial fragment attributed to *Homunculus* has a relatively short snout with procumbent incisors and moderate-size orbits (indicating diurnal habits) with complete postorbital closure. The lacrimal bone is well within the orbit margin, as in most higher primates. The cranium appears relatively gracile. with no sagittal crest.

The limb elements resemble those of a callitrichid and suggest that *Homunculus* was predominantly quadrupedal. In some details of its limbs, such as the size of the lesser trochanter on the femur, *Homunculus* resembles the parapithecids from Egypt in what are probably primitive features.

As the name *Homunculus* indicates, Ameghino (1891) originally thought the genus was in the ancestry of humans; it is not. Most later studies have noted either the unique features of the genus or dental similarities to *Aotus, Callicebus,* or *Alouatta.* In describing a new facial skeleton attributed to *Homunculus,* Tauber (1991) has noted many similarities to pitheciines.

There are several fossil platyrrhines from the slightly younger Canadon del Tordillo Formation in northern Patagonia. The newly described ***Propithecia*** has been identified as an early pitheciine on the basis of several dental features (Kay *et al.*, 1998).

The apparent contradictions in resolving the phylogenetic position of *Homunculus* apply to all the Patagonian platyrrhines: they show unusual combinations of features linking them in several directions with what today are distinct clades of platyrrhines. In phylogenetic analyses they can just as easily be placed with several different clades or as the outgroup to all modern subfamilies. There are several possible explanations for this phenomenon, which are not mutually exclusive. One possibility is that the Patagonian primates are collateral to all later evolution of platyrrhines, a conclusion that would be compatible with both their age and their geographic distinctiveness. Patagonia has a long history as a separate biogeographic area within South America, with a distinct flora and invertebrate fauna. However, without similar-age primates from elsewhere in the continent, this hypothesis cannot be tested.

It is more likely that the unusual combinations of features seen in these early platyrrhines reflect their proximity to the initial split of modern clades, before most acquired the larger suite of features that characterize their living members. *Chilecebus* and *Dolichocebus* may well be near the origin of the squirrel monkey lineage. Many of the primates from Santa Cruz Province seem to be broadly related to modern pitheciines (including *Callicebus*). Unusual combinations of "modern" features are often found among fossil taxa and are what make fossils especially valuable in resolving the evolutionary history of a group. There is no doubt that part of our inability to resolve clearly the phylogenetic position of these monkeys is that we lack a full understanding of the polarity of the features characterizing extant platyrrhines, that is, which features are primitive retentions and which are derived specializations in different groups. These early platyrrhines should provide considerable insight into the broader relationships of platyrrhine subfamilies when we can develop proper methods of analysis to understand the information they contain.

A More Modern Community

Fossil platyrrhines are known from very few places in the vast land mass of Central and South America, and there are no fossil sites as old as the Patagonian deposits from more tropical areas of South America. The oldest fossil platyrrhines from more tropical parts of the continent are those from the La Venta in Colombia (Kay *et al.*, 1997). Compared with the Patagonian fossil platyrrhines, which are difficult to place in extant platyrrhine subfamilies, most of the fossil monkeys from the later Miocene of La Venta, Colombia, are strikingly similar to modern platyrrhines and clearly belong in living subfamilies or even genera. The fossil platyrrhines from La Venta are found in two different geological formations, which preserve a very brief period of time from approximately 13 million to 12 million years ago. Comparison of the La Venta fauna with modern South American faunas indicates a tropical forest environment for this region in the late Miocene (Kay and Madden, 1997). Compared with the earlier Patagonian faunas, the La Venta fauna contains some of the same "archaic taxa," including caenolestid marsupials and astrapotheres, but also many more modern mammals, such as anteaters. There are nearly a dozen species of fossil primates described from this area (Fleagle *et al.*, 1997).

Cebupithecia sarmientoi (Fig. 14.7) was similar in size (2–3 kg) and in many aspects of skeletal morphology to the living saki, *Pithecia pithecia*. The fossil is known from parts of two skeletons, including most of the dentition, the mandible, and several cranial fragments, discovered almost forty years apart. In many aspects of its dental morphology, such as the stout canines, procumbent incisors, and flat cheek teeth with little cusp relief, this Miocene genus is very similar to the living pitheciines. Like living pitheciines, *Cebupithecia* probably ate mainly fruit and used its large

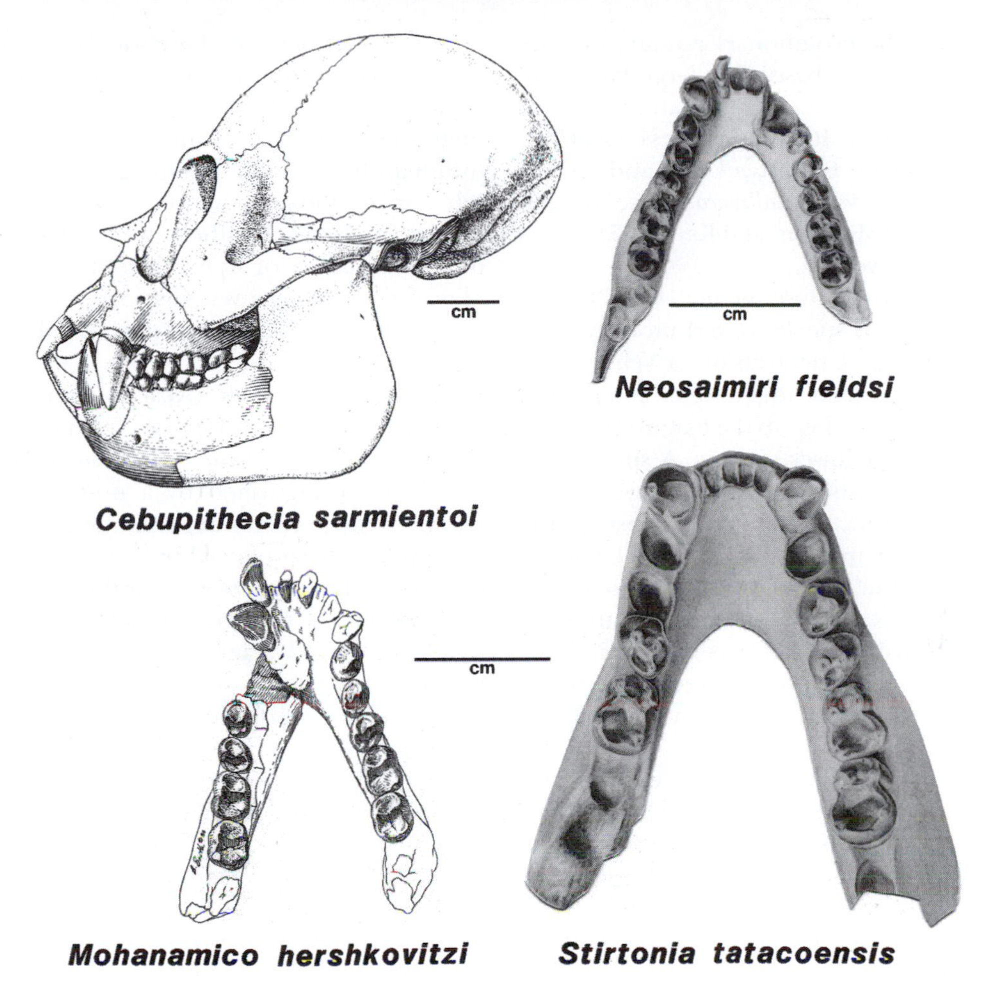

FIGURE 14.7 Cranium and dentition representing four primates from the late Miocene of La Venta, Colombia.

anterior dentition for opening seeds. The *Cebupithecia* skeleton shows more similarities to the saltatory *Pithecia* than to the more quadrupedal sakis, such as *Chiropotes.* In contrast with its uniquely pitheciine dentition, the skeleton retains many features found in other platyrrhine subfamilies while lacking some shared derived features of living pitheciines. There are also indications of vertical clinging habits in the morphology of the elbow. It is probably very near the base of the modern pitheciine radiation.

Nuciruptor rubricae is a new pitheciine from La Venta that differs from *Cebupithecia* in having a smaller canine and no canine-incisor diastema (Meldrum and Kay, 1997).

Mohanamico hershkovitzi (Fig. 14.7) is a small (1 kg) fossil monkey from La Venta that is known from a single mandible (Luchterhand *et al.,* 1986). It has also been placed near the base of the evolutionary radiation of the pitheciines on the basis of its large lateral incisor and the structure of the canine and anterior premolar. It is much less clearly a pitheciine than is *Cebupithecia,* and others have linked it with *Callimico* (Rosenberger, 1992, see also Meldrum and Kay, 1997). It was probably frugivorous.

Setoguchi and Rosenberger (1987) have described a new species of owl monkey from the later Miocene deposits of La Venta. ***Aotus dindensis*** is virtually identical in molar and premolar morphology to the extant *Aotus,* but it has narrower lower incisors. A small facial fragment suggests that the Miocene species also has large orbits similar to those of the nocturnal owl monkey. There has been debate about the affinities and similarities of *Aotus dindensis* and *Mohanamico,* and the phyletic relationships of both are uncertain. This debate illustrates mostly the conservative nature of the mandibular dentition in small-size frugivorous platyrrhines.

A fossil genus that is more clearly closely related to a modern subfamily is ***Stirtonia*** (Fig. 14.7), the largest monkey in the La Venta fauna (6–10 kg). There are two species from different stratigraphic levels. The older, larger species is *Stirtonia victoriae;* the smaller, younger species is *S. tatacoensis. Stirtonia* has many dental similarities in its upper and lower dentition to the living howling monkey (*Alouatta*). Like *Alouatta, Stirtonia* has long molars with a relatively small trigonid and large talonid and very large upper molars with well-developed shearing crests and styles. It was a folivore. Isolated molars that resemble those of *Stirtonia* (and *Alouatta*) also have been recovered from late Miocene deposits along the Rio Acre in western Brazil (Kay and Frailey, 1993).

Neosaimiri fieldsi (Fig. 14.7) is virtually identical to the living squirrel monkey in size and in all known details of dental anatomy, differing most clearly in molar proportions and the relative size of the lateral incisor. Like *Saimiri,* it was insectivorous and frugivorous. An isolated humerus from the same deposits is indistinguishable from the same bone in *Saimiri.* Although there is but a single genus of squirrel monkey today with numerous allopatric species, it seems that there was a larger taxonimic radiation of squirrel monkeys in the Miocene. *Neosaimiri* was just one of several squirrel monkey-like taxa in the later Miocene of Colombia.

Laventiana annectens (Rosenberger *et al.,* 1991b) is another fossil species from La Venta that is similar to *Saimiri* and *Neosaimiri.* This species is distinguished by a postentoconid notch on the lower molar, a characteristic not seen in any living platyrrhines. A talus recovered with the *Laventiana* mandible also resembles that of *Saimiri* and lacks the clearly derived traits characterizing the smaller callitrichines and larger atelines. Some authorities place *L. annectens* in the same genus as *Neosaimiri.*

One of the most obvious gaps in the platyrrhine fossil record for many years has been the absence of any fossil evidence of callitrichines (marmosets, tamarins and *Callimico*), even though they are the most diverse subfamily of living New World monkeys. Their evolutionary origin has been a source of endless debate (e.g., Hershkovitz, 1977; Ford, 1980; Rosenberger, 1984). In recent years several putative fossil callitrichines have been described, each with different morphologies and different implications for the origin of the group!

Micodon kiotensis is a poorly-known species from La Venta that is based on three small, isolated teeth (Setoguchi and Rosenberger, 1985). It has been described as a fossil marmoset, primarily on the basis of size. The type specimen, an upper molar, lacks any marmoset features and resembles that of a small pitheciine in occlusal morphology. Its describers noted that *Micodon* demonstrates that the small size of callitrichines preceded their distinctive dental features, such as loss of a hypocone. Any determination regarding either the validity of the species or its affinities must await more fossil remains.

Patasola magdalena is a new species from La Venta that is slightly smaller than the living squirrel monkey. It shares features with both *Saimiri* and callitrichines and was identified by its describers as a callitrichine that is more closely related to marmosets and tamarins than is *Callimico,* on the basis of simplified premolar morphology and the shape of the molar trigonid and cristid obliqua (Kay and Meldrum, 1997).

Lagonimico conclutatus (Fig. 14.8) is another new species from La Venta. It was roughly the size of an owl monkey and has been described by Kay (1994) as a giant tamarin. Based on a different set of features, mainly upper molar shape, he placed it in exactly the same phyletic position as *Patasola.* The absence of upper molar hypocones in a monkey the size of *Lagonimico* suggests that acquisition of the distinctive marmoset and tamarin molar morphology did not necessarily evolve in conjunction with small size.

The presence of a several putative fossil callitrichines at La Venta is exciting, but they clearly demand further analysis, since each offers a somewhat different picture of the origin and early evolution of the group. It is almost certain that the origin of this group involved a rather bushy phylogeny, as evidenced by the parallel and convergent features found in

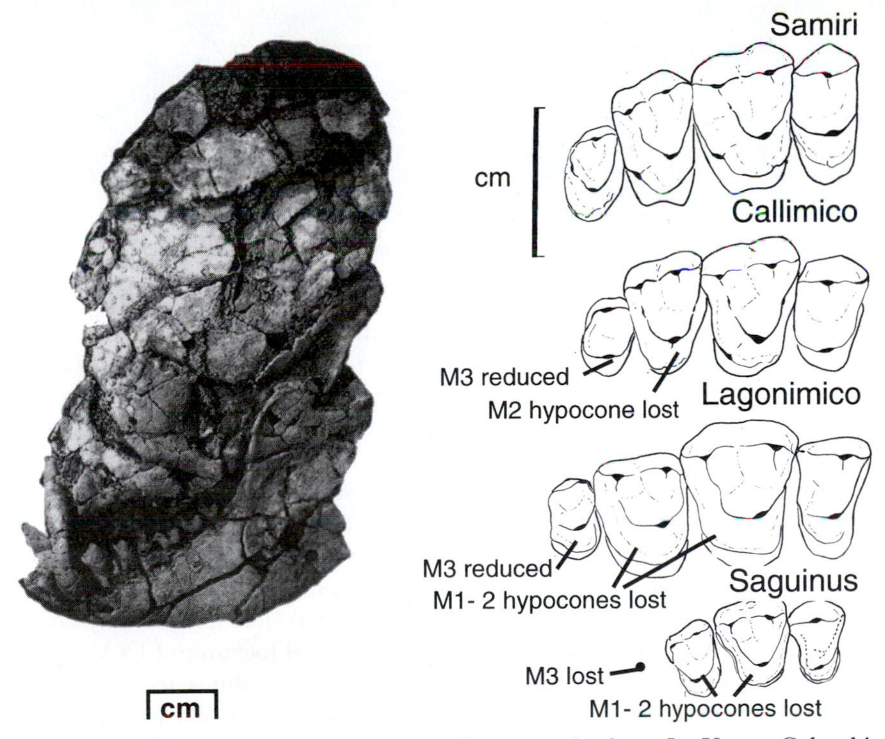

FIGURE 14.8 *Lagonimico conclutatus,* a giant tamarin from La Venta, Colombia (courtesy of R. F. Kay).

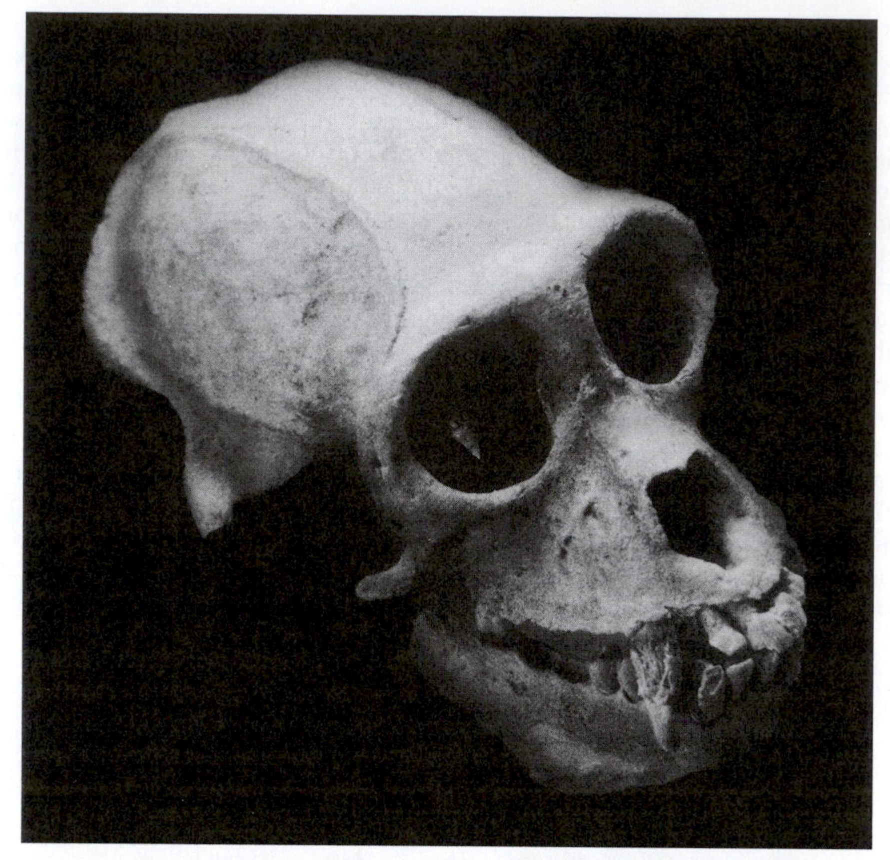

FIGURE 14.9 Cranium of *Protopithecus brasiliensis* from the Pleistocene of Brazil.

marmosets, tamarins, *Leontopithecus, Lagonimico,* and *Callimico.* Analysis of these fossils also appears to support the view that callitrichines are the sister taxon of the squirrel monkey, *Saimiri,* as suggested for many years by Rosenberger (e.g., 1981) and by recent molecular studies (Schneider *et al.,* 1996). However, we are still far from understanding many details of the origin and radiation of these most successful platyrrhines.

Despite debate over the validity of a few taxa and the proposed phylogenetic relationships of others, the fauna from La Venta indicates that many groups of extant platyrrhines were clearly differentiated and present in northern Colombia by 12 million years ago. Perhaps most notable is the absence at La Venta of any putative relatives of either *Cebus* or *Callicebus,* two of the most widespread modern genera, and the absence of any spider or woolly monkeys.

The La Venta primate fauna is clearly modern compared with other fossil platyrrhines, both taxonomically and adaptively. This modernity may reflect its relatively late age or the geographical location of La Venta closer than other fossil localities to the Amazon Basin, where living New World monkeys are most abundant. Most probably, both of these relationships contribute to the modern appearance of this fauna. Aside from a few isolated teeth from the latest Miocene of the upper reaches of the Amazon and some remains from the latest Pleistocene and Recent cave sites in Brazil (to be described shortly), La Venta is the youngest fossil deposit yielding platyrrhines from all of Central and South America. Thus, although the pre-Pleistocene fossil record of South America provides documentation of the major diversification of subfamilies and major clades, we know very little about the origin of living genera and species and have no real appreciation of platyrrhine species diversity or biogeography in the past.

Pleistocene Platyrrhines

In many respects, the most unusual fossil platyrrhines are the youngest, those from Pleistocene and Recent caves of the Caribbean and Brazil. One of the earliest fossil primates ever recovered, and the first that was recognized to be unlike any living species, was found in a Brazilian cave in the 1830s by the Danish naturalist Peter Wilhelm Lund. The numerous cave deposits of Lagoa Santa are late Pleistocene and/or Recent in age and contain a mixture of extinct and extant fauna. In 1836, Lund found a proximal femur and distal humerus of an ateline-like primate that was larger than any living platyrrhines and probably had a body weight nearly two and a half times as large (Hartwig, 1995). Although Lund's fossils have generally been placed in the same genus and species as the living muriqui, *Brachyteles arachnoides*, it is clearly a different monkey and should be recognized by the name Lund gave it 150 years ago, ***Protopithecus brasiliensis*** (Fig. 14.9).

A nearly complete skeleton of *Protopithecus* and another large ateline have been recovered recently from Pleistocene caves in northeastern Brazil (Cartelle, 1993; Hartwig and Cartelle, 1996). The locomotor skeleton of *Protopithecus* resembles that of a large, suspensory spider monkey or woolly spider monkey, whereas the skull most closely resembles that of *Alouatta*, the howling monkey. This combination of features suggests considerable parallelism in the evolution of atelines.

A new species from the same caves in northeastern Brazil, ***Caipora bambuiorum***, is also about twice as large as any living platyrrhine. In both cranial and skeletal morphology *Caipora* most closely resembles the living spider monkey, *Ateles* (Cartelle and Hartwig, 1996).

In addition to *Protopithecus* and *Caipora*, the Brazilian caves also contain remains of several extant taxa, including *Callithrix*, *Cebus*, and *Alouatta*, genera that are found in the region today.

Caribbean Primates

There are no nonhuman primates on any of the Caribbean islands today. When primate remains from these islands were described earlier this century, they were generally considered not endemic species, but "pets" brought over from the mainland by humans. Ironically, it now seems most likely that this region once harbored a diverse fauna of endemic primates and that humans were either directly or indirectly responsible for their extinction.

Xenothrix mcgregori is a latest-Pleistocene or Recent primate from the island of Jamaica. It is known from postcranial material, a mandible and newly discovered cranial remains (Horovitz *et al.*, 1997). It had a dental formula of 2.1.3.2, as in marmosets and tamarins. However, at 2 kg it was relatively much larger than marmosets and tamarins, and the molars, with large, bulbous cusps, are very different in both cusp morphology and proportions from any modern callitrichids. *Xenothrix* was probably a frugivorous species, or it may have specialized on insect larvae, like the aye-aye of Madagascar. The postcranial remains that have been attributed to *Xenothrix* evidence an unusual type of slow quadrupedal locomotion that has no counterpart among living platyrrhines. The relationship of *Xenothrix* to other platyrrhines is uncertain.

In recent years, it has become clear that *Xenothrix* is just one of several platyrrhines that lived in the Caribbean prior to the first appearance of humans several thousand years ago. Fragmentary platyrrhine fossils have been found in two other Jamaican caves. Both specimens are proximal femora and are quite different from the bones of *Xenothrix,* suggesting that there were at least three primates on that island. (Ford, 1990).

Antillothrix bernensis is a fossil monkey known from numerous, largely undescribed, dental specimens and a tibia recovered from Pleistocene or recent cave deposits in Haiti and the Dominican Republic (MacPhee *et al.,* 1995). The dental remains, which may be as much as 100,000 years old, suggest a large primate (2–3 kg) with a diet of hard fruit or seeds.

Paralouatta varonai (Fig. 14.10) is a Pleistocene platyrrhine from Cuba known from a nearly complete skull, a mandible, numerous isolated teeth, and several limb elements (Rivero and Arredondo, 1991; MacPhee *et al.,* 1995). *Paralouatta* was a large monkey, with a long, unflexed skull, a small brain, and relatively large orbits. Despite the name and the superficial similarities between the skull of *Paralouatta* and that of the living howling monkey, the most recent study suggests that *Paralouatta* is not related to *Alouatta.* Rather, it may be part of an endemic radiation of Caribbean

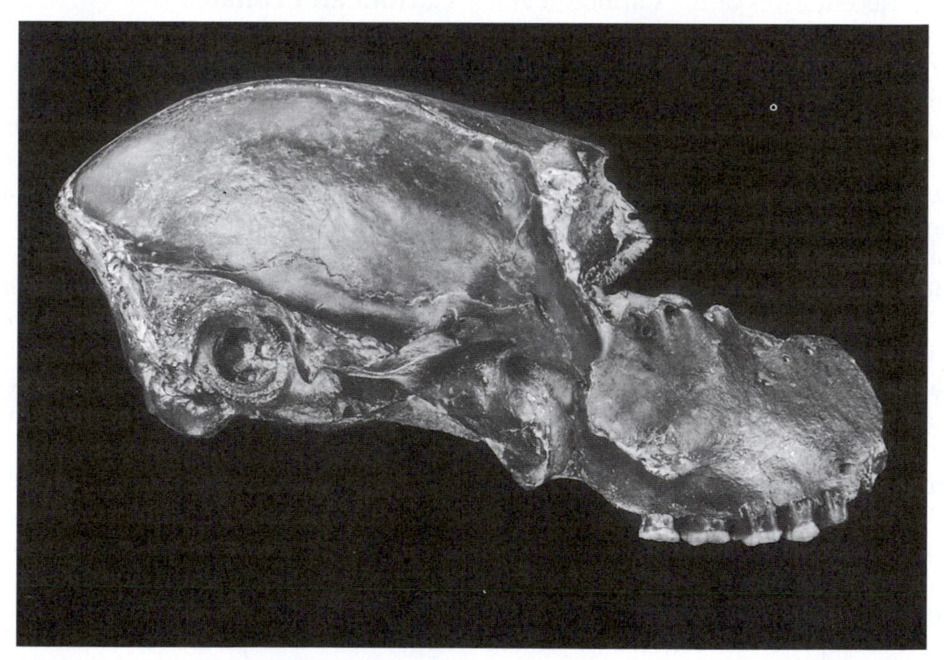

FIGURE 14.10 Cranium of *Paralouatta varonai* from the Pleistocene of Cuba (photograph by Lorraine Meeker, American Museum of Natural History).

primates related to *Callicebus* (MacPhee *et al.*, 1995; Cole *et al.*, 1997).

The recovery in recent years of numerous fossil remains of primates that are strikingly different from anything known elsewhere in the New World, often from sites that predate human colonization of the islands, demonstrates quite clearly that there was an endemic primate fauna in the Caribbean until quite recently. Interestingly, island biogeographic studies of the larger islands "predicts" the presence of more small-to-medium-size frugivores than had previously been described. However, this unveiling of an extensive Caribbean fauna raises even larger issues about the origin and ultimate extinction of these primates.

The simplest explanation for the origin of the Caribbean fauna is over-water dispersal from nearby parts of Central and South America. Cuba is only 50 miles from the Yucatan Peninsula, where *Alouatta, Ateles,* and *Cebus* are found today. In addition, the Lesser Antilles are close to Venezuela, where there is an even more extensive fauna. The fact that the primates on Cuba and Hispaniola seem most closely related to *Ateles, Alouatta,* and *Cebus* accords well with this view of a simple dispersal, perhaps relatively recently. However, the evidence from recent studies suggesting that the fossil primates from the Caribbean are quite different from extant platyrrhines on the "mainland" is more compatible with a view that the Caribbean primate may have been separated from other platyrrhines for many millions of years. This hypothesis is further supported by a Miocene primate talus from Cuba (MacPhee and Iturralde-Vinent, 1995). The distinctiveness of many other aspects of the Caribbean mammal fauna has been noted as well (Hedges, 1996). For example, the Caribbean sloths have been reported to be more similar to the Miocene sloths of Argentina than to the extant sloths of Central and South America.

SUMMARY OF FOSSIL PLATYRRHINES

Despite the scarcity of fossils from Central and South America, the available remains of fossil platyrrhines provide a number of insights into the history and timing of appearance of many modern groups of New World monkeys. Perhaps the most striking feature of the platyrrhine fossil record is the overall similarity of the extinct species to modern lineages. Although it is important to remember that our knowledge of fossil New World monkeys is based largely on fragmentary dental remains, it seems that many lineages of extant platyrrhines have been distinct since at least the middle Miocene (Fig. 14.11). Fossil species related to the extant owl monkey (*Aotus*), squirrel monkey (*Saimiri*), pitheciines, and howling monkey (*Alouatta*) were definitely present in the middle to late Miocene of Colombia. Evidence suggesting that some of these lineages can be traced back to late Oligocene (Colhuehuapian) or early Miocene (Santacrucian) times is less clear, but suggestive. Further fossil discoveries from these earlier periods should help resolve both the age of these Miocene lineages and the relationships of the modern subfamilies.

The fossil record also provides evidence that the extant platyrrhine fauna is very impoverished from that in the Pleistocene. The Jamaican *Xenothrix* and the Brazilian *Protopithecus* are very different from any extant taxa. The late Pleistocene/Recent *Protopithecus* was an animal twice the size of the largest living platyrrhines (yet was largely unrecognized for over a century and a half). The New World, like Madagascar, has clearly suffered dramatic Pleistocene extinctions of its primate fauna. We can only imagine what other discoveries may come from further field work. Perhaps someday we will find evidence of a terrestrial lineage of platyrrhines.

PLATYRRHINE ORIGINS

The most unsettled question surrounding platyrrhine origins is the geographic one: How did platyrrhines get to South America? The issue is a particularly complex one involving not only paleontological information about fossil platyrrhines but also information about paleogeography and the faunas of other continental areas. The origins question integrates our relatively recent awareness of continental drift dynamics with a long-standing awareness of monophyly, or shared common ancestry, within anthropoids. Resolving the question requires us to explore all types of evidence.

South America was an island continent throughout most of the early Cenozoic, separated from Africa by the South Atlantic and from North America by the Caribbean Sea. Debate over the origin of neotropical primates has focused on whether North America or Africa is the most likely source of the immigrating primates (Fig. 14.12).

Most geophysical studies indicate that the positions of North and South America and Africa relative to one another were much the same in the Eocene and Oligocene as they are now; the rifting of the South Atlantic had taken place much earlier, during the Mesozoic era. There was, then, a considerable body of water for migrating primates to cross, from either North America or Africa. The geological history of the Caribbean region is, however, very poorly known (Stehli and Webb, 1985), and the geophysical evidence does not seem to favor one continental source over another.

During the early Cenozoic, though, there were probably large areas of relatively shallow water in the South Atlantic, due to crustal uplift, and possibly a series of islands in the areas of the Walvis Ridge and the Sierra Leone Rise. In periods of low sea level, such as the early Oligocene, these areas and the continental shelves of Africa were probably dry land, which would appreciably shorten the open-water distances between the continents. The reconstructed currents for this time also seem to favor a crossing from Africa to South America (Tarling, 1982).

Because all available evidence indicates that the immigrant primates that rafted to South America and gave rise to living platyrrhines were anthropoids rather than prosimians, we must also consider the nature of the fossil primates known from the potential source continents. North America or Africa could be a source area for the earliest platyrrhines only if there were suitable primates on those continents to be the ancestral platyrrhines. In this respect, Africa is unquestionably the most likely source of early platyrrhines. The only undoubted Oligocene anthropoids are those of Africa. There are many similarities between the Fayum anthropoids and platyrrhines. Parapithecids seem to be basal anthropoids that preceded the evolutionary divergence of platyrrhines and catarrhines. In particular, the late Eocene *Proteopithecus* shows many features in its dentition, cranium, and postcranium that make it a suitable platyrrhine ancestor (e.g., Miller and Simons, 1997; Simons, 1997). Moreover, South American rodents, which appeared at approximately the same time or earlier than the primates (Wyss *et al.,* 1993), are most closely related to the African porcupines (Hoffstetter and Lavocat, 1970), providing further evidence of a faunal connection between South America and Africa (see George and Lavocat, 1993).

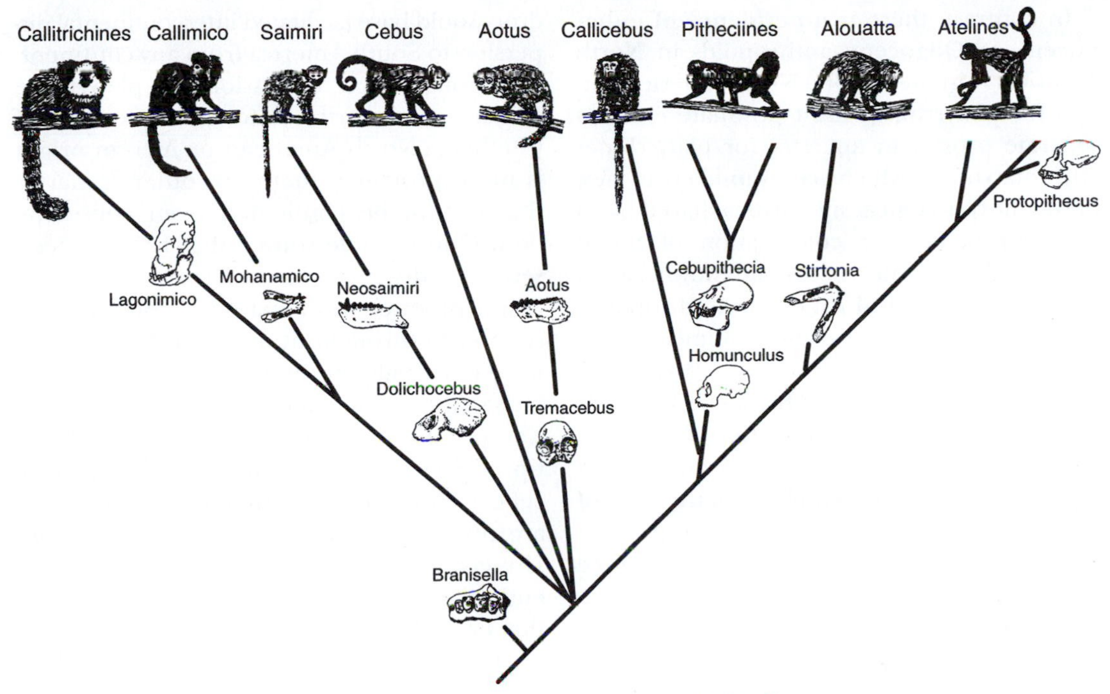

FIGURE 14.11 A platyrrhine phylogeny with fossil genera.

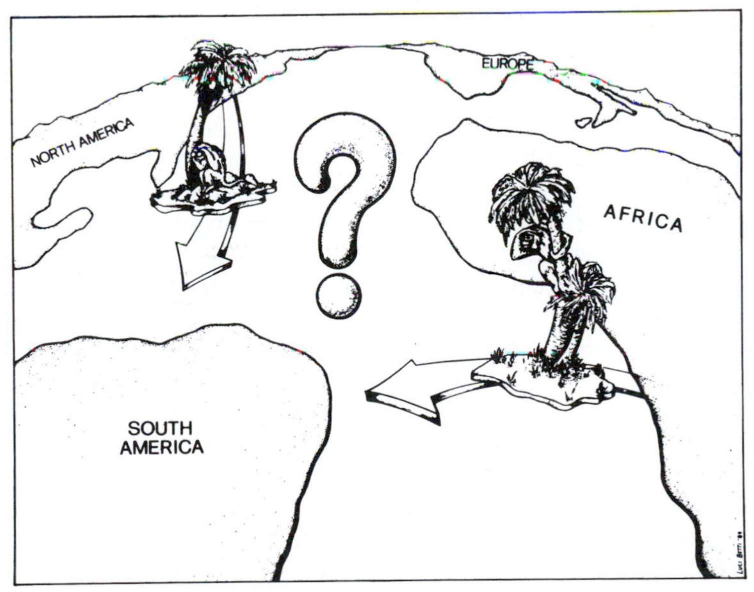

FIGURE 14.12 How did ancestral platyrrhines reach South America?

In contrast, there is no evidence of either Eocene or Oligocene anthropoids in North America. Arguments for a North American origin for platyrrhines must postulate either a separate prosimian ancestry for platyrrhines and catarrhines, which seems unlikely in view of the morphological similarities shared by all higher primates, or colonization of South America by a group of still-unknown North American or Central American anthropoids. Certainly, the discovery of prosimians with European affinities (the adapine *Mahgarita*) in the late Eocene and Oligocene of Texas demonstrates the likelihood that there were other primates in North America that remain to be discovered. Nevertheless, on the basis of the known North American Eocene primate fauna, that continent is a very unlikely source of early platyrrhines.

Finally, it must be noted that several authors (Wood and Patterson, 1970; Simons, 1976) have questioned whether long-distance rafting between any continents is a likely method for biogeographic dispersal of animals, given the dietary and climatic requirements of primates. Anthropoids seem less suited to dispersal on floating masses of vegetation than animals that can hibernate or have long periods of inactivity (such as cheirogaleids or rodents). But regardless of how unlikely rafting may seem, it is presently the only suggested mechanism for transporting terrestrial animals between continents separated by open ocean. If South America was indeed an island continent during the period in question, we must assume that primates rafted from some other continental area. Only a revision of the paleocontinental maps could eliminate the need for rafting in the origin of platyrrhines. However, rafting could be feasible; the largest drop in sea level during the entire Cenozoic occurred in the middle Oligocene, before the first appearance of *Branisella* but after the Fayum deposits were accumulated. Such a drop would have facilitated intercontinental dispersion to South America from any continent.

Although most discussions of platyrrhine origins are restricted to the question of whether a North American or African origin is more probable, there are other scenarios that cannot be eliminated from consideration. One possible source that has not been seriously discussed is Antarctica. Although South America was apparently connected to Antarctica throughout much of the early Cenozoic, virtually nothing is known about the fauna (if any) of Antarctica during this time. The presence of monkeys near the southern tip of South America during the early Miocene clearly indicates that present climates are no indicator of the past in this region. However in the absence of any evidence for eutherian mammals of any type on Antarctica, that possibly seems most unlikely.

The recent discovery of Eocene anthropoids (Beard *et al,* 1994, 1996) or even catarrhines from Asia (Chaimanee *et al.,* 1997) suggests the possibility of an Asian origin of platyrrhines. However, an Asian origin would presumably require a North American path of dispersal that seems belied by the lack of any fossil anthropoids in North America.

Finally, Szalay (1975) has suggested that perhaps anthropoids originated in the neotropics and then dispersed to the Old World. The more primitive nature and greater age of anthropoids in Africa and Asia than in South America argues (weakly) against this view. In any case, this theory generates the same problem of dispersion between South America and other continents (in reverse) and does not seem to be supported by evidence from the biogeography of other mammalian groups (in the way that the distribution of rodents suggests migration of primates from Africa to South America).

At present there is no convincing explanation for the origin of South American monkeys, but dispersal across the South Atlantic from Africa seems to be the least unlikely method (Hartwig, 1994). Future discoveries of early anthropoids should help clarify this issue.

BIBLIOGRAPHY

PLATYRRHINE FOSSIL RECORD

Ameghino, E. (1891). Nuevos restos de mamiferos fosiles descubiertos por Carlos Ameghino en al Eoceno inferior de la Patagonia Austral. *Rev. Argentina Hist. Nat.* **1**: 289–328.

______ (1893). New discoveries of fossil mammalia of southern Patagonia. *Am. Nat.* (May) pp. 439–449.

______ (1906). Les formations sedimentaires du Cretace superieur et du Tertiaire de Patagonie, avec un parallele entre leurs faunes mammalogiques et celles de l'ancien continent. *An. Mus. Nac. Hist. Buenas Aires, ser. 3,* **8**:1–568.

Ciochon, R. L., and Chiarelli, A. B. (1980). *Evolutionary Biology of the New World Monkeys and Continental Drift.* New York: Plenum Press.

Delson, E., and Rosenberger, A. L. (1980). Phyletic perspectives on platyrrhine origins and anthropoid relations. In *Evolutionary Biology of the New World Monkeys and Continental Drift,* ed. R. L. Ciochon and A. B. Chiarelli, pp. 445–458. New York: Plenum Press.

Fleagle, J. G., and Kay, R. F. (1997). Platyrrhines, catarrhines, and the fossil record. In *New World Primates: Ecology, Evolution, and Behavior,* ed. W. G. Kinzey, pp. 3–23. Hawthorne, N. Y.: Aldine DeGruyter.

Fleagle, J. G., and Rosenberger, A. L., eds. (1990). *The Platyrrhine Fossil Record.* New York: Academic Press.

Fleagle, J. G., Kay, R. F., and Anthony, M. R. L. (1997). Fossil New World monkeys. In *Vertebrate Paleontology in the Neotropics: The Miocene Fauna of La Venta, Colombia,* ed. R. F. Kay, R. H. Madden, R. L. Cifelli, and J. L. Flynn, pp. 473–495. Washington, D. C.: Smithsonian Institution Press.

Garber, P. H., and Kinzey, W. G., eds. (1992). Feeding adaptations in New World primates: An evolutionary perspective. *Am. J. Phys. Anthropol.* **88**:411–562.

Gregory, W. K. (1922). *The Origin and Evolution of the Human Dentition.* Baltimore: Williams and Wilkins.

Hartwig, W. C. (1994). Patterns, puzzles and perspectives on platyrrhine origins. In *Integrative Paths to the Past: Paleoanthropological Advances in Honor of F. Clark Howell,* ed. R. S. Corruccini and R. L. Ciochon, pp. 69–93. New York: Prentice-Hall.

Hershkovitz, P. (1977). *Living New World Monkeys* (Platyrrhini). Chicago: University of Chicago Press.

Hoffstetter, R. (1982). Les primates simiiformes (Anthropoidea) (Comprehension, Phylogenie, Histoire Biogeographique). *Ann. Paleontol. (Vert.–Invert.)* **68**(3):241–290.

Horovitz, I., and Meyer, A. (1997). Evolutionary trends in the ecology of New World monkeys inferred from a combined phylogenetic analysis of nulcear, mitochondiral and morphological data. In *Molecular Evolution and Adaptive Radiations,* ed. T. J. Givnish and K. J. Sytsma, pp. 189–224. New York: Cambridge University Press.

MacFadden, B. J. (1990). Chronology of Cenozoic primate localities in South America. *J. Human Evol.* **19**:23–60.

MacPhee, R. D. E., and Iturralde-Vinent, M. A. (1995). Earliest monkey from Greater Antilles. *J. Human. Evol.* **28**:197–200.

Meldrum, D. J. (1993). Postcranial adaptations and positional behavior in fossil platyrrhines. In *Postcranial Adaptation in Nonhuman Primates,* ed. D. L. Gebi, pp. 235–251. DeKalb: Northern Illinois University Press.

Pascual, R., and Jaureguizar, E. O. (1990). Evolving climates and mammal faunas in Cenozoic South America. *J. Human Evol.* **19**:23–60.

Patterson, B., and Pascual, R. (1972). The fossil mammal fauna of South America. In *Evolution, Mammals and Southern Continents,* ed. A. Keast, F. C. Erk, and B. Glass, pp. 247–309. Albany, N. Y.: SUNY Press.

Rosenberger, A. L. (1984). Platyrrhines contradict the molecular clock. *J. Human Evol.* **13**:737–742.

______ (1992). The evolution of feeding niches in New World monkeys. *Am. J. Phys. Anthropol.* **88**:525–562.

Simpson, G. G. (1980). *Splendid Isolation.* New Haven, Conn.: Yale University Press.

Stehli, F., and Webb, S. D., eds. (1985). *The Great American Biotic Interchange.* New York: Plenum Press.

Tarling, D. H. (1980). The geologic evolution of South America during the last 200 million years. In *Evolutionary Biology of the New World Monkeys and Continental Drift,* ed. R. L. Ciochon and A. B. Chiarelli, pp. 1–41. New York: Plenum Press.

Branisella and *Szalatavus*

Hoffstetter, R. (1969). Un primate de l'Oligocene inferieur sud-Americain: *Branisella boliviana* gen. et sp. nov. *C. R. Acad. Sci. (Paris) ser. D* **269**:434–437.

Rosenberger, A. L., Hartwig, W. C., and Wolff, R. G. (1991). *Szalatavus attricuspis* and early platyrrhine primate. *Folia Primatol.* **56**:225–233.

Takai, M., and Anaya, F. (1996). New specimens of the oldest fossil platyrrhine, *Branisella boliviana,* from Salla, Bolivia. *Am. J. Phys. Anthropol.* **99**:301–317.

Wolff, R. G. (1984). New specimens of the primate *Branisella boliviana* from the early Oligocene of Salla, Bolivia. *J. Vert. Paleontol.* **4**(4):570–574.

Dolichocebus

Bordas, A. F. (1942). Anotaciones sobre un "Cebidae" fosil de Patagonia. *Physis.* **19**:265–269.

Chopra, S. R. K. (1957). The cranial sutures in monkeys. *Proc. Zool. Soc. London* **128**:67–112.

Fleagle, J. G., and Bown, T. M. (1983). New primate fossils from late Oligocene (Colhuehuapian) localities of Chubut Province, Argentina. *Folia Primatol.* **41**:240–266.

Fleagle, J. G., and Kay, R. F. (1989). The dental morphology of *Dolichocebus gaimanensis,* a fossil monkey from Argentina. *Am. J. Phys. Anthropol.* **78**:221

Hershkovitz, P. (1970). Notes of Tertiary platyrrhine monkeys and description of a new genus from the late Miocene of Colombia. *Folia Primatol.* **12**.:1–37.

——— (1982). Supposed squirrel monkey affinities of the late Oligocene *Dolichocebus gaimanensis. Nature (London)* **298**:201–202.

Kraglievich, J. L. (1951). Contribuciones al concimienio de los primates fosilles de la Patagonia. I. Diagnosis previa de un nuevo primate fosil de Oligoceno superior (Colhuehuapiano) de Gaiman, Chubut. *Comm. Inst. Nac. Cient. Nat.* **2**:57–82.

Rosenberger, A. L. (1979). Cranial anatomy and implications of *Dolichocebus,* a late Oligocene ceboid primate. *Nature (London)* **279**:416–418.

——— (1982). Supposed squirrel monkey affinities of the late Oligocene *Dolichocebus gaimanensis. Nature (London)* **298**:202.

Tremacebus

Fleagle, J. G., and Bown, T. M. (1983). New primate fossils from late Oligocene (Colhuehuapian) localities of Chubut Province, Argentina. *Folia Primatol.* **41**:240–266.

Hershkovitz, P. (1974). A new genus of late Oligocene monkey (Cebidae, Platyrrhine) with notes on postorbital closure and platyrrhine evolution. *Folia Primatol.* **21**:1–35.

Rosenberger, A. L. (1984). Fossil New World monkeys dispute the molecular clock. *J. Human Evol.* **13**:737–742.

Rusconi, C. (1933). Nuevos restos de monos del terciario antiquo de la Patagonia. *Anal. Soc. Cient. Argentina* **116**: 286–289.

Chilicebus

Flynn, J. J., Wyss, A. R., and Swisher, C. C. (1995). An early Miocene anthropoid skull from the Chilean Andes. *Nature* **373**:603–607.

Soriacebus and *Carlocebus*

Bown, T. M., and Larriestra, C. N. (1990). Sedimentary paleoenvironments of fossil platyrrhine localities, Miocene Pinturas Formation, Santa Cruz Province, Argentina. *J. Human Evol.* **19**:87–119.

Fleagle, J. G. (1990). New fossil platyrrhines from the Pinturas Formation, southern Argentina. *J. Human Evol.* **19**:61–86.

Fleagle, J. G., Powers, D. W., Conroy, G. C., and Watters, J. P. (1987). New fossil platyrrhines from Santa Cruz Province, Argentina. *Folia Primatol.* **48**:65–77.

Fleagle, J. G. Bown, T. M., Swisher, C., and Buckley, G. (1995). Age of the Pinturas and Santa Cruz Formations. *VI Cong.Argentin de Paleont Y Bioestrat. Actas* pp. 129–135.

Meldrum, D. J. (1990). New fossil platyrrhine tali from the early Miocene of Argentina. *Am. J. Phys. Anthropol.* **83**:403–418.

——— (1993). Postcranial adaptations and positional behavior in fossil platyrrhines. In *Postcranial Adaptation in Nonhuman Primates,* ed. D. L. Gebo, pp. 235–251. DeKalb: Northern Illinois University Press.

Homunculus

Ameghino, F. (1891). Nuevos restos de mamiferos fosiles descubiertos por Carlos Ameghino en al Eoceno inferior de la Patagonia Austral. *Rev. Argentina Hist. Nat.* **1**: 289–328.

Bluntschili, H. (1931). *Homunculus patagonicus* und die ihm zugereihten Fossil funde aus den Santa-Cruz-Schichten Patagoniens. *Morphol. Jahr.* **67**:811–892.

Ciochon, R. L., and Corrucini, R. (1975). Morphometric analysis of platyrrhine femora with taxonomic implications and notes on two fossil forms. *J. Human Evol.* **4**: 193–217.

Hershkovitz, P. (1970). Notes of Tertiary platyrrhine monkeys and description of a new genus from the late Miocene of Colombia. *Folia Primatol.* **12**:1–37.

——— (1981). Comparative anatomy of platyrrhine mandibular cheek teeth dpm_4, pm_4, m_1 with particular reference to those of *Homunculus* (Cebidae) and comments on platyrrhine origins. *Folia Primatol.* **35**: 179–217.

——— (1984). More on the *Homunculus* dpm_4 and m_1 and comparisons with *Alouatta* and *Stirtonia* (Primates, Platyrrhini, Cebidae). *Am. J. Primatol.* **7**:261–283.

Tauber, A. (1991). *Homunculus patagonicus* Ameghino, 1891 (Primates, Ceboidea), Mioceno temprano, de la costa Atlantica austral, Prov. de Santa Cruz., Republica Argentina. *Acad. Nac.Ciencias Misc.* **82**:1–32.

Propithecia

Kay, R. F., Johnson, D. D., and Meldrum, D. J. (1998). A new pitheciin primate from the middle Miocene of Argentina. *Amer. J. Primatol.* In press.

La Venta Primates

Fleagle, J. G., Kay, R. F., and Anthony, M. R. L. (1997). Fossil New World monkeys. In *Vertebrate Paleontology in the Neotropics: The Miocene Fauna of La Venta, Colombia,* ed. R. F. Kay, R. H. Madden, R. L. Cifelli, and J. L. Flynn, pp. 473–495. Washington, D. C.: Smithsonian Institution Press.

Kay, R. F., and Madden, R. H. (1997). Paleogeography and paleoecology. In *Vertebrate Paleontology in the Neotropics: The Miocene Fauna of La Venta, Colombia,* ed. R. F. Kay, R. H. Madden, R. L. Cifelli, and J. L. Flynn, pp.520–550. Washington, D. C.: Smithsonian Institution Press.

Kay, R. F., Madden, R. H., Cifelli, R. L., and Flynn, J. J., eds. (1997). *Vertebrate Paleontology in the Neotropics: The Miocene Fauna of La Venta, Colombia.* Washington, D. C.: Smithsonian Institution Press.

Cebupithecia

Meldrum, D. J. (1993). Postcranial adaptations and positional behavior in fossil Platyrrhines. In *Postcranial Adaptation in Nonhuman Primates,* ed. D. L. Gebo, pp. 235–251. DeKalb: Northern Illinois University Press.

Meldrum, D. J., and Kay, R. F. (1997). Postcranial skeleton of Laventan Platyrrhines. In *Vertebrate Paleontology in the Neotropics: The Miocene Fauna of La Venta, Colombia,* ed. R. F. Kay, R. H. Madden, R. L. Cifelli, and J. L. Flynn, pp.459–472. Washington, D. C: Smithsonian Institution Press.

Meldrum, D. J., and Lemelin, P. (1991). Axial skeleton of *Cebupithecia sarmientoi* (Pitheciinae, Platyrrhini) from the middle Miocene of LaVenta, Colombia. *Am. J. Phys. Anthropol.* **25**:69–89.

Stirton, R. A. (1951). Ceboid monkeys from the Miocene of Colombia. *Bull. Univ. Calif. Pub. Geol. Sci.* **28**(11): 315–356.

Stirton, R. A., and Savage, D. E. (1951). A new monkey from the La Venta Miocene or Colombia. *Compilacion de los Estisdios Geol. Oficiales en Columbia, Serv. Geol. Nac. Bogota* **7**:345–356.

Nuciruptor

Meldrum, D. J., and Kay, R. F. (1997). *Nuciruptor rubricae,* a new pitheciin seed predator from the Miocene of Columbia. *Amer. J. Phys. Anthropol.* **102**:407–427.

Mohanamico

Fleagle, J. G., Kay, R. F., and Anthony, M. R. L. (1997). Fossil New World monkeys. In *Vertebrate Paleontology in the Neotropics: The Miocene Fauna of La Venta, Colombia,* ed. R. F. Kay, R. H. Madden, R. L. Cifelli, and J. L. Flynn, pp. 473–495. Washington, D. C.: Smithsonian Institution Press.

Kay, R. F. (1990). The Phyletic relationships of extant and fossil Pitheciinae (Platyrrhini, Anthropoidea). *J. Human Evol.* **19**:175–208.

Luchterhand, K., Kay, R. F., and Madden, R. H. (1986). *Mohanamico hershkovitzi,* gen, et so, nov. un primate du Miocene moyen d'Amerique du Sud. *C. R. Acad. Sci. (Paris) ser. 3* **303**(19) 1753–1758.

Rosenberger, A. L. (1992). The evolution of feeding niches in New World monkeys. *Am. J. Phys. Anthropol.* **88**:525–562.

Aotus

Setoguchi, T., and Rosenberger, A. L. (1987). A fossil owl monkey from La Venta, Colombia. *Nature (London)* **326**:692–694.

Stirtonia

Hershkovitz, P. (1970). Notes on Tertiary platyrrhine monkeys and description of a new genus from the late Miocene of Colombia. *Folia Primatol.* **12**.:1–37.

______ (1984). More on the *Homunculus* dPM4 and M1 and comparisons with *Alouatta* and *Stirtonia* (Primates, Platyrrhini, Cebidae). *Am. J. Primatol.* **7**:261–283.

Kay, R. F., and Frailey, C. D. (1993). Large fossil platyrrhines from the Rio Acre local fauna, late Miocene, western Amazonia. *J. Human Evol.* **25**:319–327.

Kay, R. F., Madden, R., Plavcan, J. M., Cifelli, R. L., and Diaz, J. G. (1987). *Stirtonia victoriae,* a new species of Miocene Colombian primate. *J. Human Evol.* **16**:173–283.

Setoguchi, T. (1985). *Kondous laventicus,* a new ceboid primate from the Miocene of La Venta, Colombia, South America. *Folia Primatol.* **44**:96–101.

Setagouchi, T., Watanabe, T., and Mouri, T. (1981). The upper dentition of *Stirtonia* (Ceboidea, Primates) from the Miocene of Colombia, South America, and the origin of the postero-internal cusps of upper molars of howler monkeys (*Alouatta*). *Kyoto Univ. Reports of New World Monkeys,* pp. 51–60.

Stirton, R. A., and Savage, D. E. (1951a). A new monkey from the La Venta of Miocene of Colombia. *Compilacion de los Estudios Geol. Oficiales en Columbia, Serv. Geol. Nac. Bogota* **7**:345–356.

——— (1951b). A new monkey of Colombia. *Bull. Univ. Calif. Pub. Geol. Sci.* **28**(11):315–356.

Neosaimiri and *Laventiana*

Ford, S. M. (1980). Callitrichids as phyletic dwarfs and the place of the Callitrichidae in Platyrrhini. *Primates* **21**:31–34.

Hershkovitz, P. (1977). *Living New World Monkeys* (Platyrrhini). Chicago: University of Chicago Press.

Meldrum, D. J., and Kay, R. F. (1997). Postcranial skeleton of laventan platyrrhines. In *Vertebrate Paleontology in the Neotropics: The Miocene Fauna of La Venta, Colombia,* ed. R. F. Kay, R. H. Madden, R. L. Cifelli, and J. L. Flynn, pp.459–472. Washington, D. C.: Smithsonian Institution Press.

Rosenberger, A. L., Hartwig, W. C., Takai, M., Setoguchi, T., and Shigehara, N. (1991a). Dental variability in *Saimiri* and the taxonomic status of *Neosaimiri fieldsi,* an early squirrel monkey from Colombia, South America. *Int. J. Primatol.* **12**:291–301.

Rosenberger, A. L., Setoguchi, T., and Hartwig, W. C. (1991b). *Laventiana annectens:* New fossil evidence for the origin of callitrichine New World monkeys. *Proc. Natl. Acad. Sci. USA* **88**:2137–2140

Stirton, R. A. (1951). Ceboid monkeys from the Miocene of Colombia. *Burl. Univ. Calif. Pub. Geol. Sci.* **28**(11): 315–356.

Takai, M. (1994). New specimens of *Neosaimiri fieldsi* from La Venta, Colombia: A middle Miocene ancestor of the living squirrel monkeys. *J. Human Evol.* **27**:329–360.

Micodon, Patasola, and *Lagonimico*

Fleagle, J. G., Kay, R. F., and Anthony, M. R. L. (1997). Fossil New World monkeys. In *Vertebrate Paleontology in the Neotropics: The Miocene Fauna of La Venta, Colombia,* ed. R. F. Kay, R. H. Madden, R. L. Cifelli, and J. L. Flynn, pp. 473–495. Washington, D. C.: Smithsonian Institution Press.

Ford, S. M. (1980). Callitrichids as phyletic dwarfs and the place of the Callitrichidae in Platyrrhini. *Primates* **21**:31–34.

Hershkovitz, P. (1977). *Living New World Monkeys* (Platyrrhini). Chicago: University of Chicago Press.

Kay, R. F. (1994). Giant tamarin from the Miocene of Colombia. *Am. J. Phys. Anthropol.* **95**:333–353.

Kay, R. F., and Meldrum, D. J. (1997). A new small platyrrhine and the phyletic position of Callitrichidae. In *Vertebrate Paleontology in the Neotropics: The Miocene Fauna of La Venta, Colombia,* ed. R. F. Kay, R. H. Madden, R. L. Cifelli, and J. L. Flynn, pp. 435–458. Washington, D. C.: Smithsonian Institution Press.

Rosenberger, A. L. (1981). Systematics: The higher taxa. In *Ecology and Behavior of Neotropical Primates,* vol. 1, ed. A. F. Coimbra-Filho, and R. A. Mittermeier, pp. 9–28. Rio de Janeiro: Academia Brasileira de Cincias.

——— (1984). Aspects of the systematics and evolution of the marmosets. In *A Primatologica No Brasil,* ed. M. T. de Mello, pp. 159–189. Angis do I Congresso Brasiliero de Primatologia. Sociedad de Primatologica.

Schneider, H., Sampaio, J., Harada, M. L., Barroso, C. M. L., Schneider, M. D. C., Czelusniak, J., and Goodman, M. (1996). Molecular phylogeny of the New World monkeys (Platyrrhine, Primates) based on two unlinked nuclear genes: 1RBP Intron 1 and ϵ-Globin sequences. *Amer. J. Phys. Anthropol.* **100**:153–179.

Setoguchi, T., and Rosenberger, A. L. (1985). Miocene marmosets: First fossil evidence. *Int. J. Primatol.* **6**:615–625.

Protopithecus and *Caipora*

Cartelle, C. (1993). Achado de *Brachyteles* do Pleistoceno Final. *Neotropical Primates* **1**:8

Cartelle, C, and Hartwig, W. C. (1996). A new extinct primate among Pleistocene megafauna of Bahia, Brazil. *Proc. Natl. Acad. Sci. USA* **93**:6405–6409.

Hartwig, W. C. (1995). A giant New World monkey from the Pleistocene of Brazil. *J. Human Evol.* **28**:189–196.

Hartwig, W. C. (1995) *Protopithecus:* Rediscovering the first fossil primate. *Hist. Philos. Life Sciences* **17**:447–460.

Hartwig, W. C., and Cartelle, C. (1996). A complete skeleton of the giant South American primate *Protopithecus. Nature* **381**:307–311.

Lund, P. W. (1838). Blik paa Brasiliens dyreverden for sidste jordomvaeltning. *Det Kong. Danske Viden. Selsk. Natur. Matem. Afhand.* **8**:61–144.

Antillean Fossil Platyrrhines

Cole, T. M. III, Richtsmeier, J. T., Horovitz, I., and MacPhee, R. D. E. (1997). Morphometric affinities of *Paralouatta varonai. Am. J. Phys. Anthropol. Supp.* **24**:91–92.

Ford, S. (1990). Platyrrhine evolution in the West Indies. *J. Human Evol.* **19**:237–254.

Ford, S., and Morgan, G. S. (1986). A new ceboid femur from the late Pleisiocene of Jamaica. *J. Vert. Paeleontol.* **6**:281–289.

Hedges, S. B. (1996). Historical biogeography of West Indian vertebrates. *Ann. Rev. Syst. Ecol.* **27**:163–196.

Horovitz, I., MacPhee, R. D. E., Fleming, C., and McFarlane, D. A. (1997). Cranial remains of *Xenothrix* and their bearing on the question of Antillean monkeys origins. *Journal Vert. Paleo.* **17**:54A.

MacPhee, R. D. E. (1995). Earliest monkey from Greater Antilles. *J. Human Evol.* **28**:197–200.

MacPhee, R. D. E., and Fleagle, J. G. (1991). Postcranial remains of *Xenothrix mcgregori* (Primates, Xenotrichidae) and other late Quaternary mammals from Long Mile Cave, Jamaica. Bull. *Am. Mus. Nat. Hist.* **206**:287–321.

MacPhee, R. D. E., and Iturralde-Vincent, M. A. (1995). Earliest monkey from Greater Antilles. *J. Human. Evol.* **28**:197–200.

MacPhee, R. D. E., and Woods, C. A. (1982). A new fossil cebine from Hispaniola. *Am. J. Phys. Anthropol.* **58**:419–436.

MacPhee, R. D. E., Horovitz, I., Arredondo, O., and Vasquez, O. J. (1995). A new genus for the extinct Hispaniolan monkey *Saimiri bernensis* Rimoli, 1977, with notes on its systematic position. *Am. Mus. Novitates* **3134**:1–21.

Rimoli, K. (1977). Una nueva especie de Monos (Cebidae: Saimirinae: *Saimiri*) de la Hispaniola. *Cuadernos de Cendia, Univ. Autonoma de Santo Domingo* **242**:1–14.

Rivero (de la Calle), M., and Arredondo, O. (1991). *Paralouatta varonai,* a new Quaternary platyrrhine from Cuba. *J. Human Evol.* **21**:1–11.

Rosenberger, A. L. (1977). *Xenothrix* and ceboid phylogeny. *J. Human Evol.* **6**:461–481.

Williams, E. E., and Koopman, K. E. (1952). West Indian fossil monkeys. *Am. Mus. Nov.* **1546**:1–16.

PLATYRRHINE ORIGINS

Aiello, L. .C. (1993). The Origin of the New World monkeys. In *The Africa–South America Connection,* ed. W. George and R. Lavocat, pp. 100–118. Oxford: Clarendon Press.

Beard, K. C., Qi, T., Dawson, M. R., Wang, B., and Li, C. (1994). A diverse new primate fauna from the middle Eocene fissure-fillings in southeastern China. *Nature* **368**:604–609.

Beard, K. C., Tong, Y., Dawson, M. R., Wang, J., and Huang, X. (1996). Earliest complete dentition of an anthropoid primate from the late middle Eocene of Shanxi Province, China. *Science* **272**:82–85.

Chaimanee, Y., Suteethorn, V., Jaeger, J.-J., and Ducrocq, S. (1997). A new late Eocene anthropoid from Thailand. *Nature* **385**:429–431.

Delson, E., and Rosenberger, A. L. (1980). Phyletic perspectives on Platyrrhine origins and anthropoid relations. In *Evolutionary Biology of the New World Monkeys and Continental Drift,* ed. R. L. Ciochon and A. B. Chiarelli, pp. 445–458. New York: Plenum Press.

George, W., and Lavocat, R., eds. (1993). *The Africa–South America Connection.* Oxford: Clarendon Press.

Hartwig, W. C. (1994) Patterns, puzzles and perspectives on platyrrhine origins. In *Integrative Paths to the Past: Paleoanthropological Advances in Honor of F. Clark Howell,* ed. R. S. Corruccini and R. L. Ciochon, pp. 69–93. New York: Prentice-Hall.

Hoffstetter, R. (1972). Relationships, origins and history of the ceboid monkeys and caviomorph rodents: A modern reinterpretation. In *Evolutionary Biology,* ed. T. Dobzhansky, M. K. Hecht, and W. C. Steere, pp. 323–347. New York: Appleton-Century-Crofts.

______ (1980). Origin and deployment of New World monkeys, emphasizing the southern continents route. In *Evolutionary Biology of the New World Monkeys and Continental Drift,* ed. R. L. Ciochon and A. B. Chiarelli, pp. 103–138. New York: Plenum Press.

Hoffstetter, R., and Lavocat, R. (1970). Decouverie dans le Deseadien de Bolivie de genres pentalophodenies appuyant les affiniies africaines des Rongeurs Caviornorphes. *C. R. Acad. Sci. (Paris), ser. D* **271**:172–175.

Lavocat, R. (1980). The implication of rodent palaeontology and biogeography to the geographical sources and origins of platyrrhine primates. In *Evolutionary Biology of the New World Monkeys and Continental Drift,* ed. R. J. Ciochon and A. B. Chiarelli, pp. 93–102. New York: Plenum Press.

Miller, E. R., and Simons, E. L. (1997). Dentition of *Proteopithecus sylviae,* an archaic anthropoid from the Fayum, Egypt. *Proc. Natl. Acad. Sci. USA* **94**:13760–13764.

Rand, H. M., and Mabesoone, J. M. (1982). Northeast Africa. *Palaeogeogr., Palaeoclimatol., Palaeoecol.* **38**:163–183.

Rosenberger, A. L., and Szalay, F. S. (1981). On the tarsiform origins of Anthropoidea. In *Evolutionary Biology of the New World Monkeys and Continental Drift,* ed. R. J. Ciochon and A. B. Chiarelli, pp. 139–157. New York: Plenum Press.

Simons, E. L. (1976). The fossil record of primate phylogeny. In *Molecular Anthropology,* ed. M. Goodman, R. E. Tashian, and J. H. Tashian, pp. 35–62. New York: Plenum Press.

Simons, E. L. (1997). Preliminary description of the cranium of *Proteopithecus sylviae,* an Egyptian late Eocene anthropoidean primate. *Proc. Natl. Acad. Sci. USA* **94**: 14970–14975.

Stehli, F. G., and Webb, S. D., eds. (1985). *The Great American Biotic Exchange. Topics in Geobiology,* vol. 4. New York: Plenum Press.

Szalay, F. S. (1975). Phylogeny, adaptations and dispersal of the tarsiiform primates. In *Phylogeny of the Primates: A Multidisciplimary Approach,* ed. W. C. Luckett and F. S. Szalay, pp. 357–404. New York: Plenum Press.

Tarling, D. H. (1980). The geologic evolution of South America during the last 200 million years. In *Evolutionory Biology of the New World Monkeys and Continental Drift,* ed. R. L. Ciochon and A. B. Chiarelli, pp. 1–41. New York: Plenum Press.

——— (1982). Land bridges and plate tectonics. In *Phylogenie et paleobiogeographie,* ed. E. Buffetaut, P. Janvier, J. C. Rage, and P. Tassy. *Geobios, Mem. Spec.* **6**: 361–374.

Wood, A. E., and Patterson, B. (1970). Relationships among hystricognathous and hystricomorphous rodents. *Mammalia* **34**:628–639.

Wyss, A. R., Flynn, J. J., Norell, M. A., Swisher, C. C. III, Charrier, R., Novacek, M. J., and McKenna, M. C. (1993). South America's earliest rodent and the recognition of a new interval of mammalian evolution. *Nature* **365**:434–437.

FIFTEEN

Fossil Apes

MIOCENE EPOCH

The Miocene is a relatively long epoch that began approximately 23 million years ago and ended about 5 million years ago. In the early Miocene, world temperatures warmed appreciably from the cooler Oligocene, and there were minor fluctuations of warming and cooling periods throughout much of the epoch (see Fig. 10.4) as well as increasing aridity in Africa. Several major geophysical events took place during this epoch that affected both global climate and the biogeography of mammals throughout the Old World. The Tethys Sea contracted and was cut off from the Indian Ocean by the emergence of the Arabian peninsula. On at least one occasion in the late Miocene, the Mediterranean remnant of the Tethys dried up completely. Farther east, India continued to crash into Asia, leading to the rise of the Himalayas.

In East Africa, the Miocene was characterized by considerable volcanic activity in conjunction with the developing rift system. It is here, in the early Miocene sediments of Kenya and Uganda, that we find the earliest fossil Old World monkeys and an impressive array of primitive apes. The monkeys of the early Miocene are not very diverse, and fossil evidence for their major radiation appears only in the later part of this epoch and in the succeeding Pliocene (see Chapter 16). In contrast, the Miocene deposits of Africa and Eurasia hold an extraordinary abundance and diversity of fossil apes (Fig. 15.1).

Latest Oligocene to Middle Miocene Apes from Africa

In the latest Oligocene and early to middle Miocene sediments of Kenya and Uganda (25 million to 15 million years ago) we find evidence of an extensive radiation of primitive apes, the proconsulids (Fig. 15.2, Table 15.1). Although cranial and postcranial remains are available for only a few of the genera and species, these indicate that proconsulids were more advanced than *Aegyptopithecus* and *Propliopithecus* from the early Oligocene. They seem to have all of the anatomical features that characterize living catarrhines, not just a few as found in the early Oligocene taxa, and some show features that link them exclusively with living apes. These Miocene apes ranged in size from the small, capuchin-size (3.5 kg) *Micropithecus clarki* to the female gorilla-size (50 kg) *Afropithecus* and *Proconsul major.* Fossil apes have been found in association with a variety of paleoenvironments, ranging from tropical rain forests to open woodlands (Pickford,

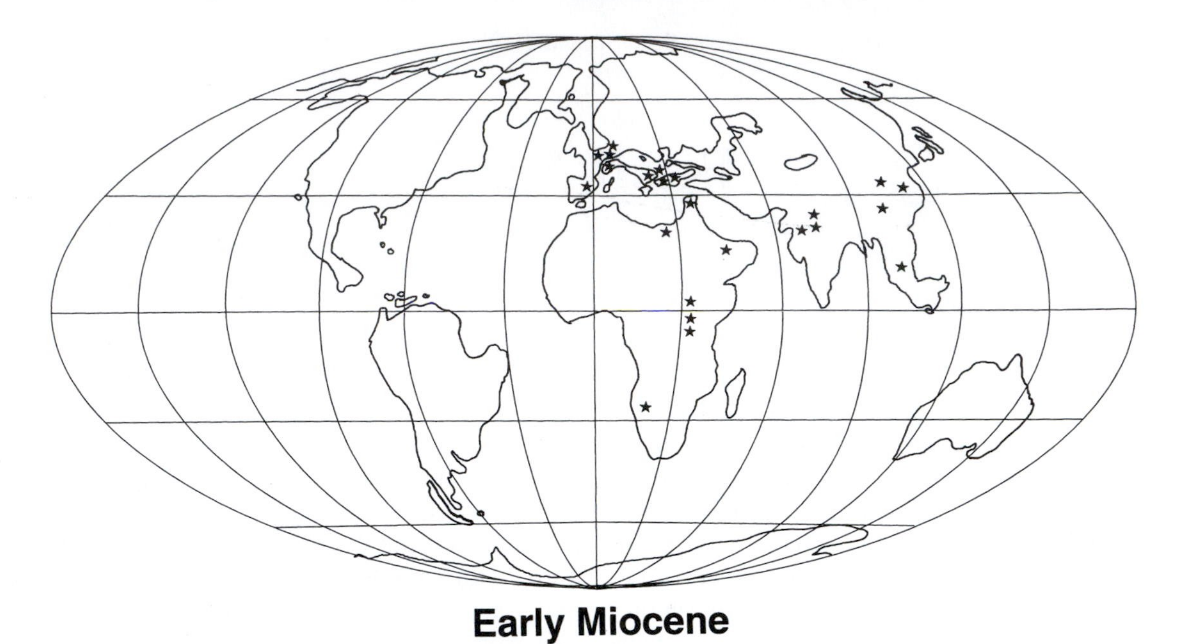

FIGURE 15.1 Map of the early Miocene world showing Miocene fossil ape localities.

1983; Andrews, 1992), and they seem to have spanned a range of ecological niches comparable to those occupied today by both Old World monkeys and apes.

The proconsulids from East Africa are clearly derived from Oligocene propliopitheids but they also have several features that distinguish them from the earlier propliopithecids (Figs. 15.3, 15.4). Like the earlier propliopithecids, they have a dental formula of 2.1.2.3, with broad upper central incisors and smaller upper laterals. The lower incisors of most species are taller and narrower than those of living apes. All species have relatively large, sexually dimorphic canines that shear against the lower anterior premolar. The upper premolars are relatively broad and bicuspid; the posterior lower premolar is a broad semimolariform tooth.

The upper molars distinguish proconsulids from the earlier propliopithecids and are characterized by their quadrate shape with a relatively larger hypocone, a pronounced, often beaded, lingual cingulum, and some details of the conules (Kay, 1977). The lower molars have a broad talonid basin surrounded by five prismlike cusps, including a large hypoconulid. The major dental differences between the many genera and species are in overall size, in the relative proportions of the anterior dentition, and in the development of shearing crests on the molars, features that are often related to dietary differences.

There are parts of the facial skeleton and other cranial bones known for many of the Miocene apes of East Africa (Figs. 15.5, 15.6; Rae, 1997). The shape of the face varies considerably, being short in some (*Micropithecus*), moderately long and broad in others (*Turkanapithecus*), and long and narrow in still others (*Afropithecus*). In most species, the nasal opening has been described as tall and relatively narrow, as in cercopithecoid monkeys, rather than broad and rounded, as in living apes. Or-

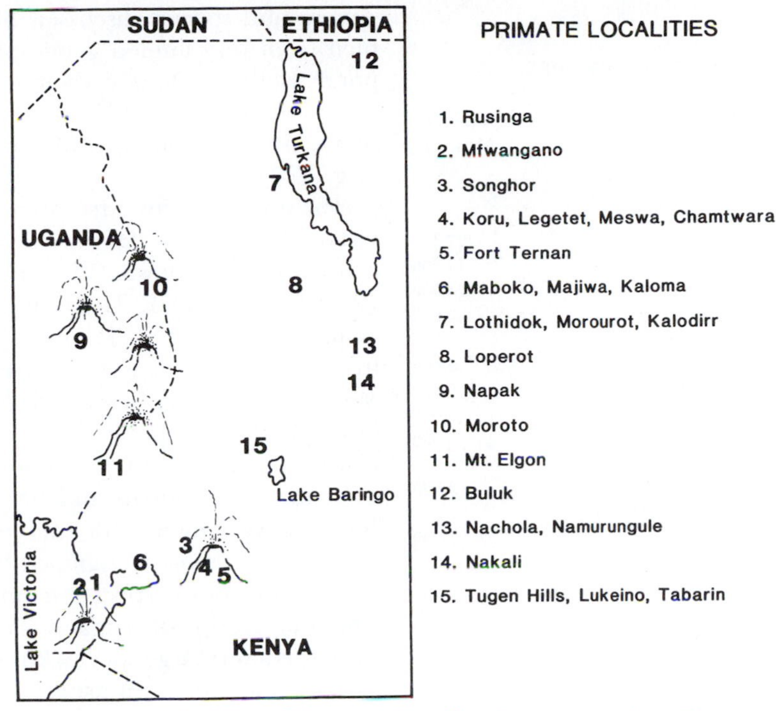

FIGURE 15.2 Map of Eastern Africa showing fossil ape localities.

bit size suggests diurnal habits for all species, but the orbits are relatively larger in the smaller species. The only known auditory region, that for *Proconsul heseloni* (formerly a part of *P. africanus*), is identical to that in living catarrhines (but unlike that in *Aegyptopithecus*) in having a tubular tympanic extending laterally from the side of the bulla. Relative brain size in *P. heseloni* and also in *Turkanapithecus* seems to have been similar to that of living Old World monkeys or perhaps a little larger, but in *Afropithecus* the brain seems to have been relatively small. The external surface of the cerebrum retains a number of primitive features lacking in extant apes.

Hundreds of isolated skeletal elements are known for these primitive apes, and relatively complete skeletons are available for several individuals of *P. heseloni* (Fig. 15.7). In limb proportions and many skeletal details these apes resemble living platyrrhines, and they lack many specialized skeletal features of the elbow or wrist, that characterize either Old World monkeys or living hominoids. They have a more primitive, in some ways more behaviorally versatile, locomotor skeleton. The interspecific skeletal differences indicate considerable locomotor diversity (Rose, 1993).

The systematics of the early and middle Miocene apes from East Africa has been in a state of flux for many years, and remains so today (see Andrews, 1992; Conroy, 1994; Begun *et al.*, 1997). Part of this is the result of attempts to break the radiation into unnatural groups on the basis of size, but mostly it reflects the continued discovery and recognition of new species, and the lack of comparable material for different taxa. The many

TABLE 15.1
Infraorder Catarrhini
EARLY AND MIDDLE MIOCENE APES

Species	Estimated Mass (g)
Family PROCONSULIDAE	
Proconsul (early Miocene, Africa)	
P. africanus	27,400
P. heseloni	17,000
P. nyanzae	28,000
P. major	50,000
Rangwapithecus (early Miocene, Africa)	
R. gordoni	15,000
Limnopithecus (early Miocene, Africa)	
L. legetet	5,000
L. evansi	6,000
Dendropithecus (early Miocene, Africa)	
D. macinnesi	9,000
Simiolus (early to middle Miocene, Africa)	
S. enjiessi	7,000
S. leakeyorum	
Micropithecus (early Miocene, Africa)	
M. clarki	3,500
Kalepithecus (early Miocene, Africa)	
K. songhorensis	5,000
Kamoyapithecus (late Oligocene, Africa)	
K. hamiltoni	
Dionysopithecus (?early Miocene, Asia)	
D. shuangouensis	3,300
Platydontopithecus (?early Miocene, Asia)	
P. janghuaiensis	15,000
Family OREOPITHECIDAE	
Mabokopithecus (middle Miocene, Africa)	
M. clarki	10,000
Nyanzapithecus (early to middle Miocene, Africa)	
N. vancouveringi	9,000
N. pickfordi	10,000
Family *Incertae sedis*	
Afropithecus (early to ?middle Miocene, Africa, Saudi Arabia)	
A. turkanensis	50,000
A. leakeyi	—
Morotopithecus (early Miocene, Africa)	
M. bishopi	40,000
Turkanapithecus (early Miocene, Africa)	
T. kalakolensis	10,000
Kenyapithecus (middle to late Miocene, Africa)	
K. africanus	41,000
K. wickeri	27,000
Otavipithecus (late Miocene, Africa)	
O. namibiensis	17,500
Samburupithecus (late Miocene, Africa)	
S. kiptalami	60,000

genera and species discussed here are sampled from very limited geographic and temporal windows. Despite their numbers, they undoubtedly underestimate the true diversity of fossil apes from the Miocene of Africa (Fig. 15.8).

Proconsul was the first Miocene ape described from Africa (Hopwood, 1933) and remains the best known (Walker and Teaford, 1989; Walker 1997). There are four species generally recognized. ***P. africanus*** (>20 kg), from the locality of Koru in Western Kenya, was the first species described. ***Proconsul heseloni,*** a similar-size species from Rusinga Island (formerly included in *P. africanus*), is known from many individuals and has become the comparative standard for the genus (Walker *et al.*, 1993). ***Proconsul nyanzae*** (20–30 kg) is a larger, well-known species from Rusinga Island and nearby Mfwangano. ***Proconsul major*** (50 kg) is a very large species known from only a few remains. In many aspects of their dentition, these species differ mainly in size (Ruff *et al.*, 1989). All have sexually dimorphic canines and a molar morphology indicating a predominantly frugivorous diet (Ungar and Kay, 1995; Kay and Ungar, 1996).

Many cranial parts are known for the smallest species, *P. heseloni* (Fig. 15.6). It has a pronounced snout with prominent canine jugae and a relatively robust zygomatic bone. The brain is similar in size to that of a large monkey. As already noted, the auditory region is identical to that of extant apes and cercopithecoid monkeys. The external surface of the brain has a primitive sulcal morphology similar to that seen in gibbons and cercopithecoids, but it lacks many features seen in the brain of living great apes (Falk, 1983).

There is a nearly complete juvenile skeleton known for *P. heseloni* (Fig. 15.7). The limb proportions are monkeylike, with an intermembral index of 89. Compared with living catarrhines, it has short limbs for its estimated

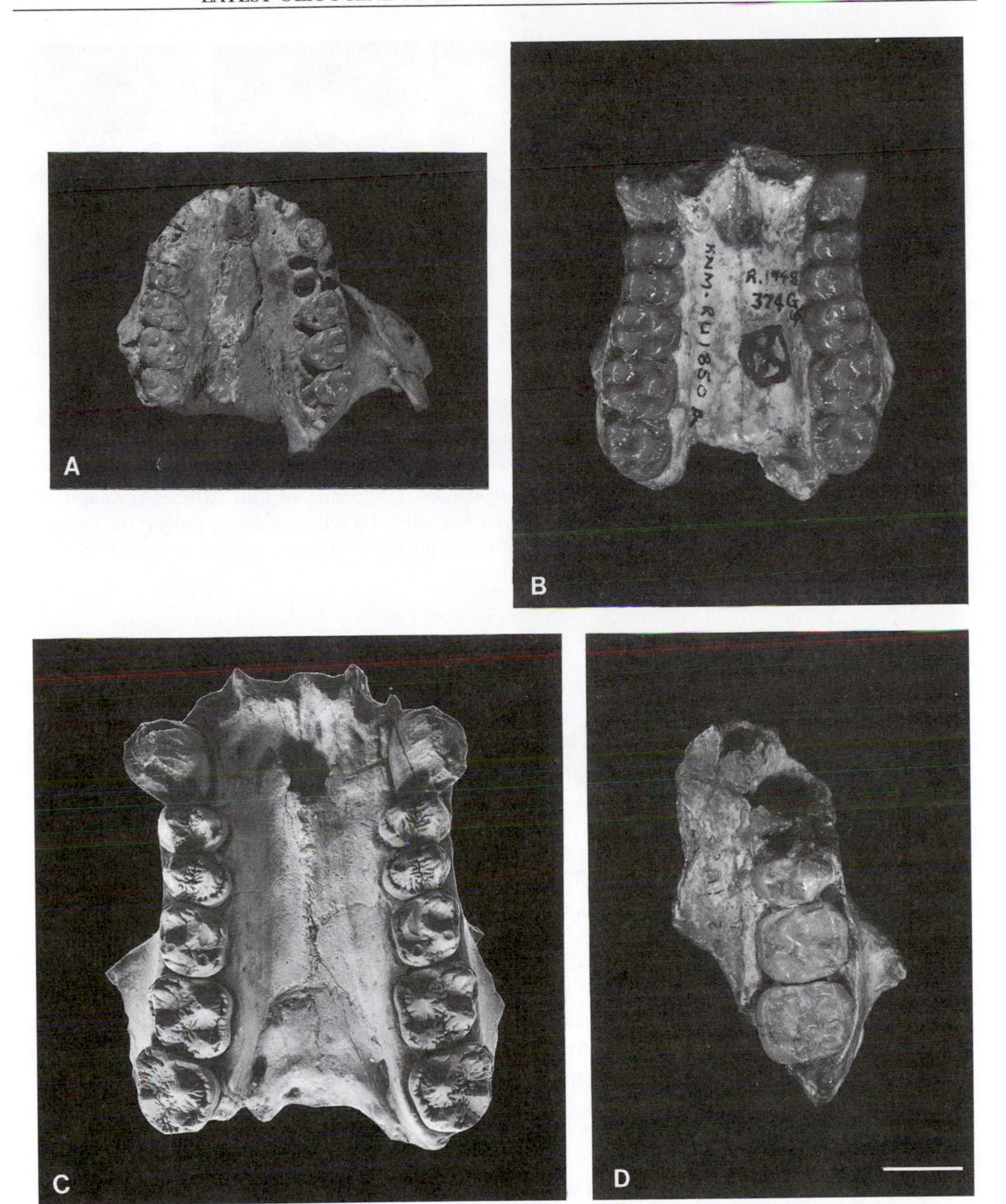

FIGURE 15.3 Upper dentitions of fossil apes from the early Miocene of East Africa: A, *Micropithecus clarki;* B, *Dendropithecus macinnesi;* C, *Rangwapithecus gordoni;* D, *Proconsul heseloni.* Notice the large lingual cingulum on the upper molars of all except *Micropithecus* and the very long *Rangwapithecus* molars. Scale = 1 cm (photographs courtesy of R. L. Ciochon and P. Andrews).

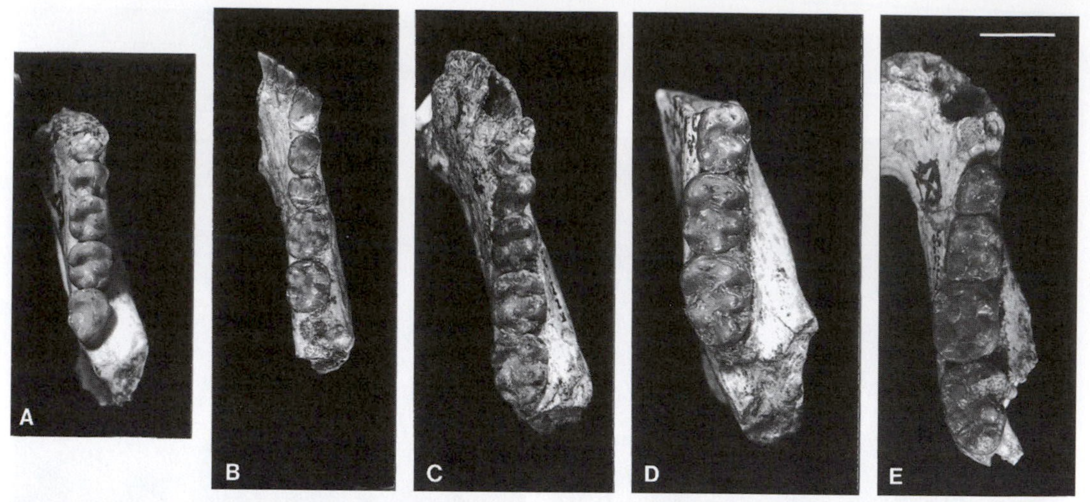

FIGURE 15.4 Lower dentitions of fossil apes from the early Miocene of East Africa: A, *Kalepithecus songhorensis;* B, *Limnopithecus legetet;* C, *Dendropithecus macinnesi;* D, *Rangwapithecus gordoni;* E, *Proconsul heseloni.* Notice the high cusps and well-developed crests on the *Rangwapithecus* molars. Scale = 1 cm (photographs courtesy of R. L. Ciochon and P. Andrews).

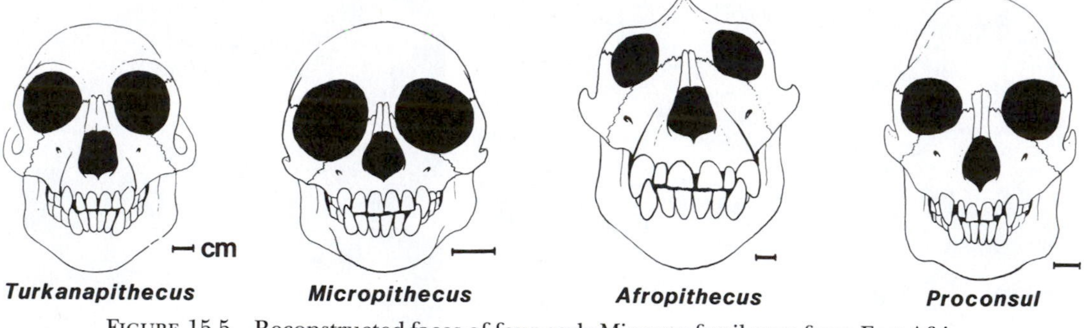

FIGURE 15.5 Reconstructed faces of four early Miocene fossil apes from East Africa.

body size and a mixture of apelike and more primitive monkeylike features throughout the skeleton (e.g. Beard *et al.,* 1986). It resembles living apes in such features as the shape of the distal part of the humerus (Rose, 1988), the robustness of the fibula, the conformation of the tarsal bones, and the absence of a tail (Ward *et al.,* 1991). It also lacks many characteristic features of living apes, such as a reduced ulnar styloid process, a short ulnar olecranon, and long curved digits. At the same time, *Proconsul* has none of the detailed skeletal features, such as a narrow elbow region, that characterize cercopithecoid monkeys. The skeleton indicates that *P. heseloni* was quadrupedal and probably arboreal but lacked the suspensory abilities of living apes (Rose, 1993).

Proconsul nyanzae resembles *P. heseloni* in many general aspects of its skeleton and lacks

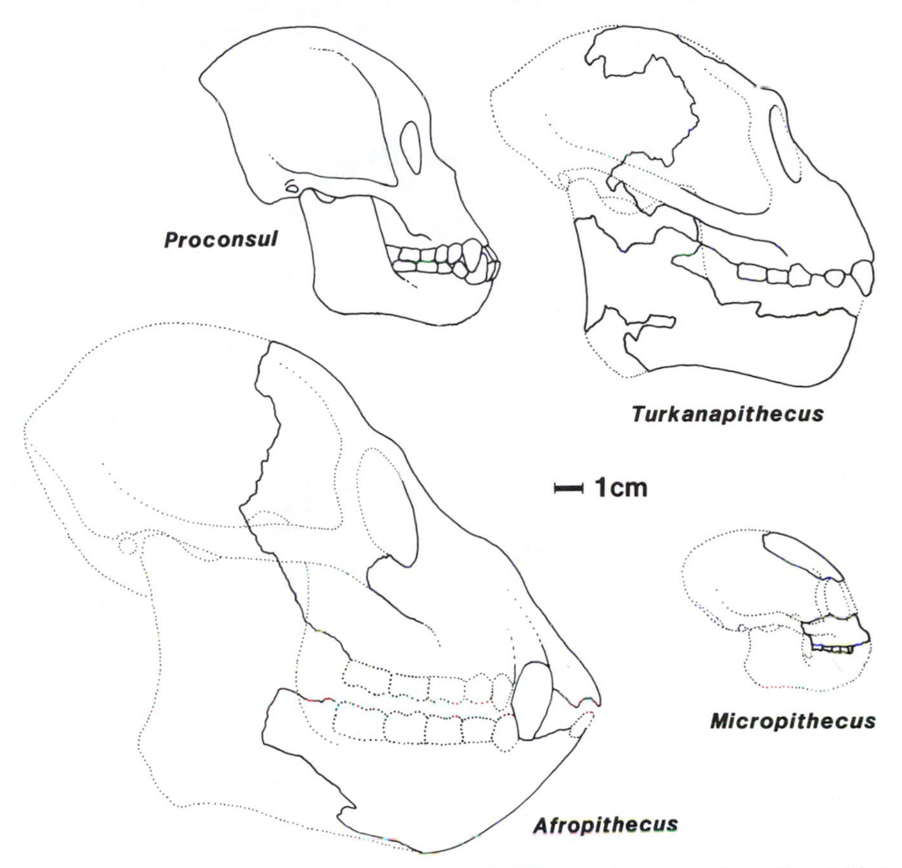

FIGURE 15.6 Reconstructed skulls of four early Miocene fossil apes from East Africa.

any clear similarities to living hominoids in its trunk and hindlimb (Ward *et al.,* 1993). However, the olecranon process in *P. nyanzae* extends posteriorly rather than proximally, suggesting more terrestrial locomotion than in *P. heseloni.*

The canine sexual dimorphism of all *Proconsul* species suggests that they did not live in monogamous social groups. However, in view of the diverse types of social groups found among extant apes, we cannot confidently speculate about many details of the social behavior of these Miocene species.

Afropithecus turkanensis is a very large fossil ape from the early to middle Miocene of Kenya (Leakey and Leakey, 1986a; Leakey and Walker, 1997). Compared to *Proconsul, Afropithecus* has a long, narrow snout, small orbits, and a broad interorbital area. The dentition is characterized by robust, procumbent incisors, short, round, tusklike canines, and extremely broad upper premolars. A maxilla from Saudi Arabia classified as "*Heliopithecus*" is probably the same genus (Andrews and Martin, 1987b). The few postcranial elements associated with *Afropithecus* from Kenya are similar to those of *Proconsul* and suggest arboreal and quadrupedal habits (Rose, 1993).

There are very diverse views about the affinities of *Afropithecus.* The long snout, straight

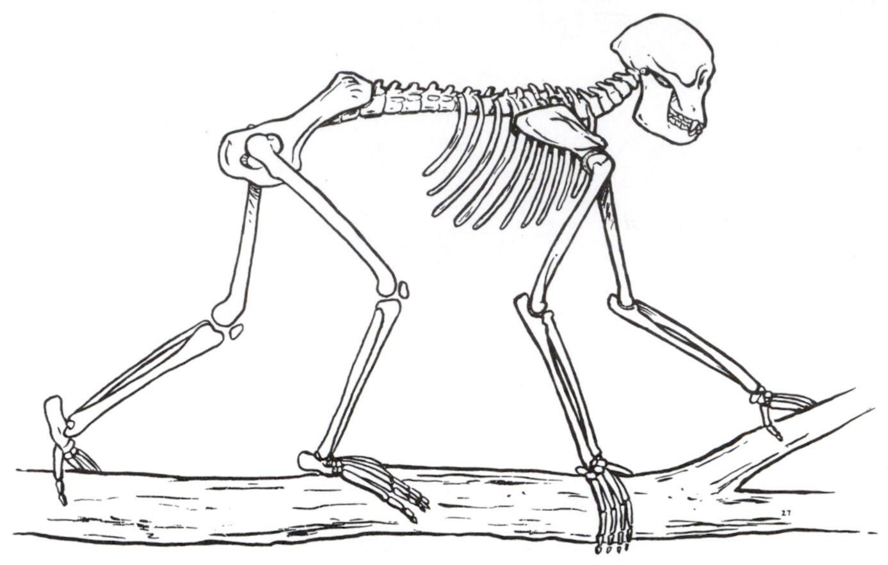

FIGURE 15.7 Juvenile skeleton of *Proconsul heseloni*.

facial profile, and small frontal bone have suggested to some (Simons, 1987) that *Afropithecus* is more closely related to primitive catarrhines such as *Aegyptopithecus*. However, the premolar morphology seems to indicate that this genus is more advanced than other proconsulids and closer to the origins of great apes and humans (Andrews and Martin, 1987; Andrews, 1992).

Morotopithecus bishopi (Gebo *et al*, 1997) is a newly named species from the early Miocene site of Moroto in Uganda. The type specimen is a large palate (Pilbeam, 1969) that was originally assigned to *Proconsul major* but is strikingly similar to that of *Afropithecus*. In contrast to the primitive nature of the limb elements of most early Miocene hominoids from Africa, including *Afropithecus*, several skeletal elements from Moroto are much more similar to those of extant hominoids. These include a lumbar vertebra with transverse processes that arise from the neural arch, as in living apes (Sanders and Bodenbender, 1994), and a rounded glenoid fossa of the scapula (Gebo *et al.*, 1997). Nevertheless, the striking similarities of the dental remains and the lack of any overlap in the known postcranial remains of the two taxa make the distinction between *Morotopithecus* and *Afropithecus* very much unresolved.

Kamoyapithecus is a poorly known genus from the latest Oligocene locality of Lothidok in Northern Kenya (Leakey *et al.*, 1995). The single species, *K. hamiltoni*, was formerly placed in the genus *Xenopithecus* (Madden, 1980). *Kamoyapithecus* differs from other East African fossil apes and resembles propliopithecids in having relatively broad upper molars with a restricted trigon, but it also shows similarities to *Afropithecus*. Its phyletic position is uncertain at present.

Turkanapithecus kalakolensis (Figs. 15.5, 15.6) is a medium-size fossil ape from Kalodirr, the same site in northern Kenya that yielded *Afropithecus* (Leakey and Leakey, 1986b). *Turkanapithecus* has relatively long upper molars with many extra cusps and relatively large anterior upper premolars. The mandible is relatively

FIGURE 15.8 Reconstruction of a fossil ape community from Rusinga Island, Kenya, approximately 18 million years ago: upper left, *Proconsul heseloni;* upper right, *Dendropithecus macinnesi;* center, *Limnopithecus legetet;* lower, *Proconsul nyanzae.*

shallow with a broad ascending ramus. The cranium shows a broad, square snout, a broad interorbital region, large, rimmed orbits, and flaring zygomatic arches. Like the other proconsulids, *Turkanapithecus* was primarily an arboreal quadruped, but some aspects of the limbs, including a reduced olecranon process, suggest that it was more suspensory than *Proconsul* (Rose, 1993).

One of the most distinctive of the early Miocene apes is ***Rangwapithecus gordoni.*** This medium-size (15 kg) species has relatively long and narrow molar teeth with numerous shearing crests that indicate a more folivorous diet than that of other early Miocene apes (Kay, 1977). It also has a very deep mandible (Hill and Odhiambo, 1987). *Rangwapithecus gordoni* seems to be found primarily in rain forest environments.

Nyanzapithecus is a small fossil ape known almost exclusively from dental remains (Harrison, 1987). There are two species, *N. vancouveringi,* from the early Miocene of Rusinga Island, and *N. pickfordi,* from the middle Miocene of Maboko Island. *Nyanzapithecus* is characterized by long upper premolars with similar-size buccal and lingual cusps, long upper molars, and lower molars with deep notches. It was a folivorous primate. Compared with the other fossil apes from the Miocene of East Africa, *Nyanzapithecus* shows greatest similarities to *Rangwapithecus* and is almost certainly derived from that genus. More interesting, however, are the distinctive dental features in molar and premolar anatomy that *Nyanzapithecus* shares with the European *Oreopithecus.* Although these dental similarities seem to indicate an African origin for *Oreopithecus* (discussed later in this chapter), such an interpretation is complicated by the fact that a humeral fragment from Maboko, assigned to *Nyanzapithecus,* shows none of the modern skeletal features found in *Oreopithecus* and extant hominoids (McCrossin, 1992; Benefit and McCrossin, 1997).

Mabokopithecus clarki is another oreopithecid from the middle Miocene of Maboko Island, Kenya. It is similar to *Nyanzapithecus,* but is only known from a few remains (see Benefit *et al.,* 1998).

The two species of ***Limnopithecus,*** *L. legetet* and *L. evansi,* were each about the size of a living gibbon (4–5 kg). They had a frugivorous diet. The few skeletal elements of these species indicate that they were arboreal quadrupeds.

Dendropithecus macinnesi is known from numerous jaws and teeth and much of a skeleton. It has tall, narrow incisors and broad molars with numerous crests, suggesting a frugivorous-folivorous diet. It was a medium-size (9 kg) animal with long, slender limbs similar to those of the neotropical spider monkey (*Ateles*). It was probably mainly quadrupedal but was the most suspensory of the earlier Miocene apes. Although there is striking canine dimorphism, both sexes have relatively long, sharp canines, suggesting that the species was possibly monogamous.

Kalepithecus songhorensis is another small hominoid that was formerly included in *Dendropithecus,* but it differs from that taxon in numerous dental and gnathic features, including broad lower premolars, a deep mandible, and a broad palate. It was probably frugivorous.

Simiolus is a small ape (7000 g) from the locality of Kalodirr in northern Kenya (*S. enjiessi*) and the middle Miocene locality of Maboko Island (*S. leakeyorum;* see Gitau and Benefit, 1995). *Simiolus* differs from other small early Miocene apes in its very narrow upper canines, triangular P_3, and long upper molars. It has a mosaic of characteristics found in various other genera but shows greatest similarities to *Dendropithecus* and *Rangwapithecus.* The limbs of *Simiolus* are very similar to those of *Dendropithecus,* indicating that it was one of the more suspensory proconsulids (Rose *et al.,* 1992).

Micropithecus clarki is the smallest known ape, with an estimated body weight of 3–4 kg. *Micropithecus clarki* is from the early Miocene of Napak and Koru. A second species, *M. leakeyorum,* from the middle Miocene of Maboko Island (Harrison, 1989), is now placed in *Simiolus* (Gitau, 1995). *Micropithecus* has distinctive dental proportions compared with the other early Miocene apes. The dentition is characterized by relatively large incisors and canines and relatively small cheek teeth. There is a reduced cingulum on the upper molars. The face of *M. clarki* (Figs. 15.5, 15.6) has a very short snout, a broad nasal opening, and large orbits, giving it a very gibbonlike appearance (Fleagle, 1975; Rae, 1993). A frontal bone attributed to this species has a smooth cranial surface and lacks any brow ridges. The endocast of the brain indicates a gibbonlike sulcal pattern.

Micropithecus is virtually identical to another small ape, ***Dionysopithecus,*** from the Miocene of China and other parts of Asia (Bernor *et al.,* 1988). Both have been frequently identified as possible gibbon ancestors because of their reduced upper molar cingulum and geographic distribution. Although there is nothing in the known dental and cranial anatomy of *Micropithecus* to preclude such a phyletic relationship, there are also few derived anatomical features of either *Micropithecus* or *Dionysopithecus* that would strongly support a unique link with living gibbons.

Most of the African fossil apes we have discussed so far are known primarily from the early Miocene. In contrast, ***Kenyapithecus*** is from the middle Miocene and seems to come from deposits representing drier, more open, woodland environments. There are two described species, the slightly older *K. africanus,* from Maboko Island and the Baringo basin, and the younger, less well-known *K. wickeri,* from Fort Ternan. *Kenyapithecus* has thicker molar enamel, a more robust mandible, and large upper premolars compared with the early Miocene proconsulids. However, the lower incisors are relatively large, suggesting possible seed-eating habits (McCrossin and Benefit, 1994).

Many aspects of the limb skeleton show that *Kenyapithecus,* like the proconsulids, lacks distinctive hominoid features of the elbow and humerus. However, *Kenyapithecus* shows greater development of terrestrial adaptations than the early Miocene species, such as a prominent greater tuberosity and a posteriorly directed medial epicondyle on the humerus (McCrossin and Benefit, 1994).

The phylogenetic position of *Kenyapithecus* has been debated for many years. Originally it was thought to be an early hominid, similar to the Asian *Ramapithecus.* More recently, several authorities have argued that it was a part of the radiation of modern hominoids because of the thick enamel and large premolars (e.g., Andrews, 1992). New skeletal elements from several sites indicate that it was probably the most terrestrial of the Miocene hominoids of Africa (McCrossin and Benefit, 1997). Most phylogenetic analyses of its dentition and limbs have failed to show that it was more closely related to living apes and humans than other proconsulids (e.g., Begun *et al.,* 1997), but newer finds indicate ape-like hand and wrist features (McCrossin *et al.,* 1998).

Otavipithecus namibiensis, from the middle Miocene (approximately 13 million years ago) of Namibia, is the only Miocene ape presently known from southern Africa (Conroy *et al.,* 1992). It is a medium-size ape with an estimated body weight of 14–20 kg. The type specimen is a right mandible with four teeth (Fig. 15.9). It shares a number of dental and

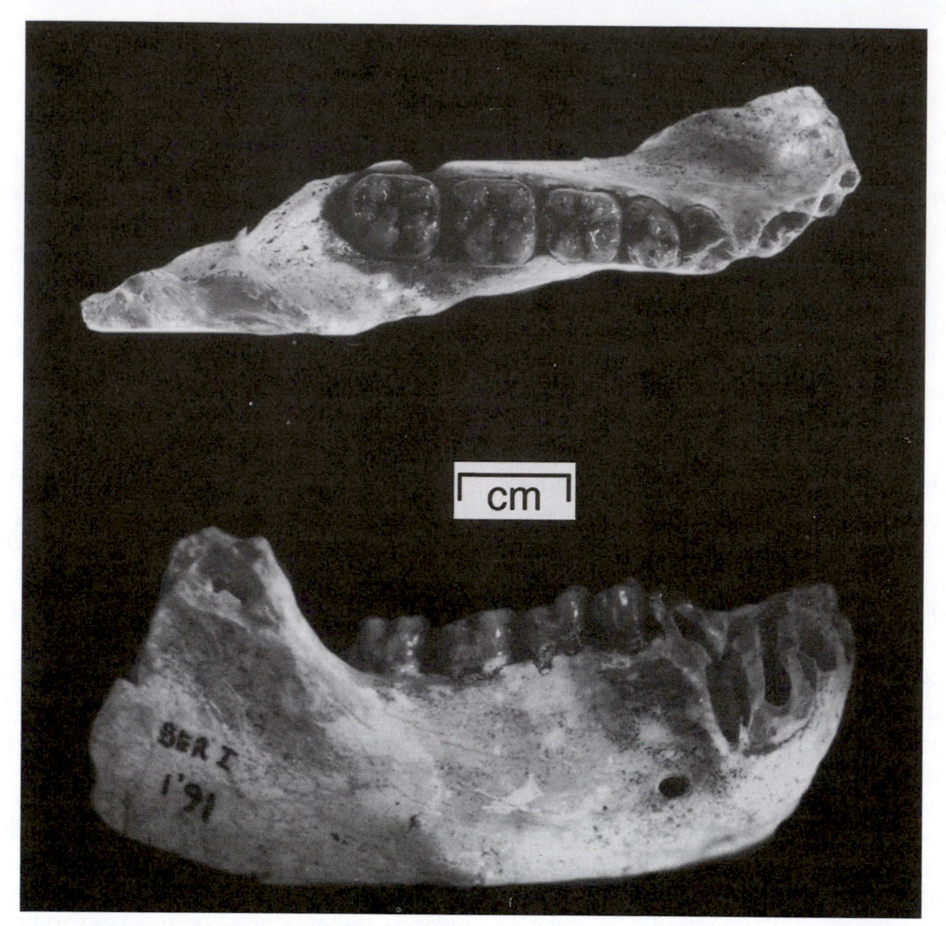

FIGURE 15.9 Lateral and occlusal views of the lower jaw of *Otavipithecus namibiensis* (courtesy of Glenn Conroy).

mandibular similarities with *Kenyapithecus* and *Sivapithecus,* but there is not enough material to determine the phylogenetic affinities of this taxon (e.g., Conroy, 1994). There is also a middle phalanx, which suggests that *Otavipithecus* was an arboreal quadruped (Conroy *et al.,* 1993).

Samburupithecus is a large late Miocene ape from Northern Kenya that is known from a single maxillary fragment (Ishida and Pickford, 1997). It resembles modern African apes and differs from most other Miocene genera in having molar teeth that are long mesiodistally. *Samburupithecus* is almost certainly a member of the clade including gorillas, chimpanzees, and hominids, but its position within the group is not clear at present.

ADAPTIVE RADIATION OF PROCONSULIDS

The fossil apes from the earlier Miocene of East Africa exhibit a diversity in size comparable to that of the living Old World monkeys

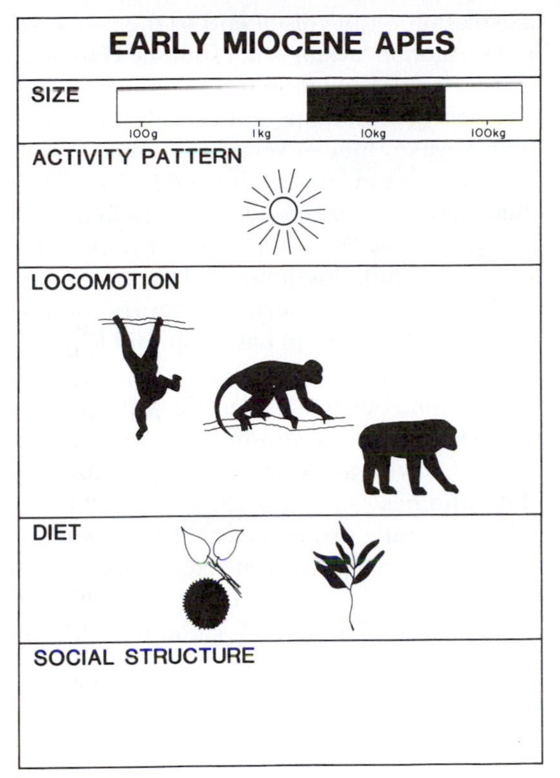

FIGURE 15.10 Adaptive diversity of early Miocene apes.

and most living apes (Fig. 15.10). Although their dental morphology indicates that most species were predominantly frugivorous, there were some genera with more folivorous tendencies, such as *Rangwapithecus* and probably also *Limnopithecus* and *Dendropithecus* (Ungar and Kay, 1995). The skeletal anatomy of the early Miocene apes is well known for only a few species, but the many isolated skeletal elements indicate that this radiation included arboreal quadrupeds (*Proconsul heseloni*), suspensory species (*Dendropithecus, Simiolus*), and more terrestrial species (*Proconsul nyanzae, Kenyapithecus*). There is no evidence of either the fast running or leaping abilities of living cercopithecoids or the brachiating habits of the lesser apes; rather, the skeletal anatomy suggests less specialized but more versatile locomotor abilities, such as those of the living spider monkeys or chimpanzees.

There is also evidence of diversity in the habitat preferences of different East African species (Pickford, 1983; Andrews, 1992). On the basis of the associated mammals, gastropods, and sediments, it seems that *Limnopithecus, Micropithecus,* and *Proconsul major* were more common in rain forest environments, whereas *Dendropithecus, Proconsul heseloni,* and *P. nyanzae* were more common in woodland localities. In the overall breadth of their ecological adaptations, the early Miocene apes seem to have filled most of the locomotor and dietary niches found among extant catarrhines, more so than did the early Oligocene propliopithecids from Egypt (see Fig. 12.16). Compared with the Egyptian fauna, the early Miocene catarrhines were larger and had more terrestrial, suspensory, and folivorous species. Differences in size dimorphism and also differences between species in the type of sexual canine dimorphism suggest a diversity of social structures.

PHYLETIC RELATIONSHIPS OF PROCONSULIDS AND OTHER AFRICAN MIOCENE APES

The proconsulids were much more similar to extant catarrhines in their dentition, cranium, and skeleton than were the early Oligocene propliopithecids. The one species that is well known, *Proconsul heseloni,* has a tubular tympanic, a larger braincase, and more modern skeletal anatomy than *Aegyptopithecus.* Although cranial and skeletal remains are rare for most other species, there is little evidence from the available remains to indicate that any of the other genera were less advanced. All seem to be full-fledged catarrhines compared with the propliopithecids.

While all of the proconsulids seem more advanced than propliopithecids in some features, their relationship to the radiation of extant apes is more difficult to ascertain. Only a few demonstrate derived features that would clearly identify them as part of a lineage uniquely related to apes rather than as primitive catarrhines that precede the monkey-ape divergence.

In most dental and skeletal features, these early Miocene "apes" are primitive compared with all living hominoids (Fig. 15.11), but nevertheless show various mosaics of primitive and derived features linking them with extant hominoids. In the same way that *Aegyptopithecus* seems to be an incipient catarrhine, the proconsulids seem to be incipient apes. Many seem to show facial features linking them with extant great apes (Rae, 1997) but exhibit only a few apelike postcranial features. *Proconsul heseloni,* for example, resembles living hominoids in having an incipiently spool-shaped articulation on the distal end of the humerus and possibly in lacking a tail (but see Harrison, 1998). At the same time it retains an articulation between the ulna and carpal bones and "monkeylike" lumbar vertebrae. Thus, *Proconsul* possesses some of the derived features that characterize the ape lineage, but it lacks others. Likewise, *Afropithecus* has relatively large premolars that link this genus with extant hominoids, but it lacks derived features of the skeleton. *Morotopithecus* has hominoid features in its premolars, lumbar vertabrae and glenoid articulation, but it has a more primitive femur. *Otavipithecus* from southern Africa is too poorly know to allow an accurate assessment of its affinities.

It is generally assumed that the common anatomical specializations of extant apes characterized their last common ancestor and that the radiation of living apes came from a type of hominoid more advanced than any of the proconsulids, even though several taxa may

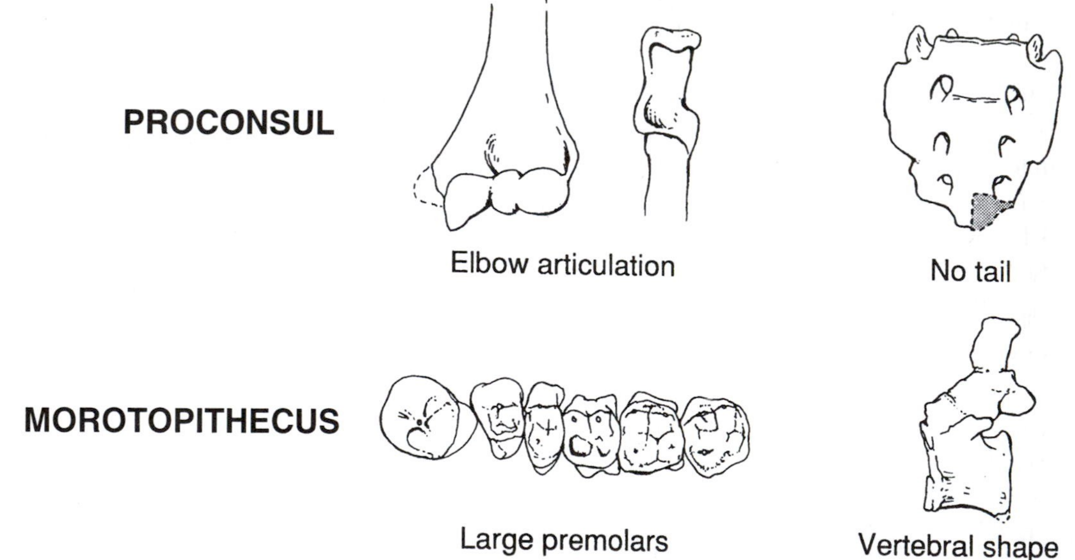

FIGURE 15.11 Features linking early Miocene apes with living hominoids.

be in the common ancestry of later apes (e. g., Begun *et al.,* 1997). Nevertheless, there has undoubtedly been considerable parallel evolution in the history of modern hominoid musculoskeletal anatomy, so the actual pathway of modern ape limb specializations is far from resolved (e.g., Larson, 1998).

Although these Miocene catarrhines are more primitive than living apes, they are also quite different from any cercopithecoid monkeys. The morphology of both these early Miocene apes and the earlier Oligocene hominoids, such as *Aegyptopithecus,* demonstrate that Old World monkeys are a very specialized group of higher primates. In their dentition, skull, and some aspects of their skeleton, the living apes have retained many more features from the early catarrhines than have cercopithecoids.

Although the fossil apes from the early and middle Miocene of East Africa are treated as a single radiation and a single taxonomic group, the proconsulids, this is undoubtedly a gradistic classification. As mentioned earlier, different taxa show different mosaics of primitive and derived features of the dentition, cranium, and limb skeleton (e.g., Begun *et al.,* 1997). It is also quite possible that several more derived lineages of higher primates can be traced back to one of these Miocene apes. The lineage leading to *Oreopithecus,* for example, may have originated via *Nyanzapithecus* from a genus such as *Rangwapithecus.* It has also been suggested that the lineage leading to great apes and humans can be traced back to *Morotopithecus.* The position of the proconsulid radiation relative to the origin of the gibbon lineage is more difficult to determine from present evidence (see Walker, 1997; Begun, 1997; Rae, 1997; Gebo *et al.,* 1997). Many authors feel that all proconsulids precede the origin of gibbons. However, if the great ape and human lineage can be traced to a genus from the Miocene of East Africa, as others believe, the more primitive gibbon lineage originated somewhere within this radiation as well.

GEOGRAPHICAL DIVERSITY OF PROCONSULIDS It has become clear in recent years that the radiation of proconsulids extended far beyond East Africa (see Fig. 15.1). Fossil apes from the Miocene of China are remarkably similar to several African genera (Gu and Lin, 1983). *Dionysopithecus* (Li, 1978) is virtually identical in its molar morphology to *Micropithecus,* despite the 10,000 km separating them, and a similar small ape has been found in the Miocene of Pakistan (Bernor *et al.,* 1988). *Platydontopithecus* (Gu and Lin, 1983) has been described as similar to *Proconsul,* and a small ape from Thailand has been placed tentatively in the genus *Dendropithecus* (Suteethorn *et al.,* 1990) but also shows similarities to *Micropithecus.* Unfortunately, these Asian proconsulids are known only from a few dental specimens, so the similarities to the East African apes, although striking, are based on very few features. However, they clearly document the radiation of primitive hominoids out of Africa and into Asia in the early Miocene (e.g., Harrison, 1996). Further discoveries in the Miocene of Asia will surely increase our knowledge of early ape evolution and bring many surprises.

Eurasian Fossil Apes

It is ironic that although fossil primates (including fossil apes) were first discovered in Europe and Asia over 150 years ago, for most of this century European fossil apes have remained very poorly known compared with the abundant African fossils. However, the fossil record of ape evolution in both Europe and Asia has increased dramatically in recent years, so many genera, such as *Dryopithecus,* are only just becoming well documented (e.g., Moya-Sola and Kohler, 1993, 1996). The earliest fossil catarrhines in Europe and Asia date from the early middle Miocene about 17 million years ago, and most are much younger (e.g., Andrews *et al.,* 1996). Even now, most remains are fragmentary or badly crushed, and many come from sites for which there are no absolute dates. Nevertheless, it has become increasingly apparent that the Miocene apes from the northern latitudes are far more distinct in their phyletic relationships than are the proconsulids from Africa. Rather than documenting different dimensions of a single broad evolutionary radiation like the proconsulids, the European and Asian apes seem to include representatives of several distinct radiations of catarrhines. In addition to the few proconsulid remains from Asia already discussed, the fossil apes from the middle and late Miocene of Europe and Asia include members of at least four families—pliopithecids, oreopithecids, dryopithecids, and pongids—as well as numerous taxa whose affinities are unresolved and hotly debated.

TABLE 15.2
Infraorder Catarrhini
Family PLIOPITHECIDAE

Species	Estimated Mass (g)
Pliopithecus (middle to late Miocene, Europe, Asia)	
P. antiquus	9,700
P. vindobonensis	7,000
P. platyodon	—
P. priensis	—
P. zhanxiangi	15,000
Plesiopliopithecus (= *Crouzelia*) (middle Miocene, Europe)	
P. lockeri	5,000
P. auscitanensis	5,000
P. rhodanica	—
Anapithecus (middle Miocene, Europe)	
A. hernyaki	13,500
Laccopithecus (late Miocene, Asia)	
L. robustus	12,000

Pliopithecids

Pliopithecus and its close relatives (Table 15.2) are the oldest and the most primitive of the fossil "apes" from Europe. This gibbon-size primate has been found in fossil sites spanning much of the middle and late Miocene of Europe and Asia (Harrison *et al.,* 1991; Andrews *et al.,* 1996). The dozen or more species of *Pliopithecus, Plesiopliopithecus, Anapithecus,* and *Laccopithecus* ranged in size from about 6 to 15 kg. The best-known species, *Pliopithecus vindobonensis,* was the size of a siamang.

The teeth of *Pliopithecus* (Fig. 15.12) are quite primitive compared with those of other Eurasian apes. The anterior dentition has broad upper central incisors, smaller upper laterals, and tall, narrow lower incisors. The canines are very sexually dimorphic, long and daggerlike in some individuals and short in others. The lower anterior premolar is similarly variable—narrow and sectorial in some individuals (presumably males) and broad in others (presumably females). The upper molars are broad and have a large lingual cingulum. The lower cheek teeth, including the posterior premolar, usually have a long and narrow occlusal surface and a prominent buccal cingulum. All have a characteristic triangle on the anterior buccal aspect of the talonid basin. Analysis of both shearing crests and microwear on the molars indicate that this radiation included both frugivorous and folivorous species (Ungar and Kay, 1995).

Parts of the skull and much of the skeleton from several individuals of *Pliopithecus vindobonensis* are known from a fissure-fill at Neudorf an der Marche, Czech Republic. The lower jaw is shallow with a broad ascending

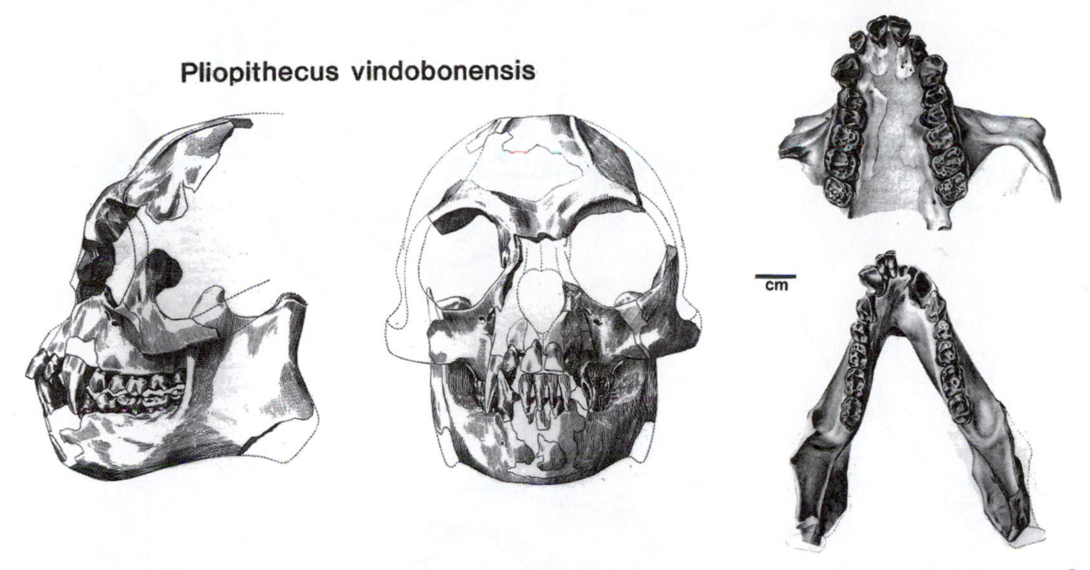

FIGURE 15.12 Cranial and dental remains of *Pliopithecus vindobonensis,* from the middle Miocene of the Czech Republic. Notice the gibbonlike face (from Zapfe, 1961).

ramus, similar to that in extant gibbons. The skull is similar to that of a living gibbon in overall appearance but is more primitive in many details. The face has a short, narrow snout. The interorbital region is very broad and the orbits are large and circular. The zygomatic region is relatively gracile. The frontal bone is high and rounded, suggesting a relatively large brain. Posteriorly the temporal lines converge to form a sagittal crest in some individuals. The structure of the ear region is intermediate between that of *Aegyptopithecus* and that found in modern catarrhines (and *Proconsul*). The tympanic bone forms a ring at the lateral surface of the bulla, but it does not form a complete tube as in living catarrhines. The inferior half of the tube is not ossified, suggesting a more primitive morphological condition than that found in either Old World monkeys or apes.

The skeleton of *Pliopithecus* (Fig. 15.13) is much like that of a large living platyrrhine such as *Ateles* or *Lagothrix.* The intermembral index is 94. Both the forelimb and the hindlimb show adaptations in the joint surfaces that are characteristic of suspensory primates. Like the Fayum propliopithecids, *Pliopithecus* lacks the distinguishing skeletal features of either living apes or living cercopithecoids, and it has many primitive skeletal features, such as an entepicondylar foramen on the humerus, a long ulnar styloid process, and a prehallux bone in the foot. Although the relatively small size of the sacral bodies would seem to suggest that *Pliopithecus* was tailless, Ankel (1965) has demonstrated that the sacral canal has proportions indicating that *Pliopithecus* had a small tail. *Pliopithecus* was an arboreal quadruped with suspensory abilities like those of the larger platyrrhines.

PHYLETIC RELATIONSHIPS Because of its size and the gibbonlike features in its face, *Pliopithecus* has traditionally been considered an ancestral gibbon. However, in many details of its dentition, skull (particularly the auditory re-

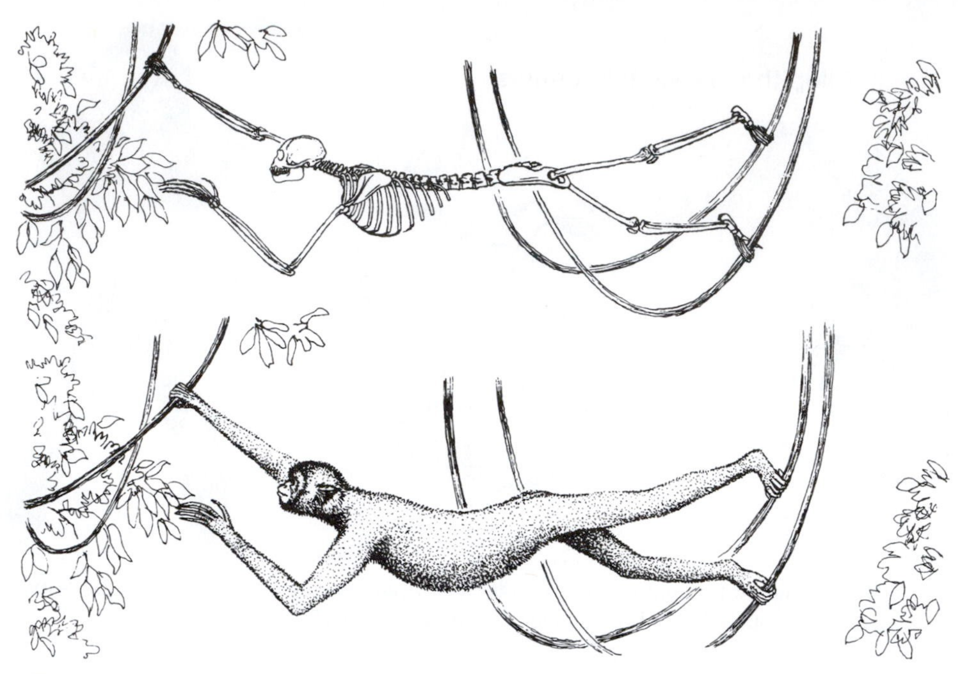

FIGURE 15.13 The skeleton of *Pliopithecus* and a reconstruction of its locomotor habits.

gion), and skeleton, *Pliopithecus,* like *Aegyptopithecus,* was more primitive than any living catarrhine (or *Proconsul*) and lacked the specializations that would be expected in the last common ancestor of Old World monkeys and apes. If gibbons evolved from *Pliopithecus,* then the anatomical features that are shared by living hominoids and Old World monkeys, such as a tubular tympanic and absence of the entepicondylar foramen on the humerus, must be the result of parallel evolution rather than inheritance from a common ancestor. At the same time, *Pliopithecus* shares no unique features with the extant lesser apes that are not also found in several other Miocene or Oligocene fossil anthropoids. Like the early Oligocene *Aegyptopithecus,* the middle Miocene *Pliopithecus* does not appear to be a full-fledged catarrhine and is more primitive than *Proconsul* from the early Miocene. *Pliopithecus* and the other pliopithecids seem to be a late member of the early catarrhine radiation that preceded the divergence of Old World monkeys and apes.

Laccopithecus robustus (Fig. 15.14), from the latest Miocene site of Lufeng, China, is a large fossil ape (12 kg) that is very similar to *Pliopithecus* in dental morphology. It has sexually dimorphic canines and anterior premolars. Like *Pliopithecus, Laccopithecus* has large orbits and a short snout, but the zygomatic region is more robust. The many dental similarities to *Pliopithecus* suggest that *Laccopithecus* is a late-surviving member of the pliopithecids. Like *Pliopithecus, Laccopithecus* has been considered a fossil gibbon (Wu and Pan, 1984, 1985), and its Asian location is compatible with such a relationship. Moreover, the parts of the anatomy in which *Pliopithecus* is more primitive than modern catarrhines (the auditory region and the skeleton) are not known for *Laccopithecus.* Should *Laccopithecus* turn out

Laccopithecus robustus

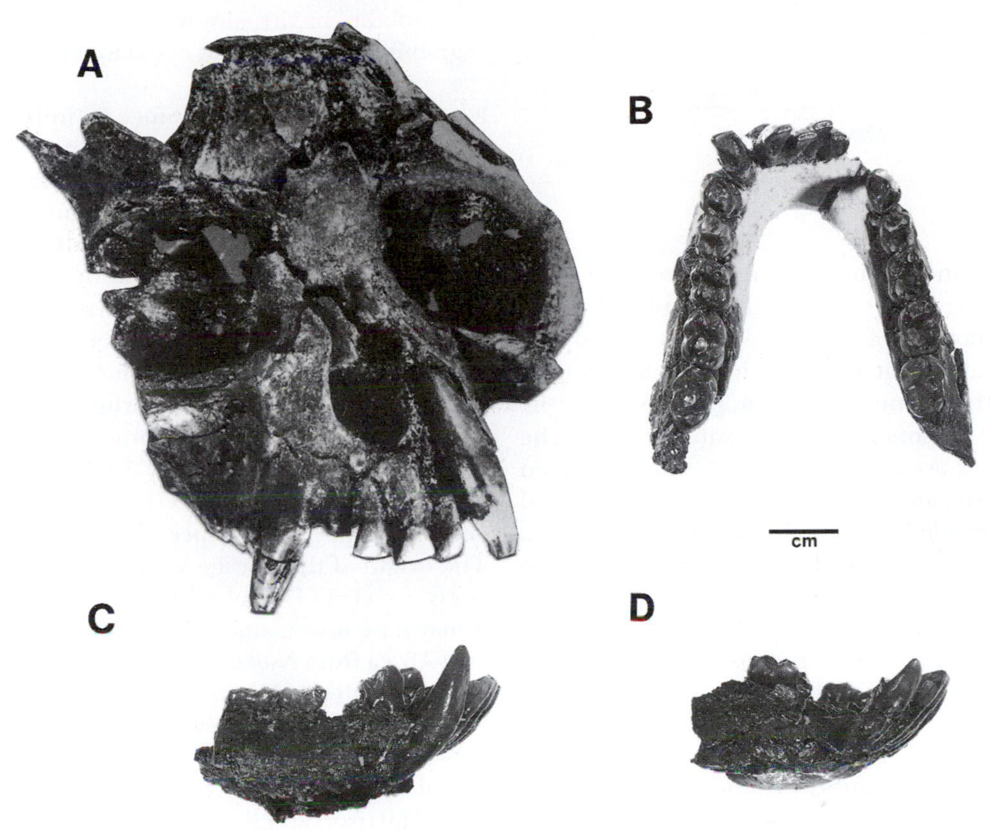

FIGURE 15.14 Cranial and dental remains of *Laccopithecus robustus*, from the latest Miocene of Lufeng, China: A, cranium; B, mandible; C and D, lateral views of male and female mandibles, showing sexual dimorphism (courtesy of Pan Yuerong).

to be more like modern gibbons in further aspects of cranial and skeletal anatomy, this would necessitate revision of the position of the pliopithecids and reconsideration of the amount of parallel evolution in hominoid evolution.

Oreopithecus

This fossil ape from the late Miocene of Europe has been an enigma to paleontologists since its initial discovery in the latter part of the last century. The single species, ***Oreopithecus bambolii*** (Table 15.3), is known from several sites in northern Italy, particularly from coal mines. Numerous remains, including cranial and skeletal elements, have been recovered (Figs. 15.15, 15.16), but the most complete remains are crushed, making interpretation of their morphology rather difficult.

Oreopithecus has a dental formula of 2.1.2.3, like all catarrhines, but many aspects of its

Table 15.3
Infraorder Catarrhini
Family OREOPITHECIDAE

Species	Estimated Mass (g)
Oreopithecus (late Miocene, Europe)	
O. bambolii	30,000

dentition are quite unique—hence the longstanding difficulties in determining its phyletic position among catarrhines. The upper central incisor is relatively large and round, and the lateral is a smaller, peglike tooth; the lower incisors are narrow, spatulate teeth. The canines are quite dimorphic, with presumed males having tall upper and lower canines and the females very small canines. In the males, the upper canine shears against the anterior surface of the anterior lower premolar; in females, the lower premolars are more semimolariform. Upper premolars are characterized by two relatively tall cusps of similar size. The upper molars are long and narrow, with a well-formed trigon, a large hypocone, and a lingual cingulum as in other Miocene catarrhines. The paraconule is particularly well developed. The lower molars have the characteristic basic cusps found in all noncercopithecoid catarrhines but also have an additional sixth cusp, the centroconid. The well-developed shearing crests clearly indicate a folivorous diet. Overall, the dentition is a more specialized version of that found in the African middle Miocene hominoid *Nyanzapithecus.*

The skull has a relatively short snout, a small brain, and a pronounced sagittal crest in some individuals. The auditory region indicates the presence of a tubular ectotympanic as in extant catarrhines.

The skeleton of *Oreopithecus* has many indications of suspensory locomotor habits, including a relatively short trunk, a broad thorax, relatively long forelimbs, short hindlimbs, long, slender manual digits, and evidence of extensive mobility in virtually all joints. The elbow region is identical to that of extant great apes.

PHYLETIC RELATIONSHIPS Since its initial discovery, *Oreopithecus* has been identified by various authorities as being closely related to parapithecids, cercopithecoids, pongids, hominids, or as an ancient higher primate lineage not closely related to any modern group of anthropoids. Many of these diverse interpretations are still championed by one or more authorities. Nevertheless, in its limb structure, *Oreopithecus* shows more similarities to extant hominoids than does any other fossil ape (Harrison, 1986; Sarmiento, 1987; Andrews *et al.,* 1996; Begun *et al.,* 1997), and it is almost certainly some type of specialized hominoid. The origin of the lineage leading to *Oreopithecus* is not clear. Dental evidence suggests that it may have arisen among the proconsulids in East Africa from *Nyanzapithecus* (e.g., Harrison, 1986; Benefit and McCrossin, 1997). However, recent discoveries of skeletal remains of *Dryopithecus* have led some authors to suggest that *Oreopithecus* is closely related to *Dryopithecus* (e.g., Harrison and Rook, 1997).

Pongidae

The systematics and evolutionary relationships of the other, more widespread, large fossil apes from the middle and late Miocene of Eurasia are very unsettled (e.g., Kelley, 1988; Begun, 1995; Andrews *et al.,* 1996; Andrews and Pilbeam, 1996). Most current authorities divide the various species of middle and late Miocene apes into four phyletic groups: *Griphopithecus,* from Europe and western Asia; *Dryopithecus,* from Europe, and *Lufengpithecus* from Asia; *Sivapithecus* and related genera, from western and southern Asia; and *Ouranopithecus* (or *Graecopithecus*) from Greece and

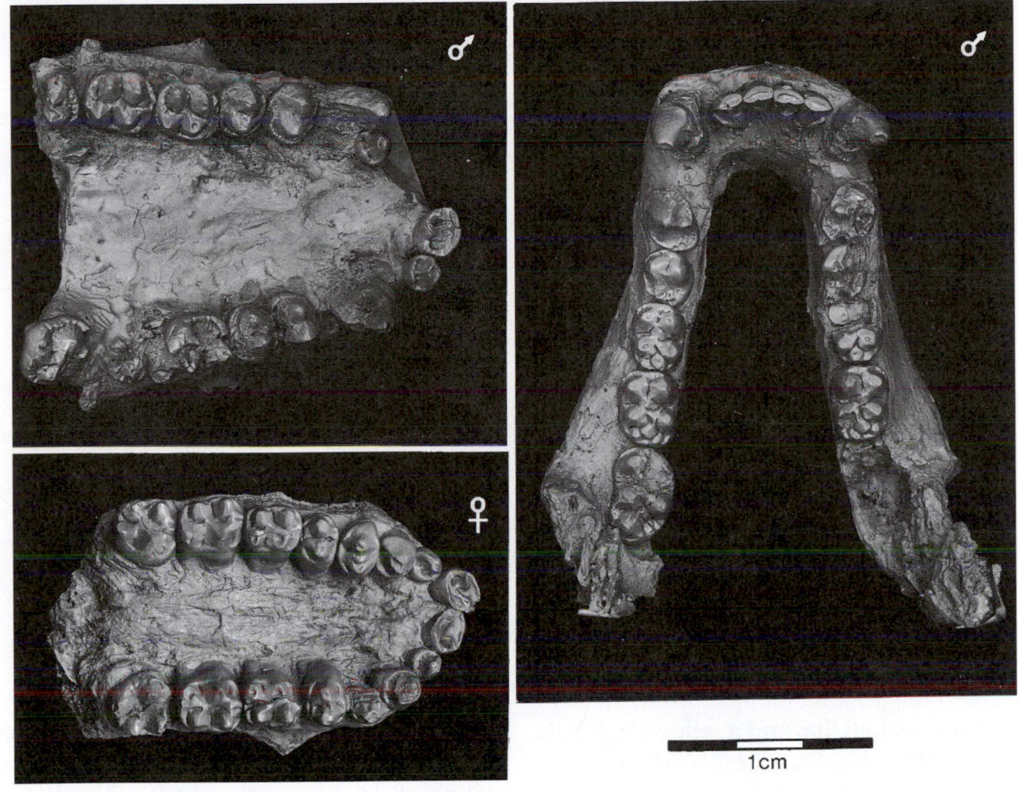

FIGURE 15.15 Upper dentition (left) and lower dentition (right) of *Oreopithecus bambolii,* from the Pliocene of Europe. (Courtesy of Eric Delson)

Turkey (Table 15.4). The exact allocation of species to one or the other of these groups varies somewhat from authority to authority. Because teeth and jaws are all that is known for many species, the division is usually made on the basis of the relative thickness of dental enamel, development of the lingual cingulum on the upper molars, premolar proportions, mandible shape, and subnasal morphology.

Dryopithecus is the best-known European fossil ape; it was first described nearly 150 years ago on the basis of a jaw and a humerus from St. Gaudens in southern France. The four species of *Dryopithecus* are known only from Europe, with three of the four taxa coming from southern France and Northern Spain, adjacent to the Pyrenees. The species of *Dryopithecus* range in size from that of a siamang to that of a chimpanzee.

The lower premolars of *Dryopithecus* are broader than those of either primitive catarrhines or extant gibbons, and the upper premolars are longer. The molar morphology is roughly intermediate between that of the early Miocene *Proconsul* from Africa and the later *Sivapithecus* from Asia. The upper molars of *Dryopithecus* are not as broad as those of the early Miocene apes or *Pliopithecus,* and they often have only a partly formed lingual cingulum. *Dryopithecus* differs from other Eurasian apes in having thin rather than thick enamel on the cheek teeth, gracile canines, a rela-

FIGURE 15.16 Skeleton of *Oreopithecus* and a reconstruction of its locomotor habits.

tively short premaxilla, and a relatively gracile mandible (Fig. 15.17). The broad, rounded cusps on the cheek teeth indicate a predominantly frugivorous diet.

Cranial remains are known for two species of *Dryopithecus, D. brancoi* from Hungary and *D. laietanus* from Spain. *Dryopithecus* was more advanced than any of the Miocene apes from East Africa (including *Proconsul, Afropithecus,* and *Turkanapithecus*). It resembles modern hominoids, or specifically great apes, in many cranial features, including development of the brow ridges, a prominent glabella, and especially the lack of a subarcuate fossa (Moya-Sola and Kohler, 1995; Begun, 1995).

The skeletal elements indicate a postcranial anatomy that is more similar to that of living hominoids than that of any of the proconsulids or *Pliopithecus* on the basis of their reduced olecranon process of the ulna, deep humeral trochlea, long hands, and short lumbar vertebrae with the transverse processes arising from the pedicle. (Morbeck, 1983; Begun, 1992; Moya-Sola and Kohler, 1996). Relative limb proportions of *D. laietanus* from Spain appear to be more like orangutans than African apes, with very long forelimbs and relatively short femora. These limbs suggest that some species were suspensory.

Griphopithecus is a poorly know large ape from Austria and Turkey. It is from slightly older deposits than *Dryopithecus,* and it differs in having thicker molar enamel and more primitive limb bones (Andrews *et al.,* 1996). It has been linked with the African *Kenyapithecus.*

Lufengpithecus is a large ape from the latest Miocene of southern China, known from more than a thousand dental remains and several skulls (Wu *et al.,* 1985; Kelley and Etler, 1989). Although the large ape fossils from Lufeng were originally placed in two taxa, it is now widely accepted that they belong to a single species, *Lufengpithecus lufengensis,* with slightly greater sexual dimorphism than any

TABLE 15.4
Infraorder Catarrhini
Family PONGIDAE

Species	Estimated Mass (g)
Dryopithecus (middle to late Miocene, Europe)	
D. brancoi (= *carinthiacus*)	—
D. crusafronti	26,000
D. fontani	35,000
D. laietanus	20,000
Lufengpithecus (late Miocene, Asia)	
L. lufengensis	50,000
Griphopithecus (middle Miocene, Europe, Asia)	
G. alpani	28,000
G. darwini	48,000
G. sp.	—
Sivapithecus (late Miocene, Europe, Asia)	
S. sivalensis (= *indicus*)	75,000
S. punjabicus	40,000
S. parvada	90,000
Ankarapithecus (late Miocene, Turkey)	
A. meteai	82,000
Gigantopithecus (late Miocene to Pleistocene, Asia)	
G. giganteus (= *bilaspurensis*)	190,000
G. blacki	225,000
Graecopithecus (late Miocene, Europe)	
G. freybergi	—
Ouranopithecus (late Miocene, Europe)	
O. macedoniensis	110,000

living ape (Kelley and Qinghua, 1991; Kelley, 1993). The dentition of *Lufengpithecus* shows many features characteristic of European *Dryopithecus,* including tall, narrow incisors and narrow molars without extremely thick enamel. The relatively complete, but crushed, cranial remains show greatest similarities to *Dryopithecus* in the broad interorbital region, vertical frontal, and lack of a broad interorbital torus. It lacks many of the derived cranial features linking *Sivapithecus* with *Pongo* and is usually considered a primitive sister group to that radiation (see Schwartz, 1997; Begun *et al.,* 1997).

Sivapithecus is from the Siwalik Hills of northern India and Pakistan and dates from approximately 12 million years ago to 8 million years ago. There are three species, *S. punjabicus, S. indicus,* and *S. parvada,* ranging in estimated body size between 40 and 90 kg (Kelly, 1993). In addition, *Gigantopithecus giganteus* (= *bilaspurensis*), discussed shortly, is often recognized as a species of *Sivapithecus.* The dentition of *Sivapithecus* is characterized by cheek teeth with thick enamel and low cusp relief, the common absence of any cingulum on the molars, very broad lower premolars, robust canines, and a thick mandible. The upper central incisors are much broader than the laterals (Kelley *et al.,* 1995). On the basis of the available material, *Sivapithecus* shows relatively little canine dimorphism compared with that found among living apes and monkeys. The combination of thick-enameled molars and low cusp relief is characteristic of living primates that eat seeds and nuts. It has been suggested that *Sivapithecus* had a diet of hard nuts, bark, or fruits with hard pits.

The skulls of *Sivapithecus* show a striking resemblance to the living orangutan in such features as a narrow snout with a very large procumbent premaxilla, a small incisive foramen, broad zygomatic arches, a tall, narrow nasal aperture, and high orbits (Fig. 15.18).

There are only a few skeletal remains of *Sivapithecus,* including phalanges, much of the humerus, and part of a femur. Although some features are similar to those of living orangutans, most indicate that *Sivapithecus* lacked the extreme adaptations for suspensory locomotion found in *Pongo.* Rather, it was a more quadrupedal ape. In view of the size range, there was probably considerable locomotor diversity within the genus, but present evidence is too scanty to determine the exact nature of such differences.

Ankarapithecus meteai, a large fossil ape from the site of Sinap in Turkey, has generally

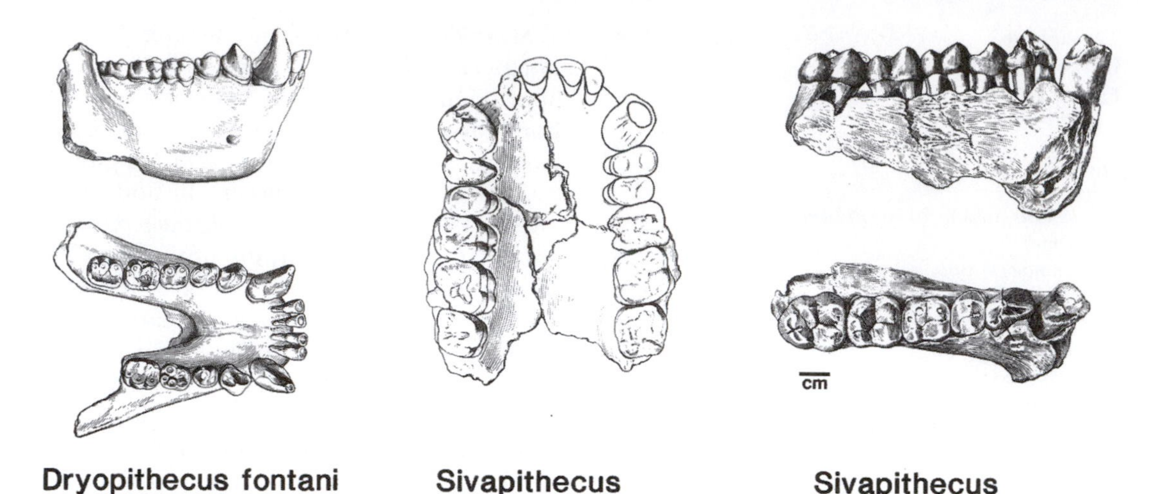

FIGURE 15.17 Dental remains of two large middle and late Miocene fossil apes from Eurasia, *Dryopithecus* and *Sivapithecus.*

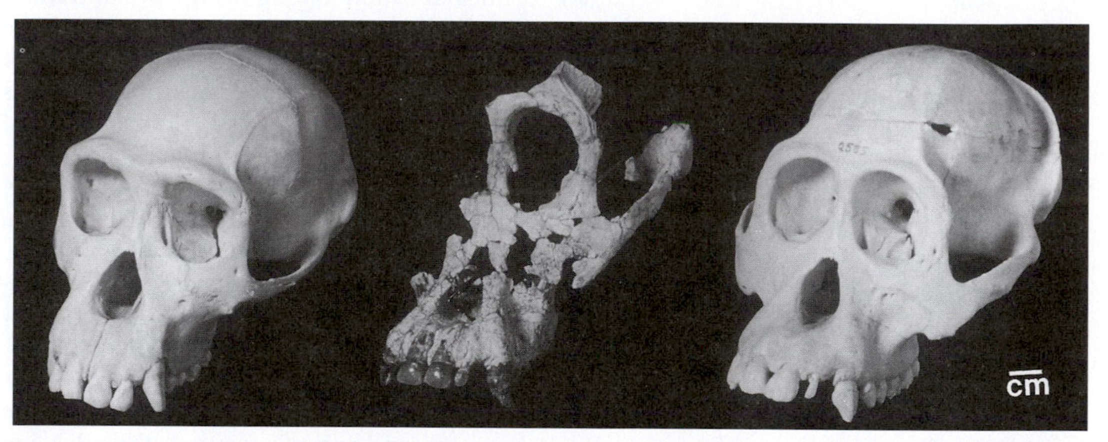

FIGURE 15.18 Cranial remains of *Sivapithecus* (center) and crania of *Pan* (left) and of *Pongo* (right) (photograph courtesy of David Pilbeam).

been considered a close relative of *Sivapithecus* (Andrews and Tekkaya, 1980). However, recent discoveries have demonstrated significant differences in the skull morphology of the two. The skull of *Ankarapithecus* (Fig. 15.19) is more flexed than that of *Sivapithecus,* and the face is relatively straighter with a broader interorbital dimension and rounder orbits (Alapagut *et al.,* 1996). It seems most likely that *Ankarapithecus* retains the more primitive facial morphology from which the specializations characteristic of *Sivapithecus* and *Pongo* were derived (Alapagut *et al.,* 1996).

A close relative of *Sivapithecus* is ***Gigantopithecus,*** the largest primate that ever lived (Figs. 15.20, 15.21). The two species of *Gigan-*

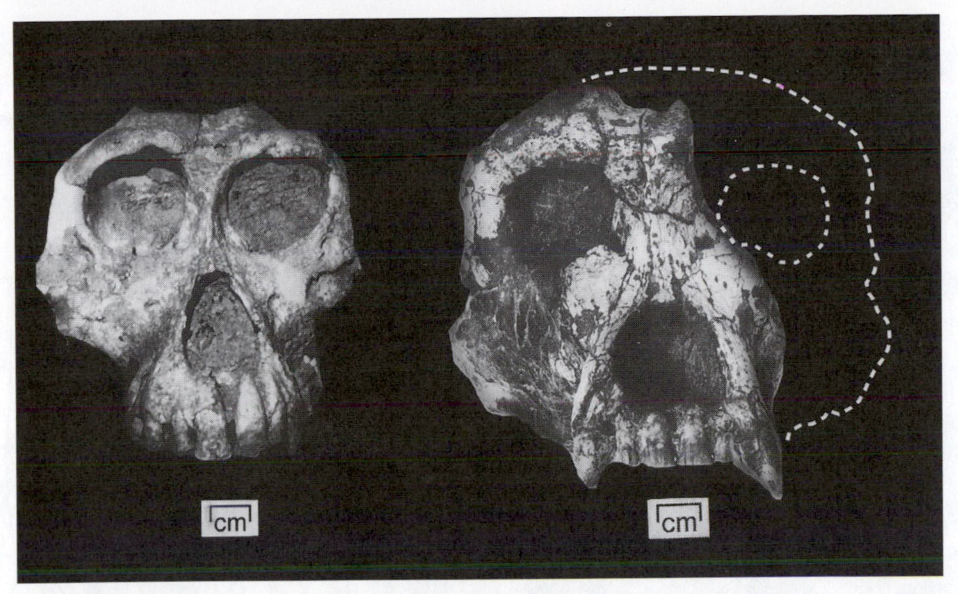

FIGURE 15.19 Facial skeleton of *Ankarapithecus meteai* (left) and *Ouranopithecus macedoniensis* (right) (courtesy of John Kappelman and L. deBonis).

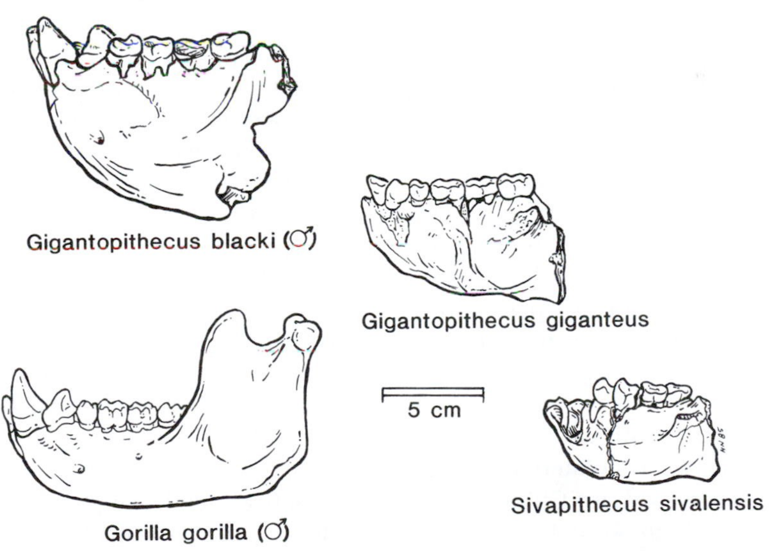

FIGURE 15.20 Lower jaws of *Gigantopithecus* and of *Sivapithecus* compared with that of a male mountain gorilla, the largest living primate.

FIGURE 15.21 A reconstruction of *Gigantopithecus blacki* from the Pleistocene of China.

topithecus were almost certainly derived from a large Asian species of *Sivapithecus.* The earlier, smaller *G. giganteus* (= *G. bilaspurensis*) is from the latest Miocene of India and Pakistan; the larger *G. blacki* is from Pleistocene caves in China and Vietnam. The smaller species probably weighed as much as a living gorilla (150+ kg), and the Pleistocene species has an estimated weight several times that (perhaps as much as 300 kg, based primarily on the large mandible).

These extraordinary primates are known only from lower jaws and isolated teeth. They were initially discovered in Chinese drugstores, where the teeth were being sold as medicine (von Koenigswald, 1983). The lower incisors are very small and vertical. The canines are thick but relatively short. The lower anterior premolar is relatively broad, as in *Homo sapiens,* rather than elongated. Like those of *Sivapithecus,* the teeth of *Gigantopithecus* have thick enamel and low, flat cusps. In *G. blacki* there are often accessory cusps. In both species, the mandible is very thick and extremely deep compared with the jaws of living apes (Fig. 15.20). The dental proportions, cheek tooth morphology, and robust mandibles indicate that *Gigantopithecus* ate some type of hard, fibrous material. One worker has suggested that they ate bamboo, like the living panda, but phytoliths recovered from the surface of fossil teeth suggest a more diverse diet including various fruits (Ciochon *et al.,* 1990). Their enormous size would seem to have precluded anything except a folivorous diet and terrestrial locomotion.

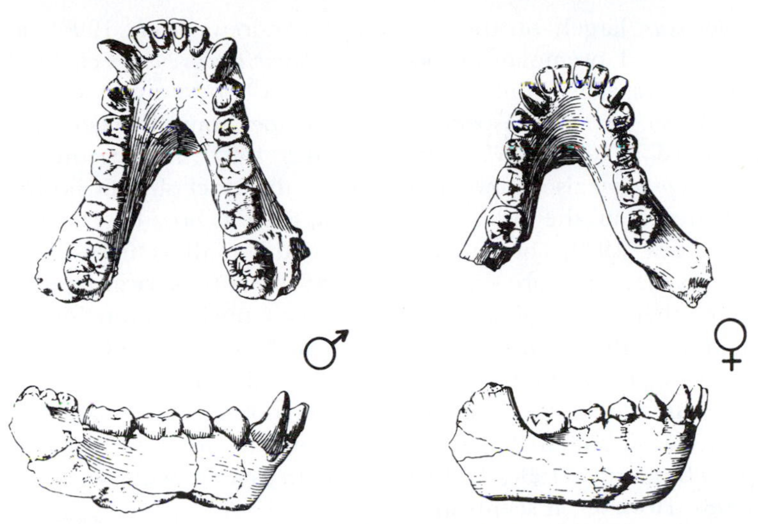

FIGURE 15.22 Male and female lower jaws of a late Miocene fossil ape from Greece, *Ouranopithecus macedoniensis.*

Ouranopithecus is a large ape from several late Miocene (9–10 million years ago) sites in Greece, all placed in a single species *O. macedoniensis* (?= *Graecopithecus freybergi;* see Andrews *et al.,* 1996). From several rich sites in northern Greece have come many jaws and a nearly complete face (Figs. 15.19, 15.22). The associated fauna suggests a mosaic environment of woodland and savannah habitats (e.g., DeBonis and Koufos, 1995).

Ouranopithecus was a large ape with an estimated body size of over 70 kg. The dentition of *Ouranopithecus* is characterized by relatively small canines and short, broad anterior lower premolars in both sexes. The molar enamel is extraordinarily thick. Studies of both cusp morphology and microwear suggest a diet for *Ouranopithecus* of hard or gritty objects, such as nuts or tubers (Ungar and Kay, 1995).

The face of *Ouranopithecus* has a broad nasal opening, a large incisive foramen, pronounced brow ridges, and a broad interorbital region with a prominent glabella (Fig. 15.19). The cranial morphology of *Ouranopithecus* has been interpreted as showing derived similarities to many different clades of living apes, including orangutans (Moya-Sola and Kohler, 1995), gorillas (Dean and Delson, 1992), and hominids (DeBonis and Koufos, 1993, 1994, 1995). Despite these differing views (e.g., Begun, 1995), most authorities agree that the facial morphology of this genus (particularly the brow ridges) link it with the African ape and human clade. Any further resolution of these alternatives must await additional remains as well as clarification of the dental and cranial morphology of the earliest hominids.

PHYLETIC RELATIONSHIPS OF LARGE EURASIAN APES Although *Griphopithecus, Dryopithecus, Ankarapithecus, Sivapithecus,* and *Ouranopithecus* are all roughly contemporary and all show some features linking them with the extant great apes, it is clear that they exhibit a considerable diversity of morphology and likely occupy very different positions in hominoid phylogeny. The poorly known *Griphopithecus* seems to be the most primitive genus and lies near the base of the radiation of extant apes. It has also been linked with the African *Kenya-*

pithecus and *Afropithecus*, largely on the basis of enamel thickness and premolar proportions. *Sivapithecus* shows striking similarities to the orangutan (*Pongo*) in many aspects of its dental and cranial anatomy (Ward and Brown, 1986) *Ankarapithecus* also shows some orangutan-like features in the lower face (Andrews and Tekkaya, 1980), but a more complete skull shows that it is more primitive than *Sivapithecus* in other aspects of its cranial anatomy (Alapagut *et al.*, 1996). The discovery of skeletal remains suggesting more quadrupedal locomotor habits for some species of *Sivapithecus* have led a few authors to reconsider this relationship (Pilbeam *et al.*, 1990; Andrews and Pilbeam, 1996). However, it seems more reasonable to view this as evidence that the orangutan clade contained a greater diversity of apes in the past and that some locomotor features shared by living great apes undoubtedly reflect parallel evolution (Larson, 1992, 1997). *Gigantopithecus* also seems to be a derived member of this radiation based on its dentition. Most authorities agree that *Ouranopithecus* (or *Graecopithecus*) is the most derived of the large Eurasian fossil apes and that it shows morphological features linking it with African apes and humans. However, some of these features are probably primitive retentions, compared with the specializations found in the *Sivapithecus*-orangutan clade, and some of the similarities to hominids (very thick enamel) may be parallelism, since they do not seem to characterize the earliest hominids (White *et al.*, 1994). Thus, the position of *Ouranopithecus* within this clade remains unresolved (see Begun, 1995). Of all the large Eurasian Miocene apes, *Dryopithecus* is the genus about which there is least agreement (Andrews and Pilbeam, 1996). Although this genus shows both dental and cranial features linking it with extant great apes, there is no consensus on where its closest affinities lie. It has been placed as the most primitive great ape (Andrews *et al.*, 1996), as a member of the *Sivapithecus-Pongo* clade (Moya-Sola and Kohler, 1995), or as the sister taxon of the African ape/hominid clade (e.g., Begun and Kordos, 1997). Despite the lack of consensus about the exact phyletic position of individual taxa, there is broad agreement that the middle and late Miocene of Eurasia document a great diversity of large apes representing an extensive phyletic radiation, only a small part of which is present in the extant genera. At present, this diversity is most evident in cranial anatomy, but a comparable diversity in locomotor anatomy will undoubtedly become evident as new fossils are recovered.

THE EVOLUTION OF LIVING HOMINOIDS

In the preceding pages we have reviewed the fossil apes from the latest Oligocene and Miocene epochs. Like the early Oligocene anthropoids from Egypt, the Miocene genera and species can be ordered on the basis of a suite of dental, cranial, and postcranial features into more primitive and more advanced species (Fig. 15.23), though not without debate over individual taxa. It is quite evident that the radiation of hominoids during the Miocene was much more extensive and produced many more lineages than many earlier workers imagined. A corollary of this increasingly complex picture of ape evolution during the Miocene is that the identification of unique lineages leading to particular extant genera is far more difficult than was previously thought. Attempts to find the ancestry of unique lineages leading to gibbons, to the great apes, or to hominids have been complicated repeatedly by the discovery of more complete fossils with unsuspected primitive features or mosaic combinations of derived features, by more careful consideration of comparative anatomy, and by more refined understanding of strati-

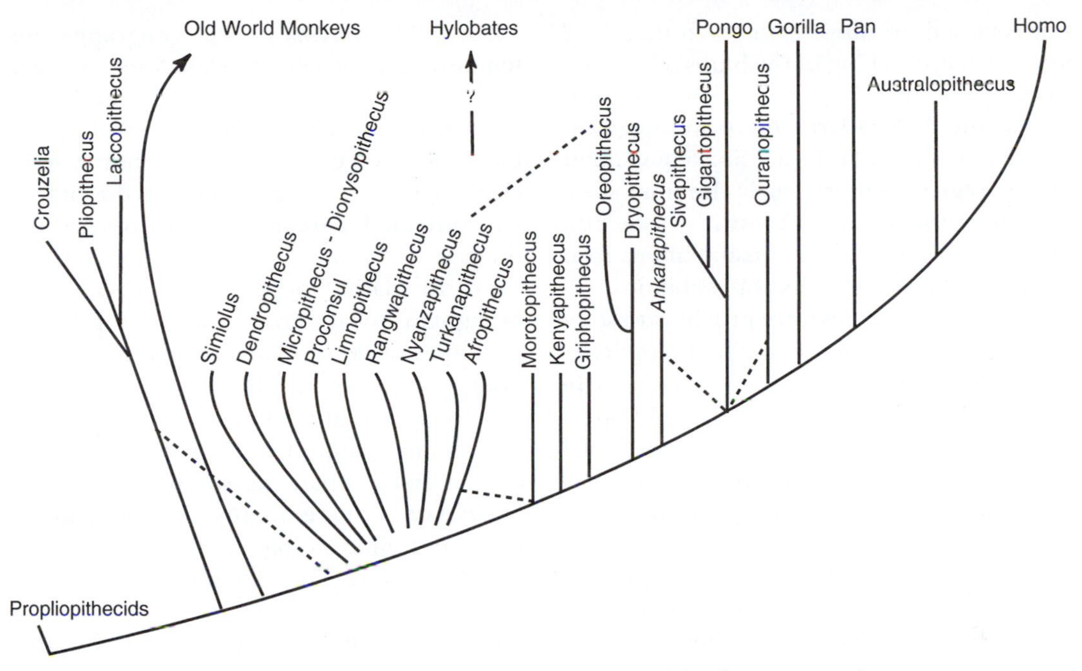

FIGURE 15.23 Phyletic radiation of fossil apes.

graphic relationships. Until very recently, the great temporal expanse of the Miocene epoch, the diversity of Miocene environments and faunas, and the morphological diversity of the fossil apes from that epoch were all unknown and largely unsuspected. This extraordinary diversity of fossil apes from both Africa (e.g., Conroy *et al.*, 1992) and Europe (Andrews *et al.*, 1996) reminds us how little we really know about ape evolution and how many early apes are yet to be uncovered. To put our current understanding of ape evolution into perspective and to contrast it with earlier views, we now examine the fossil evidence specifically for what it tells about the evolution of extant hominoids, consider the alternative ways in which they could have evolved from the diverse radiations of Miocene apes, and compare these results with predictions about ape and human evolution derived from biomolecular studies.

EVOLUTION OF GIBBONS In each of the successive radiations of Oligocene and Miocene hominoids there were small apes that at one time or another have been identified as fossil gibbons. *Propliopithecus* (= *Aeolopithecus*), *Pliopithecus, Dendropithecus, Micropithecus,* and *Dionysopithecus* all show various features (such as small size, short snouts, or large orbits) that cause them to resemble living lesser apes. As already discussed, however, most of these supposed fossil gibbons were extremely primitive in many detailed aspects of their cranial and skeletal anatomy, more so than we would expect in an ancestral gibbon based on the comparative anatomy of extant higher primates. For example, although *Propliopithecus* and *Pliopithecus* were similar to living gibbons in their size and (in some species) had simple, gibbonlike lower molars, they lacked such features as the tubular ectotympanic bone found in all living apes and Old World monkeys, and they retained primitive features in their limb bones that are lacking in the limbs of all living catarrhines. For other genera, such as *Micropithecus* and *Laccopithecus,* there are suggestive cranial remains, but critical information about the ear region and especially the limb skeleton is not available. As a result, we have little unassailable evidence for fossil gibbons from the Miocene, only a series of possibilities.

All of the small apes were probably to some extent ecological vicars of the living lesser apes, but none can be clearly shown to be uniquely related to the living Asian gibbon genus, which has a fossil record extending back only to the middle Pleistocene of China and Indonesia. Molecular estimates of the date of the divergence of gibbons from the hominoid lineage suggest an origin of the lineage leading to the lesser apes between 16 million and 18 million years ago (e.g., Pilbeam, 1996). This time range includes virtually all of the gibbonlike primitive catarrhines from the early Miocene, as well as the fossil apes that seem to mark the appearance of the great ape and human clade, coincident with the origin of gibbons (Fig. 15.23). In the absence of more definitely gibbonlike fossils from the Miocene of Asia, we cannot resolve the question of their ancestry.

EVOLUTION OF THE ORANGUTAN The one living ape whose evolutionary history is now generally considered to be well established is the orangutan. The Asian *Sivapithecus* gave rise to at least two lineages, one leading to *Gigantopithecus,* the other to the orangutan. The late Miocene specimens of *Sivapithecus* and the living *Pongo* are so similar in many details of dental and facial morphology that the latter is almost certainly derived from the former (e.g., Ward, 1997). However, this link has been questioned by some on the basis of differences between the skeletal anatomy of *Sivapithecus* and that of *Pongo* (Pilbeam *et al.,* 1990; Pilbeam, 1996). The geographic and temporal gap between the late Miocene fossils and the living great ape of Borneo and Sumatra is partly bridged by fossil teeth from the Pleistocene of China and Java, but it seems clear that the lineage leading to the orangutan contained a greater diversity of species in the past.

Precise dating of the divergence of the orangutan lineage from that leading to African apes and humans is complicated by doubts as to whether the similarities between the living orangutan and *Sivapithecus* are specializations unique to only the latest species of that genus, characteristic of all species of *Sivapithecus,* or remnants of the primitive hominoid morphology that also characterizes the ancestors of all living apes and hominids. The *Sivapithecus* fossils that show the greatest similarity to orangutans are from the late Miocene, 12 million to 9 million years ago (Kappelman *et al.,* 1991). Molecular studies have indicated dates of 10 million to 12 million years ago for the orangutan divergence, all concordant with the fossil data (Andrews, 1986; Miyamoto *et al.,* 1988; Pilbeam, 1996).

EVOLUTION OF AFRICAN APES The evolutionary history of gorillas and chimpanzees is one of the most notable gaps in our current understanding of ape and human evolution. All of the fossil apes from the Miocene of Africa seem too primitive to be uniquely ancestral to living African apes, with the exception of *Samburupithecus* from Northern Kenya (Ishida and Pickford, 1997). It is most likely a member of the African ape–human clade, but its position in that clade is debated. The other fossil ape that shows many similarities to extant African apes is *Ouranopithecus* from the late Miocene of Greece (Dean and Delson, 1992). This genus is commonly identified as a generalized ancestor for the entire

African ape and hominid clade or uniquely ancestral to hominids (e.g., Begun, 1995; Debonis and Koufos, 1994). It clearly lacks the thin enamel characteristic of gorillas and chimpanzees.

Both the primitive nature of the early Miocene apes and molecular predictions of the timing of hominoid evolution indicate that the evolutionary divergence of the lineages leading to the African great apes and to humans was probably some time in the later part of the Miocene, between 6 million and 5 million years ago. African fossil apes are extremely rare from this period, and beginning in the early Pliocene most fossil hominoids appear to be hominids, leaving the evolutionary history of chimpanzees and gorillas undocumented. This absence of fossil apes probably reflects, in part, our lack of any substantial fossil record from western and central Africa, where these apes live today.

HOMINID ORIGINS We can be almost certain that the earliest hominids evolved from some type of Miocene ape, but the identification of hominids among the various genera and species of fossil apes from that epoch has proved a fruitless exercise thus far. As discussed in a later chapter, there is a good record of hominids from approximately 4 million years ago, with suggestions of several contemporaneous taxa at this time (e.g., Kappelman and Fleagle, 1995). Earlier fragmentary remains cannot be clearly identified to taxon (e.g., Hill, 1994). Identification of hominid origins is further complicated by the rarity of limb or skeletal remains for fossil apes and the realization that the features that distinguished the earliest hominid from apes are not the small teeth and large brain that are so distinctive of ourselves, but rather the postcranial differences, particularly bipedal features of the pelvis. (We discuss hominid evolution in detail in Chapter 17.)

BIBLIOGRAPHY

GENERAL

Adams, C. G. (1981). An outline of Tertiary paleogeography. In *The Evolving Earth,* ed. L. R. M. Cocks, pp. 221–235. Cambridge: Cambridge University Press.

Andrews, P. (1992). Evolution and environment in the Hominoidea. *Nature* **360**:641–646.

——— (1996). Palaeoecology and hominoid palaeoenvironments. *Biol. Rev.* **71**:257–300.

Benefit, B. R., and McCrossin, M. L. (1995). Miocene hominoids and hominid origins. *Ann. Rev. Anthropol.* **24**:237–256.

Bernor, R., and Fahlbusch, V. (1996). *Evolution of Neogene Continental Biotypes in Central Eruope and the Eastern Mediterranean.* New York: Columbia University Press.

Begun, D. R., Ward, C. V., and Rose, M. D. (1997). *Function, Phylogeny, and Fossils: Miocene Hominoid Evolution and Adaptations.* New York: Plenum Press

Ciochon, R. L., and Corruccini, R. S. (1983). *New Interpretations of Ape and Human Evolution.* New York: Plenum Press.

Drake, R. L., Van Couvering, J. A., Pickford, M., Curtis, G. H., and Harris, J. A. (1988). New chronology for the early Miocene mammalian faunas of Kisingiri, western Kenya. *J. Geol. Soc. London* **145**:479–491.

Fleagle, J. G. (1995). The origin and radiation of anthropoid primates. In *Biological Anthropology: The State of the Science,* ed. N. T. Boaz and L. D. Wolfe, pp. 1–21. Bend, Ore.: International Institute for Human Evolutionary Research.

Pickford, M. (1986). The geochronology of Miocene higher primate faunas of East Africa. In *Primate Evolution,* ed. J. G. Else and P. C. Lee, pp. 19–33. Cambridge: Cambridge University Press.

Rose, M. D. (1993). Locomotor anatomy of Miocene hominids. In *Postcranial Adaptation in Nonhuman Primates,* ed. D. L. Gebo, pp. 252–272. DeKalb, IL: Northern Illinois Univ. Press.

AFRICAN LATEST OLIGOCENE TO MIDDLE MIOCENE APES

Andrews, P. J. (1978). A revision of the Miocene Hominoidea of East Africa. *Bull. Br. Mus. Nat. Hist. (Geol.)* **30**(2):85–224.

——— (1981). Species diversity and diet in monkeys and apes during the Miocene. In *Aspects of Human Evolution,* ed. C. B. Stringer, pp. 25–61. London: Taylor and Frances.

——— (1992). Evolution and environment in the Hominoidea. *Nature* **360**:641–646.

Andrews, P. J., and Martin, L. (1987a). Cladistic relationships of extant and fossil hominoids *J. Human Evol.* **16**: 101–118.

——— (1987b). The phyletic position of the Ad Dabtiyah hominoid. *Bull. Br. Mus. Nat. Hist. (Geol.)* **41**: 383–393.

Beard, K. C., Teaford, M. E, and Walker, A. (1986). New wrist bones of *Proconsul africanus* and *P. nyanzae* from Rusinga Island, Kenya. *Folia Primatol.* **47**:97–118.

Begun, D. R., Ward, C. V., and Rose, M. D., eds. (1997). *Function, Phylogeny, and Fossils: Miocene Hominoid Evolution and Adaptations.* New York: Plenum Press.

Benefit, B. R., and McCrossin, M. L. (1997). New fossil evidence bearing on the relationship of *Nyanzapithecus* and *Oreopithecus. Am. J. Phys. Anthropol. Suppl.* **24**:74.

Benefit, B. R., Gitau, S. N., McCrossin, M. L., and Palmer, A. K. (1998). A mandible of *Mabokopithecus clarki* sheds new light on oreopithecid evolution. *Am. J. Phys. Anthropol. Suppl.* **26**:109.

Bernor, R. L., Flynn, L. J., Harrison, T., Hussain, S. T., and Kelley, J. (1988). *Dionysopithecus* from southern Pakistan and the biochronology and biogeography of early Eurasian catarrhines. *J. Human Evol.* **17**:339–358.

Bishop, W. W. (1967). The later Tertiary in East Africa—Volcanics, sediments and faunal inventory. In *Background to Evolution in Africa,* ed. W. W. Bishop and J. D. Clark, pp. 31–56. Chicago: University of Chicago Press.

Boschetto, H. B., Brown, F. H. and McDougall, I. (1992). Stratigraphy of the Lothidok Range, northern Kenya, and K/Ar ages of its Miocene primates. *J. Human Evol.* **22**:47–51.

Conroy, G. C. (1994). *Otavipithecus:* or How to build a better hominid—Not. *J. Human Evol.* **27**:373–383.

Conroy, G. C., Pickford, M., Senut, B., and Mein, P. (1993). Diamonds in the desert: The discovery of *Otavipithecus namibiensis. Evol. Anthropol.* **2**(2):46–52.

Davis. P. R., and Napier, J. (1963). A reconstruction of the skull of *Proconsul africanus* (R. S. 51). *Folia Primatol.* **1**:20–28.

Falk, D. (1983). A reconsideration of the endocast of *Proconsul africanus:* Implications for primate brain evolution. In *New Interpretations of Ape and Human Ancestry,* ed. R. L. Ciochon and R. Corruccini, pp. 239–248. New York: Plenum Press.

Fleagle, J. G. (1975). A small gibbon-like hominid from the Miocene of Uganda. *Folia Primatol.* **24**:1–15.

——— (1983). Locomotor adaptations of Oligocene and Miocene hominoids and their phyletic implications. In *New Interpretations of Ape and Human Ancestry,* ed. R. L. Ciochon and R. Corruccini, pp. 301–324. New York: Plenum Press.

——— (1984). Are there any fossil gibbons? In *The Lesser Apes: Evolutionary and Behavioral Biology,* ed. D. J. Chivers, H. Preuschoft, N. Creel, and W. Brockelman, pp. 431–477. Edinburgh: Edinburgh University Press.

——— (1986). The fossil record of early catarrhine evolution. In *Major Topics in Primate and Human Evolution,* ed. B. A. Wood, L. B. Martin, and P. Andrews, pp. 130–139. Cambridge: Cambridge University Press.

Fleagle, J. G., and Kay, R. F. (1985). The paleobiology of catarrhines. In *Ancestors: The Hard Evidence,* ed. E. Delson, pp. 23–36. New York: Alan R. Liss.

Fleagle, J. G., and Simons, E. L. (1978). *Micropithecus clarki,* a small ape from the Miocene of Uganda. *Am. J. Phys. Anthropol.* **49**:427–440.

Gebo, D. L., MacLatchy, L., Kitio, R., Deino, A., Kingston, J., and Pilbeam, D. (1997). A new hominoid genus from the Uganda early Miocene. *Science* **276**: 401–404.

Gitau, S. N., and Benefit, B. R. (1995). New evidence concerning the facial morphology of *Simiolus leakeyorum* from Maboko Island. *Am. J. Phys. Anthropol. Suppl.* **20**:99.

Gu, Y., and Lin, Y. (1983). First discovery of Dryopithecus in east China. *Acta Anthropol. Sinica* **2**(4):305–314.

Harrison, T. (1980). New finds of small fossil apes from the Miocene locality at Koru in Kenya. *J. Human Evol.* **10**:129–137.

——— (1986). New fossil anthropoids from the Middle Miocene of East Africa and their bearing on the origin of the Oreopithecidae. *Am. J. Phys. Anthropol.* **71**: 265–284.

——— (1987). The phylogenetic relationships of the early catarrhine primates: A review of the current evidence. *J. Human Evol.* **16**:41–80.

——— (1988). A taxonomic revision of the small catarrhine primates from the early Miocene of East Africa. *Folia Primatol.* **50**:59–108.

——— (1989) A new species of *Micropithecus* from the middle Miocene of Kenya. *J. Human Evol.* **18**:537–557.

——— (1992). A reassessment of the taxonomic and phylogenetic affinities of the fossil catarrhines from Fort Ternan, Kenya. *Primates* **33**(4):501–522.

——— (1993). Cladistic concepts and the species problem in hominoid evolution. In *Species, Species Concepts, and Primate Evolution,* ed. W. H. Kimbel and L. B. Martin, pp. 345–371. New York: Plenum Press.

——— (1996). New fossil primates from the Miocene of China and their implications for the biogeography of early catarrhines in Eurasia. *Am. J. Phys. Anthropol. Suppl.* **22**:121.

Hill, A., and Odhiambo, I. (1987). New mandible of *Rangwapithecus* from Songhor, Kenya. *Am. J. Phys. Anthropol.* **72**:210.

Hopwood, A. T. (1933). Miocene primates from Kenya. *J. Linn. Soc. London, Zool.* **38**:437–464.

Ishida, H., and Pickford, M. (1997). A new late Miocene hominoid from Kenya: *Samburupithecus kiptalami* gen. et sp. nov. *C. R. Acad. Sci. Paris Earth Planetary Sci.* **326**: 823–829.

Ishida, H., Ishida, L., and Pickford, M. (1984). In *African Studies Monographs,* suppl. 3. Kyoto: Kyoto University Press.

Kay, R. F. (1977). Diets of early Miocene hominoids. *Nature* **268**:628–630.

Kay, R. F., and Ungar, P. S. (1996). Dental evidence for diet in some Miocene catarrhines with comments on the effects of phylogeny on the interpretation of adaptation. In *Miocene Hominoid Fossils: Functional and Phylogenetic Implications,* ed. D. Begun, C. Ward, and M. Rose, pp. 131–151. New York: Plenum Press

Kelley, J. (1986). Species recognition and sexual dimorphism in *Proconsul* and *Rangwapithecus. J. Human Evol.* **15**:461–495.

Larson, S. G. (1998). Parallel evolution in the hominoid trunk and forelimb. *Evol. Anthropol.* **6**:87–99.

Leakey, R. E., and Leakey, M. G. (1986a). A new Miocene hominoid from Kenya. *Nature* **324**:143–146.

——— (1986b). A second new Miocene hominoid from Kenya. *Nature (Lendon)* **324**:146–148.

——— (1987). A new Miocene small-bodied ape from Kenya. *J. Human Evol.* **16**:369–387.

Leakey, M. G., and Walker, A. (1997). *Afropithecus:* Function and phylogeny. In *Function, Phylogeny and Fossils: Miocene Hominoid Evolution and Adaptation,* ed. EDITORS, pp. 225–240. New York: Plenum Press.

Leakey, R. E., Leakey, M. G. and Walker, A. C. (1988). Morphology of *Afropithecus turkanensis* from Kenya. *Am. J. Phys. Anthropol.* **76**:289–307.

Leakey, M. G., Ungar, P. S., and Walker, A. (1995). A new genus of large primate from the late Oligocene of Lothidok, Turkana District, Kenya. *J. Human Evol.* **28**: 519–531.

LeGros Clark, W. E., and Leakey, L. S. B. (1951). The Miocene Hominoidea of East Africa. In *Fossil Mammals of Africa. Br. Mus. Nat. Hist.* **1**:1–117.

LeGros Clark, W. E., and Thomas, D. P. (1951). Associated jaws and limb bones of *Limnopithecus macinnesi.* In *Fossil Mammals of Africa. Br. Mus. Nat. Hist.* **3**:1–27.

Li, C.-K. (1978). A Miocene gibbon-like primate from Shihhung, Kiangsu Province. *Vertebr. Palasiat.* **16**: 187–192.

Madden, C. T. (1980). New *Proconsul* (*Xenopithecus*) from the Miocene of Kenya. *Primates* **21**:241–252.

McCrossin, M. L. (1992). An Oreopithecid proximal humerus from the middle Miocene of Maboko Island, Kenya. *Int. J. Primatol.* **13**(6):659–677.

McCrossin, M. L., and Benefit, B. R. (1993). Recently recovered *Kenyapithecus* mandible and its implications for great ape and human origins. *Proc. Natl. Acad. Sci. USA* **90**:1962–1966.

——— (1994) Maboko Island and the evolutionary history of Old World monkeys and apes. In *Integrative Paths to the Past: Paleoanthropological Advances in Honor of F. Clark Howell,* ed. R. S. Corruccini and R. L. Ciochon, pp. 95–122. Englewood Cliffs, N. J.: Prentice Hall.

——— (1997). On the relationships and adaptations of *Kenyapithecus,* a large-bodied hominoid from the middle Miocene of Eastern Africa. In *Function, Phylogeny and Fossils: Miocene Hominoid Evolution and Adaptation,* ed. D. Begun, C. Ward, and M. Rose, pp. 241–267. New York: Plenum Press.

McCrossin, M. L., Benefit, B. R., and Gitau, S. N. (1998). Functional and phylogenetic analysis of the distal radius of *Kenyapithecus,* with comments on the origin of the African great ape and human clade. *Am. J. Phys. Anthropol. Suppl.* **26**:158.

Napier, J. R., and Davis, P. R. (1959). The forelimb skeleton and associated remains of *Proconsul africanus.* In *Fossil Mammals of Africa. Br. Mus. Nat. Hist.* **16**:1–69.

Pickford, M. (1983). Sequence and environment of the lower and middle Miocene hominoids of western Kenya. In *New Interpretations of Ape and Human Ancestry,* ed. R. L. Ciochon and R. Corruccini, pp. 421–440. New York: Plenum Press.

——— (1986). Hominoids from the Miocene of East Africa and the phyletic position of *Kenyapithecus. Z. Morphol. Anthropol.* **76**:117–130.

Pilbeam, D. R. (1969). Tertiary Pongidae of East Africa: Evolutionary relationships and taxonomy. *Bull. Peabody Mus. Nat. Hist.* **31**:1–185.

Rae, T. C. (1993). *Phylogenetic Analysis of Proconsulid Facial Morphology.* Ph.D. Dissertation, State University of New York at Stony Brook.

——— (1997). The face of Miocene hominoids. In *Miocene Hominoid Fossils: Functional and Phylogenetic Implications,* ed. D. Begun, C. Ward, and M. Rose, pp. 59–77. New York: Plenum Press.

Rose, M. D. (1983). Miocene hominoid postcranial morphology: Monkey-like, ape-like, neither, or both? In *New Interpretations of Ape and Human Ancestry,* ed. R. L. Ciochon and R. Corruccini, pp. 405–420. New York: Plenum Press.

——— (1988). Another look at the anthropoid elbow. *J. Human Evol.* **17**:193–224.

——— (1993). Locomotor anatomy of Miocene hominids. In *Postcranial Adaptation in Nonhuman Primates,* ed. D. L. Gebo, DeKalb: NIU Press, pp. 252–272.

Rose, M. D., Leakey, M. G., Leakey, R. E. F., and Walker A. C. (1992). Postcranial specimens of *Simiolus enjiessi* and other small apes from the Early Miocene of Lake Turkana, Kenya. *J. Human Evol.* **22**:171–237.

Ruff, C. B., Walker, A., Teaford, M. F. (1989). Body mass, sexual dimorphism and femoral proportions of *Proconsul* from Rusinga and Mfangano Islands, Kenya. *J. Human Evol.* **18**:515–536.

Sanders, W. J., and Bodenbender, B. E. (1994) Morphometric analysis of lumbar vertebra UMP 67–28: Implications for spinal function and phylogeny of the Miocene Moroto hominoid. *J. Human Evol.* **26**: 203–238.

Simons, E. L. (1967). The earliest apes. *Sci. Am.* **217**: 28–35.

——— (1987) New faces of *Aegyptopithecus,* early human forebear from the Oligocene of Egypt. *J. Human Evol.* **16**:273–290.

Simons, E. L., and Pilbeam, D. R. (1965). Preliminary revision of the Dryopithecinae (Pongidae, Anthropoidea). *Folia Primatol.* **3**:81–152.

Suteethorn, V., Buffetaut, E., Buffetaut-Tong, H., Ducrocq, S., Helmcke-Ingavat, R., Jaeger, J.-J., and Jongkanlanasoontorn, Y. (1990). A hominoid locality in the Middle Miocene of Thailand. *C. R. Acad. Sci. Paris* **311** (Ser. II):1449–1454.

Teaford, M. F., Beard, K. C., and Walker, A. (1988). New hominoid facial skeleton from the Early Miocene of Rusinga Island, Kenya, and its bearing on the relationship between *Proconsul nyanzae* and *Proconsul africanus. J. Human Evol.* **17**:461–477.

Ungar, P. S., and Kay, R. F. (1995) The dietary adaptations of European Miocene catarrhines. *Proc. Natl. Acad. Sci. USA* **92**:5479–5481.

Walker, A. (1997). *Proconsul:* Function and phylogeny. In *Function, Phylogeny and Fossils: Miocene Hominoid Evolution and Adaptation,* ed. D. R. Begun, C. V. Ward, and M. D. Rose, pp. 209–224. New York: Plenum Press.

Walker, A. C., and Pickford, M. (1983). New postcranial fossils of *Proconsul africanus* and *Proconsul nyanzae.* In *New Interpretations of Ape and Human Ancestry,* ed. R. L. Ciochon and R. Corruccini, pp. 325–352. New York: Plenum Press.

Walker, A., and Teaford, M. (1989). The hunt for *Proconsul. Sci. Am.* **260**:76–82.

Walker, A., Teaford, M. F., Martin, L., and Andrews, P. (1993). A new species of *Proconsul* from the early Miocene of Rusinga/Mfangano Islands, Kenya. *J. Human Evol.* **25**:43–56.

Ward, C. V. (1993). Torso morphology and locomotion in *Proconsul nyanzae. Am. J. Phys. Anthropol.* **92**(3): 291–328.

Ward, C. V., Walker, A., and Teaford, M. F. (1991). *Proconsul* did not have a tail. *J. Human Evol.* **21**:215–220.

Ward, C. V., Walter, A., Teaford, M. F., and Odhiambo, I. (1993). Partial Skeleton of *Proconsul nyanzae* from Mfangano Island, Kenya. *Am. J. Phys. Anthropol.* **90**: 77–111.

EURASIAN FOSSIL APES

Alapagut, B., Andrews, P., and Martin, L. (1990). Miocene paleoecology of Pasalar, Turkey. In *European Neogene Mammal Chronology,* ed. E. H. Lindsay, *et al.,* pp. 443–459. New York: Plenum Press.

Alapagut, B., Andrews, P., Fortelius, M., Kappleman, J., Temizsoy, I., Celebi, H., and Lidsay W. (1996). A new specimen of *Ankarapithecus meteai* from the Sinap Formation of central Anatolia. *Nature* **382**:349–351.

Andrews, P. J. (1983). The natural history of *Sivapithecus.* In *New Interpretations of Ape and Human Ancestry,* ed. R. L. Ciochon and R. Corruccini, pp. 441–464. New York: Plenum Press.

Andrews, P. J., and Cronin, J. E. (1982). The relationships of *Sivapithecus* and *Ramapithecus* and the evolution of the orangutan. *Nature (London)* **297**: 541–546.

Andrews, P., and Martin, L. (1987). Cladistic relationships of extant and fossil hominoids. *J. Human Evol.* **16**: 101–118.

Andrews, P. J., and Pilbeam, D. R. (1996). The nature of the evidence. *Nature* **379**:123–124.

Andrews, P. J., and Tekkaya, I. (1980). A revision of the Turkish Miocene hominoid *Sivapithecus meteai. Palaeontology* **23**:85.

Andrews, P., Harrison, T., Delson, E., Bernor, R., and Martin, L. (1996). Distribution and biochronology of European and Southwest Asian Miocene Catarrhines. In *Evolution of Neogene Continental Biotypes in Central Europe and the Eastern Mediterranean,* ed. R. Bernor and V. Fahlbusch, pp. 167–207. New York: Columbia University Press.

Ankel, F. (1965). Der Canalis Sacralis als Indikator fur die Lange der Caudelregion der Primaten. *Folia Primatol.* **3**:263–276.

Begun, D. R. (1988). Catarrhine phalanges from the late Miocene (Vallesian) of Rudabánya, Hungary. *J. Human Evol.* **17**:413–438.

______ (1989). A large Pliopithecine molar from Germany and some notes on the Pliopithecinae. *Folia Primatol.* **52**:156–166.

______ (1991). European Miocene catarrhine diversity. *J. Human Evol.* **20**:521–526.

______ (1992a). Phyletic diversity and locomotion in primitive European hominids. *Am. J. Phys. Anthropol.* **87**:311–340.

______ (1992b). *Dryopithecus crusafonti* sp. nov., a new Miocene hominoid species from Can Ponsic (Northeastern Spain). *Am. J. Phys. Anthropol.* **87**:291–309.

______ (1992c). Miocene fossil hominids and the chimp-human clade. *Science* **257**:1929–1933.

______ (1993). New catarrhine phalanges from Rudabánya (Northeastern Hungary) and the problem of parallelism and convergence in hominoid postcranial morphology. *J. Human Evol.* **24**:373–402.

______ (1995) Late Micocene European orang-utants, gorillas, humans, or none of the above? *J. Human Evol.* **29**:169–180.

Begun, D. R., and Kordos, L. (1997). Phyletic affinities and functional convergence in *Dryopithecus* and other Miocene and living hominoids. In *Function, Phylogeny and Fossils: Miocene Hominoid Evolution and Adaptation,* ed. D. R. Begun, C. V. Ward, and M. D. Rose, pp. 291–316. New York: Plenum Press.

Begun, D. R., Moyá-Sola, M., and Kohler, M. (1990). New Miocene hominoid specimens from Can Llobateres (Vallés Penedés, Spain) and their geological and paleoecological context. *J. Human Evol.* **19**:255–268.

Begun, D. R., Ward, C. V., and Rose, M. D., eds. (1997). *Function, Phylogeny, and Fossils: Miocene Hominoid Evolution and Adaptations.* In *Advances in Primatology.* New York: Plenum Press.

Benefit, B. R. and McCrossin, M. L. (1997). New fossil evidence on the relationships of *Nyanzapithecus* and *Oreopithecus Amer. J. Phys. Anthrop. Suppl.* **24**:74.

Bernor, R. L., Flynn, L. J., Harrison, T., Hussain, S. T., and Kelley, J. (1988). *Dionysopithecus* from southern Pakistan and the biochronology and biogeography of early Eurasian catarrhines. *J. Human Evol.* **17**: 339–358.

Ciochon, R. E., *et al.* (1990). Opal phytoliths found on the teeth of the extinct ape *Gigantopithecus blacki:* Implications for paleodietary studies. *Proc. Natl. Acad. Sci.* **87**:8120–8124.

Dean, D. and Delson, E. (1992). Second gorilla or third chimp? *Nature* **359**:676–677.

DeBonis, L., and Koufos, G. D. (1993). The face and the mandible of *Ouranopithecus macedoniensis:* Description of new specimens and comparisons. *J. Human Evol.* **24**: 469–491.

DeBonis, L. and Koufos, G. D. (1994). Our Ancestors' Ancestor: *Ouranopithecus* is a Greek link in Human Ancestry. *Evol. Anthropol.* **3**:75–83.

DeBonis, L., and Melentis, J. (1977). Les primates hominoides du Vellesian de Macedonia (Grece). Etude de la machoire infereure. *Geobios* **10**:849–885.

Fleagle, J. G. (1984). Are there any fossil gibbons? In *The Lesser Apes: Evolutionary and Behavioral Biology,* ed. D. J. Chivers, H. Preuschoft, N. Creel, and W. Brockelman, pp. 431–477. Edinburgh: Edinburgh University Press.

Ginsburg, L. (1975). Le Pliopitheque des faluns Helvetiens de la Touraine et de I'Anjou. *Coll. Int. Cent. Nat. Rech. Sci.* **218**:877–885.

______ (1986). Chronology of the European pliopithecids. In *Primate Evolution,* ed J. G. Else and P. C. Lee, pp. 47–58. Cambridge: Cambridge University Press.

Ginsburg, L., and Mein, P. (1980). *Crouzelia rhodanica,* nouvelle espece de primate Catarrhinien ei essai sur la position systematique des Pliopithecidae. *Bull. Mus. Hist. Nat., Paris,* pp. 57–85.

Gu, Y., and Lin, Y. (1983). First discovery of *Dryopithecus* in east China. *Acta Anthropol. Sinica* **2**(4):305–314.

Harrison, T. (1986). A reassessment of the phylogenetic relationships of *Oreopithecus bambolii* Gervais. *J. Human Evol.* **15**:541–584.

______ (1991). Some observations on the Miocene hominoids from Spain. *J. Human Evol.* **20**:515–520.

______ (1996). New fossil primates from the Miocene of China and their implications for the biogeography of early catarrhines in Eurasia. *Am. J. Phys. Anthropol. Suppl.* **22**:121.

Harrison, T., and Rook, L. (1997). Enigmatic anthropoid or misunderstood ape? The phylogenetic status of *Oreopithecus bamboli* reconsidered. In *Function, Phylogeny and Fossils: Miocene Hominoid Evolution and Adaptation,* ed. D. R. Begun, C. V. Ward, and M. D. Rose, pp.327–362. New York: Plenum Press.

Harrison, T., Delson, E., and Jian, G. (1991). A new species of *Pliopithecus* from the middle Miocene of China and its implications for early catarrhine zoogeography. *J. Human Evol.* **21**:329–361.

Huxley, T. H. (1863). *Evidence as to Man's Place in Nature.* London: Williams and Norgate.

Jungers, W. L. (1987). Body size and morphometric affinities of the appendicular skeleton in *Oreopithecus bambolii* (IGF 11778). *J. Human Evol.* **16**:445–456.

Kappelman, J., Kelley, J., Pilbeam, D., Sheikh, K. A., Ward, S., Anwar, M., Barry, J. C., Brown, B., Hake, P., Johnson, N. M., Raza, S. M., and Shah, S. M. I. (1991). The earliest occurrence of *Sivapithecus* from the middle Miocene Chinji Formation of Pakistan. *J. Human Evol.* **21**:61–73.

Kay, R. F., and Simons, E. L. (1983). A reassessment of the relationships between later Miocene and subsequent Hominoidea. In *New Interpretations of Ape and Human Ancestry,* ed. R. L. Ciochon and R. Corruccini, pp. 577–624. New York: Plenum Press.

Kay, R. F., and Ungar, P. S. (1996). Dental evidence for diet in some Miocene catarrhines with comments on the effects of phylogeny on the interpretation of adaptation. In *Miocene Hominoid Fossils; Functional and Phylogenetic Implications,* ed. D. Begun, C. Ward, and M. Rose, pp. 131–151. New York: Plenum Press.

Kelley, J. (1988). A new large species of *Sivapithecus* from the Siwaliks of Pakistan. *J. Human Evol.* **17**:305–324.

——— (1992). Evolution of apes. In *The Cambridge Encyclopedia of Human Evolution,* ed. S. Jones, R. Martin and D. Pilbeam, pp. 223–230. Cambridge: Cambridge University Press.

——— (1993). Taxonomic implications of sexual dimorphism in *Lufengpithecus.* In *Species, Species Concepts, and Primate Evolution,* ed. W. H. Kimbel and L. B. Martin, pp. 429–458. New York: Plenum Press.

Kelley, J., and Etler, D. (1989). Hominoid dental variability and species number at the late Miocene site of Lufeng, China. *Am. J. Primatol.* **18**:15–34.

Kelley, J., and Pilbeam, D. R. (1986). The dryopithecines: Taxonomy, comparative anatomy, and phylogeny of Miocene large hominoids. In *Comparative Primate Biology,* vol. 1: *Systematics, Evolution, and Anatomy,* ed. D. R. Swindler and J. Erwin, pp. 361–411. New York: Alan R. Liss.

Kelly, J., and Qinghua, X. (1991). Extreme sexual dimorphism in a Miocene hominoid. *Nature* **352**:151–153.

Kelley, J., Anwar, M., McCollum, M. A., and Ward, S. C. (1995). The anterior dentition of *Sivapithecus parvada,* with comments on the phylogenetic significance of incisor heteromorphy in Hominoidea. *J. Human Evol.* **28**:503–517.

Kondopoulou, D., Sen, S., Koufos, G. D., and deBonis, L. (1992). Magneto-and biostratigraphy of the late Miocene mammalian locality of Prochoma (Macedonia, Greece). *Paleont. Evol.* **24–25**:135–139.

Koufos, G. D. (1995). The first female maxilla of the hominoid *Ouranopithecus macedoniensis* from the late Miocene of Macedonia, Greece. *J. Human Evol.* **29**: 385–399.

Larson, S. G. (1992). Parallel evolution of the ape forelimb. *Evol. Anthropol.* **6**.

Lipson, S., and Pilbeam, D. R. (1982). *Ramapithecus* and hominoid evolution. *J. Human Evol.* **11**:545–548.

Martin, L. (1985). Significance of enamel thickness in hominid evolution. *Nature (London)* **314**:260–263.

Martin, L., and Andrews, P. (1993). Renaissance of Europe's ape. *Nature* **365**:494.

Mein, P. (1986). Chronological succession of hominoids in the European Neogene. In *Primate Evolution,* ed. J. G. Else and P. C. Lee, pp. 59–70. Cambridge: Cambridge University Press.

Morbeck, M. E. (1983). Miocene hominoid discoveries from Rudabanya: Implications from the postcranial skeleton. In *New Interpretations of Ape and Human Ancestry,* ed. R. L. Ciochon and R. Corruccini, pp. 369–404. New York: Plenum Press.

Moya-Sola, S., and Köhler, M. (1993). Recent discoveries of *Dryopithecus* shed new light on evolution of great apes. *Nature* **365**:543–545.

——— (1995). New partial cranium of *Dryopithecus* Lartet, 1863 (Hominoidea, Primates) from the upper Miocene of Can Llobateres, Barcelona, Spain. *J. Human Evol.* **29**:101–140.

——— (1996). A *Dryopithecus* skeleton and the origins of great-ape locomotion. *Nature* **379**:156–159.

Pilbeam, D. R. (1996). Genetic and morphological records of the Hominoidea and hominid origins: A synthesis. *Mol. Phylogenet. Evol.* **5**:155–168.

Pilbeam, D. R., Meyer, G. E., Badgley, C., Rose, M. D., Pickford, M. H. L., Behrensmeyer, A. K., and Shah, S. M. I. (1977). New hominoid primates from the Siwaliks of Pakistan and their bearing on hominoid evolution. *Nature (London)* **270**:689–695.

Pilbeam, D., Rose, M. D., Barry, J. C., and Shah, S. M. I. (1990). New *Sivapithecus* humeri from Pakistan and the relationship of *Sivapithecus* and *Pongo. Nature* **348**: 237–239.

Runestad, J. A., Ruff, C. B., Neih, J. C., Thorington, R. W., Jr., and Teaford, M. F. (1993). Radiographic estimation of long bone cross-sectional geometric properties. *Am. J. Phys. Anthropol.* **90**:207–213.

Sarmiento, E. E. (1987). The phyletic position of *Oreopithecus* and its significance in the origin of the Hominoidea. *Am. Mus. Nov., no. 2881,* pp. 1–44.

Schwartz, J. H. (1986). *The Red Ape.* Boston: Houghton Mifflin.

——— (1997). *Lufengpithecus* and hominoid phylogeny. In *Function, Phylogeny and Fossils: Miocene Hominoid Evolution and Adaptation,* ed. D. R. Begun, C. V. Ward, and M. D. Rose, pp. 363–388. New York: Plenum Press.

Simons, E. L., and Ettel, P. C. (1970). *Gigantopithecus. Sci. Am.* **222**(1):76–85.

Simons, E. L., and Fleagle, J. G. (1973). The history of extinct gibbon-like primates. *Gibbon and Siamang* **2**: 121–148.

Simons, E. L., and Pilbeam, D. R. (1965). Preliminary revision of the Dryopithecinae (Pongidae, Anthropoidea). *Folia Primatol.* **3**:81–152.

Suteethorn, V., Buffetaut, E., Buffetaut-Tong, H., Ducrocq, S., Helmcke-Ingavat, R., Jaeger, J.-J., and

Jongkanlanasoontorn, Y. (1990). A hominoid locality in the middle Miocene of Thailand. *C. R. Acad. Sci. Paris* **311 (Ser. II)**:1449–1454.

Thenius, E. (1981). Bemerkungen zur taxonomischen und stammesgeschichtlichen Position der Gibbons (Hylobatidae, Primates). *Z. Saugetierkunde* **46**: 232–241.

Ungar, P. S., and Kay, R. F. (1995). The dietary adaptations of European Miocene catarrhines. *Proc. Natl. Acad. Sci.* **92**:5479–5481.

von Koenigswald, G. H. R. (1983). The significance of hitherto undescribed Miocene hominoids from the Siwaliks of Pakistan in the Senchenberg Museum, Frankfurt. In *New Interpretations of Ape and Human Ancestry,* ed. R. L. Ciochon and R. S. Corruccini, pp. 517–526. New York: Plenum Press.

Ward, S. (1997). The taxonomy and phylogenetic relationships of *Sivapithecus* revisited. In *Function, Phylogeny and Fossils: Miocene Hominoid Evolution and Adaptation,* ed, D. R. Begun, C. V. Ward, and M. D. Rose, pp. 269–290. New York: Plenum Press.

Ward, S., and Brown, B. (1986). The facial skeleton of Sivapithecus indicus. In *Comparative Primate Biology,* vol. 1: *Systematics, Evolution, and Anatomy,* ed. D. R. Swindler and J. Erwin, pp. 413–452. New York. Alan R. Liss.

Ward, S. C., and Kimbel, W. H. (1983). Subnasal alveolar morphology and the systematic position of *Sivapithecus. Am. J. Phys. Anthropol.* **61**:157–171.

Ward, S. C., and Pilbeam, D. R. (1983). Maxillofacial morphology of Miocene hominoids from Africa and Indo-Pakistan. In *New Interpretations of Ape and Human Ancestry,* ed. R. L. Ciochon and R. Corruccini, pp. 211–238. New York: Plenum Press.

Welcomme, J-L., Aguilar, J-P., and Ginsburg, L. (1991). Découverte d'un nouveau Pliopithèque (Primates, mammalia) associé à des rongeurs dans les sables du Miocène supérieur de Priay (Ain, France) et remarques sur la paléogéographie de la Bresse au Vallésien. *C. R. Acad. Sci. Paris* **313**:723–729.

White, T. D. (1975). Geomorphology to paleoecology: *Gigantopithecus* reappraised. *J. Human Evol.* **4**:219–233.

White, T. D., Suwa, G., and Asfaw, B. (1994). *Australopithecus ramidus,* a new species of early hominid from Aramis, Ethiopia. *Nature* **371**:306–312.

Wolpoff, M. H. (1983). *Ramapithecus* and human origins: An anthropologist's perspective of changing interpretations. In *New Interpretations of Ape and Human Ancestry,* ed. R. L. Ciochon and R. Corruccini, pp. 651–676. New York: Plenum Press.

Wu, R. (1983). Hominid fossils from China and their bearing on human evolution. *Can. J. Anthropol.* **3**(2): 207–214.

——— (1987). A revision of the classification of the Lufeng great apes. *Acta Anthropol. Sinica* **6**:265–271.

Wu, R., and Pan, Y. (1984). A late Miocene gibbon-like primate from Lufeng, Yunnan Province. *Acta Anthropol. Sinica* **3**:193–200.

——— (1985). Preliminary observation on the cranium of *Laccopithecus robustus* from Lufeng, Yunnan with reference to its phylogenetic relationship. *Acta Anthropol. Sinica* **4**:7–12.

Wu, R., Xu, Q., and Lu, Q. (1983). Morphological features of *Ramapithecus* and *Sivapithecus* and their phylogenetic relationships—morphology and comparisons of the cranium. *Acta Anthropol. Sinica* **2**(1):1–10.

——— (1985). Morphological features of *Ramapithecus* and *Sivapithecus* and their phylogenetic relationships—morphology and comparison of the teeth. *Acta Anthropol. Sinica* **4**:197–204.

Zapfe, H. (1958). The skeleton of *Pliopithecus* (*Epipliopithecus*) *vindobonensis* Zapfe and Hurzeler. *Am. J. Phys. Anthropol.* **16**:441–458.

——— (1960). Die Primatenfunde aus der Miozanen Spaltenfullung von Neudorf an der march (Devinzka nova ves), Tschechoslowakev. Mit. Anhang: Er Primatenfund aus dem Miozan von klein Hadersdorf in Niederoesterreich. *Schweiz. Pal. Abh.* **78**:4–293.

——— (1961). Ein primaten Fun aus der Miozanen Molasse von Oberosterreich. *Z. Morphol. Anthropol.* **51**(3): 247–267.

EVOLUTION OF LIVING HOMINOIDS

Andrews, P. (1986). Fossil evidence on human origins and dispersal. *Cold Spring Harbor Symposia on Quantitative Biology* **51**:419–428.

Andrews, P., and Martin, L. (1987). Cladistic relationships of extant and fossil hominoids. *J. Human Evol.* **16**: 101–118.

Andrews, P. J., and Pilbeam, D. R. (1996). The nature of the evidence *Nature* **379**:123–124.

Andrews, P., Harrison, T., Delson, E., Bernor, R., and Martin, L. (1996). Distribution and biochronology of European and Southwest Asian Miocene catarrhines. In *Evolution of Neogene Continental Biotypes in Central Europe and the Eastern Mediterranean,* ed. R. Bernor and V. Fahlbusch, pp. 167–207. New York: Columbia University Press.

Ciochon, R. L. (1983). Hominoid cladistics and the ancestry of modern apes and humans: A summary statement. In *New Interpretations of Ape and Human Ancestry,* ed. R. J. Ciochon and R. Corruccini, pp. 781–843. New York: Plenum Press.

Conroy , G. C., Senut, B., Gommery, D., Pickford, M., and Mein, P. (1992). Brief communication: New primate remains from the Miocene of Namibia, southern Africa. *Am. J. Phys. Anthropol.* **99**:487–492.

Fleagle, J. G. (1976). Locomotion and posture of the Malayan siamang and implications for hominoid evolution. *Folia Primatol.* **26**:245–269.

Gebo, D. L., MacLatchy, L., Kitio, R., Deino, A., Kingston, J., and Pilbeam, D. (1997). A new hominoid genus from the Uganda early Miocene. *Science* **276**:401–404.

Pilbeam, D. R. (1996). Genetic and morphological records of the Hominoidea and hominid origins: A synthesis. *Mol. Phylogenet. Evol.* **5**:155–168.

Ward, S., and Brown, B. (1986). The facial skeleton of *Sivapithecus indicus.* In *Comparative Primate Biology,* vol. 1: *Systematics, Evolution, and Anatomy,* ed. D. R. Swindler and J. Erwin, pp. 413–452. New York. Alan R. Liss.

Evolution of Gibbons

Fleagle, J. G. (1976). Locomotion and posture of the Malayan siamang and implications for hominoid evolution. *Folia Primarol.* **26**:245–269.

——— (1984). Are there any fossil gibbons? In *The Lesser Apes: Evolutiomary and Behavioral Biology,* ed. D. J. Chivers, H. Preuschoft, N. Creel, and W. Brockelman, pp. 431–477. Edinburgh: Edinburgh University Press.

Pilbeam, D. R. (1996). Genetic and morphological records of the Hominoidea and hominid origins: A synthesis. *Mol. Phylogenet. Evol.* **5**:155–168.

Evolution of the Orangutan

Andrews, P. (1986). Fossil evidence on human origins and dispersal. *Cold Spring Harbor Symposia on Quantitative Biology* **51**:419–428.

Andrews, P. J., and Cronin, J. E. (1982). The relationships of *Sivapithecus* and *Ramapithecus* and the evolution or the orangutan. *Nature (London)* **297**:541–546.

Kappelman, J., Kelley, J., Pilbeam, D., Sheikh, K. A., Ward, S., Anwar, M., Barry, J. C., Brown, B., Hake, P., Johnson, N. M., Raza, S. M., and Shah, S. M. I. (1991). The earliest occurrence of *Sivapithecus* from the middle Miocene Chinji Formation of Pakistan. *J. Human Evol.* **21**:61–73.

Pilbeam, D. R. (1996). Genetic and morphological records of the Hominoidea and hominid origins: A synthesis. *Mol. Phylogenet. Evol.* **5**:155–168.

Pilbeam, D. R., Rose, M. D., Barry, J. C., and Shah, S. M. I. (1990). New *Sivapithecus* humeri from Pakistan and the relationship of *Sivapithecus* and *Pongo. Nature* **348**: 237–239.

Preuss, T. M. (1982). The face of *Sivapithecus indicus:* Description of a new, relatively complete specimen from the Siwaliks of Pakistan. *Folia Primatol.* **38**:141–157.

Ward, S. (1997). The taxonomy and phylogenetic relationships of *Sivapithecus* revisited. In *Function, Phylogeny and Fossils: Miocene Hominoid Evolution and Adaptation,* ed. D. R. Begun, C. V. Ward, and M. D. Rose, pp. 269–290. New York: Plenum Press.

Ward, S., and Brown, B. (1986). The facial skeleton of *Sivapithecus indicus.* In *Comparative Primate Biology,* vol. 1: *Systematics, Evolution, and Anatomy,* ed. D. R. Swindler and J. Erwin, pp. 413–452. New York. Alan R. Liss.

Evolution of African Apes

Begun, D. R. (1995). Late Micocene European orangutants, gorillas, humans, or none of the above? *J. Human Evol.* **29**:169–180.

Dean, D., and Delson, E. (1992). Second gorilla or third chimp? *Nature* **359**:676–677.

DeBonis, L., and Koufos, G. D. (1994). Our ancestors' ancestor: *Ouranopithecus* is a Greek link in human ancestry. *Evol. Anthropol.* **3**(3):75–83.

Ishida, H., Ishida, L., and Pickford, M. (1984). In *African Studies Monographs,* suppl. 3. Kyoto: Kyoto University Press.

Hominid Origins

Benefit, B. R., and McCrossin, M. L. (1995). Miocene hominoids and hominid origins. *Ann. Rev. Anthropol.* **24**:237–256.

Hill, A. (1994). Late Miocene and early Pliocene hominoids from Africa. In *Integrative Paths to the Past: Paleoanthropological Advances in Honor of F. Clark Howell,* ed. R. S. Corruccini and R. L. Ciochon, pp. 123–145. Englewood Cliffs, N. J.: Prentice Hall.

Kappleman, J., and Fleagle, J. G. (1995). Age of early hominids. *Nature* **376**:558–559.

Martin, L. B. (1986). Relationships among great apes and humans. In *Major Topics in Primate and Human Evolution,* ed. B. Wood, L. Martin, and P. Andrews, pp. 161–187. Cambridge: Cambridge University Press.

SIXTEEN

Fossil Old World Monkeys

CERCOPITHECOID EVOLUTION

Old World monkeys are the modern success story of catarrhine evolution. Although they first appear in the fossil record at approximately the same time as apes, the beginning of the Miocene, they are quite rare throughout that epoch, and the major radiation of the group appears to have taken place much later. From the Pliocene to the Recent, Old World monkeys have an extensive fossil record from Africa, Europe, and Asia, including many parts of the world where they are absent or rare today (Fig. 16.1). With the recent discovery of many new fossils from the time of the earliest Old World monkeys and from the period of their great diversity in the Plio-Pleistocene of Africa, China, and Southeast Asia, Old World monkey evolution has become one of the most exciting areas in primate evolution.

Victoriapithecids: The Earliest Old World Monkeys

The first record of cercopithecoid monkeys comes from early Miocene deposits in northern and eastern Africa. Monkeys are relatively uncommon from this time period; they are absent from many localities, and only a few, very similar species are known. Early monkeys are known primarily from teeth only; recovery of additional cranial and postcranial material may indicate greater taxonomic and adaptive diversity than is currently recognized. However, the absence of cercopithecoids from many early Miocene localities is a real phenomenon. At some early and middle Miocene localities, early monkeys are as common as fossil apes; at other localities in the same time period and geographical area, monkeys are absent. It has been suggested that early monkeys are more abundant in drier, more open habitats (Pickford and Senut, 1988), but the Wadi Moghara locality in Egypt was probably very wet. The fauna includes many water adapted species, including a duck (Miller, 1998).

Early Old World monkeys are much more primitive than extant cercopithecids and cannot be placed conveniently in either Colobinae or Cercopithecinae. Rather, they form a separate family of more primitive monkeys, the victoriapithecids, which preceded the divergence of colobines and cercopithecines (Table 16.1). These early monkeys are generally placed in two genera: ***Prohylobates*** and *Victoriapithecus* (Fig. 16.2). *Prohylobates* is known from the early Miocene of Egypt and Libya. Fossils from northern Kenya and southern Ethiopia have also been referred to this genus (Leakey, 1985; Fleagle *et al.*, 1997). ***Victoriapithecus*** is known primarily from the middle Miocene site of Maboko

FIGURE 16.1 The modern Old World, showing fossil monkey localities from the Miocene, Pliocene, and Pleistocene.

TABLE 16.1
Infraorder Catarrhini
Family Victoriapithecidae
Subfamily VICTORIAPITHECINAE

Species	Estimated Mass (g)
Prohylobates (early Miocene, North and East Africa)	
P. tandyi	7,000
P. simonsi	25,000
Victoriapithecus (?early to middle Miocene, Kenya)	
V. macinnesi	7,000

Island in Lake Victoria in Kenya. A single victoriapithecid tooth from the early Miocene site of Napak, Uganda, has not been assigned to any taxon. *Prohylobates* and *Victoriapithecus* are small to medium-size monkeys (4–25 kg). Like all later cercopithecoids, they have bilophodont lower molars, but their teeth are more primitive than those of later Old World monkeys and more like those of hominoids in several features. The upper molars frequently have a crista obliqua linking the metacone with the protocone, and the lower molars often have a small hypoconulid (Fig. 16.3).

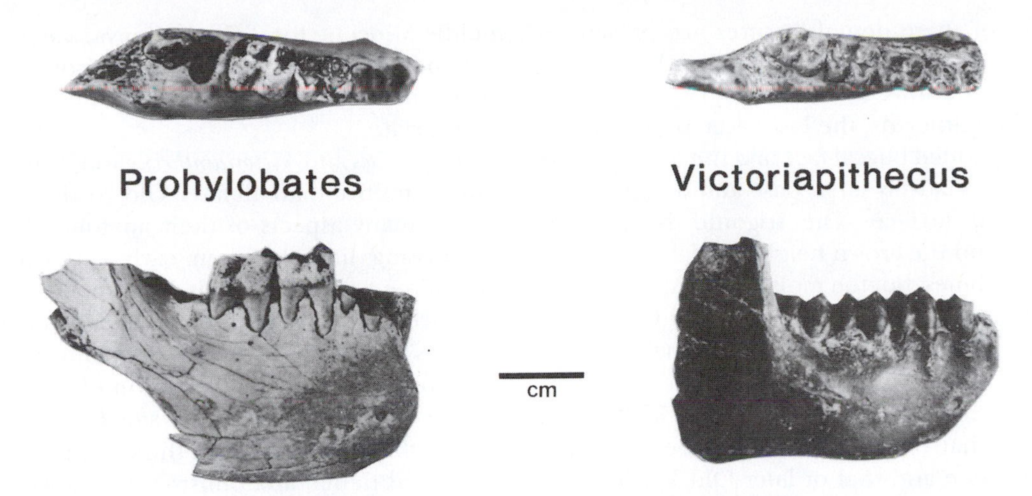

FIGURE 16.2 Lower jaws of cf. *Prohylobates* and *Victoriapithecus,* from the early and middle Miocene of Kenya (courtesy of M. G. Leakey).

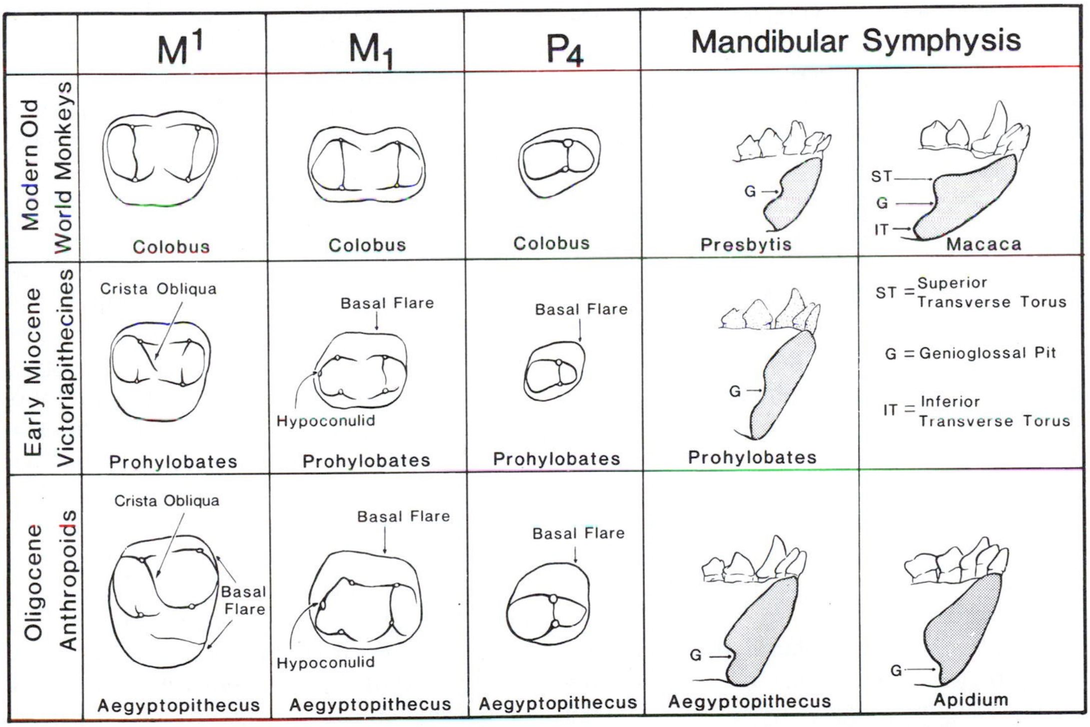

	M^1	M_1	P_4	Mandibular Symphysis	
Modern Old World Monkeys	Colobus	Colobus	Colobus	G; Presbytis	ST; G; IT; Macaca
Early Miocene Victoriapithecines	Crista Obliqua; Prohylobates	Basal Flare; Hypoconulid; Prohylobates	Basal Flare; Prohylobates	G; Prohylobates	ST = Superior Transverse Torus; G = Genioglossal Pit; IT = Inferior Transverse Torus
Oligocene Anthropoids	Crista Obliqua; Basal Flare; Aegyptopithecus	Basal Flare; Hypoconulid; Aegyptopithecus	Basal Flare; Aegyptopithecus	G; Aegyptopithecus	G; Apidium

FIGURE 16.3 Dental and mandibular features of Oligocene anthropoids, early cercopithecoids, and modern cercopithecoids, showing the intermediate morphological features of *Victoriapithecus* and *Prohylobates.*

Both of these dental features are present in primitive catarrhines and in apes but are absent in extant Old World monkeys. As in the propliopithecids, the last lower premolar has an expanded buccal face and the lower molars have a relatively large base and a constricted occlusal surface. The trigonid is relatively short and the crown height is relatively low, as in colobines, but the molar cusps are relatively low, as in cercopithecines. Like all Old World monkeys, *Victoriapithecus* has sexually dimorphic canines. Overall, the dentition of these basal Old World monkeys is intermediate between that of the early catarrhines from the Oligocene and that of later Old World monkeys. The mandible is relatively deep, and the symphysis resembles later Old World monkeys in the position of the genioglossal pit, but the mandible lacks other characteristic features of either colobines or cer-copithecines (Fig. 16.3; Benefit, 1993; Benefit and McCrossin, 1991, 1993; Leakey, 1985).

The cranial anatomy of *Victoriapithecus* (Fig. 16.4) shows a narrow interorbital pillar, a deep maxilla, tall orbits, and a small frontal trigon. *Victoriapithecus* had a relatively small brain compared with modern cercopithecoids. Overall the facial morphology of this early monkey resembles early catarrhines, such as *Aegyptopithecus* and living cercopithecines, and is strikingly different from that of colobines and gibbons (Benefit and McCrossin, 1993; 1997). Colobines and gibbons had previously been suggested to represent the ancestral cercopithecoid facial morphology.

Postcranial remains of *Victoriapithecus* show the narrow articulation in the distal end of the humerus and the narrow and deep ulnar notch characteristic of living cercopithecoids. The limb bones are most similar to those of a small cercopithecine, such as the vervet monkey, suggesting that *Victoriapithecus* was quadrupedal and adept on either arboreal or terrestrial substrates. The Maboko Island site, like most middle Miocene localities in Kenya, seems to have been an open woodland environment, where such locomotor abilities would be most appropriate.

Prohylobates and *Victoriapithecus* are distinctly more primitive than all later Old World monkeys in many aspects of their anatomy. They are missing links between early catarrhines and modern cercopithecids and provide clear evidence of the way in which characteristic features of both subfamilies of extant cercopithecoids evolved. The retention of a trigon on the upper molars and a small hypoconulid on the lower molars in these genera confirms what dental anatomists have known for years—that the bilophodont teeth of Old World monkeys are derived from an ancestor with more apelike teeth. Likewise, the cranial anatomy of *Victoriapithecus* shows many similarities to that of primitive catarrhines and to that of some other fossil monkeys (Benefit and McCrossin, 1997). The limb skeleton suggests that Old World monkeys were probably partly terrestrial from their origin (see also Strasser, 1988; Gebo, 1989).

Although we now know much about the early evolution of this group, the identity of their closest catarrhine relatives is less clear. In the past, many authorities have argued that parapithecids, especially *Parapithecus grangeri,* are ancestral cercopithecoids, on the basis of the bilophodont appearance of the lower molars in some species and some similarities in ankle structure. However, we now know that parapithecids are much more primitive than any other Old World anthropoids. If parapithecids are uniquely ancestral to Old World monkeys, then many characteristic catarrhine (and anthropoid) features must have evolved independently in Old World monkeys and apes. In addition, the genus with the most monkeylike molars, *Parapithecus,* lacks permanent incisors, precluding it from ancestry of any later catarrhine, and also lacks the ankle

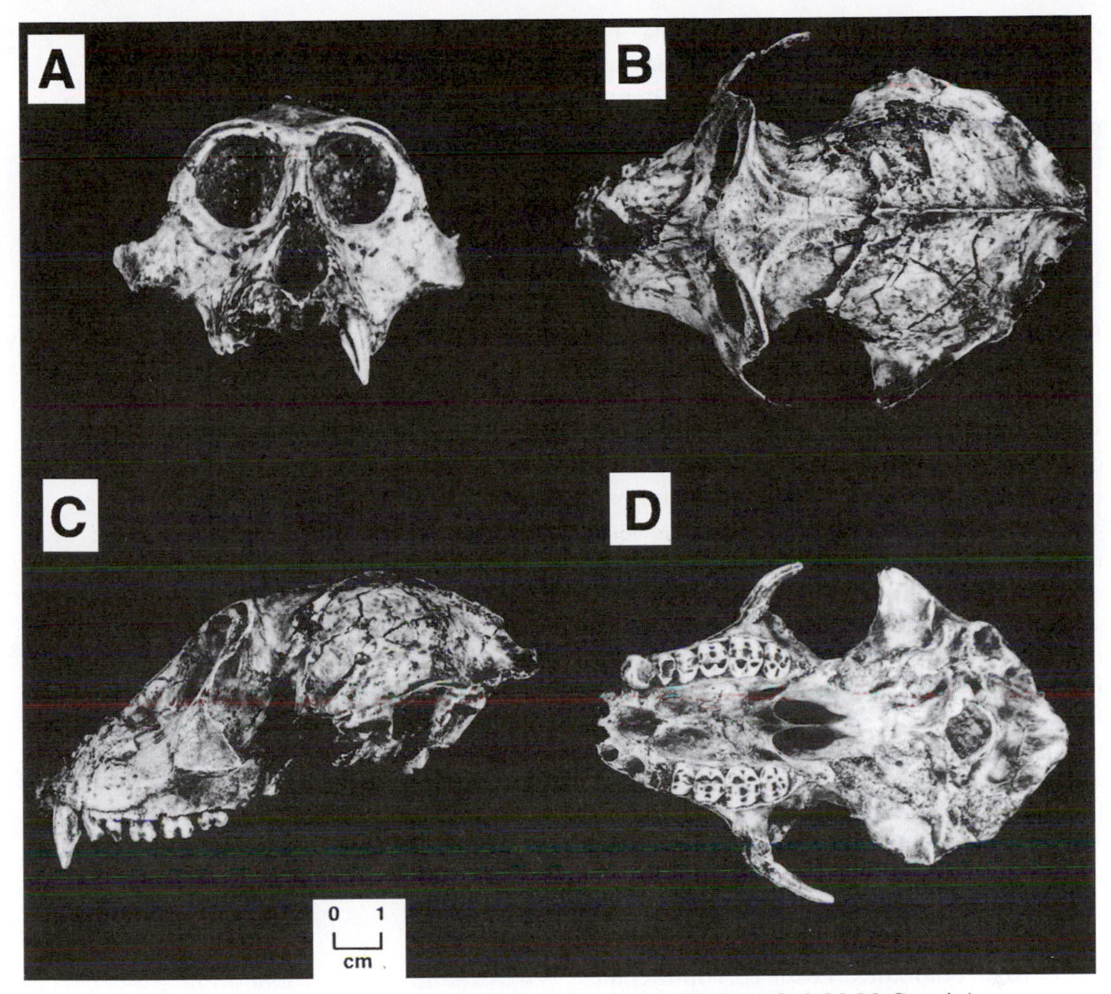

FIGURE 16.4 The skull of *Victoriapithecus* (courtesy of B. Benefit & M. McCrossin).

similarities to Old World monkeys. It therefore seems most unlikely that Old World monkeys evolved directly from parapithecids. Rather, Old World monkeys and apes were probably derived from an early catarrhine that was more advanced than the parapithecids and similar to either the propliopithecids from Egypt or the early proconsulids from East Africa.

Fossil Cercopithecids

After *Victoriapithecus* and *Prohylobates* in the early and middle Miocene there are very few fossil monkeys from the remainder of that epoch. However, approximately 10 million years later, in the latest Miocene and continuing through the Pliocene and Pleistocene, fossil monkeys are relatively abundant in deposits throughout Africa and Eurasia. This radiation of monkeys was, for the most part, the same one that dominates living higher primate communities today, most of the extinct forms can be readily grouped into the same subfamilies as living Old World monkeys.

Fossil Cercopithecines

Because many of the features that most clearly distinguish the living subfamilies of Old World monkeys are soft tissues, such as the sacculate stomachs of colobines or the cheek pouches of cercopithecines, there are potential hazards in assigning fossil monkeys to one family or another solely on the basis of dental and cranial remains. Nevertheless, extant cercopithecines can be distinguished from colobines by several dental and cranial features, including molars with long trigonids, higher crowns and relatively lower molar cusps, and skulls with longer snouts and narrower interorbital dimensions (Fig. 6.3). While these same dental and cranial features seem to sort fossil members of the two subfamilies, the postcranial differences that characterize the living taxa do not so readily distinguish fossil cercopithecids, except for the tendency of cercopithecines to have longer thumbs and shorter digits than colobines. Fossil cercopithecines (Table 16.2) can be readily divided into three major groups: macaques; mangabeys, baboons and geladas; and guenons.

TABLE 16.2
Infraorder Catarrhini
Family Cercopithecidae
Subfamily CERCOPITHECINAE

Species	Estimated Mass (g)
Macaca (latest Miocene to Recent, North Africa, Europe, Asia)	
M. sylvanus	13,000
M. prisca	12,000
M. majori	9,500
M. libyca	14,000
M. anderssoni	?
M. palaeindica	?
M. jiangchuanensis	?
Procynocephalus (Pliocene, Asia)	
P. wimani	22,000
P. subhimalayensis	23,000
Paradolichopithecus (Pliocene, Europe)	
P. arvernensis	23,000
P. sushkini	35,000
Papio (Plio-Pleistocene to Recent, Africa)	
P. robinsoni	
P. izodi	?
P. quadratirostris	?
Dinopithecus (Pleistocene, Africa)	
D. ingens	77,000
Cercocebus (Plio-Pleistocene, Africa)	
Parapapio (late Miocene to early Pleistocene, Africa)	
P. broomi	23,000
P. jonesi	19,000
P. whitei	30,000
P. antiquus	?
P. ado	17,000
Gorgopithecus (Pleistocene, southern Africa)	
G. major	41,000
Theropithecus (Plio-Pleistocene, Africa, ?Asia)	
(*Simopithecus*)	
T. oswaldi	96,000
T. delsoni	—
T. darti	?
(*Omopithecus*)	
T. brumpti	50,000
?T. baringensis	?
Cercopithecus (Pliocene to Recent, Africa)	
unnamed species	?

Macaques

The genus *Macaca* has the widest distribution of any nonhuman primate, extending from North Africa and Gibraltar in the west to Japan and the Philippines in Asia. Fossil macaques were even more widespread, especially in Europe and North Africa. Although they are quite abundant and widespread, most fossil macaques are strikingly similar to the extant genus, indicating that *Macaca* has retained a very conservative morphology over the last 5 million years (Delson and Rosenberger, 1984).

There are numerous, mostly dental, remains of macaques from latest Miocene or earliest Pliocene localities in Algeria, Libya, and Egypt. ***Macaca prisca,*** from the early Pliocene of southern France, is the earliest fossil cercopithecine in Europe. In the later Pliocene, macaques were widespread throughout much of North Africa and Europe (including Spain, France, Germany, Italy, the Netherlands, and Yugoslavia), and during the middle Pleistocene their range extended into Great Britain, southern Russia, and the Middle East. Most of these fossil populations cannot be distinguished in dental features from the living Barbary macaque, *M. sylvanus,* of Gibraltar and North Africa (Delson, 1980). The most distinctive fossil macaque, the Pliocene "dwarf macaque," ***Macaca majori,*** from the island of Sardinia, was about 5–10 percent smaller in dental dimensions than living species.

In Asia, the earliest macaque is ***Macaca palaeindica,*** from the late Pliocene of northern India and Pakistan. Although exact ages are not available for most specimens, it seems likely that the first appearance was between 3.0 and 2.0 million years ago, probably in association with a faunal change at 2.5 million years ago (Barry, 1987). Macaques were also relatively common throughout the Pleistocene of China and Southeast Asia. A larger species, ***Macaca anderssoni*** or ***M. robusta,*** is found in early and middle Pleistocene localities in northern China (including Zhoukoudian), and there are many younger, smaller remains from middle to late Pleistocene and Recent deposits in more southern, subtropical areas that are generally identified only as *Macaca sp.* There are also abundant remains of macaques from many late Pleistocene and Recent deposits in other parts of Asia, including Japan, Korea, Vietnam, Java, Sumatra, and Borneo. Many of these are similar to species currently living in the same region. For example, the fossil macaques from Vietnam have been attributed to three of the five species currently living in that country. A more precise biochronology of fossil macaques offers the possibility to test theories regarding the colonization of Asia by successive radiations of different species groups during the Pleistocene (Jablonski and Pan, 1988).

In addition to fossil representatives of the living *Macaca,* there are two genera of larger macaquelike monkeys from the late Pliocene and Pleistocene of Asia and Europe. ***Procynocephalus*** is a late Pliocene, Asian genus with one species from northern India and one from southern China. It has a macaquelike dentition and skull, and its baboonlike skeleton suggests that its locomotor habits resembled those of the more terrestrial macaques, such as *M. nemestrina.* ***Paradolichopithecus,*** a similar baboonlike macaque from the Pliocene of Europe, seems to have sexually dimorphic canines like most cercopithecines, but it lacks any evidence of dimorphism in the cheek teeth or skull.

Mangabeys, Baboons, and Geladas

Macaques are the only cercopithecines to colonize Europe and Asia successfully. The other members of the subfamily are known almost totally from sub-Saharan Africa, where they remain abundant today. Recent biomolecular studies have demonstrated that the phylogenetic relationships among mangabeys, baboons, mandrills, and geladas are much more complex than was previously realized (Disotell, 1996; Chapter 6). Mandrills are derived from one group of living mangabeys and baboons and geladas from another. Thus far there have been no attempts to verify these relationships in the fossil record, although mangabeys, baboons, and geladas all have fossil records from the Pliocene and Pleistocene of Africa.

Mangabeys ***Parapapio,*** from the late Miocene to early Pleistocene of eastern and southern

Africa, is the most primitive member of the baboon-mangabey group and is probably near the ancestry of both living genera. *Parapapio* is intermediate in size between mangabeys and savannah baboons, and it seems to have little sexual size dimorphism. It is known from numerous species. There are also numerous fragmentary remains attributed to *Cercocebus*, from the late Pliocene and early Pleistocene of eastern and southern Africa, that might ultimately bear on the phylogeny of this group.

Baboons Although the extant savannah baboons of the genus ***Papio*** are widespread and abundant in Africa today, the fossil record of this genus is very sparse and controversial. The oldest specimens attributed to *Papio* are from Laetoli at 3.7 million years ago, and dental and cranial remains attributed to *Papio* have been described from the Pliocene through much of the Pleistocene in eastern and southern Africa. However, many of these fossils may actually belong to other papionins, such as *Theropithecus* (discussed shortly) or *Parapapio*, and the genus *Papio* may actually be quite rare in the fossil record. ***Papio robinsoni***, from several South African cave deposits, and ***P. izodi***, from Taung, are recognized by virtually all authorities as members of the modern radiation of savannah baboons. However, there are several other large baboons from the past 3 million years in many parts of Africa that are linked with *Papio* by some authorities and with *Theropithecus* by others.

Papio quadratirostris is a large baboon from the Pliocene of Kenya and Ethiopia that is considered by some to be the earliest baboon (Delson and Dean, 1993) and by others as the earliest member of the *Theropithecus brumpti* lineage (Jablonski, 1993; see the next subsection). There are also fossil baboons from Angola placed in this taxon. ***Dinopithecus*** is a very large (70–80 kg), sexually dimorphic baboon known mainly from the Swartkrans cave deposits (early Pleistocene) of South Africa that is frequently grouped with *Papio quadratirostris* (Delson and Dean, 1993). There are no skeletal remains assigned to the genus.

Gorgopithecus is a medium-size (40 kg) baboon from South African Pleistocene deposits. *Gorgopithecus* seems to have little sexual dimorphism in the size of the cheek teeth, but otherwise it is probably very much like living savannah baboons.

Geladas *Theropithecus gelada*, from the Ethiopian highlands, is the only living representative of this group of large papionins that had an extensive radiation during the Pliocene and Pleistocene. Fossil *Theropithecus* remains have been recovered from many parts of Africa and also from India. Fossil species of *Theropithecus* resemble the living gelada in their complex cheek teeth, presumably related to a dietary specialization on grass blades, seeds, and tubers. *Theropithecus* is the only predominantly folivorous cercopithecine. Unlike folivorous colobines, however, the living *Theropithecus* exploits this dietary niche on the ground by specializing on grass. All *Theropithecus* have long forelimbs and short phalanges, indicating terrestrial quadrupedalism. The extinct species seem to have the same digital proportions as extant geladas, with relatively long thumbs compared with the size of the index finger, and were probably manual foragers. The extinct species are all much larger than the living gelada, and they show many more extreme dental, cranial, and skeletal specializations.

Fossil remains of *Theropithecus* are known from the early Pliocene, with some unpublished remains reported from slightly earlier, and they extend throughout most of the Plio-Pleistocene. There is a general consensus that *Theropithecus* can be divided into three long-lived lineages that have been separated since the beginning of the Pliocene. There is less agreement how many species should be recognized in each lineage and how the three are related to one another (Jablonski, 1993).

Theropithecus* (*Omopithecus*) *brumpti is an early species from the late Pliocene of East Africa that has a large anterior dentition (as in the living gelada) and extraordinary development of the zygomatic arches that must have given its face an extremely imposing appearance (Fig. 16.5). Its molars have the greatest development of shearing crests of any known cercopithecine, suggesting even more folivorous habits than the extant gelada (Benefit and McCrossin, 1990). This species has been recovered from deposits indicating

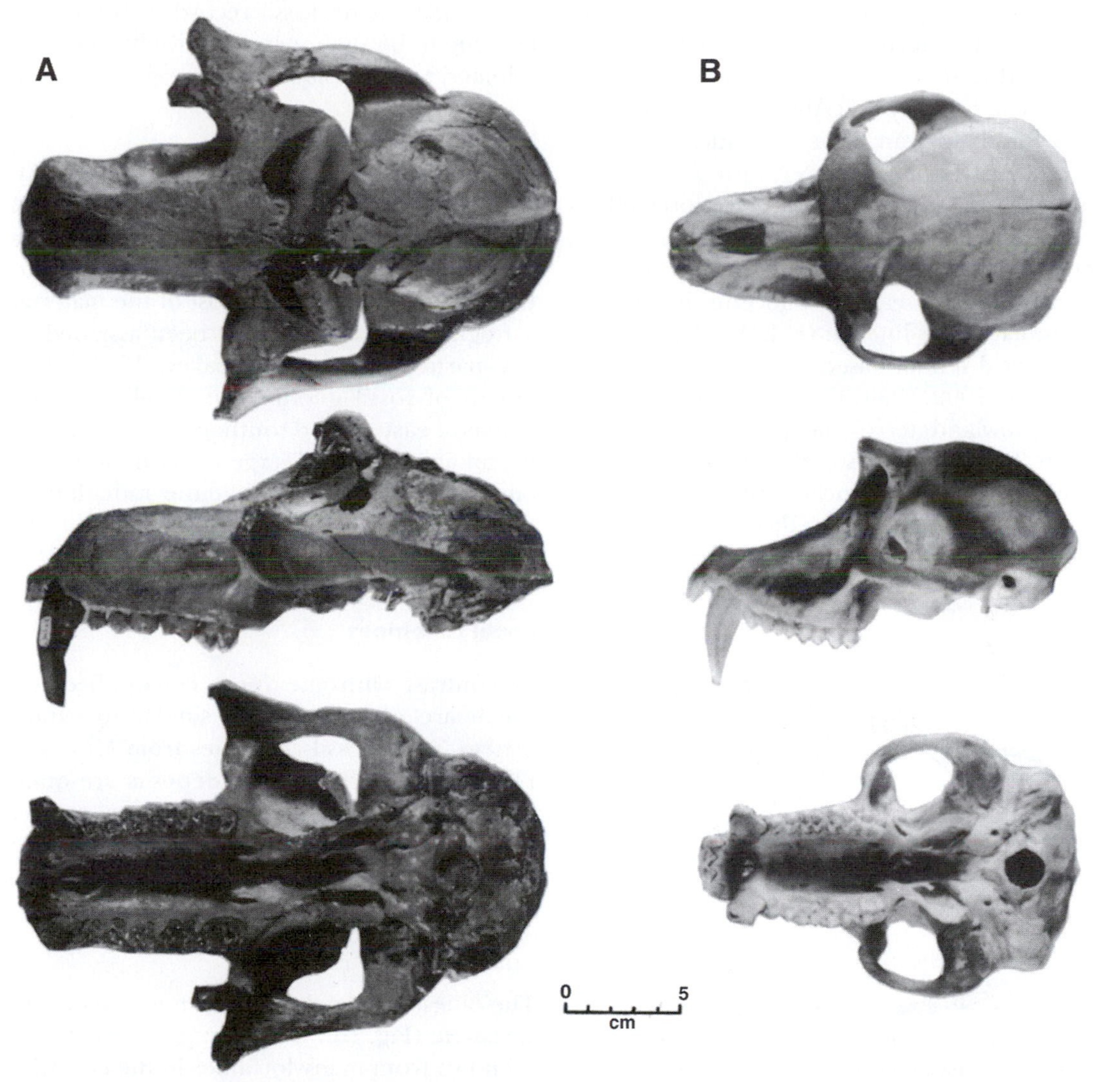

FIGURE 16.5 Skulls of (A) *Theropithecus brumpti* and (B) *Theropithecus gelada* (courtesy of Gerald Eck).

forested environments. Its limbs also show greater similarities to the limbs of the forest-living mandrills than do the limbs of other *Theropithecus* species or baboons (Ciochon, 1993). The early Pliocene ***T. baringensis*** is recognized by many authorities as a primitive member of this lineage, along with *Papio quadratirostris* (e.g., Jablonski, 1993, 1994).

Theropithecus* (*Simopithecus*) *oswaldi (with numerous temporal and geographical subspecies) was an enormous monkey that probably weighed as much as 100 kg. It was extremely abundant in many East African, North African, and South African Pliocene and Pleistocene sites (Fig. 16.6). Compared with *T. brumpti* it has greatly reduced, laterally compressed incisors and canines, large molar teeth, a short, deep face, and very long limbs. ***Theropithecus darti,*** from the earliest Pliocene, is an early member of this lineage (Eck, 1993). From an ecological model based on the living gelada, Dunbar (1992, 1993) has suggested that these large lowland terrestrial grazers must have been dependent on standing water and especially vulnerable to local extinction through climatic fluctuation, a prediction that accords well with their sporadic distribution in the fossil record, especially in South Africa (Foley, 1993; Lee and Foley, 1993).

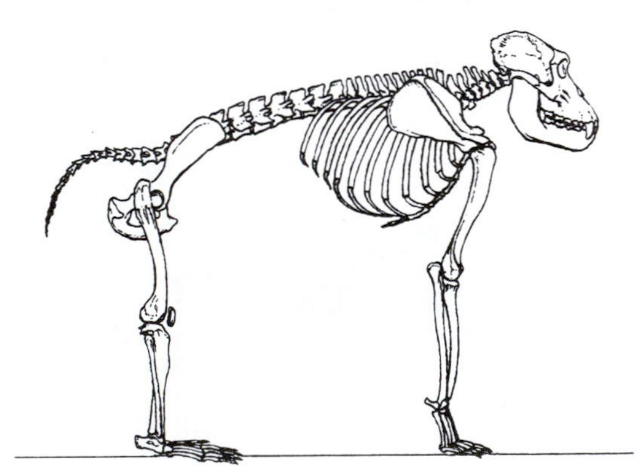

FIGURE 16.6 Skeleton of *Theropithecus oswaldi.* Scale: .05 × natural size

Theropithecus delsoni is a fossil gelada from the Pleistocene of northern India that seems related to the *Simopithecus* lineage. It is the only record of geladas outside Africa and indicates how much we do not know about the biogeography of this radiation.

The living gelada *Theropithecus* (*Theropithecus*) *gelada* has no fossil record. It is thought by some to be the most primitive lineage and is linked to the *Simopithecus* lineage by others.

Guenons

Despite their abundance in sub-Saharan Africa today, guenons are very rare in the fossil record. There are guenonlike teeth from several late Pliocene and Pleistocene localities in Kenya and Ethiopia, but most of the material is fragmentary and has not been assigned to any particular species (Leakey, 1988). The dearth of fossil guenons in Plio-Pleistocene faunas of eastern and southern Africa containing an abundance of large baboons and colobines probably means that the radiation of this group took place in other parts of the continent.

Fossil Colobines

In contrast with the fossil cercopithecines, which are often relatively similar to extant genera, many fossil colobines from Miocene, Pliocene, and Pleistocene deposits are quite different from any living taxa. These extinct colobines had both a broader geographic range and more diverse ecological adaptations than the extant species (Table 16.3).

European Colobines

The oldest fossil colobine from Eurasia is ***Mesopithecus*** (Fig. 16.7). This langur-size monkey is known from many localities in the late Mio-

TABLE 16.3
Infraorder Catarrhini
Family Cercopithecidae
Subfamily COLOBINAE

Species	Estimated Mass (g)
Mesopithecus (late Miocene to Pliocene, Europe, West Asia)	
M. pentelici	8,000
M. monspessulanus	5,000
Dolichopithecus (Pliocene, Europe)	
D. ruscinensis	18,000
D. hanaumani	—
cf. *Semnopithecus* (late Miocene to Recent, Asia)	
S. sivalensis	8,000
Presbytis (Pleistocene-Recent, Asia)	7,000
Trachypithecus (Pleistocene, Asia)	7,000
Rhinopithecus (e. Pleistocene-Recent, Asia)	
R. lantainensis	15,000
Colobus (late Miocene to Recent, Africa)	
many undescribed species	
C. flandrini	16,000
Libypithecus (late Miocene to Pliocene, North Africa)	
L. markgrafi	8,400
Microcolobus (late Miocene, Africa)	
M. tugenensis	4,000
Cercopithecoides (Pliocene, Africa)	
C. williamsi	33,000
C. kimeui	—
Paracolobus (Plio-Pleistocene, Africa)	
P. chemeroni	35,000
P. mutiwa	—
Rhinocolobus (Plio-Pleistocene, Africa)	
R. turkanensis	21,000

cene through Pliocene of southern and central Europe. The genus ranged as far west as England and as far east as Iran. There are two species, *M. pentelicus* (about 8 kg) and a younger, smaller species, *M. monspessulanus* (5 kg). *Mesopithecus* resembles living colobines in most dental and cranial features, including relatively small incisors, high-crowned cheek teeth, a deep mandible, a short face with large

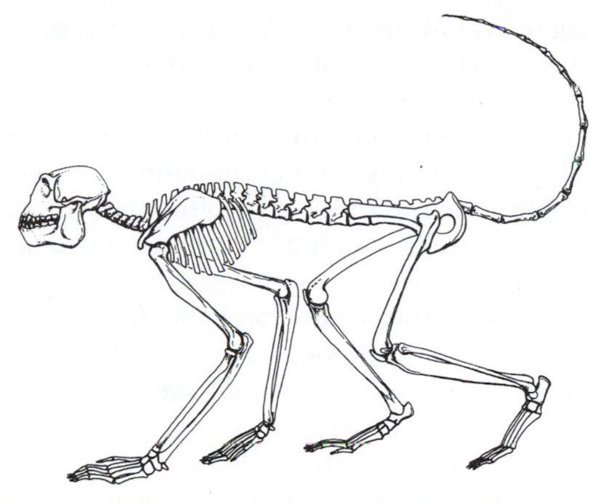

FIGURE 16.7 Skeleton of *Mesopithecus*. Scale: .15×

orbits, a narrow nasal opening, and a broad interorbital distance. It was probably a relatively folivorous monkey.

The limb skeleton of *Mesopithecus* resembles that of living colobines in having a relatively short thumb and a long tail. However, in the older species, *M. pentelicus,* the limbs are more robust than in those of most living colobines and the digits relatively shorter, suggesting that it was partly terrestrial, like the Hanuman langur of India. The localities that have yielded remains of this species seem to be characterized by woodland savannah environments (Delson, 1975). The later species, *M. monspessulanus,* is more like living colobines in its limb skeleton and also has been found in more wooded environments. Both species are sexually dimorphic and presumably lived in polygynous social groups.

Dolichopithecus is a larger Pliocene colobine that seems to be related to *Mesopithecus. Dolichopithecus* was previously known only from Europe, but several new colobines have recently been reported from northeastern Asia that seem to belong to this genus (Hasagawa, 1993; Delson, 1994). Dentally it is similar to *Mesopithecus,* but it has a longer snout and a larger overall size (15–20 kg). It also is sexually dimorphic in tooth and skull size.

In its skeleton, *Dolichopithecus* has more extensive adaptations for terrestrial quadrupedalism than any other colobine. Its limb proportions and many of its joint articulations are baboonlike, and it has short, stout phalanges. The genus seems to have been associated with humid forests and probably foraged on the forest floor, a habitus that would have separated it ecologically from the sympatric, more arboreal *Mesopithecus*.

It is not clear whether *Mesopithecus* and *Dolichopithecus* are related more closely to the living colobines of Africa or to those of Asia. There are few diagnostic features to link them unequivocally with either group, but their Eurasian distribution suggests affinities with the Asian langurs (Simons, 1970; Jablonski, 1998).

Asian Colobines

The fossil record of Asian colobines is relatively poor considering the diversity and abundance of leaf-eating monkeys on that continent today (Jablonski, 1993). The earliest Asian colobine is ?*Presbytis sivalensis* from late Miocene deposits about 6 million years old in the Siwaliks of Pakistan (Barry, 1987) However, there is an undescribed cranial specimen several million years older that may also belong to this species (Delson, 1994). The allocation of these early Asian colobines to the genus *Presbytis* is only a general identification, and they are probably more appropriately assigned to *Semnopithecus* or a new genus. The actual phyletic and taxonomic relationships to the modern Asian leaf monkeys have never been investigated carefully. Other langurs in the broadest sense are known from Pleistocene and Recent deposits in India, Burma, Vietnam, Java, Sumatra, and Borneo. *Rhinopithecus* is known from several Pleistocene localities in China (Jablonski and Pan, 1988; Jablonski, 1993). Understanding the radiation of Asian colobines in light of the geomorphological and climatic changes of that region during the past 6 million years is a major challenge for paleoanthropology (e. g., Jablonski, 1993; Brandon-Jones, 1996).

African Colobines

In contrast with Asia, which has a diverse colobine fauna today and a meager fossil record, Africa has a relatively low diversity of colobines today but an abundant record of fossil colobines, beginning in the late Miocene and extending into the Pliocene and Pleistocene. During this time there was an extensive radiation of African leaf-eating monkeys, many of which were unlike anything living today.

Microcolobus tugenensis was a small (about 4 kg) fossil colobine from the later Miocene of Kenya and one of the very few fossil monkeys from sub-Saharan Africa between 15 and 6 million years ago. It differs from later colobines and resembles *Mesopithecus* in having slightly lower molar cusps and more crushing surfaces on the lower premolars. It is also unusual among colobines in the shape of the mandibular symphysis. Both *Microcolobus* and *Mesopithecus* seem more primitive than all later colobines and probably preceded the modern radiations in Africa and Asia. In view of its small body size and less-developed shearing crests, it has been suggested that it was probably less folivorous than many later colobines.

Libypithecus markgrafi (Fig. 16.8), from Wadi Natrun in Egypt, was another small, late Miocene colobine. The species is known from a relatively complete skull and an isolated molar. The skull has a long snout, compared with most extant colobines, and well-developed sagittal and nuchal crests. In many aspects of its cranial morphology, *Libypithecus* resembles *Victoriapithecus* and probably retains the primitive cercopithecoid condition (Benefit and McCrossin, 1997). Its relationship to later colobines is less clear. Some authors have sug-

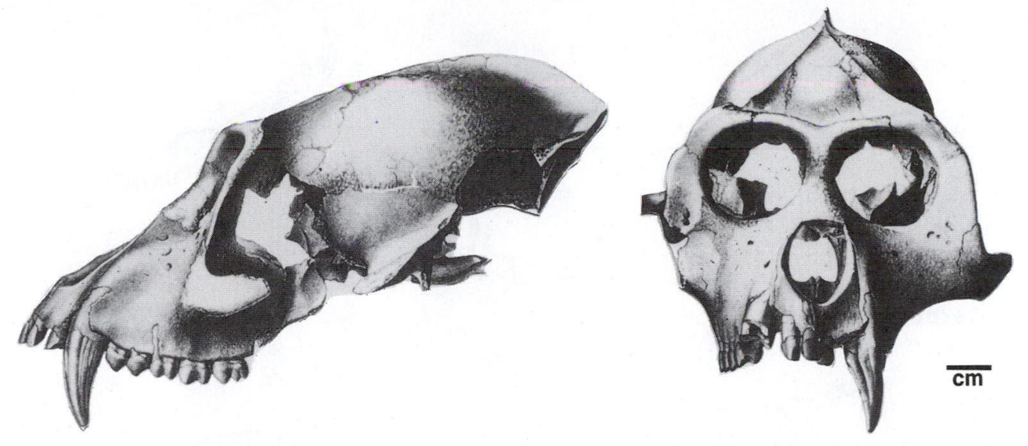

FIGURE 16.8 Skull of *Libypithecus markgrafi*, a Pliocene colobine from Egypt (from Stromer, 1913). Scale: approximately 0.5×

gested that it is closely allied with the European *Mesopithecus;* others have argued that it shows similarities to *Colobus* from sub-Saharan Africa. Because *Libypithecus* is known only from a skull, there is not suitable material for a direct comparison with *Microcolobus.*

From latest Miocene and early Pliocene deposits in Algeria, Libya, and Kenya there are a few fossils that are loosely assigned to the genus ***Colobus.*** Most of these monkeys are known only from isolated teeth or single jaws, and both their habits and their affinities with later forms are indeterminate at present. There are also many isolated teeth or jaws from the Pliocene and Pleistocene of East Africa that have been attributed to the living genus *Colobus* but have not been assigned to any particular species.

In addition to these *Colobus* fossils there is an impressive array of large extinct colobines from the Pliocene and earliest Pleistocene of southern and eastern Africa (Fig. 16.9). ***Cercopithecoides,*** from the Pliocene and Pleistocene, has two species: *C. williamsi* (about 15 kg), from both southern and eastern Africa, and the larger *C. kimeui,* from eastern Africa. Both have relatively broad molars and a short-snouted skull associated with a relatively shallow, cercopithecine-like mandible (Fig. 16.9). Aside from canine differences, *Cercopithecoides* shows no evidence of sexual dimorphism in either the dentition or the skull. In the larger species, the broad molars have an inflated, baboonlike appearance and are heavily worn on all of the individuals, suggesting a soft but perhaps gritty diet compared with that of most extant colobines.

The most striking adaptations of *Cercopithecoides* are in its limbs, which (if properly associated) resemble a terrestrial cercopithecine more than any other colobine. *Cercopithecoides* was presumably a terrestrial forager and was particularly common in grassland environments.

Paracolobus is the largest colobine known and probably weighed over 30 kg. There are several species from the Pliocene of eastern Africa. *Paracolobus* has a longer face and deeper jaw than *Cercopithecoides* (Fig. 16.9). Dentally it is similar to living colobines, sug-

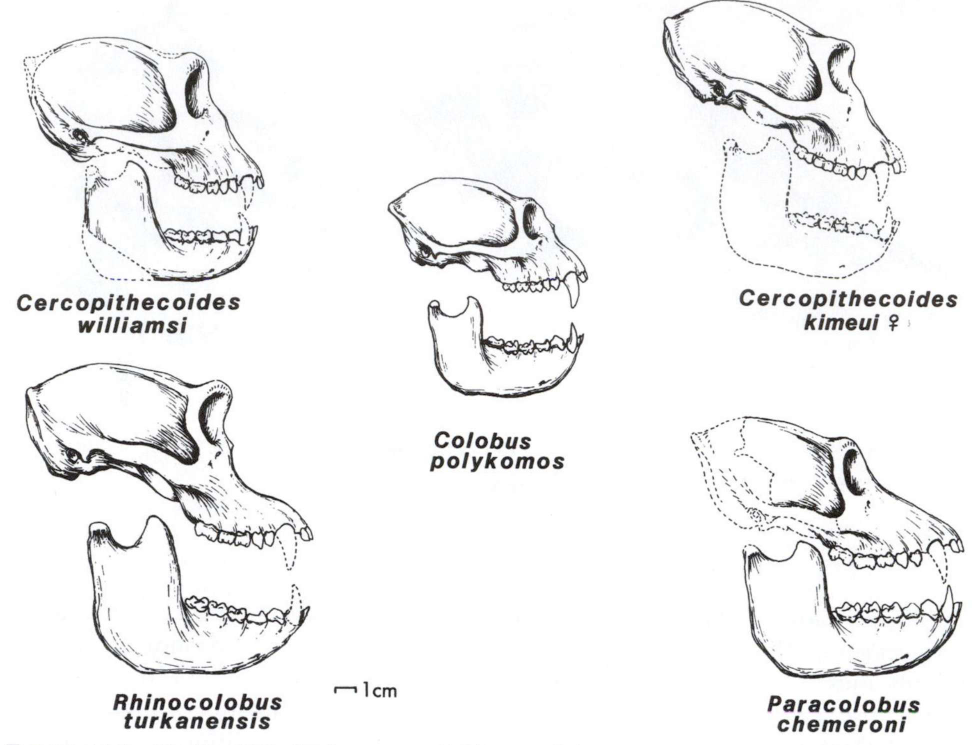

FIGURE 16.9 Skulls of Plio-Pleistocene colobines and the extant *Colobus polykomos.* Notice the greater size of the fossil monkeys.

gesting a largely folivorous diet. It has an intermembral index of 92, similar to that of the living proboscis monkey and red colobus. The skeleton indicates that *Paracolobus* was probably an arboreal quadruped.

Rhinocolobus turkanensis is another large monkey from the later Pliocene and early Pleistocene of eastern Africa. It was slightly smaller than *Cercopithecoides* or *Paracolobus* and probably weighed about 20 kg. As the name indicates, *Rhinocolobus* has a pronounced snout on its relatively deep face (Fig. 16.9). Its dentition indicates a folivorous diet, and the few skeletal remains suggest that it was an arboreal monkey. It was common in woodland and gallery forest environments.

SUMMARY OF FOSSIL CERCOPITHECOIDS

The fossil record of Old World monkeys is quite different from that of apes, the other major catarrhine group. For apes, we have abundant remains in the early and middle Miocene until about 10 million years ago and virtually nothing from the latest Miocene to Recent. In contrast, Old World monkeys have a moderate fossil record of the early victoriapithecids from the early and middle Miocene and increasing numbers of fossil monkeys in the latest Miocene through early Pleistocene. For apes, there are far more extinct genera and species than there are living taxa, and the many extinct species suggest diverse adaptive and phyletic radiations in the past that have left few if any descendants. In contrast, living monkeys far outnumber the extinct taxa, and many of the fossil monkeys seem to be part of the present-day radiation (Fig. 16.10).

Many authors have argued that the temporal pattern of change in the relative abundance of monkeys and apes during the last 20 million years (Fig. 16.11) indicates an ecological replacement of early apes by Old World monkeys. It is equally likely, however, that this apparent change in the primate fauna reflects climatic changes during the Miocene of Africa and Europe rather than simply competition between monkeys and apes in a stable environment (see also Kelley, 1998).

The earliest fossil monkeys, like the earliest fossil apes, provide evidence of intermediate stages in catarrhine evolution. The victoriapithecids and *Libypithecus* demonstrate that both colobines and cercopithecines preserve a mosaic of both primitive features from the earliest monkeys and also derived features unique to their respective subfamilies. The fossils expand our knowledge of Old World monkey evolution in several ways. They show that both colobines (*Mesopithecus* and *Dolichopithecus*) and cercopithecines (*Macaca* and *Paradolichopithecus*) ranged over much of Europe during the last 5 million years and that *Theropithecus* was once found in Asia. The fossil record also suggests that the arboreal nature of most living colobines has not characterized all members of that subfamily. Both *Dolichopithecus* and *Cercopithecoides* were very terrestrial colobines.

A particularly striking feature of the cercopithecoid fossil record is the size difference between extinct and living monkeys. Many extinct colobines and cercopithecines were

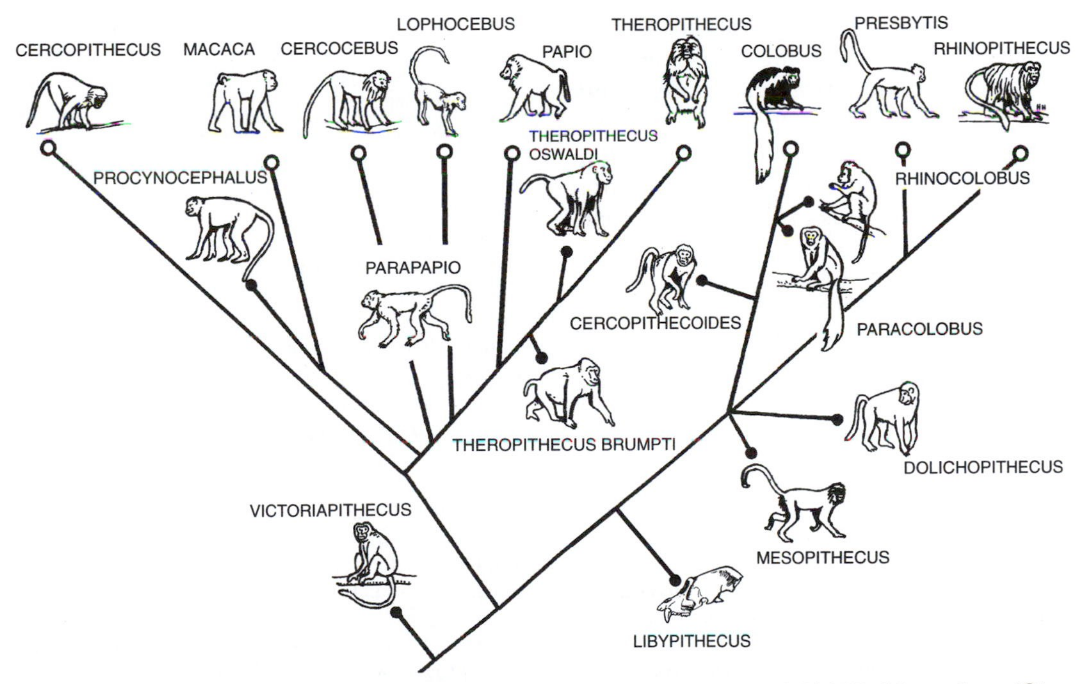

FIGURE 16.10 Cladogram of living Old World monkeys (O) and fossil Old World monkeys (●).

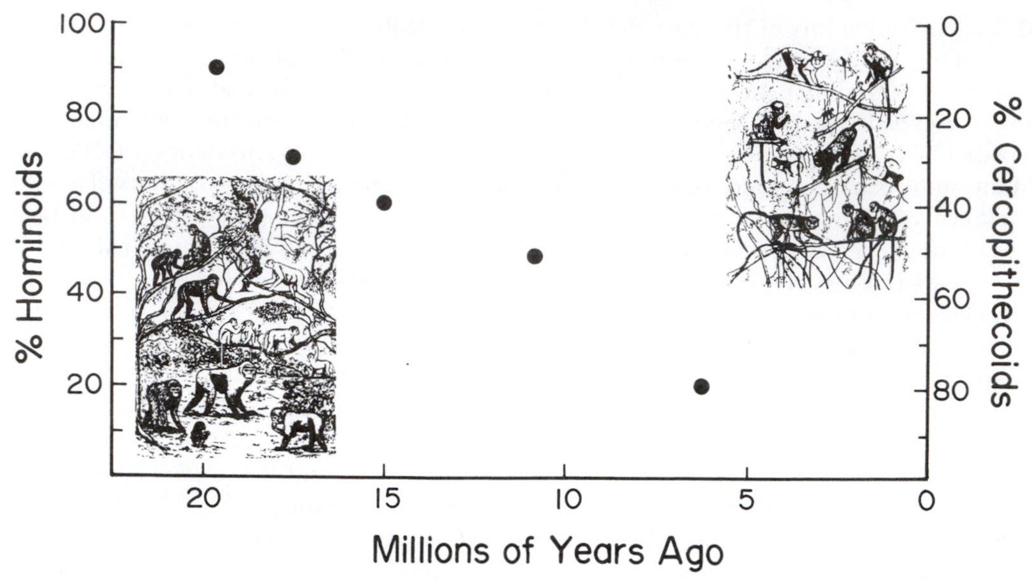

FIGURE 16.11 Relative species diversity of hominoids and cercopithecoids during the past 20 million years in Africa. The diversity of monkeys has increased as the diversity of hominoids has decreased (after Andrews, 1986).

larger than related living genera. Like the extant Malagasy fauna, the living cercopithecoids are the smaller genera from the Pliocene and early Pleistocene. This Pleistocene extinction of relatively large species is a common phenomenon around the world that cannot clearly be attributed exclusively to either climatic changes or hominid hunting (Martin and Klein, 1984).

BIBLIOGRAPHY

EARLIEST OLD WORLD MONKEYS

Benefit, B. R.(1993). The permanent dentition and phylogenetic position of *Victoriapithecus* from Maboko Island, Kenya. *J. Human Evol.* **25**:83–172.

Benefit, B. R., and McCrossin, M. L. (1991). Ancestral facial morphology of Old World higher primates. *Proc. Natl. Acad. Sci.* **88**:5267–5271.

——— (1993). Facial anatomy of *Victoriapithecus* and its relevance to the Ancestral Cranial Morphology of Old World Monkeys and apes. *Am. J. Phys. Anthropol.* **92**: 329–370.

——— (1997). Earliest known Old World monkey skull. *Nature* **388**:368–371.

Benefit, B. R., and Pickford, M. (1986). Miocene fossil cercopithecoids from Kenya. *Am. J. Phys. Anthropol.* **69**: 441–464.

Butler, P. M. (1986). Problems of dental evolution in the higher primates. In *Major Topics in Primate and Human Evolution,* ed. B. Wood, L. Martin, and P. Andrews, pp. 89–106. Cambridge: Cambridge University Press.

Delson, E. (1975). Evolutionary history of the Cercopithecidae. In *Approaches to Primate Paleobiology. Contributions to Primatology,* vol. 5, ed. F. S. Szalay, pp. 167–217. Basel, Switzerland: Karger.

——— (1979). *Prohylobates* (Primates) from the early Miocene of Libya: A new species and its implication for cercopithecid origin. *Geobios* **12**:725–733.

Fleagle, J. G., Bown, T. M., Harris, J. M., Watkins, R. W., and Leaket, M. G. (1997). Fossil monkeys from northern Kenya. *Am. J. Phys. Anthropol. Suppl.* **24**: 111.

Fourtau, R. (1918). *Contribution a l'etude des vertebres miocenes de l'Egypte.* Survey Dept., Ministry of Finance, Cairo.

Gebo, D. L. (1989). Locomotor and phylogenetic considerations in anthropoid evolution. *Am. J. Phys. Anthropol. J. Human Evol.* **18**:201–233.

Harrison, T. (1989). New postcranial remains of *Victoriapithecus* from the middle Miocene of Kenya. *J. Human Evol.* **18**:3–54.

Leakey, M. G. (1985). Early cercopithecids from Buluk, northern Kenya. *Folia Primatol.* **44**:1–17.

McCrossin, M. L., and Benefit, B. R. (1992). Comparative assessment of the ischial morphology of *Victoriapithecus macinnesi. Am. J. Phys. Anthropol.* **87**:277–290.

Miller, E. R. (1998). Faunal correlation of Wadi Moghara, Egypt—Implications for the age of *Prohylobates tandyi. J. Human Evol.*

Pickford, M., and Senut, B. (1988). Habitat and locomotion in Miocene cercopithecoids. In *A Primate Radiation: Evolutionary Biology of the African Guenons.*, ed. A. Gautier-Hion, F. Bourlère, J-P. Gautier, pp. 35–53. New York: Cambridge University Press.

Simons, E. L. (1969). Miocene monkey (*Prohylobates*) from north Egypt. *Nature (London)* **223**:687–689.

Simons, E. L., and Delson, E. (1978). Cercopithecidae and Parapithecidae. In *Evolution of African Mammals,* ed. V. J. Maglio and H. B. S. Cooke, pp. 100–119. Cambridge, Mass.: Harvard University Press.

Strasser, E. (1988). Pedal evidence for the origin and diversification of cercopithecid clades. *J. Human Evol.* **17**: 225–245.

von Koenigswald, G. H. R. (1969). Miocene Cercopithecoidea and Oreopithecoidea from the Miocene of East Africa. *Foss. Verts. Afr.* **1**:39–51.

FOSSIL CERCOPITHECINES

Fossil Macaques

Andrews, P., Harrison, T., Delson, E., Bernor, R. L., and Martin, L. B. (1997) Distribution and biochronology of European and Southwest Asian Miocene catarrhines. In *Evolution of Neogene Continental Biotypes in Central Euurope and the Eastern Mediterranean,* ed. R. Bernor, V. Fajlbusch, and H.-W. Mittman, pp. 168–207. New York: Columbia University Press.

Barry, J. C. (1987). The history and chronology of Siwalik cercopithecids. *Human Evol.* **2**(1):47–58.

Delson, E. (1980). Fossil macaques, phyletic relationships and a scenario of deployment. In *The Macaques—Studies in Ecology, Behavior and Evolution,* ed. D. G. Lindberg, pp. 10–30. New York: Van Nostrand.

Delson, E., and Rosenberger, A. L. (1984). Are there any anthropoid primate living fossils? In *Living Fossils,* ed. N. Eldridge and S. M. Stanley, pp. 50–61. New York: Springer Verlag.

Hooijer, D. A. (1962). Quaternary langurs and macaques from the Malay Archipelago. *Zool. Verhandl. Mus. Leiden* **55**:3–64.

______ (1963). Miocene mammalia of Congo. *Ann. Mus. Roy. Afr. Cent., ser. 8, Sci. Geol.* **46**:1–71.

Jablonski, N. G. (1993). Quaternary environments and the evolution of primates in East Asia, with notes on two new specimens of fossil cercopithecidae from China. *Folia Primatol.* **60**:118–132.

Jablonski, N. G., and Pan, Y. (1988). The evolution and palaeobiogeography of monkeys in China. In *The Palaeoenvironment of East Asia from the Mid-Tertiary,* vol. II: *Oceanography, Paleozoology and Paleoanthropology,* ed. E. K. Y. Chen, pp. 849–867. Hong Kong: Center of Asian Studies.

Nisbett, R. A., and Chiochon, R. L. (1993). Primates in Northern Viet Nam: A review of the ecology and conservation status of extant species, with notes on Pleistocene localities. *Int. J. Primatol.* **14**(5):765–795.

Pan, Y.-R., and Jablonski, N. G. (1987). The age and geographical distribution of fossil cercopithecids in China. *Human Evol.* **2**(1):59–69.

Fossil Mangabeys, Baboons, and Geladas

Benefit, B., and McCrossin, M. L . (1990). Diet, species diversity, and distribution of African Fossil baboons. *Kroeber Anthropological Papers* **71/72**:77–92.

Ciochon, R. L. (1993). *Evolution of the Cercopithecoid Forelimb.* Berkeley: University of California Press.

Cronin, J. E., and Meikle, W. E. (1982). Hominid and gelada baboon evolution: Agreement between molecular and fossil time scales. *Int. J. Primatol.* **3**(4):469–482.

Delson, E. *Theropithecus* fossils from Africa and India and the taxonomy of the genus. In *Theropithecus. The Rise and Fall of a Primate Genus,* ed. N. G. Jablonski, pp. 15–83. Cambridge: University of Cambridge Press.

Delson, E., and Dean, D. (1993). Are *Papio baringensis* R. Leakey, 1969, and *P. quadradirostris* Iwamoto 1982, species of *Papio* or *Theropithecus*? In *Theropithecus: The Rise and Fall of a Primate Genus,* ed. N. G. Jablonski, pp. 125–156. Cambridge: University of Cambridge Press.

Disotell, T. R. (1996). The phylogeny of Old World monkeys. *Evol. Anthropol.* **5**:18–24.

Dunbar, R. I. M. (1992). Behavioural ecology of the extinct papionines. *J. Human Evol.* **22**:407–421.

______ (1993). Socioecology of the extinct theropiths: A modelling approach. In *Theropithecus: The Rise and Fall of a Primate Genus,* ed. N. G. Jablonski, pp. 465–486. Cambridge: University of Cambridge Press.

Eck, G. (1977). Diversity and frequency distribution of Omo group Cercopithecoidea. *J. Human Evol.* **6**: 55–63.

Eck. G. G. (1993). *Theropithecus darti* from the Hadar Formation, Ethiopia. In *Theropithecus. The Rise and Fall of a Primate Genus,* ed. N. G. Jablonski, pp. 15–83. Cambridge: University of Cambridge Press.

Eck, G. G., and Jablonski, N. (1984). A reassessment of the taxonimic status and phyletic relationships of *Papio baringensis* and *Papio quadratirostris* (Primates: Cercopithecidae). *Am. J. Phys. Anthropol.* **65**:109–134.

Foley, R. (1993). African terrestrial primates: The comparative evolutionary biology of *Theropithecus* and the Hominidae. In *Theropithecus: The Rise and Fall of a Primate Genus,* ed. N. G. Jablonski, pp. 245–272.

Freedman, L. (1957). The fossil Cercopithecoidea of South Africa. *Ann. Transvaal Mus.* **23**:121–262.

——— (1965). Fossil and subfossil primates from the limestone deposits at Taung, Bolt's Farm and Witkrans, South Africa. *Paleontol. Afr.* **9**:19–48.

——— (1976). South African fossil Cercopithecoidea: A re-assessment including a description of new material from Makapandsgat, Sterkfontein and Taung. *J. Human Evol.* **5**:297–315.

Freedman, L., and Brain, C. K. (1972). Fossil cercopithecoid remains from the Kromdraai australopithecine site (Mammalia, Primates). *Ann. Transvaal Mus.* **28**(1): 1–17.

Jablonski, N. G.(ed) (1993a). *Theropithecus. The Rise and Fall of a Primate Genus.* Cambridge: University of Cambridge Press. 536 pp.

——— (1993b). The phylogeny of *Theropithecus.* In *Theropithecus: The Rise and Fall of a Primate Genus,* ed. N. G. Jablonski, pp. 209–224. Cambridge: University of Cambridge Press.

Jolly, C. J. (1967). The evolution of the baboons. In *The Baboon in Medical Research,* vol. 2, ed. H. Vagtborg, pp. 427–457. Austin: University of Texas Press.

——— (1970). The large African monkeys as an adaptive array. In *Old World Monkeys,* ed. J. P. Napier and P. H. Napier, pp. 141–174. New York: Academic Press.

——— (1972). The classification and natural history of *Theropithecus* (*Simopithecus*) (Andrews, 1916), baboons of the African Plio-Pleistocene. *Bull. Brit. Mus. Nat. Hist. Geol.* **22**:1–122.

Leakey, M. G. (1993). Evolution of *Theropithecus* in the Turkana Basin. In *Theropithecus: The Rise and Fall of a Primate Genus,* ed. N. G. Jablonski, pp. 85–123. Cambridge: University of Cambridge Press.

Leakey, M. G., and Leakey, R. E. F. (1973). Further evidence of *Simopithecus* (Mammalia, Primates) from Olduvi and Olorgesailie. *Foss. Verts. Afr.* **3**:101–120.

——— (1976). Further Cercopithecinae (Mammalia, Primates) from the Plio-Pleistocene of East Africa. *Foss. Verts. Afr.* **4**:121–146.

Lee, P. C., and Foley, R. A. (1993). Ecological energetics and extinction of giant gelada baboons. In *Theropithecus: The Rise and Fall of a Primate Genus,* ed. N. G. Jablonski, pp. 487–498. Cambridge: University of Cambridge Press.

Maier, W. (1970a). Neue Ergebnisse der Systematik und der Stammesge schichte der Cercopithecoidea. *Z. Saugerticrk.* **35**:193–217.

——— (1970b). New fossil Cercopithecoidea from the lower Pleistocene cave deposits of the Makapansgat limeworks. *South Africa. Paleontol. Afr.* **13**:69–108.

——— (1971a). The first complete skull of *Simopithecus darti* from Makapansgat, South Africa, and its systematic position. *J. Human Evol.* **1**:395–405.

——— (1971b). Two new skulls of *Parapapio antiquus* from Taung and a suggested phylogenetic arrangement of the genus *Parapapio. Ann. Sth. Afr. Mus.* **59**: 1–17.

McKee, J. K. (1993). Taxonomic and evolutionary affinities of *Papio izodi* fossils from Taung and Sterkfontein Paleonto. *Africana* **30**:43–49.

McKee, J. K., and Keyser, A. W. (1994). Cranio-dental remains of *Papio angusticeps* from the Haasgat Cave Site, South Africa. *Int. J. Primatol.* **15**:823–841.

Pickford, M. (1993). Climatic change, biogeography, and *Theropithecus.* In *Theropithecus: The Rise and Fall of a Primate Genus,* ed. N. G. Jablonski, pp. 227–243. Cambridge: University of Cambridge Press.

Shipman, P., Bosler, W., and Davis, K. L. (1981). Butchering of giant geladas at an Acheulian site. *Curr. Anthropol.* **22**(3):257–268.

Simons, E. L., and Delson, E. (1978). Cercopithecidae and Parapithecidae. In *Evolution of African Mammals,* ed. V. J. Maglio and H. B. S. Cooke, pp. 100–119. Cambridge, Mass.: Harvard University Press.

Fossil Guenons

Eck, G., and Howell, F. C. (1972). New fossil *Cercopithecus* material from the lower Omo Basin, Ethiopia. *Folia Primatol.* **13**:325–355.

Hamilton, A. C. (1988). Guenon evolution and forest history. In *A Primate Radiation: Evolutionary Biology of the African Guenons,* ed. A. Gautier-Hion, F. Bourliere, J.-P. Gautier, and J. Kingdon, pp. 13–34. Cambridge: Cambridge University Press.

Leakey, M. G. (1976). Cercopithecoidea of the East Rudolf succession. In *Earliest Man and Environment in the Lake Rudolf Basin,* ed. Y. Coppens, F. C. Howell, G. LI.

Isaac, and R. E. F. Leakey, pp. 345–350. Chicago: University of Chicago Press.

——— (1988). Fossil evidence for the evolution of the guenons. In *A Primate Radiation: Evolutionary Biology of the African Guenons,* ed. A. Gautier-Hion, F. Bourliere, J.-P. Gautier, and J. Kingdon, pp. 7–12. Cambridge: Cambridge University Press.

Simons, E. L. (1970). The deployment and history of Old World monkeys (Cercopithecidae, Primates). In *Old World Monkeys,* ed. J. R. Napier and P. H. Napier, pp. 97–137 New York: Academic Press.

Szalay, F. S., and Delson, E. (1979). *Evolutionary History of the Primates.* New York: Academic Press.

Zapfe, H. (1990). *Mesopithecus pentelicus Wagner aus dem Turolien von Pikermi bei Athen, Odontologie und Osteologie (Eine Dokumentation).* Wien, Austria: F. Berger.

FOSSIL COLOBINES

European Colobines

Andrews, P., Harrison, T., Delson, E., Bernor, R. L., and Martin, L. B. (1997). Distribution and biochronology of European and Southwest Asian Miocene catarrhines. In *Evolution of Neogene Continental Biotypes in Central Europe and the Eastern Mediterranean,* ed. R. Bernor and V. Fahlbusch, pp. 168–207 New York:Columbia University Press

Aquirre, E., and Soto, E. (1978). *Paradolichopithecus* in La Puebla de Valverde, Spain: Cercopithecoidea in European Neogene stratigraphy. *J. Human Evol* **7**:559–565.

DeBonis, L. , Bouvrain, G., Geraads, D, and Koufos, G. (1990). New remains of *Mesopithecus* (Primates, Cercopithecoidea) from the Late Miocene of Macedonia (Greece) with a description of a new species. *J. Vert. Paleont.* **10**:473–483.

Delson, E. (1975). Evolutionary history of the Cercopithecidae. In *Approaches to Primate Paleobiology. Contributions to Primatology,* vol. 5, ed. F. S. Szalay, pp. 167–217. Basel: Karger.

——— (1994). Evolutionary history of the colobine monkeys in paleoenvironmental perspective In *Colobine Monkeys: Their Ecology, Behaviour and Evolution,* ed. A. G. Davies and J. F. Oates, pp. 11–43. Cambridge: Cambridge University Press.

Hasagawa, Y. (1993). Japan's oldest 2,500,000-year-old monkey fossil discovery. *Kagaku Asahi (Monthly Journal of Science)* **53**(6):136–139.

Hohenegger, J., and Zapfe, H. (1990). Craniometric investigations on *Mesopithecus* in comparison with two recent colobines. *Beitr. Paläont. Österr.* **16**:111–143.

Jablonski, H. G. (1998). The Evolution of the Doucs and Snub-nosed Monkeys and the Question of the Phyletic Unity of the Odd-nosed Colobines. In *The Natural History of the Doucs and Snub-nosed Monkeys,* ed. N. G. Jablonski, pp. 13–41. Singapore: World Scientific.

Kalmykov, N. P., and Maschenko, E. N. (1992). The most northern representative of early Pliocene Cercopithecoidea from Asia. *Paleontologichesky Zhurnal (Moscow)* **2**: 136–138.

Asian Colobines

Barry, J. C. (1987). The history and chronology of Siwalik cercopithecids. *Human Evol.* **2**(1):47–58.

Brandon-Jones, D. (1996). The Asian Colobinae (Mammalia: Cercopithecidae) as indicators of Quaternary climatic change. *Biol. J. Linnean Soc.* **59**:327–350.

Delson. E. (1994). Evolutionary history of the colobine monkeys in paleoenvironmental perspective. In *Colobine Monkeys: Their Ecology, Behavior and Evolution,* ed. A. G. Davies and J. F. Oates, pp. 11–43. Cambridge: Cambridge University Press.

Gu, Y. M., and Hu, C. K. A fossil cranium of *Rhinopithecus* found in Xinan, Henan province. *Vert. Paleontol. Asia* **29**:55–58.

Hooijer, D. A. (1962). Quaternary langurs and macaques from the Malay Archipelago. *Zool. Verhandl. Mus. Leiden* **55**:3–64.

——— (1963). Prehistoric bone: The gibbons and monkeys of Niah Great Cave. *Sarawak Mus. J.* **10**:428–449.

Jablonski, N. G. (1993). Quaternary environments and the evolution of primates in East Asia, with notes on two new specimens of fossil cercopithecidae from China. *Folia Primatol.* **60**:118–132.

Jablonski, N. G., and Gu, Y. (1991). A reassessment of *Megamacaca lantianensis,* a large monkey from the Pleistocene of north-central China. *J. Human Evol.* **20**:51–66.

Jablonski, N., and Pan, Y. R. (1988). The evolution and paleobiogeography of monkeys in China. In *The Paleoenvironment of East Asia from the Mid-Tertiary,* ed. P. Whyte, J. Aigner, N. G. Jablonski, G. Taylor, D. Walker, and P. X. Wag, pp. 849–867. Hong Kong: Centre of Asian Studies.

Jablonski, N. G., and Tyler, D. E. (1994). The oldest fossil monkey from southeast Asia. *Am. J. Phys. Anthropol.* **18**: 113–117.

Matthew, W. D., and Granger, W. (1923). New fossil mammals from the Pliocene of Szechuan, China. *Bull. Am. Mus. Nat. Hist.* **48**:563–598.

African Colobines

Benefit, B. (1995). Earliest Old World monkey skull. *Am. J. Phys. Anthropol. Suppl.* **20**:64.

Birchette, M. G., Jr. (1982). The postcranial skeleton of *Parocolobus chemeroni*. Ph.D. Dissertation, Harvard University.

Freedman, L. (1957). The fossil Cercopithecoidea of South Africa. *Ann. Transvaal Mus.* **23**:121–262.

Leakey, M. G. (1976). Cercopithecoidea of the East Rudolf succession. In *Earliest Man and Environment in the Lake Rudolf Basin*, ed. Y. Coppens, F. C. Howell, G. Ll. Isaac, and R. E. F. Leakey, pp. 345–350. Chicago: University of Chicago Press.

——— (1982). Extinct large colobines from the Plio-Pleistocene of Africa. *Am. J. Phys. Anthropol.* **58**: 153–172.

Leakey, M. G., and Leakey, R. E. F. (1973). New large Pleistocene Colobinae from East Africa. *Foss. Verts. Afr.* **3**:121–138.

Leakey, R. E. F. (1969). New Cercopithecidae from the Chemeron beds of Lake Baringo, Kenya. *Fass. Verts. Afr.* **1**:53–69.

Simons, E. L., and Delson, E. (1978). Cercopithecidae and Parapithecidae. In *Evolution of African Mammals*, ed. V. J. Maglio and H. B. S. Cooke, pp. 100–119. Cambridge, Mass.: Harvard University Press.

FOSSIL CERCOPITHECOIDS: GENERAL

Andrews, P. (1981). Species diversity and diet in monkeys and apes during the Miocene. In *Aspects of Human Evolution*, ed. C. B. Stringer, pp. 25–61. London: Taylor and Francis.

Delson, E. (1984). Cercopithecid biochronology of the African Plio-Pleistocene :Correlation among eastern and southern hominid-bearing localities. *Courier Forschungs-Institut Senckenberg* **69**:199–218.

Delson, E., and Rosenberger, A. L. (1984). Are there any anthropoid primate living fossils? In *Living Fossils*, ed. N. Eldridge and S. M. Stanley, pp. 50–61. New York: Springer Verlag.

Disotell, T. R. (1996). The phylogeny of Old World monkeys. *Evol. Anthropol.* **5**:18–24.

Fleagle, J. G., and Kay, R. F. (1985). The paleobiology of catarrhines. In *Ancestors: The Hard Evidence*, ed. E. Delson, pp. 23–36. New York: Alan R. Liss.

Kelley, J. (1998). Noncompetitive replacement of apes by monkeys in the late Miocene of Eurasia. *Am. J. Phys. Anthrop. Suppl.* **26**:137–138.

Martin, P. S., and Klein, R. G., eds. (1984). *Quaternary Extinctions—A Prehistoric Revolution*. Tucson: University of Arizona Press.

SEVENTEEN

Hominids, the Bipedal Primates

PLIOCENE EPOCH

The short Pliocene epoch was a time of considerable faunal change in many parts of the world in association with major geographical and climatic events. The most significant tectonic event was the completion of the Panama land bridge between North and South America, which led to the exchange of faunas between those two previously separated continents. In the Old World, the Mediterranean Sea refilled at the beginning of the Pliocene after drying up in the late Miocene. In general, sea levels were higher and temperatures were warmer in the early Pliocene than in the late Miocene. In primate evolution, the Pliocene is characterized by two major events: the spread of cercopithecoid monkeys throughout many parts of the Old World (see Chapter 16) and the first appearance of hominids.

The separation of the lineages leading to the living African apes on the one hand and to humans on the other took place sometime in the late Miocene, between 10 million and 5 million years ago. There are, however, only a few hominoid fossils from this period, and those that have been recovered are so fragmentary that it is usually difficult to determine if they are apes or hominids (Hill, 1994). At present, the earliest unequivocal hominids come from the early Pliocene, between 4 million and 4.4 million years ago.

Australopithecines

The most primitive hominids, subfamily Australopithecinae, are divided into three genera: *Ardipithecus, Australopithecus,* and *Paranthropus.*

Ardipithecus

The single species of this genus, ***Ardipithecus ramidus,*** is from deposits approximately 4.4 million years old at the locality of Aramis near the Awash River in Ethiopia (White *et al.*, 1994, 1995). It has relatively smaller cheek teeth with thin enamel and larger canines than later hominids. Very little is actually known about this taxon at present, but informal reports suggest that it has more primitive skeletal morphology than any later hominid. Phylogenetic relationships of *Ardipithecus* among fossil and living hominoids are very uncertain, but it is most probably near the base of the hominid lineage.

Australopithecus

Australopithecus (southern ape), from central, eastern, and southern parts of Africa, is the best-known and most widespread genus of early hominid (Table 17.1). The earliest species, *A. anamensis* and *A. afarensis,* have been found in sites over 4 million years old, and some specimens of this genus persist until the late Pliocene, between 3 million and 2 million years ago (Figs. 17.1, 17.2). *Australopithecus* species have big teeth and small brains compared with modern humans. They were relatively short in stature, but their size range is much greater than that of any living human population and more comparable to our entire species, with estimated body weights between 30 kg (the size of an Ituri pygmy) for the smallest individuals and 85 kg (the size of a small college football player) for the largest. Most species were highly sexually dimorphic in body size.

Compared to living apes, all *Australopithecus* species have small incisors and canines relative to their body weight (Kay, 1985). The lower anterior premolar does not function as a sharpening blade for the upper canine. The molars of *Australopithecus* are large, and are characterized by thick enamel and bulbous cusps, features shared with some Miocene apes. The mandible is thick and has a high ascending ramus.

Cranially, *Australopithecus* is more apelike than humanlike in proportions, with a large face and relatively small brain. Details of brain morphology in *Australopithecus* have been debated since the first discovery of the genus. Our understanding of early hominid brain evolution is hindered by problems in estimat-

TABLE 17.1
Infraorder Catarrhini
Family HOMINIDAE

	Estimated Mass (g)	
	F	M
Subfamily Australopithecinae		
Ardipithecus (early Pliocene, Ethiopia)		
A. ramidus	—	—
Australopithecus (Pliocene, Africa)		
A. anamensis	—	—
A. afarensis	29,300	44,600
A. africanus	30,200	40,800
A. bahrelghazali	—	—
Paranthropus (late Pliocene to early Pleistocene, Africa)		
P. aethiopicus	38,000	38,000
P. boisei	34,000	48,600
P. robustus	31,900	40,200
Subfamily Homininae		
Homo (late to Recent, Cosmopolitan)		
H. habilis	31,500	51,600
H. rudolfensis	45,000	45,000
H. erectus	52,300	63,000
H. heidelbergensis	10,000	10,000
H. neanderthalensis	76,000	76,000
H. sapiens	60,000	60,000

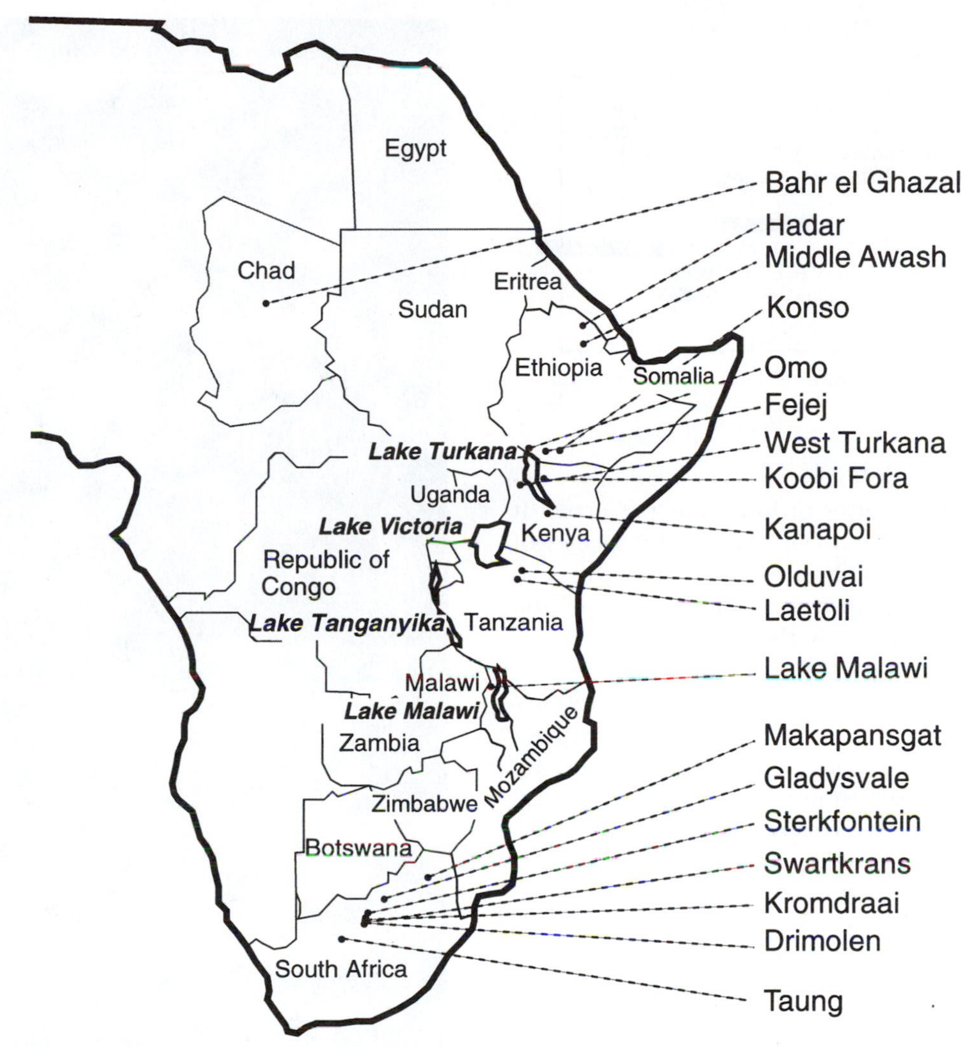

FIGURE 17.1 Map of sites yielding fossils of *Australopithecus, Paranthropus,* and early *Homo.*

ing the body size of the different species and by the lack of clear impressions on the internal surface of the cranium. In general, it appears that their brains were relatively larger than those of nonhuman primates but much smaller than those of later fossil hominids or living humans. In external morphology their brains are generally apelike, with few human features (Falk, 1991).

There are isolated skeletal elements for several species of *Australopithecus,* and a relatively complete associated skeleton for *A. afarensis* (Fig. 17.3). Like all later hominids, *Australopithecus* was bipedal. This is evident from many aspects of its skeleton, including the relatively long legs, the short, broad ilium, and the angulation of the knee joint. The reconstruction of bipedal habits for *Australopithecus* based on

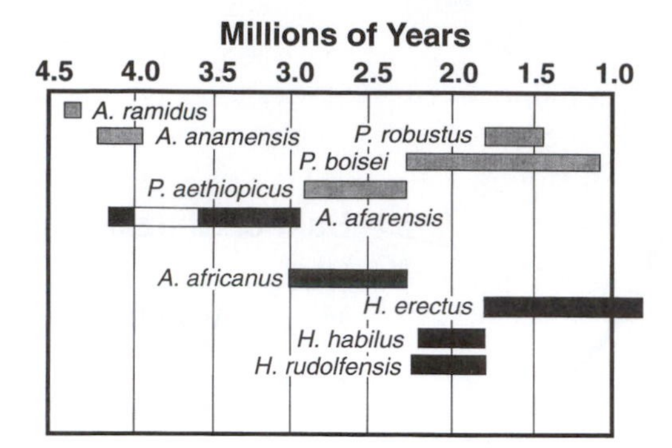

FIGURE 17.2 Temporal span of fossil hominids.

skeletal morphology was dramatically confirmed by a series of footprints preserved at Laetoli, Tanzania (Fig. 17.4). The species of *Australopithecus* show as many similarities to living apes as to humans in features of the skeleton, including the shoulder, the hand, the foot, and even details of the pelvis, femur, and tibia. The skeletal anatomy of *Australopithecus* is in many ways intermediate between that of living apes and humans, suggesting that these early hominids were both arboreal climbers and terrestrial bipeds (Figs. 17.5, 17.6). The diversity in locomotor abilities among *Australopithecus* species is difficult to determine because of a lack of skeletal material for some species and because of our weakness in creating models for interpreting a locomotor radiation of bipeds with only one extant analogue, ourselves. All species of *Australopithecus* are found associated with faunas suggesting wooded habitats (Reed, 1997).

The systematics and biogeography of *Australopithecus* species are complicated by the same factors that cause confusion in the systematics of most other groups of fossil primates: inadequate dating of sites, fragmentary remains, sexual dimorphism, and differing taxonomic philosophies. Four species of *Australopithecus* have been recognized, and more are probably

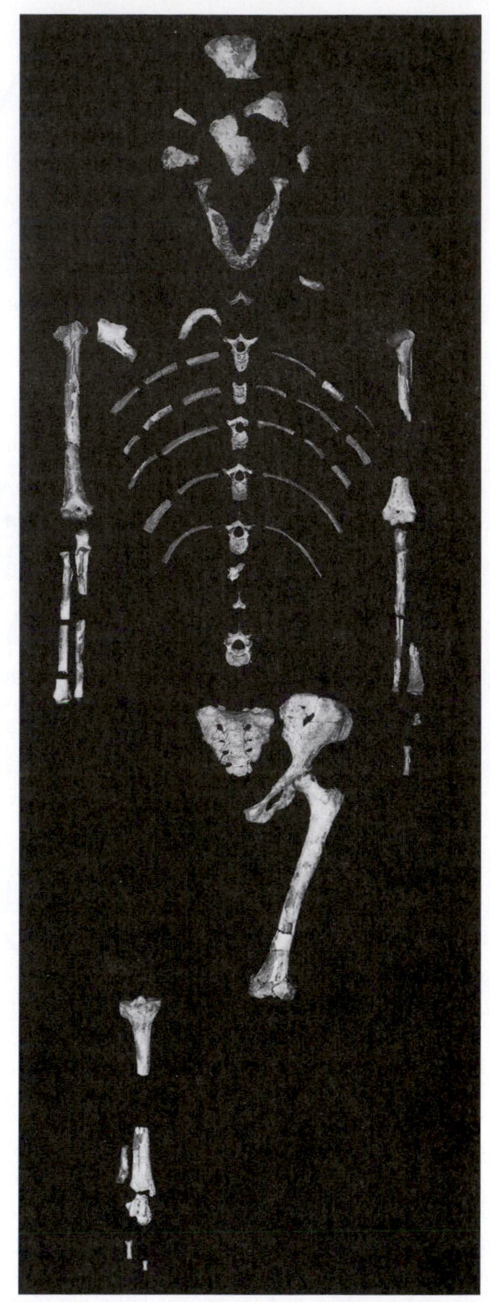

FIGURE 17.3 The skeleton of *Australopithecus afarensis*, "Lucy" (AL-288), from Hadar, Ethiopia. This is the most complete skeleton of an early hominid.

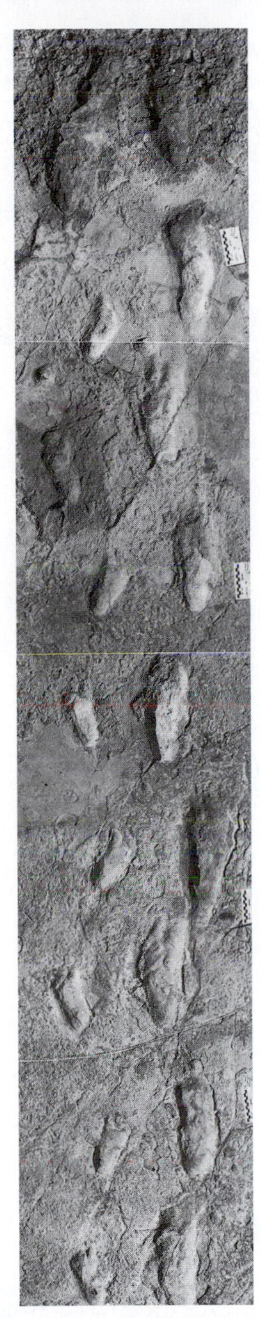

FIGURE 17.4 The 3.5-million-year-old footprints from Laetoli, Tanzania, presumably made by *Australopithecus afarensis* (photograph by P. Jones and T. White).

waiting to be uncovered. *Australopithecus anamensis* and *A. afarensis* are from eastern Africa (Ethiopia, Kenya, and Tanzania); *A. africanus* is from South Africa; and *A. bahrelghazali* has recently been described from Chad in Central Africa. The robust species of early hominids are placed in a separate genus, *Paranthropus,* described in an upcoming section.

Australopithecus anamensis is the oldest species of *Australopithecus,* found at several sites in Northern Kenya that date between 3.9 million and 4.2 million years old. It is known primarily from dental remains and a few limb elements, including a humerus and a tibia (Leakey *et al.,* 1995, 1998). The dental remains indicate considerable sexual dimorphism in this species, and the tibia suggests that it was bipedal.

Australopithecus afarensis (Figs. 17.3, 17.5, 17.6), is the best-known species of *Australopithecus.* It has been recovered from Pliocene deposits in many parts of East Africa (Ethiopia, Tanzania, and possibly Kenya). This species has a long temporal span. Although it is best known from 3.4- to 3.0-million-year-old deposits at Hadar in Ethiopia, older remains (3.7 million years old) have been described from Laetoli in Tanzania and from several sites in Ethiopia, including Omo and Fejej (as much as 4.2 million years old). This early hominid was extremely sexually dimorphic in body size, with the smallest individuals weighing no more than 30 kg and the largest probably twice as much.

In dental proportions, *A. afarensis* is similar to a chimpanzee, with larger canines and incisors than more modern hominids. The molars, however, are larger than those of living apes and have the low cusps and thick enamel also found in several Miocene apes. The relatively large anterior dentition suggests that this species was frugivorous, and the thick enamel indicates that nuts, grains, or hard fruit pits may have been part of its diet. Although in size *A. afarensis* is one of the most sexually

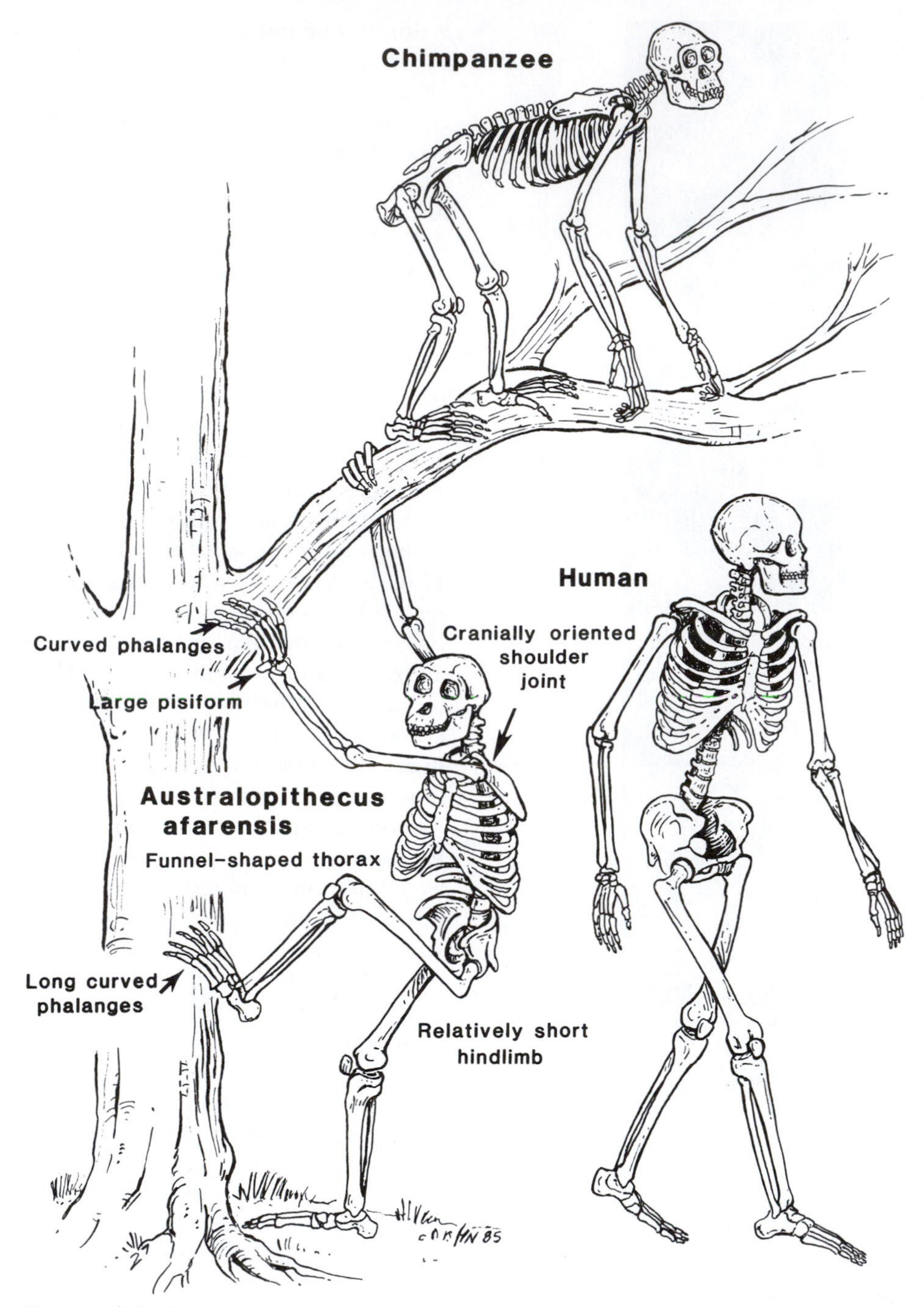

FIGURE 17.5 The skeletons of *Australopithecus afarensis, Pan troglodytes,* and *Homo sapiens.* Note the apelike features in *A. afarensis* that suggest climbing behavior.

FIGURE 17.6 A small *Australopithecus afarensis* group.

dimorphic primate species, it has little canine dimorphism compared with living great apes—but more than modern humans.

The cranial anatomy of *A. afarensis,* known from several relatively complete specimens (Kimbel *et al.,* 1994), is similar to that of living chimpanzees in many aspects. This species has a longer snout and shallower face than later hominids and an apelike nuchal region. The brain was small, the size of an orange. There is a sagittal crest both anteriorly, as in other australopithecines, and posteriorly, as in apes (Asfaw, 1987).

The skeletal anatomy of *A. afarensis* is better known than that of any other early hominid. One individual fossil from Hadar, "Lucy" (AL-288), is known from 40 percent of a skeleton, including large portions of almost all long bones (Fig. 17.3). In limb proportions, Lucy is intermediate between living chimpanzees and humans. Based on an estimated body weight of 30 kg, she has relatively short hindlimbs but forelimbs similar in length to those of a small human. Compared with a pygmy chimpanzee of the same size, she has relatively short arms but similar hindlimbs.

The forelimb remains of Lucy, although humanlike in proportions, are more chimpanzee-like in other features (Fig. 17.5). The curved phalanges, large pisiform bone, and cranially oriented shoulder joint all suggest suspensory abilities for this early hominid, as do other chimplike features of the humerus and ulna (Susman *et al.,* 1984). The pelvis of *A. afarensis,* like that of all later hominids, has a short, broad ilium and a relatively short ischium, resembling that of bipedal humans more than that of any living ape in these features, but the iliac blade faces posteriorly as in nonhuman primates rather than laterally as in humans. The distal part of the femur is strikingly humanlike in its valgus (knock-kneed) angulation, yet it lacks the enlarged lateral lip of the patellar groove found in bipedal humans. Many details of the ankle and foot, however, such as the relatively long, curved pedal phalanges, are more chimpanzee-like and suggest grasping behavior. Although clearly a biped, *A. afarensis* was of a rather nonhuman variety, for even the more humanlike hindlimb elements are different in detail from those of all later hominids, suggesting that the bipedal locomotion of *A. afarensis* was different from that of extant humans (Fig. 17.6).

As with the dental and cranial remains of *A. afarensis,* there is considerable variability in both size and morphological detail among the skeletal remains attributed to this species. In general (but not always), the smaller bones tend to be more chimpanzee-like, while the larger ones tend to be more similar to those of living humans. The size differences in the limb elements attributed to this species are comparable to those of the most dimorphic living apes, gorillas and orangutans (Lockwood *et al.,* 1996). The most widely held interpretation for this variation is that *A. afarensis* was probably characterized by sexual dimorphism in locomotor abilities. Perhaps the larger (male?) individuals were more terrestrial than the smaller (female?) ones.

Overall, *A. afarensis* is very much a missing link between the living African apes and later hominids in its dental, cranial, and skeletal morphology. Although intermediate in many aspects of its behavior, *A. afarensis* was also uniquely different from any living primate in many ways. It was probably frugivorous and could also eat very hard objects, such as seeds and nuts. It traveled bipedally on the ground but probably slept in and perhaps foraged in the trees.

The social structure of *A. afarensis* is difficult to reconstruct, since the combination of little canine dimorphism with considerable body size dimorphism is unique among living primates but most like the pattern found among modern humans. Several authors have inferred monogamous habits, based on the canines, and Lovejoy (1981) has suggested that the size dimorphism reflects different foraging patterns and antipredator strategies for the two sexes. Others (Hrdy and Bennett, 1981) have theorized that the size dimorphism indicates a polygynous social structure for early hominids. The canine reduction might have nothing to do with social structure but could be related to dietary adaptations. In view of the diversity in the social groups found among both extant apes and humans, it seems unlikely that we will ever be able to reconstruct with confidence the social habits of early hominids by arguments from analogy (see Plavcan and van Schaik, 1997).

Originally described by Raymond Dart (1925) from the limeworks at Taung in the Cape Province of South Africa, ***Australopithecus africanus*** is best known from the caves at Sterkfontein and Makapansgat. Because the limestone caves are not amenable to radiometric dating, the absolute age of *A. africanus* is not precisely known, but faunal associations suggest that this species comes from deposits between 3.0 million and 2.3 million years old (see Fig. 17.2).

Compared with *A. afarensis, A. africanus* has more similar-sized central and lateral upper incisors and larger cheek teeth (Fig. 17.7). The relatively smaller anterior dentition resulted in a shorter snout in *A. africanus.* The occipital region and the tympanic bones of the South African species are more like those of later hominids. There is considerable variability in cranial and dental morphology of

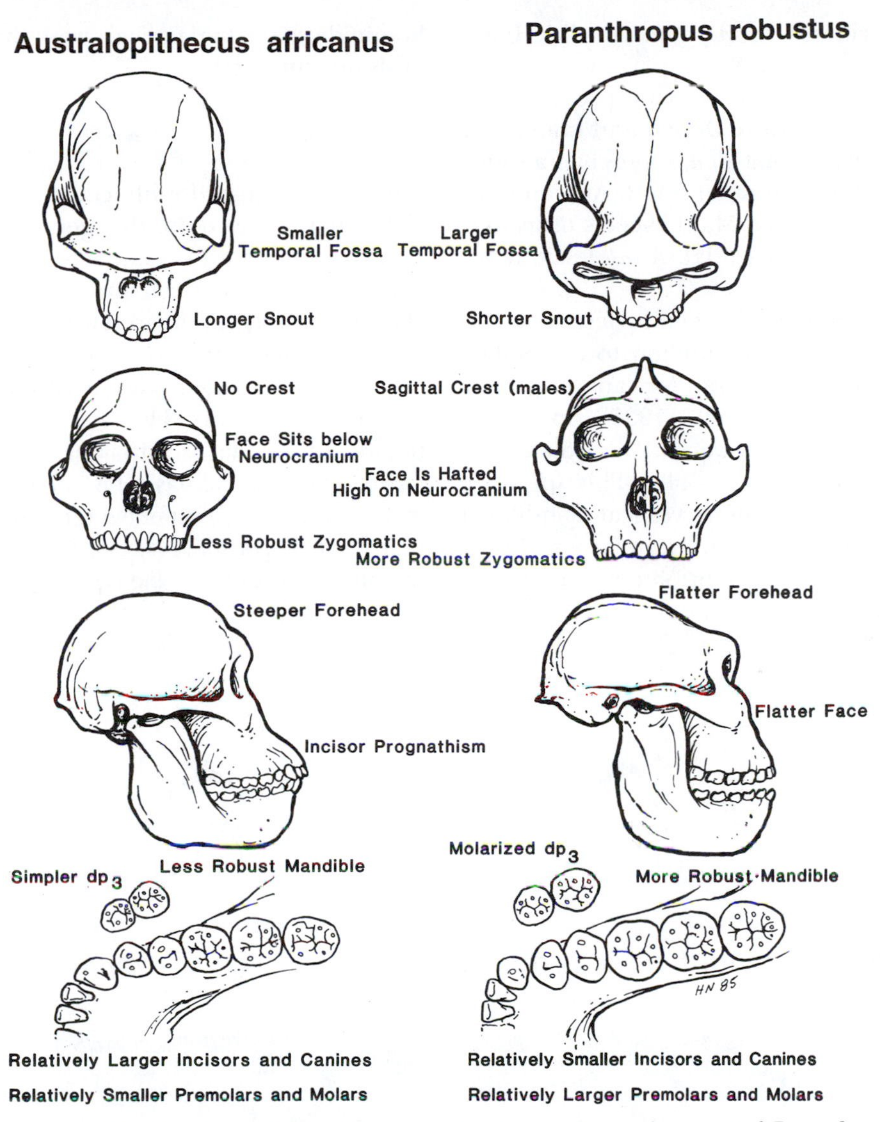

FIGURE 17.7 Cranial and dental features of *Australopithecus africanus* and *Paranthropus robustus.*

the many specimens assigned to *A. africanus* (both published and unpublished), and it is possible that several taxa are included in this sample.

The skeleton of *A. africanus* is similar to that of *A. afarensis* in many features, including relatively large upper limb elements and relatively short legs. A recently recovered foot has been suggested to indicate that this species had an abducted big toe for climbing (Clarke and Tobias, 1995). As in *A. afarensis,* there is considerable size dimorphism in limb elements attributed to *A. africanus* (McHenry, 1994).

Behaviorally, *A. africanus* was probably very similar to *A. afarensis*. Dental and cranial differences suggest that *A. africanus* had a softer, less gritty diet than other South African hominids (see Fig. 17.7). Many aspects of its limb skeleton suggest that, like *A. afarensis*, it was an adept climber.

A. bahrelghazali is a recently described species from deposits 3.0 million to 3.5 million years old in the central African country of Chad (Brunet *et al.*, 1995, 1996). So far it is known only from a lower jaw, an isolated upper premolar, and a maxilla. This species is unusual in having a more vertical mandibular symphysis than other species of *Australopithecus*. It extends the geographic range of early hominids 2500 km farther west than previously documented.

Paranthropus

The robust australopithecines (Figs. 17.7, 17.9), here placed in the genus *Paranthropus*, are from late Pliocene and early Pleistocene deposits in eastern and southern Africa (Figs 17.1, 17.2). Compared with *Australopithecus*, *Paranthropus* species have larger molars and premolars combined with relatively smaller canines and incisors. They have much flatter, broader, dished faces, and later species have a more flexed cranial base (Fig. 17.7). The skeletal anatomy of *Paranthropus* is less well known than that of *Australopithecus*, but the species was almost certainly some type of primitive bi-

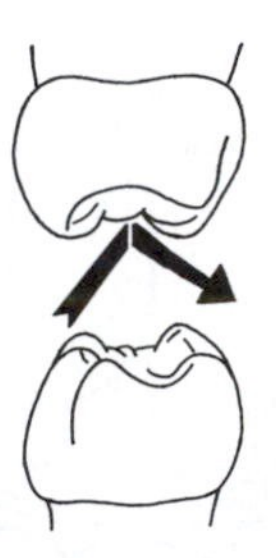

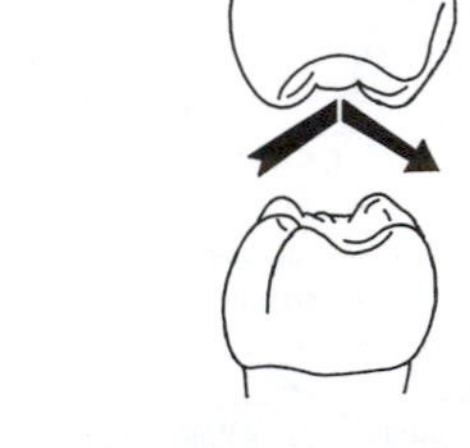

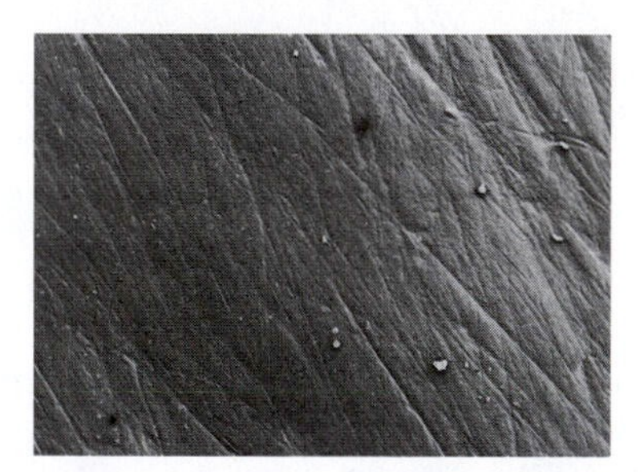

FIGURE 17.8 Differences in dental wear between *Australopithecus africanus* and *Paranthropus robustus*. Wear facets on the molar teeth indicate that in *P. robustus* the chewing stroke was flatter than in *A. africanus*. The molar teeth of *P. robustus* also show more pits than do the molars of *A. africanus* (courtesy of F. Grine).

FIGURE 17.9 Reconstruction of a group of *Paranthropus robustus* from Swartkrans in southern Africa. Faunal evidence suggests a wet habitat (Reed, 1997). Dental studies suggest a herbivorous, gritty diet, and anatomical studies of the hands and feet indicate bipedal behavior and possible tool use (courtesy of F. Grine).

ped. Despite the large teeth, body size estimates based on limbs suggest body weights similar to those of *Australopithecus* and similarly high levels of sexual dimorphism. There are three species currently recognized. All seem to have been most common in relatively wet habitats characterized by edaphic grasslands (Reed, 1997).

Paranthropus aethiopicus is the oldest and most primitive species of *Paranthropus,* from deposits between 2.7 million and 2.3 million years old in southern Ethiopia and northern Kenya. The best specimen, KNM-WT 17,000, "the Black Skull" (Fig. 17.10), has a massive face combined with a relatively long snout, a primitive cranial base like *A. afarensis,* and very large sagittal and nuchal crests.

Paranthropus robustus (Fig. 17.9) is from the cave sites of Swartkrans and Kromdraai in South Africa, estimated at between 2 million and 1 million years old, and was contemporaneous with members of the genus *Homo. Paranthropus robustus* was probably similar in size to *A. africanus* and was also sexually dimorphic in size.

FIGURE 17.10 *Paranthropus aethiopicus,* represented by "the Black Skull," KNM-WT 17,000 (photo by Alan Walker © National Museums of Kenya).

Paranthropus robustus has smaller incisors and canines, larger cheek teeth with thicker enamel, and a thicker mandible than *Australopithecus* species (Fig. 17.7). These differences in dental morphology are associated with differences in both gross and microscopic tooth wear (Fig. 17.8). Individuals of *P. robustus* wore down their teeth flatter and used more crushing than shearing. Their teeth are more heavily scratched and pitted than those of *A. africanus,* suggesting that their herbivorous diet contained harder, more resistant, and perhaps smaller food objects (Grine, 1981, 1986).

Paranthropus robustus has a shorter, broader face with deeper zygomatic arches and a larger temporal fossa than seen in the skull of *A. africanus* (Fig. 17.7). The larger individuals (males?) have sagittal and nuchal crests. Like the molar differences, the cranial differences seem related to more powerful chewing in *P. robustus.*

The skeletal differences between *P. robustus* and *Australopithecus* are difficult to assess, because there are few well-associated complete limb elements. Susman (1988) suggests that *P. robustus* was more humanlike in both hands and feet than was *A. afarensis.* The hand bones show evidence of manipulative abilities, suggesting that this species was capable of using and making tools (Susman, 1994). The foot bones indicate that it was bipedal and less arboreal than *A. afarensis.* Habitats for *P. robustus* have been reconstructed as secondary grasslands associated with rivers and wetlands (Reed, 1997).

The dental and cranial features that characterize the South African *P. robustus* were developed even further by the "hyper-robust" ***P. boisei,*** from East Africa. This species, known from deposits in Ethiopia, Kenya, and Tanzania between approximately 2.2 million and 1.2 million years old, was contemporaneous with members of our own genus, *Homo*. *Paranthropus boisei* was similar in size to *P. robustus*, with an estimated body weight of 30–50 kg. This species is sexually dimorphic in both size and cranial shape. Compared with *P. robustus*, *P. boisei* has smaller incisors and canines, absolutely larger cheek teeth, and a heavier mandible. The skull has an extremely broad, short face with a large temporal fossa between the flaring zygomatic arches and the relatively small brain. There are pronounced sagittal and nuchal crests in large males.

Although there are few limb bones that can be definitely attributed to *P. boisei*, several very large forelimb bones from East African sites are often assigned to the species (McHenry, 1994). These bones suggest suspensory abilities. Tools are often found in association with *P. boisei* in East Africa, but the existence of more advanced hominids (*Homo* sp.) from the same time span precludes unequivocal determinations as to which or how many species made or used tools. Like *P. robustus*, *P. boisei* seems to have become extinct before 1 million years ago (Fig. 17.2).

AUSTRALOPITHECINE ADAPTATIONS AND HOMINID ORIGINS

As the earliest hominids, australopithecines provide important clues to the adaptations associated with the origin of the human lineage and its divergence from that leading to living apes. Many people have attempted to reconstruct the habits of this basal hominid, and often these interpretations have been heavily colored by theoretical or personal views about human origins. Many authors have been unduly influenced by the fact that there is only a single hominid living today, and have reconstructed australopithecines as "Pliocene people" little different from modern humans. Others seem to have been overly influenced by their own views of the primitive aspects of human behavior and have seen early hominids as vicious killers bearing little resemblance to any living primate. With increased knowledge of nonhuman primate behavior, more complete and better-dated fossil remains, and a better appreciation of early hominid diversity, current reconstructions of the behavior of Pliocene hominids are probably more realistic, but also more vague, than those put forth earlier in the century.

Australopithecines were medium-size hominoids with relatively small incisors and canines, large cheek teeth, very robust jaws, and extremely large chewing muscles. In other mammals this combination of features is associated with a herbivorous diet of tough plant material. Early hominids of the genus *Paranthropus* evolved even smaller incisors and canines with even larger cheek teeth and thicker jaws, suggesting even more extreme adaptations for tough foods. All of the dental and cranial evidence indicates that these early hominids were herbivores. Contrary to earlier suggestions, recent estimates of australopithecine body size based on the limb skeleton suggest that although all species show considerable sexual dimorphism, their mean body size was very similar, between 35 and 45 kg (see Table 17.1; McHenry, 1994). This high sexual dimorphism is most probably the primitive condition retained from the common ancestor of living African apes and humans (Lockwood *et al.*, 1996). The unusual combination of low canine dimorphism and high body size dimorphism among australopithecines makes reconstruction of social behavior by analogy with living primates difficult (Plavcan and van Schaik, 1997).

The teeth of australopithecines provide interesting information about growth and development in early hominids. One of the most characteristic features distinguishing humans from other primates is our slow development (see Figs. 2.26, 2.27) and long period of growth and maturation, usually associated with our reliance on learning. Studies of dental development indicate that in timing and sequence of dental development *Australopithecus* was quite different from modern humans and more similar to African apes (Smith, 1986, 1994), whereas *Paranthropus* had a developmental pattern that resembled modern humans in many ways.

Relative to body size, the brains of australopithecines were no larger than the brains of living apes (Falk, 1991). There is also no evidence from either the external morphology of the brain or the shape of the skull base that they had any greater capabilities for articulate speech than do living apes.

Australopithecines share several distinctive features of their locomotor skeleton with *Homo sapiens,* indicating that they were bipedal primates. However, many of their differences from living humans and similarities to living pongids indicate that they retained considerable abilities for arboreal locomotion, such as climbing, and that their bipedal gaits were noticeably different from those of humans. They probably climbed trees for foraging, sleeping, and escaping predators but traveled bipedally on the ground. Arboreal foraging and terrestrial travel between food sources is a common behavioral pattern in a variety of living primates, including chimpanzees and pig-tailed macaques. A striking difference from quadrupedal monkeys and apes is that the bipedal hominids would have been able to use their hands for transporting food from place to place.

As many authors have emphasized, all early hominids may have made and used some type of perishable tools, since such tools are used by both chimpanzees and later hominids. However, the fossils suggest more specific associations. The anatomy of the hand and the lack of associated stone tools would seem to indicate that *Australopithecus* did not make or use stone tools. However, there is some evidence that *Paranthropus* was a tool user (Fig. 17.9). Bone and stone tools are often found in deposits with species of *Paranthropus,* and *P. robustus* from South Africa has anatomical features of its hand consistent with tool use (Susman, 1991, 1994).

Overall, australopithecines were rather nonhuman hominids. Many of their characteristic dental features, such as small canines and large, flat molar teeth, were also present in earlier fossil apes. There are weak trends towards increasing brain size throughout early hominid evolution, but australopithecine brains are small compared with those of later hominids. Only the anatomical features indicative of bipedal walking and perhaps some manipulative abilities in *Paranthropus* seem to separate these early hominids from other nonhuman primates. Most morphological features that characterize modern humans, such as an arched foot with an adducted hallux and a large brain, as well as our characteristically slow rates of growth and development, were not present in these primitive hominids. Australopithecines are very much missing links between apes and people.

Humans differ from living apes in numerous morphological and behavioral features, and there has been a tendency in the study of human evolution to see all human features, including bipedalism, large brains, manipulative hands, tool use, and language, as integrally related to a single adaptive complex extending back to the origin of the hominid lineage (Darwin, 1871; Washburn, 1963). Such an approach was reasonable when there was only one known hominid, *Homo sapiens,* and no fossil record of more primitive ancestors lacking this complete suite of features. But such completely integrated models provide no insight into the evolution of the group. The fossil record of australopithecines provides direct evidence that the cluster of features characterizing living humans are not inseparably linked but, rather, evolved one by one (e.g., McHenry, 1975).

Early hominids were bipedal well before they evolved brains that were appreciably larger than those of living apes and before they regularly made stone tools. *Australopithecus* appears to have thrived for 2 million years as a biped without ever evolving, or for that matter needing, these novelties of later hominids. Like all other members of our order, each species of australopithecine was a primate with a particular suite of adaptations that enabled it to make a living in its particular time and place. From the beginning, the one feature that seems clearly to have separated our lineage from other primates is bipedal locomotion. What, then, is the ecological context in which our ancestors evolved this unique primate adaptation (Fig. 17.11)?

One of the most striking advantages of bipedal locomotion is that it frees the hands for carrying things, and freeing of the hands has long been considered the major selective advantage of bipedalism (Hewes, 1961). Because stone tools do not normally seem to be associated with *Australopithecus,* some authors have suggested that the earliest hominids must have used some type of wooden tools, such as spears or clubs.

Other authors have leaned toward other, more primatelike foraging adaptations associated with bipedalism. Zihlman and Tanner (1978) see considerable similarity between australopithecines and living chimpanzees in many aspects of their behavior. They suggest that many of the traditionally accepted differences between hominids and apes are based on a misguided (and male-oriented) overemphasis on hunting as a characteristic hominid behavior (Fedigan, 1986). In modern hunter-gatherer communities, it is the gathering of plants by females that provides most of the food for subsistence, just as plants provide most of the food for living apes. In their view, we should pay more attention to the function of characteristic hominid features in a gathering rather than a hunting context. In such a context, bipedalism would enable females to carry hominid infants more easily (although it is unlikely that these infants would have been as helpless as our own). Free hands would also enable gatherers to carry extra water, thus extending foraging range into dry areas, and to carry surplus food in a way that living apes cannot. Zihlman and Tanner suggest that, if the earliest hominids used tools, those tools were probably digging sticks or natural containers for transporting food and water, not clubs or spears.

Another common argument for the origin of bipedalism is that it is related to life in open habitats and a need to see over tall grass. Although it is true that many primates living in open habitats, such as patas monkeys, frequently stand to look for conspecifics or look out for predators, none seem to have adopted bipedalism as a normal pattern of movement. Moreover, hominid origins seem unlikely to have taken place in open habitats.

Rodman and McHenry (1980) also associate bipedalism with foraging in an open woodland habitat. However, in their view the earliest hominids may well have fed in trees but adopted bipedal postures for traveling between trees. They suggest that bipedal walking would be a more efficient method of locomotion than quadrupedal travel and cite studies showing that modern human bipedalism is indeed a very efficient pattern of movement. However, as Steudel (1994) has emphasized, bipedal and quadrupedal chimps show no differences in cost of locomotion (Taylor and Rowntree, 1973), so bipedalism per se is not inherently a cheap way of moving. Because

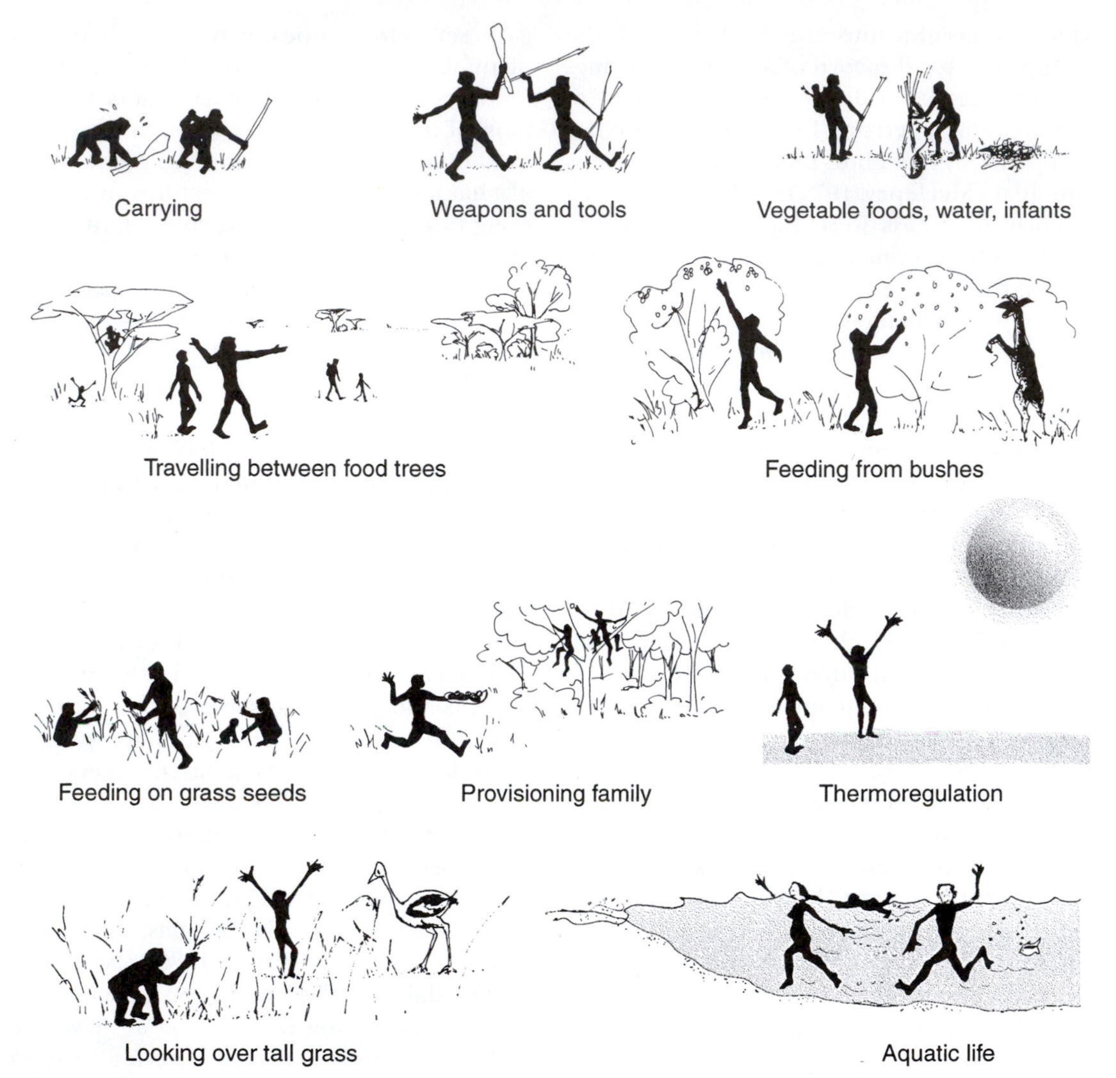

FIGURE 17.11 Various theories on the adaptive significance of the origin of hominid bipedalism (courtesy of Jeanne Sept and L. Betti).

the earliest hominids lacked many of the distinctive anatomical features associated with the striding bipedalism of modern humans, an energetic argument for the origin of this pattern of locomotion seems unrealistic.

There are several theories suggesting that bipedalism arose as a postural adaptation for foraging, either on the ground or in the trees. Several authors have suggested that the evolution of hominid bipedalism began as some sort of feeding posture enabling early hominids to feed on tall bushes or small trees, and Hunt (e.g., 1996) has argued from comparisons with living chimps that early hominids began their bipedal behavior by standing during feeding in an arboreal setting. Like the open-habitat hypothesis these suggestions fail to explain why the other primates that regularly use bipedal postures have never adopted bipedal locomotion.

Several other theories address this problem of hominid uniqueness directly. Jolly (1970) has argued that bipedalism in a herbivorous primate is advantageous only when the animal is feeding on small, evenly distributed objects such as nuts, grains, or small seeds. In this type of feeding situation, an individual's foraging efficiency is linked directly to the speed with which it can pluck and ingest food items. A squatting or partly bipedal animal with both hands free for foraging is best adapted to this type of diet. Furthermore, small, hard objects such as seeds seem to be the type of food that the *Australopithecus* dentition, with its broad, flat, thick-enameled molars, was designed to chew. Finally, Jolly argues that many foods of this nature are common in open habitats, in which bipedal locomotion seems most appropriate. Jolly's model of hominid origins, the "seed-eater" model, is based largely on the habits of the gelada baboon, an open savannah, small-object-feeding baboon. However, there is increasing evidence that the earliest hominids did not inhabit open habitats but, rather, were most commonly associated with woodland-dwelling mammals (Reed, 1997).

In a highly provocative paper on hominid origins, Lovejoy (1981) suggests that the major adaptive change distinguishing early hominids from their more apelike ancestors was in their reproductive capabilities. Living great apes, he argues, normally give birth at intervals of three to five years, largely because of the difficulties of caring for large, slow-growing offspring. Humans, he suggests, generally have much shorter spacing between successive offspring. This increased reproductive efficiency was made possible through a monogamous social system in which males provisioned their mates and offspring. Bipedalism, by freeing the hands, allowed the males to bring extra food back to the less widely foraging members of their family, a behavior unknown in other primates. Although Lovejoy notes that the teeth of *Australopithecus* clearly indicate a herbivorous diet, the advantage of bipedalism is not linked to any particular type of food, only to the ability to transport it. Lovejoy's suggestion that *Australopithecus* lived in a monogamous social system (in which male parental investment would be expected) is based on his view that there was little canine dimorphism. Other authors have suggested that the evidence of considerable sexual dimorphism in body size argues for a more polygynous social structure. The evidence for any major differences between early hominids and living apes in patterns of maturation or in birth spacing is also questionable.

Peter Wheeler (e.g., 1984, 1991, 1993) has advanced a quite different, physiological theory for the origin of bipedalism. In his view bipedalism would provide an early hominid considerable thermoregulatory advantages in limiting direct exposure to sun during the middle of the day and facilitating convective heat loss in a stressful, open equatorial habitat.

Most recently, Jablonski and Chaplin (1993) have argued that human bipedalism arose as an extension of bipedal displays of threat and gestures of appeasement found among extant African apes. In their view it is this aspect of bipedalism that links human bipedal locomotion with the bipedal behavior of our closest relatives and that provided the selective advantage for this behavior in the earliest hominid bipeds, prior to any energetic or physiological benefits that later hominids may have experienced. Finally there is the fanciful aquatic ape theory championed most vigorously by Morgan (1994).

Although no current theory seems satisfactory on all counts, most of these theories attempt to explain the evolution of a bipedal herbivorous primate with small canines based on the everyday parameters of a nonhuman

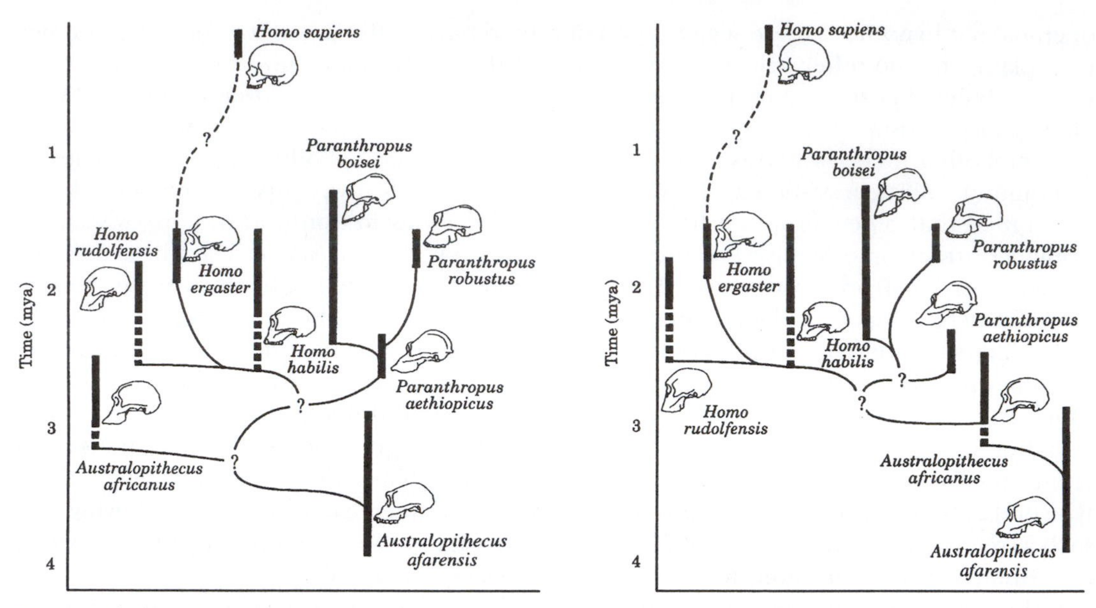

FIGURE 17.12 Alternative phylogenies for the relationships among australopithecines and early *Homo* (from Strait *et al.*, 1997).

primate way of life (feeding, travel, and reproduction) rather than on modern human behavior. All are necessarily speculative, but they do represent a changing perspective on human origins. Today we see hominids as one of many peculiar radiations in primate history, the evolution of which should be explicable in terms of adaptation to ecological surroundings. This is in significant contrast to the more traditional views, in which nonhuman primates were, at best, stepping stones or, at worst, failed experiments on the road to humanity. This new view of hominid origins has come in part from an increased appreciation that *Australopithecus* was a more primitive, apelike hominid than earlier workers had suspected and one that was also uniquely specialized in ways that were adaptively distinct from both living apes and later hominids.

Although it is important to see early hominids in the context of hominoid evolution, it is equally important to realize that in the same way that they were not little people, they also were not just bipedal chimps, but the beginning of a new radiation of very different hominoids. It is this uniqueness that makes reconstructing hominid origins so difficult. Thus, although early hominids and their bipedal adaptations are certainly derived from an African apelike ancestry, human bipedalism is morphologically and physiologically different from the occasional facultative bipedal behaviors occasionally seen in other primates (Rose, 1991). The morphological and behavioral commitment to bipedalism that characterized early hominids suggests unique ecological and historical circumstances as well. Identifying the ecological and phylogenetic conditions surrounding hominid origins will undoubtedly challenge paleoanthropologists for many years to come. Ultimately, those factors can be identified only through a more complete fossil record that documents better the actual transformations and ecological conditions during early hominid evolution.

PHYLETIC RELATIONSHIPS OF EARLY HOMINIDS

There are several long-standing debates, and a few very new ones, concerning the phyletic relationships among australopithecine species and the genus *Homo* (Fig. 17.12). One rapidly changing new issue is the phylogeny of the very earliest hominids. Until very recently, most authorities agreed that *A. afarensis* was the most primitive hominid species, ancestral to all later forms. Although *A. afarensis* is the best known of the earliest hominids, *Ardipithecus ramidus* and *Australopithecus anamensis* appear to be slightly older and more primitive in many features. Unfortunately, very little is known at present of either the anatomy or the distribution of these early species, so it is not possible to determine whether these three species are best interpreted as a single time-successive lineage or as evidence of a bushy phylogeny at the base of our family tree (Fig. 17.2; Kappelman and Fleagle, 1995).

Although the robust australopithecine species have traditionally been considered a single clade, the discovery of the Black Skull (KNM-WT 17,000) led many researchers to argue that the robust species were not a natural group, but rather they evolved their dental and cranial features in parallel (Grine, 1988). There is still a diversity of views on this topic, but the most recent analyses indicate that the robust species are a natural group (Strait *et al.*, 1997).

Perhaps the most unsettled issue in early hominid evolution is the relationship of different australopithecine genera and species to our own genus, *Homo*. Both *Australopithecus afarensis* and *A. africanus* have been advanced by various authorities (e.g., Johanson and White, 1979; Grine, 1988, 1993) as the species closest to the ancestry of *Homo*. However, phylogenetic analyses (e.g., Skelton and McHenry, 1992; Strait *et al.*, 1997) have repeatedly identified all or part of the robust clade as the sister group of *Homo*. These phylogenetic analyses suggest the existence of an ancestor common to the robust clade and the *Homo* clade that has yet to be identified in the fossil record but may well be hidden somewhere amidst the diversity of forms currently identified as *Australopithecus africanus* (Strait *et al.*, 1997). Obviously, our understanding of early hominid phylogeny will continue to be modified as new fossils are recovered and new taxa are identified through more detailed analyses. At present there is perhaps no group of primate fossils more complex and poorly sorted out than those identified as early members of the genus *Homo*.

PLEISTOCENE EPOCH

The Pleistocene epoch, from approximately 1.7 million years ago until recent times, was characterized geologically by repeated glaciations of the northern hemisphere. The initial onset of dramatic cooling seems to have begun in the latest Pliocene (around 2.5 million years ago), and there is evidence of another extreme cooling after 1 million years ago. Some researchers have argued for major shifts in the flora and fauna in Africa at 2.5 million years ago, and have attributed these to worldwide changes from relatively warm, wet climates to cooler, drier climates (Vrba *et al.*, 1996). Others find more gradual trends in faunas and reconstructed environments until approximately 1.8 million years ago, at the beginning of the Pleistocene, when there was a major appearance of secondary grassland habitats (Cerling, 1992; de Menocal, 1995; Reed, 1997).

In hominid evolution, the Pleistocene is characterized by the radiation and geographical expansion of the genus *Homo,* which began in the late Pliocene and continues to the present, and by the extinction of the robust australopithecines approximately 1 million years ago. The correlation between these major events in human evolution and patterns of global climate change is a topic of considerable interest and controversy (e.g., Vrba *et al.,* 1996; Potts, 1996; Reed, 1997; Spencer, 1997; Stanley, 1992).

Early *Homo*

The fossil record of our own genus begins in the late Pliocene, just before 2 million years ago. Until most recently, it was widely agreed that the genus *Homo* consisted of three or fewer species—*H. habilis, H. erectus,* and *H. sapiens*—all very human creatures and all drawn from a single, continuously evolving lineage that has been characterized by moderate geographic variation throughout its history. The major disagreements have been over the timing and nature of the transitions between species. Views of human diversity in the Pleistocene have changed dramatically in the past few years. Few researchers currently agree on how many species of our genus should be recognized at any time period before about 30,000 years ago, but the number is undoubtedly growing, with many more to be discovered or recognized (Fleagle, 1995). A relatively conservative assessment suggests at least seven species, three of which are found in the earliest Pleistocene deposits of Africa (Table 17.1).

Compared with *Australopithecus* and *Paranthropus, Homo* is characterized by smaller molars and premolars and a more slender mandible. Throughout the evolution of the genus, there has been a trend toward reduction in the size of the cheek teeth (Fig. 17.13). The anterior teeth, canines and incisors, are larger than those of *Paranthropus.* The cranium of *Homo* is characterized by a relatively larger brain size and a smaller face than that of *Australopithecus.* The genus *Homo* resembles later species of *Paranthropus* in having a flexed skull base.

There is evidence of considerable diversity in limb proportions and femoral anatomy among the earliest species of the genus *Homo* (e.g., Walker and Shipman, 1996). Nevertheless, overall our genus seems to be characterized by a less beaked ilium and a larger femoral head than *Australopithecus.* The foot of *Homo* has shorter digits than the feet of more primitive hominids.

The earliest specimens confidently attributed to the genus *Homo* are dated at approximately 2.3 million years ago (Kimbel *et al.,* 1996). Until very recently, all of the early remains of this genus were attributed to *Homo habilis.* However, it is now evident that there was a considerable diversity of advanced hominids present in eastern and southern Africa. Three species are commonly recognized: *Homo habilis, Homo rudolfensis,* and *Homo erectus* (or *Homo ergaster*). In addition, there is evidence of an additional species from southern Africa (Grine *et al.,* 1993). Although the details of both the systematics of this genus and the morphological and taxonomic diversity present among early members of the genus *Homo* are far from resolved (see Lieberman *et al.,* 1996; Rightmire, 1993), it seems clear that the origin of this genus was very bushy, with

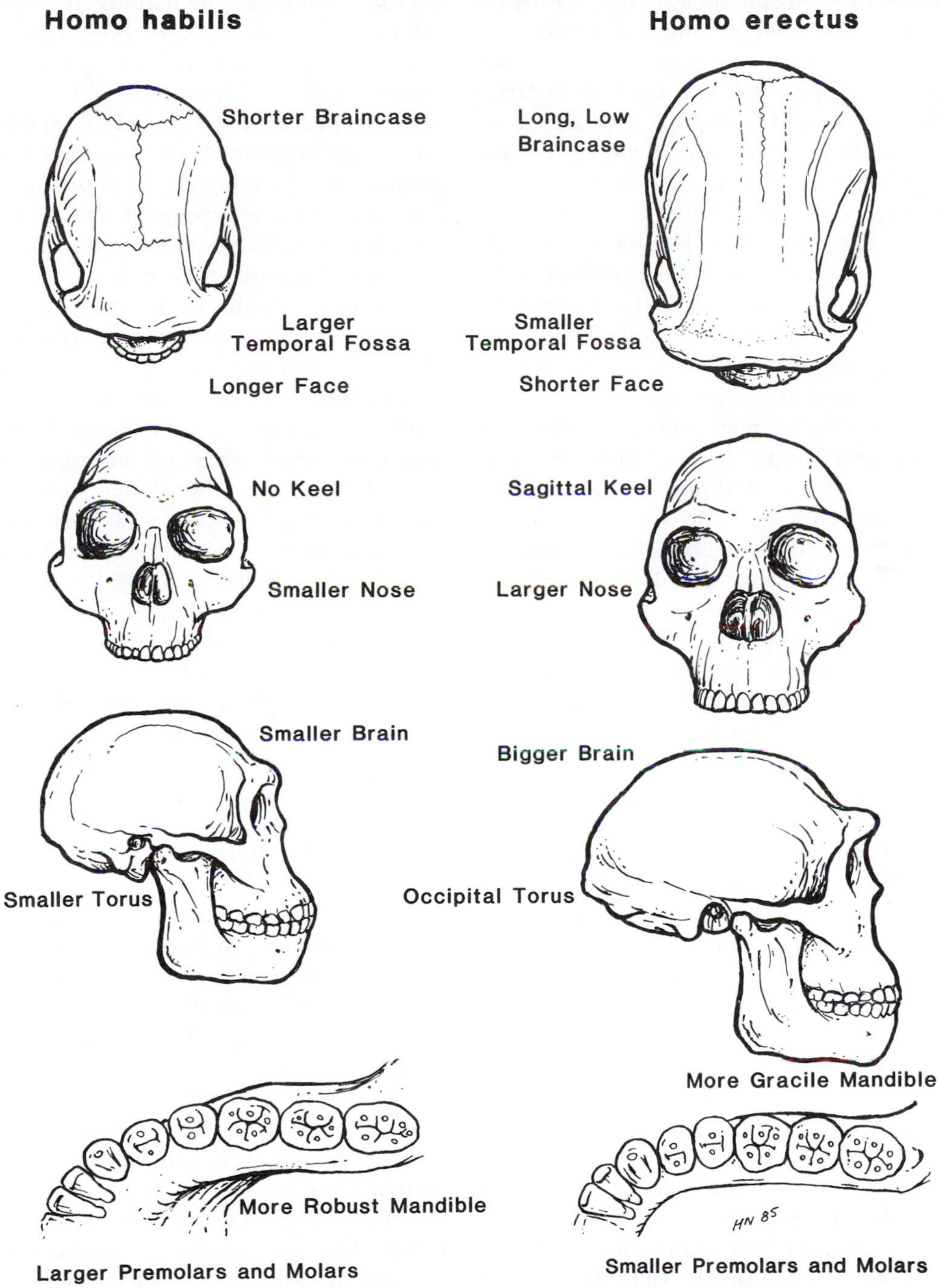

FIGURE 17.13 Cranial and dental characteristics of *Homo habilis* and *Homo erectus*.

considerable morphological and ecological diversity soon after its initial appearance, a common phenomenon in primate evolution.

The first appearance of stone tools in Africa coincides roughly with the appearance of *Homo* in the fossil record, suggesting to some that this genus used and made the artifacts (Semaw *et al.,* 1997; Kimbel *et al.,* 1996; but see also Susman, 1991). The tools are crude choppers and scrapers, collectively called the Oldowan culture because of the original discovery at Olduvai Gorge, Tanzania (Fig. 17.14). Wear on the cutting edges indicates that these tools were used in a variety of activities, including butchering of small animals, trimming of leather, and preparation of plant remains. However, experimental work involving manufacture and use of such primitive tools by present-day anthropologists indicates that many Oldowan "tools" may actually be the larger cores that were left in the process of producing smaller flakes (Toth, 1987). Although the core choppers were probably used for some activities, it is the smaller, razor-sharp flakes that are more effective in food processing.

Homo habilis

Homo habilis (Fig. 17.13) first appeared around 2.3 million years ago in the latest Pliocene and earliest Pleistocene of eastern Africa, where it is best known from Olduvai Gorge in Tanzania and the Turkana Basin of northern Kenya and southern Ethiopia. A newly discovered maxilla from Hadar is also attributed to this species (Kimbel *et al.,* 1996). Compared with *Australopithecus, H. habilis* has narrower premolars and first molars, a narrower mandible, a more coronal orientation of the petrous part of the temporal bone, and delayed eruption of the canines. The average cranial capacity is larger than that of the more primitive *Australopithecus* and smaller than that of most *H. erectus,* although individual specimens overlap with both taxa. The hand bones are more robust than those of later hominids, which suggests that this species retained some suspensory abilities. The foot is more advanced than that of *Australopithecus* and resembles the foot of extant humans in most features, suggesting a similar bipedal gait (Susman, 1988). However, in many aspects of its skeletal proportions, *H. habilis* is strikingly similar to some australopithecines, such as Lucy, in having very long forelimbs (Johanson *et al.,* 1987; Hartwig-Scherer and Martin, 1994). Although *H. habilis* has long been identified as the earliest member of our genus, there are some workers who argue that this species is perhaps best seen as an advanced australopithecine because it is much more similar to that radiation than to other species of *Homo* that are known from the same time period (Walker and Shipman, 1996).

The sites that have yielded *Homo habilis* (and *Paranthropus boisei*) regularly include Oldowan stone tools. Indeed, this species was originally named on the assumption that it was a tool user. Some of the broken animal bones found in association with Oldowan tools and *H. habilis* fossil remains show cut marks that appear to have been made by the stone tools. Furthermore, the concentrations of stone tools and broken bones suggest to some workers that animal parts were transported to the sites by hominids (Isaac, 1983); others see both the cut marks and the accumulations as possible results of geological processes (Binford, 1981). Isaac (1983) has argued that these concentrations indicate the emergence of some type of home base or "central-place foraging" and possibly food sharing among early humans. Whether the animal parts are the result of hunting or of scavenging activities cannot be determined. It seems likely that the role of meat eating and hunting in early hominids has been overemphasized in past archaeology. Unfortunately, plant foods leave few fossilized remains, so it is not possible to reconstruct the relative proportions of meat and plant material in the diet of *H. habilis.*

Homo rudolfensis

Homo rudolfensis, a species of early hominid that is roughly contemporary with *Homo habilis,* has only recently been widely recognized as a separate taxon (Wood, 1991). It is known mainly from the Turkana basin of Northern Kenya and possibly from Malawi as well (Schrenk *et al.,* 1993). *Homo rudolfensis* differs from *H. habilis* in having a flatter, broader face and broader cheek teeth with more complex crowns and thicker enamel. The holotype and best-known specimen is KNM-ER 1470. The limbs of *H. rudolfensis* are not well known, but the species seems to be characterized by longer hindlimbs and a larger femoral head than those of *H. habilis.*

An additional species of early *Homo* has recently been identified by Grine *et al.* (1993) from Swartkrans in South Africa. However, the taxon has not yet been formally named

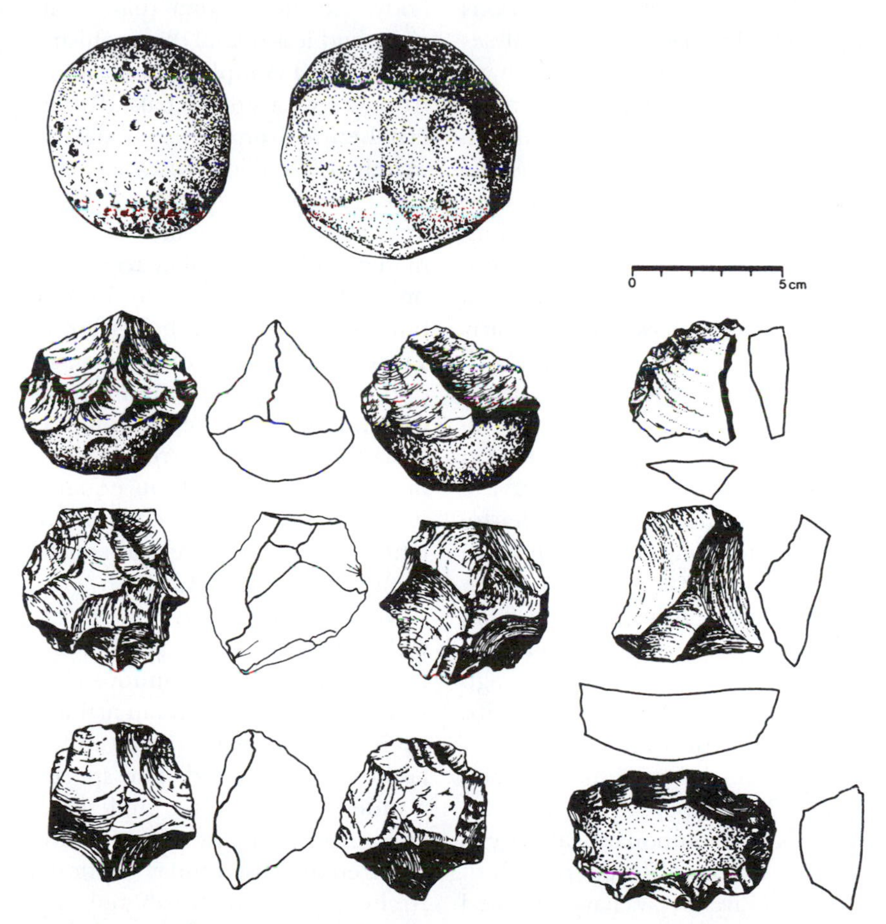

FIGURE 17.14 Primitive Oldowan tools (courtesy of Kathy Schick).

and is currently placed by most authorities in *Homo habilis.*

Homo erectus and *Homo ergaster*

For many years it has been evident that in addition to "more primitive" taxa like *H. habilis* and *H. rudolfensis,* there was a more advanced hominid in Africa at the beginning of the Pleistocene. This species is similar to ***Homo erectus,*** which has long been known from the middle Pleistocene of Asia (Rightmire 1990). The African version of *Homo erectus* is considered by some authorities as a separate species, *H. ergaster,* with distinctive cranial proportions (e.g., Wood, 1992). However, other analyses comparing African and Asian samples have not been able to find consistent differences, leading most authorities to attribute both the African and the Asian specimens to *Homo erectus* (e.g., Rightmire, 1992; Walker and Leakey, 1993; Brauer, 1994). Although it has traditionally been thought that the African fossils were substantially older than those from Asia, a number of recent analyses have reported early Pleistocene dates for early *Homo* in both China and Indonesia (Swisher *et al.,* 1994; Wanpo *et al.,* 1995). If these reports are accurate, then *Homo erectus* was well established on both continents from the beginning of the Pleistocene until well into the middle Pleistocene (Larick and Ciochon, 1996). There are no fossils clearly attributable to this taxon from Europe, but there is a mandible from the country of Georgia in western Asia with an assigned date of 1.8 million years ago (Gabunia and Vekua, 1995; Brauer and Schultz, 1996). The geographic distribution of *H. erectus* exceeds that of any other primate species prior to its time. The temporal span of *H. erectus* and its ultimate evolutionary fate are a source of continued debate. Recent restudy of fossils from the Solo River in Java have indicated that this species survived until 27,000 years ago in Southeast Asia (Swisher *et al.,* 1996).

Compared with *Australopithecus* or *H. habilis, H. erectus* (including *H. ergaster*) has smaller cheek teeth and a more slender mandible, in keeping with the general trend of tooth reduction within the genus (Fig. 17.13). Brain size is larger than in earlier hominids, with considerable variation among specimens. The cranium of *H. erectus* is characterized by very thick bones (less so in specimens attributed to *H. ergaster*), a long, low vault with sagittal keeling, projecting brow ridges, and a prominent occipital torus. The face of this species was relatively broad and had a large nasal opening. *Homo erectus* seems to have had a mean body size larger than that of australopithecines and less sexual dimorphism.

The most complete skeleton of *H. erectus* is from the west side of Lake Turkana in northern Kenya, from deposits dated at approximately 1.6 million years ago (Brown *et al.,* 1985; Walker and Leakey, 1994; Fig 17.15). The young (12-year-old) male had an estimated adult height of nearly six feet. The limb proportions are similar to those of *H. sapiens,* but most of the limb bones are more robust. The chest is more conical, as in apes. The femoral neck is long, as in *Australopithecus,* but the femoral head is large, as in modern humans. It had a relatively narrow trunk and slender limbs, associated with an equatorial climate. Populations of *Homo erectus* in more northern latitudes seem to have had shorter limbs and broader trunks (Ruff, 1993).

In Africa, *H. erectus* remains are associated with Acheulian hand axes; in Asia, the species is found with more primitive chopping tools, similar to earlier Oldowan artifacts. *Homo erectus* is the first fossil primate with a substantial archeological record. The species developed a wide range of stone implements for different purposes, many of which are still manufactured and used today by modern humans. Archeological sites attributed to *H. erectus* are widespread and diverse. Some seem to have been camps, others were sites of animal kills,

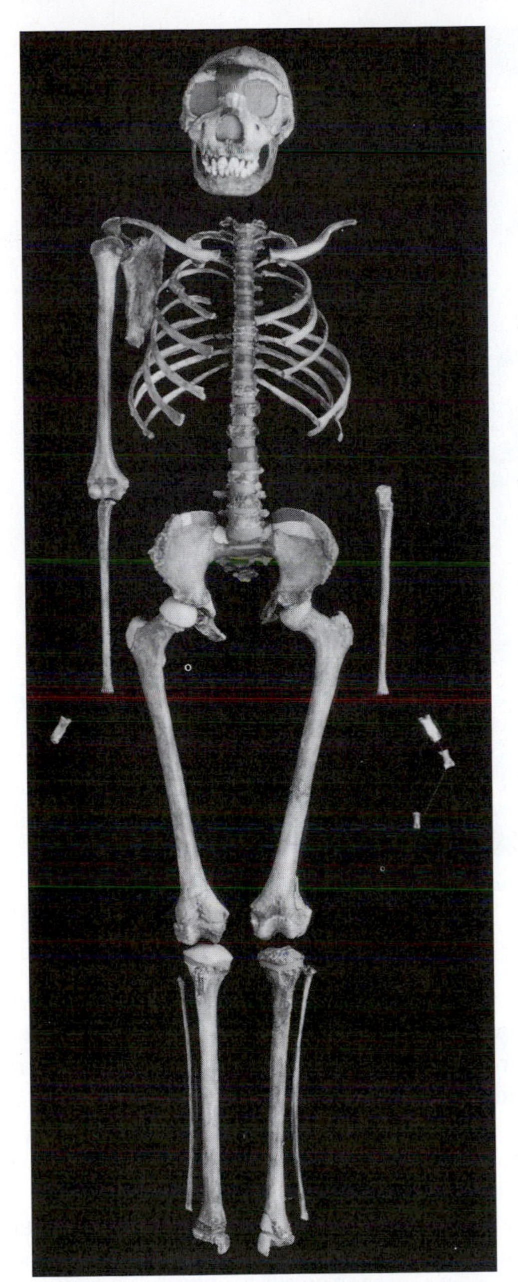

FIGURE 17.15 The 1.6-million-year-old skeleton of a *Homo erectus* boy from West Turkana, Kenya (photograph by David Brill, National Geographic Society).

and others were butchering sites. Some of the later sites show evidence of simple structures. The variation in size of the camps suggests a social organization of individual families that sometimes camped (and presumably foraged) alone and that at other times joined with other families, a social structure similar to that of living hunter-gatherers. Many researchers have argued that *Homo erectus* used fire, but this is a topic of considerable debate (Balter, 1995).

Homo erectus were hunters or scavengers that successfully preyed on a variety of medium-size and large mammals, including elephants, antelopes, horses, and deer. At later sites, the archeological evidence indicates that they exploited virtually all available animals in the area. Like both their primate forebears and living hunter-gatherers, *H. erectus* probably relied on plant parts of some sort for most of their diet. There are remains of berries at Zhoukoudian and other sites. As with other hominids, this part of the diet of *H. erectus* is very difficult to reconstruct, and our view of their subsistence behavior is certainly distorted by an overemphasis on hunting because of abundant animal bones.

Late *Homo*

Homo heidelbergensis

In deposits dating from much of the middle Pleistocene, approximately 700,000 to 130,000 years ago, we find archaic hominids from Africa and Europe that are more modern than *Homo erectus* but lacking the diagnostic features of either Neandertals or modern humans. Commonly described as "archaic *Homo sapiens*," they are now generally placed in a separate taxon, ***Homo heidelbergensis*** (Rightmire, 1998). Fossils attributed to this species are known from southern Africa (Broken Hill or Kabwe, and Berg Aukas), eastern Africa

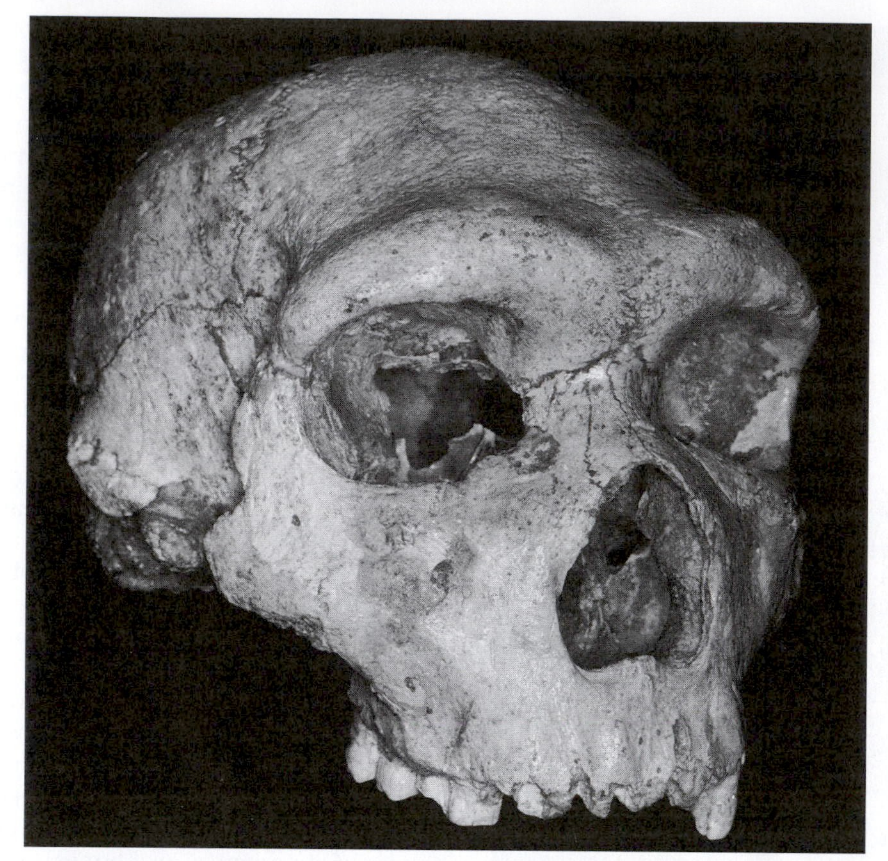

FIGURE 17.16 The skull of *Homo heidelbergensis* from Petrolona Greece (courtesy of M. F. Wolpoff).

(Bodo), Greece (Petralona; Fig. 17.16), Germany (Heidelberg), and China (Dali). Some workers (Tattersall, 1986; Groves, 1992) have argued that this group should be divided into several more species. Less complete remains probably attributable to this taxon are known from many other sites in Europe and Africa. The cranial similarities between African and European fossils are striking, and it is likely that this group of middle Pleistocene hominids is an African radiation that spread to the northern continents in the late early Pleistocene or early middle Pleistocene, as *Homo erectus* did much earlier (e.g., Klein, 1995).

Compared to *Homo erectus, H. heidelbergensis* has smaller teeth and a mandible with a broader ramus and a deeper corpus. The cranium is characterized by a greater cranial capacity, a divided brow, and an upright nasal aperture, among other features (Rightmire, 1996). Many of the better-known crania of *H. heidelbergensis* are enormous. This was a very large species of hominid, with estimated body sizes of roughly 100 kg, based on orbit dimensions (Kappelman, 1996) and postcranial remains (Grine *et al.,* 1995; Ruff *et al.,* 1997).

Postcranial elements of *H. heidelbergensis* are rare. There are several elements from Europe;

and from southern Africa there is a pelvis from Kabwe and a femur and parts of an ulna and radius, all possibly attributed to this taxon (Grine *et al.*, 1995). The femur is massive.

Fossils attributed to *H. heidelbergensis* are associated with Acheulian handaxes in both Africa and Europe. Like earlier *Homo erectus,* these archaic humans were hunters and gatherers or scavengers (Klein, 1995).

Homo neanderthalensis

Neandertals, now commonly placed in a distinct species, ***Homo neanderthalensis,*** are a distinct group of middle to early upper Pleistocene hominids from Europe and the Mideast. Neandertals seem to have originated among European populations of *Homo heidelbergensis;* and middle Pleistocene hominids, such as the 300,000 year old discoveries from Atapuerca, Spain (Arsuage *et al.*, 1993), are often considered transitional between the two species (e.g., Stringer, 1995). The latest Neandertals are contemporaneous with modern humans in Europe of approximately 35,000 years ago. Neandertals are characterized by low, crowned molars with a large pulp cavity, a sloping mandibular ramus, a prognathic face with a large and unusually formed nasal region (Schwartz and Tattersall, 1996), large brow ridges, and a large brain in a rounded cranium (Fig. 17.17). Their limbs are quite robust, with shortened distal elements (Fig. 17.18).

Neandertals were hunter-gatherers that used a wide range of Mousterian flake stone tools as well as wooden implements. They show considerable diversity in both prey selection and tool use in different geographical areas and temporally within regions. There is evidence that they used free-standing shelters as well as caves. They used fire, although cooking hearths are relatively rare. The are the first hominids known to actively bury their dead.

After a quarter millennium of considerable ecological and demographic success, Neandertals disappeared approximately 35,000 years ago. Why and how are two of the most hotly debated issues in human evolution since their disappearance roughly coincides with the initial appearance in Europe of modern humans belonging to our own species, *Homo sapiens.*

Homo sapiens

Anatomically, modern ***H. sapiens,*** similar to living populations of our species, first appear in the fossil record of Africa and the Middle East approximately 100,000 years ago with transitional forms even earlier (Bräuer *et al.*, 1997) and only about 40,000 years ago in Europe. Our species is distinguished from earlier members of the genus *Homo* by our smaller teeth, vertical mandibular ramus, development of a chin, shortened face, reduced brow ridges, and high vertical forehead (Fig. 17.17). Compared with Neandertals, we have smaller brains. Our limbs are more gracile and have much thinner bony shafts (Churchill, 1996). The markedly lower robustness of the skeleton in modern *H. sapiens* in the late Pleistocene seems to be related to the replacement of physical exertion with technological skill.

Although the earliest *H. sapiens* were probably similar to earlier hominids in being hunters and gathers, many populations specialized in fish, shellfish, small mammals, and undoubtedly many other foods before developing the sophisticated habits of agriculture and animal husbandry in the very recent past. The development of food preparation by cooking, a striking departure from our primate heritage, has been a major contributor to the continued dental reduction that characterizes our species. The technological skill of our species has enabled us to exploit virtually all available habitats on earth, from the more traditional woodlands and savannahs to tropical

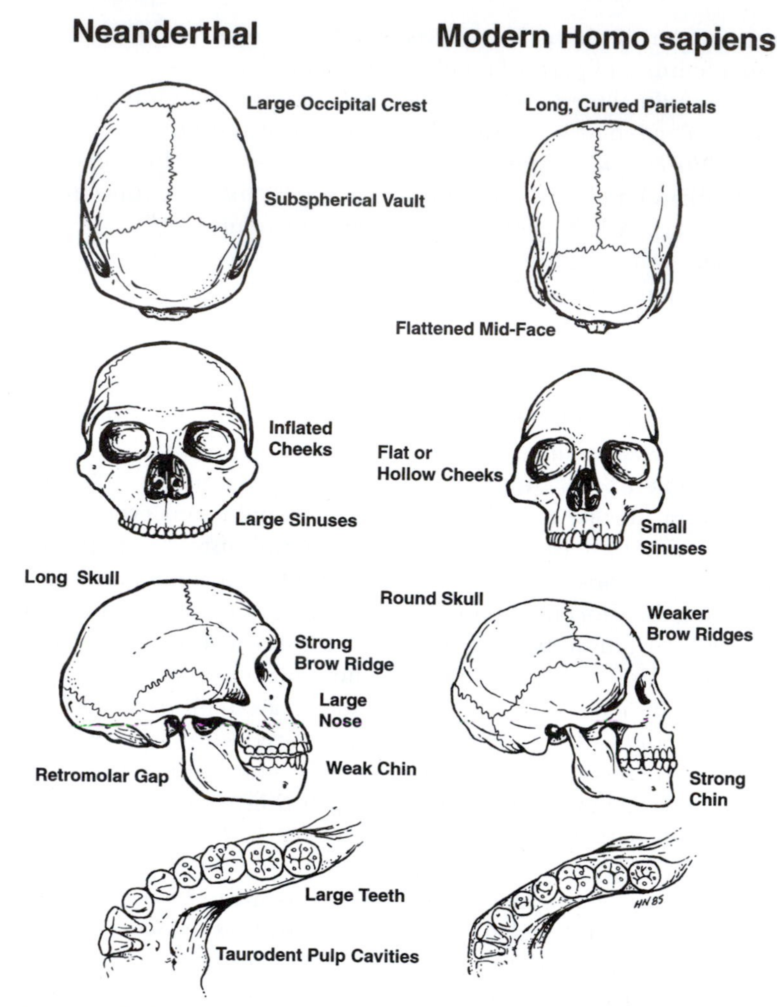

FIGURE 17.17 Comparative dental and cranial features of *Homo neanderthalensis* and *Homo sapiens.*

forests, oceanic islands, and the arctic. In addition, from as early as about 50,000 years ago, *Homo sapiens* have been distinguished from more archaic hominids in the range of cultural and artistic abilities. *Homo sapiens* considerably extended the geographic range of *H. erectus* and earlier hominids by successfully colonizing areas that had previously been beyond the range of catarrhines, such as North America and South America, as well as islands, such as Madagascar, that had never been colonized by higher primates, and Australia, which had never seen a primate. We are still expanding our range. Our own species, which has been studied more thoroughly than any other primate species, demonstrates a complex history of geographic differences (e.g., Cavalli-Sforza *et al.,* 1994).

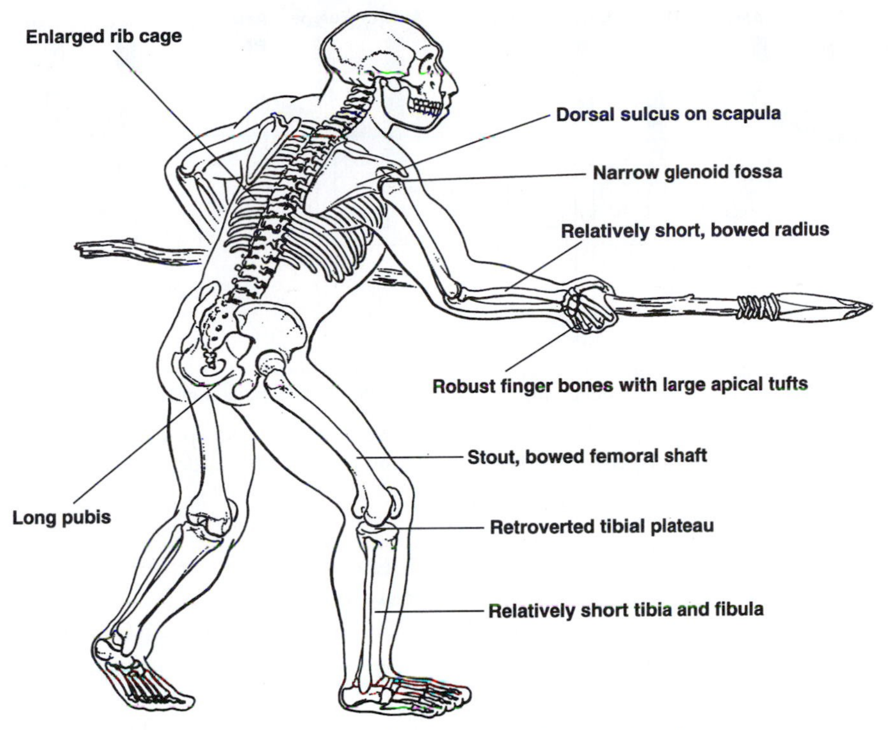

FIGURE 17.18 The skeleton of *Homo neanderthalensis,* with distinctive features indicated.

HUMAN PHYLOGENY

The major issues in human phylogeny concern whether the various populations and species of fossil humans are part of a single lineage or whether there have been multiple lineages at any one time, some of which became extinct while others gave rise to later species. This issue has been raised for virtually every potential transition between species in hominid evolution during the past 2 million years. From a paleontological perspective, these debates reflect the remarkable detail with which we can reconstruct the biochronology and biogeography. Given the limitations of the fossil record, it is not surprising that such phylogenetic details cannot be clearly resolved for a group of animals with a nearly cosmopolitan distribution for perhaps as long as 2 million years.

It is generally accepted that there were at least three, and probably more, species of *Homo* living in Africa approximately 2 million years ago (Fig. 17.20). However, there is no consensus on how these species are related to one another, except that *H. habilis* seems to be the most primitive in many aspects of its skeleton. Of these early Pleistocene hominids, most authorities feel that *Homo erectus* (or *Homo ergaster*) is the species that gave rise to later hominid species. Whether the African population should be separated as a distinct species that is more closely related to later hominids than the Asian radia-

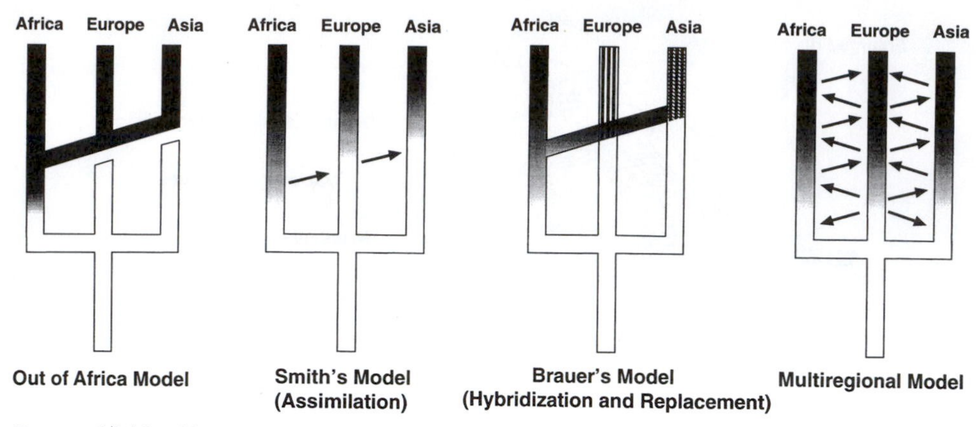

FIGURE 17.19 Alternative models of the evolutionary and biogeographical relationship between anatomically modern humans (black lines) and archaic hominids (white lines).

tion is a much debated issue (see Franzen, 1994).

The geographically widespread hominids from the early middle Pleistocene are probably the most poorly understood. In the past they have been variously considered as late *Homo erectus,* "early (archaic) *Homo sapiens,*" or several distinct species. Now they are most commonly placed in a single species, *Homo heidelbergensis.* The phylogenetic relationships implied by different systematic schemes are equally diverse. The phylogenetic relationships between these archaic middle Pleistocene hominids and later, more well-defined groups, such as Neandertals and anatomically modern *Homo sapiens,* are probably complex. In some cases there seem to be transitions between time-successive populations within a single geographical area, such as *H. heidelbergensis* and *H. neanderthalensis* in Europe and between *H. heidelbergensis* and *H. sapiens* in Africa (Bräuer *et al.,* 1997). However, in other areas, such as the Levant or Asia, it seems that very different populations or species of hominids persisted synchronously for long periods of time. The geographical and phylogenetic history of humans in the middle Pleistocene is a long way from being resolved (see Howell, 1994).

The most controversial issue in later human evolution is the geographical and phyletic origin of anatomically modern *Homo sapiens* (see Minugh-Purvis, 1996). There are two major competing theories, or models, and a host of intermediate, compromise models (Fig. 17.19). Did our species have a single African origin, with anatomically modern humans replacing all other hominid populations in the rest of the globe during the last 40,000 years? Or did populations of more archaic hominids, traditionally identified as *Homo erectus* in Asia or Neandertals in Europe, give rise to modern humans either independently or through hybridization with more modern immigrants from other regions?

The simplest view of human evolution over the past 2 million years is that of Wolpoff and colleagues (1994), who argue that there has been but a single cosmopolitan lineage of hominids since the first appearance of *Homo erectus* and that all hominids from *Homo erectus* through modern humans should be placed in a single species. In their view (e.g., Thorne and Wolpoff, 1981), many regional character-

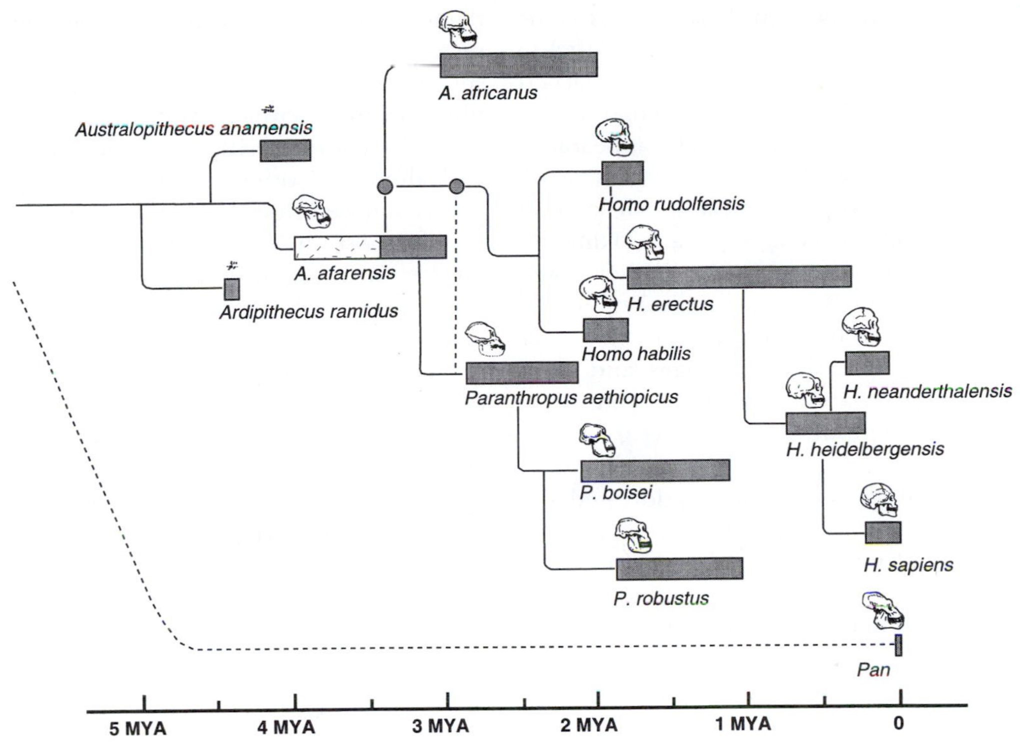

FIGURE 17.20 Summary of hominid phylogeny with the temporal range of individual species.

istics seen in *H. erectus* fossils from Africa and Asia and the Neandertals of Europe persist into later *H. sapiens* and modern populations. This is usually called the "Regional Continuity Model."

Most other authorities argue that the transition from *H. erectus* to *H. sapiens* occurred only once, probably in Africa, and then *H. sapiens* replaced archaic hominids *H. erectus, H. heidelbergensis,* and *H. neanderthalensis* throughout the world (e.g., Stringer and Andrews, 1988). This model is usually referred to as the "Out of Africa Replacement Model."

Between these extremes are numerous intermediate models that involve replacement in some regions and/or various amounts of population interbreeding in others (see reviews by Aiello, 1993; Mignuh-Purvis, 1995; Stringer and Gamble, 1993). For example, both Smith (e.g., Smith *et al.,* 1989) and Brauer (e.g., 1992) agree that modern human features originated in Africa and spread to the rest of the world, but they believe that there was nevertheless some degree of interbreeding or hybridization between the modern hominids and the archaic groups in each region. In yet another alternative, Lahr and Foley (1994) have suggested that the diversity of modern *Homo sapiens* may be the result of a succession of independent migrations from Africa.

Information from a wide range of disciplines, including paleontology, genetics, archeology, and geochronology, has been brought to bear on this debate. Indeed, the more we learn, the more complex the likely scenarios turn out to be. Although the earliest fossils of anatomically modern *Homo sapiens* come from sites in Africa and the Levant dated to approximately 100,000 years ago, the earliest *H. sapiens* had a culture similar to that of contemporary archaic hominids. The appearance of characteristic archeological features associated with our species appear much later, at 40,000–50,000 years ago, or approximately the time this group spread to other continents. This suggests major neural or behavioral changes in our lineage took place later than the original skeletal changes, and it was these later innovations that enabled our rapid spread throughout the globe (e.g., Klein, 1992, 1994).

Many genetic studies have addressed the question of the origins of *Homo sapiens.* Most show that Africa has the greatest diversity of human genetic material, suggesting a longer period of human history on that continent (e.g., Stoneking, 1993; Hammer and Zegura, 1997). However, these dates for the origin of our species rely on a wide range of assumptions, and many of the results could reflect patterns of differential Pleistocene population size rather than area of origin (e.g., Relethford, 1996). Nevertheless, the genetic studies regularly point to a genetic origin for our species between 100,000 and 200,000 years ago and recent studies of DNA from a Neandertal limb bone indicate that the neandertal lineage is very distant from modern humans with a last common ancestor over 500,000 years ago (Krings *et al.,* 1997; Ward and Stringer, 1997). Again, human evolution is remarkable in the detail of information that can be brought to bear on this well-documented and relatively recent aspect of primate evolution.

HUMANS AS AN ADAPTIVE RADIATION

In some respects it seems inappropriate to discuss a single genus (*Homo*) as an adaptive radiation. Yet, from a primate perspective, our genus is remarkable in both the geographic spread and the diversity of ecological conditions we have occupied over the past 2 million years. In contrast with nonhuman primates and almost all other organisms, our genus seems to have achieved considerable adaptive abilities, associated with the exploitation of diverse environments and geographic dispersion, with only moderate degrees of morphological change and speciation. Moreover, since the late Pleistocene, modern humans have ranged throughout Africa, Asia, and Europe with relatively little morphological differentiation in different environments—apart from minor differences, such as skin color and various blood polymorphisms, in association with different environments and perhaps other factors, such as random genetic drift and geographic isolation. Like macaques, humans are an example of what one scientist has called "specialized generalists" (Rose, 1983).

The evolutionary specializations that have permitted humans to exploit such a wide range of environments in many different ways are those that characterize the genus: a large brain that enhances learning and memory, our uniquely proportioned hand with its very mobile thumb, and our uniquely shaped vocal tract, which in conjunction with our brain permits a wide range of linguistic communication. Together, these features facilitated the technological capabilities that enabled humans to exploit diverse habitats without extensive morphological diversification.

BIBLIOGRAPHY

GENERAL

Ciochon, R. L., and Fleagle, J. G. (1993). *The Human Evolution Sourcebook.* Englewood Cliffs, N. J.: Prentice-Hall.

Conroy, G. C. (1997). *Reconstructing Human Origins.* New York: Norton.

Corruccini, R. S., and Ciochon, R. L. (1994). *Integrative Paths to the Past.* Englewood Cliffs, N. J.: Prentice-Hall.

Delson, E. (1985). *Ancestors: The Hard Evidence.* New York: Alan R. Liss.

Grine, F. E. (1988). *Evolutionary History of the Robust Australopithecines.* Hawthorne, N. Y.: Aldine.

Hill, A. (1994). Late Miocene and early Pliocene hominoids from Africa. In *Integrative Paths to the Past: Paleoanthropological Advances,* ed. R. S. Corruccini and R. L. Ciochon, pp. 123–146. Englewood Cliffs, N. J.: Prentice-Hall.

Johanson, D. C., and Edgar, B. (1996). *From Lucy To Language.* New York: Simon & Schuster

Klein, R. (1989). *The Human Career.* Chicago: University of Chicago Press.

Smith, E. H., and Spencer, F. (1984). *The Origins of Modern Humans: A World Survey of the Fossil Evidence.* New York: Alan R. Liss.

Toth, N., and Schick, K. D. (1986). The first million years: The archeology of protohuman culture. In *Advances in Archaeological Method and Theory,* vol. 9, ed. M. B. Schiffer, pp. 1–96. Orlando, Fla.: Academic Press.

Wolpoff, M. (1996). *Human Evolution.* New York: McGraw-Hill.

AUSTRALOPITHECINES

Asfaw, B. (1987). The Belohdelie frontal: New evidence of early hominid cranial morphology from the Afar of Ethiopia. *J. Human Evol.* **16**:611–624.

Asfaw, B., Beyene, Y., Suwa, G., Walter, R. C., White, T. D., WoldeGabriel, G., and Tesfaye, Y. (1992). The earliest Acheulean from Knoso-Gardula. *Nature* **360**: 732–735.

Brace, C. L. (1979). Biological parameters and Pleistocene hominid lifeways. In *Primate Ecology and Human Origins: Ecological Influences on Social Organization,* ed. I. S. Bernstein and E. O. Smith, pp. 263–289. New York: Garland STPM Press.

Brain, C. K. (1981). *The Hunters or the Hunted? An Introduction to African Cave Taphonomy.* Chicago: University of Chicago Press.

______ ed. (1993). Swartkrans: A cave's chronicle of early man. *Transvaal Mus. Monograph* **8**:1–270.

Bromage, T. G., and Dean, M. C. (1985). Re-evaluation of the age at death of immature fossil hominids. *Nature (London)* **317**:525–527.

Brunet, M., Beauvilain, A., Coppens, Y., Heintz, E., Moutaye, A. H. E., and Pilbeam, D. (1995). The first australopithecine 2,500 kilometres west of the Rift Valley (Chad). *Nature* **378**:373–374.

______ (1996). *Australopithecus bahrelghazali,* une novelle espece d'Hominide ancien de la region de Koro Toro (Tchad). *C. R. Acad. Sci. Paris* **322**:907–913.

Clark, R. J., and Tobias, P. V. (1995). Sterkfontein member 2 foot bones of the oldest South African hominid. *Science* **269**:521–524.

Conroy, G. C., and Vannier, M. W. (1987). Dental development of the Taung skull from computerized tomography. *Nature (London)* **329**:625–627

Coppens, Y., and Senut, B. (199X). *Origine(s) de la Bipédie chez les Hominidés.* Colloque Int. de la Fondation Singer-Polignac. CNRS, Paris.

Dart, R. A. (1925). *Australopithecus africanus:* The man-ape of South Africa. *Nature (London)* **115**:195–199.

Darwin, C. (1871). *The Descent of Man and Selection in Relation to Sex.* London: Murray.

de Menocal, P. B. (1995). Plio-Pleistocene African climate. *Science* **270**:53–58.

Falk, D. (1987). Hominid paleoneurology. *Ann. Rev. Anthropol.* **16**:13–30.

______ (1991). 3.5 Million years of hominid brain evolution. *Seminars in the Neurosicences* **3**:409–416.

Fedigan, L. M. (1986). The changing role of women in models of human evolution. *Ann. Rev. Anthropol.* **15**:25–66.

Fleagle, J. G., Rasmussen, D. T., Yirga, S., Bown, T. M., and Grine, F. E. (1991). New hominid fossils from Fejej, Southern Ethiopia. *J. Human Evol.* **21**:145–152.

Grine, F. E. (1981). Occlusal morphology of the mandibular permanent molars of the South African Negro and the Kalahari San (Bushman). *Ann. S. Afr. Mus.* **86**: 157–215.

______ (1986). Dental evidence for dietary differences in *Australopithecus* and *Paranthropus:* A quantitative analysis of permanent molar microwear. *J. Human Evol.* **15**: 783–822.

______ (1988). *Evolutionary History of the Robust Australopithecines.* Hawthorne, N. Y.: Aldine.

______ (1993). Australopithecine taxonomy and phylogeny: historical background and recent interpretation. In *The Human Evolution Sourcebook,* ed. R. L. Ciochon and J. G. Fleagle, pp. 196–208. Englewood Cliffs, N. J.: Prentice-Hall.

Harris, J. W. K. (1983). Cultural beginnings: Plio-Pleistocene archaeological occurrences from the Afar, Ethiopia. *Afr. Archaeol. Rev.* **1**:3–31.

Hewes, G. W. (1961). Food transport and the origin of hominid bipedalism. *Am. Anthropol.* **69**:63–67.

Hill, A. (1985). Early hominid from Baringo District, Kenya. *Nature (London)* **315**:222–224.

Holloway, H. L. (1983). Human brain evolution: A search for units, models and synthesis. *Can. J. Anthropol.* **3**:215.

Howell, F. C. (1978). Hominidae. In *Evolution of African Mammals,* ed. V. J. Maglio and H. B. S. Cooke, pp. 154–248. Cambridge, Mass.: Harvard University Press.

Hrdy, S. B., and Bennet, W. (1981). Lucy's husband: What did he stand for? *Harvard Magazine* (July–August), p. 7.

Hunt, K. D. (1994). The evolution of human bipedality: Ecology and functional morphology. *J. Human Evol.* **26**: 183–202.

——— (1996). The postural feeding hypothesis: An ecological model for the evolution of bipedalism *S. Afr. J. Sci.* **9**:77–90

Isaac, G. (1983). Aspects of human evolution. In *Evolution from Molecules to Men,* ed. D. S. Bendall, pp. 509–543. Cambridge: Cambridge University Press.

——— (1984). The archaeology of human origins: Studies of the lower Pleistocene in East Africa 1971–1981. In *Advances in World Archaeology,* vol. 3, ed. F. Wendorf and A. Close, pp. 1–87 Orlando, Fla.: Academic Press.

Jablonski, N. G. and Chaplin, G. (1993). Origin of habitual terrestrial bipedalism in the ancestor of the Hominidae. *J. Human Evol.* **24**:259–280.

Johanson, D. C., and Edey, M. (1981). *Lucy: The Beginnings of Humankind.* New York: Simon & Schuster.

Johanson, D. C., and White, T. D. (1979). A systematic assessment of early African hominids. *Science* **203**:321–330.

Johanson, D. C., White, T. D., and Coppens, Y. (1978). A new species of the genus *Australopithecus* (Primates: Hominidae) from the Pliocene of eastern Africa. *Kirtlandia,* no. 28, pp. 1–14.

Jolly, C. F. (1970). The seed-eaters: A new model of hominid differentiation based on a baboon analogy. *Man* **5**: 5–28.

Jungers, W. L. (1982). Lucy's limbs: Skeletal allometry and locomotion in *Australopithecus afarensis. Nature (London)* **297**:676–678.

Kappelman, J., and Fleagle, J. G. (1995). Age of early hominids. *Nature* **376**:558–559.

Kappelman, J., Swisher, C. C. III, Fleagle, J. G., Yirga, S., Bown, T. M., and Feseha, M. (1996). Age of *Australopithecus afarensis* from Fejej, Ethiopia. *J. Human Evol.* **30**:139–146.

Kay, R. E. (1985). Dental evidence for the diet of *Australopithecus. Ann. Rev. Anthropol.* **14**:315–342.

Kimbel, W. H., White, T. D., and Johanson, D. C. (1984). Cranial morphology of *Australopithecus afarensis*: A comparative study based on a composite reconstruction of the adult skull. *Am. J. Phys. Anthropol.* **64**: 337–388.

Kimbel, W. H., Johanson, D. C., and Rak, Y. (1994). The first skull and other new discoveries of *Australopithecus afarensis* at Hadar, Ethiopia. *Nature* **368**: 449–451.

Leakey, M. D., and Hay, R. L. (1979). Pliocene footprints in the Laetoli beds at Laetoli, northern Tanzania. *Nature (London)* **278**:317.

Leakey, M. G., Feibel, C. S., McDougall, I., and Walker, A. (1995). New four-million-year-old hominid species from Kanapoi and Allia Bay, Kenya. *Nature* **376**:565–571.

Leakey, M. G., Feibel, C. S., McDougall, I., Ward, C., and Walker, A. (1998). New specimens and confirmation of an early age for *Australopithecus anamensis. Nature* **393**:62–66.

Lockwood, C. A., Richmond, B. G., Jungers, W. L., and Kimbel, W. H. (1996). Randomization procedures and sexual dimorphism in *Australopithecus afarensis. J. Human Evol.* **31**:537–548.

Lovejoy, C. O. (1981). The origin of man. *Science* **211**: 341–350.

Lovejoy, C. O., Johanson, D. C., and Coppens, Y. (1982). Hominid lower limb bones recovered from the Hadar Formation: 1974–1977 collections. *Am. J. Phys. Anthropol.* **57**:679–700.

McHenry, H. M. (1975). Fossils and the mosaic nature of human evolution. *Science* **190**:425–431.

——— (1986). The first bipeds: A comparison of the *A. afarensis* and *A. africanus* postcranium and implications for the evolution of bipedalism. *J. Human Evol.* **15**:177–191.

——— (1994). Tempo and mode in human evolution. *Proc. Natl. Acad. Sci.* **91**:6780–6786.

Morgan, E. (1994). *The Scars of Evolution: What our bodies tell us about human origins.* Oxford: Oxford University Press.

Oxnard, C. E. (1975). *Uniqueness and Diversity in Human Evolution.* Chicago: University of Chicago Press.

Plavcan, J. M., and van Schaik, C. P. (1997). Interpreting hominid behavior on the basis of sexual dimorphism. *J. Human Evol.* **32**:345–374.

Reed, K. E. (1997). Early hominid evolution and ecological change through the African Plio-Pleistocene. *J. Human Evol.* **32**:289–322.

Rodman, P. S., and McHenry, H. M. (1980). Bioenergetics and the origin of hominid bipedalism. *Am. J. Phys. Anthropol.* **52**:103–106.

Rose, M. D. (1976). Bipedal behavior of olive baboons (*Papio anubis*) and its relevance to an understanding of the evolution of human bipedalism. *Am. J. Phys. Anthropol.* **44**:247–261.

——— (1991). The process of bipedalization in hominids. In *Origine(s) de la Bipedie chez les Hominides,* ed. Y. Coppens and B. Senut, pp. 37–48. Paris: CNRS.

Senut, B. (1981). *L'Humerus et ses articulations chez les Hominides Plio-Pleistocene.* Paris: CNRS.

Skelton, R. R., and McHenry, H. M. (1992). Evolutionary relationships among early hominids. *J. Human Evol.* **23**: 309–349.

Smith, B. H. (1986). Dental development in *Australopithecus* and early *Homo*. *Nature (London)* **323**:327–330.

______ (1994). Sequence of emergence of the permanent teeth in *Macaca, Pan, Homo,* and *Australopithecus*: Its evolutionary significance. *Am. J. Hum. Biol.* **6**:61–76.

Spencer. L. M. (1997) Dietary adaptations of Plio-Pleistocene Bovidae: Implications for hominid habitat use. *J. Human Evol.* **32**:201–228.

Spoor, F., Wood, B., and Zonneveld, F. (1994). Implications of early hominid labyrinthine morphology for evolution of human bipedal locomotion. *Nature* **369**: 645–648.

Stern, J. T., Jr., and Susman, R. L. (1983). The locomotor anatomy of *Australopithecus afarensis*. *Am. J. Phys. Anthropol.* **60**:279–317.

Steudel, K. L. (1994). Locomotor energetics and hominid evolution. *Evol. Anthropol.* **3**:42–48.

Strait, D. S., Grine, F. E., and Moniz, M. A. (1997). A reappraisal of early hominid phylogeny. *J. Human Evol.* **32**:17–82.

Susman, R. L. (1983). Evolution of the human foot: Evidence from Plio-Pleistocene hominids. *Foot Ankle* **3**: 365–376.

______ (1988). New postcranial remains from Swartkrans and their bearing on the functional morphology and behavior of *Paranthropus robustus*. In *Evolutionary History of the Robust Australopithecines,* ed. F. E. Grine, pp. 144–172. Hawthorne, N. Y.: Aldine.

______ (1991). Who made the Olduwan tools? Fossil evidence for tool behavior in Plio-Pleistocene hominids. *J. Anthropol. Res.* **47**:129–149.

______ (1994). Fossil evidence for early hominid tool use. *Science* **265**:1570–1573.

Susman, R. L., Stern, J. T., Jr., and Jungers, W. L. (1984). Arboreality and bipedality in the Hadar hominids. *Folia Primatol.* **43**:113–156.

Suwa, G., White, T. D., andHowell, F. C. (1996). Mandibular postcanine dentition from the Shungura Formation, Ethiopia: Crown morphology, taxonomic allocations, and Plio-Pleistocene hominid evolution. *Am. J. Phys. Anthropol.* **101**:247–282.

Taylor, C. R., and Rowntree, V. J. (1970). Running on two or four legs: Which consumes more energy? *Science* **179**:186–187.

Tobias, P. V. (1967). *Olduvai Gorge,* vol. 2: *The Cranium and Maxillary Dentition of Australopithecus (Zinjanthropus) boisei.* Cambridge: Cambridge University Press.

______ (1983). Hominid evolution in Africa. *Can. J. Anthropol.* **3**:163–190.

______ (1985). *Hominid Evolution.* New York: Alan R. Liss.

Vrba, E. S. (1985). Ecological and adaptive changes associated with early hominid evolution. In *Ancestors: The Hard Evidence,* ed. E. Delson, pp. 63–71. New York: Alan R. Liss.

Vrba, E. S., Denton, G. H., Partridge, T. C., and Burckle, (1996). *Paleoclimate and Evolution with Emphasis on Human Origins.* New Haven, Conn.: Yale University Press.

Walker, A., and Leakey, R. E. F. (1978). The hominids of East Turkana. *Sci. Am.* **239**(2):54–66.

Walker, A., Leakey, R. E., Harris, J. M., and Brown, E. H. (1986). 2.5 Myr *Australopithecus boisei* from west of Lake Turkana, Kenya. *Nature* **322**:517–522.

Washburn, S. L. (1963). Behavior and human evolution. In *Classification and Human Evolution,* ed. S. L. Washburn, pp. xx–yy. Chicago: Aldine.

Wheeler, P. E. (1984) The evolution of bipedality and loss of functional body hair in hominids. *J. Human Evol.* **13**:91–98.

______ (1991). The influence of bipedelism in the energy and water budgets of early hominids. *J. Human Evol.* **23**:379–388.

______ (1993). The influence of stature and body form on hominid energy and water budgets: A comparison of *Australopithecus* and early *Homo* physiques. *J. Human Evol.* **24**:13–28.

White, T. D. (1994). Ape and hominid limb length. *Nature* **369**:194.

White, T. D., Johanson, D. C., and Kimbel, W. H. (1981). *Australopithecus africanus*: Its phyletic position reconsidered. *S. Afr. J. Sci.* **77**:445–470.

White, T. D., Suwa, G., Hart, W. K., Walter, R. C., WoldeGabriel, G, de Heinzelin, J., Clark, J. D., Asfaw, B., and Vrba, E. (1993). New discoveries of *Australopithecus* at Maka in Ethiopia. *Nature* **366**:261–265.

White, T. D., Suwa, G., and Asfaw, B. (1994). *Australopithecus ramidus,* a new species of early hominid from Aramis, Ethiopia. *Nature* **371**:306–333.

______ (1995). Corrigendum: *Australopithecus ramidus,* a new species of early hominid from Aramis, Ethiopia. *Nature* **375**:88

Wood, B. (1991). *Koobi Fora Research Project,* vol. 4: *Hominid Cranial Remains.* Oxford: Clarendon Press.

Wood, B. A., and Chamberlain, A. T. (1986). *Australopithecus*: Grade or clade? In *Major Topics in Primate and Human Evolution,* ed. B. A. Wood, L. Martin, and P. Andrews, pp. 220–248. Cambridge: Cambridge University Press.

Wrangham, R. W. (1980). Bipedal locomotion as a feeding adaptation in gelada baboons and its implications for hominid evolution. *J. Human Evol.* **9**: 329–332.

Zihlman, A., and Tanner, N. M. (1978). Gathering and the hominid adaptation. In *Female Hierarchies,* ed. L. Tiger and H. Fowler, pp. 163–194. Chicago: Beresford Book Service.

EARLY *HOMO*

Asfaw, B., Beyene, Y., Suwa, G., Walter, R. C., White, T. D., WoldeGabriel, G., and Yemane, T. (1992). The earliest Acheulean from Konso-Gardula. *Nature* **360**: 732–735.

Balter, M. (1995). Did *Homo erectus* tame fire first? *Science* **268**:1570.

Binford, L. R. (1981). *Bones and Ancient Men and Modern Myths.* New York: Academic Press.

Bräuer, G. (1994). How different are Asian and African *Homo erectus*? *Courier Forschungs-Institut Senckenberg* **171**: 301–317.

Bräuer, G., and Schultz, M. (1996). The morphological affinities of the Plio-Pleistocene mandible from Dmanisi, Georgia. *J. Human Evol.* **30**:445–481.

Brown, F., Harris, J. R., Leakey, R. E. F., and Walker, A. (1985). Early *Homo erectus* skeleton from west Lake Turkana, Kenya. *Nature* **316**:788–792.

Cachel, S., and Harris, J. W. K. (1995). Ranging patterns, land-use and subsistence in *Homo erectus,* from the perspective of evolutionary ecology. In *Evolution and Ecology of* Homo erectus, vol. 1, *Pithecanthropus,* ed. J. R. F. Bower and S. Sartono, pp. 51–66. CITY: Centennial Foundation,

Cerling, T. E. (1992). Development of grasslands and savannas in East Africa during the Neogene. *Paleogeog. Paleoclimatol. Paleoecol. (Global Planetary Change Section)* **97**:241–247.

Clark, J. D., de Heinzelin, J., Schick, K. D., Hart, W. K., White, T. D., WoldeGabriel, G., Walter, R. C., Suwa, G., Asfaw, B., Vrba, E., and Selassie, Y. H. (1994). African *Homo erectus*: Old radiometric ages and young Oldowan assemblages in the Middle Awash Valley, Ethiopia. *Science* **264**:1907–1910.

de Menocal, P. B. (1995). Plio-Pleistocene African climate. *Science* **270**:53–58.

Fleagle, J. G. (1995) Too many species? *Evol. Anthropol.* **4**: 37–38.

Gabunia, L., and Vekua, A. (1995). A Plio-Pleistocene hominid from Dmanisi, East Georgia, Caucasus. *Nature* **373**:509–512.

Grine, F. E., Demes, B., Jungers, W. L., and Cole, T. M. III. (1993). Taxonomic affinity of the early *Homo* cranium from Swartkrans, South Africa. *Am. J. Phys. Anthropol.* **92**:411–426.

Harris, J. W. K. (1983). Cultural beginnings: Plio-Pleistocene archaeological occurrences from the Afar, Ethiopia. *Afr. Archaeol. Rev.* **1**:3–31.

Hartwig-Scherer, S., and Martin, R. D. (1991). Was "Lucy" more human than her "child"? Observations on early hominid postcranial skeletons. *J. Human Evol.* **21**:439–449.

Isaac, G. (1983). Aspects of human evolution. In *Evolution from Molecules to Men,* ed. D. S. Bendall, pp. 509–543. Cambridge: Cambridge University Press.

Johanson, D. C., Masao, T. T., Eck, G. G., White, T. D., Walter, R. C., Kimbel, W. H., Asfaw, B., Manega, P., Ndessokia, P., and Suwa, G. (1987). New partial skeleton of *Homo habilis* from Olduvai Gorge, Tanzania. *Nature* **327**:205–209.

Kimbel, W. H., Walter, R. C., Johanson, D. C., Aronson, J. L., Assefa, Z., Eck, G. G., Hovers, E., Marean, C. W., Rak, Y., Reed, K. E., Vondra, C., Yemane, T., and Bobe-Quinteros, R. (1996). Late Pliocene *Homo* and Oldowan tools from the Hadar Foundation (Kada Hadar Member), Ethiopia. *J. Human Evol.* **31**:549–561.

Klein, R. G. (1995). Anatomy, behavior, and modern human origins. *J. World Prehistory* **9**:167–198.

Larick, R., and Ciochon, R. L. (1996). The African emergence and early Asian dispersals of the genus *Homo. Am. Sci.* **84**:538–551.

Leakey, R. E. F., and Walker, A. (1976). *Australopithecus, Homo erectus,* and the single-species hypothesis. *Nature* **261**:572–574.

Lieberman, D. E., Wood, B. A., and Pilbeam, D. R. (1996). Homoplasy and early *Homo*: An analysis of the evolutionary relationships of *H. habilis* and *H. rudolfenosis. J. Human Evol.* **30**:97–120.

Potts, R. (1994). Variables versus models of early Pleistocene hominid land use. *J. Human Evol.* **27**:7–24.

——— (1996). *Humanity's Descent: The Consequences of Ecological Instability.* New York: William Morrow.

Potts, R., and Shipman, P. (1981). Cutmarks made by stone tools on bones from Olduvai Gorge, Tanzania. *Nature* **291**:577–580.

Reed, K. E. (1997). Early hominid evolution and ecological change through the African Plio-Pleistocene. *J. Human Evol.* **32**:289–322.

Rightmire, G. P. (1981). Patterns in the evolution of *Homo erectus. Paleobiology* **7**:241–246.

——— (1990). *The Evolution of Homo erectus.* Cambridge: Cambridge University Press.

——— (1992). *Homo erectus*: Ancestor or evolutionary side branch? *Evol. Anthropol.* **1**:43–49.

——— (1993). Variation among early *Homo* crania from Olduvai Gorge and Koobi Fora region. *Am. J. Phys. Anthropol.* **90**:1–33.

Ruff, C. B. (1993). Climatic adaptation and hominid evolution: the thermoregulatory imperative. *Evol. Anthropol.* **2**:53–60.

______ (1995). Biomechanics of the hip and birth in early *Homo. Am. J. Phys. Anthropol.* **98**:527–574.

Schrenk, F., Bromage, T. G., Betzler, C. G., Ring, U., and Juwayeyl, Y. M. (1993). Oldest *Homo* and Pliocene biogeography of the Malawi Rift. *Nature* **365**: 833–836.

Semaw, S., Renne, P., Harris, J. W. K., Feibel, C. S., Bernor, R. L., Fesseha, N., and Mowbray, K. (1997). 2.5-million-year-old stone tools from Gona, Ethiopia. *Nature* **385**:333–336.

Sigmon, B. A., and Cybulski, J. S. (1981). Homo erectus: *Papers in Honor of Davidson Black.* Toronto: University of Toronto Press.

Spencer. L. M. (1997). Dietary adaptations of Plio-Pleistocene Bovidae: Implications for hominid habitat use. *J. Human Evol.* **32**:201–228.

Stanley, S. M. (1992). An ecological theory for the origin of *Homo. Paleobiology* **18**(3):237–257.

Susman, R. L. (1983). Evolution of the human foot: Evidence from Plio-Pleistocene hominids. *Foot Ankle* **3**(6): 365–376.

______ (1988). Hand of *Paranthropus robustus* from Member 1, Swartkrans: Fossil evidence for tool behavior. *Science* **240**:781–784.

______ (1991). Who made the Olduwan tools? Fossil evidence for tool behavior in Plio-Pleistocene hominids. *J. Anthropol. Res.* **47**:129–149.

Susman, R. L., and Creel, N. (1979). Functional and morphological affinities of the subadult hand (O. H. 7) from Olduvai Gorge. *Am. J. Phys. Anthropol.* **51**: 311–332.

Susman, R. L., and Stern, J. T. (1982). Functional morphology of *Homo habilis. Science* **217**:931–934.

Swisher, C. C. III, Curtis, G. H., Jacob, T., Getty, A. G., Suprijo, A., Widiasmoro, (1994). Age of the earliest known hominids in Java, Indonesia. *Science* **263**: 1118–1122.

Swisher, C. C. III, Rink, W. J., Anton, S. C., Schwarcz, H. P., Curtis, G. H., Soprijo, A., Widiasmoro, . (1996). Latest *Homo erectus* of Java: Potential contemporaneity with *Homo sapiens* in Southeast Asia. *Science* **274**: 1870–1874.

Tobias, P. V. (1991). The environmental background of hominid emergence and the appearance of the genus *Homo. Human Evol.* **6**(2):129–142.

______ (1992). The species *Homo habilis*: Example of a premature discovery. *Ann. Zool. Fennici* **28**:371–380.

Toth, N. (1987). The first technology. *Sci. Am.* **256**: 112–121.

Toth, N., and Schick, K. D. (1986). The first million years: The archeology of protohuman culture. In *Advances in Archaeological Method and Theory,* vol. 9, ed. M. B. Schiffer, pp. 1–96. Orlando, Fla.: Academic Press.

Vrba, E. S., Denton, G. H., Partridge, T. C., and Burckle, L. H. (1996). *Paleoclimate and Evolution with Emphasis on Human Origins.* New Haven, Conn.: Yale University Press.

Walker, A. C., and Leakey, R. E. F. (1994). *The Nariokotome* Homo erectus *Skeleton.* Cambridge, Mass.: Harvard University Press.

Walker, A. C., and Shipman, P. (1996). *The Wisdom of the Bones.* New York: A. A. Knopf.

Wanpo, H., Ciochon, R., Yumin, G., Larick, R., Qiren, F., Schwarcz, H., Yonge, C., de Vos, J., and Rink, W. (1995). Early *Homo* and associated artifacts from Asia. *Nature* **378**:275–278.

Wolpoff, M. H. (1971). Competitive exclusion among lower Pleistocene hominids: The single species hypothesis. *Man* **6**:601–614.

______ (1984). Evolution in *Homo erectus*: The question of stasis. *Paleobiology* **10**(4):389–406.

Wood, B. A. (1985). Early *Homo* in Kenya, and its systematic relationships. In *Ancestors: The Hard Evidence,* ed. E. Delson, pp. 206–214. New York: Alan R. Liss.

______ (1991). *Koobi Fora Research Project,* vol. 4: *Hominid Cranial Remains.* Oxford: Oxford University Press.

______ (1992). Origin and evolution of the genus *Homo. Nature* **355**:783–790.

LATE *HOMO*

Aiello, L. C. (1993). The fossil evidence for modern human origins in Africa: A revised view. *Am. Anthropol.* **95**: 73–96.

Aitken, M. J., Stringer, C. B., and Mellars, P. A. (1993). *The Origin of Modern Humans and the Impact of Chronometric Dating.* Princeton, N. J.: Princeton University Press.

Akazawa, T., Aoki, K., and Kimura, T. (1992). *The Evolution and Dispersal of Modern Humans in Asia.* Tokyo: Hokusen-Sha.

Andrews, P., and Franzen, J. L., eds. (1984). *The Early Evolution of Man with Special Emphasis on Southeast Asia and Africa. Cour. Forschr. Inst. Senckenberg* **69**.

Arsuaga, J.-L., Martinez, I., Gracia, A., Carretero, J.-M., and Carbonell, E. (1993). Three new human skulls from the Sima de los Huesos middle Pleistocene site in Sierra de Atapuerca, Spain. *Nature* **362**:534–537.

Bräuer, G. (1984). The "Afro-European sapiens hypothesis," and hominid evolution in East Asia during the late Middle and Upper Pleistocene. In *The Early Evolu-*

tion of Man with Special Emphasis on Southeast Asia and Africa, ed. P. Andrews and J. L. Franzen, *Cour. Forschr. Inst. Senckberg* **69**:145–165.

——— (1992). Africa's place in the evolution of *Homo sapiens.* In *Continuity or Replacement: Controversies in* Homo sapiens *Evolution,* ed. G. Brauer and F. H. Smith, pp.83–98. New York: A. R. Liss.

Bräuer, G., and Smith, F. H. (1992). *Continuity or Replacement: Controversies in* Homo sapiens *Evolution.* Rotterdam: Balkema.

Bräuer, G., Yokoyama, Y., Falguères, C., and Mbua, E. (1997). Modern human origins backdated. *Nature* **386**: 337–338.

Cann, R. L., Stoneking, M., and Wilson, A. C. (1987). Mitochondrial DNA and human evolution. *Nature (London)* **325**:31–36.

Cavalli-Sforza, L. L., Menozzi, P., and Piazza, A. (1994). *The History and Geography of Human Genes.* Princeton, N. J.: Princeton University Press.

Chen, T., Yang, Q., and Wu En. (1994). Antiquity of *Homo sapiens* in China. *Nature* **368**:55–56.

Churchill, S. E. (1996). Particulate versus integrated evolution of the upper body in late Pleistocene humans: A test of two models. *Am. J. Phys. Anthropol.* **100**: 559–583.

Delson, E. (1985). Late Pleistocene human fossils and evolutionary relationships. In *Ancestors: The Hard Evidence,* ed. E. Delson, pp. 296–300. New York: Alan R. Liss.

Eldredge, N., and Tattersall, I. (1975). Evolutionary models, phylogenetic reconstruction and another look at hominid phylogeny. In *Approaches to Primate Paleobiology,* ed. F. S. Szalay, pp. 218–242. Basel, Switzerland: Karger.

Franzen, J. L., ed. (1994). 100 years of *Pithecanthropus*: The *Homo erectus* problem. *Cour. Forsch. Inst. Senckenberg* **171**:1–361.

Grine, F. E., Jungers, W. L., Tobias, P. V., and Person, O. M. (1995). Fossil *Homo* femur from Berg Aukus, Northern Namibia. *Am. J. Phys. Anthropol.* **97**:151–185.

Groves, C. P. (1992). *A Theory of Human and Primate Evolution,* 2nd ed. Oxford: Oxford University Press.

Hammer, M. and Zegura, S. (1997). The role of the Y chromosome in human evolutionary studies. *Evol. Anthropol.* **5**:116–133.

Howell, F. C. (1994). A chronostratigraphic and taxonomic framework of the origins of modern humans. In *Origins of Anatomically Modern Humans,* ed. M. H. Niteki and D. V. Niteki, pp. 253–319. New York: Plenum Press.

Howells, W. W. (1976). Explaining modern man: Evolutionists versus migrationists. *J. Human Evol.* **5**:577–596.

Kappelman, J. (1996). The evolution of body mass and relative brain size in fossil hominids. *J. Human Evol.* **30**: 243–276.

Klein, R. G. (1992). The archeology of modern human origins. *Evol. Anthropol.* **1**:5–14.

——— (1994). The problem of human origins. In *Origins of Anatomically Modern Humans,* ed. M. H. Nitecki and D. V. Nitecki, pp. 3–22. New York: Plenum Press.

——— (1995). Anatomy, behavior and modern human origins. *J. World Prehistory* **9**:167–198.

Krings, M., Stune, A., Schmitz, R. W., Krainitzkijlt, ?., Stoneking, M., and Pääbo, S. (1997). Neandertal DNA Sequences and the Origin of Modern Humans. *Cell* **90**: 14–30.

Lahr, M. M. (1996). *The Evolution of Modern Human Diversity: A Study in Cranial Variation.* Cambridge: Cambridge University Press.

Lahr, M. M., and Foley, R. (1994). Multiple dispersals and modern human origins. *Evol. Anthropol.* **3**:48–60.

Mellars, P. (1996). *The Neanderthal Legacy: An archeological Perspective from Western Europe.* Princeton: Princeton University Press.

Mellars, P., and Stringer, C. B. (1989). *The Human Revolution: Behavioral and Biological Perspectives on the Origins of Modern Humans.* Princeton, N. J.: Princeton University Press.

Minugh-Purvis, N. (1996). The modern human origins controversy: 1984–1994. *Evol. Anthropol.* **4**:140–146.

Niteki, M. H., and Niteki, D. V. (1994). *Origins of Anatomically Modern Humans.* New York: Plenum Press.

Rak, Y. (1986). The Neanderthal: A new look at an old face. *J. Human Evol.* **15**:151–164.

Relethford, J. (1996). Genetics and modern human origins. *Evol. Anthropol.* **4**:53–63.

Rightmire, G. P. (1985). The tempo of change in the evolution of mid-Pleistocene *Homo.* In *Ancestors: The Hard Evidence,* ed. E. Delson, pp. 255–264. New York: Alan R. Liss.

——— (1996). The human cranium from Bodo, Ethiopia: Evidence for speciation in the middle Pleistocene? *J. Human Evol.* **31**:21–39.

——— (1998). Human evolution in the middle Pleistocene: The role of *Homo heidelbergensis. Evol. Anthropol.* **6**: 218–227.

Rose, M. D. (1983). Miocene hominoid postcranial morphology: Monkey-like, ape-like, neither or both? In *New Interpretation of Ape and Human Ancestry,* ed. R. S. Ciochon and R. S. Corruccini, pp. 405–420. New York: Plenum Press.

Ruff, C. B., Trinkaus, E., and Holliday, T. W. (1997). Body mass and encephalization in Pleistocene *Homo. Nature* **387**:173–176.

Schwartz, J. H., and Tattersall, I. (1996). Significance of some previously unrecognized apomorphies in the nasal region of *Homo neanderthalensis*. *Proc. Natl. Acad. Sci. USA* **93**:10852–10854.

Shipman, P. (1986). Scavenging or hunting in early hominids: Theoretical framework and tests. *Am. Anthropol.* **88**:27–43.

Smith, F. H., and Spencer, F. I., eds. (1985). *The Origins of Modern Humans.* New York: Alan R. Liss.

Smith, F. H., Falsetti, A. B., and Donnelley, S. M. (1989). Modern Human Origins. Yearbook of Physical Anthropology **32:**35–68.

Stoneking, M. (1993). DNA and recent human evolution. *Evol. Anthropol.* **2**:60–74.

Stringer, C. B. (1985). Middle Pleistocene hominid variability and the origin of late Pleistocene humans. In *Ancestors: The Hard Evidence,* ed. E. Delson, pp. 289–295. New York: Alan R. Liss.

______ (1995). The evolution and distribution of later Pleistocene human populations. In *Paleoclimate and Evolution: With Emphasis on Human Origins,* pp.524–531. New Haven, Conn.: Yale University Press.

Stringer, C. B., and Andrews, P. (1988). Genetic and fossil evidence for the origin of modern humans. *Science* **239**:1263–1268.

Stringer, C. B., and Gamble, C. S. (1993). In *Search of the Neanderthals.* London: Thames and Hudson.

Tattersall, I. (1986). Species recognition in human paleontology. *J. Human Evol.* **15**:131–143.

Tattersall, I., and Eldredge, N. (1977). Fact, theory and fantasy in human paleontology. *Am. Sci.* **65**:204–211.

Thorne, A. G., and Wolpoff, M. H. (1981). Regional continuity in Australasian Pleistocene hominid evolution. *Am. J. Phys. Anthropol.* **55**:337–349.

Trinkaus, E. (1986). The Neandertals and modern human origins. *Ann. Rev. Anthropol.* **15**:193–217.

______ (1989). *The Emergence of Modern Humans: Biocultural Adaptations in Later Pleistocene.* New York: Cambridge University Press. 300p

Valladas, H., Reyss, J. L., Joron, J. L., Valladas, G., Bar-Yosef, D., and Vandermeersch, B. (1988). Thermoluminescence dating of Mousterian "Proto-Cro-Magnon" remains from Israel and the origin of modern man. *Nature (London)* **331**:614–616.

Vrba, E. S., Denton, G. H., Partridge, T. C., and Burckle, L. H. (1996). *Paleoclimate and Evolution with Emphasis on Human Origins.* New Haven, Conn.: Yale University Press.

Waddle, D. M. (1994). Matrix correlation tests support a single origin for modern humans. *Nature* **368**:452–454.

Ward, R., and Stringer, C. (1997). A molecular handle on the Neanderthals. *Nature* **388**:225–226.

Weidenreich, E. (1946). *Apes, Giants and Man.* Chicago: University of Chicago Press.

Wolpoff, M. H., Thorne, A. G., Jelinek, J., and Yinyun, Z. (1994). The case for sinking *Homo erectus.* 100 years of *Pithecanthropus* is enough! In *100 Years of* Pithecanthropus: *The* Homo erectus *Problem,* ed. J. L. Franzen. *Cour. Forsch. Inst. Senckenberg,* **171**:341–361.

Wu, R., and Olsen, J. W., eds. (1985). *Palaeoanthropology and Palaeolithic Archaeology in the People's Republic of China.* New York: Academic Press.

EIGHTEEN

Patterns in Primate Evolution

PRIMATES AND EVOLUTIONARY THEORY

Throughout the preceding chapters we have examined primates—their phyletic relationships and ecological and behavioral adaptations—in more or less chronological and taxonomic order, family by family. Now that we have outlined and described the primate radiations, we are in a position to look for general trends. How can we characterize primate evolution as a whole? Are there repetitive patterns in the evolution of this order? With a good account of primate history and phylogeny at hand, we can also begin to examine theoretical questions about evolutionary processes. How do the various theories of evolutionary mechanisms, of speciation and of species extinction, fit the primate evidence? The fossil record of primates is particularly appropriate for such investigations, since it is as complete as that of any group of mammals, and it has certainly been more thoroughly studied.

Primate Adaptive Radiations

The primate fossil record documents an extraordinary diversity of extinct forms, not just isolated species and genera but major radiations of families. There are over fifty genera and two hundred species of living primates. Roughly twice as many fossil species have been discovered and described, and many more remain to be uncovered. The vast majority of primate taxa that have ever lived are now extinct.

Unfortunately, comparing the numbers of living species with the total number of extinct species cannot give us a good indication about how the present diversity of primates compares with that in the past. Living primates are from a single slice in time over a broad geographic area, whereas our knowledge of the fossil record is derived from samples of very restricted geographic areas and relatively long periods of time. More significant, there is virtually no paleontological record from the areas in which primate diversity is greatest today—the Amazon Basin, the Zaire Basin, and Southeast Asia.

As an alternative to comparing the taxonomic diversity of primates in the past with that of today, we can compare the morphological and reconstructed ecological diversity of selected primate faunas from the past with what we find among primate faunas today. In this way we can at least speculate about how the types of adaptations exploited by the primate faunas of the past were similar to or different from those characterizing the living radiations.

Body Size Changes

As we discussed in Chapter 9, body size is closely correlated with many aspects of a species' ecology, including diet and locomotion, and is also an easy parameter by which to compare species. Comparing the size distributions of living and fossil primate faunas gives us some indication of the adaptive diversity of the groups and the extent to which they seem to occupy similar ecological niches in a very broad sense. Figure 18.1 compares the range of body sizes occupied by various extinct primate groups with that found among living primates.

Several patterns are evident in the comparisons. First, there has been considerable diversity in size among primates throughout their evolution. This size diversity is certainly associated with a considerable ecological diversity, as we discussed in earlier chapters.

Like many other groups of mammals, primates seem to have increased in size during the past 55 million years. This overall size increase for primates during their evolution is

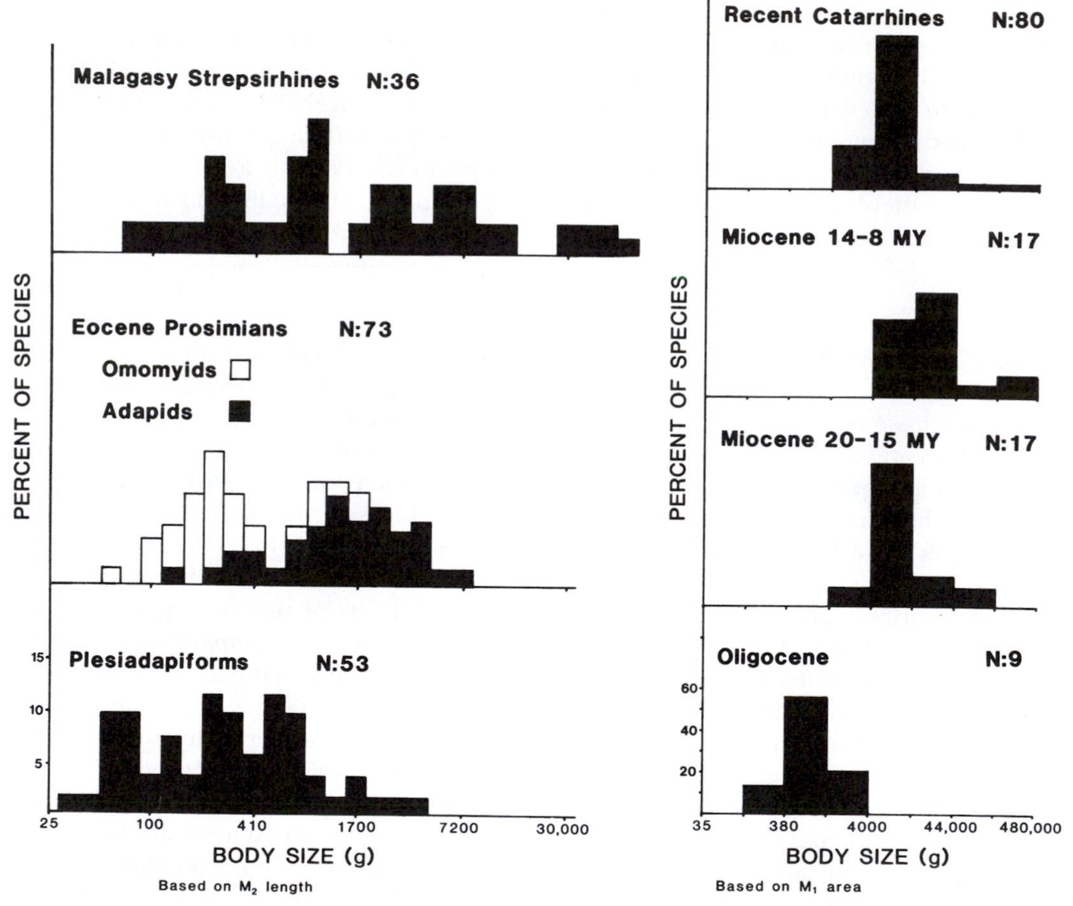

FIGURE 18.1 Body size distribution of prosimians and Old World anthropoids through time (redrawn and modified from Covert, 1986; Fleagle and Kay, 1985).

reflected in two different aspects of the distributions. First, very tiny primates or primate-like physical adaptations were relatively common in the Paleocene and Eocene, but they are less common in most Oligocene to Recent faunas (Fleagle, 1978; Covert, 1986). Second, Miocene through Recent primates include very large species that are unknown from earlier periods (Fleagle and Kay, 1985). Both of these size changes suggest that the adaptive space occupied by primates has shifted through time and that primates of the past showed different ecological adaptations from those found among living primate species. In this regard, the size changes corroborate the indications from specific morphological features, such as teeth and limbs (discussed in the following section).

Dietary Diversity

Although there is evidence of ecological diversity (especially in diet) throughout primate evolution, the expression of adaptations among different groups has probably varied considerably. A frugivorous plesiadapid was probably not much like a frugivorous adapid in many of its nondental adaptations, just as a frugivorous gibbon is very different in many aspects of its biology and foraging strategy from a frugivorous macaque. The detailed differences in foraging strategy that have been documented for extant primates in earlier chapters of this book are, of course, far beyond the realistic scope of paleontological studies, but we can see some general trends in the adaptive diversity of primate radiations through time. On the basis of dental morphology, it seems most likely that the plesiadapiforms were predominantly insectivorous and frugivorous, probably with some gum specialists. There are few species that show indications of extensive adaptation for folivory. In contrast, the fossil prosimians from the Eocene of North America and Europe include many folivorous species (especially among the adapoids), as well as others adapted for insectivorous (especially the omomyoids) and frugivorous diets (see Covert, 1986; Strait, 1997). We know relatively little about the radiation of modern groups of prosimians, except that the ancestors of galagos and lorises appear to have been relatively diverse by middle Miocene times (McCrossin, 1992) and that the diversity of Malagasy primates has been dramatically reduced in just the last few thousand years (e.g., Godfrey *et al.*, 1996).

Old World anthropoids have undergone dramatic changes in dietary diversity during the past 35 million years (Fig. 18.2). Early Oligocene higher primates, from the rich Fayum deposits of Egypt, are predominantly frugivorous anthropoids. There are no species that show dental adaptations to folivory comparable to those of many modern leaf eaters. In the early part of the Miocene epoch there are a few species with dentitions suggesting folivory, but it seems that most of the species were frugivorous (Kay and Ungar, 1997). The contemporaneous early Miocene lorises and galagos were presumably frugivores, insectivores, and gum eaters like their modern relatives. By the latter part of the Miocene there were more higher primates with teeth suggesting folivorous habits (e.g., *Mesopithecus, Microcolobus,* and *Oreopithecus*), and the proportions of frugivores to folivores was comparable to that among living Old World anthropoids. Moreover, there are indications that the fossil apes developed increasing specializations for shearing of leafy vegetation throughout the Miocene, perhaps as a result of competition with cercopithecoids (Kay and Ungar, 1997).

Less is known about the evolution of dietary diversity among fossil platyrrhines. Although the record is relatively poor for the Oligocene and earliest Miocene, the late middle Miocene fauna from La Venta in Colombia shows

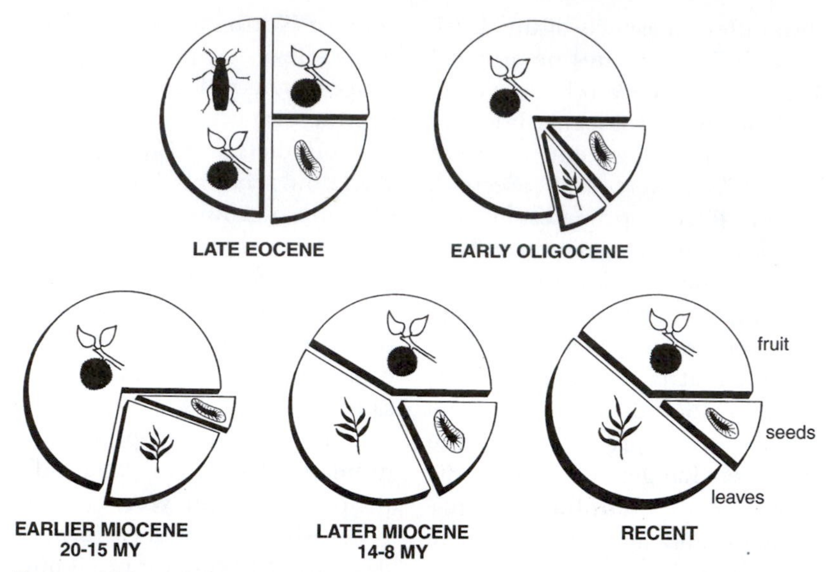

FIGURE 18.2 Changes in dietary diversity of Old World higher primates over the past 35 million years. Late Eocene anthropoids included insectivorous and frugivorous species; early Oligocene anthropoids were predominantly frugivorous; and in subsequent epochs folivorous and seed-eating species have become more common (modified from Fleagle and Kay, 1985).

a diversity comparable to that of modern platyrrhines, with specialized seed-eaters, folivores, frugivores, and insectivores (Fleagle *et al.*, 1997).

Locomotor Diversity

Temporal changes in primate locomotor habits are very difficult to document in the paleontological record because of the rarity of fossil skeletons. Still, there seem to be some general patterns (Figs. 18.3, 18.4). The few skeletal remains of plesiadapiforms, particularly the ankle and the claws, suggest arboreal habits, but the lack of an opposable hallux and the presence of long, curved claws indicate that their arboreal behavior was qualitatively different from that of Eocene to Recent primates. They were probably not leapers. In contrast, the Eocene fossil prosimians are similar to extant prosimians in general skeletal anatomy. Most seem to have been arboreal quadrupeds and quadrupedal leapers similar to cheirogaleids (Covert, 1995). Many of the diagnostic features of modern primates that first appear in the early Eocene seem to reflect an adaptation to arboreal leaping (Dagosto, 1993). On the other hand, there are few indications among the Eocene prosimians of the specialized vertical clinging behaviors that characterize living indriids or tarsiers. There is also no evidence of either terrestrial species or large species with extreme suspensory abilities, such as those found in *Palaeopropithecus*.

Old World anthropoids show locomotor change over the past 30 million years. Late Eocene Oligocene higher primates were arboreal, platyrrhine-like species, and all of the skeletal remains indicate leaping and quadru-

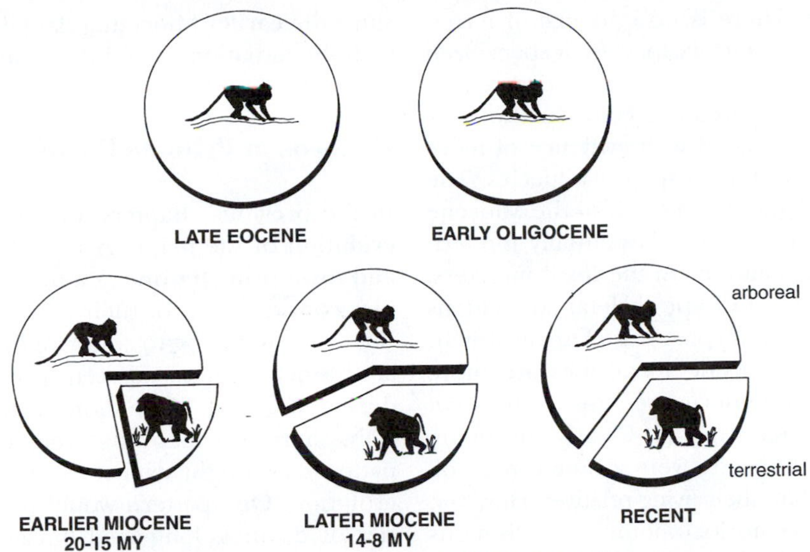

FIGURE 18.3 Changes in substrate use of Old World higher primates during the past 35 million years. Late Eocene and early Oligocene anthropoids were arboreal, but in later epochs terrestrial species have become common (modified from Fleagle and Kay, 1985).

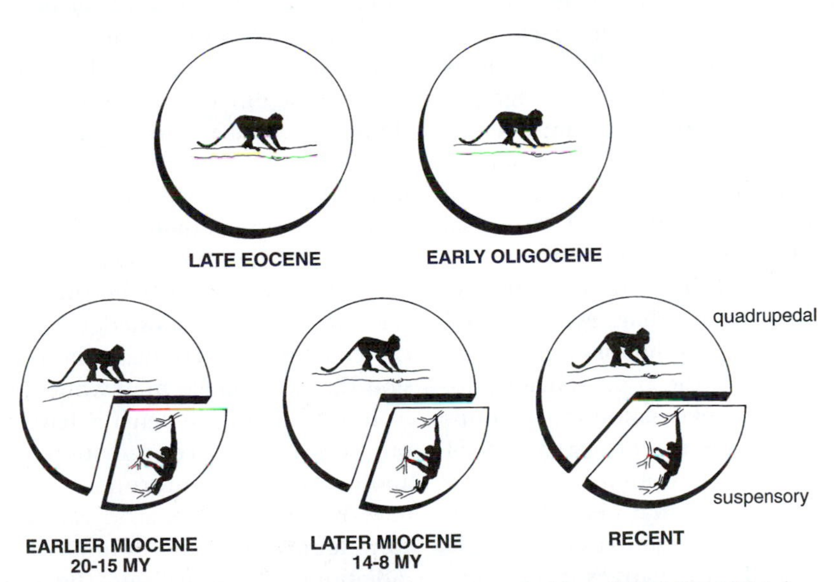

FIGURE 18.4 Changes in arboreal locomotor habits of Old World higher primates during the past 35 million years. Late Eocene and early Oligocene anthropoids were all arboreal quadrupeds and leapers. Since the early Miocene, there have been a variety of suspensory species (modified from Fleagle and Kay, 1985).

pedal habits. There is no evidence of terrestrial species, suspensory species, or specialized clingers. However, the first appearance of hominoids and cercopithecoids in the early Miocene is associated with evidence of more terrestrial and more suspensory species. The few remains from the last half of the Miocene indicate the presence of essentially modern locomotor adaptations in the fossil monkeys, but the evolution of ape skeletal adaptations are difficulty to interpret (e.g., Larson, 1998). Most middle Miocene fossil apes are more primitive than modern apes, and many seem to be more quadrupedal. Clearly the hominoids of the Miocene were a much more diverse group than their living relatives. However, the most extreme locomotor specializations found among living hominoids seem to have evolved more recently. The skeletal specializations of gibbons associated with brachiation are unknown prior to the Pleistocene, and human bipedal locomotion seems to have evolved during the past 4 million years.

The patterns in body size, diet, and locomotion described in the previous paragraphs are interrelated. It is not surprising that the first appearance of relatively large higher primates in the early Miocene is associated with evidence of more folivorous and terrestrial habits, since folivory and terrestriality are functionally linked with relatively large size. In addition, many of the changes in size and adaptation that we see in the fossil record are clearly associated with the appearance or disappearance of particular taxonomic groups (Fig. 18.5). The major taxonomic groups of living and fossil primates often have characteristic adaptive features that permit them to exploit a unique array of resources (see Fleagle and Reed, 1996). Thus, changes in the "primate" adaptive zone through time are linked with the appearance or disappearance of particular groups of primates. For example, the increase in folivory among higher primates since the earlier Miocene is largely associated with the radiation of Old World monkeys.

Patterns in Primate Phylogeny

In the previous chapters, we considered the evolution of the major groups of both living and fossil primates one at a time, with particular consideration of their adaptive diversity. It is also interesting to compare the evolutionary history of these different primate groups during the past 65 million years. Theoretically, there are many different evolutionary patterns we might expect to find in primate evolution. One pattern would be a whole series of distinct, long-lived lineages. Alternatively, we might find a series of evolutionary radiations succeeding one another in time, or maybe one slowly replacing another.

Not surprisingly, the record of primate evolution shows evidence of all of these evolutionary patterns in various groups at various times. At a gross level, the major pattern seems to be one of succeeding adaptive radiations: an initial radiation of plesiadapiforms in the Paleocene, followed by a radiation of prosimians at the beginning of the Eocene, and finally the radiation of anthropoids beginning in the late Eocene and early Oligocene (Fig. 18.6). However, this sequence is a summary of our knowledge of all primates rather than an account of the faunal succession that took place in any one locality. Only from Eurasia do we actually have a fossil record in which all three radiations succeed one another, and the anthropoids never seem to have been very diverse in Europe. In other continental areas, one or more of these major radiations is absent from the fossil record, either because the animals were never there or because we have not uncovered the fossils of some particular period. Higher primates other than humans have never successfully in-

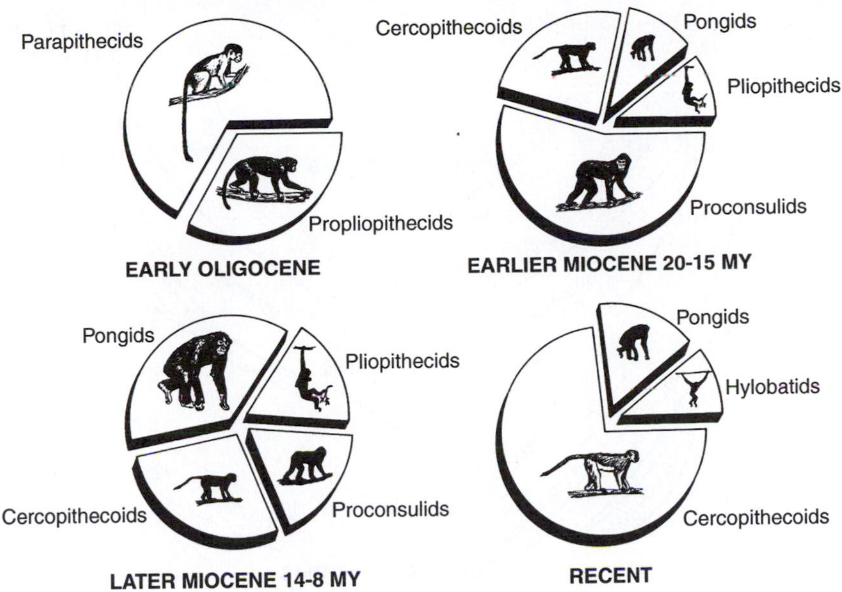

FIGURE 18.5 Changes in the taxonomic abundance of different groups of Old World higher primates during the past 30 million years. In the early Oligocene, parapithecids were the most common anthropoids and propliopithecids were slightly less common. The earlier Miocene was characterized by an abundance of fossil apes; the proconsulids were the most diverse group. In the later Miocene, cercopithecoids and pongids became more common. Recent Old World higher primate communities are dominated by cercopithecoid monkeys (modified from Fleagle and Kay, 1985).

vaded North America as far as we know; only higher primates are known from South America; and on Madagascar, prosimians were the only primates until the recent arrival of humans. Many of the global patterns that we see in the primate fossil record reflect our available sample of fossil primates more than the actual timing or biogeography of the evolution of particular taxa. Regional changes through time have been much more diverse.

If we examine evolutionary radiations of subfamilies and families of primates within restricted geographic areas, we obtain a less distorted but more parochial view of evolutionary changes in primate history. Again, we find a diversity of evolutionary patterns. Among the platyrrhine monkeys of South America we see evidence of many distinctive, relatively old lineages, none of which has ever been very diverse (see, e.g., Rosenberger, 1984, 1992). The history of higher primate evolution in the Old World seems to have been very different, with a succession of very different anthropoids in the late Eocene (many small species of unclear affinities, including oligopithecines and parapithecids); Oligocene (parapithecids and propliopithecids); the early Miocene (proconsulids), and the late Miocene to Recent (cercopithecoid monkeys) (Fig. 18.5).

An analysis at the generic level shows similar heterogeneity. There are a few primate

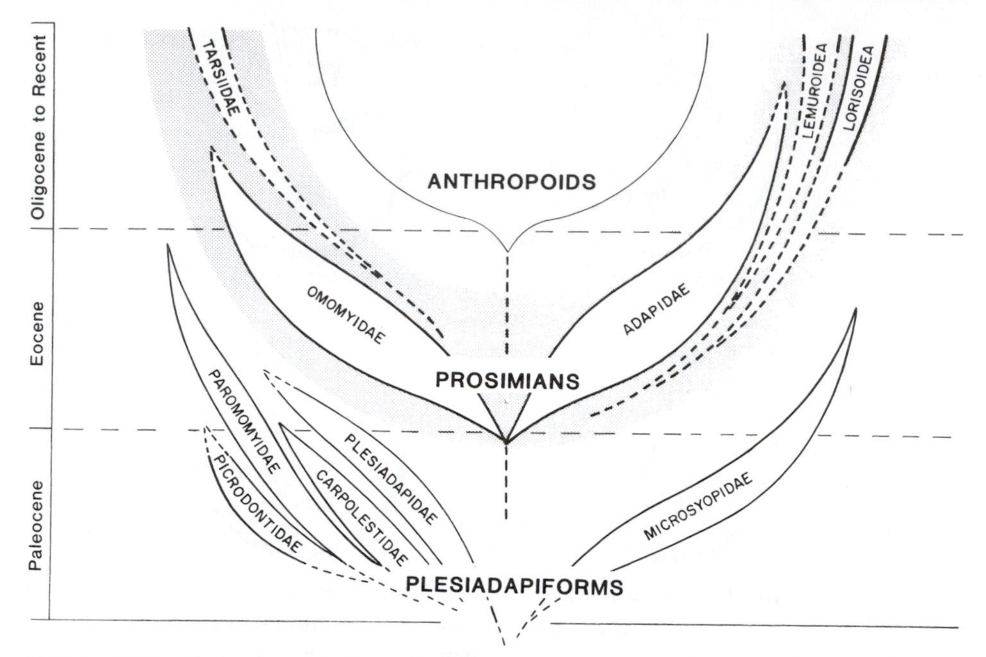

FIGURE 18.6 The major primate radiations of the Cenozoic era, showing the successive appearance of different radiations in the Paleocene, the Eocene, and the Oligocene through Recent (modified from Gingerich, 1986).

genera that seem to have persisted for tens of millions of years with very little change: *Tarsius* (Beard *et al.*, 1993; Ginsburg and Mein, 1987), *Aotus* (Setoguchi and Rosenberger, 1987), and *Macaca* (Delson and Rosenberger, 1984). Other lineages have undergone dramatic morphological changes in a relatively short time, the most notable being *Homo.*

Much of our perception of the patterns in primate evolution is determined by the density of the fossil record (e.g., Howells, 1997). When only a few fossils are know from widely separated time periods, the simplest scenario is a single transformational lineage through time. However, as more fossils are recovered, phyletic trees become more complex, defying any simple linear sequence. In general, it seems that the initial appearance of most taxa is characterized by a rapid proliferation of forms, yielding an evolutionary tree with a bushy base. This has been clearly demonstrated with increasing discoveries of omomyid prosimians, proconsulid apes, the earliest hominids (*Australopithecus* and *Paranthropus*), and early humans of the genus *Homo.*

Primate Evolution at the Species Level

One of the most hotly debated issues in evolutionary biology today is the same one that preoccupied Darwin: the origin of species. In Darwin's day the major issue was over the mechanism leading to evolutionary change and the appearance of new species. Darwin (1859) resolved this issue with his "discovery" and description of natural selection. Current debate is over the tempo of evolutionary change—whether new species appear by gradual modification of earlier types or through rapid changes in form (e.g., Carroll, 1997; Eldredge and Gould, 1972; Gould and Eldredge, 1993).

There are several periods in primate evolution for which the fossil record is sufficiently well sampled over a long period of time that questions of this nature can be fruitfully examined. Two of the best examples come from western North America, where P. D. Gingerich, T. M. Bown, and K. D. Rose have carefully documented the evolutionary history of fossil mammals through a long, continuous series of late Paleocene and early Eocene sediments in northern Wyoming. Gingerich (e.g., Gingerich, 1976, 1979, 1985) has studied evolutionary change in the late Paleocene plesiadapids and the early Eocene notharctines. He found extensive evidence of gradual morphological change through time in both lineages, and most of the "new" species are most likely the result of the gradual modification of earlier forms. For the few instances in which a new species appears abruptly, it is impossible to determine whether this abrupt appearance is the result of rapid, discontinuous change from another local form, immigration from another area, or an absence of linking forms because of missing fossils. This is one of the major difficulties in testing theories of evolutionary change with fossil evidence. A record of gradual change is positive evidence for gradual evolution, but a record of discontinuous change can be interpreted as the result of several very different phenomena.

More recently, Bown and Rose (1987) have produced extraordinarily detailed documentation of evolutionary change within lineages of early Eocene omomyoid and adapoid prosimians in northern Wyoming (Figs. 18.7, 18.8). Charting gradual change in many aspects of dental morphology, including reduction in size and loss of teeth, changes in the size and shape of cusps, and changes in the number and size of tooth roots, they show that the transition from one paleospecies to another within a lineage is characterized by changing frequencies of the diagnostic features within intermediate populations. However, the morphological differences that characterize the end products rarely change at the same rate. Thus species-specific differences in morphology are a dynamic phenomenon and are easiest to characterize in the absence of intermediate forms. For a very good fossil record the identification of species boundaries becomes arbitrary, depending on which morphological feature is being used. *Tetonius matthewi,* for example, gradually evolves into *Pseudotetonius ambiguus* in the early Eocene deposits of northern Wyoming. However, the diagnostic features of the latter appear at different levels in the stratigraphic section and at different rates. Loss of P_2 occurs relatively low in the geological section, reduction of the roots of P_3 takes place in a series of steps, and changes in the size of P_3/P_4 take place gradually. Depending on which criterion or combination of criteria is used to define *Pseudotetonius ambiguus,* one can place the taxonomic transition at almost any time within the evolutionary lineage. There is no abrupt appearance of a new species; rather, there is a mosaic appearance of morphological changes that can be used, in retrospect, to define a different species. It seems likely that this is a common pattern of evolutionary change, and the distinctiveness of many species in the fossil record is to a large degree an artifact of the incompleteness of the record.

One period of primate evolution for which there has been extensive debate over the tempo and pattern of evolutionary change is the past 4 million years of hominid evolution (e.g., Conroy, 1997). For virtually all aspects

FIGURE 18.7 Changes through time in the lower dentition of a lineage of fossil prosimians from the early Eocene of Wyoming. Note the loss of P_2 and the gradual change in the size and shape of P_3 and P_4 (modified from Bown and Rose, 1987).

of hominid evolution there are ongoing debates over the distinctiveness of individual taxa (e.g., Howell, 1996) and whether new taxa arise through gradual change from earlier forms as part of a continual transformation (e.g., Wolpoff *et al.,* 1994) or whether new taxa arise through rapid transitions at critical periods of speciation, perhaps due to climate shifts (Vrba *et al.,* 1995). Rightmire (1981) has argued that for over a million years during the first half of the Pleistocene human evolution was essentially static. During this period, he claims, *Homo erectus* showed no significant morphological change, a conclusion that has been challenged by several workers (Levinton, 1982; Wolpoff, 1984). Overall, hominid evolution seems to show episodes of all of these phenomena (e.g., Kimbel, 1995), but

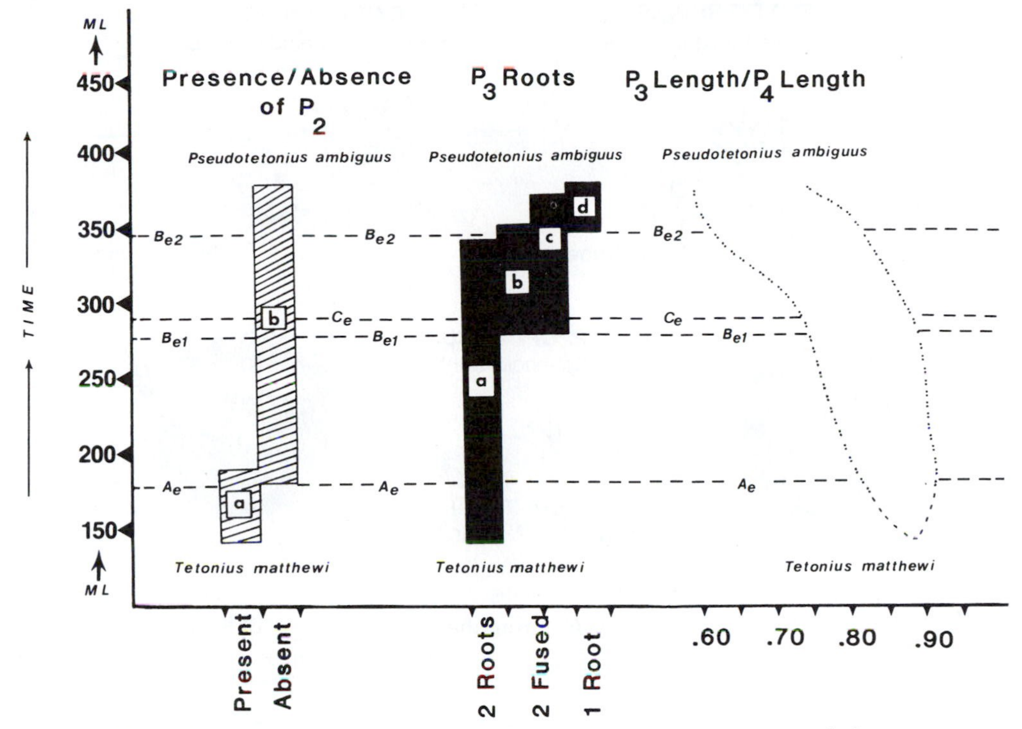

FIGURE 18.8 Changes in the size of P_3/P_4, the number of roots on P_3, and the presence or absence of P_3 in an evolving lineage of early Eocene omomyids. The morphological changes take place at different rates and at different times in the stratigraphic sequence. Because of the mosaic nature of morphological change in this lineage, identification of the boundary between the older species, *Tetonius matthewi,* and the younger species, *Pseudotetonius ambiguus,* must be based on an arbitrary choice of criteria (modified from Bown and Rose, 1987).

the debates continue, and the alternatives become more complex with an increasingly dense fossil record.

Mosaic Evolution

Perhaps the clearest pattern evident in the fossil record of primates is that the evolution of living primates has not taken place through major transformations of one group into another, but rather through a pattern of mosaic evolution (McHenry, 1975). The characteristic anatomical features of living species, genera, families, and subfamilies have appeared one by one over long expanses of time rather than in sudden bursts of several coordinated changes. The large anatomical differences we see today between many living taxa are clearly the result of the extinction of intermediate forms. A major contribution of the fossil record is in providing us with intermediate forms that document the sequence through which modern differences appeared and the timing of the appearances of the successive novelties (Figs. 18.9, 18.10). For example, all living catarrhines are distinguished from platyrrhines by a suite of characters, including reduction

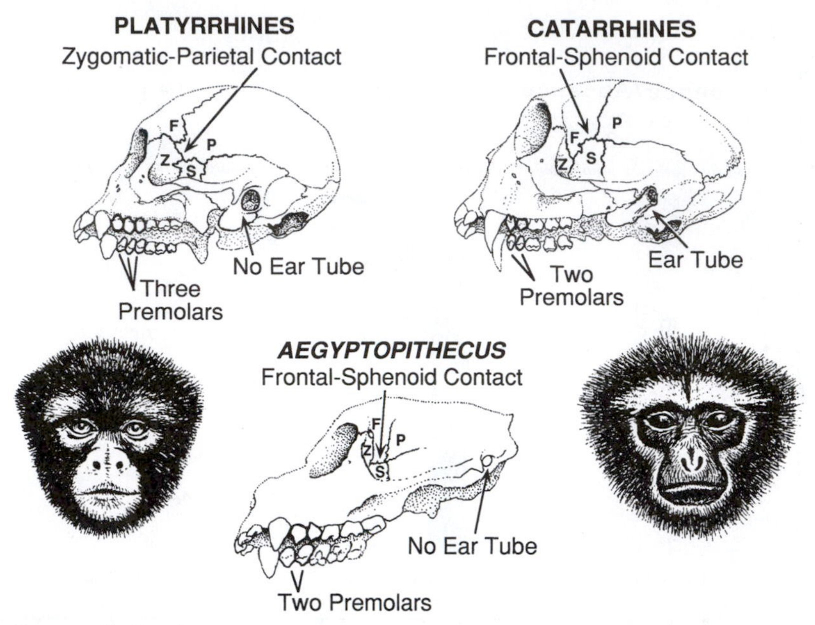

FIGURE 18.9 *Aegyptopithecus zeuxis,* from the early Oligocene of Egypt, displays an intermediate mixture of cranial features when compared with living platyrrhines and catarrhines.

of the premolar count from three to two in each quadrant, a connection of the frontal and sphenoid bone on the side wall of the skull, and a bony tube extending laterally from the ectotympanic ring (Fig. 10.9). However, the early Oligocene *Aegyptopithecus zeuxis* shows an intermediate condition with two premolars, the catarrhine skull wall, but no ear tube and demonstrates both the sequence in which modern catarrhine features appeared and also provide a minimum age for the first appearance of the premolar loss and appearance of the bony pattern on the skull wall. *Aegyptopithecus* is indeed a missing link as the name *zeuxis* suggests.

Thus humans (Fig. 18.10) and all other organisms are made up of a mosaic of features which have been acquired at various times and for various purposes during their evolutionary past. For example, our fingernails, which we share with all primates first appeared approximately 55 million years ago with the earliest record of primates in the early Eocene. The posterior wall of our orbit, which we share with all anthropoids first appeared approximately 35 million years ago, while our tailless condition which we share with apes is a much later development, and probably dates to around 20 million years ago. Although living humans can be distinguished from chimpanzees by a whole suite of anatomical differences, we know that these were not evolved in a single change, but slowly over millions of years. On the basis of the current record, it seems that our shortened ilium was the first distinctive feature to appear, approximately 3.5 million years ago. The characteristic arch of our foot and our unusual knee cartilages appeared much later, about 2 million years ago. Features such as our promi-

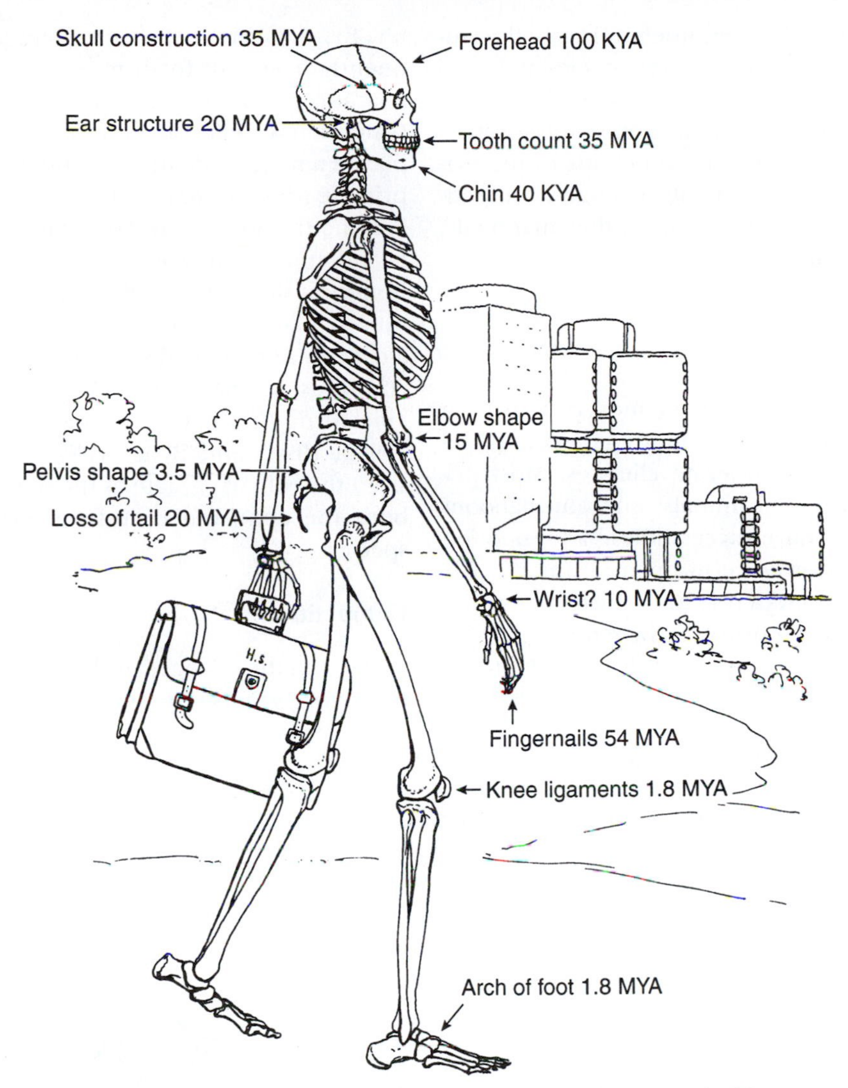

FIGURE 18.10 The timing of the appearance of distinctive anatomical features of the human skeleton that we share with various other groups of primates.

nent chin and high forehead are much more recent, dating to the last one hundred thousand years. Obviously other primates and other organisms are likewise the result of many millions of years of mosaic evolution, the details of which can only be elucidated by a well-dated fossil record.

Primate Extinctions

Over the past 55 million years, many new primate species have appeared and slightly fewer have become extinct. In many cases, a species disappeared by evolving into an animal with a different morphology that we recognize as a new species; this phenomenon is called **pseudoextinction.** In other cases, species and lineages disappeared, leaving no descendants. Three primary reasons are commonly given to account for the major extinctions in the primate fossil record: climatic changes, competition from other primates or other mammals, and predation.

Climatic Changes

Living primates are for the most part tropical animals, with only a few hardy genera and species found in temperate climates. Most primates are arboreal animals; only some baboons and humans have successfully abandoned forested areas for a life in open savannahs. Climate clearly plays a role in the distribution and diversity of living primates. In general, primate species are more common at lower latitudes and in areas with increased rainfall (see Chapter 8). Likewise, climatic change has frequently been put forth to explain the disappearance of a primate group (Fig. 18.11). Gingerich (1986) suggests that climate has played a major role in the extinction of many primate groups from northern continents, including the adapids and omomyids of Europe at the end of the Eocene and the European hominoids at the end of the Miocene. In addition, increased aridity has been suggested as a contributor to the disappearance of the proconsulids in the middle Miocene of Africa (Pickford, 1983). On a broader scale, Vrba (1985, 1995) and Boaz (1997) have emphasized the role of climate as the major factor in both the appearance and the extinction of species.

Competition

Since climatic changes in the fossil record are invariably associated with faunal change re-

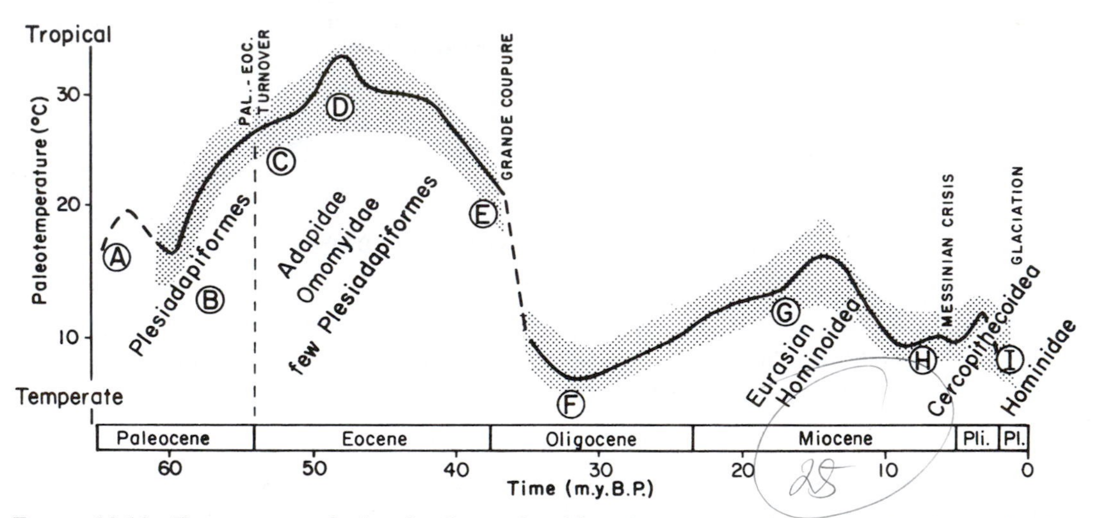

FIGURE 18.11 Temperatures during the Cenozoic, with major events in the primate fossil record of the northern hemisphere marked: A, appearance of *Purgatorius*; B, radiation of plesiadapiforms; C, appearance of adapids and omomyids; D, height of primate diversity in Europe and North America; E, decline and extinction of primates in Europe and North America; F, nadir of primate diversity in northern continents; G, appearance of hominoids in Europe and Asia; H, disappearance of northern hominoids, extensive ecological radiation of cercopithecoids, and emergence of hominids; I, evolution and dispersal of humans (from Gingerich, 1986).

sulting from the appearance or spread of new groups of primates and other mammals, we must attempt to distinguish the effects of climate per se from the effects of ecological competition with other species. The extinction of the plesiadapiforms, for example, coincides roughly with the radiation of both rodents and early prosimians (Fig. 18.12), and the decline of proconsulids in the middle and late Miocene of Africa is associated with an increased abundance of cercopithecoid monkeys (see Fig. 16.11). Unfortunately, It is very difficult to determine whether a new group contributes to the extinction of earlier groups through direct competition for resources or merely fills the gap left by their extinction, and careful analysis is required to sort out the alternatives (see Maas *et al.,* 1988). Although there is some evidence for ecological competition among extant primates (e.g., Ganzhorn, 1988, 1989), it has been suggested that the fossil record may not provide adequate temporal resolution to permit the identification of ecological competition (Sepkoski, 1996; but see Gingerich, 1996). In the case of the extinction of plesiadapiforms, it appears that competition with rodents may well have contributed to the demise of the group, whereas the radiation of the prosimians occurred after the decline or disappearance of most plesiadapiforms and so seems a less likely cause of extinction.

Predation

A cause of primate extinctions that seems well documented, in at least one case, is human predation. The disappearance of the large diurnal lemurs from the fauna of Madagascar clearly postdates the appearance of humans on the island and is most likely the result of human hunting and habitat destruction (Dewar, 1984, 1997). The lemur extinction on Madagascar was a relatively recent event, dating to the last 5000 years, but it is possible that hominid predation was also responsible for the extinction of most of the large monkeys from East Africa during the early and middle Pleistocene for the disappearance of the orangutan from mainland Asia (Ciochon, 1988), and for the extinction of New World monkeys in the Caribbean.

Limiting Primate Extinctions

We often have difficulty reconstructing the events that led to the extinction of particular species, genera, and lineages of primates from earlier periods. In contrast, the factors that will probably lead to the extinction of many primate species alive today are relatively easy to identify: habitat destruction, hunting, and live capture for the pet or research markets (Mittermeier and Cheney, 1986). Of these, habitat destruction due to our expanding human population is the greatest threat. More than 90 percent of all primate species live in the tropical forests of Asia, Africa, and South and Central America, and their fate is intimately linked with the future of these forests. There are endangered primates in virtually every part of the tropical world, but the greatest threats are in Madagascar and eastern Brazil, where faunas of unique primates with limited geographic ranges are under extreme pressure from rapidly growing populations that need the same resources of forests and land (Fig. 18.13).

With an expanding global human population and the developmental demands of many parts of the tropical world, it is inevitable that many primate populations and species will disappear in the coming decades. The goal of conservationists is to limit the losses as much as possible. This involves "1) protecting areas for particularly endangered and vulnerable species; 2) creating large national parks and reserves in areas of high primate diversity or abundance; 3) maintaining parks and reserves that already exist and enforcing protective

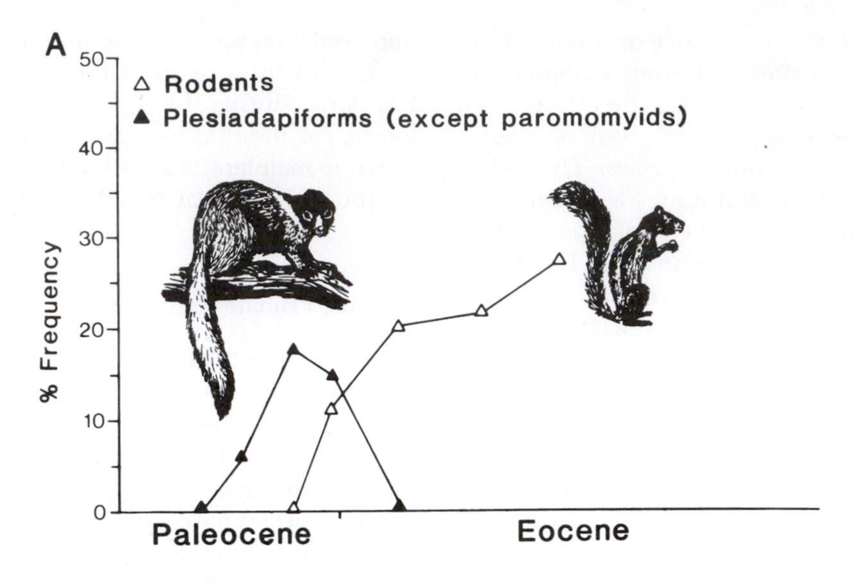

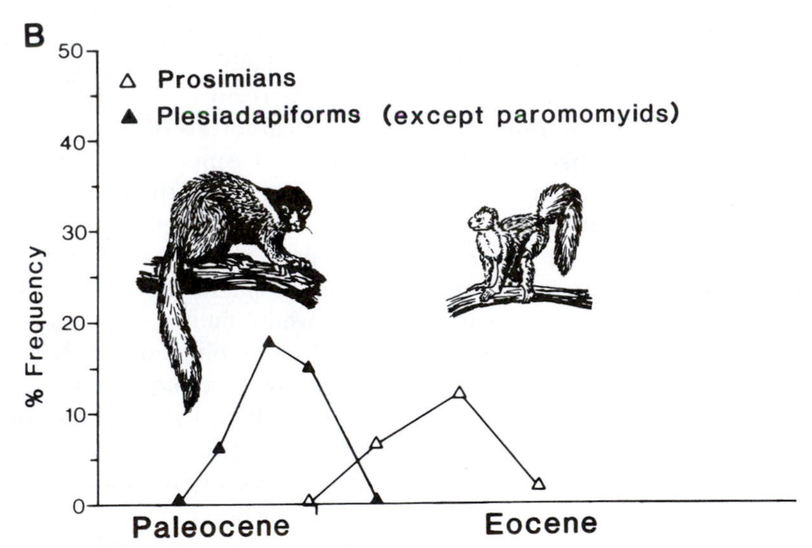

FIGURE 18.12 Abundance of plesiadapiforms in the fossil record from the Paleocene and Eocene of North America compared with the abundance of early rodents (A) and fossil prosimians (B). Note that the decline of plesiadapiform abundance and dramatic explosion of fossil rodents in this period, suggesting that the appearance of rodents was primarily responsible for the extinction of plesiadapiforms. In contrast, the radiation of fossil prosimians occurred primarily after the major decline in plesiadapiforms, suggesting that the prosimians may have exploited some aspect of the ecological space occupied by plesiadapiforms without being primarily responsible for the extinction of the plesiadapiforms. Percent frequency based on the minimum number of individuals (MNI) (modified from Maas *et al.*, 1988).

FIGURE 18.13 Eight of the primates closest to extinction today. Left to right: top row, the indris from Madagascar, the mountain gorilla from central Africa, and the aye-aye from Madagascar; middle row, the muriqui from Brazil and the lion-tail macaque from India; bottom row, the golden monkey from China, the golden lion tamarin from Brazil, and the pygmy chimpanzee from the Democratic Republic of the Congo (courtesy of Stephen Nash, World Wildlife Fund-U.S.).

legislature in them; 4) creating public awareness of the need for primate conservation and the importance of primates as both a national heritage and a resource; . . . 5) determining ways in which people and other primates can coexist in multiple-use areas" (Mittermeier and Cheney, 1986, p. 488). Only such a broad approach will ensure that future generations are able to appreciate the living remnants of 65 million years of primate adaptation and evolution.

BIBLIOGRAPHY

PRIMATE ADAPTIVE RADIATIONS

Covert, H. H. (1986). Biology of early Cenozoic primates. In *Comparative Primate Biology,* vol. 1: *Systematics, Evolution, and Anatomy,* ed. D. W. Swindler and J. Erwin, pp. 335–359. New York: Alan R. Liss.

——— (1995). Locomotor adaptations of Eocene primates: Adaptive diversity among the earliest prosimians. In *Creatures of the Dark; the Nocturnal Prosimians,* ed. L. Alterman, G. A. Doyle, and M. K. Izard, pp. 495–509. New York: Plenum Press.

Dagosto, M. (1993). Postcranial anatomy and locomotor behavior in Eocene primates. In *Postcranial Adaptation in Nonhuman Primates,* ed. D. L. Gebo, pp. 199–219. DeKalb: Northern Illinois University Press.

Fleagle, J. G. (1978). Size distributions of living and fossil primate faunas. *Paleobiology* **4**:67–76.

Fleagle, J. G., and Kay, R. F. (1985). The paleobiology of catarrhines. In *Ancestors: The Hard Evidence,* ed. E. Delson, pp. 23–36. New York: Alan R. Liss.

Fleagle, J. G., and Reed, K. E. (1996). Comparing primate communities: A multivariate approach. *J. Human Evol.* **30**:489–510.

Fleagle, J. G., Kay, R. F. and Anthony, M. (1997). Fossil New World monkeys. In *Vertebrate Paleontology in the Neotropics: The Miocene Fauna of La Venta, Colombia,* ed. R. Kay, R. Madden, R. Cifelli, and J. Flynn, pp. 473–495. Washington, D. C.: Smithsonian Institution Press.

Gautier-Hion, A., Bourliere, F., Gautier, J.-P. and Kingdon, J., eds. (1988). *A Primate Radiation: Evolutionary Biology of the African Guenons.* Cambridge: Cambridge University Press.

Godfrey, L. R., Jungers, W. L., Reed, K. E., Simons, E. L., and Chatrath P. S. (1997), Subfossil lemurs: Inferences about past and present primate communities in Madagascar. In *Natural Change and Human Impact in Madagascar,* ed. S. M. Goodman and B. D. Patterson, pp. 218–256. Washington, D. C.: Smithsonian Institution Press.

Kay, R. F., and Ungar, P. S. (1997). Dental evidence for diet in some Miocene catarrhines with comments on the effects of phylogeny on the interpretation of adaptation. In *Function, Phylogeny, and Fossils: Miocene Hominoid Evolution and Adaptations,* ed. D. R. Begun, C. V. Ward, and M. D. Rose, pp. 131–151. New York: Plenum Press.

Larson, S. G. (1998). Parallel evolution in the hominoid trunk and forelimb. *Evol. Anthropol.* **6**:87–99.

McCrossin, M. (1992). New species of bushbaby from the middle Miocene of Maboko Island, Kenya. *Am. J. Phys. Anthropol.* **89**:215–234.

Strait, S. G. (1997). Tooth use and the physical properties of food. *Evol. Anthropol.* **5**:199–211.

PATTERNS IN PRIMATE PHYLOGENY

Beard, K. C., Tao, Q., Dawson, M. R., Wang, B., and ChuanKuei, L. (1994). A diverse new primate fauna from the middle Eocene fissure-fillings in southeastern China. *Nature* **368**:604–609.

Delson, E., and Rosenberger, A. L. (1984). Are there any anthropoid primate living fossils? In *Living Fossils,* ed. N. Eldredge and S. M. Stanley, pp. 50–61. New York: Springer-Verlag.

Gingerich, P. D. (1984). Primate evolution. In *Mammals: Notes for a Short Course,* ed. T. W. Broadhead, pp. 167–181. Knoxville: University of Tennessee, Dept. of Geological Sciences.

Ginsburg, L., and Mein, P. (1987). *Tarsius thailandica nov. sp.,* premier Tarsiidae (Primates, Mammalia) fossile d'Asie. *C. R. Acad. Sc. (Paris), t. 304, Serie II,* no. 19, pp. 1213–1215.

Howells, W. W. (1997). *Getting Here: The Story of Human Evolution.* Washington, D. C.: Compass Press.

Rosenberger, A. L. (1984). Fossil New World monkeys dispute the molecular clock. *J. Human Evol.* **13**: 737–742.

——— (1992). Evolution of feeding niches in New World monkeys. *Am. J. Phys. Anthropol.* **88**:525–562.

Schneider, H., and Rosenberger, A. L. (1997). Molecules, morphology, and platyrrhine systematics. In *Adaptive Radiations of Neotropical Primates,* ed. M. A. Norconk, A. L. Rosenberger, and P. A. Garber, pp. 3–19. New York: Plenum Press.

Setoguchi, T., and Rosenberger, A. L. (1987). A fossil owl monkey from La Venta, Colombia. *Nature (London)* **326**:692–694.

EVOLUTION OF PRIMATE SPECIES

Bown, T. M., and Rose, K. D. (1987). Patterns of dental evolution in early Eocene anaptomorphine primates (Omomyidae) from the Bighorn Basin, Wyoming. Paleontological Society Memoir no. 23. *J. Paleontol.* **61**: 1–62.

Carroll, R. L. (1997). *Patterns and Processes of Vertebrate Evolution.* Cambridge Paleobiology Series. Cambridge: Cambridge University Press.

Conroy, G. C. (1997). *Reconstructing Human Origins.* New York: Norton.

Darwin, C. (1859). *On the Origin of Species by Means of Natural Selection, or the Presentation of Favoured Races in the Struggle for Life.* London: John Murray.

Eldredge, N., and Gould, S. J. (1972). Punctuated equilibria: An alternative to phyletic gradualism. In *Models in Paleobiology,* ed. T. J. M. Schopf, pp. 82–115. San Francisco: Freeman, Cooper.

Gingerich, P. D. (1976). Cranial anatomy and evolution of early Tertiary Plesiadapidae (Mammalia, Primates). *Contr. Mus. Paleontol., Univ. Michigan* **15**.

——— (1979). Paleontology, phylogeny, and classification: An example from the mammalian fossil record. *Systematic Zool.* **28**:451–464.

______ (1985). Species in the fossil record: Concepts, trends, and transitions. *Paleobiology* **11**:27–41.

Gould, S. J., and Eldredge, N. (1993). Punctuated equilibrium comes of age. *Nature* **366**:223–227.

Howell, F. C. (1996). Thoughts on the study and interpretation of the human fossil record. In *Contemporary Issues in Human Evolution,* ed. W. E. Meikle, F. C. Howell, and N. G. Jablonski, pp. 1–45. San Francisco: California Academy of Sciences.

Kimbel, W. H. (1995). Hominid speciation and Pliocene climatic change. In *Paleoclimate and Evolution, with Emphasis on Human Origins,* ed. E. S. Vrba, G. H. Denton, T. C. Partridge, and L. H. Burckle, pp. 425–437. New Haven, Conn.: Yale University Press.

Levinton, J. S. (1982). Estimating stasis: Can a null hypothesis be too null? *Paleobiology* **8**:307.

Rightmire, G. P. (1981). Patterns in the evolution of *Homo erectus. Paleobiology* **7**:241–246.

Rose, K. D., and Bown, T. M. (1986). Gradual evolution and species discrimination in the fossil record. In *Vertebrates, Phylogeny, and Philosophy,* ed. K. M. Flanagan and J. A. Lillegraven, pp. 119–130. *Contr. Geol., Univ. Wyoming, spec. paper no. 3.*

Vrba, E. S., Denton, G. H., Partridge, T. C., and Burckle, L. H., eds. (1995). *Paleoclimate and Evolution with Emphasis on Human Origins.* New Haven, Conn.: Yale University Press.

Wolpoff, M. H. (1984). Evolution in *Homo erectus*: The question of stasis. *Paleobiology* **10**:389–406.

Wolpoff, M. H., Thorne, A. G., Jelinek, J., and Yinyun, Z. (1994). The case for sinking *Homo erectus,* 100 years of *Pithecanthropus* Is *Enough! Courier Forschungs-Institut Senckenberg* **171**:341–361.

MOSAIC EVOLUTION

Fleagle, J. G. (1986). The fossil record of early catarrhine evolution. In *Major Topics in Primate and Human Evolution.* eds. B. Wood, L. Martin, and P. Andrews. pp. 130–139. Cambridge: Cambridge University Press.

Fleagle, J. G. (1994). Anthropoid Origins. In: *Integrative paths to the Past: Paleoanthropological Advances in Honor of F. Clark Howell.* eds. R. S. Corruccini and R. L. Ciochon. pp. 17–35. New York: Prentice-Hall.

McHenry, H. M. (1975). Fossils and the mosaic nature of human evolution. *Science* **190**:425–431.

PRIMATE EXTINCTIONS

Boaz, N. T (1997). *Eco Homo.* New York: Basic Books.

Ciochon, R. L. (1988). *Gigantopithecus*: The king of all apes. *Animal Kingdom* **91**:32–39.

Dewar, R. E. (1984). Extinctions in Madagascar: The loss of the subfossil fauna. In *Quaternary, Extinctions: A Prehistoric Revolution,* ed. P. S. Martin and R. G. Klein, pp. 574–593. Tucson: University of Arizona Press.

______ (1997). Were people responsible for the extinction of Madagascar's subfossils, and how will we ever know? In *Natural Change and Human Impact in Madagascar,* ed. S. M. Goodman and B. D. Patterson, pp. 364–380. Washington, D. C.: Smithsonian Institution Press.

Ganzhorn, J. U. (1988). Food partitioning among Malagasy primates. *Oecologia* **75**:436–450.

______ (1989). Primate species separation in relation to secondary plant chemicals. *Human Evol.* **4**:125–132.

Gingerich, P. D. (1986). *Plesiadapis* and the delineation of the order Primates. In *Major Topics in Primate and Human Evolution,* ed. B. Wood, L. Martin, and P. Andrews, pp. 32–46. Cambridge: Cambridge University Press.

______ (1996). Rates of evolution in divergent species lineages as a test of character displacement in the fossil record: Tooth size in Paleocene *Plesiadapis* (Mammalia, Proprimates). *Paleovertebrata* **25**:193–204.

Maas, M. C., Krause, D. W., and Strait, S. G. (1988). Decline and extinction of *Plesiadapiformes* (Mammalia: ?Primates) in North America: Displacement or replacement? *Paleobiology* **14**(4):410–431.

Marsh, C. W., and Mittermeier, R. A., eds. (1987). *Primate Conservation in the Tropical Rain Forest. Monographs in Primatology,* vol. 9. New York: Alan R. Liss.

Mittermeier, R. A., and Cheney, D. L. (1986). Conservation of primates and their habitats. In *Primate Societies,* ed. B. B. Smuts, D. L. Cheney, R. M. Seyfarth, R. W. Wrangham, and T. T. Struhsaker, pp. 477–490. Chicago: University of Chicago Press.

Nitecki, M. H., ed. (1984). *Extinctions.* Chicago: University of Chicago Press.

Pickford, M. (1983). Sequence and environments of the lower and middle Miocene hominoids of western Kenya. In *New Interpretations of Ape and Human Ancestry,* ed. R. L. Ciochon and R. S. Corruccini, pp. 421–439. New York: Plenum Press.

Sepkoski, J. J. (1996). Competition in macroevolution: The double wedge revisited, In *Evolutionary Paleobiology,* ed. D. Jablonski, D. H. Erwin, and J. H. Lipps, pp. 211–255 Chicago: University of Chicago Press.

Walker, A. (1967). Patterns of extinction among the subfossil Madagascan lemuroids. In *Pleistocene Extinctions: The Search for a Cause,* ed. P. S. Martin and

H. E. Wright, Jr., pp. 425–432. New Haven, Conn.: Yale University Press.

Vrba, E. S. (1985). Environment and evolution: Alternative causes of the temporal distribution of evolutionary events. *S. Afr. J. Sci.* **81**:229–236.

——— (1995). On the connections betweeen paleoclimate and evolution. In *Paleoclimate and Evolution with Emphasis on Human Evolution,* ed. E. S. Vrba, G. H. Denton, T. C. Partridge, and L. H. Burckle, pp. 24–45. New Haven, Conn.: Yale University Press.

Vrba, E. S., Denton, G. H., Partridge, T. C., and Burckle, L. H., eds. (1995). *Paleoclimate and Evolution with Emphasis on Human Origins.* New Haven, Conn.: Yale University Press.

GLOSSARY

abduction movement of a limb or part of a limb away from the midline of the body.

absolute dating determination of the age, in years, of a fossil or fossil site, usually on the basis of the amount of change in radioactive elements in rocks.

adaptation process whereby an organism changes in order to survive in its given environment; or, a specific new characteristic that enables survival.

adaptive radiation a group of closely related organisms that have evolved morphological and behavioral features enabling them to exploit different ecological niches.

adduction movement of a limb or part of a limb toward the midline of the body.

age-graded group a social group with several adult females and several adult males who differ in social and reproductive status according to age. Thus, with time, an age-graded group can change from a one-male reproductive system to a multimale system.

allometry the relationship between the size and shape of an organism, or, more broadly, the relationship between an organisms's size and various aspects of its biology, such as morphology, ecology, or behavior; also, the study of such relationships.

alloparenting (aunting) assistance in care of infants and juveniles by individuals other than parents.

allopatry the absence of overlap in the geographical range of two species or populations.

arboreal living in trees.

arboreal quadrupedalism mode of locomotion in which the animal moves along horizontal branches with a regular gait pattern involving all four limbs.

articulation a joint between two or more bones.

basal metabolism the energy requirements of an animal at rest.

bilateral symmetry a type of developmental shape in which right and left sides of the organism are mirror images of one another.

bilophodonty a condition of the molar teeth in which the mesial and distal pairs of cusps form ridges or lophs.

biomass the sum of the weights of the organisms in a particular area.

bipedalism mode of locomotion using only the hindlimbs, usually alternately rather than together.

brachiation arboreal locomotion in which the animal progresses below branches by using only the forelimbs.

buccal the cheek side of a tooth.

bunodont (teeth) have low, rounded cusps.

canopy a layer of forest foliage that is laterally continuous and usually distinct vertically from other layers. A tropical forest often has one or more distinct canopies.

carnivore an animal that eats primarily the flesh of other animals; also often used to refer to the mammalian order Carnivora (which includes cats, dogs, skunks, raccoons, and bears).

cathemeral active intermittently throughout the twenty-four-hour day rather than active only during the day (diurnal) or only during the night (nocturnal).

clade a group composed of all the species descended from a single common ancestor; a holophyletic group.

cladogenesis division of an ancestral species into two separate descendent species.

cladogram branching tree diagram used to represent phyletic relationships.

conspecific belonging to the same species.

convergent evolution the independent evolution of similar morphological features from different ancestral conditions. The wings of bats and birds are an example of convergent evolution.

core area the part of a group's home range that is used most intensively.

cranial capacity the volume of the brain, usually determined by measuring the volume of the inside of the neurocranium.

crepuscular active primarily during the hours around dawn and dusk.

cryptic hidden; not normally visible.

day range the distance a group of animals travels during a single day.

deciduous dentition the milk teeth or first set of teeth in the mammalian jaw. The deciduous dentition is replaced by the permanent dentition.

dental eruption sequence the order in which the different teeth erupt or come into use.

dental formula a notation of the number of incisors, canines, premolars, and molars in the upper and lower dentition of a species. In humans, the adult dental formula is $\frac{2.1.2.3.}{2.1.2.3.}$.

derived feature a specialized morphological (or behavioral) characteristic that departs from the condition found in the ancestors of a species or group of species.

diagnostic distinguishing or characteristic, as the diagnostic features of a group of organisms.

diurnal active primarily during daylight hours.

dorsal toward the back side of the body; the opposite of ventral.

dizygotic twins (fraternal twins) twins that develop from two fertilized eggs or zygotes. This contrasts with monozygotic (or identical) twins, which develop from a single fertilized egg or zygote.

eclectic coming from many sources, as an eclectic diet.

ecological niche the complex of features (such as diet, forest type preference, canopy preference, activity pattern) that characterize the position a species occupies in the ecosystem.

ecology the study of the relationship between an organism and all aspects of its environment; or, all aspects of the environment of an organism which affect its way of life.

emergent trees the trees in a tropical forest which extend above the relatively continuous canopy.

endocast an impression of the inside of the cranium, often preserving features of the surface of the brain.

extant living, as opposed to extinct.

extension a movement in which the angle of a limb joint increases.

exudate a substance, such as gum, sap, or resin, which flows from the vascular system of a plant.

faunal correlation determination of the relative ages of different geological strata by comparing the fossils within the strata and assigning similar ages to strata with similar fossils; a method of relative dating.

faunivore an animal that eats primarily other animals (includes insectivores and carnivores).

fission-fusion society a type of social organization in which individuals regularly form small subgroups for foraging but from time to time also join together in larger groups; the variation in grouping usually depends on the type of food.

flexion a movement in which the angle of a limb joint decreases; the opposite of extension.

folivore an animal that feeds primarily on leaves.

foraging strategy the behavioral adaptations of a species related to its acquisition of food items.

founder effect changes in an allele-frequency that are the result of a small initial population.

frugivore an animal that feeds primarily on fruit.

gallery forest a forest along a river or stream.

genetic drift change in allele frequencies in a population due to chance rather than selection.

genotype the genetic make-up of an organism.

gracile relatively slender or delicately built.

grade a level or stage of organization, or a group of organisms sharing a suite of features (either primitive or derived) that distinguishes them from more advanced or more primitive animals but does not necessarily define a clade.

gradistic classification a classification in which organisms are grouped according to grade or level of organization rather than according to ancestry or phylogeny.

graminivore an animal that eats primarily grains; often also used to describe an animal that eats seeds.

gregarious living in regular social groups; contrasted to solitary living.

grooming the cleaning of the body surface by licking, biting, picking with fingers or claws, or other kinds of manipulation.

growth allometry the relationship between size and shape during the growth (or ontogeny) of an organism.

gummivore an animal that eats **exudates**—gum, saps, or resins.

holophyletic group a taxonomic group of organisms which has a single common ancestor and which includes all descendants of that ancestor.

home range the area of land that is regularly used by a group of animals for a year or longer.

homologous having the same developmental and evolutionary origin. The bones in the hands of primates and the wings of bats are homologous.

homoplasy morphological similarity in two species that is not the result of common ancestry, including **convergent evolution** and **parallel evolution**.

infanticide the killing of infants.

insectivore an animal that eats primarily insects (and other invertebrates); also used to refer to the mammalian order Insectivora (which includes shrews, moles, and hedgehogs).

insertion the attachment of a muscle or ligament farthest from the trunk or center of the body.

intermembral index a measure of the relative length of the forelimbs and hindlimbs of an animal: humerus plus radius length × 100 divided by femur plus tibia length.

interspecific allometry the relationship between size and shape among a range of different species; for example, a comparison between mouse and elephant.

ischial callosity a fatty sitting pad on the ischium of all Old World monkeys and gibbons.

Kay's threshold the body weight (approximately 500 grams) that is roughly the upper size limit of predominantly insectivorous primates and the lower size limit of predominantly folivorous primates.

keystone resources critical elements in a species' ecology that limit its population size and distribution.

knuckle-walking a type of quadrupedal walking, used by chimpanzees and gorillas, in which the upper body is supported by the dorsal surface of the middle phalanges of the hands.

kyphosis dorsally convex curvature of the back.

life history parameter a characteristic of the growth and development of an organism such as the length of gestation, timing of sexual maturity, length of reproductive period, or lifespan.

locomotion movement from one place to another.

lordosis ventrally convex curvature of the back.

mandible jawbone.

mandibular symphysis the joint between the right and left halves of the mandible. In human and other higher primates, this joint is fused.

monogamy a social system based on mated pairs and their offspring.

monophyletic group a taxonomic group of organisms which has a single common ancestor.

morphology the shape of anatomical structures.

multi-male group a group of animals in which several adult males and several adult females are reproductively active.

natural selection a nonrandom differential preservation of genotypes from one generation to the next which leads to changes in the genetic structure of a population.

nectivore an animal that eats nectar.

neoteny the retention of the features of a juvenile animal of one species in the adult form of a different species.

neotropics the tropical regions of North America, Central America, and South America.

nocturnal active primarily during the night.

noyau a type of social organization in which adult individuals have separate home ranges; ranges of individuals do not overlap with those of other individuals of the same sex, but they do overlap with ranges of individuals of the opposite sex.

olfaction the sense of smell.

one-male group a social group containing several reproductively active females but only one reproductively active male.

ontogeny the development of an organism from conception to adulthood.

organ of Jacobson an organ for chemical reception found in the anterior part of the roof of the mouth of many vertebrates.

paleomagnetism study of the magnetism of rocks that were formed in earlier time periods. More broadly, the study of changes in the earth's magnetic fields during geological time.

palmar pertaining to the palm side of the hand.

parallel evolution independent evolution of similar (and homologous) morphological features in separate lineages.

paraphyletic classification a classification in which a taxonomic group contains some, but not all, of the members of a clade.

phyletic classification a classification in which taxonomic groups correspond to monophyletic groups.

phyletic gradualism a model of evolution in which change takes place slowly in small steps, in contrast to the punctuated equilibrium model.

phylogeny the evolutionary or genealogical relationships among a group of organisms.

plantar pertaining to the sole of the foot.

polyandry a type of social organization in which there are two or more reproductively active males and a single reproductively active female.

polygyny any type of social organization in which one male mates with more than one female.

prehensile capable of grasping; for example, the prehensile tail of some platyrrhine monkeys.

primary rain forest rain forest characterized by the later stages of the vegetational succession cycle.

primitive feature a behavioral or morphological feature that is characteristic of a species and its ancestors.

procumbent inclined forward, protruding, as in the procumbent incisors of some primates.

prognathism prominence of the snout.

pronation rotation of the forearm so that the palm faces dorsally or downward; the reverse movement from supination.

punctuated equilibrium a model of evolution in which change takes place primarily by abrupt genetic shifts, in contrast to phyletic gradualism.

quadrumanous four-handed; as in quadrumanous climbing, in which many suspensory primates use their feet in the same manner that they use their hands.

ramus the vertical part of the mandible, often called the ascending ramus.

Rapaport's Rule a biogeographical pattern in which species found closer to the equator have smaller geographical ranges than those found farther from the equator.

relative dating A determination of whether a fossil or fossil site is younger or older than other fossils or sites, usually through study of the stratigraphic position or evolutionary relationships of the fauna; contrasts with absolute dating.

reproductive strategy an organism's complex of behavioral and physiological features concerned with reproduction. The reproductive strategy of oysters, for example, is characterized by the production of large numbers of offspring and no parental care, whereas the reproductive strategy of humans is often characterized by production of relatively few offspring and investment of large amounts of parental care by both parents.

reproductive success the contribution of an individual to the gene pool of the next generation.

sagittal crest a bony ridge on the top of the neurocranium formed by the attachment of the temporalis muscles.

saltation leaping; either a type of locomotion, or a description of rapid evolutionary change characterized by a lack of intermediate forms, i.e., "leaping" from one distinct species to another.

savannah a type of vegetation zone characterized by grasslands with scattered trees.

schizodactyly grasping between the second and third digits of the hand rather than between the pollex (thumb) and second digit.

secondary compounds poisons produced by plants which exist in leaves, flowers, etc., and deter animals from eating them.

secondary rain forest rain forest characterized by immature stages of the succession cycle, commonly found on the edges of forests, along rivers, and around tree falls.

sexual dichromatism the condition in which males and females of a species differ in color.

sexual dimorphism any condition in which males and females of a species differ in some aspect of their nonreproductive anatomy such as body size, canine tooth size, or snout length.

single-species hypothesis the theory that there has never been more than one hominid lineage at any time because all hominids are characterized by culture and thus all occupy the same ecological niche.

speciation appearance of new species.

Species–Area Relationship a biogeographical rule that larger geographical areas contain more species.

subfossil recently extinct, often from historical time periods. Some prosimians from Madagascar, for example, have become extinct in the past thousand years.

supination rotation of the forearm such that the palmar surface faces anteriorly or upward; the reverse movement from pronation.

suspensory behavior locomotor and postural habits characterized by hanging or suspension of the body below or among branches rather than walking, running, or sitting on top of branches.

suture a joint between two bones in which the bones interdigitate and are separated by fibrous tissue. The joints between most of the bones of the skull are sutures.

sympatry overlap in the geographical range of two species or populations.

systematics the science of classifying organisms and the study of their genealogical relationships.

taphonomy study of the processes that affect the remains of organisms from the death of the organism through its fossilization.

taxonomy the science of describing, naming, and classifying organisms.

terrestrial on the ground.

terrestrial quadrupedalism four-limbed locomotion on the ground.

territory part of a home range that is exclusive to a group of animals and is actively defended from other groups of the same species.

tooth comb a formation of the lower incisors into a comblike structure for grooming.

tympanic bone the bone that forms the bony ring for the eardrum.

type specimen a single designated individual of an organism which serves as the basis for the original name and description of the species.

understory the part of a forest that lies below the canopy layers.

valgus an angulation of the femur such that the knees are closer together than the hip joints; "knock-kneed."

ventral toward the belly side of an animal; the opposite of dorsal.

vertical clinging and leaping a type of locomotion and posture in which animals cling to vertical supports and move by leaping between these vertical supports.

woodland a vegetation type characterized by discontinuous stands of relatively short trees separated by grassland.

CLASSIFICATION OF THE ORDER PRIMATES

Genera in boldface contain extant members

ORDER PRIMATES

SUBORDER PROSIMII

Family *Incertae sedis*
Altanius
Altiatlasius

INFRAORDER LEMURIFORMES

Superfamily Adapoidea
Family Adapidae
Subfamily Adapinae
Adapis
Adapoides
Cryptadapis
Leptadapis
Microadapis

Family Notharctidae
Subfamily Notharctinae
Cantius
Copelemur
Hesperolemur
Notharctus
Pelycodus
Smilodectes
Subfamily Cercamoniinae
Aframonius
Agerinia
Anchomomys
Buxella
Caenopithecus
Cercamonius
Djebelemur
Donrussellia
Europolemur
Huerzeleria
Mahgarita
Omanodon
Panobius
Periconodon
Pronycticebus
Protoadapis
Shizarodon
Wadilemur

Family Sivaladapidae
Subfamily Sivaladapinae
Hoanghonius
Indraloris
Rencunius
Sinoadapis
Sivaladapis
Wailekia

Family *Incertae sedis*
Azibius
Lushius

Superfamily Lemuroidea
Family Cheirogaleidae
Allocebus
Cheirogaleus
Microcebus
Mirza
Phaner

Family Daubentoniidae
Daubentonia

Family Indriidae
Subfamily Archaeolemurinae
Archaeolemur
Hadropithecus
Subfamily Indriinae
Avahi
Indri

Propithecus
Subfamily Palaeopropithecinae
Archaeoindris
Babakotia
Mesopropithecus
Palaeopropithecus

Family Lemuridae
Eulemur
Hapalemur
Lemur
Pachylemur
Varecia

Family Lepilemuridae (Megaladapidae)
Subfamily Lepilemurinae
Lepilemur
Subfamily Megaladapinae
Megaladapis

Superfamily Lorisoidea
Family Galagidae
Euoticus
Galago
Galagoides
Komba
Otolemur
Progalago

Family Lorisidae
Arctocebus
Loris
Mioeuoticus
Nycticeboides
Nycticebus
Peodicticus
Pseudopotto

Family Plesiopithecidae
Plesiopithecus

INFRAORDER TARSIIFORMES

Superfamily Omomyoidea
Family Microchoeriidae
Microchoerus
Nannopithex
Necrolemur
Pseudoloris

Family Omomyidae
Subfamily Anaptomorphinae
Tribe Anaptomorphini
Absarokius
Acrossia
Anaptomorphus
Gazinius
Pseudotetonius
Strigorhysis
Tatmanius
Teilhardina
Tetonius
Tribe Trogolemurini
Anemorhysis
Arapohovius
Chlororhysis
Sphacorhysis
Tetonoides
Trogolemur
Tribe Washakiini
Dyseolemur
Loveina
Shoshonius
Washakius

Subfamily Omomyinae
Tribe Macrotarsiini
Hemiacodon
Macrotarsius
Yaquius
Tribe Omomyinae
Chumashius
Omomys
Steinius
Tribe Ourayini
Ageitodendron
Asiomomys
Chipetaia
Ourayia
Stockia
Wyomomys
Tribe Uintaniini
Jemezius
Uintanius
Tribe *Incertae sedis*
Ekgmowechashala

Family *Incertae sedis*
Kohatius
Rooneyia

Superfamily Tarsioidea
Family Tarsiidae
Afrotarsius
Tarsius
Xantorhysis

SUBORDER ANTHROPOIDEA

INFRAORDER *INCERTAE SEDIS*

Family Eosimiidae
Eosimias

Family Parapithecidae
Apidium
Biretia
Parapithecus
Qatrania
Serapia

Family *Incertae sedis*
Algeripithecus
Amphipithecus
Arsinoea
Pondaungia
Proteopithecus
Siamopithecus
Tabelia

INFRAORDER CATARRHINI

Superfamily Propliopithecoidea
Family Oligopithecidae
Catopithecus
Oligopithecus

Family Pliopithecidae
Anapithecus
Laccopithecus
Plesiopliopithecus
Pliopithecus

Family Propliopithecidae
Aegyptopithecus
Propliopithecus

Superfamily Cecopithecoidea
Family Voctiriapithecidae
Subfamily Victoriapithecinae
Prohylobates
Victoriapithecus

Family Cercopithecidae
Subfamily Cercopithecinae
Allenopithecus
Cercocebus
Cercopithecus
Chlorocebus
Dinopithecus
Erythrocebus
Gorgopithecus
Lophocebus
Macaca
Mandrillus
Miopithecus
Papio
Paradolichopithecus
Parapapio
Procynocephalus
Simopithecus
Theropithecus
Subfamily Colobinae
Cercopithecoides
Colobus
Dolichopithecus
Kasi
Libypithecus
Mesopithecus
Microcolobus
Nasalis
Paracolobus
Piliocolobus
Presbytis
Procolobus
Pygathrix
Rhinocolobus
Rhinopithecus
Semnopithecus
Simias
Trachypithecus

Superfamily Hominoidea
Family Hominidae
Subfamily Australopithecinae
Ardipithecus
Australopithecus
Paranthropus
Subfamily Homininae
Homo

Family Hylobatidae
Hylobates

Family Oreopithecidae
Mabokopithecus
Nyanzapithecus
Oreopithecus

Family Pongidae
Ankarapithecus
Dryopithecus
Gigantopithecus
Gorilla
Graecopithecus
Griphopithecus
Lufengpithecus
Ouranopithecus
Pan
Pongo
Sivapithecus

Family Proconsulidae
Dendropithecus
Dionysopithecus
Kalepithecus
Kamoyapithecus
Limnopithecus
Micropithecus
Platydontopithecus
Proconsul
Rangwapithecus
Simiolus

Family *Incertae sedis*
Afropithecus
Kenyapithecus
Morotopithecus
Otavipithecus
Samburupithecus
Turkanapithecus

INFRAORDER PLATYRRHINI

Superfamily Ceboidea
Family Atelidae
Subfamily Atelinae
Alouatta
Ateles
Brachyteles
Caipora
Lagothrix
Protopithecus
Stirtonia
Subfamily Callicebinae
Callicebus
Subfamily Pitheciinae
Cacajao
Carlocebus
Cebupithecia
Chiropotes
Homunculus
Nuciruptor
Pithecia
Propithecia
Soriacebus

Family Cebidae
Subfamily Aotinae
Aotus
Tremacebus
Subfamily Callithrichinae
Callimico
Callithrix
Cebuella
Lagonimico
Leontopithecus
Micodon
Patasola
Saguinus
Subfamily Cebinae
Cebus
Chilecebus
Dolichocebus
Laventiana
Neosaimiri
Saimiri
Subfamily *Incertae sedis*
Antillothrix
Branisella
Mohanamico
Paralouatta
Szalatavus
Xenothrix

INDEX

Note: Page numbers in italic indicate illustrations.